SYNTHESIS, PROPERTIES, AND APPLICATIONS OF OXIDE NANOMATERIALS

The Wiley Bicentennial–Knowledge for Generations

Each generation has its unique needs and aspirations. When Charles Wiley first opened his small printing shop in lower Manhattan in 1807, it was a generation of boundless potential searching for an identity. And we were there, helping to define a new American literary tradition. Over half a century later, in the midst of the Second Industrial Revolution, it was a generation focused on building the future. Once again, we were there, supplying the critical scientific, technical, and engineering knowledge that helped frame the world. Throughout the 20th Century, and into the new millennium, nations began to reach out beyond their own borders and a new international community was born. Wiley was there, expanding its operations around the world to enable a global exchange of ideas, opinions, and know-how.

For 200 years, Wiley has been an integral part of each generation's journey, enabling the flow of information and understanding necessary to meet their needs and fulfill their aspirations. Today, bold new technologies are changing the way we live and learn. Wiley will be there, providing you the must-have knowledge you need to imagine new worlds, new possibilities, and new opportunities.

Generations come and go, but you can always count on Wiley to provide you the knowledge you need, when and where you need it!

William J. Pesce
President and Chief Executive Officer

Peter Booth Wiley
Chairman of the Board

SYNTHESIS, PROPERTIES, AND APPLICATIONS OF OXIDE NANOMATERIALS

José A. Rodríguez
Brookhaven National Laboratory
Upton, New York

Marcos Fernández-García
Instituto de Catálisis y Petroleoquímica CSIC
Madrid, Spain

WILEY-INTERSCIENCE
A JOHN WILEY & SONS, INC., PUBLICATION

Published by John Wiley & Sons, Inc., Hoboken, New Jersey
Published simultaneously in Canada

Wiley Bicentennial Logo: Richard J. Pacifico

Library of Congress Cataloging-in-Publication Data:

Synthesis, properties, and applications of oxide nanomaterials/[edited by]
José A. Rodríguez, Marcos Fernández-García.
p. cm.
Includes bibliographical references and index.
ISBN 978-0-471-72405-6
1. Nanostructured materials. 2. Oxides. I. Rodríguez, José A. II. Fernández García, Marcos.
TA418.9.N35S978 2007
661′.0721—dc22 2006024024

Printed in the United States of America

10 9 8 7 6 5 4 3 2 1

CONTENTS

CONTRIBUTORS

James A. Anderson, Department of Chemistry, King's College, University of Aberdeen, Aberdeen AB24 3UE

Jonas Baltrusaitis, Departments of Chemistry and Chemical and Biochemical Engineering, University of Iowa, Iowa City IA 52242

Carolina Belver-Coldeira, Instituto de Catálisis y Petroleoquímica CSIC, Campus Cantoblanco 28049 Madrid, Spain

Juan Bisquert, Departament de Ciències Experimentals, Universitat Jaume I, 12071 Castelló, Spain

S. Buzby, Department of Materials Science & Engineering, University of Delaware Newark, DE 19716

Gerardo Colón-Ibáñez, Instituto de Ciencia de Materiales de Sevilla CSIC, C/Américo Vespucio, 49 41092-Sevilla, Spain

Lawrence D'Souza, International University Bremen, Bremen, Germany

Marcos Fernández-García, Instituto de Catálisis y Petroleoquímica CSIC, Campus Cantoblanco, 28049 Madrid, Spain

Lucio Forni, Dipartimento di Chimica Fisica ed Elettrochimica, Università degli Studi di Milano, v. C. Golgi 19, I-20133 Milano, Italy

S. Franklin, Department of Materials Science & Engineering, University of Delaware Newark, DE 19716

Vicki H. Grassian, Departments of Chemistry and Chemical and Biochemical Engineering, University of Iowa, Iowa City IA 52242

Jonathan C. Hanson, Chemistry Department, Brookhaven National Laboratory, Upton, NY 11973, USA

Russell F. Howe, Department of Chemistry, King's College, University of Aberdeen, Aberdeen AB24 3UE

Francesc Illas, Departament de Química Física and Centre especial de Recerca en Química Teòrica, Universitat de Barcelona and Parc Científic de Barcelona, C/Martí i Franquès 1, E-08028 Barcelona, Spain

Pethaiyan Jeevanandam, 111, Willard Hall, Department of Chemistry, Kansas State University, Manhattan, KS, 66506, USA

Kenneth J. Klabunde, 111, Willard Hall, Department of Chemistry, Kansas State University, Manhattan, KS, 66506, USA

Doina Lutic, School of Technology and Design, Växjo University, 35195 Växjö, Sweden

Arturo Martínez-Arias, Instituto de Catálisis y Petroleoquímica, CSIC, C/Marie Curie 2, Campus Cantoblanco, 28049 Madrid, Spain

Glenn C. Mather, Instituto de Cerámica y Vidrio, CSIC, C/Kelsen 5, Campus Cantoblanco, 28049 Madrid, Spain

Gianfranco Pacchioni, Dipartimento di Scienza dei Materiali, Università di Milano-Bicocca, via R. Cozzi, 53-I-20125 Milano, Italy

Javier Pérez-Ramírez, Institut Catalá d'Investigació Química Av. Paisos Catalans 16 Campus Univesitary, Tarragona, Spain

John M. Pettibone, Departments of Chemistry and Chemical and Biochemical Engineering, University of Iowa, Iowa City IA 52242

Ryan Richards, International University Bremen, Bremen, Germany

Vicente Rives, Departamento de Química Inorgánica, Universidad de Salamanca, Spain

José A. Rodríguez, Chemistry Department, Brookhaven National Laboratory, Upton, NY 11973, USA

Ilenia Rossetti, Dipartimento di Chimica Fisica ed Elettrochimica, Università degli Studi di Milano, v. C. Golgi 19, I-20133 Milano, Italy

F. Ruette, Laboratorio. de Química Computacional, IVIC, Apartado. 21827, Caracas 1020-A, Venezuela

Mehri Sanati, School of Technology and Design, Växjo University, 35195 Växjö, Sweden

M. Sánchez, Laboratorio. de Química, Computacional, IVIC, Apartado. 21827, Caracas 1020-A, Venezuela

D.C. Sayle, DEOS, Cranfield University, Defense Academy of the United Kingdom Shrivenham, Swindon, UK

T.X.T. Sayle, DEOS, Cranfield University, Defense Academy of the United Kingdom Shrivenham, Swindon, UK

William F. Schneider, Department of Chemical and Biomolecular Engineering, and Department of Chemistry and Biochemistry, University of Notre Dame, Notre Dame, IN 46556

S. Ismat Shah, Department of Materials Science & Engineering, University of Delaware Newark, DE 19716

William A. Shelton, Computer Science and Mathematics Division, Oak Ridge National Laboratory, Oak Ridge, TN 37831

Anita Lloyd Spetz, S-SENCE and Div. of Applied Physics, Linköping University, 58183 Linköping, Sweden

Chang Q. Sun, School of Electrical and Electronic Engineering, Nanyang Technological University, Singapore 639798 and Institute of Advanced Materials Physics and Faculty of Science, Tianjin University, 300072, P. R. China

Raquel Trujillano, Departamento de Química Inorgánica, Universidad de Salamanca, Spain

Lionel Vayssieres, International Center for Young Scientists, National Institute for Materials Science, Tsukuba, Japan 305-0044

Miguel A. Vicente, Departamento de Química Inorgánica, Universidad de Salamanca, Spain

Xianqin Wang, Institute for Interfacial Catalysis, Pacific Northwest National Laboratory, Richland, WA, 99352 (USA)

Ye Xu, Computer Science and Mathematics Division, Oak Ridge National Laboratory, Oak Ridge, TN 37831

INTRODUCTION

The World of Oxide Nanomaterials

JOSÉ A. RODRÍGUEZ
Brookhaven National Laboratory, Upton, New York
MARCOS FERNÁNDEZ-GARCÍA
Instituto de Catálisis y Petroleoquímica CSIC, Madrid, Spain

Metal oxides play a very important role in many areas of chemistry, physics, and materials science (1–6). The metal elements can form a large diversity of oxide compounds (7). These elements can adopt many structural geometries with an electronic structure that can exhibit metallic, semiconductor, or insulator character. In technological applications, oxides are used in the fabrication of microelectronic circuits, sensors, piezoelectric devices, fuel cells, coatings for the passivation of surfaces against corrosion, and as catalysts. For example, almost all catalysts used in industrial applications involve an oxide as active phase, promoter, or "support." In the chemical and petrochemical industries, products worth billions of dollars are generated every year through processes that use oxide and metal/oxide catalysts (8). For the control of environmental pollution, catalysts or sorbents that contain oxides are employed to remove the CO, NO_x, and SO_x species formed during the combustion of fossil-derived fuels (9,10). Furthermore, the most active areas of the semiconductor industry involve the use of oxides (11). Thus, most of the chips used in computers contain an oxide component.

In the emerging field of nanotechnology, a goal is to make nanostructures or nanoarrays with special properties with respect to those of bulk or single-particle species (12–16). Oxide nanoparticles can exhibit unique physical and chemical properties due to their limited size and a high density of corner or edge surface sites. Particle size is expected to influence three important groups of properties in any material. The first one comprises the structural characteristics, namely, the lattice symmetry and cell parameters (17). Bulk oxides are usually robust and stable systems with well-defined crystallographic structures. However, the growing importance of surface-free energy

Synthesis, Properties, and Applications of Oxide Nanomaterials, Edited by José A. Rodríguez and Marcos Fernández-Garcia

and stress with decreasing particle size must be considered: Changes in thermodynamic stability associated with size can induce modification of cell parameters and/or structural transformations (18–21), and in extreme cases, the nanoparticle can disappear because of interactions with its surrounding environment and a high surface-free energy (22). To display mechanical or structural stability, a nanoparticle must have a low surface-free energy. As a consequence of this requirement, phases that have a low stability in bulk materials can become very stable in nanostructures. This structural phenomenon has been detected in TiO_2, VO_x, Al_2O_3, or MoO_x oxides (21–24).

Size-induced structural distortions associated with changes in cell parameters have been observed in nanoparticles of Al_2O_3 (22), NiO (24), Fe_2O_3 (25), ZrO_2 (26), MoO_3 (24), CeO_2 (27), and Y_2O_3 (28). As the particle size decreases, the increasing number of surface and interface atoms generates stress/strain and concomitant structural perturbations (29). Beyond this "instrinsic" strain, there may be also "extrinsic" strain associated with a particular synthesis method that may be partially relieved by annealing or calcination (30). On the other hand, interactions with the substrate on which the nanoparticles are supported can complicate the situation and induce structural perturbations or phases not observed for the bulk state of the oxide (23,31).

The second important effect of size is related to the electronic properties of the oxide. In any material, the nanostruture produces the so-called quantum size or confinement effects, which essentially arise from the presence of discrete, atom-like electronic states. From a solid-state point of view, these states can be considered as being a superposition of bulk-like states with a concomitant increase in oscillator strength (32). Additional general electronic effects of quantum confinement experimentally probed on oxides are related to the energy shift of exciton levels and optical bandgap (33,34). An important factor to consider when dealing with the electronic properties of a bulk oxide surface are the long-range effects of the Madelung field, which are not present or limited in a nanostructured oxide (35–37). Theoretical studies for oxides show a redistribution of charge when going from large periodic structures to small clusters or aggregates, which must be roughly considered to be relatively small for ionic solids while significantly larger for covalent ones (38–43). The degree of ionicity or covalency in a metal-oxygen bond can, however, strongly depend on size in systems with partial ionic or covalent character; an increase in the ionic component to the metal–oxygen bond in parallel to the size decreasing has been proposed (20).

Structural and electronic properties obviously drive the physical and chemical properties of the solid, and the third group of properties is influenced by size in a simple classification. In their bulk state, many oxides have wide band gaps and a low reactivity (44). A decrease in the average size of an oxide particle does in fact change the magnitude of the band gap (37,45), with strong influence on the conductivity and chemical reactivity (46,47). Surface properties are a somewhat particular group included in this subject because of their importance in chemistry. Solid–gas or solid–liquid chemical reactions can be mostly confined to the surface and/or subsurface regions of the solid. As mentioned, the two-dimensional (2D) nature of surfaces has (*1*) notable structural consequences, typically a rearrangement or reconstruction of bulk geometries (3,15,48), and (*2*) electronic consequences, such as the presence of mid-gap states (47,49). In the case of nanostructured oxides, surface properties are strongly modified with respect to 2D-infinite surfaces, producing solids with unprecedent sorption or acid/base characteristics (50). Furthermore, the presence of under-coordinated atoms (like corners or edges) or O vacancies in an oxide nanoparticle should produce specific geometrical arrangements as well as occupied electronic states located above the valence band of the corresponding bulk material (51–53), enhancing in this way the chemical activity of the system (48,50,54,55).

This book is organized into five parts that cover different aspects associated with the world of oxide nanomaterials. The first part of the book deals with fundamental aspects quantum–mechanical (56) and thermodynamic (57) that determine the behavior and growth mode of nanostructured oxides. To prepare these materials, novel synthethic procedures have been developed that can be described as physical and chemical methods (58,59). In general, they use top-down and bottom-up fabrication technologies, which involve liquid–solid or gas–solid transformations that are reviewed in the second part of the book. To characterize the structural and electronic properties of nanostructured oxides, one needs an array of sophisticated experimental techniques (6) and state-of-the-art theory (37,41,44,60,61). This theory is the subject of the third part of the book. In the fourth part, the emphasis is on systematic studies examining the physicochemical properties of oxide nanoparticles (6,62,63). The fifth and final part of the book is focused on technological applications of nanostructured oxides as sorbents, sensors, ceramic materials, photo-devices, and catalysts for reducing environmental pollution, transforming hydrocarbons, and producing H_2 (64–68). We feel that it is important to stress the need for a fundamental understanding of the properties of nanostructured oxides, particularly for sizes in which the atoms directly affected in their properties are a significant percentage of the total number of atoms present in the solid particle; this usually implies a dimension limited to about or below 10 nm. When this fact occurs exclusively in one dimension, we are dealing with a surface or film, whereas in two dimensions, nanotubes, nanowires, and other interesting morphologies can be formed. Finally, when the three dimensions are limited to the nanoscale, nanoparticles are formed.

REFERENCES

(1) Noguera, C. *Physics and Chemistry at Oxide Surfaces*; Cambridge University Press: Cambridge, UK, 1996.

(2) Kung, H.H. *Transition Metal Oxides: Surface Chemistry and Catalysis*; Elsevier: Amsterdam, 1989.

(3) Henrich, V.E.; Cox, P.A. *The Surface Chemistry of Metal Oxides*; Cambridge University Press: Cambridge, UK, 1994.

(4) Wells, A.F. *Structural Inorganic Chemistry, 6th ed*; Oxford University Press: New York, 1987.

(5) Harrison, W.A. *Electronic Structure and the Properties of Solids*; Dover: New York, 1989.

(6) Fernández-García, M.; Martínez-Arias, A.; Hanson, J.C.; Rodríguez, J.A. *Chem. Rev.* **2004**, *104*, 4063.

(7) Wyckoff, R.W.G. *Crystal Structures, 2nd ed*; Wiley: New York, 1964.

(8) *Handbook of Heterogeneous Catalysis*; Ertl, G.; Knozinger, H.; Weitkamp, J. (Editors); Wiley-VHC: Weinheim, 1997.

(9) Shelef, M.; Graham, G.W.; McCabe, R.W. In *Catalysis by Ceria and Related Materials*; Trovarelli, A. (Editor); Imperial College Press: London, 2002; Chapter 10.

(10) Stirling, D. *The Sulfur Problem: Cleaning Up Industrial Feedstocks*; Royal Society of Chemistry: Cambridge, UK, 2000.

(11) Sherman, A. *Chemical Vapor Deposition for Microelectronics: Principles, Technology and Applications*; Noyes Publications: Park Ridge, NJ, 1987.

(12) Gleiter, H. *Nanostruct. Mater.* **1995**, *6*, 3.

(13) Valden, M.; Lai, X.; Goodman, D.W. *Science* **1998**, *281*, 1647.

(14) Rodriguez, J.A.; Liu, G.; Jirsak, T.; Hrbek, Chang, Z.; Dvorak, J.; Maiti, A. *J. Am. Chem. Soc.* **2002**, *124*, 5247.

(15) Baumer, M.; Freund, H.-J. *Progress in Surf. Sci.* **1999**, *61*, 127.

(16) Trudeau, M.L.; Ying, J.Y. *Nanostruct. Mater.* **1996**, *7*, 245.

(17) Ayyub, P.; Palkar, V.R.; Chattopadhyay, S.; Multani, M.; *Phys. Rev. B.* **1995**, *51*, 6135.

(18) Millot, N.; Aymes, D.; Bernard, F.; Niepce, J.C.; Traverse, A.; Bouree, F.; Cheng, B.L.; Perriat, P. *J. Phys. Chem. B.* **2003**, *107*, 5740.

(19) Schoiswohl, J.; Kresse, G.; Surnev, S.; Sock, M.; Ramsey, M.G.; Netzer, F.P. *Phys. Rev. Lett.* **2004**, *92*, 206103.

(20) McHale, J.M.; Auroux, A.; Perrota, A.J.; Navrotsky, A. *Science* **1997**, *277*, 788.

(21) Zhang, H.; Bandfield, J.F. *J. Mater. Chem.* **1998**, *8*, 2073.

(22) Samsonov, V.M.; Sdobnyakov, N.Yu.; Bazulev, A.N. *Surf. Sci.* **2003**, *532–535*, 526.

(23) Song, Z.; Cai, T.; Chang, Z.; Liu, G.; Rodriguez, J.A.; Hrbek, J. *J. Am. Chem. Soc.* **2003**, *125*, 8060.

(24) Rodriguez, J.A.; Rodriguez, L.; Ruette, F.; Gonzalez, L. In preparation.

(25) Ayyub, P.; Multani, M.; Barma, M.; Palkar, V.R.; Vijayaraghavan, R. *J. Phys. C: Solid State Phys.* **1988**, *21*, 229.

(26) Garvie, R.C.; Goss, M.F. *J. Mater. Sci.* **1986**, *21*, 1253.

(27) Hernández-Alonso, M.D.; Hungría, A.B.; Coronado, J.M.; Martínez-Arias, A.; Conesa, J.C.; Soria, J.; Fernández-García, M. *Phys. Chem. Chem. Phys.* **2004**, *6*, 3524.

(28) Skandan, G.; et al. *Nanostruct. Mater.* **1992**, *1*, 313.

(29) Cammarata, R.C.; Sieradki, K. *Phys. Rev. Lett.* **1989**, *62*, 2005.

(30) Fernández-García, M.; Wang, X.; Belver, C.; Hanson, A.; Iglesias-Juez, J.C.; Rodriguez, J.A. *Chem. Mater.* **2005**, *17*, 4181.

(31) Surnev, S.; Kresse, G.; Ramsey, M.G.; Netzer, F.P. *Phys. Rev. Lett.* **2001**, *87*, 86102.

(32) Moriarty, P. *Rep. Prog. Phys.* **2001**, *64*, 297.

(33) Yoffre, A.D. *Advances in Phys.* **1993**, *42*, 173.

(34) Brus, L. *J. Phys. Chem.* **1986**, *90*, 2555.

(35) Pacchioni, G.; Ferrari, A.M.; Bagus, P.S. *Surf. Sci.* **1996**, *350*, 159.

(36) Mejias, J.A.; Marquez, A.M.; Fernandez-Sanz, J.; Fernández-García, M.; Ricart, J.M.; Sousa, C.; Illas, F. *Surf. Sci.* **1995**, *327*, 59.

(37) Fernández-García, M.; Conesa, J.C.; Illas, F.; *Surf. Sci.* **1996**, *349*, 207.

(38) Albaret, T.; Finocchi, F.; Noguera, C. *Faraday Discuss.* **2000**, *114*, 285.

(39) Rodriguez, J.A.; Maiti, A. *J. Phys. Chem. B.* **2000**, *104*, 3630.

(40) Rodriguez, J.A.; Jirsak, T.; Chaturvedi, S. *J. Chem. Phys.* **1999**, *111*, 8077.

(41) Bredow, T.; Apra, E.; Catti, M.; Pacchioni, G. *Surf. Sci.* **1998**, *418*, 150.

(42) Casarin, M.; Maccato, C.; Vittadini, A. *Surf. Sci.* **1997**, *377–379*, 587.

(43) Scamehorn, C.A.; Harrison, N.M.; McCarthy, M.I. *J. Chem. Phys.* **1994**, *101*, 1547.

(44) Rodriguez, J.A. *Theor. Chem. Acc.* **2002**, *107*, 117.

(45) Rodriguez, J.A.; Chaturvedi, S.; Kuhn, M.; Hrbek, J. *J. Phys. Chem. B.* **1998**, *102*, 5511.

(46) Hoffmann, R. *Solids and Surfaces: A Chemist's View of Bonding in Extended Structures*; VCH: New York, 1988.

(47) Albright, T.A.; Burdett, J.K.; Whangbo, M.H. *Orbital Interactions in Chemistry*; Wiley-Interscience: New York, 1985.

(48) Luth, H.; *Surface and Interface of Solid Materials*; Springer: Berlin, 1997.

(49) Bardeen, J. *Phys. Rev.* **1947**, *71*, 717.

(50) Lucas, E.; Decker, S.; Khaleel, A.; Seitz, A.; Futlz, S.; Ponce, A.; Li, W.; Carnes, C.; Klabunde, K.J. *Chem. Eur. J.* **2001**, *7*, 2505.

(51) Anchell, J.L.; Hess, A.C. *J. Phys. Chem.* **1996**, *100*, 18317.

(52) Rodriguez, J.A.; Hrbek, J.; Dvorak, J.; Jirsak, T.; Maiti, A. *Chem. Phys. Lett.* **2001**, *336*, 377.

(53) Ferrari, A.M.; Pacchioni, G. *J. Phys. Chem.* **1995**, *99*, 17010.

(54) Richards, R.; Li, W.; Decker, S.; Davidson, C.; Koper, O.; Zaikovski, V.; Volodin, A.; Rieker, T. *J. Am. Chem. Soc.* **2000**, *122*, 4921.

(55) Rodriguez, J.A.; Wang, X.; Liu, G.; Hanson, J.C.; Hrbek, J.; Peden, C.H.F.; Iglesias-Juez, A.; Fernández-García, M. *J. Molec. Catal. A: Chemical* **2005**, *228*, 11.

(56) Sun, C.Q. *Prog. Mater. Sci.* **2003**, *48*, 521.

(57) Vayssieres, L. *Int. J. Nanotechnol.* **2005**, *2*, 411.

(58) *Nanomaterials; Synthesis, Properties and Applications*; Edelstein, A.S.; Cammarata, R.C. (Editors); Institute of Physics Publishing: London, 1998.

(59) Khaleel, A.; Richards, R.M. In *Nanoscale Materials in Chemistry*; Kladunde, K.J. (Editor); Wiley: New York, 2001.

(60) Ruette, F.; Sanchez, M.; Sierraalta, A.; Mendoza, C.; Añez, R.; Rodriguez, L.; Lisboa, O.; Daza, J.; Manrique, P.; Perdomo, Z.; Rosa-Brussin, M. *J. Molec. Catal. A.: Chemical* **2005**, *228*, 211.

(61) Sayle, D.C.; Doig, J.A.; Parker, S.C.; Watson, G.W. *J. Materials Chem.* **2003**, *13*, 2078.

(62) Li, G.; Larsen, S.C.; Grassian, V.H. *J. Molec. Catal. A.* **2005**, *227*, 25.

(63) Anderson, J.A.; Martínez-Arias, A.; Fernández-García, M.; Hungría, A.B.; Iglesias-Juez, A.; Gálvez, O.; Conesa, J.C.; Soria, J.; Munuera, G. *J. Catal.* **2003**, *214*, 261.

(64) Gratzel, M. *Nature* **2001**, *414*, 338.

(65) Fu, Q.; Saltsburg, H.; Flytzani-Stephanopoulos, M. *Science* **2003**, *301*, 935.

(66) Klabunde, K.J.; Stark, J.; Koper, O.; Mobs, C.; Park, D.G.; Decker, S.; Jiang, Y.; Lagadic, I.; Zhang, D. *J. Phys. Chem.* **1996**, *100*, 12142.

(67) Wang, X.; Rodriguez, J.A.; Hanson, J.C.; Gamarra, D.; Martínez-Arias, A.; Fernández-García, M. *J. Phys. Chem. B.* **2005**, *109*, 19595.

(68) Gao, X.; Jehng, J.; Wachs, I.E. *J. Catal.* **2002**, *209*, 43.

PART I

BASIC CONCEPTS

CHAPTER 1

Theory of Size, Confinement, and Oxidation Effects

CHANG Q. SUN

School of Electrical and Electronic Engineering, Nanyang Technological University, Singapore 639798 and Institute of Advanced Materials Physics and Faculty of Science, Tianjin University, 300072, P. R. China

1.1. INTRODUCTION

Both the processes of oxidation and the size reduction form the essential entities that dictate the behavior of a nanostructured oxide. The electronic processes of oxidation destroy the initially metallic bonds to create new kinds of bonds with specific properties, such as polarization, localization, and transportation of charge, which determine the behavior of the oxides to vary from their parent metals (1). Oxygen interaction with atoms of metals relates to the technical processes of corrosion, bulk oxidation, heterogeneous catalysis, and so on (2). Studies of these processes laid the foundations for applications in microelectronics (gate devices and deep submicron integrated circuit technologies), photo-electronics (photoluminescence, photo-conductance, and field emission), magneto-electronics (superconductivity and colossal magneto-resistance) and dielectrics (ferro-, piezo-, and pyro-electrics) (3).

Nanoparticles are entities that appear in between extended solids and molecules or even in an isolated atom. Properties of nanosolids determined by their shapes and sizes are indeed fascinating, which form the basis of the emerging field of nanoscience and nanotechnology that has been recognized as the key significance in science, technology, and economics in the 21st century. When examining nanostructures, many concepts developed in both molecular chemistry and solid-state physics have to be considered. One may develop a "top-down" theory on the behavior of nanosystems starting from a solid and confining it to a limited size. Another chemical-like "bottom-up" approach is to start with a molecular system and expand its size. Each approach

Synthesis, Properties, and Applications of Oxide Nanomaterials, Edited by José A. Rodríguez and Marcos Fernández-Garcia

has different advantages and disadvantages. There is a challenge to bridge these two approaches and develop new concepts for nanostructures.

The size-induced property change of nanostructures has inspired tremendous theoretical efforts. For instance, several models have been developed to explain how the size reduction could induce the blue shift in the photoluminescence (PL) of nanosemiconductors. An *impurity luminescent center* model (4) assumed that the PL blue shift arises from different types of impurity centers in the solid and suggested that the density and types of the impurity centers vary with particle size. S*urface states* and the *surface alloying* mechanism (5,6) proposed that the PL blue shift originates from the extent of surface passivation that is subject to the processing parameters, aging conditions, and operation temperatures (7). The model of *inter-cluster interaction and oxidation* (8) also claimed responsibility for the PL blue shift. The most elegant model for the PL blue shift could be the "quantum confinement (QC)" theory (9–13). According to the QC theory, the PL energy corresponds to the band gap expansion dictated by electron-hole (e-h) pair (or exciton) production:

$$E_G(R) - E_G(\infty) = \pi^2\hbar^2/(2\mu R^2) - 1.786e^2/(\varepsilon_r R) + 0.284E_R \qquad (1.1)$$

where $\mu = m_h^* m_e^*/(m_h^* + m_e^*)$, being the reduced mass of the e-h pair, is an adjustable parameter. The E_G expansion originates from the addition of the kinetic energy E_K and the Coulomb interaction E_p of the e-h pairs that are separated by a distance of the particle radius R and contribution of the Rydberg or spatial correlation (electron–electron interaction) energy E_R for the bulk semiconductor. The effective dielectric constant ε_r and the effective mass μ describe the effect of the homogeneous medium in the quantum box, which is simplified as a mono-trapping central potential by extending the dimension of a single atom d_0 to that of the solid D. According to the QC theory, electrons in the conduction band and holes in the valence band are confined spatially by the potential barrier of the surface or are trapped by the potential well of the quantum box. Because of the confinement of both the electrons and the holes, the lowest energy optical transition from the valence to the conduction band increases in energy, effectively increasing the E_G. The sum of the kinetic and potential energy of the freely moving carriers is responsible for the E_G expansion, and therefore, the width of the confined E_G grows as the characteristic dimensions of the crystallites decrease.

In contrast, a free-exciton collision model (14) suggested that the E_G expansion arises from the contribution of thermally activated phonons in the grain boundaries rather than from the QC effect. During a PL measurement, the excitation laser heats the free excitons that then collide with the boundaries of the nanometer-sized fragments. The laser heating the free-excitons up to the temperature in excess of the activation energy required for the self-trapping gives rise to the extremely hot self-trapping excitons (STEs). Because the resulting temperature of the STEs is much higher than the lattice temperature, the cooling of the STEs is dominated by the emission of phonons. However, if the STE temperature comes into equilibrium with the lattice temperature, the absorption of lattice phonons becomes possible. As a result, the blue shift of the STE–PL band is suggested to originate from the activation of hot-phonon-assisted electronic transitions. The blue shift of the STE–PL band depends on the temperature of laser-heated free-excitons that in turn is determined by the size of nanofragments. This event happens because the temperature (kinetic energy) of the laser-heated free-exciton increases with the number of boundary collisions, which tends to be higher with decreasing size of the nanofragments. The energy gained from laser heating of the exciton increases with decreasing nanosolid size in an $\exp(1/R)$ way.

Another typical issue of nanostructures is their thermal stability. The melting point (T_m) of an isolated nanosolid, or a system with weakly linked nanoparticles, drops with solid size (called supercooling), whereas the T_m may rise (called superheating) for an embedded nanosystem because of the interfacial effect. The T_m is characterized by the Lindermann's criterion (15) of atomic vibration abruption or Born's criterion (16) of shear modulus disappearance at the T_m. The T_m elevation or suppression and the mode of melting in the nanometer regime have been described with the following models: (*1*) homogeneous melting and growth (17,18); (*2*) random fluctuation melting (19); (*3*) liquid shell nucleation and growth (20–23); (*4*) liquid-drop (24) formation; (*5*) lattice-vibrational instability (25,26); and (*6*) surface-phonon instability (27,28).

Often, numerous modeling arguments exist for a specific phenomenon. The challenge is how to correlate all outstanding arguments for all observations to the effect of bond order loss of the under-coordinated surface atoms or the effect of confinement. The origin for one phenomenon must intrinsically sustain to others. Understanding the mechanisms for both size reduction and oxidation and their joint contribution of the two entities is critical to understanding the behavior of nanostructured oxide materials and related devices.

This chapter describes a bond order–length–strength (BOLS) correlation mechanism (29) for the effect of physical size (bond breaking) and a chemical-bond–valence-band–potential-barrier (BBB) correlation mechanism (1) for the effect of oxidation (bond making) on the performance of a nanostructured oxide. The BBB correlation indicates the essentiality of sp-orbital hybridization of an oxygen atom upon reacting with atoms in the solid phase. In the process of oxidation, electronic holes, nonbonding lone pairs, lone-pair-induced antibonding dipoles, and hydrogen bonds are involved through charge transportation, localization, and polarization, which dictate the performance of an oxide. Charge transport from metal to oxygen creates the band gap, which turns the metal to be a semiconductor or an insulator; lone-pair-induced charge polarization lowers the work function of the surface, whereas hydrogen bond-like formation caused by overdosing with the oxygen additives restores the work function. The often-overlooked events of nonbonding and antibonding are expected to play significant roles in the functioning of an oxide.

The BOLS correlation mechanism indicates the significance of bond order loss of an atom at a site surrounding a defect or near the edge of a surface or in an amorphous phase in which the coordination (CN) reduction (deviation of bond order, length, and angle) distributes randomly. Bond order loss causes the remaining bonds of the under-coordinated atom to contract spontaneously associated with bond-strength gain or atomic potential well depression, which localizes electrons and enhances the density of charge, mass, and energy in the relaxed region. The energy density rise in the relaxed region perturbs the Hamiltonian and the associated properties such as the band gap width, core-level energy, Stokes shift (electron–phonon interaction), and dielectric susceptibility. On the other hand, bond order loss lowers the cohesive energy of the under-coordinated atom, which dictates the thermodynamic process such as self-assembly growth, atomic vibration, thermal stability, and activation energies for atomic dislocation, diffusion, and chemical reaction.

1.2. EFFECTS OF SIZE AND CONFINEMENT

1.2.1. Basic Concepts

Involvement of interatomic interaction causes the performance of a solid, or a cluster of atoms, to vary from that of an isolated atom; adjustment of the relative number of the under-coordinated surface atoms provides an additional freedom that allows one to tune the properties of a nanosolid with respect to that of its bulk counterpart. Hence, contribution from the under-coordinated atoms and the involvement of interatomic interaction could be the starting points of consideration to bridge the gap between an isolated atom and a bulk solid in chemical and physical performance.

1.2.1.1. Intra-atomic Trapping Electrons of a single atom confined by the intra-atomic trapping potential $V_{\text{atom}}(r)$ move around the central ion core in a standing-wave form inside the potential well. The $V_{\text{atom}}(r)$ describes the electron–nucleus interaction, and it takes a value that varies from several electronvolts to infinity, depending on the orbitals in which electrons are revolving. The Hamiltonian and the corresponding eigenwave functions and the eigenenergies for an electron in the isolated atom are given as

$$\hat{H}_0 = -\frac{\hbar^2 \nabla^2}{2m} + V_{\text{atom}}(r)$$

$$\phi_v(r) \propto \sin^2(k_n r)$$

and

$$E(n) = \hbar^2 k_k^2 / 2m_e; \quad k_n = 2n\pi/d_0^2, \quad n = 1, 2, 3, \ldots \tag{1.2}$$

the atomic diameter d_0 corresponds to the dimension of the potential well of the atom. The branch numbers (n) correspond to different energy levels. The energy separation between the nearest two levels depends on $(n+1)^2 - n^2 = 2n + 1$.

1.2.1.2. Interatomic Bonding and Intercluster Coupling When two atoms or more joined as a whole, interatomic interaction comes into play, which causes the performance of a cluster of atoms to be different from that of an isolated atom.

The interatomic bonding is essential to make a solid or even a liquid. Considering an assembly composed of n particles of mean size K_j and with each particle, there are N_j atoms, the total binding energy $V_{\text{cry}}(r, n, N_j)$ is (30):

$$\begin{aligned} V_{\text{cry}}(r, n, N_j) &= \sum_{n} \sum_{l \neq i} \sum_{i} v(r_{li}) \\ &= \frac{n}{2}\left[N_j \sum_{i=1} v(r_{li}) + \sum_{k \neq j} V(K_{kj}) \right] \\ &\cong \frac{n}{2}\left[N_j^2 v(d_0) + nV(K_j) \right] \end{aligned} \tag{1.3}$$

The $V_{\text{cry}}(r, n, N_j)$ sums over all N_j atoms and the n particles. The high order r_{li} is a certain fold of the nearest atomic spacing d_0. Interaction between the nearest clusters, k and j, $V(K_{kj})$, is negligible if the K_{kj} is considerably large. Normally, the intercluster interaction, $V(K_{kj})$, is much weaker than the interatomic interaction, if the cluster is treated as an electric or a magnetic dipole of which the Van der Waals or the super-paramagnetic potentials dominate.

1.2.1.3. Hamiltonian and Energy Band According to the band theory, the Hamiltonian for an electron inside a solid is in the form:

$$\hat{H} = \hat{H}_0 + \hat{H}' = -\frac{\hbar^2 \nabla^2}{2m} + V_{\text{atom}}(r) + V_{\text{cry}}(r + R_C) \tag{1.4}$$

where the $\hat{H}_0$ is the Hamiltonian for an isolated atom as given in Eq. 1.2. $\hat{H}' = V_{\text{cry}}(r) = V_{\text{cry}}(r + R_C)$ is the periodic potential of the crystal, describing the interaction between the specific electron with the ion cores of all other atoms. The term of electron–electron interaction is treated in a mean field as a constant background from the first-order approximation. R_C is the lattice constant. According to the nearly free-electron approximation, the E_G between the valence and the conduction bands originates from the crystal potential. The width of the gap depends on the integral of the crystal potential in combination with the Bloch wave of the nearly free electron $\phi(k_l, r)$:

$$E_G = 2|V_1(k_l)| \quad \text{and} \quad V_1(k_l) = \langle \phi(k_l, r)|V(r + R_C)|\phi(k_l, r)\rangle \tag{1.5}$$

where k_l is the wave-vector and $k_l = 2l\pi/R_C$. Actually, the E_G is simply twice the first Fourier coefficient of the crystal potential.

As illustrated in Figure 1.1, the energy levels of an isolated atom will evolve into energy bands when interatomic bonding is involved. When two atoms are bonded together, such as H_2 dimer-like molecules, the single energy level of the initially

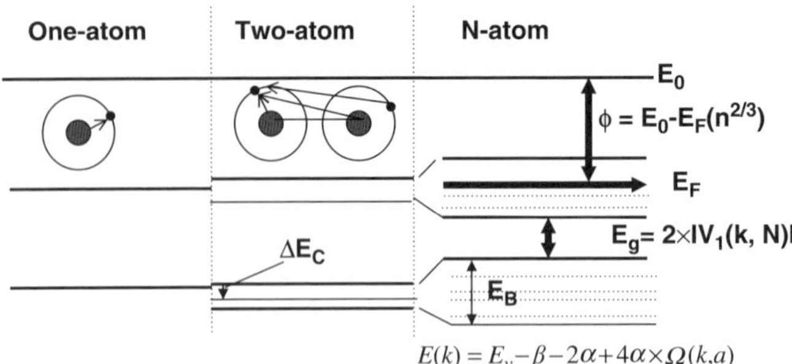

Figure 1.1. The involvement of interatomic interaction evolves a single energy level to the energy band when a particle grows from a single atom to a bulk solid that contains N_j atoms. Indicated is the work function ϕ, band gap E_G, core level shift ΔE_C, and bandwidth E_B. The number of allowed sublevels in a certain band equals the number of atoms of the solid (33).

isolated atom splits into two sublevels. The presence of interatomic interaction lowers the center of the two sublevels, which is called the core level shift. Increasing the number of atoms up to N_j, the single energy level will expand into a band within which there are N_j sublevels.

What distinguishes a nanosolid from a bulk chunk is that for the former the N_j is accountable, whereas for the latter, the N_j is too large to be accounted despite the portion of the under-coordinated atoms in the surface skin. Therefore, the classic band theories are valid for a solid over the whole range of sizes or for containing any number of atoms. As detected with X-ray photoelectron spectroscopy (XPS), the density-of-states (DOS) of the valence region for a nanosolid exhibits band-like features rather than the discrete spectral lines of a single atom. If the N_j is sufficiently small, the separation between the sublevels is resolvable. The energy level spacing between the successive sublevels in the valence band, known as the Kubo gap ($\delta_K = 4E_F/3N_j$), decreases with the increase of the number of valence electrons of the system N_j (31). Where E_F is the Fermi energy of the bulk. Because of the presence of the δ_K in an individual nanosolid, properties such as electron conductivity and magnetic susceptibility exhibit quantized features (32).

According to the tight-binding approximation, the energy dispersion of an electron in the vth core band follows the relation:

$$\begin{aligned} E_v(k) &= E_v(1) + \Delta E_v(\infty) + \Delta E_B(k_l, R_C, z) \\ &= E_v(1) - (\beta + 2\alpha) + 4\alpha\Omega(k_l, R_C, z) \end{aligned} \tag{1.6}$$

where

$E_v(1) = \langle \phi_v(r)|\hat{H}_0|\phi_v(r)\rangle$ is the vth energy level of an isolated atom.

$\beta = -\langle \phi_v(r)|V_{cry}(r)|\phi_v(r)\rangle$ is the crystal potential effect on the specific core electron at site r.

$\alpha = -\langle\phi_v(r - R_C)|V_{\text{cry}}(r - R_C)|\phi_v(r - R_C)\rangle$ is the crystal potential effect on the coordinate neighboring electrons. For an fcc structure example, the structure factor $\Omega(k_l, R_C) = \sum_z \sin^2(k_l R_C/2)$. The sum is over all contributing coordinates (z) surrounding the specific atom in the solid.

Equations 1.5 and 1.6 indicate that the E_G, the energy shift $\Delta E_v(\infty) = -(\beta + 2\alpha)$ of the $E_v(1)$ and the bandwidth ΔE_B (last term in Eq. 5.16) are all functions of interatomic interaction or crystal potential. Any perturbation to the crystal potential will vary these quantities accordingly as the change of the Block wavefunction is negligible in the first-order approximation. The band structure has nothing to do with the actual occupancy of the particular orbitals or events such as electron-hole pair creation or recombination, or the processes of PL and PA that involve the electron–phonon coupling effect. Without interatomic interaction, neither the E_G expansion nor the core-level shift would be possible; without the interatomic binding, neither a solid nor even a liquid would form.

If one intends to modify the properties of a solid, one has to find ways of modulating the crystal potential physically or chemically. Bond nature alteration by chemical reaction or bond length relaxation by size reduction, as discussed in the following sections, will be the effective ways of modulating the interatomic potential.

1.2.1.4. Atomic Cohesive Energy and Thermal Stability Another key concept is the cohesive energy per discrete atom. The binding energy density per unit volume contributes to the Hamiltonian that determines the entire band structure and related properties, whereas the atomic cohesive energy determines the activation energy for thermally and mechanically activated processes, including self-assembly growth, phase transition, solid–liquid transition, evaporation, atomic dislocation, diffusion, and chemical reaction.

The cohesive energy (E_{coh}) of a solid containing N_j atoms equals to the energy dividing the crystal into individually isolated atoms by breaking all bonds of the solid. If no atomic CN reduction is considered, the E_{coh} is the sum of bond energy over all z_b coordinates of all N_j atoms:

$$E_{\text{coh}}(N_j) = \sum_{N_j} \sum_{z_i} E_i \cong N_j z_b E_b = N_j E_B \tag{1.7}$$

The cohesive energy for a single atom E_B is the sum of the single bond energy E_b over the atomic CN, $E_B = z_b E_b$. One may consider a thermally activated process such as phase transition in which all bonds are loosened to a certain extent due to thermal activation. The energy required for such a process is a certain portion of the atomic E_B although the exact portion may change from process to process. If one considers the relative change to the bulk value, the portion will not be accounted. This approximation is convenient in practice, as one should be concerned with the origins and the trends of changes. Therefore, bulk properties such as the thermal stability of a solid could be related directly to the atomic cohesive energy—the product of the bond number and bond energy of the specific atom.

1.2.2. Boundary Conditions

1.2.2.1. Barrier Confinement vs. Quantum Uncertainty The termination of lattice periodicity in the surface normal direction has two effects. One is the creation of the surface potential barrier (SPB), work function, or contact potential, and the other is the reduction of the atomic CN. The SPB is the intrinsic feature of a surface, which confines only electrons that are freely moving inside the solid. However, the SPB has nothing to do with the *strongly localized* electrons in deeper core bands or with those form sharing electron pairs in a bond. The localized electrons do not suffer such barrier confinement at all as the localization length is far shorter than the particle size.

According to the principle of quantum uncertainty, reducing the dimension (D) of the space inside which energetic particles are moving increases the fluctuation, rather than the average value, of the momentum p or kinetic energy E_k of the moving particles:

$$\begin{aligned} \Delta p D &\geq \hbar/2 \\ p &= \bar{p} \pm \Delta p \\ \overline{E_k} &= \bar{p}^2/(2\mu) \end{aligned} \tag{1.8}$$

where $\hbar$ being the Plank constant corresponds to the minimal quanta in energy and momentum spaces and μ is the effective mass of the moving particles. The kinetic energy of a freely moving carrier is increased by a negligible amount due to the confinement effect on the fluctuation that follows the principle of quantum uncertainty.

1.2.2.2. Atomic CN Reduction The atomic CN reduction is referred to the standard value of 12 in the bulk of an fcc structure irrespective of the bond nature or the crystal structure. Atomic CN reduction is referred to an atom with a coordinate less than the standard value of 12. The CN is 2 for an atom in the interior of a monatomic chain or an atom at the open end of a single-walled carbon nanotube (CNT); while in the CNT wall, the CN is 3. For an atom in the fcc unit cell, the CN varies from site to site. The CN of an atom at the edge or corner differences from the CN of an atom in the plane or the central of the unit cell. Atoms with deformed bond lengths or deviated angles in the CNT are the same as those in amorphous states that are characterized with the band tail states (34). For example, the effective CN of an atom in diamond tetrahedron is the same as that in an fcc structure as a tetrahedron unit cell is an interlock of two fcc unit cells. The CN of an atom in a highly curved surface is even lower compared with the CN of an atom at a flat surface. For a negatively curved surface (such as the inner side of a pore or a bubble), the CN may be slightly higher than that of an atom at the flat surface. Therefore, from the atomic CN reduction point of view, there is no substantial difference in nature among a nanosolid, a nanopore, and a flat surface. This premise can be extended to the structural defects or defaults, such as voids; atoms surrounding these defects also suffer from CN reduction. Unlike a nansolid with ordered CN reduction at the surface, an amorphous solid possesses defects that are distributed randomly.

1.2.3. Surface-to-Volume Ratio

It is easy to derive the volume or number ratio of a certain atomic layer, denoted i, to that of the entire solid by differentiating the natural logarithm of the volume:

$$\gamma_{ij} = \frac{N_i}{N_j} = \frac{V_i}{V_j} = dLn(V_j) = \frac{\tau dR_j}{R_j} = \frac{\tau c_i}{K_j} \tag{1.9}$$

where $K_j = R_j/d_0$ is the dimensionless form of size, which is the number of atoms lined along the radius of a spherical dot ($\tau = 3$), a rod ($\tau = 2$), or cross the thickness of a thin plate ($\tau = 1$). The volume of a solid is proportional to R_j^τ. For a hollow system, the γ_{ij} should count both external and internal sides of the hollow structure.

With reducing particle size, the performance of surface atoms becomes dominant because at the lower end of the size limit, $(K_j \to \tau c_i)\gamma_1$ approaches unity. At $K_j = 1$, the solid will degenerate into an isolated atom. Therefore, the γ_{ij} covers the whole range of sizes and various shapes. The definition of dimensionality (τ) herein differs from the convention in transport or quantum confinement considerations in which a nanosphere is zero-dimension (quantum dot), a rod as one dimension (quantum wire), and a plate two dimension (quantum well). If we count atom by atom, the number ratio and the property change will show quantized oscillation features at smaller sizes, which varies from structure to structure (35).

1.2.4. BOLS Correlation

1.2.4.1. Bond Order-Length Correlation As the consequence of bond order loss, the remaining bonds of the under-coordinated atoms contract spontaneously. As asserted by Goldschmidt (36) and Pauling (37), the ionic and the metallic radius of the atom would shrink spontaneously if the CN of an atom is reduced. The CN reduction-induced bond contraction is independent of the nature of the specific bond or structural phases (38). For instances, a 10% contraction of spacing between the first and second atomic surface layers has been detected in the liquid phase of Sn, Hg, Ga, and In (39). A substitutional dopant of As impurity has induced 8% bond contraction around the impurity at the Te sublattice in CdTe has also been observed using extended X-ray absorption fine structure (EXAFS) and X-ray absorption near edge spectroscopy (XANES) (40). Therefore, bond order loss-induced bond contraction is universal.

Figure 1.2*a* illustrates the CN dependence of bond length. The solid curve $c_i(z_i)$ formulates the Goldschmidt premise, which states that an ionic radius contracts by 12%, 4%, and 3%, if the CN of the atom reduces from 12 to 4, 6 and 8, respectively. Feibelman (41) has noted a 30% contraction of the dimer bond of Ti and Zr, and a 40% contraction of the dimer-bond of Vanadium, which is also in line with the formulation.

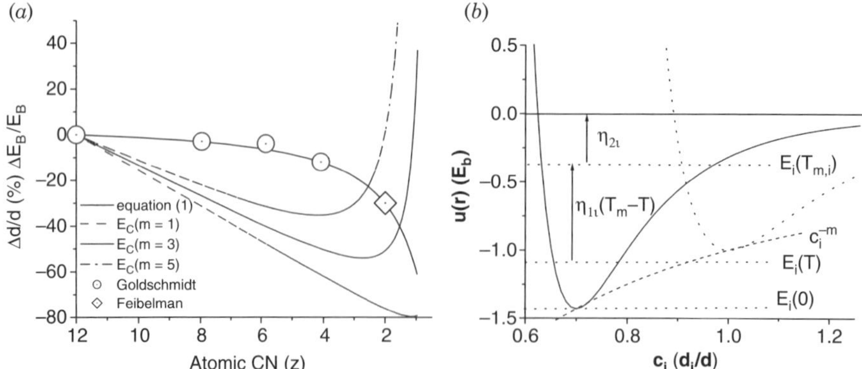

Figure 1.2. Illustration of the BOLS correlation. Solid curve in (*a*) is the contraction coefficient c_i derived from the notations of Goldschmidt (36) (open circles) and Feibelman (41) (open square). As a spontaneous process of bond contraction, the bond energy at equilibrium atomic separation will rise in absolute energy $E_i = c_i^{-m} E_b$. The m is a parameter that represents the nature of the bond. However, the atomic cohesive energy $z_i E_i$ changes with both the m and the z_i values. (*b*) Atomic CN reduction modified pairing potential energy. CN reduction causes the bond to contract from one unit (in d_0) to c_i, and the cohesive energy per coordinate increases from one unit to c_i^{-m} unit. Separation between $E_i(T)$ and $E_i(0)$ is the thermal vibration energy. Separation between $E_i(T_{m,i})$ and $E_i(T)$ corresponds to melting energy per bond at T, which dominates the mechanical strength. $T_{m,i}$ is the melting point. η_{2i} is $1/z_i$ fold energy atomizing an atom in molten state.

1.2.4.2. Bond Length-Strength Correlation As the atomic CN reduction-induced bond contraction is a spontaneous process, the system energy will be lowered with an association of bond strength gain. The contraction coefficient and the associated bond energy gain form the subject of the BOLS correlation mechanism that is formulated as

$$\begin{cases} c_i(z_i) = d_i/d_0 = 2/\{1 + \exp[(12 - z_i)/(8z_i)]\} & \text{(BOLS-coefficient)} \\ E_i = c_i^{-m} E_b & \text{(Single-bond-energy)} \\ E_{B,i} = z_i E_i & \text{(Atomic-cohesive-energy)} \end{cases} \tag{1.10}$$

Subscript i denotes an atom in the ith atomic layer, which is counted up to three from the outermost atomic layer to the center of the solid as no CN-reduction is expected for $i > 3$. The index m is a key indicator for the nature of the bond. Experience (42) revealed that for Au, Ag, and Ni metals, $m \equiv 1$; for alloys and compounds, m is around four; for C and Si, the m has been optimized to be 2.56 (43) and 4.88 (44), respectively. The m value may vary if the bond nature evolves with atomic CN (45). If the surface bond expands in cases, we simply expand the c_i from a value that is smaller than unity to greater, and the m value from positive to negative to represent the spontaneous process of which the system energy is minimized. The $c_i(z_i)$ depends on the effective CN rather than a certain order of CN. The z_i also varies with the particle size due to the change of the surface curvature. The z_i takes the following values (44):

$$z_1 = \begin{cases} 4(1 - 0.75/K_j) & \text{curved-surface} \\ 4 & \text{flat-surface} \end{cases} \tag{1.11}$$

Generally, $z_2 = 6$ and $z_3 = 8$ or 12.

Figure 1.2*b* illustrates schematically the BOLS correlation using a simple interatomic pairing potential, $u(r)$. When the CN of an atom is reduced, the equilibrium atomic distance will contract from one unit (in d_0) to c_i and the cohesive energy of the shortened bond will increase in magnitude from one unit (in E_b) to c_i^{-m}. The solid and the broken $u(r)$ curves correspond to the pairing potential with and without CN reduction. The BOLS correlation has nothing to do with the particular form of the pairing potential as the approach involves only atomic distance at equilibrium. The bond length-strength correlation herein is consistent with the trend reported by Bahn and Jacobsen (46) although the extent of bond contraction and energy enhancement therein vary from situation to situation.

Several characteristic energies in Figure 1.2*b* correspond to the following facts:

(1) $T_{m,i}$ being the local melting point is proportional to the cohesive energy $z_iE_i(0)$ (47) per atom with z_i coordinate (48).
(2) Separation between $E = 0$ and $E_i(T)$, or $\eta_{1i}(T_{m,i} - T) + \eta_{2i}$, corresponds to the cohesive energy per coordinate E_i at T, being energy required for bond fracture under mechanical or thermal stimulus. η_{1i} is the specific heat per coordinate.
(3) The separation between $E = 0$ and $E_i(T_m)$, or η_{2i}, is the $1/z_i$ fold energy that is required for atomization of an atom in molten state.
(4) The spacing between $E_i(T)$ and $E_i(0)$ is the vibration energy purely due to thermal excitation.
(5) The energy contributing to mechanical strength is the separation between the $E_i(T_m)$ and the $E_i(T)$, as a molten phase is extremely soft and highly compressible (49).

Values of η_{1i} and η_{2i} can be obtained with the known c_i^{-m} and the bulk η_{1b} and η_{2b} values that vary only with crystal structures as given in Table 1.1.

1.2.4.3. Densification of Mass, Charge, and Energy Figure 1.3 compares the potential well in the QC convention with that of the BOLS for a nanosolid. The QC convention extends the monotrapping potential of an isolated atom by

TABLE 1.1. Relation Between the Bond Energy E_b and the T_m of Various Structures (24). $\eta_{2b} < 0$ for an fcc Structure Means that the Energy Required for Breaking All Bonds of an Atom in Molten State is Included in the Term of $\eta_{1b}zT_m$, and Therefore, the η_{2b} Exaggerates the Specific Heat per CN.

$E_b = \eta_{1b}T_m + \eta_{2b}$	fcc	bcc	Diamond Structure
$\eta_{1b}(10^{-4}\ \mathrm{eV/K})$	5.542	5.919	5.736
$\eta_{2b}(\mathrm{eV})$	−0.24	0.0364	1.29

expanding the size from d_0 to D. BOLS scheme covers contribution from individual atoms that are described with multi-trapping-center potential wells and the effect of atomic CN reduction on the surface skin. Atomic CN reduction-induced bond-strength gain depresses the potential well of trapping in the surface skin. Therefore, the density of charge, energy, and mass in the relaxed surface region are higher than other sites inside the solid. Consequently, surface stress that is in

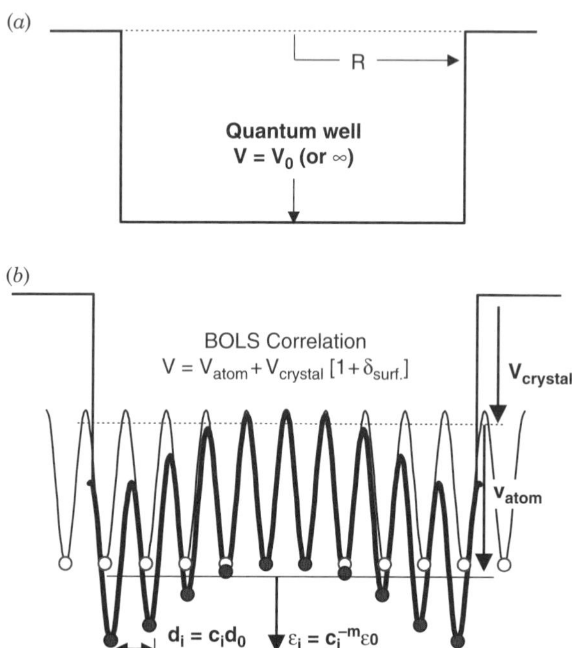

Figure 1.3. Schematic illustration of conventional quantum well (*a*) with a monotrapping center extended from that of a single atom, and the BOLS-derived nanosolid potential (*b*) with multi-trap centers and CN reduction-induced features. In the relaxed surface region, the density of charge, energy, and mass will be higher than other sites due to atomic CN reduction.

the dimension of energy density will increase in the relaxed region. Electrons in the relaxed region are more localized because of the depression of the potential well of trapping, which lowers the work function and conductivity in the surface region due to boundary scattering (50,51), but it enhances the angular momentum of the surface atoms (35).

1.2.4.4. Oxide Long-Range Interaction For an oxide nanosoild, the long-order dipole–dipole interaction is involved, which may change the Madelung potential in the ionic oxide nanosolid (52). The value of Madelung potential shifts their effective electronic levels, thus renormalizing their electronegativity. The concept of electronegativity in the oxide nanosolid is very useful for discussing the ability of under-coordinated atoms to bind to adatoms or molecules. The Madelung potential value also enters the energetic balance that fixes the local atomic arrangement, and its long-range character may induce peculiar effects, for example, the stabilization of noncompact structures especially in small clusters. These aspects are akin to insulating compounds with a non-negligible ionic character of the oxygen-cation bonding and are not met in metals or compound semiconductors in low dimensionality.

The immediate effect of long-range interaction is its contribution to the atomic cohesive energy that dominates the ferroelectric properties of oxides (53). For a spherical dot with radius $R = Kd$, we need to consider the interaction between the specific central atom and its surrounding neighbors within the critical volume $V_C = 4\pi K_C^3/3$, in addition to the BOLS correlation in the surface region, as illustrated in Figure 1.4. The ferroelectric property drops down from the bulk value to a value smaller than $5/16$ (estimated from Figure 1.4) when one goes from the central atom to the edge along the radius to the correlation radius K_C. If the surrounding volume of the central atom is smaller than the critical V_C, the ferroelectric feature of this central atom attenuates; otherwise, the bulk value remains. For an atom in the ith surface layer, the number of the lost exchange bonds is proportional to the volume V_{vac} that is the volume difference between the two caps of the V_C-sized spheres as illustrated in Figure 1.4*a*. Therefore, the relative change of the ferroelectric exchange energy of an atom in the ith atomic layer compared with that of a bulk atom becomes

$$\frac{\Delta E_{exc,i}}{E_{exc}(\infty)} = \frac{V_C - V_{vac}}{V_C} - 1 = -\frac{V_{vac}}{V_C} \tag{1.12}$$

1.2.5. Shape-and-Size Dependency

1.2.5.1. Scaling Relation Generally, the mean relative change of a measurable quantity of a nanosolid containing N_j atoms, with dimension K_j, can be expressed as $Q(K_j)$ and as $Q(\infty)$ for the same solid without contribution from bond order loss. The correlation between the $Q(K_j)$ and $Q(\infty) = N_j q_0$ and the relative change of Q

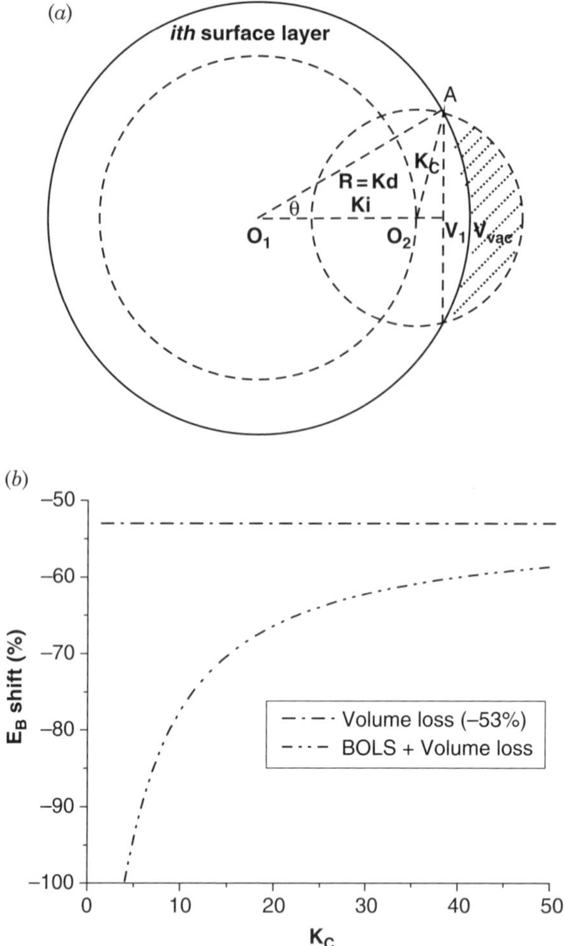

Figure 1.4. (*a*) Schematic illustration of the long-range exchange bonds lost in an atom in a spherical oxide nanosolid with radius K ($K = R/d$ is the number of atoms of size d lined along the radius R of a sphere or cross a thin plate of R thick). K_C is the critical correlation radius. The volume loss V_{vac} (the shaded portion) is calculated by differencing the volumes of the two spherical caps:

$$V_{\text{vac}} = \pi(K_C + K_i - K\cos\theta)^2 \left(K_C - \frac{K_C + K_i - K\cos\theta}{3}\right) - \pi(K - K\cos\theta)^2 \left(K - \frac{K - K\cos\theta}{3}\right)$$

where the angle θ is determined by the triangle O_1O_2A. (*b*) Correlation radius K_C dependence of the atomic cohesive energy. For the $K_C = 5$ example, the BOLS lowers the E_B by -41.1% (follows the curve in Figure 1.4*a*), and the long-range bond loss contribution that is a constant lowers the E_B by -53% (53).

caused by bond order loss is given as

$$
\begin{aligned}
Q(K_j) &= N_j q_0 + N_s(q_s - q_0) \\
\frac{\Delta Q(K_j)}{Q(\infty)} &= \frac{Q(K_j) - Q(\infty)}{Q(\infty)} = \frac{N_s}{N_j}\left(\frac{q_s}{q_0} - 1\right) \\
&= \sum_{i \le 3} \gamma_{ij}(\Delta q_i / q_0) = \Delta_{qj}
\end{aligned}
\tag{1.13}
$$

The weighting factor γ_{ij} represents the geometrical contributions from dimension (K_j) and dimensionality (τ) of the solid, which determines the magnitude of change. The quantity $\Delta q_i/q_0$ is the origin of change. The $\sum_{i \le 3} \gamma_{ij}$ drops in a K_j^{-1} fashion from unity to infinitely small when the solid grows from atomic level to infinitely large. For a spherical dot at the lower end of the size limit, $K_j = 1.5(K_j d_0 = 0.43\,\text{nm}$ for an Au spherical dot example), $z_1 = 2$, $\gamma_{1j} = 1$, and $\gamma_{2j} = \gamma_{3j} = 0$, which is identical in situation to an atom in a monatomic chain (MC) despite the orientation of the two interatomic bonds. Actually, the bond orientation is not involved in the modeling iteration. Therefore, the performance of an atom in the smallest nanosolid is a mimic of an atom in an MC of the same element without the presence of an external stimulus such as stretching or heating. At the lower end of the size limit, the property change of a nanosolid relates directly to the behavior of a single bond.

Generally, experimentally observed size-and-shape dependence of a detectable quantity follows a scaling relation. Equilibrating the scaling relation to Eq. 1.10, one has

$$
Q(K_j) - Q(\infty) = \begin{cases} bK_j^{-1} & \text{(measurement)} \\ Q(\infty) \times \Delta_{qj} & \text{(theory)} \end{cases}
\tag{1.14}
$$

where the slope $b \equiv Q(\infty) \times \Delta_{qj} \times K_j \cong$ constant is the focus of various modeling pursues. The $\Delta_j \propto K_j^{-1}$ varies simply with the $\gamma_{ij}(\tau, K_j, c_i)$ if the functional dependence of $q(z_i, c_i, m)$ on the atomic CN, bond length, and bond energy is given.

1.2.5.2. Cohesive Energy Modification The heat energy required for loosening an atom is a certain portion of the atomic E_B that varies with not only the atomic CN but also the bond strength. The variation of the mean E_B with size is responsible for the fall (supercooling) or rise (superheating) of the T_C (critical temperature for melting, phase transition, or evaporation) of a surface and a nanosolid. The E_B is also responsible for other thermally activated behaviors such as phase transition, catalytic reactivity, crystal structural stability, alloy formation (segregation and diffusion), and stability of electrically charged particles (Coulomb explosion). The cohesive energy also determines crystal growth and atomic diffusion and atomic gliding displacement that determines the ductility of nanosolids.

Considering both the BOLS correlation in the surface region and the long-range bond loss, we have a universal form for the cohesive energy suppression for an oxide nanosolid:

$$\frac{\Delta E_B(K_j)}{E_B(\infty)} = \begin{cases} \sum_{i\leq 3} \gamma_{ij}(z_{ib}c_i^{-1} - 1) = \Delta_B & \text{(BOLS)} \\ \sum_{i\leq K_C} \gamma_{ij}\left(\frac{-V_{\text{vac}}}{V_C}\right) + \Delta_B = \Delta_{\text{COH}} & \text{(BOLS + long − range)} \end{cases} \tag{1.15}$$

For the short spin–spin interaction, it is sufficient to sum over the outermost three atomic layers; for a ferroelectric and a superconductive solid, the sum should be within the sphere of radius K_C. The BOLS correlation considers only contribution from atoms in the shells of the surface skin, whereas the long-range contribution involves the concept of correlation radius that is used in ferroelectric systems. The BOLS contribution is obtained by considering the shell structures:

$$\begin{aligned} \langle E_{\text{coh}}(N_j)\rangle &= N_j z_b E_b + \sum_{i\leq 3} N_i(z_i E_i - z_b E_b) \\ &= N_j E_B(\infty) + \sum_{i\leq 3} N_i z_b E_b(z_{ib}E_{ib} - 1) \\ &= E_{\text{coh}}(\infty)\left[1 + \sum_{i\leq 3}\gamma_{ij}(z_{ib}c_i^{-m} - 1)\right] = E_{\text{coh}}(\infty)(1 + \Delta_B) \end{aligned}$$

or

$$\Delta E_B(K_j)/E_B(\infty) = \sum_{i\leq 3}\gamma_{ij}(z_{ib}c_i^{-m} - 1) = \Delta_B \tag{1.16}$$

where $E_{\text{coh}}(\infty) = N_j z_b E_b$ represents the ideal situation without *CN reduction*. The $z_{ib} = z_i/z_b$ is the normalized *CN*, and the $E_{ib} = E_i/E_b \cong c_i^{-m}$ is the normalized binding energy per coordinate of a surface atom. For an isolated surface, $\Delta_B < 0$; for an intermixed interface, Δ_B may be positive depending on the strength of interfacial interaction. Therefore, the relative change of $T_C(K_j)$ and activation energy $E_A(K_j)$ for thermally and mechanically activated process can be expressed as

$$\frac{\Delta T_C(K_j)}{T_C(\infty)} = \frac{\Delta E_A(K_j)}{E_A(\infty)} = \frac{\Delta E_B(K_j)}{E_B(\infty)} = \Delta_B(K_j) \tag{1.17}$$

Interestingly, the critical temperature for sensoring operation could be lowered from 970 to 310 K of $SrTiO_3$ by ball milling to obtain 27 nm-sized powders (54). The resistivity of the $SrTiO_3$ increases when the $SrTiO_3$ particle size is decreased (55). Decreasing the particle sizes of ferroelectric $BaTiO_3$ could lower the T_C to 400 K and the refractive index (dielectric constant) and, hence, the transmittance of $BaTiO_3$ infilled SiO_2 photonic crystals, as a consequence (56,57). The suppression of the critical temperatures for sensoring and phase transition results and the modulation of resistivity and refractive index could be consequences of energy densification and cohesive energy suppression in the surface skin.

1.2.5.3. Hamiltonian Perturbation The perturbation to the energy density in the relaxed region that contributes to the Hamiltonian upon assembly of the nanosolids is

$$\begin{aligned}\Delta_H(K_j) &= \frac{V_{\text{cry}}(r, n, N_j)}{V_{\text{cry}}(d_0, n, N_j)} - 1\\ &= \sum_{i\leq 3} \gamma_{ij} \frac{\Delta v(d_i)}{v(d_0)} + \delta_{kj}\\ &= \sum_{i\leq 3} \gamma_{ij}(c_i^{-m} - 1) + \delta_{kj}\end{aligned}$$

where

$$\delta_{kj} = \frac{nV(K_j)}{N_j^2 v(d_0)} \tag{1.18}$$

With the perturbation, the $\hat{H}'$ in Eq. 1.4 becomes $\hat{H}'(\Delta_H) = V_{\text{cry}}(r)[1 + \Delta_H(K_j)]$, which dictates the change of not only the E_G width, but also the core-level energy:

$$\frac{\Delta E_G(K_j)}{E_G(\infty)} = \frac{\Delta E_v(K_j)}{E_v(\infty)} = \Delta_H(K_j) \tag{1.19}$$

where $\Delta E_v(K_j) = E_v(K_j) - E_v(1)$. This relation also applies to other quantities such as the bandwidth and band tails (33).

Most strikingly, without triggering electron–phonon interaction or electron–hole generation, the scanning tunneling microscopy/spectroscopy (STM/S) measurement at low temperature revealed that the E_G of Si nanorod varies from 1.1 to 3.5 eV with decreasing the rod diameter from 7.0 to 1.3 nm associated with ~12% Si–Si bond contraction from the bulk value (0.263 nm) to ~0.23 nm. The STS findings concur excitingly with the BOLS premise: CN reduction shortens the remaining bonds of the under-coordinated atoms spontaneously with an association of E_G expansion.

1.2.5.4. Electron–Phonon Coupling The electron–phonon (e–p) interaction contributes to the processes of photoemission, photoabsorption, photoconduction, and electron polarization that dominates the static dielectric constant. Figure 1.5 illustrates the effect of e–p coupling and crystal binding on the energy of photoluminescence and absorbance E_{PL} and E_{PA}. The energies of the ground state (E_1) and the excited

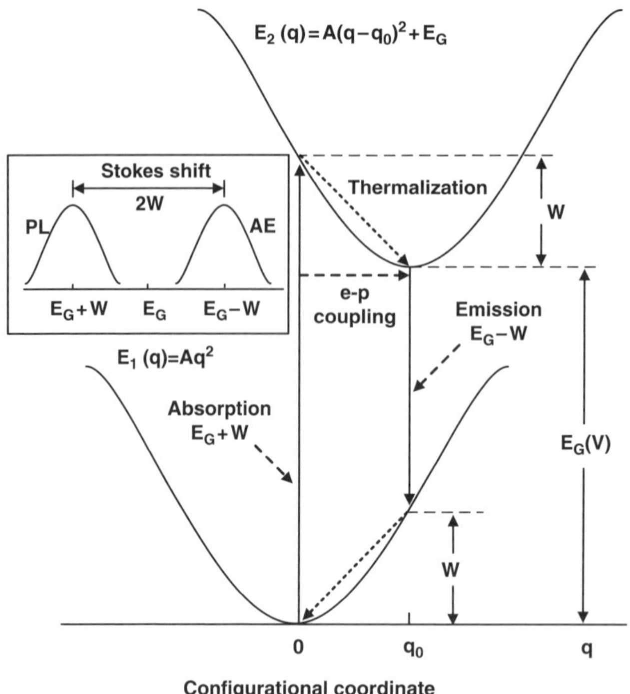

Figure 1.5. Mechanisms for E_{PA} and E_{PL} of a nano-semiconductor, involving crystal binding (E_G) and electron–phonon coupling (W). Insertion illustrates the Stokes shift from E_{PA} to E_{PL}. Electron is excited by absorbing a photon with energy $E_G + W$ from the ground minimum to the excited state, then undergoes a thermalization to the excited minimum, and then transmits to the ground emitting a photon with energy $E_G - W$ (58).

state (E_2) are expressed in parabola forms (34):

$$\begin{cases} E_1(q) = Aq^2 \\ E_2(q) = A(q - q_0)^2 + E_G \end{cases} \tag{1.20}$$

Constant A is the slope of the parabolas. The q is in the dimension of wave-vector. The vertical distance between the two minima is the true E_G that depends uniquely on the crystal potential. The lateral displacement (q_0) originates from the e–p coupling that can be strengthened by lattice contraction. Therefore, the blue shift in the E_{PL} and in the E_{PA} is the joint contribution from crystal binding and e–p coupling.

In the process of carrier formation, or electron polarization (34), an electron is excited by absorbing a photon with $E_G + W$ energy from the ground minimum to the excited state with creation of an electron-hole pair. The excited electron then undergoes a thermalization and moves to the minimum of the excited state, and eventually it transmits to the ground and combines with the hole. The carrier recombination is associated with emission of a photon with energy $E_{PL} = E_G - W$. The transition processes (e-h pair production and recombination) follow the rule of momentum and energy conservation, although the conservation law may be subject to relaxation for the short-ordered nanosolid. Relaxation of the conservation law is responsible for the broad peaks in the PA and PL.

The insertion illustrates the Stokes shift, $2W = 2Aq_0^2$, or the separation from E_{PL} to E_{PA}. The q_0 is inversely proportional to atomic distance d_i, and hence, $W_i = A/(c_i d_i)^2$, in the surface region. Based on this premise, the blue shift of the E_{PL}, the E_{PA}, and the Stokes shift can be correlated to the CN reduction-induced bond contraction (58).

$$\left.\begin{array}{l}\dfrac{\Delta E_{PL}(K_j)}{E_{PL}(\infty)}\\[2ex] \dfrac{\Delta E_{PA}(K_j)}{E_{PA}(\infty)}\end{array}\right\} = \frac{\Delta E_G(K_j) \mp \Delta W(K_j)}{E_G(\infty) \mp W(\infty)} \cong \sum_{i\leq 3} \gamma_i[(c_i^{-m} - 1) \mp B(c_i^{-2} - 1)]$$

$$= \Delta_H \mp B\Delta_{e-p}$$

$$\left(B = \frac{A}{E_G(\infty)d^2}; \quad \frac{W(\infty)}{E_G(\infty)} \approx \frac{0.007}{1.12} \approx 0\right) \qquad (1.21)$$

Compared with the bulk $E_G(\infty) = 1.12$ eV for silicon, the $W(\infty) \sim 0.007$ eV obtained using tight-binding calculations (59) is negligible. One can easily calculate the size-dependent E_{PL}, E_{PA}, and $E_G = (E_{PL} + E_{PA})/2$ using Eq. 21. Fitting the measured data gives the values of m and A for a specific semiconductor.

1.2.5.5. Mechanical Strength The mechanical yield strength is the strain-induced internal energy deviation that is proportional to energy density or the sum of bond energy per unit volume (17). Considering the contribution from heating, the strength (stress, flow strength), the Young's modulus, and the compressibility (under compressive stress) or extensibility (under tensile stress) at a given temperature can be expressed by

$$P_i(z_i, T) = -\left.\frac{\partial u(r, T)}{\partial V}\right|_{d_i, T} \sim \frac{N_i \eta_{1i}(T_{m,i} - T)}{d_i^\tau}$$

$$\beta_i(z_i, T) = -\left.\frac{\partial V}{V\partial P}\right|_T = [Y_i(z_i, T)]^{-1} = \left[-V\left.\frac{\partial u^{2^{-1}}(r, T)}{\partial V^2}\right|_T\right]^{-1}$$

$$= \frac{d_i^\tau}{N_i \eta_{1i}(T_{m,i} - T)} = [P_i(z_i, T)]^{-1} \qquad (1.22)$$

β is an inverse of dimension of the Young's modulus or the hardness. N_j is the total number of bonds in d^τ volume. If calibrated with the bulk value at T and using the size-dependent specific heat, melting point, and lattice parameter, the temperature, bond nature, and size-dependent strength and compressibility of a nanosolid will be

$$\frac{P(K_j,T)}{P(\infty,T)} = \frac{Y(K_j,T)}{Y(\infty,T)} = \frac{\beta(\infty,T)}{\beta(K_j,T)} = \frac{\eta_1(K_j)}{\eta_1(\infty)}\left(\frac{d(\infty)}{d(K_j)}\right)^3 \times \frac{T_m(K_j,m)-T}{T_m(\infty)-T} \quad (1.23)$$

The bond number density between the circumferential neighboring atomic layers does not change upon relaxation ($N_i = N_b$). Equation 1.23 indicates that the mechanical strength is dictated by the value of $T_m(K_j)-T$ and the specific heat per bond. At T far below the T_m, a surface or a nanostructure is harder than the bulk interior. However, the T_m drops with size K_j, and therefore, the surface or nanosolid become softer when the $T_m(K_j)-T$ value becomes smaller. This relation has led to quantification of the surface mechanical strength, the breaking limit of a single bond in monatomic chain, and the anomalous Hall–Petch relationship in which the mechanical strength decreases with size in a fashion of $D^{-0.5}$ and then deviates at the order of 10 nm from the Hall–Petch relationship (49).

1.2.6. Summary

If one could establish the functional dependence of a detectable quantity Q on atomic separation or its derivatives, the size dependence of the quantity Q is then certain. One can hence design a nanomaterial with desired functions based on such prediction. The physical quantities of a solid can be normally categorized as follows:

(1) Quantities that are directly related to bond length, such as the mean lattice constant, atomic density, and binding energy.

(2) Quantities that depend on the cohesive energy per discrete atom, $E_{B,i} = z_i E_i$, such as self-organization growth; thermal stability; Coulomb blockade; critical temperature for liquidation, evaporation, and phase transition of a nanosolid; and the activation energy for atomic dislocation, diffusion, and bond unfolding (60).

(3) Properties that vary with the binding energy density in the relaxed continuum region such as the Hamiltonian that determine the entire band structure and related properties such as band gap, core-level energy, photo-absorption, and photo-emission.

(4) Properties that are contributed from the joint effect of the binding energy density and atomic cohesive energy such as mechanical strength, Young's modulus, surface energy, surface stress, extensibility and compressibility of a nanosolid, as well as the magnetic performance of a ferromagnetic nanosolid.

Using the scaling relation and the BOLS correlation, we may derive solutions to predict the size and shape dependence of various properties. Typical samples are given in Table 1.2 and Figure 1.6.

TABLE 1.2. Summary of Functional Dependence of Various Quantities on Particle Size and Derived Information.

Quantity Q	$\Delta Q(K_j)/Q(\infty) = \Delta_q(K_j)$	Refs.	Comments
Lattice constant (d)	$\sum_{i\le 3} \gamma_{ij}(c_i - 1)$	42	Only outermost three atomic layers contribute
Bond energy (E_i) Band gap (E_G) Core-level shift (ΔE_v)	$\sum_{i\le 3} \gamma_{ij}(c_i^{-m} - 1) = \Delta_H$	29, 33	Δ_H—Hamiltonian perturbation
Electron–phonon coupling energy (Stokes shift, W)	$B\sum_{i\le 3} \gamma_{ij}(c_i^{-2} - 1) = B\Delta_{e-p}$	58	B-constant
Photoemission and photoabsorption energy (E_{PL}, E_{PA})	$\Delta_H \mp B\Delta_{e-p}$	58	$E_G = (E_{PA} + E_{PL})/2$
Critical temperature for phase transition (T_C); activation energy for thermally and mechanically activated processes	$\begin{cases} \sum_{i\le 3} \gamma_{ij}(z_{ib}c_i^{-1} - 1) = \Delta_B & \text{(Ferromagnetic)} \\ \sum_{i\le K_C} \gamma_{ij}\left(\frac{-V_{vac}}{V_C}\right) + \Delta_B = \Delta_{COH} & \text{(else)} \end{cases}$	47, 53	Δ_B—atomic cohesive perturbation V_{vac}—volume loss V_C—correlation volume

(Continued)

TABLE 1.2. *Continued.*

Quantity Q	$\Delta Q(K_j)/Q(\infty) = \Delta_q(K_j)$	Refs.	Comments
Mechanical strength and Yung's modulus of monatomic bond (P, Y)	$\frac{\eta_{1i} d^3 (T_{m,i} - T)}{\eta_{1b} d_i^3 (T_m(\infty) - T)} - 1$	61	η_1—specific heat per bond T_m—melting point
Inverse Hall–Petch relation; solid–semisolid–liquid transition	$\frac{\eta_1(K_j) d^3 (T_m(K_j) - T)}{\eta_{1b} d^3 (K_j)(T_m(\infty) - T)} - 1$	49	
Optical phonon frequency (ω)	$\sum_{i \le 3} \gamma_{ij} (z_{ib} c_i^{-(m/2+1)} - 1)$	62	
Fermi level	$\sum_{i \le 3} \gamma_{ij} (c_i^{-2\tau/3} - 1)$	63, 64	
Dielectric permittivity $(\chi = \varepsilon_r - 1)$	$\Delta_d - (\Delta_H - B\Delta_{e-p})$	65	

Typical samples of consistency are shown in Figure 1.6.

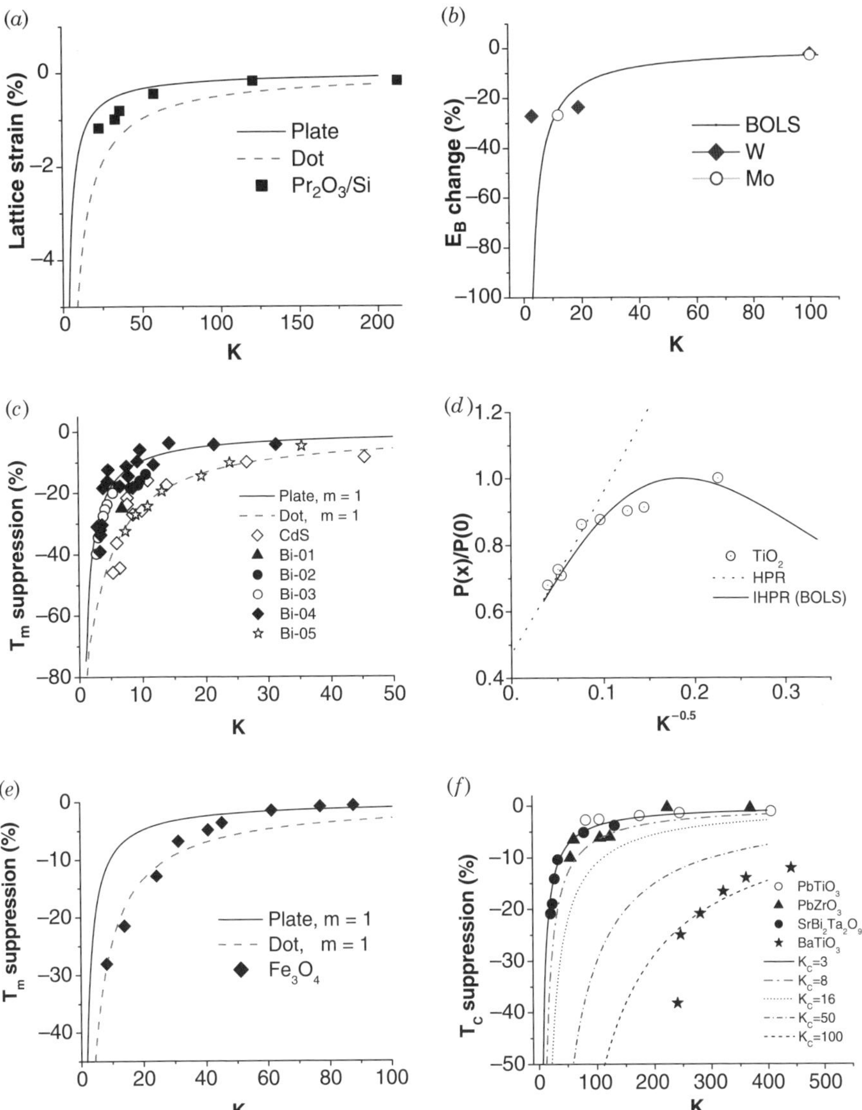

Figure 1.6. Comparison of BOLS predictions with measured size dependence of (*a*) Lattice contraction of Pr_2O_3 films on Si substrate (66). (*b*) Atomic cohesive energy of Mo and W (67). (*c*) Mechanical strength (Inverse Hall–Petch relationship, IHPR) of TiO_2 (68) nanosolids; straight line is the traditional Hall–Petch relationship (HPR). (*d*) T_m suppression of Bi (69–73) and CdS (74); (*e*) T_C suppression of ferromagnetic Fe_3O_4 nanosolids (75). (*f*) T_C suppression of ferroelectric $PbTiO_3$ (76), $SrBi_2Ta_2O_9$ (77), $BaTiO_3$ (78), and anti-ferroelectric $PbZrO_3$ (79) nanosolids. High-order CN reduction is considered for dipole–dipole interaction.

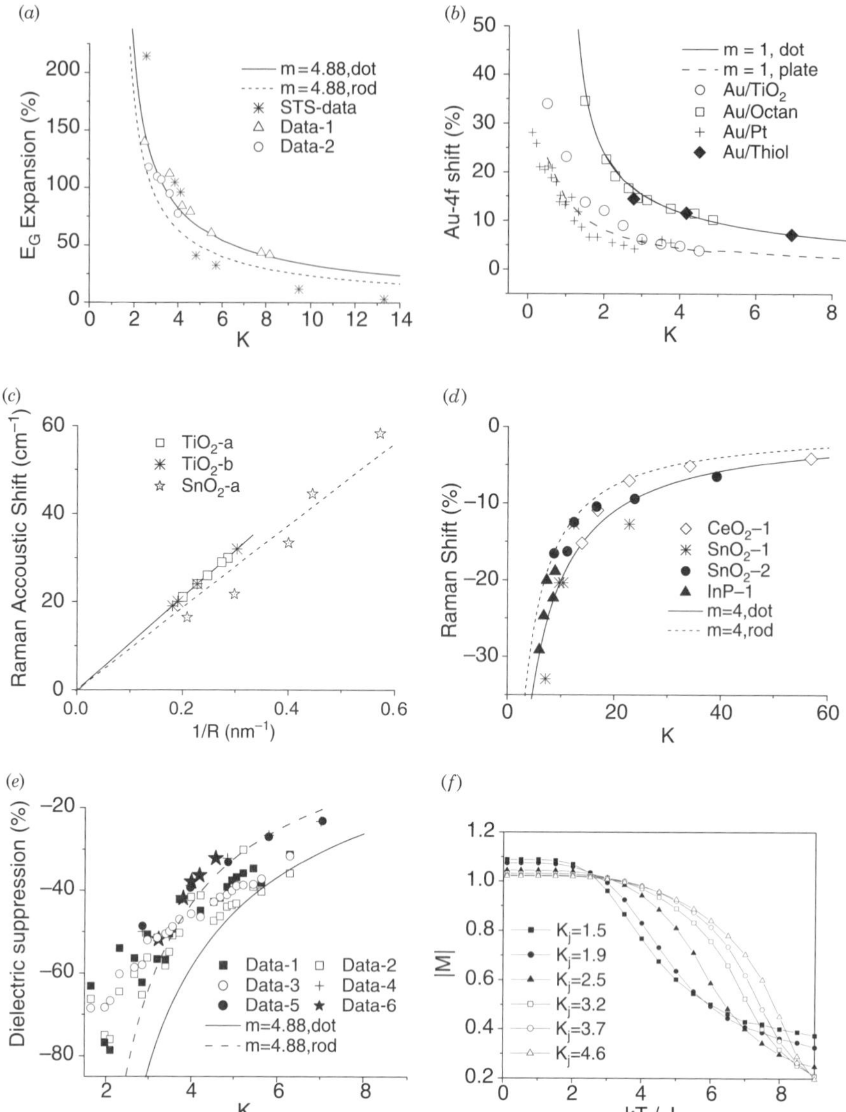

Figure 1.7. Comparison of BOLS predictions with measured size dependence of (*a*) E_G-expansion measured using STS (80) and optical method, Data-1 ($E_G = E_{PA} - W$) (81), Data-2 ($E_G = (E_{PL} + E_{PA})/2$) (82). (*b*) Core-level shift of Au capped with Thiol (83) and deposited on Octan (84) shows three-dimensional features, whereas core-level shift of Au deposited on TiO_2 (85) and Pt (86) show one-dimensional pattern. (*c*) Raman acoustic frequency shift of TiO_2-a and TiO_2-b (87) SnO_2-a (88) nanostructures caused by interparticle interaction. (*d*) Raman optical frequency shift of CeO_2 (89), SnO_2-1 (90), SnO_2-2 (88), InP (91). (*e*) Dielectric suppression of nanosolid silicon with Data 1, 2, and 3 (92); Data 4 and 5 (93); and Data-6 (94). (*f*) Temperature and size dependence of magnetization.

1.3. EFFECT OF OXIDATION

1.3.1. Bond–Band–Barrier (BBB) Correlation

The BBB correlation mechanism indicates that it is necessary for an atom of oxygen, nitrogen, and carbon to hybridize its *sp* orbitals upon interacting with atoms in the solid phase. Because of tetrahedron formation, nonbonding lone pairs, anti-bonding dipoles, and hydrogen-like bonds are produced, which add corresponding features to the DOS of the valence band of the host, as illustrated in Figure 1.8 (95). Bond forming alters the sizes and valences of the involved atoms and causes a collective dislocation of these atoms. Alteration of atomic valences roughens the surface, giving rise to corrugations of surface morphology. Charge transportation not only alters the nature of the chemical bond but also produces holes below the E_F and thus creates or enlarges the E_G (96). In reality, the lone-pair-induced metal dipoles often direct into the open end of a surface because of the strong repulsive forces among the lone pairs and among the dipoles. This dipole orientation leads to the surface dipole layer with lowered Φ. For a nitride tetrahedron, the single lone pair may direct into the bulk center, which produces an ionic layer at the surface. The ionic surface network deepens the well depth, or increases the Φ, as the host surface atoms donate their electrons to the electron acceptors. For carbide, no lone pair is produced, but the weak antibonding feature exists because of the ion-induced polarization. However, hydrogen adsorption neither adds DOS features to the valence band nor expands the E_G as hydrogen adsorption terminates the dangling bond at a surface, which minimizes the midgap impurity DOS of silicon, for instance (34).

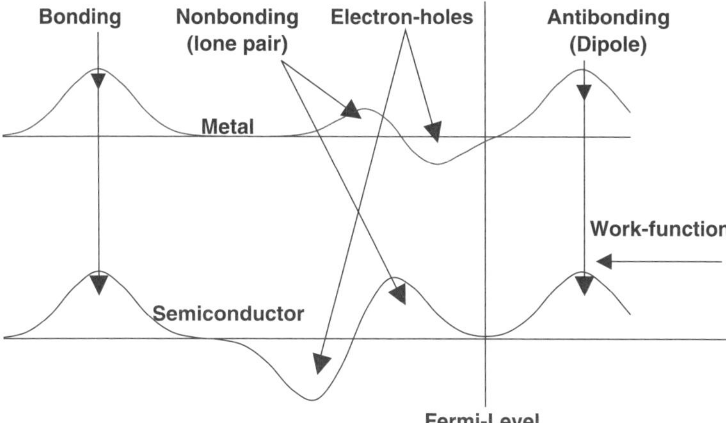

Figure 1.8. Oxygen-induced DOS differences between a compound and the parent metal (upper) or the parent semiconductor (lower). The lone-pair polarized anti-bonding state lowers the Φ, and the formation of bonding and anti-bonding generates holes close to E_F of a metal or near the valence band edge of a semiconductor. For carbide, no lone-pair features appear, but the ion-induced anti-bonding states will remain.

1.3.2. Experimental Evidence

1.3.2.1. Surface Potential Barrier and Bond Geometry The work function is expressed as $\Phi = E_0 - E_F(\rho(E)^{2/3})$ (97), which is the energy separation between the vacuum level E_0 and the Fermi energy E_F. The Φ can be modulated by enlarging the charge density ($\rho(E)$) through lattice contraction or by raising the energy where the DOS is centered via dipole formation (63). Dipole formation could lower the Φ of a metal surface by ~1.2 eV. (1) The work function varies from site to site with strong localized features. However, if additional oxygen atoms are adsorbed to the surface, the Φ will restore to the original value or even higher because the metal dipoles donate the polarized electrons to the additional electronegative additives to form a "+/dipole" at the surface (1).

Figure 1.9 shows a typical STM image and the corresponding model for the Cu(001)–(2 × 1)–O^{2-} phase. The round bright spots of 0.8 ± 0.2 Å in height correspond to the lone-pair-induced Cu dipoles. In contrast, the protrusion for the clean Cu(110) surface is about 0.15 Å (98). Single O–Cu–O strings are formed along the [010] direction associated with every other row missing because of the tetrahedron bond saturation.

Dynamic XRD and very-low-energy electron diffraction (VLEED) optimization have led to quantification of atomic positions that are determined by the bond geometry, such as the bond length and bond angles. A simple conversion between

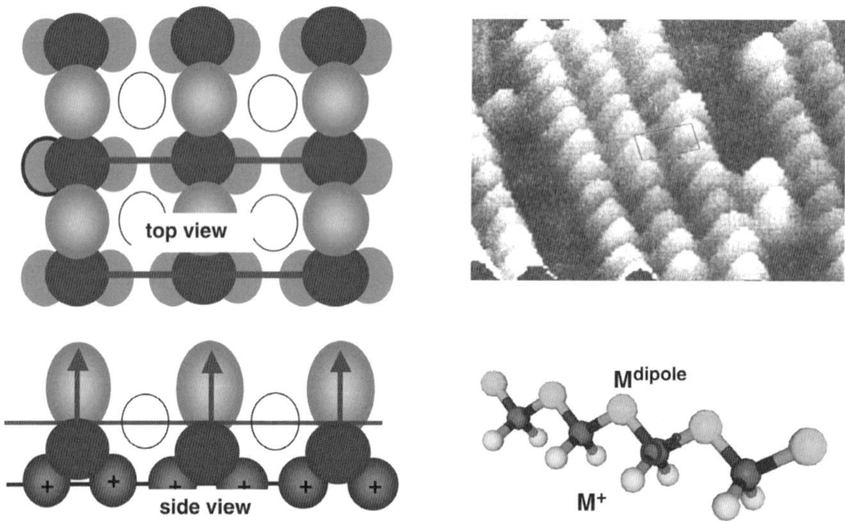

Figure 1.9. STM image (98) and the corresponding models (1) for the surface atomic valencies of Cu(1 1 0)–(2 × 1)–O^{2-}. The STM gray scale is 0.85 Å, much higher than that of metallic Cu on a clean (1 1 0) surface (0.15 Å). The single "O^{2-}: Cu^{dipole}: O^{2-}" chain zigzagged by the nonbonding lone pairs and composed of the tetrahedron. See color insert.

TABLE 1.3. Geometrical Parameters for the Cu_2O Tetrahedron Deduced from the XRD Data of O–Cu(1 1 0) Phase (99) and the Cu_3O_2 Paired Tetrahedra Derived from VLEED Calculation of O–Cu(0 0 1) Surface (1 0 0). Bond Contraction with Respect to the Ideal Length of 1.85 Å Results from the Effect of Bond Order Loss. The ":" Represents the Lone-Pair Interaction.

Parameters	O–Cu(110)	O–Cu(001)	Conclusion
Ionic bond length (Å)	1.675 (Cu^+–O^{2-})	1.628 (Cu^{2+}–O^{2-})	<1.80
Ionic bond length (Å)	1.675 (Cu^+–O^{2-})	1.776 (Cu^+–O^{2-})	<1.80
Lone-pair length (Å)	1.921 (Cu^p:O^{2-})	1.926 (Cu^p:O^{2-})	~1.92
Ionic bond angle (°)	102.5	102.0	<104.5
Lone-pair angle (°)	140.3	139.4	~140.0

atomic position and bond parameters could give the bond geometry for O–Cu examples as shown in Table 1.3.

1.3.2.2. Valence Density of States Oxygen-derived DOS features can be detected using STS and ultraviolet photoelectron spectroscopy (UPS). The STS spectra in Figure 1.10 for an O–Cu(110) surface (98) revealed the lone-pair and dipole features. Spectrum A was taken from the clean Cu(110) surface, whereas B and C were taken from, respectively, the site above the bright spot (dipole) and the site between two bright spots along a "O^{2-} : Cu^{dipole} : O^{2-}" chain at the Cu(110)–(2 × 1)–O^{2-} surface. On the clean surface, empty DOS at 0.8 ~ 1.8 eV above E_F are resolved and no extra DOS structures are found below E_F. The STS spectra recorded from the Cu(110)–(2 × 1)–O^{2-} islands reveal that the original empty-DOS above E_F are partially occupied by electrons upon chemisorption, which result in a slight shift of the empty DOS to higher energy. Additional DOS features are generated around −2.1 eV below the E_F. The sharp features around −1.4 eV have been detected with angular-resolved photoelectron spectroscopy (ARPES) (101) and with the de-excitation spectroscopy of metastable atoms (102). The DOS for Cu-3d electrons are between −2 and −5 eV (103), and the O–Cu bonding derivatives are around the 2p-level of oxygen, −5.6 ~ −7.8 eV below E_F (104). The DOS features for Cu-3d and O–Cu bonding are outside the energy range of the STS ($E_F \pm 2.5$ eV). The band-gap-expansion mechanism implies that it is possible to discover or invent new sources for light emission with a desired wavelength by controlling the extent of the catalytic reaction. Intense blue-light emission from $P(Zr_xTi_{1-x})O_3$ ceramics under Ar^+ ultraviolet (UV) irradiation could be direct evidence for this mechanism (105).

It has been found that the crystal geometry and the surface morphology may vary from surface to surface and from material to material; the oxygen-derived DOS features are substantially the same in nature, as summarized in Table 1.4. The O-derived DOS features include oxygen-metal bonding (−5 ~ −8 eV), nonbond lone-pair of oxygen (−1 ~ 2 eV), holes of metal ions (~E_F) and antibonding metal-dipole states (>E_F).

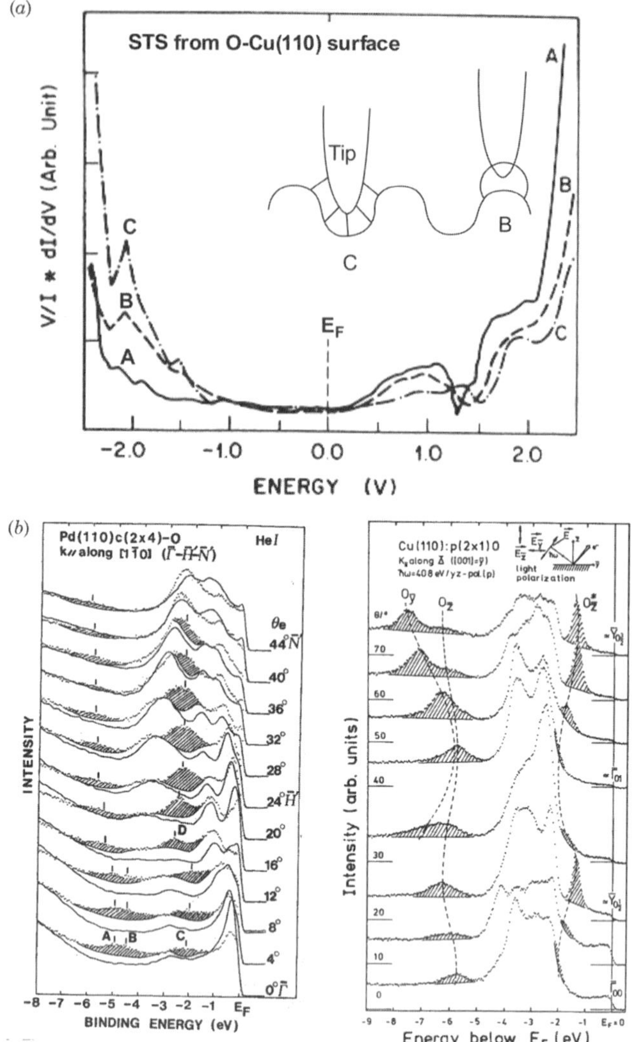

Figure 1.10. (*a*) STS profiles of a Cu(110) surface (98) with and without chemisorbed oxygen. Spectra in panel (*a*) were obtained (A) at a metallic region, (B) on top of, and (C) between protrusions of the "O^{2-}: Cu^{dipole}: O^{2-}" chain (95) on the Cu(110)–(2 × 1)–O^{2-} surface. (*b*) Oxygen-derived DOS features (shaded areas) in the valence band of O–Pd(110) (106) and O–Cu(110) (107) surfaces. Although the microscopy and crystallography of these two systems are different, the PES features are substantially the same. A slight difference in the feature positions results from the difference in electronegativity. Features around $-1 \sim -2$ eV and $-5 \sim -8$ eV correspond to the nonbonding lone pairs and the O sp-hybrid bond states, respectively.

TABLE 1.4. Oxygen-Derived DOS Features Adding to the Valence Band of Metals (unit in eV). Holes are Produced Below E_F. All the Data were Probed with Angular-Resolved Photoelectron Spectroscopy Unless Otherwise Indicated.

Oxide Surfaces	Refs.	Methods	Anti-Bond Dipole > E_F	Lone Pair ≮ E_F	O–M Bond < E_F
O–Cu(001)	108,109	VLEED (1)		-1.5 ± 0.5	-6.5 ± 1.5
			1.2	−2.1	
O–Cu(110)	101,107, 110,111	STS (98)	∼ 2.0	-1.5 ± 0.5	-6.5 ± 1.5
			1.3 ± 0.5	-2.1 ± 0.5	
O–Cu(poly)	103			−1.5	-6.5 ± 1.5
O–Cu/Ag(110)	112			−1.5	−3.0; −6.0
O–Rh(001)	113	DFT	1.0	−3.1	−5.8
O–Pd(110)	114			-2.0 ± 0.5	-4.5 ± 1.5
O–Gd(0001)	115			−3.0	−6.0
O–Ru(0001)	116			-1.0 ± 1.0	-5.5 ± 1.5
O–Ru(0001)	117			−0.8	−4.4
O–Ru(0001)	118	Ab initio	1.5	−4	−5.5, −7.8
O–Ru(0001)	119		1.7	−3.0	−5.8
O–Ru(1010)	120	DFT	2.5	−2 ∼ −3.0	−5.0
O–Co(Poly)	121			−2.0	−5.0
O–diamond (001)	122			−3.0	
(O, S)–Cu(001)	123			−1.3	−6.0
(N, O, S)–Ag	124			−3.4	−8.0

1.3.2.3. Lone-Pair Interaction A direct determination of the lone-pair interaction is to measure the frequency of vibration in metal oxide surfaces using Raman and high-resolution electron energy loss spectroscopy (EELS) (125–127) in the frequency range below 1000 cm^{-1} or the shift energy of ∼50 meV. Typical Raman spectra in Figure 1.11 show the lone pair vibration features in Al_2O_3 and TiO_2 powders (128). The energy of the stretching vibration of O–M in EELS around 50 meV coincides with the energy of hydrogen bond detected using infrared and Raman spectroscopy from H_2O, protein and DNA (129). The energy for an ionic bond is normally around 3.0 eV, and the energy for a Van der Waals bond is about 0.1 eV. The ∼0.05-eV vibration energies correspond to the weak nonbonding interaction between the host dipole and the oxygen adsorbate.

A nuclear inelastic scattering of synchrotron radiation measurement (130) revealed that additional vibrational DOS present at energies around 18 meV and 40–50 meV for the oxide-caped nanocrystalline α-Fe (6–13-nm sizes) compared with that of the coarse-grained α-Fe. The 50-meV DOS corresponds apparently to the lone-pair states, whereas the 18-meV modes could be attributed to intercluster interaction that should increase with the inverse of particle size (Figure 1.7*c*).

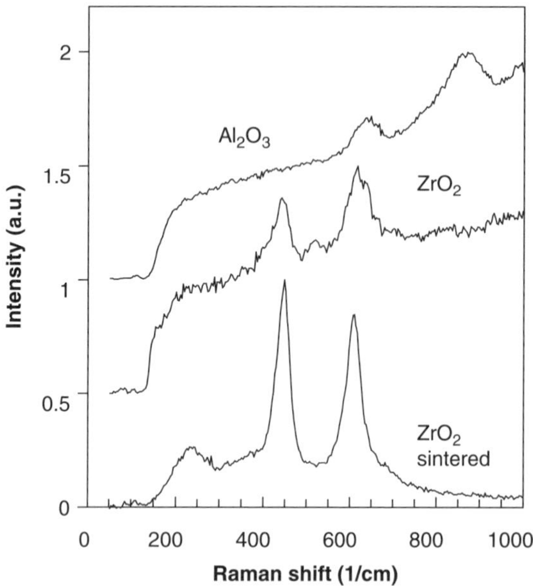

Figure 1.11. Low-frequency Raman shifts indicate that weak bond interaction exists in Ti and Al oxides, which correspond to the nonbonding electron lone pairs generated during the *sp*-orbital hybridization of oxygen.

1.3.2.4. Bond-Forming Kinetics The spectral signatures of LEED, STM, PES/STS, TDS, and EELS can be correlated to the chemical bond, surface morphology, valence DOS, and the bond strength, which enables the kinetics of oxide tetrahedron formation to be readily understood. It has been found generally that an oxide tetrahedron forms in four discrete stages: (*1*) O^{1-} dominates initially at very low oxygen dosage; (*2*) O^{2-} hybridization begins with lone-pair and dipole formation upon second bond formation; (*3*) interaction develops between lone pairs and dipoles; and finally, (*4*) H-bond-like forms at higher dosages. These processes give rise to the corresponding DOS features in the valence band and modify the surface morphology and crystallography, accordingly. Therefore, the events of *sp*-hybrid bonding, nonbonding lone pair, anti-bonding dipole, and the H-like bonding are essential in the electronic process of oxidation, which should dominate the performance of an oxide.

1.3.3. Summary

It is essential that an oxygen atom hybridizes its *sp*-orbitals upon reacting with atoms in the solid phase. In the process of oxidation, electronic holes, nonbonding lone pairs, anti-bonding dipoles, and hydrogen-like bonds are involved, which add corresponding density-of-states features to the valence band of the host. Formation of

TABLE 1.5. Summary of Special Bonding Events and Potential Applications of Oxide Nanomaterials.

Events	Functions	Potential Applications
Anti-bonding(dipole) > E_F	Work function-reduction ($\Delta\phi$)	Cold-cathode Field emission
Holes < E_F	Band gap expansion	PL blue-shift UV detection
Nonbonding (Lone pair) < E_F	Polarization of metal electrons	High-elasticity Raman and Far-infrared low-frequency activity
H- or CH-like bond	$\Delta\phi$-recovery	Bond network stabilization
Bond order loss	BOLS correlation charge, mass, and energy densification Cohesive energy Hamiltonian	Origin for the tunability of nanosolids

the basic oxide tetrahedron and, consequently, the four discrete stages of bond-forming kinetics and the oxygen-derived DOS features are intrinsically common for all analyzed systems, although the patterns of observations may vary from situation to situation. What differs one oxide surface from another in observations are as follows: (*1*) the site selectivity of the oxygen adsorbate, (*2*) the order of the ionic bond formation, and (*3*) the orientation of the tetrahedron at the host surfaces. The valencies of oxygen, the scale and geometrical orientation of the host lattice, and the electronegativity of the host elements determine these specific differences extrinsically.

Knowing the bonding events and their consequences would help us to scientifically design the synthesis of oxide nanostructures with desired functions. The predictions of the functions and the potential applications of the bonding events at a surface with chemisorbed oxygen are summarized in Table 1.5. Oxidation modifies directly the occupied valence DOS by charge transportation or polarization, in particular the band gap and work function. The involvement of the often-overlooked events of lone pair and dipoles may play significant roles in many aspects of the performance of an oxide. The bond contraction is not limited to an oxide surface, but it happens at any site, where the atomic CN reduces.

1.4. CONCLUSION

The impact of the often-overlooked event of atomic CN reduction is indeed tremendous, which unifies the performance of a surface, a nanosolid, and a solid in amorphous state consistently in terms of bond relaxation and its consequences on bond energy. The unusual behavior of a surface and a nanosolid has been consistently understood and systematically formulated as functions of atomic CN reduction and its derivatives on the atomic trapping potential, crystal binding intensity, and electron–phonon coupling. The properties include the lattice contraction (nanosolid densification and surface relaxation), mechanical strength (resistance to both elastic and plastic deformation), thermal stability (phase transition, liquid–solid transition, and evaporation), and lattice vibration (acoustic and optical phonons). They also cover photon emission and absorption (blue shift), electronic structures (core-level disposition and work function modulation), magnetic modulation, dielectric suppression, and activation energies for atomic dislocation, diffusion, and chemical reaction. Structural miniaturization has indeed given a new freedom that allows us to tune the physical properties that are initially nonvariable for the bulk chunks by simply changing the shape and size to make use of the effect of atomic CN reduction.

The effect of size reduction and the effect of oxidation enhance each other in an oxide in many aspects such as the charge localization, band gap expansion, and work function reduction. For instance, the enhancement of the energy density in surface skin enlarges the band gap intrinsically through the Hamiltonian modification, whereas charge transport from metal to oxygen enlarges the E_G extrinsically by hole production at the upper edge of the valence band (131). A theoretical calculation (132) suggested that the band gap of nanosolid Si decreases with increasing dot size. Furthermore, the band gap increases as much as 0.13 eV and 0.35 eV on passivating the surface of the dot with hydrogen and oxygen, respectively. So both size and surface passivation contribute to the optical and electronic properties of Si nanosolids. The lone-pair-induced dipoles in oxidation will add a new DOS above the E_F, whereas charge densification due to bond contraction will intensify the DOS. Therefore, both oxidation and size reduction could lower the work function.

We need to note that the bond contraction for under-coordinated metal atoms and oxygen-metal bonds is general. In particular, in cases such as V, Mo, W, and other oxides, atoms at surfaces will have double bonds and the average M–O distance will decrease. However, at surfaces terminated by hydroxyl groups, the bond contraction is not so clear. For example, earlier EXAFS measurements suggested an average distance decreases in TiO_2, but new results using cumulant expansion (a more sophisticated analysis) give the opposite result (3). The bond expansion in the latter case and the hydroxyl-terminated surface may be connected with the fact that the TiO_2, CeO_2, and other "reducible" oxides have a certain quantity of oxygen vacancies and reduced (Ti^{3+}, Ce^{3+}) cation states; the corresponding radii are larger than the ones of fully oxidized states and, typically, display larger average M–O bond distances. This latter is also observed in α- and γ-Fe_2O_3, but several "synthesis" details concerning the coexistence of Fe^{2+}/Fe^{3+} are still unclear. Alternative explanations indicate that M–OH bond distances would be larger than M–O and dominate the first coordinating distance behavior. Therefore, we could not exclude the situation of bond expansion as observed in compound nanostructures. Both situations, e.g., distance increase and decrease, are covered by the BOLS correlation as consequences on the bond energy disregarding the possible origins for the bond expansion.

It is the practitioner's view that we are actually making use of the effect of bond order loss in size reduction and the effect of bond nature alteration in oxidation in dealing with oxide nanomaterials. Grasping with the factors controlling the process of bond making, breaking and relaxing would be more interesting and rewarding.

ACKNOWLEDGMENT

Helpful discussions and valuable input by Dr. Fernandez-Garcia and Dr. Martinez-Arias are gratefully acknowledged.

REFERENCES

(1) Sun, C.Q. Oxidation electronics, bond-band-barrier correlation and its applications. *Prog. Mater. Sci.* **2003**, *48*, 521–685.

(2) Wintterlin, J.; Behm, R.J. In *The Scanning Tunnelling Microscopy I*; Günthert, H.J.; Wiesendanger, R. (Editors); Springer-Verlag: Berlin, 1992; Chapter 4.

(3) Fernandez-Garcia, M.; Martinez-Arias, A.; Hanson, J.C.; Rodriguez, J.A. Nano-structured oxides in chemistry. Characterization and properties. *Chem. Rev.* **2004**, *104(9)*, 4063–4104.

(4) Qin, G.G.; Song, H.Z.; Zhang, B.R.; Lin, J.; Duan, J.Q.; Yao, G.Q. Experimental evidence for luminescence from silicon oxide layers in oxidized porous silicon. *Phys. Rev. B.* **1996**, *54*, 2548–2555.

(5) Koch, F.; Petrova-Koch, V.; Muschik, T.; Nikolov, A.; Gavrilenko, V. In *Micrystalline Semiconductors, Materials Science and Devices*; Materials Research Society: Pittsburgh, PA; **1993**, *283*, 197.

(6) Wang, X.; Qu, L.; Zhang, J.; Peng, X.; Xiao, M. Surface-related emission in highly luminescent CdSe quantum dots. *Nanolett* **2003**, *3*, 1103–1106.

(7) Prokes, S.M. Surface and optical properties of porous silicon. *J. Mates. Res.* **1996**, *11*, 305–319.

(8) Iwayama, T.S.; Hole, D.E; Boyd, I.W. Mechanism of photoluminescence of Si nanocrystals in SiO2 fabricated by ion implantation, the role of interactions of nanocrystals and oxygen. *J. Phys. Condens. Matter.* **1999**, *11*, 6595–6604.

(9) Trwoga, P.F.; Kenyon, A.J.; Pitt, C.W. Modeling the contribution of quantum confinement to luminescence from silicon nanoclusters. *J. Appl. Phys.* **1998**, *83*, 3789–3794.

(10) Efros, A.L.; Efros, A.L. Interband absorption of light in a semiconductor sphere. *Sov. Phys. Semicond.* **1982**, *16*, 772–775.

(11) Brus, J.E. On the development of bulk optical properties in small semiconductor crystallites. *J. Lumin.* **1984**, *31*, 381–384.

(12) Kayanuma, Y. Quantum-size effects of interacting electrons and holes in semiconductor microcrystals with spherical shape. *Phys. Rev.* **1988**, *38*, 9797–9805.

(13) Ekimov, A.I.; Onushchenko, A.A. Quantum size effect in the optical spectra of semiconductor microcrystals. *Sov. Phys. Semicon.* **1982**, *16*, 775–778.

(14) Glinka, Y.D.; Lin, S.H.; Hwang, L.P.; Chen, Y.T.; Tolk, N.H. Size effect in self-trapped exciton photoluminescence from SiO2-based nanoscale materials. *Phys. Rev. B.* **2001**, *64*, 085421.

(15) Lindemann, F.A. *Z. Phys.* **1910**, *11*, 609.

(16) Born, M. *J. Chem. Phys.* **1939**, *7*, 591.

(17) Buffat, P.; Borel, J.P. Size effect on the melting temperature of gold particles. *Phys. Rev. A.* **1976**, *13*, 2287–2298.

(18) Pawlow, P. *Z. Phys. Chem. Munich* **1909**, *65*, 1.

(19) Vekhter, B.; Berry, R.S. Phase coexistence in clusters, an "experimental" isobar and an elementary model. *J. Chem. Phys.* **1997**, *106*, 6456–6459.

(20) Reiss, H.; Mirabel, P.; Whetten, R.L. Capillarity theory for the "coexistence" of liquid and solid clusters. *J. Phys. Chem.* **1988**, *92*, 7241–7246.

(21) Sakai, H. Surface-induced melting of small particles. *Surf. Sci.* **1996**, *351*, 285–291.

(22) Ubbelohde, A.R. *The Molten State of Materials*; Wiley: New York; 1978.

(23) Vanfleet, R.R.; Mochel, J.M. Thermodynamics of melting and freezing in small particles. *Surf. Sci.* **1995**, *341*, 40–50.

(24) Nanda, K.K.; Sahu, S.N.; Behera, S.N. Liquid-drop model for the size-dependent melting of low-dimensional systems. *Phys. Rev. A.* **2002**, *66*, 013208.

(25) Jiang, Q.; Liang, L.H.; Li, J.C. Thermodynamic superheating and relevant interface stability of low-dimensional metallic crystals. *J. Phys. Condens. Matt.* **2001**, *13*, 565–571.

(26) Jiang, Q.; Liang, L.H.; Zhao, M. Modelling of the melting temperature of nano-ice in MCM-41 pores. *J. Phys. Condens. Matt.* **2001**, *13*, L397–L401.

(27) Jiang, Q.; Zhang, Z.; Li, J.C. Superheating of nanocrystals embedded in matrix. *Chem. Phys. Lett.* **2000**, *322*, 549–552.

(28) Vallee, R.; Wautelet, M.; Dauchot, J.P.; Hecq, M. Size and segregation effects on the phase diagrams of nanoparticles of binary systems. *Nanotechnology* **2001**, *12*, 68–74.

(29) Sun, C.Q. Surface and nanosolid core-level shift: impact of atomic coordination number imperfection. *Phys. Rev.* **2004**, *69*, 045105.

(30) Sun, C.Q.; Gong, H.Q.; Hing, P. Behind the quantum confinement and surface passivation of nanoclusters. *Surf. Rev. Lett.* **1999**, *6*, L171–L176.

(31) Rao, C.N.R.; Kulkarni, G.U.; Thomas, P.J.; Edwards, P.P. Size-dependent chemistry. Properties of nanocrystals. *Chem. Euro. J.* **2002**, *8*, 28–35.

(32) Aiyer, H.N.; Vijayakrishnan, V.; Subbanna, G.N.; Rao, C.N.R. Investigations of Pd clusters by the combined use of HREM, STM, high-energy spectroscopies and tunneling conductance measurements. *Surf. Sci.* **1994**, *313*, 392–398.

(33) Sun, C.Q.; Chen, T.P.; Tay, B.K.; Li, S.; Huang, H.; Zhang, Y.B.; Pan, L.K.; Lau, S.P.; Sun, X.W. An extended 'quantum confinement' theory, surface-coordination imperfection modifies the entire band structure of a nanosolid. *J. Phys. D.* **2001**, *34*, 3470–3479.

(34) Street, R.A. *Hydrogenated Amorphous Silicon*; Cambridge University Press: Cambridge, UK, 1991.

(35) Zhong, W.H.; Sun, C.Q.; Li, S.; Bai, H.L.; Jiang, E.Y. Impact of bond order loss on surface and nanosolid magnetism. *Acta. Materialia.* **2005**, *53*, 3207–3214.

(36) Goldschmidt, V.M. *Ber. Deut. Chem. Ges.* **1927**, *60*, 1270.

(37) Pauling, L. Atomic radii and interatomic distances in metals. *J. Am. Chem. Soc.* **1947**, *69*, 542–553.

(38) Sinnott, M.J. *The Solid State for Engineers*; Wiley and Sons: New York, 1963.

(39) Shpyrko, O.G.; Grigoriev, A.Y.; Steimer, C.; Pershan, P.S.; Lin, B.; Meron, M.; Graber, T.; Gerbhardt, J.; Ocko, B.; Deutsch, M. *Phys. Rev. B.* **2004**, *70*, 224206.

(40) Mahnke, H.-E.; Haas, H.; Holub-Krappe, E.; Koteski, V.; Novakovic, N.; Fochuk, P.; Panchuk, O. Lattice distortion around impurity atoms as dopants in CdTe. *Thin Solid Films* **2005**, *480–481*, 279–282.

(41) Feibelman, P.J. Relaxation of hcp(0001) surfaces. A chemical view. *Phys. Rev.* **1996**, *53*, 13740–13746.

(42) Sun, C.Q.; Tay, B.K.; Zeng, X.T.; Li, S.; Chen, T.P.; Zhou, J.; Bai, H.L.; Jiang, E.Y. Bond-order-length-strength (BOLS) correlation mechanism for the shape and size dependency of a nanosolid. *J. Phys. Condens. Matt.* **2002**, *14*, 7781–7795.

(43) Sun, C.Q.; Bai, H.L.; Tay, B.K.; Li, S.; Jiang, E.Y. Dimension, strength, and chemical and thermal stability of a single C-C bond in carbon nanotubes. *J. Phys. Chem. B.* **2003**, *107*, 7544–7546.

(44) Chen, T.P.; Liu, Y.; Sun, C.Q. et al. Core-level shift of Si nanocrystals embedded in SiO2 matrix. *J. Phys. Chem. B.* **2004**, *108*, 16609–16611.

(45) Sun, C.Q.; Li, C.M.; Bai, H.L.; Jiang, E.Y. Bond nature evolution in atomic clusters. *J. Phys. Chem. B.* in press.

(46) Bahn, S.R.; Jacobsen, K.W. Chain formation of metal atoms. *Phys. Rev. Lett.* **2001**, *87*, 266101.

(47) Sun, C.Q.; Wang, Y.; Tay, B.K.; Li, S.; Huang, H.; Zhang, Y.B. Correlation between the melting point of a nanosolid and the cohesive energy of a surface atom. *J. Phys. Chem. B.* **2002**, *106*, 10701–10705.

(48) Kocks, U.F.; Argon, A.S.; Ashby, A.S. *Prog. Mater. Sci.* **1975**, *19*, 1.

(49) Sun, C.Q.; Li, S.; Li, C.M. Impact of bond-order loss on surface and nanosolid mechanics. *J. Phys. Chem. B.* **2005**, *109*, 415–423.

(50) Hensel, J.C.; Tung, R.T.; Poate, J.M.; Unterwald, F.C. Specular boundary scattering and electrical transport in single-crystal thin films of CoSi2. *Phys. Rev. Lett.* **1985**, *54*, 1840–1843.

(51) Tesanovic, Z.; Jaric, M.V. Quantum transport and surface scattering. *Phys. Rev. Lett.* **1986**, *57*, 2760–2763.

(52) Nogera, C. Insulating oxides in low dimensionality: A theoretical review. *Surf. Rev. Lett.* **2001**, *8*, 167–211.

(53) Sun, C.Q.; Zhong, W.H.; Li, S.; Tay, B.K.; Bai, H.L.; Jiang, E.Y. Coordination imperfection suppressed phase stability of ferromagnetic, ferroelectric, and superconductive nanosolids. *J. Phys. Chem.* **2004**, *108*(3), 1080–1084.

(54) Hu, Y.; Tan, O.K.; Cao, W.; Zhu, W. Fabrication and characterization of nano-sized SrTiO3-based oxygen sensor for near-room temperature operation. *IEEE Sensors J.* **2004**, in press.

(55) Hu, Y.; Tan, O.K.; Pan, J.S.; Yao, X. A new form of nano-sized SrTiO3 material for near human body temperature oxygen sensing applications. *J. Phys. Chem. B.* **2004**, *108*, 11214–11218.

(56) Zhou, J.; Sun, C.Q.; Pita, K.; Lam, Y.L.; Zhou, Y.; Ng, S.L.; Kam, C.H.; Li, L.T.; Gui, Z.L. Thermally tuning of the photonic band-gap of SiO2 colloid-crystal infilled with ferroelectric BaTiO3. *Appl. Phys. Lett.* **2001**, *78*, 661–663.

(57) Li, B.; Zhou, J.; Hao, L.F.; Hu, W.; Zong, R.L.; Cai, M.M.; Fu, M.; Gui, Z.L.; Li, L.T.; Li, Q. Photonic band gap in (Pb,La)(Zr,Ti)O3 inverse opals. *Appl. Phys. Lett.* **2003**, *82*, 3617–3619.

(58) Pan, L.K.; Sun, C.Q. Coordination imperfection enhanced electron-phonon interaction in nanosolid silicon. *J. Appl. Phys.* **2004**, *95*, 3819–3821.

(59) Sanders, G.D.; Chang, Y.C. Theory of optical properties of quantum wires in porous silicon. *Phys. Rev. B.* **1992**, *45*, 9202–9213.

(60) Tománek, D.; Mukherjee, S.; Bennemann, K.H. Simple theory for the electronic and atomic structure of small clusters. *Phys. Rev. B.* **1983**, *28*, 665–673.

(61) Sun, C.Q.; Li, C.; Li, S. Breaking limit of atomic distance in an impurity-free monatomic chain. *Phys. Rev. B.* **2004**, *69*, 245402.

(62) Sun, C.Q.; Pan, L.K.; Li, C.M. Elucidating Si-Si dimer vibration from the size-dependent Raman shift of nanosolid Si. *J. Phys. Chem. B.* **2004**, *108(11)*, L3404–L3406.

(63) Zheng, W.T.; Sun, C.Q.; Tay, B.K. Modulating the work function of carbon by O and N addition and nanotip formation. *Solid State Commun.* **2003**, *128*, 381–384.

(64) Poa, C.H.P.; Smith, R.C.; Silva, S.R.P.; Sun, C.Q. Influence of mechanical stress on electron field emission of multi-walled carbon nanotube-polymer composites. *J. Vac. Sci. Technol. B.* **2005**, *23*, 698–701.

(65) Pan, L.K.; Sun, C.Q. Dielectric suppression of nanosolid Si. *Nanotechnology* **2004**, *15*, 1802–1806.

(66) Liu, J.P.; Zaumseil, P.; Bugiel, E.; Osten, H.J. Epitaxial growth of Pr2O3 on Si(111) and the observation of a hexagonal tocubic phase transition during postgrowth N2 annealing. *Appl. Phys. Lett.* **2001**, *79*, 671–673.

(67) Kim, H.K.; Huh, S.H.; Park, J.W.; Jeong, J.W.; Lee, G.H. The cluster size dependence of thermal stabilities of both molybdenum and tungsten nanoclusters. *Chem. Phys. Lett.* **2002**, *354*, 165–172.

(68) Höfler, H.J.; Averback, R.S. *Scripta. Metall. Mater.* **1990**, *24*, 2401.

(69) Peppiat, S.J. *Proc. R. Soc. London Ser. A.* **1975**, *345*, 401.

(70) Itoigawa, H.; Kamiyama, T.; Nakamura, Y. Bi precipitates in Na2O-B2O3 glasses. *J. Non-Cryst. Solids* **1997**, *210*, 95–100.

(71) Kellermann, G.; Craievich, A.F. Structure and melting of Bi nanocrystals embedded in a B2O3-Na2O glass. *Phys. Rev. B.* **2002**, *65*, 134204.

(72) Allen, G.L.; Gile, W.W.; Jesser, W.A. Melting temperature of microcrystals embedded in a matrix. *Acta. Metall.* **1980**, *28*, 1695–1701.

(73) Skripov, V.P.; Koverda, V.; Skokov, V.N. Size effect on melting of small particles. *Phys. Status Solidi. A.* **1981**, *66*, 109–118.

(74) Goldstein, N.; Echer, C.M.; Alivistos, A.P. Melting in semiconductor nanocrystals. *Science* **1992**, *256*, 1425–1427.

(75) Sadeh, B.; Doi, M.; Shimizu, T.; Matsui, M.J. Dependence of the Curie temperature on the diameter of Fe3O4 ultra-fine particles. *J. Magn. Soc. Jpn.* **2000**, *24*, 511–514.

(76) Zhong, W.L.; Jiang, B.; Zhang, P.L.; Ma, J.M.; Cheng, H.M.; Yang, Z.H.; Li, L.X. Phase transition in PbTiO3 ultrafine particles of different sizes. *J. Phys. Condens. Matt.* **1993**, *5*, 2619–2624.

(77) Yu, T.; Shen, Z.X.; Toh, W.S.; Xue, J.M.; Wang, J.J. Size effect on the ferroelectric phase transition in SrBi2Ta2O9 nanoparticles. *J. Appl. Phys.* **2003**, *94*, 618–620.

(78) Uchina, K.; Sadanaga, Y.; Hirose, T. *J. Am. Ceram. Soc.* **1999**, *72*, 1555.

(79) Chattopadhyay, S.; Ayyub, P.; Palkar, V.R.; Gurjar, A.V.; Wankar, R.M.; Multani, M. Finite-size effects in antiferroelectric PbZrO3 nanoparticles. *J. Phys. Condens. Matt.* **1997**, *9*, 8135–8145.

(80) Ma, D.D.D.; Lee, C.S.; Au, F.C.K.; Tong, S.Y.; Lee, S.T. Small-diameter silicon nanowire surfaces. *Science* **2003**, *299*, 1874–1877.

(81) Campbell, I.H.; Fauchet, P.M. Effects of microcrystal size and shape on the phonon Raman spectra of crystalline semiconductors. *Solid State Commun.* **1986**, *58*, 739–741.

(82) Pan, L.K.; Sun, C.Q.; Tay, B.K.; Chen, T.P.; Li, S. Photoluminescence of Si nanosolids near the lower end of the size limit. *J. Phys. Chem. B.* **2002**, *106*, 11725–11727.

(83) Zhang, P.; Sham, T.K. X-Ray studies of the structure and electronic behavior of alkanethiolate-capped gold nanoparticles. The interplay of size and surface effects. *Phys. Rev. Lett.* **2003**, *90*, 245502.

(84) Ohgi, T.; Fujita, D. Consistent size dependency of core-level binding energy shifts and single-electron tunneling effects in supported gold nanoclusters. *Phys. Rev. B.* **2002**, *66*, 115410.

(85) Howard, A.; Clark, D.N.S.; Mitchell, C.E.J.; Egdell, R.G.; Dhanak, V.R. Initial and final state effects in photoemission from Au nanoclusters on TiO2(110). *Surf. Sci.* **2002**, *518*, 210–224.

(86) Salmon, M.; Ferrer, S.; Jazzar, M.; Somojai, GA. Core- and valence-band energy-level shifts in small two-dimensional islands of gold deposited on Pt(100): The effect of step-edge, surface, and bulk atoms. *Phys. Rev. B.* **1983**, *28*, 1158–1160.

(87) Gotić, M.; Ivanda, M.; Sekulić, A.; Music, S.; Popović, S.; Turković, A.; Furić, K. *Mater. Lett.* **1996**, *28*, 225.

(88) Dieguez, A.; Romano-Rodrýguez, A.; Vila, A.; Morante, J.R. The complete Raman spectrum of nanometric SnO2 particles. *J. Appl. Phys.* **2001**, *90*, 1550–1557.

(89) Spanier, J.E.; Robinson, R.D.; Zhang, F.; Chan, S.W.; Herman, I.P. Size-dependent properties of CeO2 nanoparticles as studied by Raman scattering. *Phys. Rev. B.* **2001**, *64*, 245407.

(90) Shek, C.H.; Lin, G.M.; Lai, J.K.L. Effect of oxygen deficiency on the Raman spectra and hyperfine interactions of nanometer SnO2. *Nanostructured Mater.* **1999**, *11*, 831–835.

(91) Seong, M.J.; Micic, O.I.; Nozik, A.J.; Mascarenhas, A.; Cheong, H.M. Size-dependent Raman study of InP quantum dots. *Appl. Phys. Lett.* **2003**, *82*, 185–187.

(92) Lannoo, M.; Delerue, C.; Allan, G. Screening in semiconductor nanocrystallites and its consequences for porous silicon. *Phys. Rev. Lett.* **1995**, *74*, 3415–3418.

(93) Wang, L.W.; Zunger, A. Dielectric constants of silicon quantum dots. *Phys. Rev. Lett.* **1994**, *73*, 1039–1042.

(94) Pan, L.K.; Huang, H.T.; Sun, C.Q. Dielectric transition and relaxation of nanosolid silicon. *J. Appl. Phys.* **2003**, *94*, 2695–2700.

(95) Sun, C.Q.; Li, S. Oxygen derived DOS features in the valence band of metals. *Surf. Rev. Lett.* **2000**, *7*, L213–L217.

(96) Sun, C.Q. A model of bonding and band-forming for oxides and nitrides. *Appl. Phys. Lett.* **1998**, *72*, 1706–1708.

(97) Wales, D.J. Structure, dynamics, and thermodynamics of clusters, tales from topographic potential surfaces. *Science* **1996**, *271*, 925–929.

(98) Chua, F.M.; Kuk, Y.; Silverman, P.J. Oxygen chemisorption on Cu(110): An atomic view by scanning tunneling microscopy. *Phys. Rev. Lett.* **1989**, *63*, 386–389.

(99) Feidenhans'l, R.; Grey, F.; Johnson, R.L.; Mochrie, S.G.J.; Bohr, J.; Nielsen, M. Oxygen chemisorption on Cu(110): A structural determination by x-ray diffraction. *Phys. Rev. B.* **1990**, *41*, 5420–5423.

(100) Sun, C.Q.; Bai, C.L. A model of bonding between oxygen and metal surfaces. *J. Phys. Chem. Solids* **1997**, *58*, 903–912.

(101) Jacob, W.; Dose, V.; Goldmann, A. Atomic adsorption of oxygen on Cu(111) and Cu(110). *Appl. Phys. A.* **1986**, *41*, 145–150.

(102) Sesselmann, W.; Conrad, H.; Ertl, G.; Küppers, J.; Woratschek, B.; Haberland, H. Probing the local density of states of metal surfaces by de-excitation of metastable noble-gas atoms. *Phys. Rev. Lett.* **1983**, *50*, 446–450.

(103) Belash, V.P.; Klimova, I.N.; Kormilets, V.I.; Trubjtsjn, V.Y.; Finkelstein, L.D. Transformation of the electronic structure of Cu into Cu2O in the adsorption of oxygen. *Surf. Rev. Lett.* **1999**, *6*, 383–388.

(104) Courths, R.; Cord, B.; Wern, H.; Saalfeld, H.; Hüfner, S. Dispersion of the oxygen-induced bands on Cu(110)-an angle-resolved UPS study of the system p(2×1)O/Cu(110). *Solid State Commun.* **1987**, *63*, 619–623.

(105) Jin, D.; Hing, P.; Sun, C.Q. Growth kinetics and electrical properties of Pb(Zr0.9Ti0.1)O3 doped with Cerium oxide. *J. Phys. D: Appl. Phys.* **2000**, *33*, 744–752.

(106) Bondzie, V.A.; Kleban, P.; Dwyer, D.J. XPS identification of the chemical state of subsurface oxygen in the O/Pd(110) system. *Surf. Sci.* **1996**, *347*, 319–328.

(107) DiDio, R.A.; Zehner, D.M.; Plummer, E.W. An angle-resolved UPS study of the oxygen-induced reconstruction of Cu(110). *J. Vac. Sci. Technol.* **1984**, *A2*, 852–855.

(108) Warren, S.; Flavell, W.R.; Thomas, A.G.; Hollingworth, J.; Wincott, P.L.; Prime, A.F.; Downes, S.; Chen, C. Photoemission studies of single crystal CuO(100). *J. Phys. Condens. Matter.* **1999**, *11*, 5021–5043.

(109) Pforte, F.; Gerlach, A.; Goldmann, A.; Matzdorf, R.; Braun, J.; Postnikov, A. Wave-vector-dependent symmetry analysis of a photoemission matrix element: The quasi-one-dimensional model system Cu(110)(2×1)O$'$. *Phys. Rev. B.* **2001**, *63*, 165405.

(110) Dose, V. Momentum-resolved inverse photoemission. *Surf. Sci. Rep.* **1986**, *5*, 337–378.

(111) Spitzer, A.; Luth, H. The adsorption of oxygen on copper surface: I Cu(100) and Cu(110). The adsorption of oxygen on copper surface: II Cu(111). *Surf. Sci.* **1982**, *118*, 121–144.

(112) Sekiba, D.; Ogarane, D.; Tawara, S.; Yagi-Watanabe, K. Electronic structure of the Cu-O/Ag(110)(2×2)p2mg surface. *Phys. Rev. B.* **2003**, *67*, 035411.

(113) Alfe, D.; de Gironcoli, S.; Baroni, S. The reconstruction of Rh(100) upon oxygen adsorption. *Surf. Sci.* **1998**, *410*, 151–157.

(114) Bondzie, V.A.; Kleban, P.; Dwyer, D.J. XPS identification of the chemical state of subsurface oxygen in the O/Pd(110) system. *Surf. Sci.* **1996**, *347*, 319–328.

(115) Zhang, J.; Dowben, P.A.; Li, D.; Onellion, M. Angle-resolved photoemission study of oxygen chemisorption on Gd(0001). *Surf. Sci.* **1995**, *329*, 177–183.

(116) Böttcher, A.; Niehus, H. Oxygen adsorption on oxidized Ru(0001). *Phys. Rev. B.* **1999**, *60(20)*, 14396–14404.

(117) Böttcher, A.; Conrad, H.; Niehus, H. Reactivity of oxygen phases created by the high temperature oxidation of Ru(0001). *Surf. Sci.* **2000**, *452*, 125–132.

(118) Bester, G.; Fähnle, M. On the elelctronic structure of the pure and oxygen covered Ru(0001) surface. *Surf. Sci.* **2002**, *497*, 305–310.

(119) Stampfl, C.; Ganduglia-Pirovano, MV.; Reuter, K.; Scheffler, M. Catalysis and corrosion: The theoretical surface-science context. *Surf. Sci.* **2002**, *500*, 368–394.

(120) Schwegmann, S.; Seitsonen, A.P.; De Renzi, V.; Dietrich, D.; Bludau, H.; Gierer, M.; Over, H.; Jacobi, K.; Scheffler, M.; Ertl, G. Oxygen adsorption on the Ru(10^1 0) surface. Anomalous coverage dependence. *Phys. Rev. B.* **1998**, *57*, 15487–15495.

(121) Mamy, R. Spectroscopic study of the surface oxidation of a thin epitaxial Co layer. *Appl. Surf. Sci.* **2000**, *158*, 353–356.

(122) Zheng, J.C.; Xie, X.N.; Wee, A.T.S.; Loh, K.P. Oxygen induces surface states on diamond (100). *Diamond Rel. Mater.* **2001**, *10*, 500–505.

(123) Tibbetts, G.G.; Burkstrand, J.M.; Tracy, J.C. Electronic properties of adsorbed layers of nitrogen, oxygen, and sulfur on copper (100). *Phys. Rev. B.* **1977**, *15*, 3652–3660.

(124) Tibbetts, G.G.; Burkstrand, J.M. Electronic properties of adsorbed layers of nitrogen, oxygen, and sulfur on silver (111). *Phys. Rev. B.* **1977**, *16*, 1536–1541.

(125) He, P.; Jacobi, K. Vibrational analysis of the 1×1-O overlayer on Ru(0001). *Phys. Rev. B.* **1997**, *55*, 4751–4754.

(126) Moritz, T.; Menzel, D.; Widdra, W. Collective vibrational modes of the high-density (1×1)-O phase on Ru(001). *Surf. Sci.* **1999**, *427–428*, 64–68.

(127) Frank, M.; Wolter, K.; Magg, N.; Heemeier, M.; Kühnemuth, R.; Bäumer, M.; Freund, H.-J. Phonons of clean and metal-modified oxide films, an infrared and HREELS study. *Surf. Sci.* **2001**, *492*, 270–284.

(128) Sun, C.Q.; Tay, B.K.; Lau, S.P.; Sun, X.W.; Zeng, X.T.; Bai, H.L.; Liu, H.; Liu, Z.H.; Jiang, E.Y. Bond contraction and lone pair interaction at nitride surfaces. *J. Appl. Phys.* **2001**, *90*, L2615–L2617.

(129) Han, W.G.; Zhang, C.T. A theory of non-linear stretch vibration of hydrogen bonds. *J. Phys. Condens. Mat.* **1991**, *3*, 27–35.

(130) Pasquini, L.; Barla, A.; Chumakov, A.I.; Leupold, O.; Ruffer, R.; Deriu, A.; Bonetti, E. Size and oxidation effects on the vibrational properties of anocrystalline a-Fe. *Phys. Rev. B.* **2002**, *66*, 073410.

(131) Sun, C.Q.; Sun, X.W.; Gong, H.Q. Frequency shift in the photoluminescence of nanometric SiOx: Surface bond contraction and oxidation. *J Phys. Condens. MAT.* **1999**, *11*, L547–L550.

(132) Ghoshal, S.K.; Gupta, U.; Gill, K.S. Role of hydrogen and oxygen on the band gap of silicon quantum dots. *Ind. J. Pure. Appl. Phys.* **2005**, *43*, 188–191.

CHAPTER 2

On Aqueous Interfacial Thermodynamics and the Design of Metal-Oxide Nanostructures

LIONEL VAYSSIERES

International Center for Young Scientists, National Institute for Materials Science, Tsukuba, Japan 305-0044

2.1. INTRODUCTION

Among the various classes of materials, metal oxides are the most common, most diverse, and most probably the richest class of materials in terms physical, chemical, and structural properties. Such properties include, for instance, optical, optoelectronic, magnetic, electrical, thermal, photoelectrochemical, mechanical, and catalytic ones. As a result, numerous applications of metal oxides, such as ceramics, (chemical-, gas-, and bio-)sensors, actuators, lasers, waveguides, infrared (IR) and solar absorbers, pigments, photodetectors, optical switches, photochromics, refractors, electrochromics, (electro-, photo-)catalysts and support for catalysts, insulators, semiconductors, superconductors, supercapacitors, transistors, varistors, resonators, dielectrics, piezoelectrics, pyroelectrics, ferroelectrics, magnets, transducers, thermistors, thermoelectrics, protective and anticorrosion coatings, fuel cells, alkaline and lithium batteries, and solar cells, have been developed.

The diversity of such properties and applications originates from the more complex crystal and electronic structures of metal oxides compared with other classes of materials. The main reasons are related to their variety of oxidation states, coordination number, symmetry, ligand-field stabilization, density, stoichiometry, and acid–base properties, which yield fascinating compounds exhibiting, for instance, insulating, semiconducting, (supra)conducting, or magnetic behaviors with transitions among those states.

Synthesis, Properties, and Applications of Oxide Nanomaterials, Edited by José A. Rodríguez and Marcos Fernández-Garcia

Within the last decade, not only the scientific community but also the whole world has witnessed the advent of nanoscience (1–7) and its establishment as a leading science with fundamental and applied work covering all basic physical, life, and earth sciences as well as engineering and materials science (8–17). A major feature of nanoscience is its inherent ability to bridge the dimensional gap between the atomic and molecular scale of fundamental sciences and the microstructural scale of engineering and manufacturing (18). Consequently, a remarkable amount of true multidisciplinary fundamental and applied knowledge is to be explored and used (19,20). Without a doubt, such a knowledge will lead to a substantial amount of in-depth understanding of nanoscale phenomena and, thus, to the fabrication of novel high technological devices in many fields of applications (21–24). As a result, the level of technological advances should improve tremendously and at a much greater rate than human history has ever experienced. However, to unfold the full potential of nanotechnology, several crucial challenges remain. For instance, the ability and competence to hierarchically order, connect, and integrate low-dimensional building blocks (25) such as quantum dots (0D), quantum wires (1D), and nanosheets (2D) in functional network, ultrathin film coatings, and three-dimensional (3D) arrays for the fabrication and manufacturing of practical nanodevices is yet to be established and demonstrated. Future devices based on such novel and advanced building blocks should revolutionize materials science and engineering given that they possess the unique properties of nanoscale as well as the ability to connect the nanoworld to the microworld.

It goes without saying that the combinations of the variety of distinctive properties and applications of metal oxides with the unique effects of low dimensionality at nanoscale make the development of metal oxide nanostructures (26) an important challenge from both fundamental and industrial standpoints.

This chapter will address the fundamental aspect of the aqueous thermodynamic stability of transition metal oxide nanoparticles and the ability to rationally design ordered oxide nanostructures at low cost and large scale.

2.2. CHARACTERISTICS OF THE AQUEOUS METAL-OXIDE INTERFACE

2.2.1. Surface Charge and Points-of-Zero Charge

The surface charge developed at the water-oxide interface originates from the ionization of hydroxyl groups (i.e., adsorption/desorption of protons), which depends directly on the pH of the solution (27,28). Several approaches have been developed to model the proton affinity for the water-oxide interface. Such models include, for instance, the one site/1 pK, one site/2 pK, and the multisite complexation (MUSIC) models (29–35) as well as electrostatic models (33,34). Most oxides are amphoteric; that is, they can develop either positive or negative surface depending on the acidity of the solution. The surface charge depends in sign on the pH of the solution and in amplitude on the pH and the ionic strength of the solution (35).

An important characteristic of the surface of a metal oxide is its point-of-zero charge. As a general trend, the higher the formal oxidation state of the transition or post-transition metal cations, the lower the point-of-zero charge (Table 2.1 (36)). The concept of the point-of-zero charge was born in the 1950s during the studies of flocculation of hydrophobic colloids (37–39). Points-of-zero charge have been defined traditionally as a reference state from which to express the acid–base reactions of hydroxyl groups on the surfaces of oxide minerals (40,41). Using straightforward mass action descriptions of these surface reactions, it was demonstrated that the proton concentration in aqueous solution at the "zero point of charge" was equal to the geometric mean of two acid–base equilibrium constants (42). The zero point of charge was then compared with the pH value at which a colloidal particle is "electrokinetically uncharged," i.e., the isoelectric point (IEP) (43). Nowadays, points-of-zero charge are defined as pH values for which one of the categories of surface charge is equal to zero, at given ambient temperature, pressure, and aqueous solution composition (44,45). It is worth noting that variables other than pH, such as the negative logarithm of the aqueous activity of any ions that adsorb primarily through inner sphere surface complexation, can be used to define a point-of-zero charge.

TABLE 2.1. Point-of-Zero Charge of Most Common Metal Oxides. A More Complete List Can be Found in Ref. 36.

Oxidation State	Materials	PZC
+VI	WO_3	$\ll 3$
+IV	SiO_2	2–4
+IV+V	Nb_2O_5	4.3
+III	α-Cr_2O_3	4–6.7
+IV	SnO_2	4.2–6.6
+III	γ-MnO_2	5.3–5.6
+IV	TiO_2	5.2–7
+IV	CeO_2	6–11.2
+III	α-Fe_2O_3	6.3–8.5
+IV	ZrO_2	6.4–8.2
+II	CuO	6.9–8.5
+II+III	Co_3O_4	7.2
+IV	β-MnO_2	7.3–7.7
+II+III	Fe_3O_4	7.5–8
+IV	ThO_2	7.5–8.8
+II	ZnO	7.5–9.8
+II	NiO	7.8–9.5
+III	α-Al_2O_3	8–9
+III	γ-Al_2O_3	8.2–9.6
+III	In_2O_3	9
+III	Y_2O_3	7.9–11
+II	MgO	10–12.4
+II	PbO	11

Three types of points-of-zero charge are conventionally accepted (46):

- The point-of-zero net proton (PZNPC), which is the pH value at which the net proton surface charge is equal to zero (44,45).
- The point-of-zero net charge (PZNC), which is the pH value at which the intrinsic surface charge is equal to zero (44).
- The point-of-zero charge (PZC), which is the pH value at which the net total particle charge is equal to zero (44,45). At the PZC, there is no surface charge to be neutralized by ions, and any adsorbed ions that exist must be bound in surface complexes.

Official IUPAC nomenclature (47) requires the use of lower capitals separated by dots (e.g., p.z.c. for PZC).

Several experimental techniques exist for the determination of the points-of-zero charge of metal oxides. Such techniques involve the classic potentiometric titration technique (PT) (48), the mass titration technique (MT) (49,50), the electroacoustic (51) technique, and microelectrophoresis (ME) (52), which is used more frequently for determining the IEP, which is equal to the PZC in cases in which no specific adsorption takes place. Recent methodologies involve the potentiometric mass titration technique (PMT) (53,54) and the differential potentiometric titration (DPT) (55). In the former, the potentiometric curves are determined either for three different values of the mass of the oxide or by recording two potentiometric titration curves: one for the blank solution and one for a suspension of a given amount of the immersed oxide. The latter allows the determination of the PZC of metal (hydr)oxides using only one potentiometric curve. Other experimental and theoretical methods involve scanning force microscopy and crystal chemistry (56,57).

2.2.2. Interfacial Thermodynamics

When two different phases (gas/liquid, liquid/liquid, gas/solid, or liquid/solid) are in contact with each other, the atoms or molecules at the interface experience an imbalance of forces. This imbalance will lead to an accumulation of free energy at the interface. The excess energy is called surface free energy and can be quantified as the energy required to increase the surface area of the interface by a unit amount (expressed in mJ/m^2 or erg/cm^2). It is also possible to describe this situation as having a line tension or interfacial tension, which is quantified as a force/length measurement. This force tends to minimize the area of the surface, which explains the spherical shape of liquid droplets and air bubbles. The common units for interfacial tension are dynes/cm or mN/m. This excess energy exists at any interface. If one of the phases is the gas phase of a liquid being tested, the measurement is normally referred to as surface tension. If the surface investigated is the interface of two immiscible liquids, the measurement is normally referred to as interfacial tension. Solid surfaces also may be described to have an interfacial tension normally referred to as surface free energy, but direct measurement of its value is not possible through techniques used for liquids. Experimental methods to measure, yet indirectly, the interfacial tension of solids do exist; they involve mainly contact angle measurements and vapor adsorption by inverse gas chromatography as well as by atomic force microscopy (58–63).

A difference between surface free energy and surface tension (typical for solids) is shown to be the consequence of nonuniformity of chemical potential in a solid interface. The problem of using the chemical potential of a solid in interfacial thermodynamics and the anisotropy of chemical potential has been discussed by Rusanov (64), and a generalized adsorption equation is formulated for deformable solids. Gibbs (65) was first to formulate the principles of the thermodynamics of solids, including interfacial thermodynamics. However, when comparing Gibbs' thermodynamics for fluids and for solids, one may notice that a major component of interfacial thermodynamics, the Gibbs adsorption equation, is absent in Gibbs' description of solids because he did not apply the concept of chemical potential to solids. Chemical potentials were not introduced for immobile components of a solid, i.e., for the components forming the crystalline lattice of the solid, whereas chemical potentials were defined as usual for mobile components able to move freely inside the solid. Gibbs stated a difference between surface tension and surface free energy for solids, characterizing γ as the work of creation of a new surface (per unit area) by cutting off a body. The applicability of thermodynamic relations when passing from a fluid to a solid has been discussed originally by Shuttleworth (66) and Eriksson (67). The surface tension or surface free energy of solids, unlike liquids, cannot be equated with the total energy in the surface layer—the "native" specific surface free energy plus "stress" energy (64). By definition, the former is the work performed in forming a unit area of new surface, whereas the latter pertains to the work spent in stretching the surface. To see the difference between the two concepts, it is useful to visualize new surface formation as a two-step process. First, the condensed phase is divided to produce the new surface, but the molecules in the new surface are held in the exact locations (relative to the remaining bulk phases) they occupied in the bulk. In the second step, the atoms in the newly formed surface are allowed to relocate into their most stable configuration. What this means, in effect, is that some units in the original (new) surface are "pulled" into the bulk by the unbalanced forces acting on them. In a liquid system, these two steps will occur essentially simultaneously because of the mobility of the units. In solids, on the other hand, greatly reduced atomic or molecular mobility means that the rearrangement will occur much more slowly, or perhaps not at all, on a reasonable timescale. The density of units in the new solid surface will therefore be something other than the "equilibrium" value. The surface may be compressed or stretched with no coincident change in the surface unit density. What changes will be the distance between units, which means a change in their interaction forces and therefore their free energy. A mechanical model is useful for understanding exactly what is meant by surface stress in a solid, as opposed to its surface tension. Suppose that a solid surface is cleaved in a direction perpendicular to the surface. As pointed out, the solid units in the new surface will not be able to relocate to attain their equilibrium positions relative to the bulk. To "make" their positions into "equilibrium" positions, one can think of applying some external force or lateral pressure on the surface units to hold them in place. The force per unit length of a new surface needed for this equilibrium situation is the surface stress. If one takes the average surface stress for two mutually perpendicular cuts, one will obtain the surface tension of the solid. For the cases of liquids or isotropic solids, the two stresses are equal and the surface stress and surface tension are equivalent. For an anisotropic solid, however, the two will not be equal (64).

Crystallization/precipitation, that is, the formation of an "ordered" solid phase, is obtained from a liquid phase if the chemical potential of the solid phase is lower than that of the material in the solution. A solution in which the chemical potential of the dissolved component is the same as that of the solid phase is said to be in equilibrium under the given set of conditions and is called saturated solutions. For a crystallization/precipitation to occur, the equilibrium concentration must be exceeded (supersaturation). Most important is the chemical reaction, which allows the rational control of the size, shape, crystal phase and solubility, nucleation, growth, and secondary ageing, processes that all depend on the interfacial tension of the system. Control over such a pertinent parameter is the key to rational growth of nanoparticles. Indeed, nucleation and growth processes rule the generation of solid phases from solutions (68,69). A maximum is found for the first derivative of the free enthalpy of nucleation with respect to the number of precursors N (which essentially represents the overall size). This maximum corresponds to the nucleation energy barrier for the system to overcome to form a solid particle. It depends on the interfacial tension at the cubic power, therefore, reducing the interfacial tension induces a considerable reduction of the nucleation energy barrier, which leads to the decrease of the nanoparticle size and its distribution. Given that the interfacial tension depends on the composition of the interface and thus on the physical chemistry of the dispersion medium, rational control over the size of nanoparticles can be achieved by lowering the interfacial tension of a system as demonstrated in the next section.

2.3. AQUEOUS THERMODYNAMIC STABILITY OF METAL-OXIDE NANOPARTICLES

Divided and ultra-divided systems such as nanoparticle dispersions are generally unstable with regard to the size and number of their constituents because the solid-solution interfacial tension, acting as a driving force, leads to a reduction of the surface area to minimize the dispersion free enthalpy. Such phenomenon known as surface energy minimization induces an increase in average particle size as a result of the decrease of the surface area at constant volume. For such a reason, these dispersions are usually considered thermodynamically unstable. However, they can be thermodynamically stabilized if, by adsorption, the interfacial tension of the system becomes very low. This phenomenon, well known for microemulsions, is quantitatively modeled and demonstrated for transition metal-oxide nanoparticles. When the pH of precipitation is sufficiently far from the point-of-zero charge and the ionic strength sufficiently high, the ripening of nanoparticles is avoided and their size can be monitored over one order of magnitude by tailoring solution pH and ionic strength. A model based on the Gibbs adsorption equation leads to an analytical expression of the water-oxide interfacial tension as a function of the pH and the ionic strength of the dispersion/precipitation medium. The stability condition, defined by a "zero" interfacial tension, corresponds to the chemical and electrostatic saturation of the water-oxide interface. In such a condition, the density of charged surface groups reaches its maximum, the interfacial tension its minimum, and further adsorption forces the surface area to expand and, consequently, the size of nanoparticles to decrease. A general control of the metal oxide nanoparticle size when precipitated far from their PZC is thus expected.

2.3.1. Definition and Concepts

The stability of solid nanoparticles in aqueous suspensions involves two different aspects: the permanence of their dispersion state (i.e., aggregation), that is, the *kinetic stability*; and the permanence of their division state (i.e., particle size), that is, the *thermodynamic stability*. The dispersion state of nanoparticles results from an energetic barrier, which is opposed to the approach of particles due to Brownian motion. Such stability is called *kinetic stability* due to the energy barrier resulting from the balance of attractive (van der Waals) and repulsive forces (electrostatic) between the surfaces (70). The classic Dejarguin–Landau–Verwey–Overbeek (DLVO) theory (71–73) as well as more recent ones (74–81) account very well for such a stability, and thus, the domains of dispersion, aggregation, and flocculation of most nanoparticle, colloidal, and even microbial systems are correctly predicted. On the other hand, the *thermodynamic stability* originates from surface energy minimization phenomena. Indeed, nanoparticles in suspension can undergo "spontaneous" changes in size, shape, and/or crystal structure. Such transformations occur because nanoparticles are formed under kinetic control during the initial steps of the nucleation and growth processes involved in the precipitation reactions. Further changes may occur during the ripening processes to reach a thermodynamic equilibrium of minimized surface energy (82,83). The evolution in size, known as Ostwald ripening (84–87), involves dissolution–crystallization processes that lead to secondary growth processes and, thus, the overall increase of the average particle size with time. The most important contribution of such a process originates from the difference in solubility between nanoparticles of different sizes—the smaller they are, the more soluble they are. Thus, the smallest dissolve and recrystallize on the larger ones. As a result, the decrease of the total surface area of the system allows reducing the free enthalpy of formation of the system. However, it is possible to obtain dispersions consisting of very fine particles due to very slow ripening processes, e.g., by growth limitation due to surfactants. Yet, the system is not in a stable state from a thermodynamical point of view (e.g., the exposed crystal faces may not be the thermodynamically stable ones), and consequently, the final particle size cannot generally be rationally controlled or predicted.

This raises empirical questions concerning the *thermodynamic stability* of divided and ultradivided systems (88). Is it possible to avoid the ripening of nanoparticles in suspension and to control their dimension by monitoring the precipitation conditions? In other words, are highly extended water-oxide interfaces thermodynamically stable? Indeed, thermodynamically stable dispersed systems do exist and microemulsions are good examples. They owe their stability to a very low surface tension at equilibrium due to adsorption phenomena (89–95). Two nonmiscible liquids, for instance, water and oil, do not mix and minimize their surface in contact by forming two separate phases. In the presence of an amphiphilic molecule, a tensio-active agent preferring to be in contact with the two phases simultaneously, the emulsification, i.e., the dispersion of one of the phases into the other by forming nanoparticles, is spontaneous. This implies that the free energy of the dispersion formation, $\Delta Gf = \gamma \Delta A - T \Delta S$, where $\Delta A(>0)$ stands for the interfacial area expansion, is negative. Such a condition can be fulfilled because of the interfacial tension lowering due to the localization of the tensio-active at the water–oil interface. In such conditions, the enthalpic term $\gamma\Delta A$ does not exceed the positive entropic term and a state of thermodynamic stability can be reached. Such conditions are fulfilled for reversible systems with steric control of the size, but such an effect is unlikely to apply for the precipitation of metal oxides because of strong metal–oxygen chemical bonds, a rigid crystal structure, a large enthalpic contribution, and relatively slow kinetics due to the low solubility of oxides in aqueous solutions.

However, metal-oxide materials usually do not spontaneously disperse in water. The aqueous precipitation of metal cations forms clusters and ultrafine particles, which often quickly grow to minimize the extent of their interfacial area. However, such a phenomenon can be avoided, for instance, at high pH and ionic strength, when the electrostatic surface charge is at maximum during the formation/precipitation of the nanoparticles (96). This suggests that the stability results from surface thermodynamics, as for microemulsions, by a lowering of the interfacial free energy. Such an idea was theoretically proposed several decades ago by several authors, for instance, Tolman (88), Hermans (97), Overbeek (90), and de Bruyn (98). The latter did introduce the concept of a point-of-zero surface tension for slightly soluble salt, i.e., ionic solids AB resulting from the chemical reaction of $A+$ and $B-$ ions such as silver halides (e.g., AgCl). Unfortunately, no experimental evidence supported their conclusions. Our results on the precipitation of magnetite (99) gave the first experimental evidence of the existence of thermodynamic stability for metal-oxide nanoparticles in aqueous solutions. Their growth mechanism involves "real" chemical reactions such as olation (the formation of hydroxo bridges by elimination of water) and oxolation (the two-step formation of oxo bridges by elimination of water) rather than "simple" electrostatic interactions and solvation/desolvation of ions in solutions involved in pure ionic crystals, such as halides, for instance. A model and its experimental corroboration developed by the author (100) are presented to support and demonstrate that thermodynamic stability is indeed a reality for metal-oxide nanoparticles. A quantitative expression of the stability conditions is given, which allows the prediction of the stability domains of metal-oxide nanoparticles in aqueous solutions.

2.3.2. Lowering of the Water-Oxide Interfacial Tension

Generally, the surface free enthalpy of a solid is given by

$$G^{\mathrm{Surf}} = \gamma A + \sum_j \bar{\mu}_i n_j^{\mathrm{Surf}}$$

In differential forms, one obtains

$$dG^{\mathrm{Surf}} = Ad\gamma + \gamma dA + \sum_j n_j^{\mathrm{Surf}} d\bar{\mu}_j + \sum_j \bar{\mu}_j dn_j^{\mathrm{Surf}}$$

which is also given by

$$dG^{\mathrm{Surf}} = -S^{\mathrm{Surf}} dT + \gamma dA + \sum_j \bar{\mu}_j dn_j^{\mathrm{Surf}}$$

Therefore, at constant temperature, one obtains the *Gibbs–Duhem* equation:

$$Ad\gamma + \sum_j n_j^{\mathrm{Surf}} d\bar{\mu}_j = 0$$

which, by surface area, gives the *Gibbs* adsorption equation:

$$d\gamma = -\sum_j \Gamma_j d\mu_j \tag{2.1}$$

with $\Gamma_j = \dfrac{n_j^{\mathrm{Surf}}}{A}$ and where Γ_j represents the surface density of component j. Such a quantity can be positive or negative. If Γ is positive, it is called *adsorption*, and if Γ is negative, *desorption*. Equation 2.2 shows that adsorption at any surface/interface ($\Gamma > 0$) induces the lowering of the surface/interfacial tension ($d\gamma < 0$).

2.3.3. Modeling of the Water-Oxide Interfacial Tension

By considering metal-oxide nanoparticles in an aqueous solution containing a base BOH and an electrolyte AB, A^- and B^+ are single charged and noncomplexing ions, so that they do not participate in any chemical bonding (i.e., chemisorption) with the oxide surface. They are acting exclusively as counter ions involved in the electrostatic screening process. Thus, at a given ionic strength, the chemical composition of the interface solely depends on the pH through the adsorption/desorption of protons and hydroxyls. The surface composition can be expressed by the electrostatic surface charge density σ_0 defined by

$$\sigma_0 = F(\Gamma_{H+} - \Gamma_{OH-}) \tag{2.2}$$

where F is the Faraday constant; $\Gamma_i = n_i/A$ is the adsorption density (surface excess), n_i is the number of mole of adsorbed species i, and A is the surface area of the adsorbent. σ_0 depends on sign and magnitude on the pH and ionic strength of the dispersion medium, respectively. At the point-of-zero charge (101–105), $\sigma_0 = 0$ because $\Gamma_{H+} = \Gamma_{OH-}$ and/or $\Gamma_{H+} = 0$ and $\Gamma_{OH-} = 0$.

Given that the surface of metal oxides holds an electrostatic charge in water due to the presence of hydroxyl surface groups, applying Eq. 2.2 requires the introduction of the electrochemical potential of adsorbed species: $\mu_j^* = \mu_j + z_j F\Psi$, where z_j is the formal charge of species j and Ψ is the electrostatic surface potential. In addition, by considering only protons and hydroxyl ions (no adsorption of other ions) as potential determining ions, i.e., participating in the charging of the surface, Eq. 2.1 becomes

$$d\gamma = -\left(\Gamma_{H+} d\mu_{H^+}^* - \Gamma_{OH-} d\mu_{OH^-}^*\right) \tag{2.3}$$

In addition, by considering that the most commonly reported conditions of precipitation of metal oxides, hydroxides, and oxyhydroxides occurs in basic medium (i.e., alkalinization of metal cations), Γ_H can be neglected.

Introducing Eq. 2.1 into 2.3 gives

$$d\gamma = -\frac{\sigma_0}{F} d\mu_{OH} - \sigma_0 d\Psi_0 \tag{2.4}$$

Equation 2.4 shows that γ is maximum at pH = PZC, where σ_0 and Ψ are both equal to zero, and that the lowering of γ by adsorption at pH greater than PZC results from two contributions, i.e., *chemical* and *electrostatic*, respectively.

2.3.3.1. Chemical Contribution Following the argument of Stol and de Bruyn (98), the Langmuir model of adsorption can be used because, at maximum, a monolayer is involved by the proton adsorption/desorption phenomena. Such a model assumes equivalent surface sites and that the adsorption energy is independent of the density of adsorption. Consequently, it follows that

$$d\mu_{OH^-} = \mathrm{RT}\, d\left[\mathrm{Ln}\left(\frac{\theta}{1-\theta}\right)\right] \tag{2.5}$$

θ is the fraction of groups from which the proton was desorbed (pH > PZC). θ has to be evaluated as a function of σ_0.

If n^- and n^+ are the density of negatively and positively charged groups, respectively, the net surface charge is given by

$$\sigma_0 = F(n^+ - n^-) \tag{2.6}$$

The relation between the surface charge density σ_0 obtained at any pH and the maximum surface charge density σ_{max} at pH values far from the PZC can thus be easily obtained. If n_{max} is the total density of charged groups, $n_{max} = n^+ + n^-$, the fraction of negatively charged groups is given by $\theta = \frac{n^-}{n_{max}}$, and therefore, $1 - \theta = \frac{n^+}{n_{max}}$ represents the fraction of positively charged groups. Consequently, the surface charge density σ_0 is given by the following equation:

$$\sigma_0 = \frac{F}{A}\left(n^+ - n^-\right) = \frac{F}{A}(1 - 2\theta)n_{max} = (1 - 2\theta)\sigma_{max} \tag{2.7}$$

σ_0 equals to zero when $n^+ = n^- = n_{max}/2$. Such an assumption only holds if the positive and negative charges of the same surface group are equal in absolute value. It is obviously not always verified, but such an approximation facilitates the calculation and is not a real limitation, especially far from the PZC when σ_0 is very close to σ_{max}.

Equation 2.7 gives

$$\theta = \frac{\sigma_{max} - \sigma_0}{2\sigma_{max}}$$

and

$$d\mu_{OH^-} = RTd\left[\mathrm{Ln}\left(\frac{\sigma_{max} + \sigma_0}{\sigma_{max} - \sigma_0}\right)\right]$$

and Eq. 2.5 gives, finally, for the chemical contribution:

$$d\mu_{OH^-} = 2RT\left[\frac{\sigma_{max}}{\sigma_{max}^2 - \sigma_0^2}\right]d\sigma_0 \tag{2.8}$$

2.3.3.2. Electrostatic Contribution

Assuming a double-layer model for the electrostatic charge and considering that in the absence of specific adsorption (chemisorption) of ions from the solution, the surface charge density σ_0 is compensated by the charge density of the diffuse layer σ_d. With an ionic strength I set by monovalent ions, σ_d is given by the Grahame equation:

$$\sigma_0 = -\sigma_d = \frac{4FI}{\kappa}\sinh\left(\frac{F\Psi_0}{2RT}\right) \tag{2.9}$$

where Ψ_0 is the surface potential and κ^{-1} is the Debye length. In water at 25 °C, $\kappa = 0.329\sqrt{I}$ (Å^{-1}) with $I = \frac{1}{2}\sum_i c_i z_i^2 \cdot c_i$ and z_i are the concentration and the charge of the ions of the electrolyte, respectively.

The relation between the water-oxide interfacial tension and the solution pH is obtained by introducing Eqs. 2.8 and 2.9 into Eq. 2.4, which gives

$$d\gamma = -\left[\frac{2\mathrm{RT}}{\mathrm{F}}\frac{\sigma_{\max}\sigma_0}{\sigma_{\max}^2 - \sigma_0^2}\right]d\sigma_0 - \left[\frac{4\mathrm{FI}}{\kappa}\sinh\left(\frac{\mathrm{F}\Psi_0}{2\mathrm{RT}}\right)\right]d\Psi_0 \tag{2.10}$$

The integration of Eq. 2.10 from the PZC, where $\sigma_0 = 0$, $\Psi_0 = 0$, and $\gamma = \gamma_0$:

$$\int_{\gamma_0}^{\gamma} d\gamma = -\frac{\mathrm{RT}\sigma_{\max}}{F}\int_0^{\sigma_0}\frac{2\sigma_0}{\sigma_{\max}^2 - \sigma_0^2}d\sigma_0 - \frac{4\mathrm{FI}}{\kappa}\int_0^{\Psi_0}\sinh\left(\frac{\mathrm{F}\Psi_0}{2\mathrm{RT}}\right)d\Psi_0$$

gives

$$\Delta\gamma = \gamma - \gamma_0 = \frac{\mathrm{RT}\sigma_{\max}}{F}\mathrm{Ln}\left[1 - \frac{\sigma_0^2}{\sigma_{\max}^2}\right] - \frac{8\mathrm{RTI}}{\kappa}\left[\cosh\left(\frac{\mathrm{F}\Psi_0}{2\mathrm{RT}}\right) - 1\right] \tag{2.11}$$

Moreover, assuming that the water-oxide interface follows the Nernst's law, which is well established in the literature for transition metal oxides over a wide range of pH (106), the surface potential ψ_0 is given as follows:

$$\frac{\mathrm{F}\Psi_0}{2.3\mathrm{RT}} = \mathrm{PZC} - \mathrm{pH} = \Delta\mathrm{pH} \tag{2.12}$$

One obtains, finally (100),

$$\Delta\gamma = \gamma - \gamma_0 = 25.7\sigma_{\max}\mathrm{Ln}\left[1 - \mathrm{I}\left(\frac{0.117\sinh(1.15\Delta\mathrm{pH})}{\sigma_{\max}}\right)^2\right] - 6\sqrt{\mathrm{I}}[\cosh(1.15\Delta\mathrm{pH}) - 1] \tag{2.13}$$

where γ is expressed in mJ/m^2, I in mol/l, and $\sigma_{\max}$ in C/m^2.

Figure 2.1 shows the variation of the interfacial tension $\Delta\gamma$ as a function of pH, as calculated from Eq. 2.13 for different values of the ionic strength. A negative divergence of $\Delta\gamma$ is observed when the pH increases. The decrease in interfacial tension takes place for a pH variation ΔpH = pH – PZC, which decreases as the ionic strength increases. A large decrease in interfacial tension in the order of hundreds of mJ/m^2 (mN/m) is predicted due to the negative divergence of $\Delta\gamma$ as a function of ΔpH. Such behavior is supported by experimental variations obtained for Al_2O_3

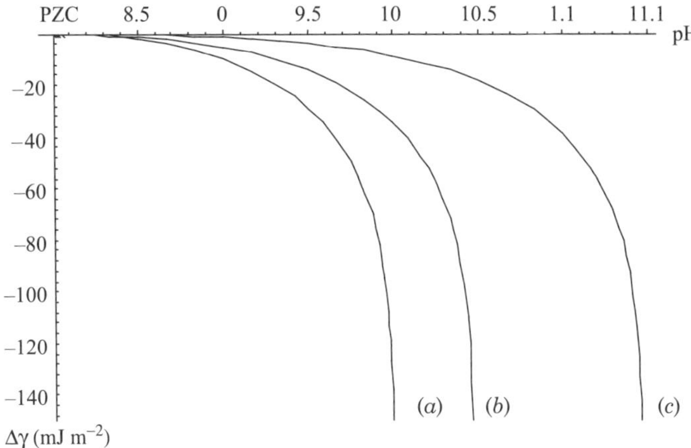

Figure 2.1. Variation of $\Delta\gamma$ as a function of pH at ionic strengths of (*a*) 3M, (*b*) 1M, and (*c*) 0.1M as calculated from Eq. 2.13 (PZC = 8 and $\sigma_{max} = 1\,Cm^{-2}$) from Ref. 100 with permission.

(106) and ThO_2 (107) from the integration of experimental σ-pH titration curves at different ionic strengths, where a similar tendency was observed (Figure 2.2). The relative magnitude of the chemical and electrostatic contributions to the interfacial tension decrease is illustrated in Figure 2.3. The electrostatic contribution initially prevails at low pH (near the PZC), but the chemical contribution quickly dominates leading to a negative divergence of the equation at higher pH values.

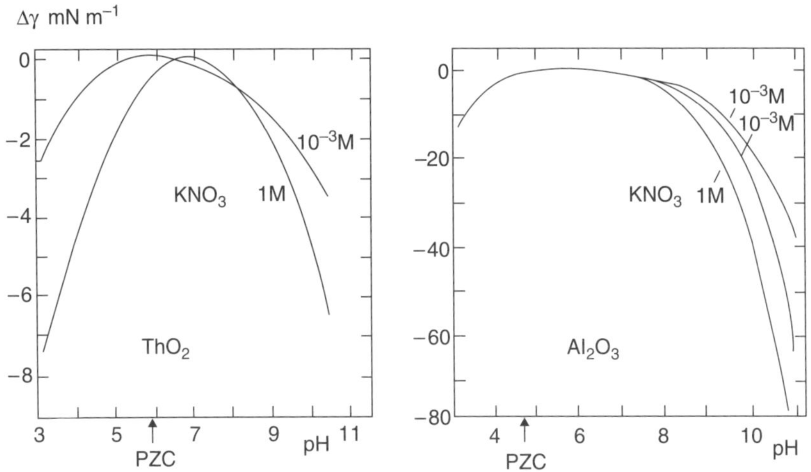

Figure 2.2. Variation of $\Delta\gamma$ as a function of pH at given ionic strengths for various metal oxides as calculated from experimental σ-pH plots from Refs. 106 and 107 permission.

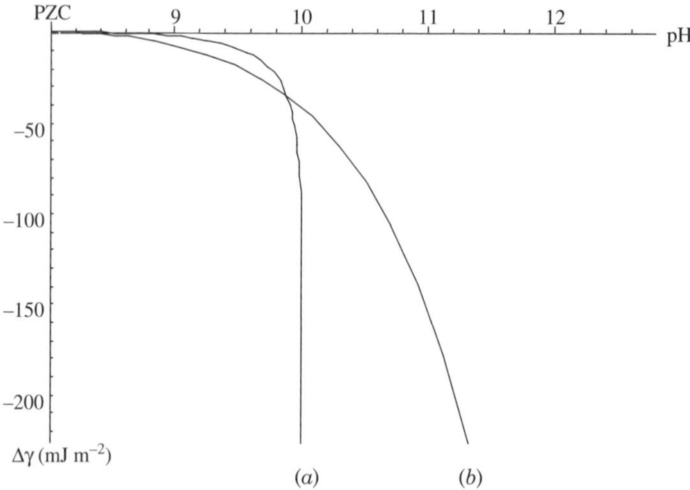

Figure 2.3. (*a*) Chemical and (*b*) electrostatic contributions to the variation of $\Delta\gamma$ as calculated from Eq. 2.13 (I = 3 M, PZC = 8, and σ_{max} = 1 Cm^{-2}) from Ref. 100 with permission.

Assuming an interfacial tension at pH = PZC, γ_0 equals to 100 mJ/m^2 (although the actual value does not bare much importance given the negative divergence of the equation), Eq. 2.13 shows that the interfacial tension reaches zero at reasonable solution pH values ranging from pH $\sim$ 9.5 at an ionic strength of 10 M to pH $\sim$ 11.5 at 0.1 M (Figure 2.4). Such results suggest a logarithmic variation of the stability conditions with the ionic strength, given that a change of two orders of magnitude in

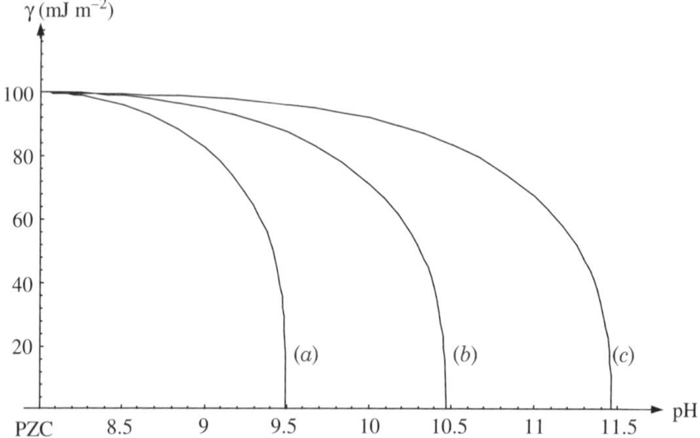

Figure 2.4. Variation of the interfacial tension γ as a function of pH at ionic strengths of (*a*) 10M, (*b*) 1M, and (*c*) 0.1M as calculated from Eq. 2.13 (γ_0 = 100 mJ/m^2, PZC = 8, and σ_{max} = 1 Cm^{-2}) from Ref. 100 with permission.

ionic strength (ca 10 to 0.1) yields a two pH units change in the intercept pH value (ca 9.5 to 11.5).

Nanoparticles in dispersions are considered thermodynamically stable (at a thermodynamic equilibrium state) if the particle size (extension of the area) does not change with time. As for microemulsions, a true thermodynamic equilibrium is also characterized by the modification of the particle size in response to a change of one parameter of the precipitation, but kinetics may be a limiting factor.

In principle, the conditions for the thermodynamic stability have to be derived from the expression of the free energy of formation (ΔG) of the system. However, an exact formulation of ΔG is unlikely to be established because numerous effects must be taken into consideration, such as the interactions between particles, the nonideal behavior of colloidal dispersions, the entropic effects, as well as the surface charge and interfacial tension dependence on the surface curvature radius (89–95, 108,109). From a qualitative point of view, a large positive value of the interfacial tension leads to the minimization of the surface area (growth) for any dispersed system, whereas a negative value of γ induces a spontaneous dispersion (dissolution). Thus, assuming a zero value of γ as the criterion of thermodynamic stability for inorganic nanoparticles appears reasonable. In fact, a low, yet positive, value of γ must correspond to the equilibrium condition because of the positive entropic contribution to ΔG due to the formation of a suspension. For instance, for microemulsions formed in NaC1 aqueous solution/heptane/AOT or in other conditions, ultra-low, yet positive, interfacial/surface tensions at equilibrium ($\approx 10^{-2}$ to 10^{-3} mJm^{-2}) have experimentally been measured (89–95).

In fact, the exact value of the interfacial tension corresponding to the equilibrium is of very little importance because of the negative divergence of γ correlated with the pH increase (Eq. 2.13, Figure 2.1). Thus, it seems sufficient to consider the pH value, for which γ equals zero, namely, the point-of-zero interfacial tension, PZIT. From Eq. 2.11, the PZIT is indeed the pH value for which $\sigma = \sigma_{max}$, that is, when the surface is chemically and electrostatically saturated.

Equation 2.13 does not have a real zero value; therefore, an analytical expression of the PZIT cannot directly be deduced. However, Eq. 2.13 can be simplified by neglecting the electrostatic contribution because, as pointed out, the chemical contribution induces a negative divergence of the equation. Therefore by solving the *Grahame* equation (Eq. 2.9), the PZIT can be established analytically:

$$\sigma_{max} = 0.117\sqrt{I}\sinh(1.15\Delta pH^*) \tag{2.14}$$

with $\Delta pH^* = PZC - PZIT$.

Given that $x = \sinh y \Rightarrow y = Ln\left(x + \sqrt{x^2 + 1}\right)$, Eq. 2.14 gives

$$\Delta pH^* = \frac{2.3}{1.15}\log\left(\frac{\sigma_{max}}{0.117\sqrt{I}} + \sqrt{\frac{\sigma_{max}^2}{0.117^2 I} + 1}\right) \tag{2.15}$$

As the argument of the square root is always positive and much larger than 1, one obtains

$$\sqrt{\frac{\sigma_{\max}^2}{0.117^2 \mathrm{I}}+1} \approx \frac{\sigma_{\max}}{0.117\sqrt{\mathrm{I}}}$$

and thus, Eq. 2.15 gives

$$\Delta \mathrm{pH}^* = 2\log\left(\frac{2\sigma_{\max}}{0.117\sqrt{\mathrm{I}}}\right)$$

and finally

$$\mathrm{PZIT} = \mathrm{PZC} + 2.46 + 2\log\sigma_{\max} - \log \mathrm{I} \tag{2.16}$$

Equation 2.16 shows that the PZIT depends on the PZC and on the maximum surface charge $\sigma_{\max}$, which are two characteristics that are easily experimentally measurable (cf. Section 2.2.1), and on the ionic strength, which is set experimentally by the precipitation conditions. Accordingly, the higher the ionic strength, the lower is the PZIT and, consequently, the greater the stability domain will be. The PZIT values calculated from Eq. 2.16 satisfy Eq. 2.13, and are in very good agreement with those deduced from the $\Delta\gamma$-pH curves (Figure 2.4) for the same set of parameters. Such an agreement justifies the approximations used in the calculations. Moreover, the variation is indeed a logarithmic function of the ionic strength as suggested from the calculated results reported in Figure 2.4.

Equation 2.16 shows that it is possible, with realistic and practical experimental conditions, to reach $\gamma \to 0$ for metal oxides in aqueous solutions by significantly raising the pH and the ionic strength of the precipitation medium from the PZC. Physically, the interfacial tension cannot become negative given that the precipitation does occur. In fact, for pH $\geq$ PZIT at a given ionic strength, the surface is already saturated and the interfacial tension cannot decrease anymore. γ must be constant with a very low, yet positive, value. An analogous behavior arises for microemulsions. The interfacial tension falls during the adsorption of surfactant at the interface. When the interface is saturated, the surfactant starts to aggregate, and, because the aggregates are not surface-active, γ remains constant. A further lowering is, however, possible if a co-surface-active component is added (89–95).

Thus, when the precipitation of an oxide is performed at pH $>$ PZIT, one can expect nanoparticles to form at the thermodynamic equilibrium condition, $\gamma = (\partial G/\partial A) = 0$, which leads to stable nanoparticle sizes with time and where a direct correlation of the size with the synthesis conditions. On the other hand, if pH $<$ PZIT, the interface is not saturated; therefore, γ will remain at a significant positive value. Thus, ripening and secondary growth processes would lead to the reduction of the surface area (due to the surface energy minimization) and, therefore, to an increase of the average particle size with time. Table 2.2 summarizes the expectations of the model.

TABLE 2.2. Scheme of the Model Expectation for the Stability of Metal Oxide Nanoparticles as a Function of the pH of Precipitation at a Given Ionic Strength.

	Domain of Precipitation	
	Unstable	Stable
Experimental condition	PZC $\leq$ pH $<$ PZIT	pH $\geq$ PZIT
Surface charge	$\sigma < \sigma_{\text{max}}$	$\sigma = \sigma_{\text{max}}$
Interfacial tension	$\gamma > 0$	$\gamma \approx 0$

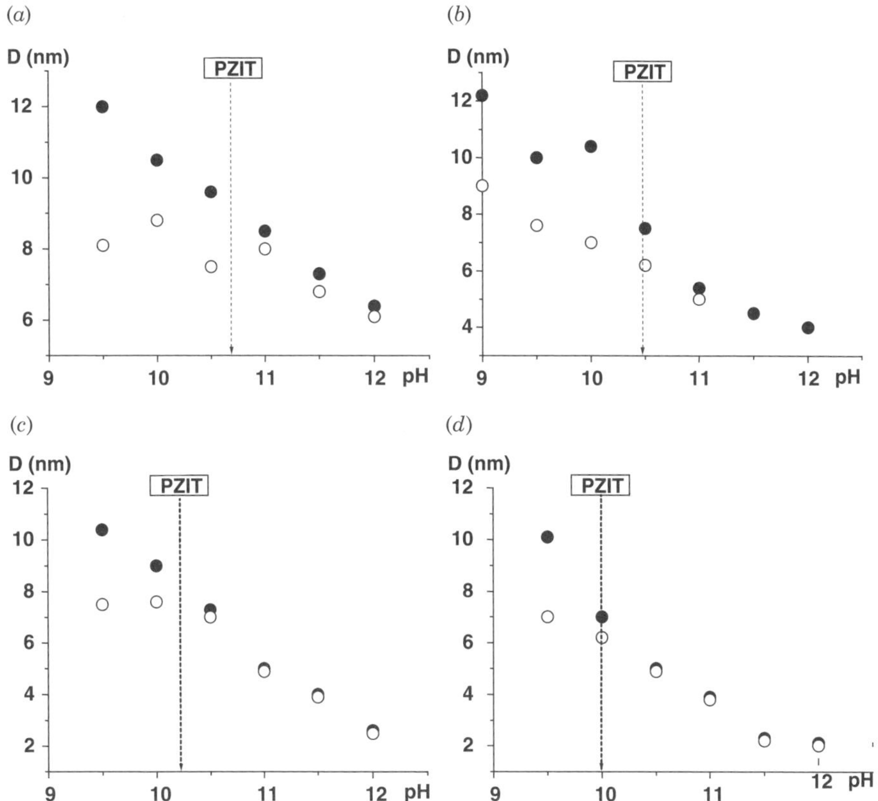

Figure 2.5. Influence of the precipitation medium on the average particle size of magnetite nanoparticles observed by TEM immediately after the synthesis (○) and after 8 days of ageing in suspension (•) for precipitations performed at pH 12.0 and (*a*) I = 0.5 M, (*b*) I = 1.0 M, (*c*) I = 1.5 M, and (*d*) I = 3.0 M. The PZIT is displayed as dotted lines and indicates the limit between the stable and unstable domains as calculated from Eq. 2.16 with PZC = 8.1 and $\sigma_{\text{max}} = 0.84\,\text{C/m}^2$ (such values were obtained by titration methods) from Ref. 100 with permission.

2.3.4. Experimental Evidence

To probe and clearly establish the thermodynamic stabilization and investigate its action on ageing phenomena, a series of experiments was conducted. A set of experiments where the average particle size of magnetite was measured by TEM immediately after the synthesis and after one week of ageing in suspension at room temperature were conducted. Figure 2.5 clearly shows that two distinct domains of pH over a large range of ionic strength can clearly be identified. A pH range where the particle size strongly increases with time (unstable domain) and another range where no evolution occurs (stable domain) are observed experimentally. The stable domain clearly extends as the ionic strength increases as suggested from Eq. 2.16. The PZIT value calculated from Eq. 2.16 corresponds to the limit between these two domains. The model seems to account well for a stabilization effect of the nanoparticle size, which is not from a kinetic origin. Such conclusions imply that the particle size within the stable domain of precipitation is directly related to the synthesis conditions as suggested by the model and confirmed in Figure 2.5 from the study of controlled precipitation of magnetite (99,100).

2.4. AQUEOUS DESIGN OF METAL-OXIDE NANOSTRUCTURES

2.4.1. Experimental Method

A novel experimental technique has emerged recently as a simple and powerful tool to fabricate, at low cost and mild temperatures, large areas of metal-oxide nano- to microparticulate thin films (110). Three-dimensional arrays consisting of oriented anisotropic nanoparticles are easily generated with enhanced control over orientations and dimensions. The synthesis involves the controlled heteronucleation of metal oxides onto substrates from the hydrolysis-condensation of metal salts in aqueous solutions. The most pertinent parameter to control the nucleation and growth and, therefore, the overall design and architecture of a thin film is the interfacial free energy of the system (111).

By applying the above-mentioned thermodynamic concepts to the thin film processing technology, an inexpensive and effective aqueous growth technique at mild temperatures has been developed to produce functionalized coating of metal-oxide materials onto various substrates. Such a technique allows the generation of advanced nano/micro particulate thin films without any template, membrane, surfactant, applied external fields, or specific requirements in substrate activation, thermal stability, or crystallinity. Given that the crystallites grow from the substrate, a large choice of thin film/substrate combinations is offered, which provides consequently a better flexibility and a higher degree of materials engineering and design.

This method consists of heating an aqueous solution of metal salts (or complexes) in the presence of a substrate at moderated temperatures (below 100 °C) in a closed vessel. Therefore, such technique does not require high-pressure containers and is entirely recyclable, safe, and environmental-friendly because only water is used as solvent. Such a process avoids the safety hazards of organic solvents and their eventual evaporation and potential toxicity. In addition, because no organic solvents or surfactants are present, the purity of the materials is substantially improved. The residual salts are easily washed out by water due to their high solubility. In most cases, no additional heat or chemical treatments are necessary, which represent a significant improvement compared with surfactant-, template-, or membrane-based synthesis methods.

A full coverage of the substrate is obtained within a few hours provided that the heat capacity of water and surface coverage control are achieved by monitoring the synthesis time in the early stages of the thin film growth. Partial coverage is obtained within the first hours, which may be necessary for certain applications to adjust and tune the overall physical properties of devices (e.g., optical properties of multi-band-gap thin films). Such a technique is a multideposition technique, and the growth of layer-by-layer of thin films is readily obtainable. The development of multilayer thin films of various morphologies and/or of various chemical compositions—i.e., composite, multi-band gap, and doped thin films—is reached. The complete thin film architecture may thus be modeled, designed, and monitored to match the application requirements. In most cases, it should improve the physical and chemical properties of the devices. It also gives the capacity to create novel and/or improved thin film integrated devices. Growing thin layers directly from the substrate does substantially improve the adherence and the mechanical stability of the thin film compared with the standard solution and colloidal deposition techniques such as spin-and-dip coating, chemical bathing, screen printing, or doctor blading. Moreover, given that such materials do precipitate in homogeneous solution from molecular scale compounds (i.e., condensed metal complex), they will virtually grow on any substrate. It goes without saying that the overall mechanical stability of the thin films does vary from substrate to substrate, but in most cases, strong adhesion is observed. Scale-up is potentially easily feasible, and this concept and the synthesis method is theoretically applicable to all water-soluble metal ions likely to precipitate in solution. Large-scale manufacturing at low cost is therefore achievable with such a technique. In addition to all these industrial-related advantages, such a technique is also potentially very interesting due to the compatibility of water and aqueous solution to biological compounds. For instance, 3D arrays of composite bio-nanomaterials have been obtained using the aqueous chemical growth. Such concepts and thin film processing technique have been applied successfully to oxides and oxyhydroxides of transition metals.

2.4.1.1. Structural Control In addition to the control of particle size, precipitation of nanoparticles at low interfacial tension allows the stabilization of metal-oxide and oxyhydroxide metastable crystal structures. An insight into structural control can thus be provided. Crystal phase transitions in solution generally operate through a dissolution-recrystallization process to comply with the surface energy minimization requirement of the system. Indeed, when a solid offers several allotropic phases and polymorphs, it is typically the one with the highest solubility and consequently the lowest stability (i.e., the crystallographic metastable phase), that precipitates first. This is understood by considering the nucleation kinetics of the solids. At a given supersaturation ratio, the germ size is as small (and the nucleation rate as high) as the interfacial tension of the system is low. Thus, given that the solubility is inversely proportional to the interfacial tension, the precipitation of the most soluble phase, and consequently the less stable thermodynamically, is therefore kinetically promoted. Due to its solubility and metastability, this particular phase is more sensitive to secondary growth and ageing, which leads to crystallographically more stable phases, essentially by heteronucleation. Secondary growth and ageing processes are delicate to control, and the phase transformations appear within a few hours to a few days in solution, whence the resulting undesired mixing of allotropic phases and polymorphs. However, by careful consideration of the precipitation conditions (i.e., at thermodynamically stable conditions), such phenomena may be avoided when nanosystems are precipitated at low interfacial tension.

2.4.1.2. Morphological Control Moreover, when the thermodynamic stabilization is achieved, not only the size and the crystal structure can be tailored but also the shape may be tailored to a certain extent. For instance, at low interfacial tension, the shape of nanoparticles does not necessarily require being spherical; indeed, very often nanoparticles develop a spherical morphology to minimize their surface energy because the sphere represents the smallest surface for a given volume. However, if the synthesis and dispersion conditions are suitable (i.e., yielding to the thermodynamic stabilization of the system), the shape of the crystallites will be driven by the symmetry of the crystal structure as well as by the chemical environment, and various morphologies may therefore be developed. Manipulating and controlling the interfacial tension enables nanoparticles to grow with sizes and shapes tailored for their applications. Applying the appropriate solution chemistry (precursors and precipitation/dispersion conditions) to the investigated transition metal ion along with the natural crystal symmetry and anisotropy or by forcing the material to grow along a certain crystallographic direction by controlling chemically the specific interfacial adsorptions of ions and/or ligand and crystal-field stabilization, one can reach the ability to develop purpose-built crystal morphology without the use of surfactants and other growth-limiting agents.

2.4.1.3. Orientation Control To develop the capability of growing nano- to micro-particulate thin films from aqueous solution and controlling the orientation of large arrays of anisotropic nanoparticles onto a substrate, one has to consider the differences between homogeneous and heterogeneous nucleation phenomena. In most cases, homogenous nucleation of solid phases from solutions requires a higher activation energy barrier, and therefore, heteronucleation is promoted and energetically more favorable. Indeed, the interfacial energy between two solids is generally smaller than the interfacial energy between a solid and a solution, and therefore, nucleation may take place at a lower saturation ratio onto a substrate than in solution. Nuclei will grow by heteronucleation onto the substrate, and various morphologies and orientation monitoring can be obtained by experimental control of the chemical composition of the precipitation medium. To illustrate and demonstrate such capabilities, the example of an anisotropic building block is taken and the possibilities of experimentally monitoring its orientations onto a substrate using the aqueous

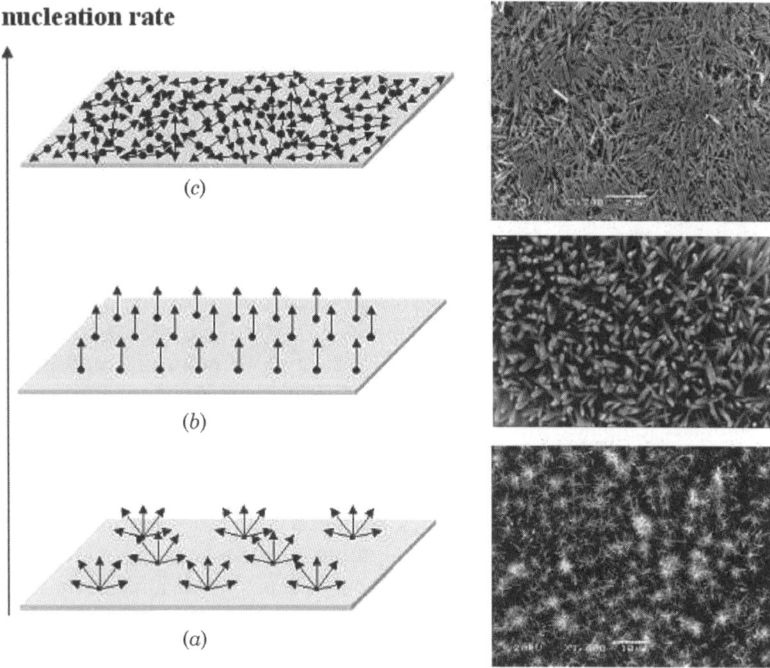

Figure 2.6. Schematic representation and scanning electron microscope (SEM) images of the effect of heteronucleation rate on the orientation of anisotropic crystalline building blocks of ZnO (wurtzite structure) grown onto substrates by the aqueous chemical growth technique at 90 °C.

chemical growth thin film processing technique are exposed hereafter and illustrated in Figure 2.6.

- Anisotropic building blocks with multi-angular orientation onto substrates may be generated when the number of nuclei is exceedingly limited by the precipitation conditions (through the chemical and electrostatic monitoring of the interfacial tension). The system will be forced to promote twinning and the preferential epitaxial three-dimensional growth of the rods along their easy axis from a very limited number of nuclei, which will induce, a type of star/flower-shape morphology (Figure 2.6*a*)
- Anisotropic building blocks with a perpendicular orientation onto substrates may be obtained when the number of nuclei is slightly limited by the precipitation conditions. The slow appearance of a limited number of nuclei will allow the slow growth (according to the crystal symmetry and relative face velocities) along the easy direction of crystallization. As a result, a condensed phase of anisotropic nanorods parallel to the substrate normal will be generated (Figure 2.6*b*)

- Anisotropic building blocks with a parallel orientation onto substrates can be obtained when the nucleation rate is enhanced by the precipitation conditions. Indeed, the fast appearance of a large number of nuclei will result in a rapid two-dimensional growth. The stacking of anisotropic nanoparticles with random orientation between each other but with an overall parallel orientation with respect to the substrate normal is therefore promoted (Figure 2.6*c*)

The ability to design materials consisting of anisotropic nanoparticles of different orientations enables the experimental studies of the angular dependence influences on the physical/chemical properties and electronic structure of materials and gives further opportunities for materials design and engineering (e.g., increased dimensionality). The outcome allows the fabrication of innovative and functional nano- to micro-particulate thin films of metal oxides directly onto substrates without membrane, template, surfactant, undercoating, or applied external fields. Such purpose-built nanomaterials are prepared in such a way that particle nucleation is separated from growth, giving exquisite control of particle size. In many cases, the differences between interfacial energies among crystal facets can be exploited to produce controlled anisotropic shapes. In addition, because the purpose-built nanomaterials are fabricated at low temperature and can be deposited on almost any substrate, it is possible to deposit one phase on another. This opens up the possibility of using the anisotropy of one material to generate a high surface area substrate upon which a second, active, material is deposited. In effect, this produces nanocomposites—materials combining properties and architectures from multiple phases to achieve results not available from any single phase.

In summary, by adjusting the experimental conditions to reach the thermodynamic stability of a system, the nanoparticle size, shape, and crystal structure may be tuned and optimized. It allows the functionalized design of nanomaterials and the ability to quantitatively probe the influence of such parameters on the physical and chemical properties of metal-oxide nanoparticles and nanoparticulate materials.

2.4.2. Major Achievements

The theoretical concepts and experimental method described in this chapter have been successfully applied to develop a new generation of functional materials, the so-called *purpose-built materials* (112–114). Indeed, when thermodynamic stabilization is applied to heterogeneous nucleation rather than homogeneous nucleation, not only the size of spherical nanoparticles can be controlled but also the shape and the orientation of anisotropic building blocks onto various substrates can be tailored to build smart nanostructures (115). Indeed, the fabrication of highly oriented crystalline arrays of large physical area consisting of nanorods of iron oxides (116), iron-chromium sesqui-oxide nanocomposites (117), iron oxyhydroxides, and iron metal (118) have been successfully designed by following such ideas. In addition, ZnO nanorods and nanowires (119), microrods (120), microtubes (121), along with other morphologies (122), nanowires and nanoparticles of manganese oxides (123), and arrays consisting of sub-monolayers of nonaggregated mesoparticles of α-Cr_2O_3 (124) have also been obtained by such method. More recently, ordered arrays of *c*-axis elongated nanorods of rutile SnO_2 (125) with a squared cross section have also been successfully fabricated following the same concept and aqueous thin film processing method. Such purpose-built nanomaterials have been designed, for instance, to improve the photovoltaic (126–128), photoelectrochemical (129,130), gas-sensing (131) properties of large band gap semiconductors (132). Better fundamental understanding of the electronic structure (133) and orbital symmetry contribution (134,135) of II–VI semiconductor nanomaterials as well as quantum confinement effects in TiO_2 quantum dots (136) and α-Fe_2O_3 ultra-fine nanorods (137) have also been obtained with such materials. Finally, the direct application of such a model allowed the demonstration of the thermodynamic stabilization of metastable crystal phases such as hydroxides and oxyhydroxides of transition metals in aqueous solutions (138) as well as the fabrication of novel bio-nanocomposites by tailoring the conformation of bioactive molecules adsorbed onto designed metal-oxide nano-arrays (139).

2.5. CONCLUSIONS AND PERSPECTIVES

In conclusion, an insight on aqueous interfacial thermodynamics of metal oxides has been presented. Thermodynamic stabilization of metal-oxide nanoparticles was demonstrated. A thermodynamic model illustrates the effect of acidity and ionic strength on protonation–deprotonation equilibria of surface hydroxylated groups, on the electrostatic charge, and consequently, on the water-oxide interfacial tension. The resulting change in interfacial chemical composition induces a decrease of the interfacial tension as stated by the Gibbs adsorption equation. The surface contribution to the free enthalpy of formation of metal-oxide nanoparticles from aqueous solutions is substantially lowered, allowing the surface area of the system to increase and therefore the creation of thermodynamically stable dispersion of metal oxide nanoparticles. The stability condition is satisfied when the interfacial tension gamma is close to zero. Such equilibrium is reversible, and thus, the average particle size is indeed directly related to the precipitation conditions. The nanoparticle size can be stabilized and controlled over an order of magnitude by precipitating the oxide nanoparticles far from their PZC, that is, at a condition where the interface is electrostatically saturated, for instance, at high pH and high ionic strength for spinel iron oxides.

Combining such thermodynamical phenomena occurring at the water-oxide interface with the direct heteronucleation growth of metal oxide, the rational fabrication of advanced nanostructures is achievable by experimental monitoring of the interfacial thermodynamics and kinetics of nucleation, growth, and ageing processes. Such ability is reached by the chemical and electrostatic control of the interfacial free energy as predicted by a model based on the Gibbs adsorption equation. Aqueous chemical growth exploits such concepts and enables the design and the fabrication of functional purpose-built metal-oxide thin films with well-defined textures onto various substrates at low temperature and at low cost. Therefore, the economical large-scale manufacturing of a new generation of smart metal-oxide materials with extended functionality and surface nanoengineered morphology is readily available.

Such purpose-built materials should contribute to develop novel and optimized devices for high technological applications as well as for biological- and medical-related ones. They also constitute chemically and structurally well-defined samples that enable reaching better fundamental knowledge and understanding of nanostructured metal oxides and their fascinating physical, chemical, and structural properties as well as their unique electronic structure.

REFERENCES

(1) Drexler, E.K. *Engines of Creation: The Coming Era of Nanotechnology*; Doubleday & Company: New York, 1987.

(2) Drexler, E.K. *Nanosystems: Molecular Machinery, Manufacturing, and Computation*; Wiley: New York, 1992.

(3) Wilson, M. *Nanotechnology: Basic Science and Emerging Technologies*; Simmons, M.; Smith, G.; Kannangara, K. (Editors); CRC Press: Boca Raton, FL, 2002.

(4) Regis, E.; Chimsky, M. *Nano: The Emerging Science of Nanotechnology, Vol. 1*; Little, Brown & Company: New York, 1996.

(5) *Nanotechnology: Research and Perspectives*; Crandall, B.C.; Lewis, J. (Editors); MIT Press: Cambridge, MA, 1992.

(6) *Prospects in Nanotechnology: Toward Molecular Manufacturing*; Krummenacker, M.; Lewis, J. (Editors); Wiley: New York, 1994.

(7) *Nanotechnology*; Timp, G. (Editor); AIP Press: New York, 1999.

(8) *Nanoscale Materials in Chemistry*; Klabunde, K.J. (Editor); Wiley-Interscience: New York, 2001.

(9) Bard, A.J. *Integrated Chemical Systems: A Chemical Approach to Nanotechnology*; Wiley: New York, 1994.

(10) Kawata, S. *Nano-Optics*; Irie, M.; Ohtsu, M. (Editor); Springer-Verlag: New York, 2002.

(11) *Nano-optoelectronics: Concepts, Physics, and Devices*; Grundmann, M. (Editor); Springer: Berlin, 2002.

(12) Nalwa, H.S. *Handbook of Nanostructured Materials and Nanotechnology*; Academic Press: San Diego, CA, 1999.

(13) Mitura, S. *Nanomaterials*; Elsevier Science: New York, 2000.

(14) Nalwa, H.S. *Nanostructured Materials and Nanotechnology*; Academic Press: San Diego, CA, 2001.

(15) *Nanomaterials: Synthesis, Properties, and Applications*; Edelstein, A.S.; Cammarata, R.C. (Editors); Institute of Physics Publishing: Bristol, PA, 1998.

(16) *Nanostructured Materials*; Hofmann, H.; Rahman, Z.; Schubert, U. (Editors); Springer-Verlag: Vienna, 2002.

(17) *Nanostructured Materials: Processing, Properties and Potential Applications*; Koch, C.C. (Editor); Noyes Publications/William Andrew Publishing Norwich: New York, 2002.

(18) *Fundamentals of Tribology and Bridging the Gap Between the Macro and Micro/Nanoscales*; Bhushan, B. (Editor); Kluwer Academic Publishers: Dordrecht, the Netherlands, 2001.

(19) *Nanoscale Characterization of Surfaces and Interface*; DiNardo, N.J. (Editor); Weinheim: New York, 1994.

(20) Wang, Z.L. *Characterization of Nanophase Materials*; Wang, Z.L. (Editor); Wiley: New York, 2000.

(21) *Molecular Nanoelectronics*; Reed, M.A.; Lee, T. (Editors); American Scientific Publishers. Stevenson Ranch, CA, 2003.

(22) Freitas, R.A. Jr. *Nanomedicine*; Landes Bioscience: Georgetown, TX, 1999.

(23) Robinson, D. *Nanotechnology in Medicine and the Biosciences*; Coombs, R. (Editor); Gordon & Breach Publishing Group: Amsterdam, The Netherlands, 1996.

(24) *Nanofabrication and Biosystems: Integrating Materials Science, Engineering, and Biology*; Hoch, H.C.; Jelinski, L.W.; Craighead, H.G. (Editors); Cambridge University Press: Cambridge, UK, 1996.

(25) *Quantum Dots and Nanowires*; Bandyopadhyay, S.; Nalwa, H.S. (Editors); American Scientific Publishers: Stevenson Ranch, CA, 2003.

(26) Vayssieres, L.; Manthiram, A. One-dimensional metal oxide nanostructures. In *Encyclopedia of Nanoscience and Nanotechnology, Vol. 8*; Nalwa, H.S. (Editor); American Scientific Publishers: Stevenson Ranch, CA, 2004; 147–166.

(27) Healy, T.W.; White, L.R. Ionizable surface group models of aqueous interfaces. *Adv. Colloid Interface Sci.* **1978**, *9*, 303–345.

(28) Bowden, J.W.; Posner, A.M.; Quirk, J.P. Ionic adsorption on variable charge mineral surfaces. Theoretical-charge development and titration curves. *Aust. J. Soil. Res.* **1977**, *15*, 121–136.

(29) Prelot, B.; Charmas, R.; Zarzycki, P.; Thomas, F.; Villieras, F.; Piasecki, W.; Rudzinski, W. Application of the theoretical 1-p*K* approach to analyzing proton adsorption isotherm derivatives on heterogeneous oxide surfaces. *J. Phys. Chem. B.* **2002**, *106*, 13280–13286.

(30) Tao, Z.; Dong, W. Comparison between the one p*K* and two p*K* models of the metal oxide–water interface. *J. Colloid Interface Sci.* **1998**, *208*, 248–251.

(31) Hiemstra, T.; Venema, P.; Van Riemsdijk, W.H. Intrinsic proton affinity of reactive surface groups of metal (hydr)oxides: The bond valence principle. *J. Colloid Interface Sci.* **1996**, *184*, 680–692.

(32) Hiemstra, T.; De Wit, J.C.M.; Van Riemsdijk, W.H. Multisite proton adsorption modeling at the solid/solution interface of (hydr)oxides: A new approach. II. Application to various important (hydr)oxides. *J. Colloid Interface Sci.* **1989**, *133*, 105–117.

(33) Sverjensky, D.A. Interpretation and prediction of triple-layer model capacitances and the structure of the oxide–electrolyte–water interface. *Geochim. Cosmochim. Acta.* **2001**, *65*, 3641–3653.

(34) Sahai, N.; Sverjensky, D.A. Evaluation of internally consistent parameters for the triple-layer model by the systematic analysis of oxide surface titration data. *Geochim. Cosmochim. Acta.* **1997**, *61*, 2801–2826.

(35) Sverjensky, D.A. Prediction of surface charge on oxides in salt solutions: Revisions for 1:1 (M+L−) electrolytes. *Geochim. Cosmochim. Acta.* **2005**, *69*, 225–257.

(36) Kosmulski, M. pH-dependent surface charging and points of zero charge II. Update. *J. Colloid Interface Sci.* **2004**, *275*, 214–224.

(37) Kruyt, H.R. *Colloid Science, Vol. 1, Irreversible Systems*; Elsevier: Amsterdam, 1952, 80.

(38) Babcock, K.L.; Overstreet, R. On the use of half calomel cells to measure Donnan potentials. *Science*, **1953**, *117*, 686–687.

(39) Overbeek, J.Th.G. Donnan-EMF and suspension effects. *J. Colloid Sci.* **1953**, *8*, 593–605.

(40) Stumm, W.; Huang, C.P.; Jenkins, S.R. Specific chemical interaction affecting stability of dispersed systems. *Croat. Chem. Acta*. **1970**, *42*, 223–245.

(41) Schindler, P.W.; Stumm, W. The surface chemistry of oxides, hydroxides, and oxide minerals. In *Aquatic Surface Chemistry: Chemical Processes at the Particle–Water Interface*, Stumm, W. (Editor); Wiley: New York, 1987; 83–110.

(42) Stumm, W.; Hohl, H.; Dalang, F. Interactions of metal ions with hydrous oxide surfaces. *Croat. Chem. Acta*. **1976**, *48*, 491–504.

(43) Parks, G.A. The isoelectric points of solid oxides, solid hydroxides, and aqueous hydroxo complex systems. *Chem. Rev.* **1965**, *65*, 177–198.

(44) Sposito, G. In *Environmental Particles, Vol. 1*; Buffle, J.; van Leeuwen, H.P. (Editors); Lewis Publishers: Boca Raton, FL, 1992; 291–314.

(45) Stumm, W. *Chemistry of the Solid-Water Interface*; Wiley: New York, 1992.

(46) Esposito, G. On points of zero charge. *Environ. Sci. Technol.*, **1998**, *32*, 2815–2819.

(47) Mills, I.; Cvitas, T.; Homann, K.; Kallay, N.; Kuchitsu, K. *Quantities, Units and Symbols in Physical Chemistry, 2nd ed*; Blackwell Scientific Publications: Oxford, 1993; 64.

(48) Bolt, G.H. Determination of the charge density of silica sols. *J. Phys. Chem.* **1957**, *61*, 1166–1169.

(49) Zalac, S.; Kallay, N. Application of mass titration to the point of zero charge determination. *J. Colloid Interface Sci.* **1992**, *149*, 233–240.

(50) Noh, J.S.; Schwarz, J. Estimation of the point of zero charge of simple oxides by mass titration. *J. Colloid Interface Sci.* **1989**, *130*, 157–164.

(51) Gunnarsson, M.; Rasmusson, M.; Wall, S.; Ahlberg, E.; Ennis, J. Electroacoustic and potentiometric studies of the hematite/water interface. *J. Colloid Interface Sci.*, **2001**, *240*, 448–458.

(52) Zhou, X.Y.; Wei, X.J.; Fedkin, M.V.; Strass, K.H.; Lvov, S.N. Zetameter for micro-electrophoresis studies of the oxide/water interface at temperatures up to 200 degrees C. *Rev. Sci. Instr.* **2003**, *74*, 2501–2506.

(53) Vakros, J.; Kordulis, C.; Lycourghiotis, A. Potentiometric mass titrations: A quick scan for determining the point of zero charge. *Chem. Comm.* **2002**, *17*, 1980–1981.

(54) Bourikas, K.; Vakros, J.; Kordulis, C.; Lycourghiotis, A. Potentiometric mass titrations: Experimental and theoretical establishment of a new technique for determining the point of zero charge (PZC) of metal (hydr)oxides. *J. Phys. Chem. B.* **2003**, *107*, 9441–9451.

(55) Bourikas, K.; Kordulis, C.; Lycourghiotis, A. Differential potentiometric titration: Development of a methodology for determining the point of zero charge of metal (hydr)oxides by one titration curve. *Environ. Sci. Technol.* **2005**, *39*, 4100–4108.

(56) Eggleston, C.M.; Jordan, G. A new approach to pH of point of zero charge measurement: Crystal-face specificity by scanning force microscopy. *Geochim. Cosmochim. Acta*. **1998**, *62*, 1919–1923.

(57) Sverjensky, D.A. Zero-point-of-charge prediction from crystal chemistry and solvation theory. *Geochim. Cosmochim. Acta*. **1994**, *58*, 3123–3129.

(58) *Surface and Interfacial Tension: Measurement, Theory and Applications, Surfactant Science Series, Vol. 119*; Hartland, Marcel Dekker: New York, 2004.

(59) Chibowski, E.; Perea-Carpio, R. A novel method for surface free-energy determination of powdered solids. *J. Colloid Interface Sci.* **2001**, *240*, 473–479.

(60) Sun, C.; Berg, J.C. Effect of moisture on the surface free energy and acid-base properties of mineral oxides. *J. Chromatogr.* **2002**, *969*, 59–72.

(61) Lewin, M.; Mey-Marom, A.; Frank, R. Surface free energies of polymeric materials, additives and minerals. *Polym. Adv. Technol.* **2005**, 16, 429–441.

(62) Shimizua, I.; Takei, Y. Temperature and compositional dependence of solid–liquid interfacial energy: Application of the Cahn–Hilliard theory. *Physica B.* **2005**, *362*, 169–179.

(63) Drelich, J.; Tormoen, G.W.; Beach, E.R. Determination of solid surface tension from particle-substrate pull-off forces measured with the atomic force microscope. *J. Colloid Interface Sci.* **2004**, *280*, 484–497.

(64) Rusanov, A.I. Interfacial thermodynamics: Development for last decades. *Solid State Ionics* **1995**, *75*, 275–279.

(65) Gibbs, J.W. *The Scientific Papers, Vol. 1*; Longmans: New York, 1906.

(66) Shuttleworth, R. The surface tension of solids. *Proc. Phys. Soc. London A.* **1950**, *63*, 444–457.

(67) Eriksson, J.C. Thermodynamics of surface phase systems: V. Contribution to the thermodynamics of the solid-gas interface. *Surface Sci.* **1969**, *14*, 221–246.

(68) Nielsen, A.E. *Kinetics of Precipitation*; Pergamon Press: Oxford, 1964.

(69) Nielsen, A.E. *Crystal Growth*; Pergamon Press: London, 1967.

(70) Overbeek, J.T.G. Strong and weak points in the interpretation of colloid stability. *Adv. Colloid Interface Sci.* **1982**, *16*, 17–30.

(71) Van Oss, C.J.; Norde, W.; Visser, H. 50 years DLVO theory. *Colloids Surf. B.* **1999**, *14*(1–4), 1–259.

(72) Usui, S. DLVO theory of colloid stability. *Surf. Sci. Ser.* **1998**, *76*, 101–118.

(73) Verwey, E.J.W.; Overbeek, J.T.G. *Theory of the Stability of Lyophobic Colloids*; Elsevier: New York, 1948.

(74) Smalley, M.V. DLVO for macro-ionic gels and colloids. *Mol. Phys.* **1990**, *71*(6), 1251–1267.

(75) Rajagopalan, R. Stability of colloidal dispersions: A thermodynamic approach. *Water Sci. Technol.* **1993**, *27*(10), 117–129.

(76) Li, Z.; Giese, R.F.; Wu, W.; Sheridan, M.F.; Van Oss, C.J. The surface thermodynamic properties of some volcanic ash colloids. *J. Dispersion Sci. Technol.* **1997**, *18*(3), 223–241.

(77) Furukawa, J. A thermodynamic theory of suspension. *Bull. Chem. Soc. Jpn.* **2000**, *73*(7), 1469–1475.

(78) Linse, P. Structure, phase stability, and thermodynamics in charged colloidal solutions. *J. Chem. Phys.* **2000**, *113*(10), 4359–4373.

(79) Sun, Z.; Xu, S.; Dai, G.; Li, Y.; Lou, L.; Liu, Q.; Zhu, R. A microscopic approach to studying colloidal stability. *J. Chem. Phys.* **2003**, *119*(4), 2399–2405.

(80) Strevett, K.A.; Chen, G. Microbial surface thermodynamics and applications. *Res. Microbiol.* **2003**, *154*(5), 329–335.

(81) Hermansson, M. The DLVO theory in microbial adhesion. *Colloids Surf. B.* **1999**, *14*(1–4), 105–119.

(82) Sugimoto, T. Preparation of monodispersed colloidal particles. *Adv. Colloid Interface Sci.* **1987**, *28*(1) 65–108.

(83) Nielsen, A.E. *Kinetics of Precipitation*; Pergamon Press: Oxford, 1964.

(84) Kahlweir, M. Ostwald ripening of precipitates. *Adv. Colloid Interface Sci.* **1975**, *5*(1), 1–35.

(85) Sugimoto, T. General kinetics of Ostwald ripening of precipitates. *J. Colloid Interface Sci.* **1978**, *63*(1), 16–26.

(86) Madras, G.; McCoy, B.J. Temperature effects on the transition from nucleation and growth to Ostwald ripening. *Chem. Eng. Sci.* **2004**, *59*(13), 2753–2765.

(87) Finsy, R. On the critical radius in Ostwald ripening. *Langmuir* **2004**, *20*(7), 2975–2976.

(88) Tolman, R.C. The general principles of equilibria in divided systems. *J. Am. Chem. Soc.* **1913**, *35*(4), 307–316.

(89) Ruckenstein, E. Origin of thermodynamic stability of microemulsions. *Chem. Phys. Lett.* **1978**, *57*(4), 517–521.

(90) Overbeek, J.T.G. Microemulsions, a field at the border between lyophobic and lyophilic colloids. *Faraday Disc. Chem. Soc.* **1978**, *65*, 7–19.

(91) Aveyard, R. Ultralow tensions and microemulsions. *Chem. Ind.* **1987**, *14*, 474–478.

(92) Shchukin, E. Conditions of spontaneous dispersion and formation of thermodynamically stable colloid systems. *J. Dispersion Sci. Technol.* **2004**, *25*(6), 875–893.

(93) Bliznakov, G.; Delineshev, S. On the thermodynamic stability of disperse system. *Cryst. Res. Technol.* **1990**, *25*(1), 41–50.

(94) Lekkerkerker, H.N.W.; Kegel, W.K.; Overbeek, J.T.G. Phase behavior of ionic microemulsions. *Ber. Buns. Ges.* **1996**, *100*(3), 206–217.

(95) Chen, W.; Gray, D.G. Interfacial tension between isotropic and anisotropic phases of a suspension of rodlike particles. *Langmuir* **2002**, *18*(3), 633–637.

(96) Vayssieres, L. *Précipitation de Nanoparticules d'oxydes en Solution Aqueuse: Contrôle de la Croissance et de la Tension Interfaciale*; Thèse de Doctorat, Université Pierre et Marie Curie: Paris, 1995; 1–145.

(97) Hermans, J.J. In *Colloid Science, Vol. 1*; Kruyt, H.R. (Editor); Elsevier: Amsterdam, 1952, 11.

(98) Stol, R.J.; de Bryun, P.L. Thermodynamic stabilization of colloids. *J. Colloid Interface Sci.* **1980**, *75*, 185–198.

(99) Vayssieres, L.; Chaneac, C.; Tronc, E.; Jolivet, J.P. Size tailoring of magnetite particles formed by aqueous precipitation: An example of thermodynamic stability of nanometric oxide particles. *J. Colloid Interface Sci.* **1998**, *205*(2), 205–212.

(100) Vayssieres, L. On the thermodynamic stability of metal oxide nanoparticles in aqueous solutions. *Int. J. Nanotechnol.* **2005**, *2*(4), 412–432.

(101) Kruyt, H.R. *Colloid Science, Vol. 1*; Elsevier: Amsterdam, 1952; 80.

(102) Mills, I.; Cvitas, T.; Homann, K.; Kallay, N.; Kuchitsu, K. *Quantities, Units and Symbols in Physical Chemistry, 2nd ed*; Blackwell Scientific: Oxford, 1993.

(103) Yaroshchuk, A.E. A model for the shift of isoelectric points of oxides in concentrated and mixed-solvent electrolyte solutions. *J. Colloid Interface Sci.* **2001**, *238*(2), 381–384.

(104) Parks, G.A. Isoelectric points of solid oxides solid hydroxides and aqueous hydroxo complex systems. *Chem. Rev.* **1965**, *65*(2), 177–198.

(105) Parks, G.A.; de Bruyn, P.L.D. Zero point of charge of oxides. *J. Phys. Chem.* **1962**, *66*(6), 967–973.

(106) Ahmed, S.M. Studies of the double layer at oxide-solution interface. *J. Phys. Chem.* **1969**, *73*(11), 3546–3555.

(107) Ahmed, S.M. Dissociation of oxide surfaces at the liquid-solid interface. *Can. J. Chem.* **1966**, *44*(14), 1663–1670.

(108) Ruckenstein, E.; Chi, J.C. Stability of microemulsions. *J. Chem. Soc. Faraday Trans. II* **1975**, *71*(10), 1690–1707.

(109) Eriksson, J.C.; Ljunggren, S. Is it necessary that the interfacial tension of a general curved interface is constant along the interface at equilibrium? *J. Colloid Interface Sci.* **1992**, *152*(2), 575–577.

(110) Vayssieres, L.; Aqueous purpose-built metal oxide thin films. *Int. J. Mater. Prod. Technol.* **2003**, *18*, 313–337.

(111) Vayssieres, L. *Précipitation de Nanoparticules d'oxydes en Solution Aqueuse: Contrôle de la Croissance et de la Tension Interfaciale*; Thèse de Doctorat, Université Pierre et Marie Curie: Paris, 1995; 1–145.

(112) Vayssieres, L.; Hagfeldt, A.; Lindquist, S.-E. Purpose-built metal oxide nanomaterials. The emergence of a new generation of smart materials. *Pure Appl. Chem.* **2000**, *72*(1–2), 47–52.

(113) Vayssieres, L.; Hagfeldt, A.; Lindquist, S.-E. Purpose-built nanostructured Ru/RuO2 thin films on plastic substrate for electrocatalysis applications. *Electrochemical Soc. Meeting Abst.* **1999**, *99*, 2118.

(114) Vayssieres, L.; Hagfeldt, A.; Lindquist, S.-E. New purpose-built nanostructured metal oxide materials for photoelectrochemical applications. *Electrochemical Soc. Meeting Abst.*, **1998**, *98*, 747.

(115) Vayssieres, L.; Manthiram, A. One-dimensional metal oxide nanostructures. In *Encyclopedia of Nanoscience and Nanotechnology Vol. 8*; Nalwa, H.S. (Editor); American Scientific Publishers: Stevenson Ranch, CA, 2004; 147–166.

(116) Vayssieres, L.; Beermann, N.; Lindquist, S.-E.; Hagfeldt, A. Controlled aqueous chemical growth of oriented three-dimensional nanorod arrays: Application to Iron(III) oxides. *Chem. Mater.* **2001**, *13*(2), 233–235.

(117) Vayssieres, L.; Guo, J.-H.; Nordgren, J. Aqueous chemical growth of α-Fe_2O_3–α-Cr_2O_3 nanocomposite thin films. *J. Nanosci. Nanotechnol.* **2001**, *1*(4), 385–388.

(118) Vayssieres, L.; Rabenberg, L.; Manthiram, A. Aqueous chemical route to ferromagnetic 3D arrays of iron nanorods. *Nano. Lett.* **2002**, *2*(12), 1393–1395.

(119) Vayssieres, L. Growth of arrayed nanorods and nanowires of ZnO from aqueous solutions. *Adv. Mater.* **2003**, *15*(5), 464–466.

(120) Vayssieres, L.; Keis, K.; Lindquist, S.-E.; Hagfeldt, A. Purpose-built anisotropic metal oxide material: 3D highly oriented microrod-array of ZnO. *J. Phys. Chem. B.* **2001**, *105*(17), 350–3352.

(121) Vayssieres, L.; Keis, K.; Hagfeldt, A.; Lindquist, S.-E. Three-dimensional array of highly oriented crystalline ZnO microtubes. *Chem. Mater.* **2001**, *13*(12), 4395–4398.

(122) Vayssieres, L. On the design of advanced metal oxide nanomaterials. *Int. J. Nanotechnol.*, **2004**, *1*(1–2), 1–41.

(123) Rabenberg, L.; Vayssieres, L. Multiple orientation relationships among nanocrystals of manganese oxides. *Microsc. Microanal.* **2003**, *9*(2), 402–403.

(124) Vayssieres, L.; Manthiram, A. 2-D mesoparticulate arrays of α-Cr_2O_3. *J. Phys. Chem. B.* **2003**, *107*(12), 2623–2625.

(125) Vayssieres, L.; Graetzel, M. Highly ordered SnO_2 nanorod-arrays from controlled aqueous growth. *Angew. Chem. Int. Ed.* **2004**, *43*(28), 3666–3670.

(126) Keis, K.; Vayssieres, L.; Rensmo, H.; Lindquist, S.-E.; Hagfeldt, A. Photoelectrochemical properties of nano-to-microstructured ZnO electrodes. *J. Electrochem. Soc.* **2001**, *148*(2), A149–A155.

(127) Keis, K.; Vayssieres, L.; Lindquist, S.-E.; Hagfeldt, A. Nanostructured ZnO electrodes for photovoltaic applications. *Nanostruct. Mater.* **1999**, *12*(1–4), 487–490.

(128) Beermann, N.; Vayssieres, L.; Lindquist, S.-E.; Hagfeldt, A. Photoelectrochemical studies of oriented nanorod thin films of Hematite. *J. Electrochem. Soc.* **2000**, *147*(7), 2456–2461.

(129) Lindgren, T.; Wang, H.; Beermann, N.; Vayssieres, L.; Hagfeldt, A.; Lindquist, S.-E. Aqueous photoelectrochemistry of hematite nanorod-array. *Sol. Energy Mater. Sol. Cells* **2002**, *71*(2), 231–243.

(130) Lindgren, T.; Vayssieres, L.; Wang, H.; Lindquist, S.-E. Photo-oxidation of water at Hematite electrodes. In *Chemical Physics of Nanostructured Semiconductors*; Kokorin, A.I.; Bahnemann, D.W. (Editors); VSP-International Science Publishers: Zeist, The Netherlands, 2003; 83–110.

(131) Wang, J.X.; Sun, X.W.; Yi, Y., Huang, H.; Lee, Y.C.; Tan, O.K.; Vayssieres, L. Hydrothermally grown ZnO nanorod arrays for gas sensing applications. *Nanotechnology* **2006**, *17*(19), 4995–4998.

(132) Vayssieres, L. Advanced semiconductor nanostructures. *C. R. Chimie.* **2006**, *9*, 691–701.

(133) Dong, C.L.; Persson, C.; Vayssieres, L.; Augustsson, A.; Schmitt, T.; Mattesini, M.; Ahuja, R.; Chang, C.L.; Guo, J.-H. The electronic structure of nanostructured ZnO from X-ray absorption and emission spectroscopy and the local density approximation. *Phys. Rev. B.* **2004**, *70*(19), 195325.

(134) Guo, J.-H.; Vayssieres, L.; Persson, C.; Ahuja, R.; Johansson, B.; Nordgren, J. Polarization-dependent soft-X-ray absorption of highly oriented ZnO microrods. *J. Phys. Condens. Matter* **2002**, *14*(28), 6969–6974.

(135) Persson, C.; Dong, C.L.; Vayssieres, L.; Augustsson, A.; Schmitt, T.; Mattesini, M.; Ahuja, R.; Nordgren, J.; Chang, C.L.; Ferreira da Silva, A.; Guo, J.-H. X-ray absorption and emission spectroscopy of ZnO nanoparticles and highly oriented ZnO microrod arrays. *Microelectron. J.* **2006**, *37*(8), 686–689.

(136) Vayssieres, L.; Persson, C.; Guo, J.-H. Quantum size effect in anatase TiO_2 quantum dot arrays. 2006; submitted for publication.

(137) Vayssieres, L.; Saathe, C.; Butorin, S.M.; Shuh, D.K.; Nordgren, J.; Guo J.-H. One dimensional quantum confinement effect in α-Fe_2O_3 ultrafine nanorod arrays. *Adv. Mater.* **2005**, *17*(19), 2320–2323.

(138) Vayssieres, L. to be published.

(139) Vayssieres, L. 3-D bio-inorganic arrays. In *Chemical Sensors VI: Chemical and Biological Sensors and Analytical Methods*; Brukner-lea, C.; Vanysek, P.; Hunter, G.; Egashira, M.; Miura, N.; Mizutani, F. (Editors); The Electrochemical Society: Pennington, NJ, 2004; 322–343.

PART II

SYNTHESIS AND PREPARATION OF NANOSTRUCTURED OXIDES

The first requirement in any study dealing with oxide nanostructures is the synthesis and some characterization of the material of interest. The metal elements can form a large diversity of oxide compounds that can adopt many bulk crystal structures. Can we obtain similar compounds and structures at the nano range? The development of systematic methods for the synthesis of metal-oxide nanostructures is a challenge from both fundamental and industrial standpoints. Chapter 2 described criteria, based on thermodynamic principles, that can be useful for the preparation of oxide nanoparticles from aqueous solution. In this part of the book, Chapters 3 and 4 focus on synthetic methods based on liquid–solid and gas–solid transformations. The ultimate goal is to produce in a controlled manner single-phase compounds with a narrow distribution of particle sizes. Methods frequently used for the synthesis of bulk oxides may not work when aiming at the preparation of oxide nanostructures or nanoparticles. For example, a reduction in particle size by mechanically grinding a reaction mixture can only achieve a limiting level of grain diameter, at best about 0.1 μm. However, chemical methods can be used to effectively reduce particle size into the nanometer range. One of the most widely used methods for the synthesis of bulk materials involves heating the components together at a high temperature over an extended period of time. This process is the route for the synthesis of many bulk phases of mixed-metal oxides like $SrTiO_3$ or $MgAl_2O_4$. Elevated temperatures ($>800\,^{\circ}C$) can be a problem when using this approach for the generation of oxide nanostructures. A much better control of the synthesis can be achieved by direct coprecipitation of the oxide components from a liquid solution with subsequent calcination, or by using sol-gels or microemulsions in the synthesis process. Following these approaches, one can control the stoichiometry of the oxide nanostructures in a precise way. Such techniques are widely used for the synthesis of catalysts and ceramics. Chemical vapor deposition is a technique employed in various industrial applications and technologies (for example, the fabrication of sensors and electronic devices) that can be extremely helpful in the synthesis of oxide nanostructures. During the last decade, pulsed laser deposition has been established as a versatile technique for the generation of nanoparticles and thin

Synthesis, Properties, and Applications of Oxide Nanomaterials, Edited by José A. Rodríguez and Marcos Fernández-Garcia

films of oxides. It is generally easier to obtain the desired stoichiometry for multi-element materials using pulsed laser deposition than with other deposition techniques. The advantages and disadvantages of these synthetic methods will be discussed in detail in the next two chapters.

CHAPTER 3

Synthesis of Metal-Oxide Nanoparticles: Liquid–Solid Transformations

LAWRENCE D'SOUZA AND RYAN RICHARDS
International University Bremen, Bremen, Germany

3.1. INTRODUCTION

This chapter presents an overview of synthetic methods for metal oxides with special attention given to recent developments in high-surface-area (HSA) (nanoscale) metal oxides. New classes of surface-decorated or surface-modified metal oxides have also been included.

Metal oxides can be found in the form of a single crystal (pure or defective), powder (with many crystals), polycrystalline (crystals with various orientations), or thin film. Various methods are available in the literature for the synthesis of metal oxides, the selection of which is often dependent on the desired properties and final application.

Among different categories of solid-state materials like halides, hydrides, and chalcogenides (oxides, sulphides, etc.), oxides and sulphides have found wide application particularly because of their nonstoichiometry, thermal and mechanical properties, as well as catalytic activity. Early methods to synthesize solid-state materials, especially mixed-metal oxides, demanded high temperatures (1000–3000 °C), and long reaction times (1 hour to 1 month). Moreover, the material formed generally possessed a very low surface area and was not suitable for many applications, particularly for catalysis, sorption, and sensing.

Synthesis, Properties, and Applications of Oxide Nanomaterials, Edited by José A. Rodríguez and Marcos Fernández-Garcia

The field of nanotechnology[1] has stimulated extensive research focused toward gaining control of particle size, shape, and composition under mild conditions. The syntheses of nanoscale materials are generally grouped into two broad categories: "bottom-up" and "top-down." Those preparations based on building up from atomic or molecular precursors, which come together to form clusters, and subsequently nanoparticles are referred to as "bottom up." Conversely, when the nanoscale is reached by physically tearing down larger building blocks, the process is referred to as "top-down." The "bottom-up" route is particularly interesting because of the possibilities to manipulate the properties (such as size, shape, stoichiometry, surface area, pore size, and surface decoration) of the end product and will be the focus of this chapter.

The "bottom-up" method is broadly applicable for synthesizing new materials and is the basis for numerous fundamental studies. In the last few decades, literature on the synthesis of solid-state materials by "bottom-up" routes has increased dramatically. Most recently, the trend is toward preparation of an ordered, crystalline mesoporous nanoscale metal oxides (mono- and mixed-metal oxides) as well as surface decoration or modification of one metal oxide on other at the atomic scale.

The insulating oxides are made up of the metals from the far left and right sides of the Periodic Table. Typical examples of insulating oxides include MgO, CaO, Al_2O_3, and SiO_2. The oxides of the metals in the middle of the Periodic Table (Sc to Zn) make up the semiconducting or metallic oxides. Typical examples include ZnO, TiO_2, NiO, Fe_2O_3, and Cr_2O_3. Additionally, the transition metal oxides, which include the oxides of Ru, Mo, W, Pt, V, and so forth, are of particular interest for applications in catalysis, sensor materials, and other potential applications.

Because metal oxides are generally ionic, their physical and chemical properties are dictated by defects. Among the defects possible, the high proportion of surface-to-bulk atoms in nanoscale materials most dramatically influences their properties. In spherical nanoparticles, for example, at a size of 3 nm, 50% of the atoms or ions are on the surface allowing the possibility of manipulation of bulk properties by surface effects and allowing near-stoichiometric chemical reaction (1). When strong chemical bonding is present, delocalization can vary with size; this in turn can lead to different chemical and physical properties (2).

The formation of the ultra-small particles is facilitated due to the refractory nature of most metal oxides (2). The ionic nature of some materials, especially MgO, Al_2O_3, ZrO_2, and TiO_2, allows the formation of many stable defect sites, including edges, corners, and anion/cation vacancies. The materials prepared via aerogel methods usually possess very low densities, can be translucent or transparent, have low thermal conductivities, and unusual acoustic properties. They have found various applications, including as detectors for radiation, superinsulators, solar concentrators, coatings, glass precursors, catalysts, insecticides, and destructive adsorbents. The very high

[1]When the size of these materials is brought down to the nanometer regime, several size-dependent properties arise primarily due to surface chemistry. This field of nanoscience is facilitated by the fact that many systems of interest have been extensively studied in the bulk form and therefore provide ready comparisons with the nanoparticulate systems.

surface areas of nanoscale particles gives rise to several defect sites. Several techniques have been used to in an attempt to clarify the types of defect sites that can exist in the case of MgO (3–7). The most common defect is coordinatively unsaturated ions due to planes, edges, corners, anion/cation vacancies, and electron excess centers.

3.2. STRUCTURE AND BONDING

Determining the exact nature of structure and bonding in nanomaterials is particularly difficult because generally these materials are made of very small crystallites or are amorphous. Recent advances in crystallography for powders and crystals employing X-ray, electron, and neutron diffraction have provided insight into the structures of metal oxides (see Chapter 5). Metal oxides crystallize in a variety of structures, and bonding in these materials can range from ionic (MgO, $Fe_{1-x}O$) to metallic (TiO_2, ReO_3) (8).

An understanding of not only the crystal structure and bonding but also the local microstructures that are due to defects is necessary to understand the structure of complex transition metal oxides. Of course, on the nanometer scale, the number of defects due to edges, corners, "f" centers, and other surface imperfections is greatly enhanced by the large surface area (see Figure 3.1).

The pursuit of an understanding of the structure/property relationship is integral to the understanding of the unique properties observed on the nanoscale.

In bulk structures, five types of crystals can be defined based on bonding considerations: covalent, ionic, metallic, molecular (Van der Waals), and hydrogen bonded. These structures are also present on the nanoscale; however, one must also consider the number of atoms at the surface when examining the structure of nanoscale materials. Ionic crystals are formed when highly electronegative and highly electropositive

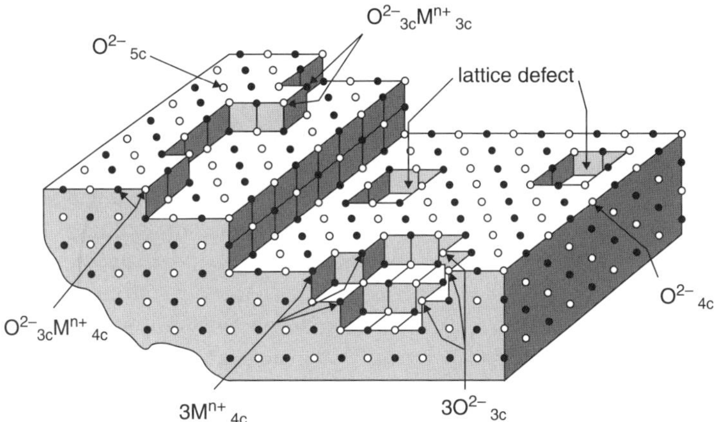

Figure 3.1. A representation of the various defects present on metal oxides (adapted from Ref. 9).

elements are combined in a lattice. It has been found that the ionic model is a poor approximation for crystals containing large anions and small cations (e.g., oxides and sulfides), where the covalent contribution to bonding becomes significant (8). Van der Walls interactions play a crucial role in many transition metal oxides, especially those with layered structures. In many oxide hydrates or hydroxy oxides, hydrogen bonding also contributes to the cohesive energy. In most transition metal oxides, the bonding is only partly ionic; however, several examples are primarily ionic such as MgO and CaO. In other words, there is a considerable overlap between the orbitals of the cations and the anions. Many transition metal oxides also exhibit metallic properties. Inorganic compounds of the formula AB can have the rock salt (B1), CsCl (B2), ZnS (B3), Wurtzite (B4), or NiAs (B8) structure (10). Alkaline earth metal oxides, such as MgO and monoxides of 3*d* transition metals, as well as lanthanides and actinides, such as TiO, NiO, EuO, and NpO, exhibit the rock salt structure with the 6 : 6 octahedral coordination.

3.3. SYNTHESIS

One of the areas of fundamental importance to the understanding and development of nanoscale materials is the development of synthetic methods, which allow scientists control over such parameters as particle size, shape, size distributions, and composition. Although considerable progress has taken place in recent years, one major challenge to scientists is the development of a "synthetic toolbox" that would afford access to size and shape control of structures on the nanoscale and conversely allow scientists to study the effects these parameters impart to the chemical and physical properties of the nanoparticles.

"Bottom-up" preparation methods are of primary interest to chemists and materials scientists because the fundamental building blocks are atoms or molecules. Gaining control over the way these fundamental building blocks come together and form particles is one of the most sought after goals of synthetic chemists. Therefore, these methods will be the focus of this section. Interest in "bottom-up" approaches to nanoscale oxides and other materials is clearly indicated by the number of reports and reviews on this subject (11–26). There are, of course, numerous "bottom-up" approaches to the preparation of nanoscale materials, and metal oxides are no exception. Gas–solid (wet chemical) and liquid–solid (physical) transformations are two different ways for synthesizing nanomaterials by "bottom-up" preparation methods. Several physical aerosol methods have been reported for the synthesis of nano-size particles of oxide materials. These methods include gas condensation techniques (27–32), spray pyrolysis (30,33–36), thermochemical decomposition of metal-organic precursors in flame reactors (32,37–39), and other aerosol processes named after the energy sources applied to provide the high temperatures during gas–particle conversion. The most common and widely used "bottom up" wet chemical method for the preparation of nanoscale oxides has been the sol-gel process. Other wet chemistry methods, including novel microemulsion techniques, oxidation of metal colloids, and precipitation from solutions, have also been used.

The methods of sample preparation are of course the determining factors in producing different morphologies (40). For example, burning Mg in O_2 (MgO smoke) yields 40–80-nm cubes and hexagonal plates, whereas thermal decomposition of $Mg(OH)_2$, $MgCO_3$, and especially $Mg(NO_3)_2$ yields irregular shapes often exhibiting hexagonal platelets. Surface areas can range from 10 m^2/g (MgO smoke) to 250 m^2/g for $Mg(OH)_2$ thermal decomposition, but surface areas of about 150 m^2/g are typical. In the case of calcium oxide, surface areas can range from 1 to 100 m^2/g when prepared by analogous methods, but about 50 m^2/g is typical. In the following discussion, different methods for the synthesis of metal oxides by liquid–solid transformation will be the focus.

3.4. COPRECIPITATION METHODS

One conventional method to prepare nanoparticles of metal oxide ceramics is the precipitation method (23,41–46). This process involves dissolving a salt precursor, usually a chloride, oxychloride, or nitrate, such as $AlCl_3$ to make Al_2O_3, $Y(NO_3)_3$ to make Y_2O_3, and $ZrOCl_2$ to make ZrO_2. The corresponding metal hydroxides usually form and precipitate in water by adding a basic solution such as sodium hydroxide or ammonium hydroxide to the solution. The resulting chloride salts, i.e., NaCl or NH_4Cl, are then washed away and the hydroxide is calcined after filtration and washing to obtain the final oxide powder. This method is useful in preparing composites of different oxides by coprecipitation of the corresponding hydroxides in the same solution. One disadvantage of this method is the difficulty to control the particle size and size distribution. Very often, fast (uncontrolled) precipitation takes place resulting in large particles.

As an example, we highlight the zirconia system for further discussion on this topic because of its interesting properties and applications.

Zirconia has very interesting properties as an acid–base catalyst, as a promoter for other catalysts, (47) as an inert support material, and so on. It has been widely employed in various heterogeneous catalytic reactions and in ceramic production. Sulfated ZrO_2 is of great interest on account of its high activity in alkylation as a solid acid catalyst, whereas ZrO_2 has also been found to act as a photocatalyst, due to its n-type semiconductor nature.

Traditionally, zirconia has been limited as a support for heterogeneous catalysis due to its thermal stability. Amorphous zirconia undergoes crystallization at around 450 °C, and hence, its surface area decreases dramatically at that temperature. At room temperature, the stable crystalline phase of zirconia is monoclinic, whereas the tetragonal phase forms upon heating to 1100–1200 °C. For several applications, it is desirable to have the tetragonal phase with a high surface area. However, preparation protocols are needed to obtain the tetragonal phase at lower temperatures. Many researchers have tried to maintain the HSA of zirconia by several means. Usually the ZrO_2 is mixed with CaO, MgO, Y_2O_3, Cr_2O_3, or La_2O_3 for stabilization of the tetragonal phase at low temperature. Zirconia and mixed-oxide zirconia have been widely studied by many methods, including sol-gel process (48–55), reverse micelle

TABLE 3.1. Literature Review: Surface Area Obtained at Different Calcination Temperatures and Techniques by Other Groups for Neat ZrO_2 Samples (Adapted from Ref. 66).

	BET Surface Area, m^2/g at Different Calcination Temperatures			
Preparation Method	400 °C	500 °C	650 °C	700 °C
Alcothermal, SCFD (67)	270	206	–	–
Alcothermal, SCFD (68)	–	–	26–85	18–62
Pre. and dig. (53)	–	54	–	–
Pre. and dig. (69, 70)	–	100–277	–	–
Pre. and dig. (55)	–	30	8	–
Pre. and dig. (71)	91–125	70–94	–	–
EISA and NH_3 treatment (72)	406	297	–	154

Abbreviations: Pre. = precipitation, dig. = digestion, calc. = calcinations, SCFD = supercritical fluid drying, EISA = evaporation-induced self-assembly.

method (56), coprecipitation (57,58) and hydrothermal synthesis (59), functionalization of oxide nanoparticles and their self assembly (60), and templating (61). Bedilo and Klabunde (62,63) have prepared HSA sulfated zirconia by a supercritical drying technique and found that the resulting material was active toward alkane isomerization reactions (64,65). Recently, Chane-Ching et al. (60) reported a general method to prepare high-surface-area materials through the self-assembly of functionalized nanoparticles. This process involves functionalizing the oxide nanoparticles with bifunctional organic anchors like aminocaproic acid and taurine. After the addition of a copolymer surfactant, the functionalized nanoparticles will slowly self-assemble on the copolymer chain through a second anchor site. Using this approach, the authors could prepare several metal oxides such as CeO_2, ZrO_2, and CeO_2–$Al(OH)_3$ composites. The method yielded ZrO_2 of surface area 180 m^2/g after calcining at 500 °C, 125 m^2/g for CeO_2, and 180 m^2/g for CeO_2–$Al(OH)_3$ composites.

Table 3.1 shows the literature data for zirconia obtained by different processes and the resulting surface area obtained at different calcination temperatures (53,55,66–72). The product obtained is referred to as a xerogel when drying is achieved by evaporation under normal conditions. Aerogel powders usually demonstrate higher porosities and larger surface areas as compared with analogous xerogel powders because the solvent is removed rapidly in a supercritical drying process that protects the porous structure. Aerogel processing has been very useful in producing highly divided powders of different metal oxides (73–79).

Table 3.2 gives the results obtained by the Richards group by the TMACl stabilization route (66). The data indicate that digesting the material in the presence of TMACl and ammonia provided higher surface areas. Moreover, surface area and thermal stability were found to depend on digestion time. In their study, digestion at 110 °C for 100 h provided high thermal stability and the highest surface area of 370 m^2/g; the same sample when calcined at 700 °C possessed a surface area of 160 m^2/g, which

TABLE 3.2. Physical Characteristics of $ZrO(OH)_2$–TMACl/NH_3 Samples at Different Temperatures (Adapted from Ref. 66).

T (hr)	T_{cal} (°C)	S_{BET} (m^2g^{-1})	D_{ave}(Å)	V_p ($cc\,g^{-1}$)[a]	%W_l	T_c(°C)
0	110[b]	167	10.93	0.083	27.85	465ex
24	110[b]	331	12.18	0.085	27	607ex
	200	258	12.38	0.08	–	–
	400	147	12.84	0.047	–	–
	500	100	12.94	0.033	–	–
	600	84	13	0.026	–	–
48	110[b]	359	13.05	0.099	22.01	635ex
	200	327	13.1	0.089	–	–
	400	271	13.17	0.088	–	–
	500	200	13.2	0.061	–	–
	600	132	13.35	0.041	–	–
100	110[b]	370	12.32	0.11	16.95	648ex
	200	353	12.51	0.107	–	–
	300	330	12.23	0.083	–	–
400	308	–	12.55	0.094	–	–
	500	226	12.72	0.059	–	–
	600	217	11.22	0.05	–	–
	700	160	11.39	0.055	–	–

[a]Total pore volume (Vp) is for pores smaller than 13.6 Å in diameter.
[b]Dried at that temperature.
Abbreviations: T_c = temperature of crystallization, %W_l = total mass loss, T_{cal} = calcination temperature, Ex = exothermic peaks, D_{ave} = pore average diameter, and T = duration of digestion.

was the best value obtained thus far. Zirconia samples that were digested more than 100 h had high thermal stability, but no increases in surface area were observed. The same samples that were dried by applying supercritical techniques did not yield better results, which indicates that the material achieved crystallinity during the digestion period, which was supported by X-ray diffraction studies. The crystallinity in the sample after 160 h of digestion was much higher than that of samples digested for 24, 48, and 100 h (diffraction patterns not shown). Obviously, the crystalline nature overcomes the stickiness of the particles irrespective of the drying method for the hydroxide samples. One possible reason this approach yields high-surface-area materials is the fact that digestion minimizes defect sites (Oswald ripening), which hinder grain growth during sintering. Hence, the crystallite remains smaller even after conversion to oxide, resulting in the high surface area. As the monoclinic phase is thermodynamically stable for zirconia, if this phase forms or predominates, surface area is usually lower because monoclinic crystallites are bigger than tetragonal (70).

It has also been demonstrated (80) that thermally stable mesoporous tetragonal zirconia with high surface area can be prepared using readily available and inexpensive materials such as zirconyl chloride and ethylene diamine (ED). ED plays multiple roles: as a precipitating agent for zirconia, a colloidal-zirconia protecting agent, as well as a binder between zirconia and Pt nanoclusters (see Figure 3.2*b*). Digesting

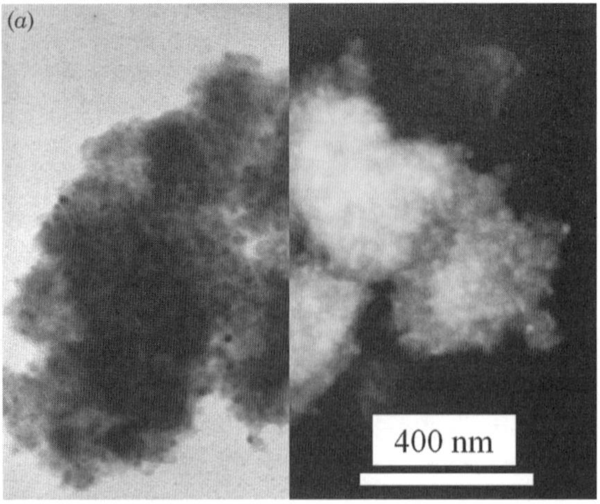

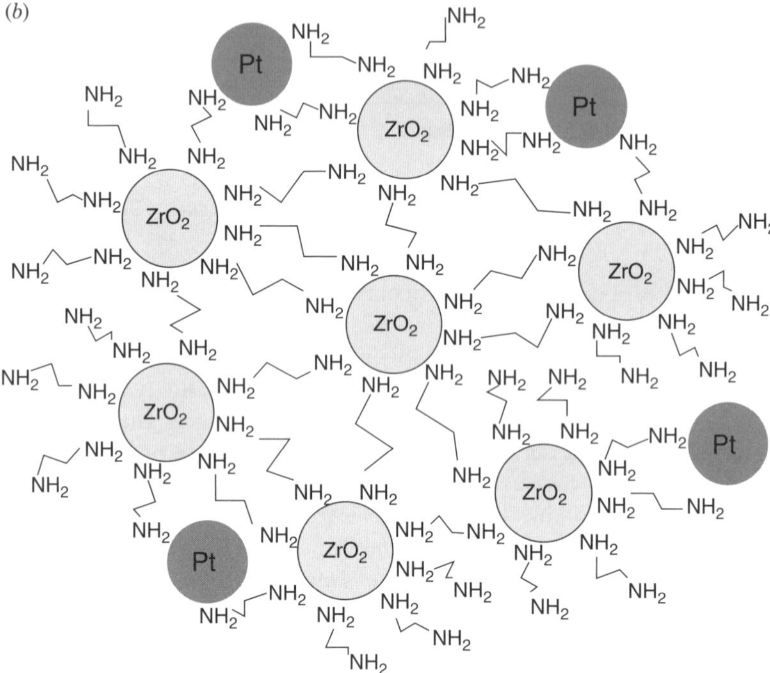

Figure 3.2. (*a*) TEM picture of sample TEM of undigested $ZrO(OH)_2$-ED matrix (half of the image was color inverted using Scion Image software for better contrast) and (*b*) schematic representation of ethylene diamine encircled zirconia matrix (adapted from Ref. 80).

the zirconia material minimizes the defect site in the crystal lattice and hence gives thermal stability. Because chlorine vapor is known to retard ZrO_2 densification, (81) it is suspected that there is some dependence of oxygen vacancy density with chloride ion concentration while digesting the sample. Elemental analysis results show the presence of nearly 0.7–0.9% of chlorine in the ZrO_2 sample. Karapetrova et al. have also observed the presence of chlorine in the ZrO_2 sample prepared from $ZrOCl_2$ precursor (82). Matsui and Ohgai observed that change in crystalline structure of ZrO_2 with the increasing chlorine content at the preparation stage (83). It is not certain whether chlorine replaces lattice oxygen or just adsorbs on the surface. If it replaces oxygen, then there should be a zirconia vacancy [Schottky defect (58)] in the resulting sample, similar to the oxygen vacancy created during alkaline-earth metals (CaO, MgO, etc.) doping to replace zirconia. Doping ZrO_2 with low valent oxides like Y_2O_3, CaO, or MgO results in stabilization of the tetragonal phase at low temperature (82). However, an oxygen deficiency has been observed in the sample sintered in air at 1000 °C for 3 h, which is in accordance with literature (82).

The data indicate that digesting the material in aqueous media results in a thermally stable material with HSA. Moreover, these factors were found to depend on digestion time (80). The crystallinity in the sample after 150 h of digestion was much higher than that of samples digested for 48 h. One possible reason why this approach yields HSA materials is the fact that digestion minimizes defect sites, which hinders grain growth during sintering. The ethylene diamine-protecting agent acts as a steric barrier for the growth of the particles during the digestion process. Hence, the crystallites maintain the original shape and size during digestion, which results in HSA after drying.

The N_2 adsorption–desorption isotherm for the zirconia obtained from the ethylene diamine-directed method as well as the calcined samples indicate a type IV isotherm with an H2 hysteresis loop based on IUPAC classifications (84). This type of hysteresis is associated with capillary condensation in mesopores; and a limiting uptake over a range of high P/P_0. The H2 loop indicates that the pores are of an inkbottle shape. Digestion of the sample for longer times did not alter the hysteresis shape, indicating that pore shape has been maintained after the digestion process.

The sample exhibits a broad pore size distribution in the mesoporous range. Figure 3.3II shows the pore size distribution for the 150 h digested, 110 °C dried, and 500 °C calcined $ZrO(OH)_2$-ED samples. The calcination in air or vacuum at a temperature of 500 °C results in substantial loss of the textural porosity of the zirconia aerogel. Further examination of the texturology by nitrogen adsorption/desorption isotherm data indicates that the average pore diameter lies between 20 and 50 Å with a mean at 35 Å in the mesoporous range (Figure 3.3II) and 10 Å in microporous region (Figure 3.3I). The results also show that pore size increases and pore distribution broadens with calcination.

A diffraction pattern for 48 h digested and 1000 °C calcined neat $ZrO(OH)_2$-ED samples exhibited only the tetragonal phase (JCPDS no. 42-1164, major peaks appear at $2\theta \approx 30, 33.9, 34.8, 49.4, 50, 58.27$, and 59.37) at all stages of calcination and there was no indication of monoclinic phase (JCPDS no. 37-1484, major peaks appears at $2\theta = 28.17, 31.46, 34.15, 34.38, 35.3$, and 40.72).

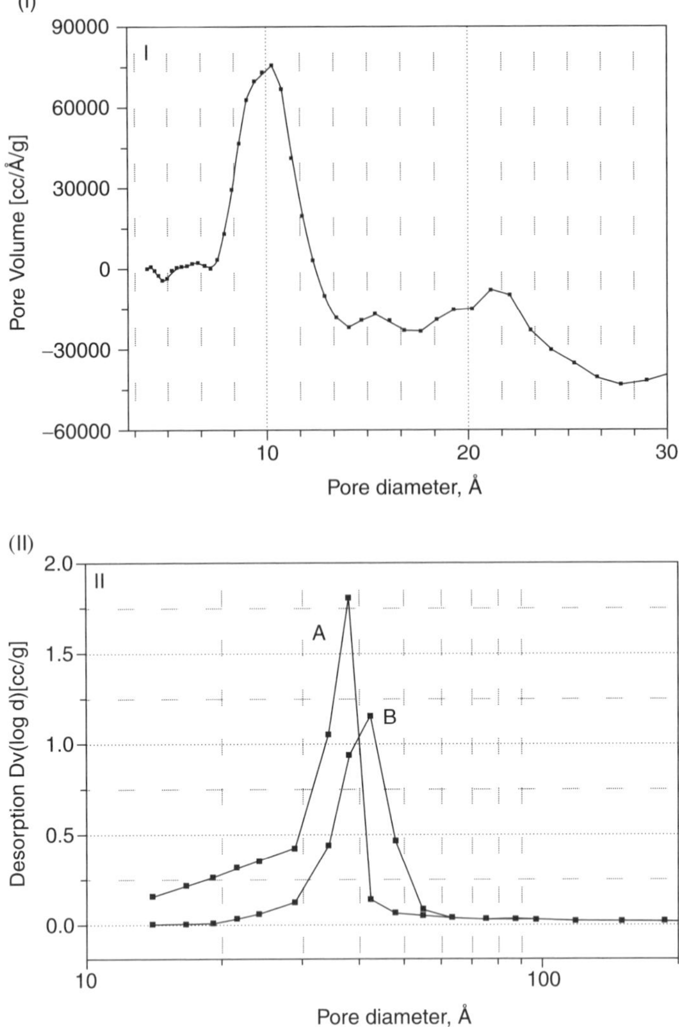

Figure 3.3. Pore size distribution of 150-hour digested $ZrO(OH)_2$-ED sample, (I) microporous region for the material calcined at 500 °C; (II) in mesoporous region (A) dried at 110 °C and (B) calcined at 500 °C (adapted from Ref. 80).

3.4.1. Crystal Structure Refinement of ZrO_2

The refinement of the crystal structure ZrO_2 was performed using powder diffraction. A set of experimental data of diffraction reflections was recorded with a Siemens, Germany D5000 diffractometer with Cu Kα radiation operated at 40 kV and 40 mA. The zirconia structure of ZrO_2 (space group $P4_2/nmc$ (137) (85)) was used as a model for the refinement. The simultaneous refinement of profile and structural parameters

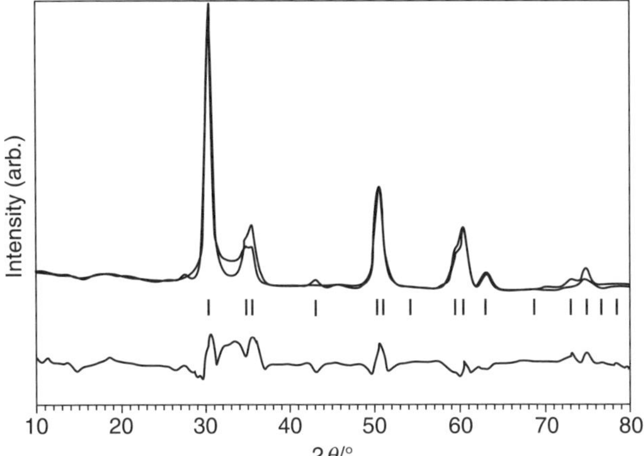

Figure 3.4. Experimental and calculated XRD patterns and corresponding difference pattern for the ZrO_2 compound (adapted from Ref. 80).

of the chosen model led to a satisfactory value of R_f factor. The X-ray diffraction diagram calculated from the obtained data agrees well with the experimental diagram (Figure 3.4).

Nanophase powders of $Y_xZr_{1-x}O_{2-x/2}$ have also been prepared from a mixture of commercially available ZrO_2 and Y_2O_3 powders (86). It was found that depending on the starting powder mixture composition, the yttrium content in the nanophases can be controlled and the tetragonal or cubic phases can be obtained. Tetragonal or a mixture of tetragonal and cubic were observed for low yttria content (3.5% mol yttria) and cubic for higher yttria contents (19, 54, and 76 mol% yttria). These powders were found to have a most probable grain radius of about 10 to 12 nm, and the grains appeared as isolated unstrained single crystals with polyhedral shapes. The grain shapes appeared to be polyhedral and not very anisotropic. Lattice fringes were parallel to the surfaces, demonstrating that (1 0 0) and (1 1 1) faces dominate.

3.4.2. Sonochemical Coprecipitation

The principle of sonochemistry is breaking the chemical bond with the application of high-power ultrasound waves usually between 20 kHz and 10 MHz. The physical phenomenon responsible for the sonochemical process is acoustic cavitation. According to published theories for the formation of nanoparticles by sonochemistry, the main events that occur during the preparation are creation, growth, and collapse of the solvent bubbles that are formed in the liquid. These bubbles are in the nanometer size range. The solute vapors diffuse into the solvent bubble, and when the bubble reaches a certain size, its collapse takes place. During the collapse, very high temperatures of 5000–25,000 K (87) are obtained, which is enough to break the chemical bonds in solute. The collapse of the bubble takes place in less than a nanosecond (88,89); hence, a high cooling rate (1011 K/s) is also obtained. This high cooling rate hinders the organization and crystallization of the products. As the breaking of bonds in the precursor occurs in the gas phase, amorphous nanoparticles are obtained. Although the reason for formation of amorphous products is well understood, the formation of nanostructures is not. The possible explanations are as follows: The fast kinetics does not stop the growth of the nuclei, and in each collapsing bubble, a few nucleation centers are formed whose growth is limited by the collapse; and the precursor is a nonvolatile compound and the reaction occurs in a 200-nm ring surrounding the collapsing bubble (90). In the latter case, the sonochemical reaction occurs in a liquid phase and the products could be either amorphous or crystalline depending on the temperature in the ring region of the bubble. Suslick has estimated the temperature of the ring region as 1900 °C. The sonochemical method has been found useful in many areas of material science starting from the preparation of amorphous products (91,92), insertion of nanomaterials into mesoporous materials (93,94), to deposition of nanoparticles on ceramic and polymeric surfaces (95,96).

Sonochemical methods for the preparation of nanoparticles were pioneered by Suslick in 1991 (87). He prepared Fe nanoparticles by sonication of $Fe(CO)_5$ in a decaline solution, which gave him 10–20-nm-size amorphous iron nanoparticles. Sonochemical decomposition methods have been further developed by Suslick et al. (97) and Gedanken et al. (98–100) and have produced Fe, Mo_2C, Ni, Pd, and Ag nanoparticles in various stabilizing environments.

If the same reaction is carried out in the presence of oxygen, the formation of the product will be an oxide. $NiFe_2O_4$ has been synthesized by sonicating the mixture of $Fe(CO)_5$ and $Ni(CO)_4$ in decaline under a 1–1.5 atm pressure of oxygen (101). ZrO_2 was synthesized using $Zr(NO)_3$ and NH_4OH in aqueous phase under ultrasonic irradiation, followed by calcinations at 800 °C to induce the tetragonal to monoclinic phase transformation (102). Nanocrystalline $La_{1-x}Sr_xMnO_3$ was prepared in a similar manner (103). Recently, the method has been successfully applied to synthesize $Ni(OH)_2$ and $Co(OH)_2$ nanoparticles (104) and iron (III) oxide (105).

Chalcogenides are commonly synthesized by this method. The metal sulphides syntheses have been carried out in ethanol (106), water (107), and ethylenediamine (108) whereas the sources of metal ions have been acetates (106,108) or the chlorides (107). The precursor for sulphur is usually thioacetamide or thiourea.

3.4.3. High-Gravity Reactive Precipitation (HGRP)

This is a new method for synthesizing nanoparticles introduced by Chen et al. (109) that was originally applied for the preparation of metal carbonates and hydroxides. Using this method, they could control and adjust the particle size of $CaCO_3$ particles in the region of 17–36 nm by adjusting the reaction parameters such as high-gravity levels, fluid flow rate, and reactant concentrations. The synthesis of nanofibrils of $Al(OH)_3$ of 1–10 nm in diameter and 50–300 nm in length as well as $SrCO_3$ with a mean size of 40 nm in diameter was also demonstrated by the same group using this technique.

HGRP is based on Higee technology (110); it consists of rotating a packed bed under a high-gravity environment. This technology is a novel technique to intensify mass transfer and heat transfer in multiphase systems. The rate of mass transfer between a gas and a liquid in a rotating packed bed is one to three orders of magnitude larger than that in a conventional packed bed. This is very helpful in the generation of higher supersaturated concentrations of the product in the gas–liquid-phase reaction and precipitation process. Due to the high shear field experienced by the passing fluid, formation of fine droplets, threads, and thin films takes place in the rotating packed bed. This effect facilitates the intense micromixing between the fluid components (109).

The experimental apparatus for synthesis by high-gravity reactive precipitation is shown schematically in Figure 3.5. The key part of the RPB (Higee machine) is a packed rotator (6). The inner diameter of the rotator is $d_{in} = 50$ mm, and the outer diameter is $d_{out} = 150$ mm. The axial width of the rotator is 50 mm. The distributor (5) consists of two pipes (10 and 1.5 mm), each having a slot of 1 mm in width and 48 mm in length, which just covers the axial length of the packing section in the rotator. The rotator is installed insidc the fixed casing and rotates at the speed of several hundred

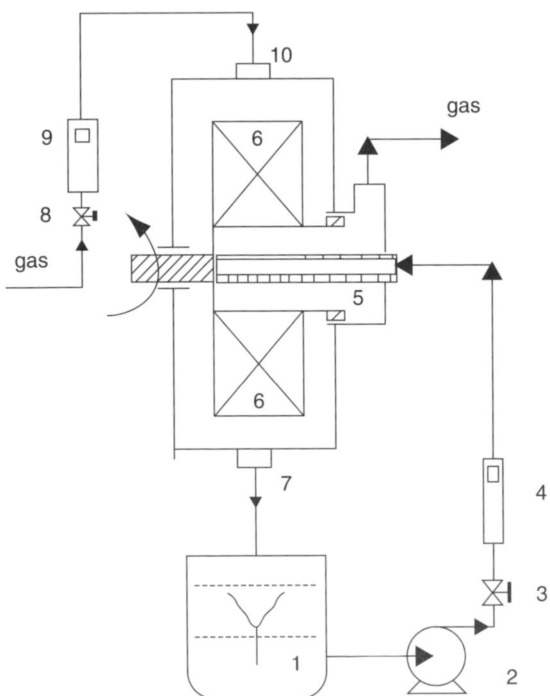

Figure 3.5. Schematic of experimental setup. 1. Stirred tank, 2. Pump, 3. Valve, 4. Rotor flow meter, 5. Distributor, 6. Packed rotator, 7. Outlet, 8. Valve, 9. Rotor flow meter, 10. Inlet (adapted from Ref. 109).

to thousands of rpm. Liquid (including slurry) is introduced into the eye space of the rotator from the liquid inlet pipe and then sprayed by the slotted pipe distributor (5) onto the inside edge of the rotator. The liquid on the bed flows in the radial direction under centrifugal force, passing through the packing, and emerges outside the space between the rotator and the shell, finally leaving the equipment through the liquid exit (7) and collects in 1. The gas is introduced from an outside source (gas cylinder) through the gas inlet (10), flows inward counter currently to the liquid in the packing of the rotator, and finally goes out through the gas exit under the force of pressure gradient.

This technique has thus far been relatively limited in application and has not been extended to any other metal-oxide systems.

3.5. SOL-GEL PROCESSING

Sol-gel techniques have long been known for the preparations of metal oxides and have been described in several books and reviews (15–19, 21, 25, 76, 111–121). The process is typically used to prepare metal oxides via the hydrolysis of metal reactive precursors, usually alkoxides in an alcoholic solution, resulting in the corresponding hydroxide. Condensation of the hydroxide molecules by giving off water leads to the formation of a network of metal hydroxide. When hydroxide species undergo polymerization by condensation of the hydroxy network, gelation is achieved and a dense porous gel is obtained. The gel is a polymer of a three-dimensional skeleton surrounding interconnected pores. Removal of the solvents and appropriate drying of the gel is an important step that results in an ultra-fine powder of the metal hydroxide. Heat treatment of the hydroxide is a final step that leads to the corresponding ultra-fine powder of the metal oxide. Depending on the heat treatment procedure, the final product may end up in the form of a nanometer scale powder, bulk material, or oxygen-deficient metal oxides.

The chemical and physical properties of the final product are primarily determined by the hydrolysis and drying steps. Hydrolysis of metal alkoxides ($M(OR)_z$) involves nucleophilic reactions with water as follows:

$$M(OR)_y + xH_2O \longleftrightarrow M(OR)_{y-x}(OH) + xROH \tag{3.1}$$

The mechanism of this reaction involves the addition of a negatively charged HO^- group to the positively charged metal center (M^+). The positively charged proton is then transferred to an alkoxy group followed by the removal of ROH. Condensation occurs when the hydroxide molecules bind together as they release water molecules and a gel/network of the hydroxide is obtained as shown in Scheme 3.1.

The rates at which hydrolysis and condensation take place are important parameters that affect the properties of the final product. Slower and more controlled hydrolysis typically leads to smaller particle sizes and more unique properties. Hydrolysis and condensation rates depend on the electronegativity of the metal atom, the alkoxy group, solvent system, and the molecular structure of the metal alkoxide. Those metals with higher electronegativities undergo hydrolysis more slowly than those with lower electronegativities. For example, the hydrolysis rate of $Ti(OEt)_4$ is about five orders of magnitude greater than that of $Si(OEt)_4$. Hence, the gelation times of silicon alkoxides are much longer (on the order of days) than those of titanium alkoxides (seconds or minutes) (34). The sensitivity of metal alkoxides toward hydrolysis decreases as the OR group size increases. Smaller OR groups lead to higher reactivity of the corresponding alkoxide toward water and, in some cases, result in uncontrolled precipitation of the hydroxide.

Because alcohol interchange reactions are possible, the choice of solvents in sol-gel processes is very important. As an example, when silica gel was prepared from $Si(OMe)_4$ and heated to 600 °C, when ethanol was used as a solvent, the surface area was 300 m^2/g with a mean pore diameter of 29 Å. However, when methanol was used, the surface area dropped to 170 m^2/g and the mean pore diameter increased to 36 Å (17).

The rate of hydrolysis also becomes slower as the coordination number around the metal center in the alkoxide increases. Therefore, alkoxides that tend to form oligomers usually show slower rates of hydrolysis and, hence, are easier to control and handle. n-Butoxide (O–n–Bu) is often preferred as a precursor to different oxides, including TiO_2 and Al_2O_3, because it is the largest alkoxy group that does not prevent oligomerization (122).

Careful handling in dry atmospheres is required to avoid rapid hydrolysis and uncontrolled precipitation because most metal alkoxides are highly reactive toward water. For alkoxides that have low rates of hydrolysis, acid or base catalysts can be used to enhance the process. The relatively negative alkoxides are protonated by acids creating a better leaving group and eliminating the need for proton transfer in the transition state. Alternatively, bases provide better nucleophiles (OH^-) for hydrolysis, however; deprotonation of metal hydroxide groups enhances their condensation rates.

Developments in the areas of solvent removal and drying have facilitated the production of nanoscale metal oxides with novel properties. When drying is achieved by evaporation under normal conditions, the gel network shrinks as a result of capillary pressure that occurs and the hydroxide product obtained is referred to as xerogel. However, if supercritical drying is applied using a high-pressure autoclave reactor at temperatures higher than the critical temperatures of solvents, less shrinkage of the gel network occurs, as there is no capillary pressure and no liquid–vapor interface, which allows the pore structure to remain largely intact. The hydroxide product obtained in this manner is referred to as an aerogel. Aerogel powders usually demonstrate higher porosities and larger surface areas as compared with analogous xerogel powders. Aerogel processing has been very useful in producing highly divided powders of different metal oxides (15,123–125).

Sol-gel processes have several advantages over other techniques for the synthesis of nanoscale metal oxides. Because the process begins with a relatively homogeneous mixture, the resulting product is a uniform ultra-fine porous powder. Sol-gel processing also has the advantage that it can also be scaled up to accommodate industrial-scale production.

Figure 3.6. TEM micrograph of the nanostructure of AP–MgO (supercritical solvent removal). Here the porosity is formed by the interconnected cubic nanocrystals of MgO (adapted from Ref. 125).

Numerous metal-oxide nanoparticles have been produced by making some modifications to the traditional aerogel method. One modification involved the addition of large amounts of aromatic hydrocarbons to the alcohol-methoxide solutions before hydrolysis and alcogel formation (Figure 3.6). This change was done to further reduce the surface tension of the solvent mixture and to facilitate solvent removal during the alcogel–aerogel transformation (1,125,126). The resulting nanoparticles exhibited higher surface areas, smaller crystallite sizes, and more porosity for samples of MgO, CaO, TiO_2, and ZrO_2 (127,128).

The mesoporous TiO_2 sphere is a better candidate for applications in catalysis, biomaterials, microelectronics, optoelectronics, and photonics. Recently, Zhang et al. (129) put forward a new method to synthesize mesoporous TiO_2 as well as hollow spheres using titanium butoxide, ethanol, citric acid, water, and ammonia. The mesoporous solid and hollow spheres are formed with the same reactants but adding them in a different order. The TiO_2 possessed a mesoporous structure with the particle diameters of 200–300 nm for solid spheres and 200–500 nm for hollow spheres. The average pore sizes and BET surface areas of the mesoporous TiO_2 solid and hollow spheres are 6.8, 7.0 nm, and 162, 90 m^2/g, respectively. Optical adsorption studies showed that the TiO_2 solid and hollow spheres possess a direct band gap structure with the optical band gap of 3.68 and 3.75 eV, respectively. The advantage of this route is that it does not use surfactants or templates as is usual for synthesizing mesoporous materials.

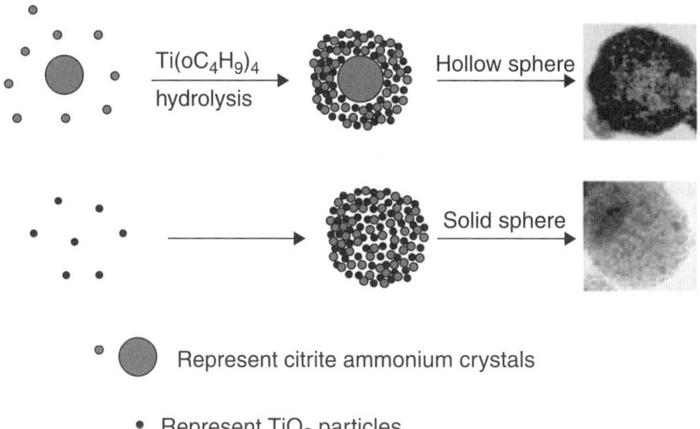

Figure 3.7. Schematic illustration of the formation mechanisms for the mesoporous TiO_2 hollow and solid spheres (adapted from Ref. 129).

The proposed mechanism is (see Figure 3.7), that along the preparation path, the ammonium citrate will form and plays a key role in the formation of the mesoporous solid (129). The formation of mesoporous TiO_2 solid or hollow spheres was highly dependent on the extent of TiO_2 condensation present at the onset of ammonium citrate crystal growth. That is, mesoporous solid spheres form when the TiO_2 condensation process takes place with the formation of ammonium citrate crystals simultaneously; mesoporous hollow spheres are produced when the nucleation and growth of ammonium citrate crystals occurs before the TiO_2 condensation process.

Only a few transition metals exhibit metal-alkoxide chemistry amenable to sol-gel synthesis. Cerda et al. prepared $BaSnO_3$ (130) by calcining a gel formed between $Ba(OH)_2$ and K_2SnO_3 at pH ≈ 11. The material possessed a particle size of 200–500 nm and contained $BaCO_3$ impurity, which can be eliminated by high-temperature calcination. O'Brien et al. (131) synthesized samples of monodisperse nanoparticles of barium titanate with diameters ranging from 6 to 12 nm by the sol-gel method. The technique was extended by Meron et al. (132) to synthesize colloidal cobalt ferrite nanocrystals. This synthesis involves the single-stage high-temperature hydrolysis of the metal alkoxide precursors to obtain crystalline, uniform, organically coated nanoparticles that are well dispersed in an organic solvent. They were also able to form Langmuir–Blodgett films consisting of a close-packed nanocrystal monolayer. The structural and magnetic properties of these nanocrystals possessed the same behavior of similarity to bulk Fe_3O_4 with a very high coercivity at low temperatures.

Other oxides such as V_2O_5 (133,134), MoO_3 (133), and MnO_2 (135,136) aerogels have also been prepared. Nanocomposites of RuO_2–TiO_2 (137) are of particular interest because of their supercapacitor properties. Nitrides and sulphides of Ti, and the sulphide of Nb have also been prepared by sol-gel methods. TiN (138) was prepared by thermal decomposition of a titanium alkoxy hydrazide precursor. TiS_2 (139) and

NbS_2 (140) were prepared by reaction of $Ti(O^iPr)_4$ or $Nb(OEt)_5$ with H_2S in benzene and acetonitrile solution, respectively. The reaction of metal propoxide with H_2S results in the formation of metal alkoxy sulfides, which was calcined at 600–800 °C subsequently to obtain the metal sulfides.

Sol-gel processes have also been used to synthesize nanocomposites like one composed of nanoparticles of transition metal(s) and nanoparticles of magnetic ferric oxide (141), Au–SiO_2, luminescent SnO_2-based supercapacitors (142), $Y_2O_3/Fe_2O_3/TiO_2$ (143), SiO_2–TiO_2 (144), Si–SiO_2 (145), Au–SiO_2 (146), FeCo–SiO_2 (147), Fe-SiO_2 (148), Ag–SiO_2 (149), Ni–SiO_2, NiO–SiO_2 (150), Ti–Si mixed oxide (151), PbE-ceramic composite materials (152), and so on.

3.6. MICROEMULSION TECHNIQUE

Microemulsions or micelles (including reverse micelles) represent an approach based on the formation of micro/nano-reaction vessels for the preparation of nanoparticles, and they have received considerable interest in recent years (153–159). A literature survey depicts that the ultra-fine nanoparticles in the size range between 2 and 50 nm can be easily prepared by this method. This technique uses an inorganic phase in water-in-oil microemulsions, which are isotropic liquid media with nanosized water droplets that are dispersed in a continuous oil phase. In general, microemulsions consists of, at least, a ternary mixture of water, a surfactant, or a mixture of surface-active agents and oil. The classic examples for emulsifiers are sodium dodecyl sulfate (SDC) and aerosol bis(2-ethylhexyl)sulfosuccinate (AOT). The surfactant (emulsifier) molecule stabilizes the water droplets, which have polar heads and nonpolar organic tails. The organic (hydrophobic) portion faces toward the oil phase, and the polar (hydrophilic) group towards water. In diluted water (or oil) solutions, the emulsifier dissolves and exists as a monomer, but when its concentration exceeds a certain limit called the critical micell concentration (CMC), the molecules of emulsifier associate spontaneously to form aggregates called micelles. These micro-water droplets then form nanoreactors for the formation of nanoparticles. The nanoparticles formed usually have monodisperse properties. One method of formation consists of mixing of two microemulsions or macroemulsions and aqueous solutions carrying the appropriate reactants to obtain the desired particles. The interchange of the reactants takes place during the collision of the water droplets in the microemulsions. The interchange of the reactant is very fast so that for the most commonly used microemulsions, it occurs just during the mixing process. The reduction, nucleation, and growth occur inside the droplets, which controls the final particle size. The chemical reaction within the droplet is very fast, so the rate-determining step will be the initial communication step of the microdroplets with different droplets. The rate of communication has been defined by a second-order communication-controlled rate constant and represents the fastest possible rate constant for the system. The reactant concentration has a major influence on the reduction rate. The rate of both nucleation and growth are determined by the probabilities of the collisions between several atoms, between one atom and a nucleus, and between two or more nuclei. Once a nucleus forms with the minimum number of atoms, the growth process starts. For the formation of monodisperse particles, all nuclei must form at the same time and grow simultaneously and with the same rate.

A typical method for the preparation of metal-oxide nanoparticles within micelles consists of forming two microemulsions, one with the metal salt of interest and the other with the reducing or oxide containing agent and mixing them together. A schematic diagram is shown in Figure 3.8 (160).

When two different reactants mix, the interchange of the reactants takes place due to the collision of water microdroplets. The reaction (reduction, nucleation, and growth) takes place inside the droplet, which controls the final size of the particles. The interchange of nuclei between two microdroplets does not take place due to the special restrictions from the emulsifier. Once the particle inside the droplets attains its full size, the surfactant molecules attach to the metal surface, thus stabilizing and preventing further growth.

Reverse micelles are used to prepare nanoparticles by using an aqueous solution of reactive precursors that can be converted to insoluble nanoparticles. Nanoparticle synthesis inside the micelles can be achieved by different methods, including hydrolysis of reactive precursors, such as alkoxides, and precipitation reactions of metal salts (161,162). Solvent removal and subsequent calcination leads to the final product. A variety of surfactants can be used in these processes, such as pentadecaoxyethylene nonylphenylether (TNP-35) (162), decaoxyethylene nonylphenyl ether (TNT-10)

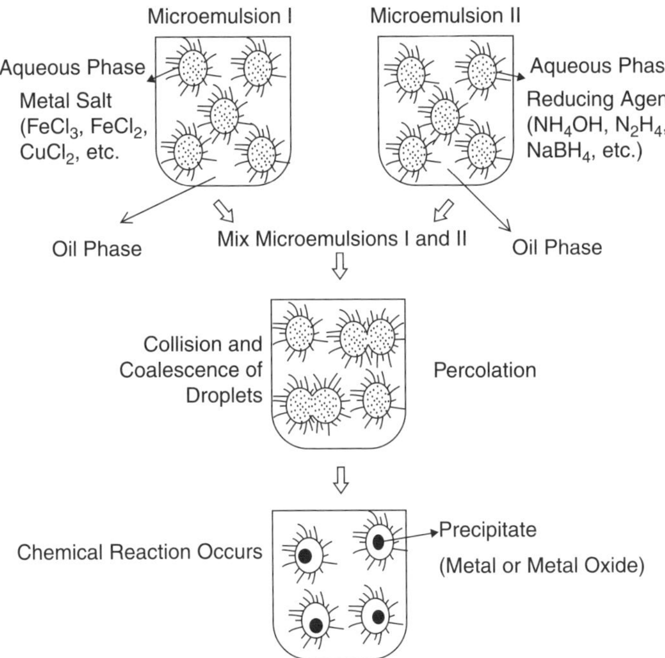

Figure 3.8. Proposed mechanism for the formation of metal particles by the microemulsions approach (adapted from Ref. 160).

(162), poly(oxyethylene)$_5$nonylphenolether (NP5) (163), and many others that are commercially available. Several parameters, such as the concentration of the reactive precursor in the micelle and the weight percentage of the aqueous phase in the microemulsion, affect the properties, including particle size, particle size distribution, agglomerate size, and phases of the final oxide powders. There are several advantages to using this method, including the preparation of very small particles and the ability to control the particle size. Disadvantages include low production yields and the need to use large amount of solvents and surfactants.

This method has been successfully applied for the synthesis of metals, metal oxides, alloys, and core-shell nanoparticles. The synthesis of metal oxides from reverse micelles is similar in most aspects to the synthesis of metal oxides in aqueous phase by the precipitation process. For example, precipitation of hydroxides is done by addition of bases like NH_4OH or NaOH to a reverse micelle solution containing aqueous metal ions at the micelle cores. This method has been also used to synthesize mixed metal iron oxides (164):

$$M^{2+} + 2Fe^{2+} + OH^-(\text{excess}) \xrightarrow{\Delta, O_2} MFe_2O_4 + xH_2O \uparrow$$

where M = Fe, Mn, or Co. The resultant particles obtained are of the order of 10–20 nm in diameter. The cation distribution in the case of spinels depends on the temperature used in the reaction (165). If the transition metal cation is unstable or insoluble in aqueous media, nanoparticles of those metals can be prepared by hydrolysis of suitable precursors; for example, TiO_2 has been prepared from tetraisopropyl titanate (166) in the absence of water. The hydrolysis occurs when a second water-containing solution of reverse micelles is added to the solution of the first micelle containing the metal precursor. Adjusting the concentration of reactants can change the final crystal form, i.e., amorphous or crystalline. The reaction can be gives as follows:

$$Ti(OiPr)_4 + 2H_2O \longrightarrow TiO_2 + 4(iPr{-}OH)$$

Microemulsions have also been employed to prepare precursors that decompose during calcinations, resulting in desired mixed metal oxides. A series of mixed-metal ferrites have been prepared by precipitating the metal precursor in the H_2O–AOT–isooctane system and calcining the products at 300–600 °C (167). A superconductor material $YBa_2Cu_3O_{7-\delta}$ with 10-nm particles has been prepared by combining the micellar solution of Y_{3+}, Ba_{2+}, and Cu_{2+} prepared in an Igepal CO-430–cyclohexene system with a second micellar solution containing oxalic acid in the aqueous cores. (Note: Igepal CO-430 is a surfactant with chemical name “Nonyl phenol 4 mole ethoxylate”). The precipitate was subsequently calcined to 800 °C to remove oxalic precursors (168). Other metal oxides, which are prepared in a similar method are tungsten and tungsten oxide nanoparticles (169), ASA $Ce_xBa_{1-x}MnAl_{11}Oy$ (170), Ni–Fe spinel (171), Al_2O_3 (172), $LaMnO_3$ (173), $BaFe_{12}O_{19}$ (174), $Cu_2L_2O_5$ (L = Ho, Er) (175), and $LiNi_{0.8}Co_{0.2}O_2$ (176), and $BaFe_{12}O_{19}$ (177).

3.7. SOLVOTHERMAL TECHNIQUE

Solvothermal techniques have also been used to synthesize metal oxide (67,178–190) and chalcogenide[2] (191–203) as well as semiconductor chalcogenides nanoparticles (197,204–209).

The metal complexes are decomposed thermally either by boiling the contents in an inert atmosphere or by using an autoclave. A suitable capping agent or stabilizer such as a long-chain amine, thiol, and trioctylphospine oxide [TOPO] is added to the reaction contents at a suitable point to hinder the growth of the particles and, hence, provide their stabilization against agglomeration. The stabilizers also help in dissolution of the particles in different solvents.

The decomposition of metal alkoxides for the synthesis of TiO_2 (210) nanoparticles, TOPO capped autoclave synthesis of TiO_2 by metathetic reaction (211), decomposition of metal N-nitroso N-phenyl hydroxylamine complex to get metal oxide (212,213), synthesis of CdE ($E = S, Se, Te$) nanoparticles using an organometallic precursor in TOPO (214) as solvent, and toluene (193) as solvent in nujol (215) are some important innovations in recent years.

Nanoscale TiO_2 was prepared using an aqueous $TiCl_4$ solution by hydrothermal synthesis; it was found that acidic conditions favored the rutile, whereas basic conditions favored anatase (216). It was also found that higher temperature favors the highly dispersed product and that grain size can be controlled by the addition of minerals such as $SnCl_4$ or NaCl, but the presence of NH_4Cl leads to agglomeration of particles. The approach was extended and revealed that phase purity of the products depends primarily on concentration, with higher concentrations of $TiCl_4$ favoring the rutile phase, whereas particle size depends primarily on reaction time (217). Yin et al. produced 2–10-nm crystallites of monodispersed, phase-pure anatase by using citric acid to stabilize the TiO_2 nanoparticles and treating the precursors hydrothermally in the presence of KCl or NaCl mineralizers (218).

Masui et al. reported the hydrothermal synthesis of nanocrystalline, monodispersed CeO_2 with a very narrow size distribution (219). They combined $CeCl_36H_2O$ and aqueous ammonia with a citric acid stabilizer and autoclaved the solution in a sealed teflon container at 80 °C. The CeO_2 nanoparticles exhibited a 3.1 nm average diameter. Inoue et al. (220) were able to reduce the particle size of CeO_2 produced by a similar approach to 2 nm by autoclaving a mixture of cerium metal and 2-methoxyethanol at 250 °C and coagulating the particles with methanol and ammonia. The higher reaction temperature resulted in increased particles size.

Metal oxides can also be synthesized by the decomposition of metal-Cupferron complexes, M^xCup^x ($Cup = C_6H_5N(NO)O^-$) (212). Seshadri and co-workers were able to replace the amine-based solvents with toluene and prepared ≈10-nm-diameter γ-Fe_2O_3 and ≈7-nm $CoFe_2O_4$ by hydrothermal processes (213). They synthesized maghemite γ-Fe_2O_3 nanoparticles from a Fe^{III}-cupferron complex and spinel

[2]Compounds containing a chalcogen element, that in-group 16 of the Periodic Table, and excluding oxides are commonly termed chalcogenides. These elements are sulfur (S), selenium (Se), tellurium (Te), and polonium (Po). Common chalcogenides contain one or more of S, Se, and Te, in addition to other elements.

$CoFe_2O_4$ nanoparticles starting from a Co^{II}-cupferron complex and Fe^{III}-cupferron complex. The authors, however, found that, the presence of at least trace quantities of a strongly coordinating amine was necessary to act as a capping agent and prevent aggregation. They could also synthesis of nanoparticulate $ZnFe_2O_4$ by a similar approach (221).

3.8. TEMPLATED/SURFACE DERIVATIZED NANOPARTICLES

In recent years, varieties of porous materials have been obtained by templated techniques. Generally two types of templates have been used in the literature, soft template (surfactants) (222) and hard template (porous solids such as carbon or silica). In the case of hard templates, the formation of porous material takes place in a confined space formed by the porosity of the template. Two types of hard templates have been made use of in the template synthesis: active carbon (223) and mesoporous silica materials (224–227). Commercial active carbons were used as a template to prepare different types of HSMO (228), monodisperse and porous spheres of oxides, and phosphates (229). However, the use of active carbons as templates has certain limitations because, at the high treatment temperature employed during the synthesis, infiltrated salts and the carbon may react with each other destroying the intended material. Moreover, if the heat treatment is performed in air, the carbon may be rapidly oxidized (ignition) even at a relatively low temperature due to the catalytic effect of infiltrated salts. On the other hand, some metallic salts may end up as metal instead of metal oxide as carbon is a good reducing agent at high temperature under inert temperature.

Mesostructured silica materials (MSMs) have been employed for the fabrication of ordered porous inorganic materials (i.e., metal oxides, sulphides) (224,225). In this method, retention of surface silanol groups is very important for effective impregnation of metal salts inside the porosity. Surface silanol groups within the silica pores are usually affected during the removal of surfactant templates used for the synthesis of MSMs. A recent and special technique called microwave digestion (Figure 3.9) helps to overcome this problem. Using this technique, Tian et al. (224) have prepared various metal oxides such as NiO, Co_2O_3, Mn_xO_y, In_2O_3, Fe_2O_3, and Cr_2O_3 (Figure 3.10). Table 3.3 give the textural properties of various nanostructured metal-oxide wires and spheres obtained by this method (224).

Although, the procedure proposed by Tian et al. (224) yields HSA metal oxides, it requires the use of expensive surfactants as templating agents for the synthesis of MSM and a sophisticated method to remove the surfactant. An inexpensive approach to synthesize high-surface-area metal oxides (HSMOs) was introduced by Fuertes (226). In this method, porous silica templates (silica xerogel) were synthesized without employing the surfactants (i.e., silica gels). The xerogel is obtained from a synthesis mixture formed exclusively by a silica source (sodium silicate), HCl, and water. Furthermore, because the templated material is obtained without the aid of structure-directing agents (i.e., surfactants) and the synthesis takes place at a moderate temperature (100 °C), the surface silanol groups can easily be preserved. This method had been demonstrated for Fe_2O_3, Cr_2O_3, NiO, CeO_2, Mn_2O_3, Co_2O_3, and Al_2O_3 (see Figure 3.11). Table 3.4 gives the physical characteristic of these samples (226).

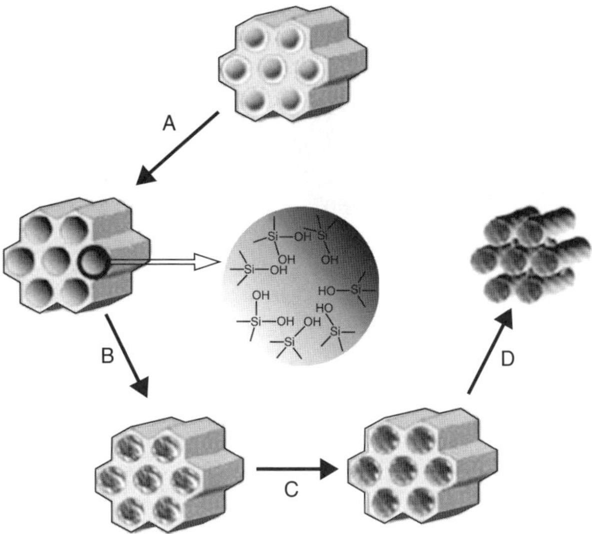

Figure 3.9. Schematic diagram of the synthesis procedure for the ordered crystallized metal-oxide nano-arrays with mesostructured frameworks. (A) microwave digestion, (B) infiltration with inorganic precursors, (C) calcination, and (D) etching by HF or NaOH. The magnified image indicates the abundance of the silanol groups on the cavity surface (adapated from Ref. 224).

Although Feurtes' method is simple and inexpensive, the product obtained is not an ordered nanorod as it is when using the SBA template (224), but it is in the form of aggregates of nanoparticles and/or three-dimensional solid-containing confined pores. The Feurtes' method can also be used to synthesize metal sulphides as well as mixed metal oxides.

A recent method proposed by Chane-Ching et al. (60) to synthesize various nanostructured metal oxides from surface-modified nanoparticle building blocks by using a liquid-crystal template is very simple and versatile. This process involves functionalizing the oxide nanoparticles with bifunctional organic anchors like aminocaproic acid and taurine. After the addition of a copolymer surfactant, the functionalized nanoparticles will slowly self-assemble on the copolymer chain through a second anchor site. By using this approach, authors could prepare several metal oxides like CeO_2 (Figure 3.12), ZrO_2, and CeO_2–$Al(OH)_3$ composites. The method yields ZrO_2 of surface area 180 m^2/g after calcining at 500 °C, 125 m^2/g for CeO_2, and 180 m^2/g for CeO_2–$Al(OH)_3$ composites. The ZrO_2 obtained by this method possessed solely the monoclinic crystal phase. For example, SAXS patterns of the as-synthesized nanostructured CeO_2 materials, after calcination at 500 °C for 6 h, showed a peak centered at 10 nm (Figure 3.13) (60). TEM images show that the material is mesoporous and possess large, well-ordered channels organized in hexagonal arrays. The d_{SAXS} peaks can be indexed to a hexagonal cell with a unit-cell dimension of 11.5 nm. The N_2 desorption curves suggested that the material possess pores with a size of 5.5 nm.

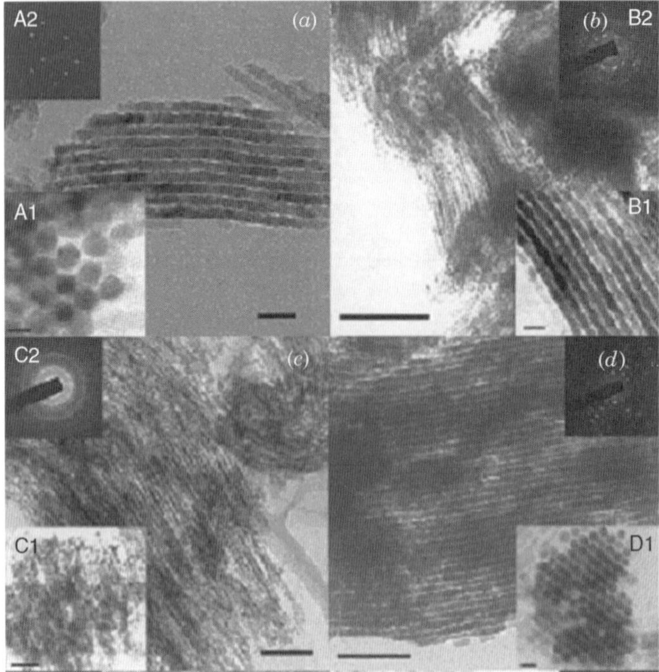

Figure 3.10. TEM images of ordered metal-oxide nanowires (obtained from MWD-SBA-15) arrays with crystallized mesoporous frameworks. (*a*) Cr_2O_3 nanowires depicted on the (100) plane and the (011) plane (inset A1); inset A2 is the SAED of A. (*b*) Mn_xO_y nanowires depicted on the (100) plane; inset B1 showing the (100) plane with larger magnification; inset B2 is the SAED of B1. (*c*) Fe_2O_3 nanowires depicted on the (100) plane and the (011) plane (inset C1); SAED of C is shown in inset C2. (*d*) Co_3O_4 nanowires, (100) plane and the (011) planes (inset D1); SAED of D is shown in D2. (adapated from Ref. 224).

TABLE 3.3. Textural Properties of Various Nanostructured Metal Oxide Wires and Spheres (Adapted from Ref. 224).

Composition	Template	Morphology	BET, m^2/g	Pore Size [nm]	Pore Volume [$cm^3\ g^{-1}$]
Cr_2O_3	MWD-SBA-15	nanowires	64.9	3.6	0.34
Mn_xO_y	MWD-SBA-15	nanowires	103.2	3.4	0.39
Fe_2O_3	MWD-SBA-15	nanowires	137.1	4	0.43
Co_3O_4	MWD-SBA-15	nanowires	82.4	3.4	0.37
	MWD-SBA-16	nanowires	92.1	6.5	0.33
NiO	MWD-SBA-15	nanowires	56.4	3.5	0.38
In_2O_3	MWD-SBA-15	nanowires	70.3	3.5	0.36
	MWD-SBA-16	nanowires	122.6	6.1	0.34

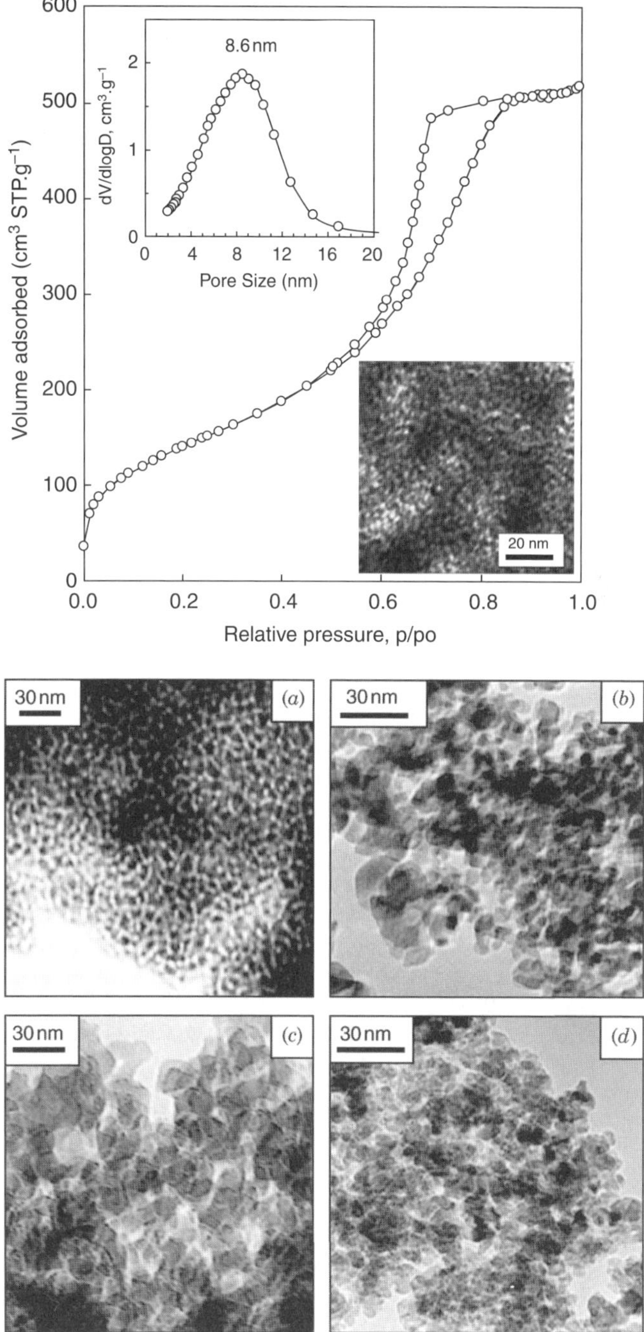

Figure 3.11. (Top) Nitrogen sorption isotherm, pore size distribution (upper left), and TEM image (lower right) for the silica xerogel used as template. (bottom) TEM images of the various synthesized metallic oxides: (*a*) Fe_2O_3 (ferrihydrite), (*b*) Co_2O_3, (*c*) Al_2O_3, and (*d*) NiO (adapted from Ref. 226).

TABLE 3.4. Structural Properties of High-Surface Area Metal Oxides (Adapted from Ref. 226).

Product (XRD)	S_{BET} ($m^2\,g^{-1}$)	V_t ($cm^3\,g^{-1}$)[a]	Structural Porosity V_s ($cm^3\,g^{-1}$)[c]	Structural Porosity Pore Size (nm)[d]	Crystallite Size (nm)[b]
Fe_2O_3	–	–			
Ferrihydrite	270	0.45	0.32	6.5	–
Maghemite	176	0.44	0.32	11	–
Cr_2O_3	118	0.57	–	–	17
NiO	96	0.39	0.12	6.9	11
CeO_2	141	0.35	0.04	4.6	7
Mn_2O_3	117	0.26	0.08	4.6	18
Co_2O_3	98	0.21	0.14	8.6	10
Al_2O_3	53	0.78	–	–	–

[a]Total pore volume from N_2 adsorption at p/po = 0.99.
[b]Estimated from XRD peak broadening by using the Scherrer formula.
[c]Volume of structural pores.
[d]Maximum PSD for the structural pores.

In a recent method (230) to synthesize one-dimentional transition metal oxide-nanocrystals, metal chlorides such as $WCl_4/MnCl_2/TiCl_4$ were dissolved in oleic acid and oleylamine and heated to 350/270/290 °C. After 60/20/2 min, the reaction was quenched using toluene and treated with acetone to precipitate dark brown flocculates of $W_{18}O_{48}/MnO_2/TiO_2$, respectively, which can be separated by centrifugation. The resulting powder was dispersed in a hydrophobic solvent like toluene and dichloromethane. By this route, the authors obtained well-defined rod-shaped $W_{18}O_{48}$ nanocrystals of 4.5 ± 0.6 nm in width and 28.3 ± 5.1 nm in length. X-ray studies revealed that these nanorods were anisotropically grown along the b-axis of

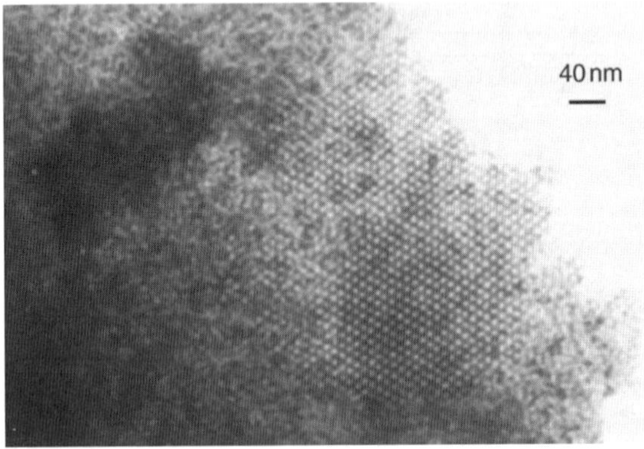

Figure 3.12. TEM image of the CeO_2 nanostructured material after calcination at 500 °C (adapted from Ref. 60).

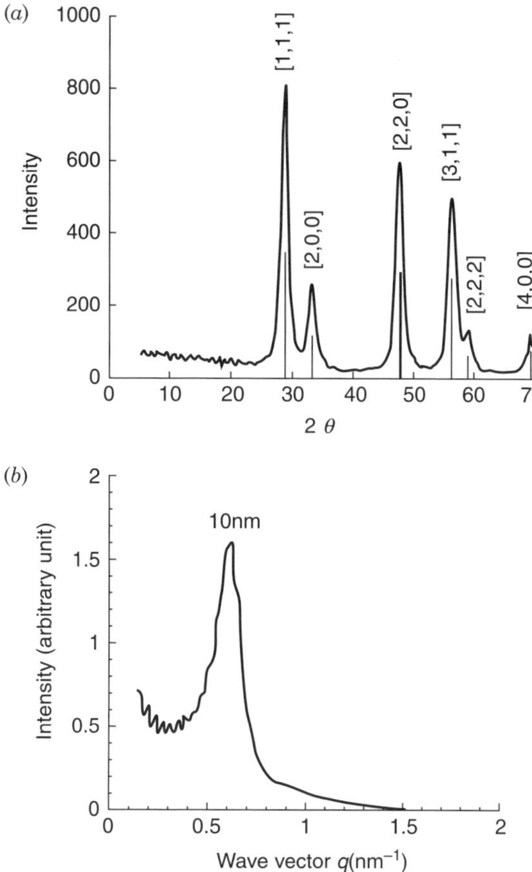

Figure 3.13. (*a*) XRD pattern of the CeO_2 nanostructured material after calcination at 500 °C for 6 h. (*b*) The SAXS pattern of the CeO_2 nanostructured material after calcination at 500 °C for 6 h (adapted from Ref. 60).

the monoclinic structure of $W_{18}O_{48}$. The HTTEM study (Figure 3.14) indicates that the $W_{18}O_{48}$ single crystallinity in the monoclinic structure and the lattice fringes that orient in the (010) direction with an interplanar distance of 3.78 Å reveal the nanorod is elongated along the (010) lattice plane (230). TEM and HRTEM images of MnO_2 nanocrystals revealed that they are one-dimensional with a width of 3.1 ± 0.3 nm and a length of 8.4 ± 1.2 nm, with the aspect ratio of ~3. It was also found that control of the aspect ratio is possible by changing the growth temperature.

Similar studies for TiO_2 nanocrystal revealed that they are highly anisotropic with the aspect ratio of ~7. XRD analysis showed that they have a tetragonal anatase structure. The HRTEM image of the TiO_2 nanorods showed that these nanocrystals have single crystallinity and are preferentially grown along the (001) direction with an interplanar distance of 2.4 Å. In summary, this method is versatile, reproducible,

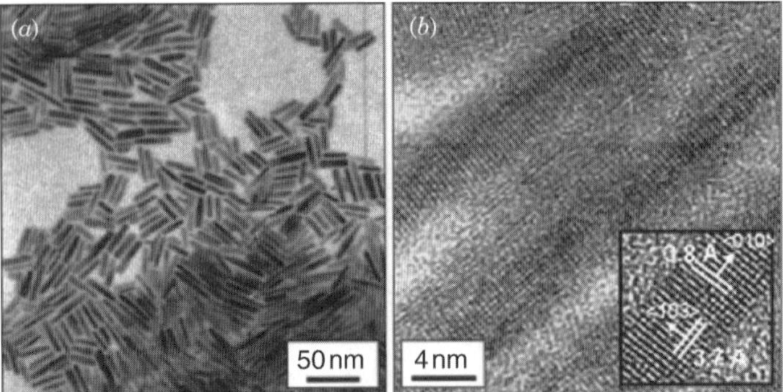

Figure 3.14. (*a*) TEM and (b) HRTEM of tungsten oxide nanorods. Inset: magnified HRTEM image showing lattice fringes of $W_{18}O_{49}$ nanorods (adapted from Ref. 230).

and simple, and therefore, it can be used as a new protocol for a one-dimensional metal-oxide nanocrystal. But applicability for other metal oxides has yet to be tested.

3.9. SPECIFIC PROPERTIES

As mentioned, the properties of nanoparticles are usually size dependent and in the case of metal oxides are primarily dictated by defects and the coordinatively unsaturated surface. When prepared in nanometer size particles, these materials exhibit unique chemical and physical properties that are remarkably different than those of the corresponding bulk materials. The study of physical and chemical properties of nanoparticles is of great interest as a way to explore the gradual transition from atomic or molecular to condensed matter systems.

As the size of a particle decreases, the percentage of atoms residing on the surface increases. As an example, a study of different samples of MgO nanoparticles has revealed that for particles ~4 nm in diameter, ~30% of the atoms are surface atoms (15). Of course, surface atoms/ions are expected to be more reactive than their bulk counterparts as a result of coordinative unsaturation. Because of this and the fact that the surface-to-volume ratio is large, it is not unusual to see unique behavior and characteristics for nanoparticles. This particle size effect is a characteristic of different nanomaterials, including metal oxides. Chapters 11–13 in this book discuss in detail some selected properties of nanophase metal oxides, which have significant size dependence.

3.10. CONCLUSION

Nanoscale metal oxides have turned out to play an important role in different fields of nanoscience from biology, chemistry, physics, and material science to electronics.

Depending on the application of metal oxides as well as the quantity of material required, one could choose different methods of synthesis as mentioned here. Although in-depth exploration of this field often requires sophisticated and expensive instrumentation, many innovative breakthroughs have been realized and the future seems to hold a great deal of potential.

Nanoscale metal oxides are of considerable importance to both the fundamental understanding of size-dependent properties and the numerous applications. Although in many cases a basic understanding of the bonding and structure present in these systems has been determined, there is still a great deal of work to be done. Additionally, the methods for the preparation of oxide systems are a continually evolving area of science. Finally, developments in the areas of synthesis, instrumentation, and modeling will aid scientists to understand the relationships between physical, electronic, and chemical properties.

REFERENCES

(1) Klabunde, K.J.; Stark, J.V.; Koper, O.; Mohs, C.; Park, D.G.; Decker, S.; Jiang, Y.; Lagadic, I.; Zhang, D.J. *Phys. Chem.* **1996**, *100*, 12142–12153.

(2) Klabunde, K.J.; Mohs, C. *Nanoparticles and Nanostructural Materials*; Wiley-VCH: New York, 1998; 317.

(3) Liu, H.; Feng, L.; Zhang, X.; Xue, O. *J. Phys. Chem.* **1995**, *99*, 332.

(4) Tanabe, K. *Solid Acids and Bases*; Academic Press: San Diego, CA, 1970.

(5) Utiyama, M.; Hattori, H.; Tanabe, K. *J. Catal.* **1978**, *53*, 237.

(6) Morris, R.M.; Klabunde, K.J. *Inorg. Chem.* **1983**, *22*, 682.

(7) Richards, R.M.; Volodin, A.M.; Bedilo, A.F.; Klabunde, J.K. *Phys. Chem. Chem. Phys.* **2003**, *19*, 4299–4305.

(8) Rao, C.N.R.; Raveau, B. *Transition Metal Oxides, Structure, Properties, and Synthesis of Ceramic Oxides, 2nd ed*; Wiley-VCH: New York, 1998.

(9) Dyrek, K.; Che, M. *Chem. Rev.* **1997**, *97*, 305–331.

(10) Shriver, D.F.; Atkins, P.W. *Inorganic Chemistry, 3rd ed*; W.H. Freeman & Company: New York, 1999.

(11) Larcher, D.; Sudant, G.; Patrice, R.; Tarascon, J.-M. *Chem. Mater.* **2003**, *15*(18), 3543–3551.

(12) Zemski, K.A.; Justes, D.R.; Castleman, A.W., Jr. *J. Phys. Chem. B.* **2002**, *106*(24), 6136–6148.

(13) Vetrivel, S.; Pandurangan, A. *Ind. Eng. Chem. Res.* **2005**, *44*(5), 692–701.

(14) Ranjit, K.T.; Klabunde, K.J. *Inorg. Mater.* **2005**, *17*(1), 65–73.

(15) Itoh, H.; Utamapanya, S.; Stark, J.V.; Klabunde, K.J.; Schlup, J.R. *Chem. Mater.* **1993**, **5**, 71.

(16) Palkar, V.R. *Nanostruct. Mater.* **1999**, *11*(3), 369.

(17) Interrante, L.V.; Hampden-Smith, M.J. *Chemistry of Advanced Materials: An Overview*; Wiley-VCH: New York, 1998.

(18) Gesser, H.D.; Gosswami, P.C. *Chem. Rev.* **1989**, *89*, 765.

(19) Bourell, D.L.J. *J. Am. Ceram. Soc.* **1993**, *76*, 705.
(20) Chatry, M.; Henry, M.; Livage, J. *Mater. Res. Bull.* **1994**, *29*, 517.
(21) Kumazawa, H.; Inoue, T.; Sasa, E. *Chem. Eng. J. (Amsterdam, Neth.)* **1994**, *55*, 93.
(22) Trueba, M.; Trasatti, S.P. *Eur. J. Inorg. Chem.* **2005**, *17*, 3393–3403.
(23) Reddy, B.M.; Khan, A. *Catalysis Rev. Sci. Eng.*, **2005**, *47*(2), 257–296.
(24) Carreon, M.A.; Guliants, V.V. *Catal. Today* **2005**, *99*(1–2), 137–142.
(25) Carreon, M.A.; Guliants, V.V. *Eur. J. Inorg. Chem.* **2005**, (1), 27–43.
(26) Taguchi, A.; Schüth, F. *Microporous Mesoporous Mater.* **2005**, *77*(1), 1–45.
(27) Siegel, R.W.; Ramasamy, S.; Hahn, H.; Zongquan, L.; Ting, L.; Gronsky, R. *J. Mater. Res.* **1998**, *3*, 1367.
(28) El-Shall, M.S.; Slack, W.; Vann, W.; Kane, D.; Hanely, D. *J. Phys. Chem.* **1994**, *98*(12), 3067.
(29) Baraton, M.I.; El-Shall, M.S. *Nanostruct. Mater.* **1995**, *6*, 301.
(30) Hadjipanayis, G.C.; Siegel, R.W. *Nanophase Materials*; Kluwer Academic Publishers: Dordrecht, the Netherlands, 1994.
(31) Huh, M.Y.; Kim, S.H.; Ahn, J.P.; Park, J.K.; Kim, B.K. *Nanostruct. Mater.* **1999**, *11*(2), 211.
(32) Cow, G.M.; Gonsalves, K.E. *Nanotechnology, Molecularly Designed Materials*; American Chemical Society: Washington, DC, 1996; 77–99.
(33) Kodas, T.T. *Adv. Mater.* **1998**, *6*, 180.
(34) Janackovic, D.; Jokanovic, V.; Kostic-Gvozdenovic, L.; Uskokovic, D. *Nanostruct. Mater.* **1998**, *10*(3), 341.
(35) Messing, G.L.; Zhang, S.C.; Jayanthi, G.V. *J. Am. Ceram. Soc.* **1993**, *76*(11), 2707.
(36) Kodas, T.T.; Datye, A.; Lee, V.; Engler, E. *J. Appl. Phys.* **1989**, *65*, 2149.
(37) Ulrich, G.D.; Riehl, J.W. *J. Colloid Interface Sci.* **1982**, *87*, 257.
(38) Lindackers, D.; Janzen, C.; Rellinghaus, B.; Wassermann, E.F.; Roth, P. *Nanostruct. Mater.* **1998**, *10*(8), 1247.
(39) Skanadan, G.; Chen, Y.-J.; Glumac, N.; Kear, B.H. *Nanostruct. Mater.* **1999**, *11*(2), 149.
(40) Klabunde, J.K.; Cardenas-Trivino, G. *Active Metals*; VCH: Weinheim, 1996; 237–278.
(41) Perez-Alonso, F.J.; Melian-Cabrera, I.; Granados, M.L.; Kapteijn, F.; Fierro, J.L.G. *J. Catal.* **2006**, *239*(2), 340–346.
(42) Zhou, L.P.; Xu, J.; Li, X.Q.; Wang, F. *Mater. Chem. Phys.* **2006**, *97*(1), 137–142.
(43) Choi, H.J.; Moon, J.; Shim, H.B.; Han, K.S.; Lee, E.G.; Jung, K.D. *J. Am. Ceram. Soc.* **2006**, *89*(1), 343–345.
(44) Gao, L.; Wang, H.Z.; Hong, J.S.; Miyamoto, H.; Miyamoto, K.; Nishikawa, Y.; Torre, S.D.D.L. *Nanostruct. Mater.* **1999**, *11*(1), 43.
(45) Qian, Z.; Shi, J.L. *Nanostruct. Mater.* **1998**, *10*(2), 235.
(46) Rao, R.M.; Rao, K.; Prasada, A.V. *Mater. Lett.* **1996**, *1*(28), 463.
(47) Tom, R.T.; Nair, S.A.; Singh, N.; Aslam, M.; Nagendra, C.L.; Philip, R.; Vijayamohanan, K.; Pradeep, T. *Langmuir* **2003**, *19*(8), 3439–3445.
(48) Pârvulescu, V.; Coman, N.S.; Grange, P.; Pârvulescu, V.I. *Appl. Catal., A* **1999**, *176*(1), 27–43.

(49) Pârvulescu, V.I.; Pârvulescu, V.; Endruschat, U.; Lehmann, C.W.; Grange, P.; Poncelet, G.; Bönnemann, H. *Microporous Mesoporous Mater.* **2001**, *44*, 221–226.

(50) Pârvulescu, V.I.; Bönnemann, H.; Pârvulescu, V.; Endruschat, U.; Rufinska, A.; Lehmann, C.W.; Tesche, B.; Poncelet, G. *Appl. Catal., A* **2001**, *214*(2), 273–287.

(51) Ward, D.A.; Ko, E.I. *J. Catal.* **1995**, *157*(2), 321–333.

(52) Mamak, M.; Coombs, N.; Ozin, G.A. *Chem. Mater.* **2001**, *13*(10), 3564–3570.

(53) Li, Y.; He, D.; Yuan, Y.; Cheng, Z.; Zhu, Q. *Energy Fuels* **2001**, *15*(6), 1434–1440.

(54) Xu, W.; Luo, Q.; Wang, H.; Francesconi, L.C.; Stark, R.E.; Akins, D.L. *J. Phys. Chem. B* **2003**, *107*, 497–501.

(55) Navio, J.A.; Hidalgo, M.C.; Colon, G.; Botta, S.G.; Litter, M.I. *Langmuir* **2001**, 17, 202–210.

(56) Sun, W.; Xu, L.; Chu, Y.; Shi, W. *J. Colloid Interface Sci.* **2003**, *266*, 99–106.

(57) Stichert, W.; Schüth, F. *J. Catal.* **1998**, *174*, 242–245.

(58) Chuah, G.K.; Jaenicke, S.; Pong, B.K. *J. Catal.* **1998**, *175*, 80–92.

(59) Tani, E.; Yoshimura, M.; Somiya, S. *J. Am. Ceram. Soc.* **1983**, *66*, 11.

(60) Chane-Ching, J.-Y.; Cobo, F.; Aubert, D.; Harvey, H.G.; Airiau, M.; Corma, A. *Chem. Eur. J.* **2005**, *11*, 979–987.

(61) Kristof, C.; Thierry, L.; Katrien, A.; Pegie, C.; Oleg, L.; Gustaaf, V.G.; René, V.G.; Etienne, F.V. *J. Mater. Chem.* **2003**, *13*, 3033–3039.

(62) Bedilo, A.F.; Klabunde, K.J. *J. Catal.* **1998**, *176*, 448–458.

(63) Bedilo, A.F.; Klabunde, K.J. *Nanostruct. Mater.* **1997**, *8*(2), 119–135.

(64) Bedilo, A.F.; Volodin, A.M. *J. Mol. Catal. A: Chem.* **2000**, *158*, 405.

(65) Bedilo, A.F.; Timoshok, A.M.; Volodin, A.M. *Catal. Lett.* **2000**, *68*, 209.

(66) D'Souza, L.; Saleh-Subaie, J.; Richards, R. *J. Colloid Interface Sci.* **2005**, *292*, 476–485.

(67) Cao, Y.; Hu, J.-C.; Hong, Z.-S.; Deng, J.-F.; Fan, K.-N. *Catal. Lett.* **2002**, *81*(1), 107–112.

(68) Xu, B.-Q.; Wei, J.-M.; Ying-Tao Y.U; Li, Y.; Li, J.-L.; Zhu, Q.-M. *J. Phys. Chem. B* **2003**, *107*, 5203–5207.

(69) Srdic, V.V.; Winterer, M. *Chem. Mater.* **2003**, *15*, 2668–2674.

(70) Chuah, G.K. *Catal. Today* **1999**, *49*, 131–139.

(71) Kaddouri, A.; Mazzocchia, C.; Tempesti, E.; Anouchinsky, R. *J. Therm. Anal.* **1998**, *53*, 97–109.

(72) Cassiers, K.; Linssen, T.; Aerts, K.; Cool, P.; Lebedev, O.; Van Tendeloo, G.; Van, G.; Vansant, E.F. *J. Mater. Chem.* **2003**, *13*(12), 3033–3039.

(73) Yang, P.; Zhao, D.; Margolese, D.I.; Chmelka, B.F.; Stucky, G.D. *Nature* **1998**, *396*, 152–155.

(74) Richards, R.; Khaleel, A. *Ceramics Nanoparticle in 'Nanostructured Materials in Chemistry'*; Wiley VCH: New York, 2001.

(75) Bönnemann, H.; Richards, R. *Catalysis and Electrocatalysis at Nanoparticle Surfaces*; Marcel Dekker: New York, 2002.

(76) Brinker, J.C.; Scherer, G. *Sol-Gel Sci*; Elsevier Science: New York, 1989.

(77) Thomas, J.M.; Thomas, W.J. *Heterogeneous Catalysis*; Wiley VCH: New York, 1997.

(78) Fenelonov, V.B.; Mel'gunov, M.S.; Mishakov, I.V.; Richards, R.; Chesnokov, V.V.; Volodin, A.; Klabunde, J.K. *J. Phys. Chem. B* **2001**, *105*, 3937–3941.

(79) Mishakov, I.V.; Bedilo, A.F.; Richards, R.; Chesnokov, V.V.; Volodin, A.; Buyanov, R.A.; Klabunde, K.J. *J. Catal.* **2002**, *206*, 40–48.

(80) D'Souza, L.; Suchopar, A.; Zhu, K.; Balyozova, D.; Devadas, M.; Richards, R.M. *Microporous Mesoporous Mater.* **2005**, *88*(1–3), 22–30.

(81) Readey, M.J.; Readey, D.W. *J. Am. Ceram. Soc.* **1986**, *69*(7), 580–582.

(82) Karapetrova, E.; Platzer, R.; Gardner, J.A.; Torne, E.; Sommers, J.A.; Evenson, W.E. *J. Am. Ceram. Soc.* **2001**, *84*(1), 65–70.

(83) Matsui, K.; Ohgai, M. *J. Ceram. Soc. Jpn.* **1999**, *107*(10), 949–954.

(84) Sing, K.S.W. *Pure Appl. Chem.* **1982**, *54*(11), 2201–2218.

(85) Joo, J.; Yu, T.; Kim, Y.W.; Park, H.M.; Wu, F.; Zhang, J.Z.; Hyeon, T. *J. Am. Chem. Soc.* **2003**, *125*, 6553–6557.

(86) Rouanet, A.; Pichelin, G.; Roucau, C.; Snoeck, E.; Monty, C. *Nanophase Materials*; Kluwer Academic Publishers: Dordrecht, the Netherlands, 1994; 85–88.

(87) Suslick, K.S.; Choe, S.B.; Cichowlas, A.A.; Grinstaff, M.W. *Nature* **1991**, *353*, 414.

(88) Hiller, R.; Putterman, S.J.; Barber, B.P. *Phy. Rev. Lett.* **1992**, *69*, 1182.

(89) Barber, B.P.; Putterman, S.J. *Nature* **1986**, *352*, 414.

(90) Suslick, K.S.; Hammerton, D.A.; Cline, R.E. *J. Am. Chem. Soc.* **1986**, *108*, 5641.

(91) Livage, J. *J. Phys.* **1981**, *42*, 981.

(92) Sugimoto, M. *J. Magn. Magn. Mater.* **1994**, *133*, 460.

(93) Landau, M.V.; Vradman, L.; Herskowitz, M.; Koltypin, Y.; Gedanken, A. *J. Cata.* **2001**, *201*, 22.

(94) Perkas, N.; Wang, Y.; Koltypin, Y.; Gedanken, A.; Chandrasekhar, N. *Chem. Commun.* **2001**, 988.

(95) Ramesh, S.; Koltypin, Y.; Prozorov, R.; Gedanken, A. *Chem. Mater.* **1997**, *9*, 546.

(96) Pol, V.G.; Reisfeld, R.; Gedanken, A. *Chem. Mater.* **2002**, *14*, 3920.

(97) Suslick, K.S.; Hyeon, T.; Fang, M.; Cichowlas, A. *Advanced Catalysts and Nanostructured Materials*; Academic Press: San Diego, CA, 1996; 197–212.

(98) Dhas, A.; Gedanken, A. *J. Mater. Chem.* **1998**, *8*, 445–450.

(99) Koltypin, Y.; Fernandez, A.; Rojas, C.; Campora, J.; Palma, P.; Prozorov, R.; Gedanken, A. *Chem. Mater.* **1999**, *11*, 1331–1335.

(100) Salkar, R.A.; Jeevanandam, P.; Aruna, S.T.; Koltypin, Y.; Gedanken, A. *J. Mater. Chem.* **1999**, *9*, 1333–1335.

(101) Shafi, K.V.P.M.; Koltypin, Y.; Gedanken, A.; Prozorov, R.; Balogh, J.; Lendavi, J.; Felner, I. *J. Phys. Chem. B* **1997**, *101*, 6409.

(102) Liang, J.; Jiang, X.; Liu, G.; Deng, Z.; Zhuang, J.; Li, F.; Li, Y. *Mater. Res. Bull.* **2003**, *38*, 161.

(103) Pang, G.; Xu, X.; Markovich, V.; Avivi, S.; Palchik, O.; Koltypin, Y.; Gorodetsky, G.; Yeshurun, Y.; Buchkremer, H.P.; Gedanken, A. *Mater. Res. Bull.* **2003**, *38*, 11.

(104) Vidotti, M.; van Greco, C.; Ponzio, E.A.; de Torresi, S.I.C. *Electrochem. Commun.* **2006**, *8*(4), 554–560.

(105) Fulton, J.L.; Matson, D.W.; Pecher, K.H.; Amonette, J.E.; Linehan, J.C. *J. Nanosci. Nanotechnol.* **2006**, *6*(2), 562–567.

(106) Wang, G.Z.; Geng, B.Y.; Huang, Y.W.; Wang, Y.W.; Li, G.H.; Zhang, L.D. *Appl. Phys. A-Mater. Sci. Process.* **2003**, *77*, 933–936.

(107) Mukaibo, H.; Yoshizawa, A.; Momma, T.; Osaka, T. *Power Sources* **2003**, *119*, 60–63.

(108) Li, Q.; Ding, Y.; Shao, M.W.; Wu, J.; Yu, G.H.; Qian, Y.T. *Mater. Res. Bull.* **2003**, *38*, 539–543.

(109) Chen, J.-F.; Wang, Y.-H.; Guo, F.; Wang, X.-M.; Zheng, C. *Ind. Eng. Chem. Res.* **2002**, *39*, 948–954.

(110) Ramshaw, C.; Mallinson, R. Mass Transfer Process; European Patent 2568B, 1979; U.S. Patent 4263255, 1981.

(111) Sanchez, C.; Boissiere, C.; Coupe, A.; Goettmann, F.; Grosso, D.; Julian, B.; Llusar, M.; Nicole, L. Design of functional nano-structured inorganic and hybrid materials. In *Nanoporous Materials IV, Vol. 156*; 2005; 19–36.

(112) Livage, J.; Ganguli, D. *Sol. Energy Mater. Sol. Cells* **2001**, *68*(3-4), 365–381.

(113) Vioux, A. *Chem. Mater.* **1997**, *9*(11), 2292–2299.

(114) Ogoshi, T.; Chujo, Y. *Composite Interfaces* **2005**, *11*(8–9), 539–566.

(115) Cushing, B.L.; Kolesnichenko, V.L.; O'Connor, C.J. *Chem. Rev.* **2004**, *104*(9), 3893–3946.

(116) Hench, L.L.; West, J.K. *Chem. Rev.* **1990**, *90*, 33.

(117) Yi, G.; Sayer, M. *Ceram. Bull.* **1991**, *70*, 1173.

(118) Avnir, D. *Acc. Chem. Res.* **1995**, *28*, 328.

(119) Chandler, C.D.; Roger, C.; Hampden-Smith, M.J. *Chem. Rev.* **1993**, *93*, 1205.

(120) Klein, L. *Sol-Gel Optics: Processing and Applications*; Kluwer: Boston, MA, 1993.

(121) Narula, C.K. *Ceramic Precursor Technology and Its Applications*; Marcel Decker: New York, 1995.

(122) Barringer, E.A.; Bowen, H.K. *J. Am. Ceram. Soc.* **1982**, *65*, C-199.

(123) Hatakeyama, F.; Kanzaki, S. *J. Am. Ceram. Soc.* **1990**, *73*(7), 2107.

(124) White, D.A.; Oleff, S.M.; Fox, J.R. *Adv. Ceram. Mater.* **1987**, *2*(1), 53.

(125) Richards, R.; Li, W.; Decker, S.; Davidson, C.; Koper, O.; Zaikovski, V.; Volodin, A.; Rieker, T.; Klabunde, J.K. *J. Am. Chem. Soc.* **2000**, *122*, 4921–4925.

(126) Utampanya, S.; Klabunde, K.J.; Schlup, J.R. *Chem. Mater.* **1991**, *3*, 175.

(127) Klabunde, K.J.; Stark, J.V.; Koper, O.; Mohs, C.; Khaleel, A.; Glavee, G.; Zhang, D.; Sorensen, C.M.; Hadjipanayis, G.C. In Nanophase Materials; Hadjipanayis, G.C.; Siegel, R.W. (Editiors); Kluwer Academic Publishers: Dordrecht, the Netherlands, 1994.

(128) Bedilo, A.; Klabunde, K.J. *Chem. Mater.* **1993**, *5*, 500.

(129) Zhang, Y.; Li, G.; Wu, Y.; Luo, Y.; Zhang, L. *J. Phys. Chem. B* **2005**, *109*, 5478–5481.

(130) Cerda, J.; Arbiol, J.; Diaz, R.; Dezanneau, G.; Morante, J.R. *Mater. Lett.* **2002**, *56*, 131.

(131) O'Brien, S.; Brus, L.; Murray, C.B. *J. Am. Chem. Soc.* **2001**, *123*, 12085–12086.

(132) Meron, T.; Rosenberg, Y.; Lereah, Y.; Markovich, G. *J. Magn. Magn. Mater.* **2005**, *292*, 11–16.

(133) Harreld, J.H.; Dong, W.; Dunn, B. *Mater. Res. Bull.* **1998**, *33*, 561.

(134) Dong, W.; Rolison, D.R.; Dunn, B. *Electrochem. Solid State Lett.* **2000**, *3*, 457.

(135) Long, J.W.; Swider, K.E.; Stroud, R.M.; Rolison, D.R. *Electrochem. Solid State Lett.* **2000**, *3*, 453.

(136) Long, J.W.; Young, A.L.; Rolison, D.R. *J. Electrochem. Soc.* **2003**, *150*, A1161.

(137) Swider, K.E.; Hagans, P.L.; Merzbacher, C.I.; Rolison, D.R. *Chem. Mater.* **1997**, *9*, 1248.

(138) Kim, I.; Kumta, P.N. *J. Mater. Chem.* **2003**, *13*, 2028.

(139) Kumta, P.N.; Kim, J.Y.; Sriram, M.A. *Ceram. Trans.* **1999**, *94*, 163.

(140) Sriram, M.A.; Kumta, P.N. *J. Mater. Chem.* **1998**, *8*, 2441.

(141) Wang, Y.; Zhang, J.; Liang, M.; Wang, X.; Wei, Y.; Gui, L. A composite material composed of nanoparticles of transition metal and magnetic ferric oxide, a method of preparing the same, and uses of the same. Publication number WO 2006042453.

(142) Wu, N.-L. *Mater. Chem. Phys.* **2002**, *75*(1–3), 6–11.

(143) Ismail, A.A. *Appl. Catal., B* **2005**, *58*(1–2), 115–121.

(144) Kim, Y.K.; Kim, E.Y.; Whang, C.M.; Kim, Y.H.; Lee, W.I. *J. Sol-Gel Sci. Technol.* **2005**, *33*(1), 87–91.

(145) Kozuka, H.; Sakka, S. *Chem. Mater.* **1993**, *5*, 222.

(146) Liz-Marzan, L.M.; Giersig, M.; Mulvaney, P. *Langmuir* **1996**, *12*, 4329.

(147) Corrias, A.; Casula, M.F.; Ennas, G.; Marras, S.; Navarra, G.; Mountjoy, G. *J. Phys. Chem. B* **2003**, *107*, 3030.

(148) Kim, H.-J.; Ahn, J.-E.; Haam, S.; Shul, Y.-G.; Song, S.-Y.; Tatsumi, T. *J. Mater. Chem.* **2006**, *16*(17), 1617–1621.

(149) Wu, P.-W.; Dunn, B.; Doan, V.; Schwartz, B.J.; Yablonovitch, E.; Yamane, M. *J. Sol-Gel Sci. Technol.* **2000**, *19*, 249.

(150) Casula, M.F.; Corrias, A.; Paschina, G. *Mater. Res. Soc. Symp. Proc.* **2000**, *581*, 363.

(151) Sun, D.; Huang, Y.; Han, B.; Yang, G. *Langmuir* **2006**, *22*(10), 4793–4798.

(152) Moore, J.T. *Abstracts of Papers, 231st ACS National Meeting, Atlanta, GA, United States, March 26–30, 2006*, INOR–768.

(153) He, P.; Shen, X.; Gao, H. *J. Colloid Interface Sci.* **2005**, *284*(2), 510–515.

(154) Wallin, M.; Cruise, N.; Palmqvist, A.; Skoglundh, M.; Klement, U. *Colloids Surf., A* **2004**, *238*(1-3), 27–35.

(155) Lu, C.-H.; Wang, H.-C. *J. Eur. Ceram. Soc.* **2004**, *24*(5), 717–723.

(156) Koetz, J.; Bahnemann, J.; Lucas, G.; Tiersch, B.; Kosmella, S. *Colloids Surf., A* **2004**, *250*, (1-3 S.I.), 423–430.

(157) Chen, D.; Gao, L. *J. Colloid Interface Sci.* **2004**, *279*(1), 137–142.

(158) Zhang, X.; Chan, K.-Y. *Chem. Mater.* **2003**, *15*(2), 451–459.

(159) Holzinger, D.; Kickelbick, G. *Chem. Mater.* **2003**, *15*(26), 4944–4948.

(160) Capek, I. *Adv. Colloid Interface Sci.* **2004**, *110*(1–2), 49–74.

(161) Bruch, C.; Kruger, J.K.; Unruh, H.G. *Ber. Bunsen-Ges. Phys. Chem.* **1997**, *101*(11), 1761.

(162) Hartl, W.; Beck, C.; Roth, M.; Meyer, F.; Hempelmann, R. *Ber. Bunsen-Ges. Phys. Chem.* **1997**, *101*(11), 1714.

(163) Fang, J.; Wang, J.; Ng, S.C.; Chew, C.H.; Gan, L.M. *Nanostruct. Mater.* **1997**, *8*(4), 499.

(164) O'Connor, C.J.; Seip, C.T.; Carpenter, E.E.; Li, S.; John, V.T. *Nanostruct. Mater.* **1999**, *12*, 65.

(165) Zhang, Z.J.; Wang, Z.L.; Chakoumakos, B.C.; Yin, J.S. *J. Am. Chem. Soc.* **1998**, *120*, 1800.

(166) Moran, P.D.; Bartlett, J.R.; Bowmaker, G.A.; Woolfrey, J.L.; Cooney, R.P. *J. Sol-Gel Sci. Technol.* **1999**, *15*, 251.

(167) Yener, D.O.; Giesche, H. *Ceram. Trans.* **1999**, *94*, 407.

(168) Kumar, P.; Pillai, V.; Bates, S.R.; Shah, D.O. *Mater. Lett.* **1993**, *16*, 68.

(169) Xiong, L.; He, T. *Chem. Mater.* **2006**, *18*(9), 2211–2218.

(170) Teng, F.; Xu, P.; Tian, Z.; Xiong, G.; Xu, Y.; Xu, Z.; Lin, L. *Green Chem.* **2005**, *7*(7), 493–499.

(171) Calbo, J.; Tena, M.A.; Monros, G.; Llusar, M.; Badenes, J.A. *J. Sol-Gel Sci. Technol.* **2006**, *38*(2), 167–177.

(172) Pang, Y.-X.; Bao, X.J. *Mater. Chem.* **2002**, *12*, 3699.

(173) Hayashi, M.; Uemura, H.; Shimanoe, K.; Miura, N.; Yamazoe, N. *Electrochem. Solid State Lett.* **1998**, *1*, 268.

(174) Pillai, V.; Kumar, P.; Multani, M.S.; Shah, D.O. *Colloids Surf. A* **1993**, *80*, 69.

(175) Porta, F.; Bifulco, C.; Fermo, P.; Bianchi, C.L.; Fadoni, M.; Prati, L. *Colloids Surf., A* **1999**, *160*, 281.

(176) Lu, C.-H.; Wang, H.-C. *J. Mater. Chem.* **2003**, *13*, 428.

(177) Palla, B.J.; Shah, D.O.; Garcia-Casillas, P.; Matutes-Aquino, J. *J. Nanopart. Res.* **1999**, *1*, 215.

(178) Niederberger, M.; Garnweitner, G.; Pinna, N.; Neri, G. *Prog. Solid State Chem.* **2005**, *33*(2–4), 59–70.

(179) Pinna, N.; Garnweitner, G.; Antonietti, M.; Niederberger, M. *J. Am. Chem. Soc.* **2005**, *127*(15), 5608–5612.

(180) Yu, D.B.; Sun, X.Q.; Bian, J.T.; Tong, Z.C.; Qian, Y.T. *Physica E-Low-Dimensional Sys. Nanostructures* **2004**, *23*(1–2), 50–55.

(181) Ghosh, M.; Seshadri, R.; Rao, C.N.R. *J. Nanosci. Nanotechnol.* **2004**, *4*(1–2), 136–140.

(182) Zhang, W.X.; Yang, Z.H.; Liu, Y.; Tang, S.P.; Han, X.Z.; Chen, M. *J. Cryst. Growth* **2004**, *263*(1–4), 394–399.

(183) Kim, C.S.; Moon, B.K.; Park, J.H.; Choi, B.C.; Seo, H.J. *J. Cryst. Growth* **2003**, *257*(3–4), 309–315.

(184) Chen, D.H.; Chen, D.R.; Jiao, X.L.; Zhao, Y.T. *J. Mater. Chem.* **2003**, *13*(9), 2266–2270.

(185) Praserthdam, P.; Phungphadung, J.; Tanakulrungsank, W. *Mater. Res. Innovations* **2003**, *7*(2), 118–123.

(186) Zhan, J.H.; Zhang, Z.D.; Qian, X.F.; Wang, C.; Xie, Y.; Qian, Y.T. *J. Solid State Chem.* **1998**, *141*(1), 270–273.

(187) Niederberger, M.; Pinna, N.; Polleux, J.; Antonietti, A. *Angew. Chem., Int. Ed.* **2004**, *43*(17), 2270–2273.

(188) Su, C.-Y.; Goforth, A.M.; Smith, M.D.; Pellechia, P.J.; Zur Loye, H.-C. *J. Am. Chem. Soc.* **2004**, *126*(11), 3576–3586.

(189) Michailovski, A.; Patzke, G.R. *Chimia* **2004**, *58*(4), 228–231.

(190) Inoue, M. *J. Phys.: Condens. Matter* **2004**, *16*(14), S1291–S1303.

(191) Li, B.; Xie, Y.; Huang, J.X.; Qian, Y.T. *Adv. Mater.* **1999**, *11*(17), 1456–1459.

(192) Zhan, J.H.; Yang, X.G.; Zhang, W.X.; Wang, D.W.; Xie, Y.; Qian, Y.T. *J. Mater. Res.* **2000**, *15*(3), 629–632.

(193) Gautam, U.K.; Rajamathi, M.; Meldrum, F.; Morgan, P.; Seshadri, R. *Chem. Commun.* **2001**, *7*, 629–630.

(194) Hai, B.; Tang, K.B.; Wang, C.R.; An, C.H.; Yang, Q.; Shen, G.Z.; Qian, Y.T. *J. Cryst. Growth* **2001**, *225*(1), 92–95.

(195) Yu, D.B.; Yu, W.C.; Wang, D.B.; Qian, Y.T. *Thin Solid Films* **2002**, *419*(1–2), 166–172.

(196) Chen, S.J.; Chen, X.T.; Xue, Z.L.; Li, L.H.; You, X.Z. *J. Cryst. Growth* **2002**, *246*(1–2), 169–175.

(197) Li, Q.; Shao, M.W.; Zhang, S.Y.; Liu, X.M.; Li, G.P.; Jiang, K.; Qian, Y.T. *J. Cryst. Growth* **2002**, *243*(2), 327–330.

(198) Hu, J.Q.; Deng, B.; Wang, C.R.; Tang, K.B.; Qian, Y.T. *Solid State Commun.* **2002**, *121*(9–10), 493–496.

(199) Mo, M.S.; Shao, M.W.; Hu, H.M.; Yang, L.; Yu, W.C.; Qian, Y.T. *J. Cryst. Growth* **2002**, *244*(3–4), 364–368.

(200) Hu, H.M.; Yang, B.J.; Zeng, J.H.; Qian, Y.T. *Mater. Chem. Phys.* **2004**, *86*(1), 233–237.

(201) Batabyal, S.K.; Basu, C.; Das, A.R.; Sanyal, G.S. *Cryst. Growth Des.* **2004**, *4*(3), 509–511.

(202) Kim, K.H.; Chun, Y.G.; Park, B.O.; Yoon, K.H. In *Synthesis of CuInSe$_2$ and CuInGaSe$_2$ Nanoparticles by Solvothermal Route. Designing, Processing and Properties of Advanced Engineering Materials, Pts 1 and 2*; Kang, S.G.; Kobayashi, T. (Editors); Trans Tech Publications: Zurich, Switzerland, 2004; 273–276.

(203) Gautam, U.K.; Seshadri, R. *Mater. Res. Bull.* **2004**, *39*(4–5), 669–676.

(204) Hou, Y.L.; Gao, S. *J. Alloys Compd.* **2004**, *365*(1–2), 112–116.

(205) Rosemary, M.J.; Pradeep, T. *J. Colloid Interface Sci.* **2003**, *268*(1), 81–84.

(206) Wei, G.D.; Deng, Y.; Nan, C.W. *Chem. Phys. Lett.* **2003**, *367*(3–4), 512–515.

(207) Zhao, X.B.; Ji, X.H.; Zhang, Y.H.; Lu, B.H. *J. Alloys Compd.* **2004**, *368*(1–2), 349–352.

(208) Wei, G.D.; Nan, C.W.; Deng, Y.; Lin, Y.H. *Chem. Mater.* **2003**, *15*(23), 4436–4441.

(209) Gao, S.; Zhang, J.; Zhu, Y.F.; Che, C.M. *New J. Chem.* **2000**, *24*(10), 739–740.

(210) Chemseddine, A.; Moritz, T. *Eur. J. Inorg. Chem.* **1999**, *2*, 235–245.

(211) Trentler, T.J.; Denler, T.E.; Bertone, J.F.; Agrawal, A.; Colvin, V.L. *J. Am. Chem. Soc.* **1999**, *121*, 1613.

(212) Rockenberger, J.; Scher, E.C.; Alivisatos, A.P. *J. Am. Chem. Soc.* **1999**, *121*, 11595.

(213) Thimmaiah, S.; Rajamathi, M.; Singh, N.; Bera, P.; Meldrum, F.; Chandrasekhar, N.; Seshadri, R. *J. Mater. Chem.* **2001**, *11*(12), 3215–3221.

(214) Murray, C.B.; Norris, D.J.; Bawendi, M.G. *J. Am. Chem. Soc.* **1993**, *115*, 8706.

(215) Mitchell, P.W.D.; Morgan, P.E.D. *J. Am. Ceram. Soc.* **1974**, *57*, 278.

(216) Cheng, H.; Ma, J.; Zhao, Z.; Qi, L. *Chem. Mater.* **1995**, *7*, 663.

(217) Yanquing, Z.; Erwei, S.; Zhizhan, C.; Wenjun, L.; Xingfang, H. *J. Mater. Chem.* **2001**, *11*, 1547.

(218) Yin, H.; Wada, Y.; Kitamura, T.; Sumida, T.; Hasegawa, Y.; Yanagida, S. *J. Mater. Chem.* **2002**, *12*, 378.

(219) Masui, T.; Hirai, H.; Hamada, R.; Imanaka, N.; Adachi, G.; Sakata, T.; Mori, H. *J. Mater. Chem.* **2003**, *13*, 622.

(220) Inoue, M.; Kimura, M.; Inui, T. *Chem. Commun.* **1999**, 957.

(221) Gautam, U.K.; Ghosh, M.; Rajamathi, M.; Seshadri, R. *Pure Appl. Chem.* **2002**, *74*(9), 1643–1649.

(222) Zhang, H.F.; Cooper, A.I. *Soft Matter* **2005**, *1*(2), 107–113.

(223) Li, W.-C.; Lu, A.-H.; Weidenthaler, C.; Schuth, F. *Chem. Mater.* **2004**, *16*(26), 5676–5681.

(224) Tian, B.Z.; Liu, X.Y.; Yang, H.F.; Xie, S.H.; Yu, C.Z.; Tu, B.; Zhao, D.Y. *Adv. Mater.* **2003**, *15*(15), 1370.

(225) Tian, B.Z.; Liu, X.Y.; Solovyov, L.A.; Liu, Z.; Yang, H.F.; Zhang, Z.D.; Xie, S.H.; Zhang, F.Q.; Tu, B.; Yu, C.Z.; Terasaki, O.; Zhao, D.Y. *J. Am. Chem. Soc.* **2004**, *126*(3), 865–875.

(226) Fuertes, A.B. *J. Phys. Chem. Solids* **2005**, *66*, 741–747.

(227) Laha, S.S.; Ryoo, R. *Chem. Commun.* **2003**, 2138.

(228) Schwickardi, M.; Johann, T.; Schmidt, W.; Schuth, F. *Chem. Mater.* **2002**, *14*, 3913.

(229) Dong, A.G.; Ren, N.; Tang, Y.; Wang, Y.J.; Zhang, Y.H.; Hua, W.M.; Gao, Z. *J. Am. Chem. Soc.* **2003**, *125*(17), 4976–4977.

(230) Seo, J.-W.; Jun, Y.-W.; Ko, S.J.; Cheon, J. *J. Phys. Chem. B* **2005**, *109*(12), 5389–5391.

CHAPTER 4

Synthesis of Metal-Oxide Nanoparticles: Gas–Solid Transformations

S. BUZBY, S. FRANKLIN, and S. ISMAT SHAH

This chapter describes the formation of nanoparticle and thin-film oxides by gas–solid transformations. The first part of the chapter will review the chemical vapor deposition (CVD) technique where solid particles are created from the reaction of volatile chemical precursors with reactive gases. Part two of the chapter will focus on the technique called pulsed laser deposition (PLD) in which a solid precursor target is ablated to form vapors by a high-energy laser pulse.

4.1. PART ONE—CHEMICAL VAPOR DEPOSITION

4.1.1. Introduction

CVD is a general term applied to the deposition of solid materials from chemical precursors in the vapor phase. The CVD process for compound formation involves reaction between a volatile precursor of the material to be deposited with other reactive gases. The process may employ various gaseous, liquid, or solid chemicals as sources of the elements of which the film is to be made. The reaction results in the production of a nonvolatile solid that deposits on an appropriately placed substrate to form either particles or a thin film. CVD processes have been used in various industrial applications and technologies such as the fabrication of electronic devices, the manufacture of cutting tools, and the formation of nanoparticles.

In this section, several topics related to the CVD process will be explained, including reaction types, system design, growth mechanisms, and examples of reactions.

Synthesis, Properties, and Applications of Oxide Nanomaterials, Edited by José A. Rodríguez and Marcos Fernández-Garcia

This section is divided into the following subsections:

1. Types of CVD processes
2. CVD systems and requirements
3. Particle growth mechanism

4.1.2. Types of CVD Processes

Due to the great variety of materials that can be created by CVD, a large number of processes and systems have been designed and constructed. The various processes can be characterized and described by such terms as low or high temperature, atmospheric or low pressure, gaseous or liquid precursors, and hot-wall or cold-wall reactors and by incorporation of physical techniques such as plasma or light sources to assist in ion formation. The following CVD processes are commonly used for the formation of nanoparticles (1,2):

CVD—chemical vapor deposition (thermally activated or pyrolytic)
MOCVD—Metalorganic CVD
PACVD—Plasma-assisted CVD
PCVD—Photo CVD

4.1.3. Thermally Activated CVD

Thermal CVD processes are initiated only with thermal energy (resistance heating, RF heating, or infrared radiation). They are usually performed at normal or low pressure down to ultra-low pressure. High-temperature CVD is the best way to achieve the growth of high-quality epitaxial thin films. It is used extensively for metallurgical coatings and for in situ calcinations to obtain the desired crystalline phase. The reactors used can be divided into two main types: hot-wall and cold-wall reactors. Hot-wall reactors are usually tubular in shape and are heated externally by use of resistance-heating elements. Multiple temperature zones are essential for the efficient transport of matrix as well as dopant atoms. By carefully controlling gas flow rates and temperatures, the composition, doping levels, and particle size of the nanoparticles can be manipulated. In a cold-wall reactor, the substrate is inductively heated while the non-conductive reaction chamber walls are either air or water cooled. Cold-wall reactors are used almost exclusively for the deposition of epitaxial Si films (1–3).

4.1.4. MOCVD

MOCVD is also a thermal CVD processes, but the sources differ in that they are metalorganic gases or liquids. Metalorganic compounds are organic compounds that contain a metal atom, particularly compounds in which the metal atom has a direct bond with a carbon atom (e.g., Titanium tetraisopropoxide, $Ti[OCH(CH_3)_2]_4$). With the availability of pure metalorganic precursors, their use offers significant advantages in the practical aspects of delivering the reactants to the substrate surface. The greatest advantage of using metalorganics is that they are volatile at a relatively low temperature, which makes it easier to transfer the precursor to the substrate. As all of the reactants are in the gas phase, it is possible to achieve precise control of the flow rates and partial pressures of the gases allowing for efficient reproducible depositions. However, due to the organic nature of the precursors, carbon contamination becomes a problem, making it difficult to produce high-purity samples (4–8).

4.1.5. PACVD and PCVD

PACVD uses plasma to ionize and dissociate the gases used for the deposition. The most common method for creating the plasma discharge is the use of a radio-frequency (RF) field. The reactor can be coupled either inductively with a coil or capacitively with electrode plates to form plasma. The plasma formed has sufficient energy to decompose gas molecules into electrons, ions, radicals, and so on. PCVD uses light (lasers) to enhance the reaction rate. The effect of the radiation is either a local heating of the substrate to decompose the gases above it or a photochemical reaction where the energetic photons directly decompose the gases. The generated species are highly reactive and their use enhances the particle growth rate, allowing for deposition at much lower temperatures than in thermal CVD. Therefore, previously impossible reactions, such as high-temperature reactions on temperature-sensitive substrates, may now be carried out (9,10). The CVD processes are summarized below in Table 4.1.

4.1.6. Comparison of CVD with Other Deposition Techniques

The advantages of the CVD technique consist of producing uniform, pure, reproducible particles, and films at low or high flow rates. However, when compared with other deposition techniques, the CVD processes are perhaps the most complex to use. Unlike growth by physical deposition techniques, such as evaporation, this method requires multiple test runs to establish the growth parameters to be used, especially for single-crystal growth. The complexity of this method results from several factors. First, CVD generally uses multiple reactant species in the chemical reactions

TABLE 4.1. Summary of CVD Processes.

Process	Energy Used	Pressure Used (torr)	Source	Advantages
CVD	Thermal	10^{-5}–760	Inorganic	High-quality epitaxial growth
MOCVD	Thermal	10–760	Metalorganic	Precise control; Reproducible
PACVD	Plasma	0.01–10	Both	Lower temperature
PCVD	Light/Laser	10–760	Both	Lower temperature; Selected control

compared with single sources for PVD or PLD. Second, CVD reactions generally produce intermediate products, with the growth occurring in more consecutive steps rather than directly forming the product as in physical methods (i.e., evaporation). Finally, the growth has numerous independent variables (flow rate, partial pressures, system pressure, etc.), and all must be monitored and controlled. Other disadvantages to the CVD process are the chemical hazards associated with the toxic, explosive, inflammable, or corrosive gases used and that not all materials have a precursor that can be economically and commercially viable (1,2).

4.1.7. CVD System Requirements

Each process described above uses a different system. However, regardless of the process type, most CVD systems have many components in common. A CVD system must have the ability to deliver and control the reactant gases into the reactor, usually accomplished by using mass flow controllers. A reaction chamber is equipped with a substrate holder where the reactant gases combine to form particles that are collected on the substrate. The reaction chamber is typically tubular in shape and constructed of stainless steel, glass, pyrex, and quartz, depending on the reaction conditions (temperature, pressure, reactivity, etc.). The system must also contain a heating arrangement with temperature control to supply heat to the reactor and substrate. The heat is necessary to ensure that the reaction can proceed efficiently and provides the energy to dissociate the gases in CVD and MOCVD processes. For PACVD or PCVD, there is in addition to the heat source a plasma generator or light source (laser) present to assist in the dissociation. Finally, vacuum pumps are used for maintaining the system pressure and for the removal of reaction byproducts and depleted gases (1,2,7). A schematic drawing of a MOCVD system for the synthesis of TiO_2 nanoparticles is shown in Figure 4.1.

4.1.8. Particle Growth Mechanism

The formation of oxide nanoparticles can be divided into three basic stages. The first step in the process is the introduction of the metal precursor gas or vapor and oxygen

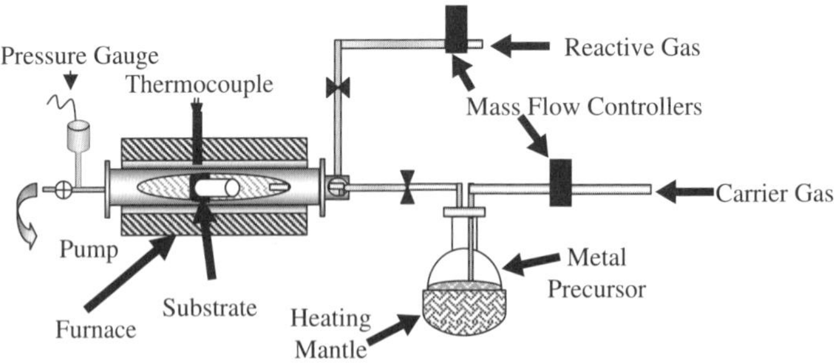

Figure 4.1. Schematic of an MOCVD system.

into the reaction chamber. The two most important steps in the formation of oxide nanoparticles by CVD are the reaction and condensation steps as they determine the composition and size of the nanoparticles formed (3,4). It is known that particle size and distribution are determined by growth temperature and deposition rate that, in turn, relies on the flow rate of the metal precursor (4,6). The selection of the precursor used determines the method and ease of introducing the precursor into the reaction chamber. Gaseous precursors may be regulated and added directly to the chamber by use of mass flow controllers. However, many of the precursors used for the formation of oxide nanoparticles are liquids, and so additional steps are required to convert them to gases before they can be used. The equilibrium concentration of a vapor over a liquid is known as the vapor pressure of the substance. The free energy change in going from the liquid to the gaseous state determines the concentration of the gas (11,12).

$$A_{\text{liq}} \longrightarrow A_{\text{gas}}$$

$$\frac{[A_{\text{gas}}]}{[A_{\text{liq}}]} = \exp(-\Delta G/RT)$$

As can be seen from the equation above, by heating the liquid precursor, it is possible to increase the concentration of precursor vapor introduced into the system resulting in higher reaction rates. The precursor transport rate may also be increased by the use of a carrier gas. An inert gas (i.e., N_2 or Ar) is bubbled through the liquid precursor carrying the precursor molecules into the reaction chamber. For oxidation reactions below 500 °C, oxygen is used as oxidant, whereas above 500 °C, carbon dioxide or nitrogen oxides are added to oxygen to reduce the flammability risk.

Second is the reaction of the precursor with oxygen to form metal-oxide monomers (4,6). The equilibrium state of any chemical system—the concentrations of reactants and products that will be present if you wait long enough—occurs when the free enthalpy G (Gibbs free energy) has reached its minimum value. This is equivalent to saying that the change in free energy due to any reaction (involving a small number of moles) at the equilibrium concentration is zero. For any single reaction,

$$aA + bB \longrightarrow cC + dD$$

We can then find the equilibrium state of the reaction from the equilibrium constant (11,12):

$$\frac{[C]^c[D]^d}{[A]^a[B]^b} = \exp(-\Delta G/RT)$$

The overall CVD reactions can be classified by examining the standard enthalpy of reaction. Reactions for which the change in free energy (ΔG) is modest are potentially reversible, with the actual reaction being a balance between deposition and etching of the products, depending on the experimental conditions used:

$$\text{SiH}_4 \longleftrightarrow \text{Si} + 2\text{H}_2 \quad \Delta G = -57\,\text{kJ/mole}$$

Reactions involving very large changes in free energy have essentially no reactants left when equilibrium is reached. They are completely irreversible and lead only to deposition. This is the case with oxide formation when the concentration of reactants is essentially zero at equilibrium:

$$SiH_4 + 2O_2 \longleftrightarrow SiO_2 + H_2O \quad \Delta G = -1307\,\text{kJ/mole}$$

Finally, consider the reaction for the reduction if titanium(IV) chloride:

$$TiCl_4 + 2H_2 \longrightarrow Ti + 4HCl \quad \Delta G = +287\,\text{kJ/mole}$$

This reaction is not thermodynamically possible at room temperature because the free energy change (ΔG) is positive (1,2). The only way this reaction can be made to proceed is by heating it to 1000 °C. This is why Ti CVD is very difficult.

Finally the monomers condense to form larger clusters by homogenous nucleation in the gas phase and the subsequent collection of the oxide nanoparticles on the substrate (4,6). As mentioned, the size and distribution of the particle formed are determined by the reaction temperature and the oxide deposition rate. The total precursor concentration can be controlled by varying the flow rate of the carrier gas, the liquid precursor temperature, or the system pressure. The reaction temperature effects the precursor concentration due to thermal expansion of the gas (6). The total concentration of the gas decreases as the temperature increases, and because the flow rate of the gas is kept constant, this reduces the time the precursor spends in the condensation region of the reactor. The lower concentration coupled with the lower residence time will cause the mean free path of the oxide monomers to increase, thereby reducing the size of the particles formed. Figure 4.2 shows the steps formation of particles in the CVD process.

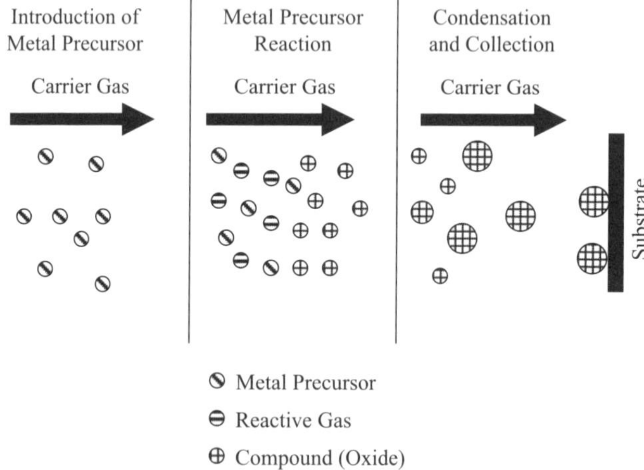

Figure 4.2. Steps of particle formation in the CVD process.

4.1.9. Conclusions

CVD techniques are vapor deposition techniques based on homogeneous and/or heterogeneous chemical reactions. These processes employ various gaseous, liquid, and solid chemicals as sources of the elements of which the film is to be made. Compared with other deposition techniques, the CVD method is perhaps the most complex. Unlike growth by physical deposition such as evaporation, this method requires numerous test runs to reach suitable growth parameters. The complexity of this method results from the facts that (1,2):

1. It generally includes multicomponent species in the chemical reactions.
2. The chemical reactions generally produce intermediate products.
3. The growth has numerous independent variables.
4. The growth includes more consecutive steps than in physical methods.

The CVD production has several advantages (7):

1. High purity particles.
2. Size control to produce uniform, reproducible particles.
3. Control over the crystal structure of the particles.
4. Particles may be produced rapidly and continuously.

4.2. PART TWO—PULSED LASER DEPOSITION

4.2.1. Introduction

During the last decade, PLD has been established as a simple, versatile, and contamination-free deposition technique. Nanoparticles and thin films of a wide variety of materials (metals, dielectrics, semiconductors, oxides, nitrides, etc.) have been deposited using the PLD technique (12). Conceptually, PLD is a very simple technique. A laser beam enters a vacuum chamber through a window and impinges on the target material that is to be deposited. The basic setup is shown in Figure 4.3.

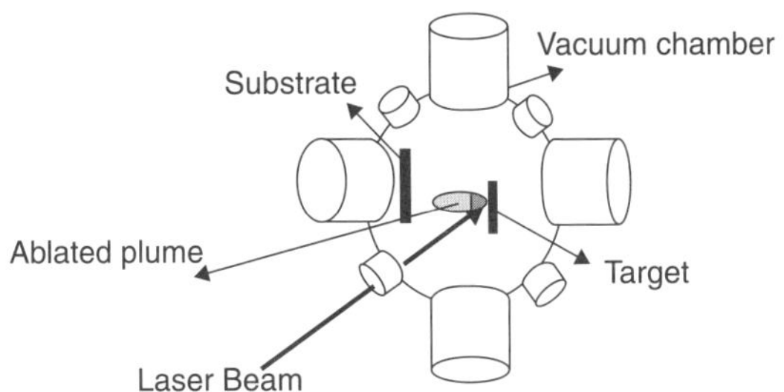

Figure 4.3. A typical setup for pulsed laser deposition of thin films.

The laser pulse with pulse durations ranging from 30 ns to a few femtoseconds is focused to an energy density of 1–10 J/cm^2 vaporizing a few hundred angstroms of surface material. The evaporated flux contains neutral atoms, positive and negative ions, electrons, molecules, molecular ions, and free radicals of the target material in both their ground and excited states. These vapors absorb incoming radiation as a result of which the particles acquire kinetic energy of 1–5 eV and move in the direction perpendicular to the target. The particles then deposit on a substrate generally heated to allow for crystalline film growth (13).

Laser beam–solid interaction is a complex phenomenon. Energy of the laser-photon beam is absorbed by the solid surface leading to electronic excitation. As thermal conductivity of the target solid is far too low to dissipate energy in the short period of the laser pulse, photon energy is converted into thermal and chemical energy. As a result, temperature of the order of 3500 °C is reached in a small volume of the target, leading to instantaneous evaporation, ionization, and decomposition of the material. The moving front of such a collection of particles constituting plasma is called a plume. The plume expands rapidly having high pressure and deposits on the substrate forming a film.

4.2.2. PLD

The laser evaporation method was first used to deposit thin films in 1965 (14), but only after successful fabrication of as-deposited 90K $YBa_2Cu_3O_7$ (YBCO) epitaxial films giving 10^6-A/cm^2 critical current during late 1987 and early 1988, the rapid development of this method (15,16) was possible.

PLD or laser ablation deposition (LAD) is now used on a worldwide basis to produce films of various kinds of materials [semiconductors, high T_c superconductors, ceramics, ferroelectrics, metal and metal compounds, polymer, biological material, and refractive material (17–20)]. In particular, PLD is used to deposit materials and combinations of materials that either cannot be deposited by other methods or their deposition is greatly problematic by other methods (17–27). Recently, PLD has been used to synthesize nanotubes (28), nanopowders (29), and quantum dots (30). The versatility of PLD is that there are almost no restrictions on the target material to be used. PLD with liquid target materials has been illustrated (31), and even PLD of solid nitrogen and methane has been probed experimentally (32). Several modifications of conventional PLD has led to the processing of thin films of not only inorganic solids but also organic polymers and biomaterials that were earlier damaged irreversibly due to interaction with lasers (33).

4.2.3. PLD Advantages

The main advantage and the reason for the enthusiastic research in LAD is the preservation of the stoichiometry during the film production process. LAD can be used to deposit films of any material irrespective of their optical properties and stoichiometry. PLD relies on a photon interaction to create an ejected plume of material from any target. The actual physical processes of material removal are quite complex; however, one can consider the ejection of material to occur due to rapid explosion of the target surface due to superheating. Unlike thermal evaporation, which produces a vapor composition dependent on the vapor pressures of elements in the target material, the laser-induced expulsion produces a plume of material with stoichiometry similar to the target. It is generally easier to obtain the desired film stoichiometry for multi-element materials using PLD than with other deposition technologies.

Another important aspect of PLD is that the introduction of background gas is easily possible, and it allows the opportunity to control the stoichiometry further. PLD has high versatility with regard over this criterion ($P \leq 300\,\text{mTorr}$), whereas other techniques generally only incorporate lower pressures of background gas. The kinetic energy of highly energetic and directed fluxes can be usefully controlled by introduction of (reactive or nonreactive) background gas (by reactions or thermalizing collisions). On the other hand, if the deposition flux does not have high kinetic energy and directionality, it will be scattered by background gas collisions. In such cases, the deposited film thickness drops precipitously as either the source-substrate distance or the pressure. When depositing from multicomponent targets, the intentional introduction of background gas can enhance the film quality or can even enable the production of the multicomponent film (oxides or nitrides) from a monocomponent target and the appropriate reactive background gas (e.g., oxygen or nitrogen). Numerous examples can be given from the field of laser ablation, for example, ablation of graphite in a nitrogen atmosphere under standard conditions gives rise to CN species (34), which are easily observed from their violet emission in the ablation plasma, and lead to production of CN films with a nitrogen content that is related to the background pressure of nitrogen. A second example is the deposition via ablation of $PbZrTiO_3$ (35).

4.2.4. Comparison of PLD with Other Deposition Techniques

Compared with conventional film deposition techniques [thermal evaporation (TE), molecular beam epitaxy (MBE), sputtering (S), chemical vapor deposition (CVD), etc.], PLD has many advantages. This technique is very reliable and permits great experimental versatility, simple, contamination-free, inexpensive, and fast. It also allows for the production of multicomponent films and low processing temperatures. Among the drawbacks are the appearances of particulates within films (22) and their nonuniformity in thickness (22), which are now being overcome by modifying the deposition techniques like movement of target and or laser beam (36,37) and off-axis deposition (38,39). Appearance of particulates can be used beneficially in the production of nano- and micrometer-sized particles, ultra-fine powders, nanocomposites, and thick coatings and films with a distribution of nanometer-sized clusters (40–42). Several optoelectronic applications require such films (43,44).

4.2.5. PLD Process/Mechanism

The fundamental processes, occurring within the laser-produced plasmas, are not fully understood; thus, deposition of novel materials usually involves a period of empirical optimization of the deposition parameters. However, this optimization requires a considerable amount of time and effort. Much of the early research into PLD concentrated on the empirical optimization of deposition conditions for individual materials and applications, without attempting to understand the processes occurring as the material is transported from target to substrate. Significant work has been done on modeling the phenomenon of laser ablation and deposition (45–50). For explaining the plume expansion dynamics in an ambient gas, a shock wave model is sometimes used. In the shock wave model, a rapidly expanding plume behaves as a piston that generates a shock wave in the ambient gas (51). The ambient gas is compressed into a thin shell between the shock and the plume front. The conditions in the compressed region (high temperature, high pressure) can be suitable for reaction between ambient gas and ablated species, and formation of oxides, nitride, and their clusters could take place in that region (52). A schematic of a PLD system is shown in Figure 4.3.

4.2.6. Reactive PLD

When PLD is conducted in the ambience of a chemically reactive gas, the ablated substance enters into a chemical reaction ending with the formation of a compound that is eventually deposited onto the surface of the collector (53). This process is usually known as reactive PLD (RPLD). RPLD has been established as a flexible, simple, and controllable method for producing stoichiometric deposition of multicomponent high-quality thin films of metallic oxide (54). RPLD is shown to be a promising technique for compound thin-film synthesis (55–58). In comparison with PLD, which consists of the stoichiometric mass transfer from the target to the substrate, the compound synthesis by reactive PLD is obtained by ablation of a target in a chemically reactive gas. This method has been used to deposit nitrides or oxides of metals, semiconductors, and other materials by ablating pure elemental targets in an appropriate atmosphere (59–61) or ablating oxides of elements in oxygen ambient to get the required stoichiometry (62). Oxide films were first deposited by Fogarassy et al. in 1990 (53). The growth of oxide thin films (PbO, Ce_2O on Si, MgO) is discussed by Beeach et al. (63). Thin films of SiO_2 (56), GeO_2 (64), and other oxides have been successfully grown. Some of the highest quality oxide layers for optical coatings have been grown by PLD (65). The optical materials ZnO and CeO_2 on glass and silicon substrates have been grown by excimer laser ablation in oxygen ambient (66). Metal films and their multilayers can be grown by the ablation of an oxide in hydrogen atmosphere. Tungsten metal film can be made by ablating WO_3 in hydrogen. Growth of metal carbides is performed by the ablation of metal carbides or metal oxides or pure metals in methane or ethane (67). Growth of binary and ternary nitride films is a distinct possibility. Besides making PLD films from stoichiometric nitrides, ablation of oxides in ammonia atmosphere can lead to metal nitride and metal oxynitride films. This process follows from the ammonolysis of oxides into nitrides. Deposition of nitrides is less efficient by PLD as only a few elements such as Ti or B allow the growth of nitride films by reactive PLD (59,68). The main problems are the nitrogen deficiency in the deposit and the oxygen contamination.

4.2.7. RPLD Mechanisms

To optimize the deposition of thin films' understanding of the fundamental mechanisms of mixing of ambient gas and ablated species is required. Depending on the conditions existing in the gas phase, formation of compounds can take place in the gas phase and the deposited thin film will be consisting of the compound. The formation of compound also depends on the ambient gas mass, the ablated particle energy, and the binding energy of the compound. For ablation of pure elemental targets in low-pressure ambient gas, it is observed that oxide formation takes place depending on the mass of the target atom. For example, SiO, SiO^+, AlO, and MgO were detected by emission spectroscopy during the ablation of pure targets in O_2 or H_2O low-pressure atmospheres (69–71). Contrary to these studies, Vega et al. (72) and Gonzalo et al. (73) did not observe any GeO or BaO and TiO emission bands during the ablation of Ge or $BaTiO_3$, respectively, in low-pressure oxygen. For reactive PLD of nitride thin films, the observation of nitride diatomics in the gas phase was mostly limited to CN during ablation of graphite in N_2 or other nitrogen-containing atmospheres (74). Most reports in the literature mention much higher densities of oxides than nitrides when ablating the same targets in O_2 and then in N_2 under otherwise identical conditions (75).

Hermann and Dutouquet (75) performed time- and space-resolved emission spectroscopic measurements during the ablation of Al, C, and Ti targets in either O_2 or N_2 low-pressure atmospheres. The spectra show that the interaction between the ablated material vapor and O_2 ambient gas leads to strong oxidation in the gas phase for all target materials, whereas nitration in an N_2 atmosphere was found to occur only for the case of carbon. These results support the hypothesis that the formation process of molecular species strongly depends on the binding energy of the reaction product with respect to the dissociation energy of the ambient gas molecule. Thus, oxidation is an exothermic reaction for the three target materials, whereas nitration is generally strongly endothermic, except for carbon. The binding energy also affects the distance from the target at which the reaction takes place. Products having higher dissociation energies are formed in plasma regions of relatively high temperature. Lighter atoms undergo reactions near the target surface, whereas heavier atoms, which have to undergo collisions to reduce their energy, react at greater distances.

As the pressure of the ambient gas increases, conditions for the formation of a shock wave are satisfied and a compressed region between the expanding plume and the ambient is created. Some plasma material is also transported inside the compressed region, and because the temperature of the shocked region is high, the particles inside the shock can undergo physical and chemical changes. Strong chemical reactions can occur at the oxygen rich boundary of the plume when the ambient gas is oxygen (52), allowing one to obtain metal oxides from ablated metal targets.

It is established that the shocked region at the edge of the plume reaches high enough temperatures to initiate chemical reactions (52). This indicates that for films to be oxygen rich, the position of the substrate relative to the target surface is an important parameter. Misra et al. (61) have discussed the PLD of Al_2O_3 thin films that are important for magnetic and optical packaging applications. They have shown that optimum deposition occurs when the substrate is placed near the shock region and that

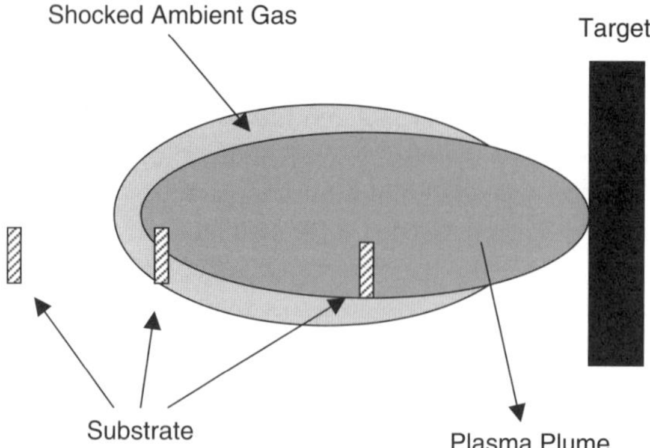

Figure 4.4. Schematic diagram showing positions of various substrates for thin-film deposition via laser ablation.

substrates not placed in the shock region show minimum deposition. Figure 4.4 shows a schematic diagram for substrate placement for optimum deposition of compounds via RPLD.

During growth of oxides, the use of oxygen is often inevitable for achieving a sufficient amount of oxygen in the growing oxide film. The amount of reactive oxygen created should be sufficient to help reduce the number of oxygen vacancies. Furthermore, the density of laser-created plasma decreases with the axial distance from the target and the energy of the particles emerging from the target becomes very low at the high pressures of ambient oxygen required to create enough reactive oxygen. For instance, for the formation of perovskite structures at high substrate temperatures in a one-step process, an oxygen pressure of about 0.3 mbar is necessary (15).

Mitra et al. (76) discussed the PLD of ZnO films by laser ablation of ZnO target in oxygen atmosphere. ZnO is a direct band gap II–VI semiconductor and a novel photonic material with a crystal structure similar to that of GaN. The size of the ZnO nanocrystallite decreases at lower ambient pressures, implying that the ambient pressure can control the size of the deposited nanocrystallite. They predicted that at higher pressure, the compressed plasma plume will lead to the nucleation of larger ZnO clusters.

4.2.8. RPLD Advantages

RPLD has a great advantage in the possibility of stabilizing transition metals in higher oxidation states. Stabilization of Cu in +2 and +3 oxidation states is the main reason for the success of PLD in obtaining as deposited 90K YBCO films. Oxides like $PrNiO_3$, $NdNiO_3$, and so on can be synthesized only at high oxygen pressure. But films of these oxides can be grown at PLD conditions (77). Furthermore, these films are grown from mixed-oxide pellets instead of stoichiometric-oxide pellets. Thin films of transition metal oxides having monodispersed nanopartices can be deposited by laser ablation of oxide targets in an inert ambient atmosphere (78). Zbroniec et al. (78) have shown that the size of the nanoparticles can be controlled by the ambient pressure.

Most oxides grown into films by PLD are perovskite-related oxides. Therefore, to grow these oxide films, lattice-matched single-crystal substrates are necessary. Commonly used substrates are $SrTiO_3$, $LaAlO_3$, MgO, ZrO_2, and sapphire, which can be cut in the [100], [110], or [111] direction. If the thermal expansion coefficients of the oxides are close to those of the substrates, the deposited oxide films on the substrates do not crack. With silicon being a covalent material, the sticking probability of ions is low due to the lack of ionic bonding between the Si and the oxides and the large thermal mismatch. Therefore, growing epitaxial films on single-crystal silicon substrates has not been successful.

The properties of thin films deviate from those of their bulk samples due to changes in the structure of the films. One important cause of such a change originates from the lattice mismatch of the growing material with the substrate. When oxide films are deposited on a substrate with a lattice parameter smaller or larger than the film materials, they grow under strain. A detailed investigation of $PrNiO_3$ films by varying the thickness has confirmed the critical thickness that undergoes strain on the substrate (78). This is analogous to the external pressure and increases the bond angle of Ni–O–Ni in turn increasing the bandwidth.

4.2.9. Conclusions

PLD is thus a versatile technique to obtain thin films of multicomponent oxide materials and, therefore, has become a standard experimental method to prepare films in general and high-quality epitaxial films in particular for research purposes. It is important to note that YBCO is grown in situ under oxygen partial pressure. Reaction between oxygen and metal ions takes place in the plume. Similar reaction between molecules such as dinitrogen, hydrogen, ammonia, methane, acetylene, and diborane can induce reactions leading to formation of compounds other than oxides. This aspect of PLD is open to research.

REFERENCES

(1) Ohring, M. *The Materials Science of Thin Films*; Academic Press: San Diego, CA, 1992.

(2) *Handbook of Thin Film Process Technology*; Glocker, D.; Shah, S.I. (Editors); Institute of Physics Publishing: Philadelphia, PA, 1997.

(3) Okuyama, K.; Kousaka, Y.; Tohge, N.; Yamamoto, S.; Wu, J.J.; Flagan, R.C.; Seinfel, J.H. *AIChE J.* **1986**, *32(12)*, 2010–2019.

(4) Li, W.; Shah, S.I. *Encyclopedia of Nanoscience and Nanotechnology*; Nalwa, H.S. (Editor); American Scientific Publishers: Stevenson Ranch, TX, 2004.

(5) Ding, Z.; Hu, X.; Yue, P.L.; Lu, G.Q.; Greenfield, P.F. *Catal. Today* **2001**, *68*, 173–182.

(6) Li, W.; Shah, S.I.; Sung, M.; Huang, C.P. *J. Vac. Sci. Technol. B* **2002**, *20(6)*, 2303–2308.

(7) Okuyama, K.; Ushio, R.; Kousaka, Y.; Flagan, R.; Seinfled, J.H. *AlChE J.* **1990**, *36(3)*, 409–419.

(8) Rajeshwar, K.; de Tacconi, N.R.; Chenthamarakshan, C.R. *Chem. Mater.* **2001**, *13*, 2765–2782.

(9) Cote, D.R.; Nguyen, S.V.; Stamper, A.K.; Armbrust, D.S.; Tobben, D.; Conti, R.A.; Lee, G.Y. *IBM J. R. D.* **1999**, *43(1–2)*, 5.

(10) Takezoe, N.; Yokotani, A.; Kurosawa, K.; Sasaki, W.; Igarashi, T.; Matsuno, H. *Appl. Surf. Sci.* **1999**, *138–139*, 340.

(11) Chawla, S. *A Text Book of Engineering Chemistry*; Chawla, T. (Editor); Dhanpat Rai & Co.: Delhi, India, 2003.

(12) West, A.R. *Solid State Chemistry and Its Applications*; Wiley: Singapore, 2003.

(13) Hubler, G.K. *Mater. Res. Soc. Bull.* **1992**, *17*, 25.

(14) Smith, H.M.; Turner, A.F. *Appl. Opt.* **1965**, *4*, 147.

(15) Dijkkamp, D.; Venkatesan, T.; Wu, X.D.; Shaheen, S.A.; Jisrawi, N.; Min-Lee, Y.H.; Mclean, W.L.; Croft, M. *Appl. Phys. Lett.* **1987**, *51*, 619.

(16) Inam, A.; Hegde, M.S.; Wu, X.D.; Venkatesan, T.; England, P.; Miceli, P.F.; Chase, E.W.; Chang, C.C.; Tarascon, J.M.; Wachtman, J.B. *Appl. Phys. Lett.* **1988**, *53*, 908.

(17) Otsubo, S.; Maeda, T.; Minamikawa, T.; Yonezawa, Y.; Morimoto, A.; Shimizu, T. *Japan. J. Appl. Phys.* **1990**, *29*, L133.

(18) Kidoh, H.; Morimoto, A.; Shimizu, T. *Appl. Phys. Lett.* **1991**, *59*, 237.

(19) Tsuboi, Y.; et al. *J. Appl. Phys.* **1999**, *85*, 4189.

(20) Tsuboi, Y.; et al. *Chem. Lett.* **1998**, 521.

(21) *Pulsed Laser Deposition of Thin Films*; Chrisey, D.B.; Hubler, G.K. (Editors); Wiley: New York, 1994.

(22) Bäuerle, D. *Laser Processing and Chemistry, 3rd edn*; Springer: Berlin, 2000.

(23) Lowndes, D.H. *Laser Ablation and Desorption*; Miller, J.C.; Haglund, R.F., Jr. (Editors); Academic Press: San Diego, CA, 1997.

(24) Willmott, P.R.; Huber, J.R. *Rev. Mod. Phys.* **2000**, *72*, 315.

(25) Gorbunoff, A. Laser-assisted synthesis of nanostructured materials. *Fortschritt-Berichte; Reihe* **2002**, *9*, 357.

(26) *Materials Research Bulletin, 2nd ed. Vol. XVII*; Hubler, G.K. (Editor); 1992.

(27) Ashfold, M.N.R.; Claeyssens, F.; Fuge, G.M.; Henley, S.J. *Chem. Soc. Rev.* **2004**, *33*, 23.

(28) Zhang, Y.; Gu, H.; Iijima, S. *Appl. Phys. Lett.* **1998**, *73*, 3827.

(29) Geohegan, D.B.; Puretzky, A.A.; Rader, D.J. *Appl. Phys. Lett.* **1999**, *74*, 3788.

(30) Goodwin, T.J.; Leppert, V.J.; Risbud, S.H.; Kennedy, I.M.; Lee, H.W.H. *Appl. Phys. Lett.* **1997**, *70*, 3122.

(31) Rong, F.X. *Appl. Phys. Lett.* **1995**, *67*, 1022.

(32) Ishiguro, T.; Shoji, T.; Inada, H. *Appl. Phys. A.* **1999**, *69*, S149.

(33) Chrisey, D.B.; Piqué, A.; McGill, R.A.; Horwitz, J.S.; Ringeisen, B.R. *Chem. Rev.* **2003**, *103*, 553.

(34) Chen, M.Y.; Murray, P.T. *J. Vac. Sci. Tech. A* **1998**, *16*, 2093.

(35) Horwitz, J.S.; Grabowski, K.S.; Chrisey, D.B.; Leuchtner, R.E. *Appl. Phys. Lett.* **1991**, *59*, 1565.

(36) Greer, J.A.; Tabat, M.D. *J. Vac. Sci. Technol. A* **1995**, *13*, 1175.

(37) Greer, J.A. *Pulsed Laser Deposition of Thin Films*; Chrisey, D.B.; Hubler, G.K. (Editors); Wiley: New York, 1994.

(38) Agostinelli, E.; Kaciulis, S.; Vittori-Antisari, M. *Appl. Surf. Sci.* **2000**, *156*, 143.

(39) Radhakrishnan, G.; Adams, P.M. *Appl. Phys. A* **1999**, *69*, S33–S38.

(40) Heitz, J.; Dickinson, J.T. *Appl. Phys. A* **1999**, *68*, 515.

(41) Ayyub, P.; Chandra, R.; Taneja, P.; Sharma, A.K.; Pinto, R. *Appl. Phys. A* **2001**, *73*, 67.

(42) Borsella, E.; Botti, S.; Giorgi, R.; Martelli, S.; Turtú, S.; Zappa, G. *Appl. Phys. Lett.* **1993**, *63*, 1345.

(43) Yamada, Y.; Orii, T.; Umezu, I.; Takeyama, S.; Yoshida, T. *Jpn. J. Appl. Phys.* **1996**, *35(Pt. 1)*, 1361.

(44) Makimura, T.; Kunii, Y.; Murakami, K. *Jpn. J. Appl. Phys.* **1996**, *35(Pt. 1)*, 4780.

(45) Ready, J.F. *Effects of High-Power Laser Radiation*; Academic Press: New York, 1971.

(46) Wood, R.F.; Giles, G.E. *Phys. Rev. B* **1981**, *23*, 2923.

(47) Singh, R.K. *Appl. Phys. Lett.* **1991**, *59*, 2115.

(48) Anisimov, S.I.; Bäuerle, D.; Luk'yanchuk, B.S. *Phys. Rev. B* **1993**, *48*, 12076.

(49) Kools, J.C.S. *J. Appl. Phys.* **1993**, *74*, 6401.

(50) Franklin, S.R.; Thareja, R.K. *Appl. Surf. Sci.* **2004**, *222*, 293.

(51) Dyer, P.E.; Sidhu, J. *J. Appl. Phys.* **1988**, *64*, 4657.

(52) Thareja, R.K.; Mishra, A.; Franklin, S.R. *Spectrochimica Acta B* **1998**, *53*, 1919.

(53) Fogarassy, E.; Fuchs, C.; Slaoui, A.; Stoquert, J.P. *Appl. Phys. Lett.* **1990**, *57*, 664.

(54) Sema, R.; Afonso, C.N. *Appl. Phys. Lett.* **1996**, *69*, 1541.

(55) Kools, J.C.S.; Nillesen, C.J.C.M.; Brongersma, S.H.; Van de Riet, E.; Dielemann, J. *J. Vac. Sci. Technol. A* **1992**, *10*, 1809.

(56) Fogarassy, E.; Slaoui, A.; Fuchs, C.; Stoquert, J.P. *Appl. Surf. Sci.* **1992**, *54*, 180.

(57) De Giorgi, M.L.; Leggieri, G.; Luches, A.; Martino, M.; Perrone, A.; Majni, G.; Mengucci, P.; Zemek, J.; Mihailescu, I.N. *Appl. Phys. A* **1995**, *60*, 275.

(58) Mihailescu, I.N.; Gyorgy, E.; Chitica, N.; Teodorescu, V.S.; Mavin, G.; Luches, A.; Perrone, A.; Martino, M.; Neamtu, J. *J. Mat. Sci.* **1996**, *31*, 2909.

(59) Gu, H.D.; Leung, K.M.; Chung, C.Y.; Han, X.D. *Surf. Coatings Technol.* **1998**, *110*, 153.

(60) He, M.; Cheng, N.; Zhou, P.; Okabe, H.; Halpern, J.B. *J. Vac. Sci. Technol. A* **1998**, *16*, 2372.

(61) Misra, A.; Bist, H.D.; Navati, M.S.; Thareja, R.K.; Narayan, J. *Mat. Sci. Eng. B* **2001**, *79*, 49.

(62) Ohshima, T.; Thareja, R.K.; Ikegami, T.; Ebihara, K. *Surf. Coatings Technol.* **2003**, *169–170*, 520.

(63) Beeach, F.; Boyd, I.W.; Amirhagi, S.; Sajjadi, A. *Laser Ablation of Electronic Materials*; Fogarassy, E.; Lazare, S. (Editors); North Holland: Amsterdam, 1992; 341.

(64) Afonso, C.N.; Vega, F.; Solis, J.; Catalina, F.; Ortega, C.; Siejka, J. *Appl. Surf. Sci.* **1992**, *54*, 175.

(65) Afonso, C.N. *Pulsed Laser Deposition of Films for Optical Applications in Materials for Optoelectronics: New Developments*; Agullo-Lopez, F. (Editor); World Scientific: Singapore, 1995.

(66) Cracium, V.; Cracium, D.; Bunescu, M.C.; Dabu, R.; Boyd, I.W. *Appl. Surf. Sci.* **1997**, *109–110*, 354.

(67) Mihailescu, I.N.; Teodorescu, V.S.; Gyorgy, E.; Luches, A.; Perrone, A.; Martino, M. *J. Phys. D: Appl. Phys.* **1998**, *31*, 2236–2240.

(68) Pfleging, W.; Klotzbücher, T.; Wesner, D.A.; Kreutz, E.W. *Diamond Rel. Mater.* **1995**, *4*, 370.

(69) Marine, W.; Gerri, M.; Scotto d'Aniello, J.M.; Sentis, M.; Delaporte, P.; Forestier, B.; Fontaine, B. *Appl. Surf. Sci.* **1992**, *54*, 264.

(70) Matsuo, Y.; Nakajima, T.; Kobayashi, T.; Takami, M. *Appl. Phys. Lett.* **1997**, *71*, 996.

(71) Ishitani, E.; Yoshimoto, S.; Higashide, H.; Kobayashi, M.; Shinohara, H.; Sato, H. *Chem. Lett.* **1993**, 1203.

(72) Vega, F.; Afonso, C.N.; Solis, J. *J. Appl. Phys.* **1993**, *73*, 2472.

(73) Gonzalo, J.; Afonso, C.N.; Madariaga, I. *J. Appl. Phys.* **1997**, *81*, 951.

(74) Vivien, C.; Hermann, J.; Perrone, A.; Boulmer-Leborgne, C.; Luches, A. *J. Phys. D: Appl. Phys.* **1998**, *31*, 1263.

(75) Hermann, J.; Dutouquet, C. *J. Phys. D: Appl. Phys.* **1999**, *32*, 2707.

(76) Mitra, A.; Thareja, R.K.; Ganesan, V.; Gupta, A.; Sahoo, P.K.; Kulkarni, V.N. *Appl. Surf. Sci.* **2001**, *174*, 232.

(77) Hegde, M.S. *Proc. Indian. Acad. Sci. (Chem. Sci.)* **2001**, *113*, 445.

(78) Zbroniec, L.; Sasaki, T.; Koshizaki, N.; Materialia, S. **2001**, *44*, 1869.

PART III

STUDY AND CHARACTERIZATION OF NANOSTRUCTURED OXIDES

As described in Chapters 3 and 4, nanostructured solids are nowadays generated by a series of different physical and chemical preparation methods. Although such synthetic methods have reached significant advances in terms of homogeneity in the structural/electronic properties of the particles forming the solid, as well as in the size distribution, much work is still needed to obtain full control of the synthesis variables and their influence in the final metal-oxide solid prepared. Problems related to oxide stoichiometry and metal oxidation state (commonly at surface layers) or the presence of impurities and amorphous phases coexisting with the crystalline one are often encountered. A glimpse into these problems would indicate that the structural and electronic properties as well as the primary particle size distribution are strongly dependent on the preparation method, and at the state-of-the-art, a first goal is to perform a detailed and full characterization of the particular solid yielded by the specific preparation method used. To do such a study in a systematic way, one needs a diverse array of experimental techniques (X-ray diffraction and scattering, microscopies, vibrational and electron spectroscopies, etc.) and theoretical methods (ab initio and semi-empirical quantum-mechanical calculations, Monte Carlo simulations, molecular dynamics, etc.). An important issue is the correct interpretation of the experimental and theoretical results. To help in reaching this goal, in this part of the book, a series of experimental techniques is presented that allows the investigation of the structural (Chapter 5) and electronic (Chapter 6) properties of oxide nanomaterials. This is followed by chapters that describe theoretical approaches that are useful for studying oxide nanoparticles and are based on post-Hartree–Fock formalisms or density-functional theory (Chapter 7), semi-empirical or parametric quantum-chemical methods (Chapter 8), and atomistic models with molecular dynamics (Chapter 9).

Synthesis, Properties, and Applications of Oxide Nanomaterials, Edited by José A. Rodríguez and Marcos Fernández-Garcia

CHAPTER 5

Techniques for the Study of the Structural Properties

JOSÉ A. RODRIGUEZ

Chemistry Department Brookhaven National Laboratory Upton, NY 11973, USA

MARCOS FERNÁNDEZ-GARCÍA

Instituto de Catálisis y Petroleoquímica CSIC, Campus Cantoblanco 28049 Madrid, Spain

ARTURO MARTÍNEZ-ARIAS

Instituto de Catálisis y Petroleoquímica CSIC, Campus Cantoblanco 28049 Madrid, Spain

JONATHAN C. HANSON

Chemistry Department Brookhaven National Laboratory Upton, NY 11973, USA

5.1. TECHNIQUES FOR THE STUDY OF STRUCTURAL PROPERTIES

The evolution of our understanding of the behavior of oxide nanostructures depends heavily on the structural information obtained from a wide range of physical methods traditionally used in solid-state physics, surface science, and inorganic chemistry (1,2). In this chapter, we describe several techniques that are useful for the characterization of the structural properties of oxide nanostructures: X-ray diffraction (XRD) and scattering, X-ray absorption fine structure (XAFS), Raman spectroscopy, transmission electron microscopy (TEM), scanning tunneling microscopy (STM), and atomic force microscopy (AFM). The ultimate goal is to obtain information about the spatial arrangement of atoms in the nanostructures with precise interatomic distances and bond angles. This may not be possible for complex systems, and one may get only partial information about the local geometry or morphology.

Synthesis, Properties, and Applications of Oxide Nanomaterials, Edited by José A. Rodríguez and Marcos Fernández-Garcia

5.2. XRD AND PDF

Since the 1910s, X-ray crystallography has been a valuable tool for obtaining structural parameters of metal oxides (1). X-ray diffraction powder patterns come from the interference pattern of elastically dispersed X-ray beams by atom cores and, in the case of materials with moderate-to-long-range order, contain information that arises from the atomic structure and the particle characteristics (for example, size and strain) (1,2). The effect of atomic structure on peak positions and intensities is described in textbooks (1), and the effect of particle characteristics on peak shape has been recently reviewed (3). This technique is frequently used for studying the structural properties of non-amorphous oxide nanoparticles (2).

Alternatively, the radial distribution of the materials can be determined by the appropriate Fourier transform of the diffraction pattern independent of any requirement of long-range coherence. This is an advantage in the study of nanoparticles or nanostructures where there is reduced long-range order. Egami et al. have applied this technique to metal oxide particles and obtained information about the structure and its defects (4). Finally, low-angle X-ray diffraction (5–7) can be used to determine the structure of nanoparticles.

5.2.1. Particle Characteristics from Whole Powder Profile Fitting

Recent analyses of particle characteristics rely on least-squares refinement of the whole powder diffraction profile, although several useful approximations give some of these characteristics from individual peaks. The pattern is represented as a superposition of peaks where the peak position depends on the cell parameters, and the peak width is a sum of instrument parameters, particle size parameters (τ), and strain related parameters (ε). The particle size peak width is given by

$$\beta_\tau = K\lambda/\tau \cos\theta \tag{5.1}$$

whereas the strain peak width varies as

$$\beta_\varepsilon = 4\varepsilon \tan\theta \tag{5.2}$$

with λ representing the wavelength and θ the diffraction angle. The fact that the strain broadening follows a $\tan\theta$ function, whereas the crystallite size broadening has a $1/\cos\theta$ dependence, allows the separation of these effects in whole-profile fitting through a range of θ (8). Here, the strain (ε) can be the result of a nonuniform application of stress in the lattice of the nanoparticles and reflects variations in cell dimension within the sample. Obvious variations are present at the surface of the material with respect to bulk because local symmetry and distances are different. Additionally, oxygen vacancies and the different types of defects (9) shown in

TABLE 5.1. Possible Imperfections in Crystals.[a]

Type of Imperfection		Description of Imperfection
Point defects	Interstitials	Extra atom in an interstitial site
	Schottky defect	Atom missing from correct site
	Frenkel defect	Atom displaced to interstitial site creating nearby vacancy
Line defects	Edge dislocation	Row of atoms marking edge of a crystallographic plane extending only part way in crystal
	Screw dislocation	Row of atoms about which a normal crystallographic plane appears to spiral
Plane defects	Lineage boundary	Boundary between two adjacent perfect regions in the same crystal that are slightly tilted with respect to each other
	Grain boundary	Boundary between two crystals in a polycrystalline solid
	Stacking fault	Boundary between two parts of closest packing having alternate stacking sequences

[a]From Ref. 9.

Table 5.1 can lead to strain in the lattice of a nanoparticle. The relative importance of each type of defect depends on the size of the nanoparticle and the chemical nature of the oxide.

Additional structural features can be obtained from diffraction. In some cases, the refinement of the powder pattern can include structural parameters for the thermal motion of the atoms, the fractional occupancy, and the positions. The quality of these refinements can be greatly improved with high-energy X-rays where the absorption correction is negligible, and a larger range of Q $(4\pi \sin(\theta)/\lambda)$ can be measured (10).

5.2.2. Local Structure from Fourier Analysis of Whole Pattern

The local structure from amorphous nanoparticles can be determined from the diffraction pattern in a manner similar to techniques that have been used to determine the structures of liquids and glasses (11). As an example, we describe the determination of the atomic pair distribution function (PDF) of amorphous quartz. Figure 5.1 shows the diffraction pattern from a quartz capillary. The local structure $g(r)$ is defined as the probability of finding an atom at a distance r from a reference atom and is derived from $G(r)$:

$$G(r) = 4\pi r[\rho(r) - o_0] \qquad \rho(r) = \rho_0 g(r) \tag{5.3}$$

where $\rho(r)$ is the pair density function and ρ_0 is the average pair density.

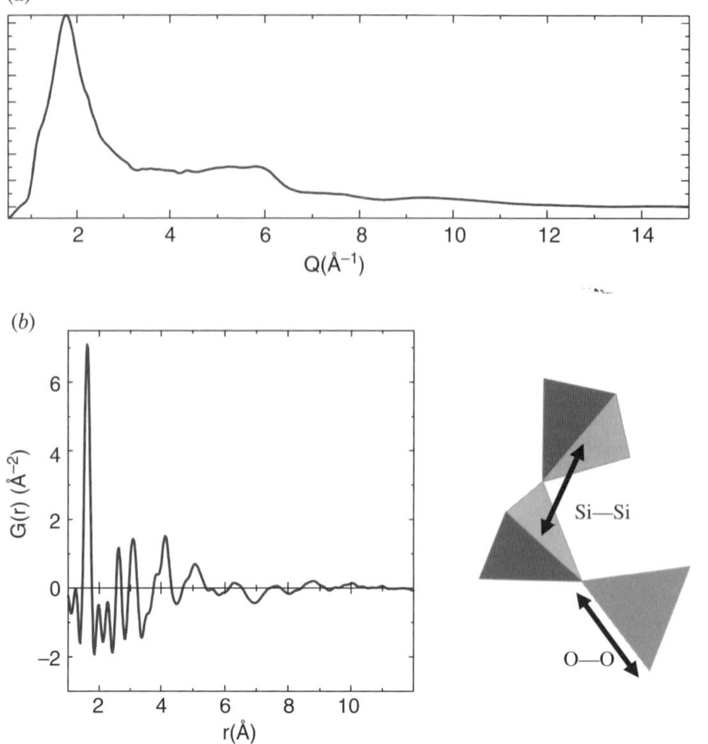

Figure 5.1. (*a*) The diffraction pattern for a quartz capillary. (*b*) The G(r) obtained from the data in (*a*).

$G(r)$ is also the Fourier transform of the total structure factor $S(Q)$:

$$G(r) = (2/\pi) \int Q[S(Q) - 1] \sin(Qr) dQ \tag{5.4}$$

$$Q = 4\pi \sin(\theta)/\lambda \tag{5.5}$$

The structure factor $S(Q)$ is related to the elastic part of the diffraction intensity (I^{el}) as follows:

$$S(Q) = 1 + \left[I^{el}(Q) - \sum c_i \, |f_i(Q)|^2\right] \Big/ \left|\sum c_i f_i(Q)\right|^2 \tag{5.6}$$

where c_i is the fraction of species i and $f_i(Q)$ is the atomic scattering factor of species i. The I^{el} is obtained from the total scattering after subtraction of background and corrections for Compton scattering, absorption, and multiple scattering. The calculated $G(r)$ is shown in Figure 5.1b. In this figure, we clearly see the Si–O, O–O, and

Si–Si distances and some longer distances. The number of interatomic distances has decreased to nearly zero by the distance of 10 Å.

The real space resolution obtained in the PDF is directly dependent on the Q range measured, and it is therefore a necessity to measure data to high Q, with adequate counting statistics, to obtain well-resolved peaks in the PDF (12). Generally, the data are collected at synchrotrons where high energy (short wavelength) and high intensity facilitate the above requirement. When the PDF technique is applied to crystalline materials, it has the advantage that the diffuse scattering data in the "background" (not under Bragg diffraction peaks) region contributes to the PDF. The obvious disadvantage of having a one-dimensional result is partly compensated for by refining three-dimensional models of the one-dimensional data (13).

The PDF analysis of nanoparticles of ceria is another good example of the application of this tool to nanostructures of oxides (14). In this case, data of neutron diffraction was used in the analysis. The neutron scattering cross sections from cerium and oxygen are nearly equal, whereas X-ray diffraction scattering by oxygen is much less than cerium, and consequently, neutron diffraction provides a more reliable determination of the oxygen structural features. However, a recent experiment on nano-ceria using high-energy X rays ($\lambda = 0.15$) obtained similar results (15). Frenkel defects of the oxygen atoms in ceria were observed using this technique. Recent PDF studies (16) of Zr-doped ceria have show that the oxygen storage capacity can be related to interfacial domains between the zirconium and the cerium-rich domains.

In addition to nano-CeO_2 and $Ce_{1-x}Zr_xO_2$, several structures of nano-crystalline metal oxides have been elucidated using the PDF technique (17–19). In these systems, the Bragg peaks were very broad and there was very limited long-range order. The short-range structures were solved by matching the observed $G(R)$ to related bulk structures. Nano-tubes of V_2O_5 have a very well-defined structure based on double layers of pyramidal/octahedral vanadium-oxygen units derived from the structure of $BaV_7O_{16}.nH_2O$ (17). In a similar way, the structure of a V_2O_5–nH_2O Xerogel was shown to consist of square pyramidal V_2O_5 at the nanometer length scale (18). In addition, the structure of hydrated lithium manganese oxide was found to contain layers made up of edge shared MnO_6 octahedra encapsulating K and Li atoms with the hydrous sample and Li and Mn atoms within the anhydrous sample (19). This result indicates that the improved electrochemical performance of the water-based nano-crystalline lithium manganese oxide could be attributed to its unique structure with well-separated MnO layers leaving Li atoms ample room to move in and out during the charge–discharge process (19).

5.2.3. In Situ Time-Resolved Powder Diffraction

Investigations at Brookhaven National Laboratory and other institutions have established the feasibility of conducting subminute, time-resolved XRD experiments under a wide variety of temperature and pressure conditions ($-190\,°C < T < 900\,°C$; $P < 45$ atm) (20–22). This is made possible by the combination of synchrotron radiation with new parallel data-collection devices. Using time-resolved XRD, one can follow structural changes during the preparation of nanoparticles and under reaction

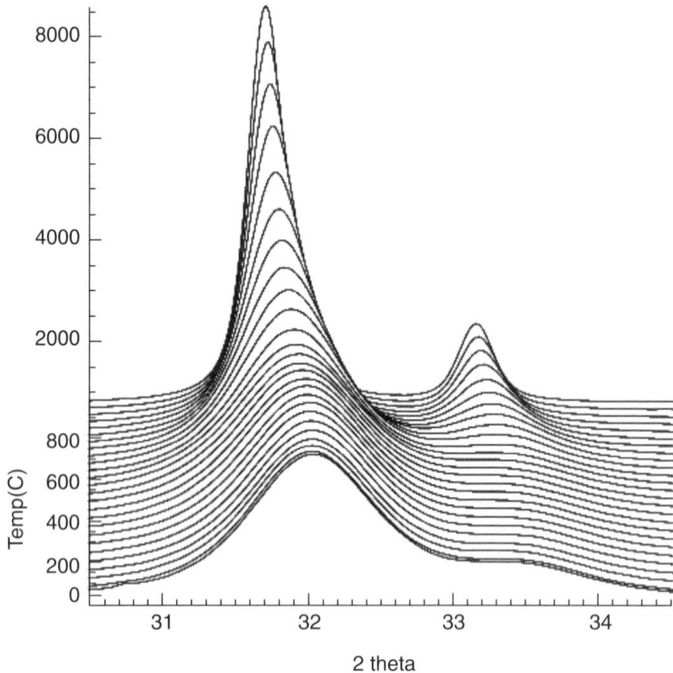

Figure 5.2. Time-resolved XRD data for the heating of $Ce_{0.9}Zr_{0.1}O_2$ nanoparticles in air. Heating rate = 4.8 °C/min, λ = 0.8941 Å (taken from Ref. 23).

conditions. Figure 5.2 shows time-resolved powder diffraction results for Zr-doped nano ceria ($Ce_{0.9}Zr_{0.1}O_2$) during heating from 25 °C to 925 °C in air (23). The sintering of the nanoparticles produces a sharpening of the diffraction peaks at temperatures above 600 °C. This type of approach can also provide information of changes in cell dimension or lattice strain with temperature.

Recently, the PDF technique has been extended/modified to allow the rapid acquisition of the necessary data. Time-resolved PDF measurements of nano-catalysts under reaction conditions are now in progress (24).

5.3. XAFS

The XAFS acronym refers to the oscillatory structure observed in the absorption coefficient [$\mu(E)$] just above or close to the absorption edge of an element constituting the sample. X-ray absorption spectroscopies are, essentially, synchrotron-based techniques that measure the absorption coefficient as a function of the X-ray energy $E = h\omega$. A typical absorption spectra [$\mu(E)$ vs. E] show three general features. The first one is an overall decrease in $\mu(E)$ with increasing energy that, sometimes, also contains well-defined peaks at specific energies called pre-edge transitions. The second is the presence of the edge, which roughly resembles a step function with a sharp rise at the corresponding edge energy and reflects the ionization of an inner shell (or core) electron of the element under study. The third component appears above the edge and corresponds to the oscillatory structure that can be roughly described as a periodical function, progressively damped as evolves from the edge energy. Being absorption techniques, XAFS spectroscopies are element-specific. On the other hand, the deep penetration of X rays in the matter makes them to probe the whole system, being only surface sensitive by performing specific detection schemes, will be briefly discussed below.

The XAFS spectrum $\chi(E)$ is defined phenomenologically as the normalized oscillatory structure of the X-ray absorption, e.g., $\chi(E) = (\mu(E) - \mu_0(E)/\mu_0(E))$, where $\mu_0(E)$ is the smooth-varying atomic-like background absorption. Essentially, the XAFS spectrum involves the quantum-mechanical transition from an inner, atomic-like core orbital electron to unoccupied bound (pre-edge transition) or unbound, free-like continuum levels. The oscillatory structure therefore reflects the unoccupied part of the electronic bands/structure of the system in the presence of a core-hole (25–29). Note that this differs from the initial, ground state by physical effects induced by the fact that the core hole is not infinitely long lived but must decay as a function of time and distance from the photoabsorber atom (22–26). The latter is a point to stress here as their short-range order character makes these techniques particularly suitable to analyze nanostructured materials that, frequently, do not posses or have a strongly disturbed long-range order.

Historically, it was controversial to recognize the local nature of the XAFS techniques [see Lytle for a recent discussion (30)] but now the theoretical description of the techniques is reasonably well settled (24,25). The absorption of X rays by matter is described in many textbooks (31). The treatment of the radiation as an electric field without practical spatial variation on a molecular/local scale and eliminating magnetic parts leads to the Fermi Golden Rule for the X-ray cross section:

$$\mu(E) = K[\langle\phi_f/e \cdot \bar{r}/\phi_i\rangle]^2 \delta_{Ef-Ei+h\gamma} \tag{5.7}$$

The intensity of the absorption process is then proportional to the square of the transition matrix element connecting the initial (ϕ_i) and final (ϕ_f) states times a delta function that ensures the fulfillment of the conservation energy theorem. The elimination of the spatial dependence of the electric field corresponds to a series expansion of its $e^{2\pi z/\lambda}$ dependence up to the first term (linear dependence in r_j, Eq. 5.7); this yields the dipole approximation of the interaction energy between the atom electronic cloud and the X-ray radiation field. Better approximations will include quadrupole, octupole, and so forth terms. However, except in a few cases, some of them here detailed, the dipole approximation gives quantitative analysis of the X-ray absorption near-edge spectroscopy (XANES) shape.

An important approximation is to assume that the matrix element can be rewritten into a single-electron matrix element. This is based in the sudden approximation that the transition element in Eq. 5.7 is reordered in terms of a overlap term of the $N - 1$ "inactive" electrons, which is roughly independent of energy, and the mentioned single-electron matrix element connecting wave functions of the unbound and inner core electrons (32). The validity of the sudden approximation depends primarily on the photo-excited electron kinetic energy and has been the subject of many studies (23,27,29). All electron rearrangements (interatomic and mainly extra-atomic relaxation of valence electrons) in response to the core-hole creation are thus neglected; the series of delta functions corresponding to different final states identifies with the local density of states (ρ) (33), and the corresponding X-ray absorption cross section becomes

$$\mu(E) = K[\langle\varphi_f/e \cdot \bar{r}/\varphi_i\rangle]^2\rho \tag{5.8}$$

where $\varphi_{i,f}$ correspond to the mentioned initial and final mono-electronic wave functions. The dipole matrix element dictates that the local density of states has an orbital moment that differs by 1 from the core state ($\Delta L = \pm 1$), whereas the spin is conserved ($\Delta S = 0$). If the electric quadrupole radiation is considered, the selection rules change and are ($\Delta L = 0, \pm 2$; $\Delta S = 0; \pm 1, \pm 2$).

There are two XAFS techniques: XANES and extended X-ray absorption fine structure (EXAFS). They differ in the energy of the final electronic state sampled, which is limited to a maximum of about 40–50 eV for XANES and above that point for EXAFS (Figure 5.3). Intuitively, it is obvious that pure electronic information (e.g., chemical bonding information) is only enclosed in the low-lying extended states and is thus confined to the XANES region. This will be discussed in Chapter 6. On the other hand, at high energy in the continuum of electrons participating in EXAFS, the effect of neighboring atoms becomes small and electron states approximate to spherical waves that are simply scattered by such atoms. The information extracted is thus of geometrical local character.

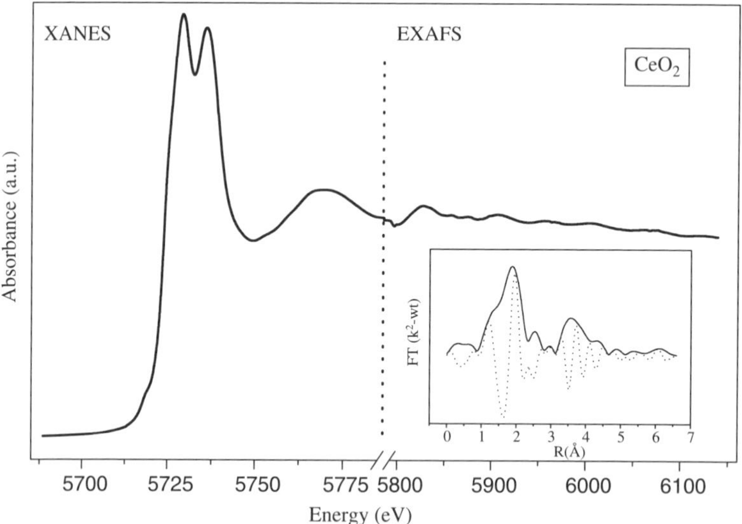

Figure 5.3. Ce L_{III} absorption spectrum of fluorite CeO_2. Inset: Fourier transform of the EXAFS spectrum.

As a general rule, Eqs. 5.7 and 5.8 can be solved by using modern quantum-mechanical methods. Two general approaches are discerned. The first one employs ab initio or DFT methods to calculate the initial and final wave functions of the electronic transition. The second reformulates the Schorodinger equation in terms of the scaterring theory and the absorption equation is expressed as a correlation function. This latter is particularly useful for EXAFS analysis of nanostructured materials as they can be treated as clusters for calculations that, in turn, can be performed in steps of growing size, showing the effect of introducing different coordination spheres. Detailed accounts for the multiple scattering EXAFS theory can be found in Refs. 34 and 35. Using this theoretical framework, the EXAFS formulas can be written as

$$\chi(E) = S_0^2 \sum_i \frac{N_i F(k)}{kR_i^2} e^{-2R/\lambda(k)} e^{-2DW} \sin(2kR + \phi(k)) \qquad (5.9)$$

This was originally proposed by Sayers et al. (36), based on a single-scattering formulism, but can be generalized to represent the contribution of N equivalent multiple-scattering contributions of path length $2R$ (33). It can be thus considered the standard EXAFS formulas, providing a convenient parameterization for fitting the local structure around the absorbing atoms. The dependence of the oscillatory structure on interatomic distance and energy is reflected in the sin($2kR$) term. The decay of the wave due to mean free path or lifetime (including intrinsic core-hole and extrinsic inelastic losses) is enclosed in the exponential $2R/\lambda$, which is largely responsible for the relatively short range probed by EXAFS in a material. As recognized in Ref. 34, the sinusoidal nature of the EXAFS phenomenon allows a Fourier analysis of the signal, yielding key information to give initial guesses for the fitting parameters. In fact, for each shell contributing to the EXAFS signal, we may have a peak in the Fourier transform spectrum. The strength of the interfering waves depends on the type and number (N) of neighboring atoms through the backscattering amplitude $[F(k)]$, which largely dictates the magnitude of the signal. The phase function $[\phi(k)]$ reflects the quantum-mechanical wave-like nature of backscattering, which depends on both the absorber and the scatterer atom properties and accounts for the difference between the measured and the geometrical interatomic distances (which is typically of 0.3–0.4 A for low atomic number (Z) backscatterers and lower for the rest of the atoms) displayed in the Fourier transform (26,28).

Another important factor is the exponential Debye–Waller (DW) term. This accounts for thermal and static disorder effects concerning the movement/position of atoms around their equilibrium/averaged position. A point to stress is that the nature of this term is different than the counterpart term in XRD (37). As vibrations increase with temperature, EXAFS spectra are usually acquired at low temperature to maximize information. Spectra at different temperatures may, on the other hand, allow us to decouple thermal and static contributions to DW. The DW term smears the sharp interference pattern of the sinusoidal term and cuts off EXAFS at sufficiently large energy beyond ca. 20 A^{-1}. Important to stress is that the DW factor is in fact a complex mathematical function and has a natural cumulant expansion in powers of k. The amplitude of the Debye–Waller term contains even moments $DW(k) = 2C_2k^2 - (2/3)C_4k^4 + \ldots$, whereas odd moments contribute to the EXAFS phase $\phi(k) = 2c_1k - (4/3)C_3k^3 + \ldots$ (38). Although the fitting of bulk systems makes use up to the squared term of the k-series expansion, the point is that the k^3 behavior (third cumulant) is important for "disordered" systems; as surface atoms have a less symmetric environment with respect to bulk ones and correspond to a significant part in nanostructured materials, the cumulant expansion needs to be considered in analyzing EXAFS data. Inclusion of the third cumulant would mainly influence coordination distance, whereas the fourth would influence coordination number and/or (second-order) Debye–Waller factor.

5.3.1. Effect of Nanostructure

EXAFS analysis of oxides as a function of size is scarce and mainly devoted to study Ti (39–41), Zn (42), Y (43), Zr (43,44), and Ce (45–47) containing systems. A point to stress is that, as mentioned throughout the text, nanocrystalline phases prepared by chemical methods are commonly accompanied by amorphous ones (45,48), and different techniques may in fact be selective to one of them (XRD), whereas others (as XAFS) inform over the whole system. Although EXAFS studies of dispersed surface species are not the main target of this review, it can be mentioned here that revisions of related papers can be found in Refs. 49 and 50. Surface species can have isolated or oligomeric structures, differing in their local order from the surface termination of nano-particulated materials by effect of the oxide anion sharing between support and surface components. In this case, there are an infinite number of possible local arrangements as a function of the support nature, oxides like alumina, silica, or carbon-containing materials (surface-modified carbon using oxygen-containing groups, nanotubes, etc.) and morphological properties (surface area and face exposed to the atmosphere) and the loading of the surface species.

Analysis of the EXAFS data of nanostructured oxides that account for the relative asymmetric (anharmonic) pair distribution of bond distances is even more scarce and exclusively devoted to Ti and Ce oxide systems (51,52). This, however, has been frequently worked out for nanostructured chalcogenides (53) and has been studied in detail for nanostructured fcc metals (54). As metal oxides display a multitude of different crystal structures, no general correlation between EXAFS coordination numbers of the three or four first shell coordination numbers and average particle size/morphology has been published. This can be found for fcc metals (55) and might be generalized to certain oxide cubic structures. It can be, however, noted that first shells display coordination numbers (CN) rather close to bulk ones, differing typically by less than one unit as under-coordinated atoms complete their first coordination shell at surfaces by using hydroxyl groups (see below). Second and further shells display CNs departing from bulk values, but their analysis is structure and shape dependent. A point to mention is that in some case, like ZrO_2 oxides, the coexistence of two crystalline (or one amorphous) phases can lead to destructive interference effects and misleading interpretation of results (56).

For titanium oxides, the studies usually identify the presence of fivefold-coordinated species in the nanostructured anatase and rutile surfaces, with a shortening of the Ti-O first shell bond distance (39,40), which in some cases is identified as a result of the presence of Ti hydroxyl groups at the surface (39). Note, however, that the inclusion of the third cumulant for studying mesoporous (amorphous by XRD but with a surface local structure similar to nanostructured anatase) titania suggests a larger Ti-O average first coordination distance with respect to the rutile/anatase phase (52). The presence of surface hydroxyl-coordinated centers has been also mentioned in ZrO_2 but detected by using other techniques (44). The case of cerium-containing nanostructured oxides is still under study because the series of samples with increasing size is obtained by calcining at progressively higher temperatures solids from Ce(III) precursors, and this influences the presence and concentration of two oxidation states Ce(III)/Ce(IV) in materials with small particle size (45,46).

Geometrical information is also enclosed in the XANES spectrum. The energy position of the peaks corresponding to electronic transitions to final quasibound, continuum states (called continuum resonances, CRs, in solid-state physics) depends on both electronic and geometrical factors (25,31,57). Electronic factors are analyzed in Chapter 6. Geometrical factors are summarized in the so-called $1/R^2$ rule, which states that the quantity $\Delta E \times R^2$ is a constant, where ΔE refers to the energy difference between the CR energy position and the zero electron kinetic energy and R is the nearest neighbor distance. This rule has been used fruitfully in solid-state studies, but an article by Kizler (58) demonstrates limitations in its applicability. Essentially, this rule can be used with generality only for CRs dominated by scattering from first or second nearest neighbors.

5.4. RAMAN SPECTROSCOPY

Raman spectroscopy is a powerful tool to analyze structural/morphological properties of solid oxides at a local level, given the strong sensitivity of the phonon characteristics to the crystalline nature of the materials (59). The theoretical basis for the application of this technique to the study of the properties of extended crystals as well as of individual molecular complexes present at the surface of oxide materials is out of the scope of this work and can be found in different textbooks or review articles (60,61). Raman spectroscopy is a bulk-sensitive technique, but the use of excitation frequencies allowing charge-transfer band absorption (resonance Raman) allows some surface sensitivity. Essentially, the theoretical background for the study of nanocrystalline (considered as a whole as crystalline materials) oxides as well as the spectral consequences observed by Raman is provided by the phonon confinement model that will be briefly analyzed in a first part; complete details of this model can be found elsewhere (62–64). The study by Raman spectroscopy of other properties that can in some cases be linked to the nanoscopic nature of the oxide materials, like the existence of a disorder associated with the presence of lattice defects (see also Section 5.1) or surface phonons, will be described later.

5.4.1. Phonon Confinement Model

The model employed for analyzing the features observed in the Raman spectra of nanocrystalline materials is the so-called "phonon confinement" (or "spatial correlation") model (66). Within such a model, nanocrystalline materials can be considered as an intermediate case between that of a perfect infinite crystal and that of an amorphous material. Thus, in a perfect crystal, conservation of phonon momentum (q) during phonon creation or decay upon interaction with the external radiation field requires that, in first-order Raman scattering, only optic phonons near the center of the Brillouin zone ($q \approx 0$) are involved. In amorphous materials, due to the lack of long-range order, the q vector selection rule does not apply and the Raman spectrum resembles the phonon density of states. In the intermediate case of nanocrystalline materials, due to the uncertainty principle, a range of q vectors of the order $\Delta q \approx 1/L$ (where L is the particle size) can be involved in the Raman process (62–65). The approach employed in the phonon confinement model follows from the work by Richter et al. (62), which shows that the relaxation of the conservation of phonon momentum selection rule arises from substitution of the phonon wave function in an infinite crystal by a wave function in which the phonon is confined in the volume of the crystallite. In their approach, which is employed to explain the first-order Raman spectrum of microcrystalline spherical silicon particles, it is used as a Gaussian function to impose the phonon localization. Later, Campbell and Fauchet extend the model by using other confinement functions and considering different shapes for the nanocrystallites (spherical, columnar, or thin films) (64). They found that the Raman spectrum of different semiconductor nanocrystals is best explained by using a Gaussian confinement function similar to the one employed by Richter et al., although with lower amplitude at the boundaries. By using this model, the first-order Raman spectrum is, considering the general case of a Gaussian confinement function with phonon amplitudes at the borders depending on a confinement parameter β (62–65), given by

$$I(\omega) \cong \int_{BZ} \frac{d^3q \exp(-q^2L^2/8\beta)}{(\omega - \omega(q))^2 + (\Gamma_0/2)^2} \tag{5.10}$$

where the integration is carried out over the entire Brillouin zone in the corresponding symmetry directions, $\omega(q)$ is the phonon dispersion curve, and Γ_0 is the natural line width. In the case of an infinite crystal ($L \to \infty$), $I(\omega)$ is a Lorentzian centered at $\omega(0)$ (the Raman frequency) with a line width of Γ_0. Application of Eq. 5.10 to any given Raman line of the perfect crystal results, as experimentally observed (62–64) in frequency shifts and asymmetric broadenings that increase with decreasing the particle size of the nanocrystal, and of a sign and magnitude that depends on the form and dispersion of the $\omega(q)$ function. Thus, a negative/positive dispersion in ω values as q increases is expected to produce asymmetric low/high-frequency broadening and a red/blue shift of the peak.

Examples on the application of the phonon confinement model to the analysis of Raman lines in nanosized TiO_2 and SnO_2 materials (62,67) are shown in Figure 5.4

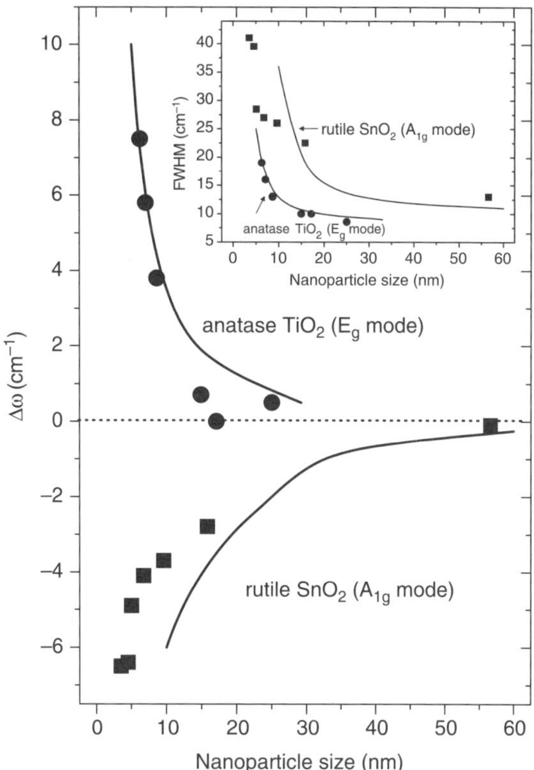

Figure 5.4. Experimental values (points) and values calculated according to the phonon confinement model (lines) for the Raman shift with respect to the large crystal value for peaks at $142\,cm^{-1}$ of anatase TiO_2 (E_g mode) and $638\,cm^{-1}$ of rutile SnO_2 (A_{1g} mode) as a function of the particle size. The evolution of the corresponding peaks widths is shown in the inset (experimental: points; calculated: lines); as inferred from the model, a high/low-frequency asymmetry was also observed with decreasing the particle size for the 142/638-cm^{-1} peaks. The figure was constructed with data extracted from Refs. 64 and 65.

similar analyses successfully applied to the study of Raman lines in ZrO_2 or ZnO nanoparticles are found in the literature (67,68). In the general case, a qualitative approach to the model is used to explain the evolutions of Raman peaks with decreasing the size of the nanoparticles. In this sense, it must be considered that in addition to the uncertainty in the choice of the confinement function and phonon dispersion curves (in the absence of precise theoretical or experimental results by neutron inelastic scattering), the presence of nonuniform particle size distributions or of imperfect crystallites (further confining the phonons by the microdomain boundaries or defects) often leads us to resort to empirical correlations for the analysis (66–74). Some examples of application of the confinement model for qualitative interpretation of Raman results in a series of nanostructured oxides like anatase TiO_2, CuO, Cr_2O_3, $ZrTiO_4$, or manganese oxides can be found elsewhere (69–75).

In some cases, refinement of the model by including effects of defects or more precise consideration of particle size distributions leads to closer approaches to the experimental results, as shown by Spanier et al. for CeO_2 nanoparticles (65). Thus, the analysis of the three-fold degenerated F_{2g} peak at 465 cm^{-1} (the only one allowed in first order) in CeO_2 nanosized samples with multidisperse size distributions requires consideration of both the effects of the presence of the vacancies (proposed to increase with decreasing particle size, at least for sizes below ca. 2 nm) (76) and the associated change in lattice dimensions. Although the former, according to experimental evidence, is expected to give rise to blue shifts of the peak, this is compensated by the red shift produced upon lattice expansion (Δa), in accordance with the equation:

$$\Delta\omega = -3\gamma\omega_0(\Delta a/a_0) \tag{5.11}$$

where ω_0 is the F_{2g} frequency in pure stoichiometric CeO_2, a_0 is the CeO_2 lattice constant, and $\gamma = (B/\omega)\, d\omega/dP$ is the Grüneisen constant with B the bulk modulus and $d\omega/dP$ the linear shift of the frequency with hydrostatic pressure (77). The result is that the red shift and broadening observed in the Raman line is significantly larger than expected from the phonon confinement model and that adequate fitting of the experimental spectra is only obtained when jointly considering the effects of inhomogeneous strain (associated with the dispersion in lattice constants as a consequence of the presence of vacancies and reduced cerium centers) (76) and phonon confinement (65). A similar approach has been employed to explain qualitatively the shifts observed in the main band of Ce–Tb or Ce–Cu mixed-oxide nanoparticles as a function of the Tb or Cu content (78,79). Another effect of the presence of oxygen vacancies in the CeO_2 nanoparticles is the appearance of a new broad peak at ca. 600 cm^{-1} that can be related to a second-order phonon or a local mode centered on the vacancy positions (65,77,80). The presence of second-order peaks associated with nanostructured configurations has been also reported for ZrO_2 (81).

5.4.2. Observation of Acoustic Modes

Another possible consequence of decreasing the particle size in the nanosized oxides is the appearance of low-frequency bands associated with acoustic modes. An example is the study performed on SnO_2, which for particles from 3.7 nm to 9.5 nm of average size, displays a peak from ca. 15 cm^{-1} to 60 cm^{-1} that shifts to higher frequency and becomes less intense and broader with decreasing the particle size (Figure 5.5 shows the spectrum obtained under a typical backscattering geometry of the spectrometer; note that correction of the Bose–Einstein phonon occupation factor required for the analysis is not included) (82). Analysis of such a peak considers the vibration of the nanoparticle as a whole. Thus, the nanoparticle is considered an elastic body of spherical shape with two types of vibrational modes: spheroidal and torsional. Then, after considering different boundary conditions (rigid or stress free), one arrives at expressions for estimating the frequency of the vibration as a function of the particle

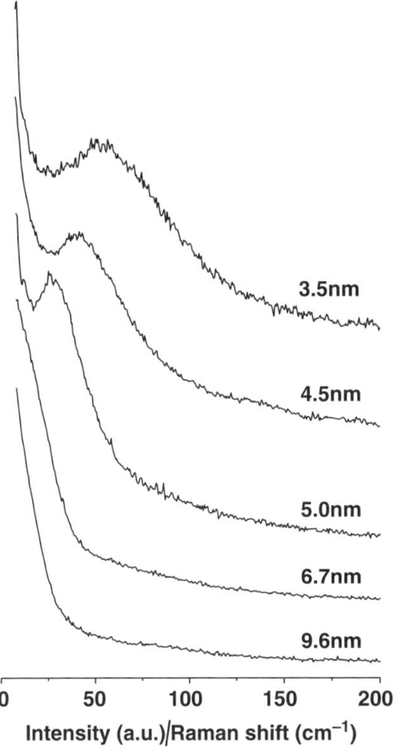

Figure 5.5. Acoustic modes observed in the Raman spectra of SnO_2 nanoparticles as a function of the particle size (courtesy of Dr. A. Diéguez).

size of the type (82):

$$\omega = \frac{S_{l,t} v_{l,t}}{Lc} \tag{5.12}$$

where $v_{l,t}$ is one of the two sound velocities (v_l for spherical or longitudinal and v_t for torsional or transversal vibrations) in the bulk of the oxide, $S_{l,t}$ is a coefficient that depends on the sound velocities, and c is the vacuum light velocity. As an interesting application of this analysis, nondestructive estimation of the particle size distribution can be done with this model after selecting the most adequate theoretical condition and applying it to deconvolute the experimental spectrum (82). Similar analyses have been applied to explain the presence of acoustical phonon modes in ZrO_2 nanoparticles (83).

5.4.3. Other Spectral Consequences of Nanostructure

An obvious consequence of decreasing the particle size is that a relatively large fraction of atoms in the nanoparticle reside on the surface, and as a result, surface optical phonons can also be observed with measurable intensity and analyzed by Raman spectroscopy (in addition to high-resolution electron energy loss spectroscopy and helium atom scattering techniques, most typically used for this purpose). An interesting application for this purpose is the use of resonant Raman spectroscopy tuned to the surface region, taking advantage of the different band structures of bulk and surface (84). The appearance of new vibrations (ascribed to surface phonon modes) different than those observed for the bulk material and increasing their relative intensity with decreasing the particle size has been observed in different oxide systems (82,85). A classic model is usually employed to analyze such surface phonons. According to the classic optical dispersion theory, each vibrational motion is characterized by an independent harmonic electric dipole oscillator with the corresponding transverse optical (TO) and longitudinal optical (LO) phonon modes. The frequency corresponding to the surface phonon modes must lie between ν_{TO} and ν_{LO} of any determinate TO–LO pair and satisfies the following condition:

$$\nu^2 = \nu_{TO}^2 \left[\frac{\varepsilon_0 + \varepsilon_m(1/L_k - 1)}{\varepsilon_\infty + \varepsilon_m(1/L_k - 1)} \right] \tag{5.13}$$

where $\varepsilon_0, \varepsilon_\infty$, and ε_m are dielectric constants for the material (static and high-frequency ones) and the surrounding medium and L_k are depolarization factors that depend on the particle shape. Full details of this theoretical approach and its successful application to the analysis of SiO_2 surface modes can be found elsewhere (86).

5.5. TEM AND STEM

TEM is one of the most powerful and versatile techniques for the characterization of nanostructured systems. Its unique characteristics allow us to achieve atomic resolution of crystal lattices as well as to obtain (with the assistance of energy dispersive X-ray spectroscopy—EDS—and electron energy-loss spectroscopy—EELS—complementary techniques) chemical and electronic information at the sub-nanometer scale. Numerous monographs have addressed technical or theoretical aspects involved in modern TEM microscopy employed for characterization of nanostructured systems as well as particular applications in different fields (see, for instance, Refs. 87–96 and references therein). The vast extension of the field precludes giving an exhaustive treatment here, and only a brief outline of the fundamental aspects of the TEM application to characterization of nanostructured systems will be thus attempted. As it is well known, the interaction of an electron beam with a solid specimen results in several elastic or inelastic scattering phenomena (backscattering or reflection, emission of secondary electrons, X rays or optical photons, and transmission of the undeviated beam along with beams deviated as a consequence of elastic—single atom scattering, diffraction—or inelastic phenomena). The TEM technique is dedicated to the analysis of the transmitted or forward-scattered beam. Such a beam is passed through a series of lenses, among which the objective lens mainly determines the image resolution, in order to obtain the magnified image. In low-resolution TEM, the objective aperture will be adjusted for selection of the central beam (containing the less-scattered electrons) or of a particular diffracted (or scattered in any form) beam to form the bright-field or dark-field image, respectively. In high-resolution TEM (HRTEM), which is usually performed in bright-field mode, the image is formed by collecting a few diffracted beams in addition to the central one.

Analysis of contrasts observed in the experimental TEM images when using the central beam by intercepting the electrons scattered through angles larger than a selected objective aperture is usually done on the basis of amplitude (or scattering) and mass-thickness or atomic number contrast theories (87,88). Figure 5.6 shows a typical example of a bright-field image collected in this mode (97,98). The higher contrast resolution usually attained in dark-field with respect to bright-field images can be most useful to analyze particle size or shape of the nanoparticles; an interesting example in this sense has been shown in a work by Yin and Wang in which the tetrahedral shape of CoO nanoparticles is assessed from contrast analysis in dark-field images (99). On the other hand, interference between the scattered and the incident wave at the image point is usually analyzed in terms of the phase contrast theoretical approach,

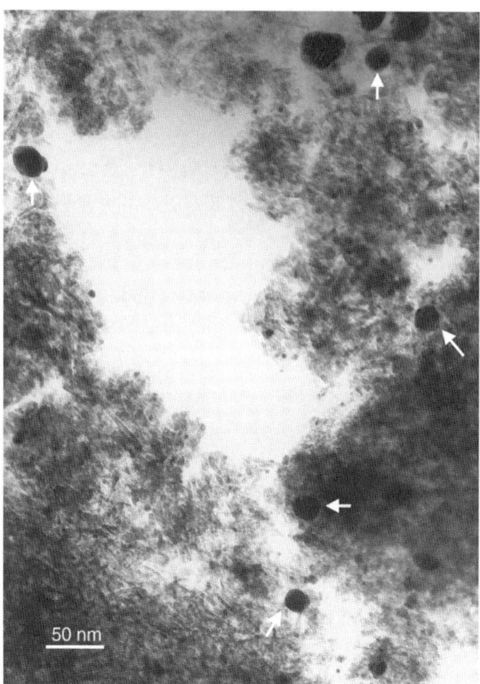

Figure 5.6. Bright-field TEM picture of a Ag/Al_2O_3 catalyst. The darkest zones (some of them marked with arrows) appear as a consequence of the higher scattering power of silver and are attributed (on the basis of analysis with other techniques) to oxidic silver nanoparticles (2). Thickness contrast is also revealed by comparing contrast in different zones in the picture.

which is employed for interpretation of HRTEM images (87–91). As a result of interaction with the specimen, the incoming electron wave is modified into a transmitted electron wave whose function $\psi(x, y)$ contains information [the electron wave exiting from the specimen can be approximated as $\psi(x, y) = \exp[i\sigma V_p(x, y)]$, with $V_p(x, y)$ being the electrostatic potential of the specimen projected in the x, y plane and σ an interaction constant) (87–91) on the structural details of the sample. Extraction of such information by formation of an image at the detector requires further interaction of the transmitted beam with the series of lenses constituting the microscope optics, in such a way that the observed image intensity distribution ($I(x, y) = |\psi_m(x, y)|^2$) is correlated with $\psi(x, y)$ through a function T that contains instrumental parameters (like apertures, wave attenuations, defocus and spherical aberration coefficient of the objective lens, etc.), according to (87–91).

$$\hat{\psi}_m(u, v) = \hat{\psi}(u, v)T(u, v) \tag{5.14}$$

where $\hat{\psi}_m(u, v)$ represents the Fourier transform of the wave function after the lenses, $\hat{\psi}(u, v)$ is the Fourier transform of $\psi(x, y)$, and u, v are coordinates of the reciprocal space vector.

To take into account the influence of the different parameters affecting the contrasts observed in the experimental images (electron scattering by the sample—particularly influences of irregular, columnar, etc. disposal of the scatterers—lens imperfections, microscope instability, etc.), HRTEM quantitative analysis is usually based on matching experimental and calculated images, for which various computational methods have been developed (87,89,93,94,100,101). Figure 5.7 shows an example of application of simulated images to resolve geometrical aspects in a supported oxide system (93).

Resolution limits of the TEM technique and schemes to achieve sub-angstrom resolution have been analyzed in detail by Spence (89). The highest structural resolution possible (point resolution) when the obtained images can be directly related to the projected structure of the specimen is given by the Scherzer expression $R_s = 0.66\lambda^{3/4}C_s^{1/4}$, where λ is the electron wavelength and C_s is the spherical aberration coefficient (89). From this expression and considering the de Broglie formula, higher resolution is achieved upon use of high-voltage instruments (acceleration voltages higher than 0.5 MeV). Enhanced radiation damage, which may have stronger effects for nanostructured materials (91), must, however, be considered in that case. Aberration correction has in turn allowed us to achieve sub-angstrom resolution with microscopes operating at lower voltages (typically 200 KeV) and to resolve oxygen atoms in oxides materials (89,93). On the other hand, considering that high resolution is achieved in TEM as the result of electron wave interference among diffracted peaks, a limitation to structural resolution can also arise from decreasing the number of atoms in the nanoparticle (101). This also can pose limitations to routinely employed determination of particle size distributions, particularly in cases when contrast differences between various sample components are not clearly resolved, as may occur for supported catalysts (101). Considering that such estimations are both

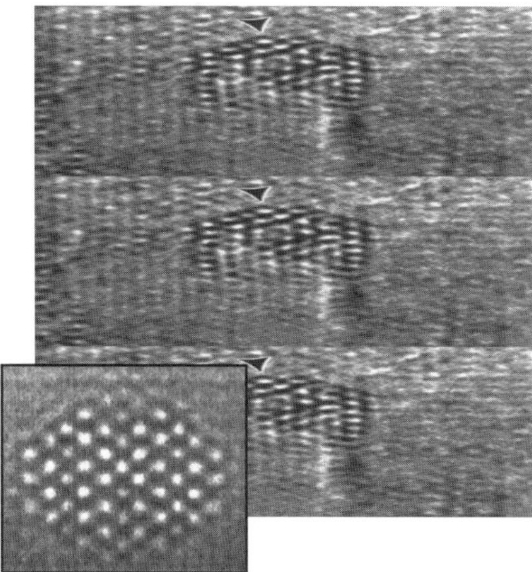

Figure 5.7. HRTEM image of a Nd_2O_3/MgO catalyst and simulation (left bottom) of the Nd_2O_3 nanoparticle appearing in the experimental image (marked with an arrow). The best match is observed to correspond to a rounded, unstrained, monolayer raft of cubic Nd_2O_3 in [001] orientation grown under parallel orientation over a (001) MgO surface (courtesy of Dr. J.J. Calvino).

microscope and sample dependent, calculations are generally required to establish limitations in contrast resolution in each particular case (101,102).

As mentioned, information on chemical and electronic characteristics of the nanocrystals at a high spatial resolution is also achievable with TEM instruments. The use of scanning transmission electron microscopy (STEM) in which a fine convergent electron probe is scanned over the sample (the resolution being related to the probe size that can be as small as ≈ 0.1 nm) (95) is particularly useful for this purpose (87,89,92,100,103). The electrons kinetic energy losses or the X rays emitted as a consequence of inelastic scattering processes during electron–specimen interaction can be analyzed by EELS and EDS techniques, which are the two most commonly used in chemical microanalysis. Due to EDS limitations in detection of light elements, EELS is most employed for the analysis of oxide materials. In addition to elemental information, the EELS spectra provide information on the electronic structure, bonding, and nearest neighbor distribution of the specimen atoms. Among the different loss regions, the one most useful for this purpose is the high-loss one, related to electrons that have interacted with inner shell or core electrons of the specimen atoms. Thus, the information obtained is similar to that given by XAS (see above); i.e., K-, L-, M-, etc. ionization edges of the elements present in the sample appear in the EELS spectrum (near edge and extended energy-loss fine structure—ELNES and EXELFS—regions being defined within them). A generally higher energy resolution is, however, achieved with EELS, which comparatively facilitates analysis of the states present in a determinate band (91). An interesting example of the high spatial resolution that can be achieved with this type of instruments is provided by a recent TEM study of ceria nanoparticles (3–20 nm in diameter) in which EELS spectra can resolve between valence states of cerium at the edge in comparison with the center of the particles (104). Thus, even though as shown in a joint XPS–XANES study, this phenomenon is still subject to some debate (105), such a TEM–EELS study nicely shows the valence reduction of cerium with decreasing the particle size and the preferential reduction of cerium at the surface of the nanoparticles (104).

In connection with this, TEM images can be also formed with electrons that have lost a specific energy with respect to a determinate energy threshold of the atomic inner shell, giving rise to energy-filtered TEM (EF-TEM). A chemical mapping at the nanoscale can be obtained in this mode, which allows us to resolve the location of the corresponding elements in the samples, being most useful in cases when small contrast differences between different sample components are observed in the bright-field images; in addition, differences observed in the EELS spectra as a function of the valence state of a determinate element can be exploited by EF-TEM for getting valence state distributions across the specimen (91,99).

On the other hand, collecting the electrons scattered at high angles in a STEM instrument (by means of a high-angle annular dark field detector—HAADF) has proven most useful to obtain images (by the so-called Z-contrast imaging) of small clusters in catalysts or even single (relatively heavy) atoms or point defects (95,96). Considering the Mott formula for the electron scattering factor of a single atom ($f_e(s) = e[Z - f_x(s)]/16\pi^2\varepsilon_0 s^2$, with $s = \sin\theta/\lambda$, f_x the X-ray scattering factor of the atom, θ the scattering angle, and λ the electron wavelength), it is inferred that the electrons collected at high angles give significant information directly related to the atomic number. Another application of high utility in the analysis of materials, and in which significant progress has been achieved during the last years, is in situ TEM (95,106,107). Recent advances in this field are collected in the review of Wang (95).

5.6. STM AND AFM

Since the 1980s, STM- and STM-derived local probe methods have experienced a remarkable development. The seminal ideas behind these microscopies can be traced back to the early 1970s with the invention of the topografiner (108,109). Ten years later, Binnig and Rohrer established the basic principles for STM (110,111). For this important contribution, they received the 1986 Nobel Prize in Physics. In an STM, the sample surface is scanned with a sharp tip located at a distance of less than a few nanometers (112). The tip is mounted on a piezoelectric positioner. A voltage is applied between the sample and the tip, generating a tunneling current that varies exponentially with separation (110,112). The tunneling current is then measured, and a feedback control system compares the actual tunneling current with a (user-adjustable) "set current." If the actual current is too large or too small, the feedback control system moves the tip back or forward (respectively). In a first approximation, the movement of the tip represents variations in the position of the atoms in the surface of the sample. The two-dimensional array of numbers representing the surface "height" at each point is recorded in a computer, displayed in the computer screen, and then stored on a magnetic or optical device for subsequent analysis (112).

Several approaches have been proposed for the analysis of STM images (113–118). In general, there is a relationship between the STM image and the geometric and electronic properties of the surface of a sample. A precise interpretation of the image at an atomic scale can be a problem (111,117,118) because the image is not at all a steric representation of the surface but is a view of its electronic structure at the Fermi level energy. In most cases, it is therefore not possible to directly and simply relate the bumps and shapes of the STM image to the actual lateral positions of atoms (118). One must establish how the structural parameters affect the tunneling current and formation of the image (119,120). A good knowledge of the electronic structure of the surface and tip is a prerequisite for the calculation of the tunneling current. To obtain this electronic structure, different quantum mechanical approaches can be used ranging from the extended Hückel method to calculations based on density functional theory (113,118). Various levels of approximation have been proposed for the calculation of the tunneling current (113–119).

Many methods used to calculate the current rely on perturbation theory (121–125). Following this formalism, the current between two electrodes (represented by μ and ν) can be described as (112).

$$I = (2\pi |e|/h) \sum_{\mu,\nu} [f(E_\nu) - f(E_\mu)] |M_{\mu\nu}|^2 \delta(E_\nu + |e|V - E_\mu) \tag{5.15}$$

In this equation, $f(E_\nu)$ represents the Fermi–Dirac distribution function (the first factor in the summation reflects the fact that electrons must tunnel from an occupied electronic state to an unoccupied electronic state); $M_{\mu\nu}$ represents the tunneling matrix element between state μ on the sample and state ν on the tip. The δ-function term reflects energy conservation during the tunneling process. The calculation of $M_{\mu\nu}$ is not trivial (112,113,118). In the early work of Tersoff and Hamann, an s-wave was assumed for the tip (114), but in subsequent studies, tunneling matrix elements have been calculated with p and i states as tip orbitals (112). A single metal atom and clusters have been used to model the tip (121–125).

When using perturbation theory, the tip and the surface are treated separately, neglecting any interaction between them. Technically, this is valid only in the limit of large tip-to-surface distances. Methods have been proposed for the calculation of the tunneling currents that take into account interactions between the tip and the sample through a scattering theory formalism (126,127). In principle, the tunnel event is viewed as a scattering process (118): Incoming electrons from the sample scatter from the tunnel junction and have a small probability to penetrate into the tip and a large one to be reflected towards the sample. Multiple scattering events can affect the current (128,129). Tip-induced localized states and barrier resonances seem to play an important role in the conductance (130,131).

Although it may not be easy to obtain structural parameters from STM images, this microscopy can be used in a straightforward way to study the shape or morphology

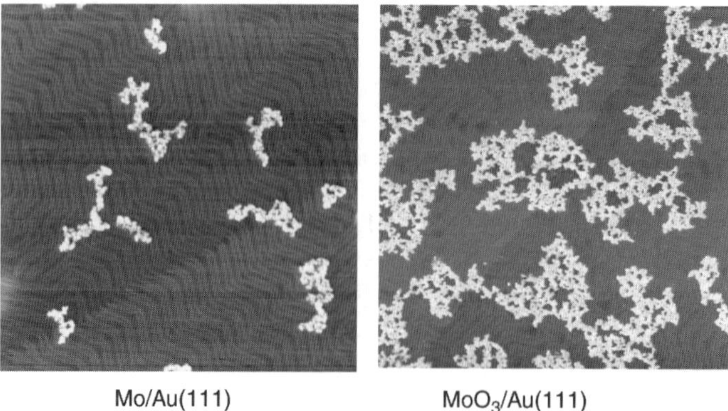

Figure 5.8. STM images recorded before and after oxidizing Mo nanoparticles on a Au(111) template. Image size: $190 \times 190\,\mathrm{nm}^2$ (taken from Ref. 138). See color insert.

of oxide nanostructures as a function of temperature and under reaction conditions (132–137). Figure 5.8 illustrates how useful this technique can be (138). Initially, Mo nanoparticles are present on a Au(111) substrate. There is Au–Mo intermixing, and the nanoparticles are tri-dimensional. Upon oxidation and formation of MoO_3, the STM image shows a radical change in the morphology of the system. There is no more Au–Mo intermixing, and two-dimensional aggregates of MoO_3 spread out over the Au(111) substrate (138).

The dependence of STM images on applied bias can be used as the basis for a true atomic-resolution tunneling spectroscopy (135). The effects of the voltage on the tunneling current depend on the electronic properties of the sample, and thus, information can be obtained about local density of states and band gaps (139). Tunneling spectroscopy data can be compared directly with area-averaging probes of electronic structure such as photoemission and inverse photoemission spectroscopies (140,141). As the tunneling data can be acquired with atomic resolution, one could be able to directly visualize the spatial distribution of individual electronic states (140,141).

In AFM, a tip is mounted on a cantilever (112,113). Forces exerted by the sample on the tip bend the cantilever, and this deflection is used in a feedback system in a manner analogous to the tunneling current in the STM. The forces relevant to AFM are of electromagnetic origin (142). In the absence of external fields, the dominant forces are van der Waals interactions, short-range repulsive interactions, adhesion, and capillary forces (142). The interaction between the tip and the sample can be operated on either the attractive or the repulsive parts of the interatomic potential curve. Most experiments for contact AFM are performed on the steeply rising repulsive part of the potential curve (112). Operation in the attractive region is intrinsically more difficult because it involves weaker interactions and requires the measuring of both the force and the force gradient (112). However, true atomic resolution is easier to achieve using attractive-mode contact AFM (112,113). Images obtained with repulsive-mode contact AFM may show atomic periodicity while not truly achieving atomic resolution.

In the last decade, ATM has experienced a significant transformation by using a vibrating probe to explore the surface morphology (142). Dynamic or noncontact AFM methods are emerging as powerful and versatile techniques for atomic and nanometer-scale characterization (112,113,142). In dynamic AFM, there are two major modes of operation (143–145): amplitude modulation and frequency modulation (AM–AFM and FM–AFM, respectively). Using FM–AFM, images with very good atomic resolution have been obtained for semiconductors and insulator samples (146–148). Because AFM does not involve the induction of a tunneling current between the probing tip and sample, it can be applied to study oxide systems that have a low conductivity (insulators or wide band-gap semiconductors) (149–150).

To obtain structural parameters from AFM images, one must model the movement of the cantilever over the sample in an accurate way (112,142). To gain insight into the tip motion, several authors have considered the cantilever-tip ensemble as a point-mass spring, and thus, the movement can be described by a nonlinear, second-order differential equation (151–153).

$$m\ddot{I} + kI + (m\omega_o/Q)\dot{I} = F_{ts} + F_o\cos(\omega t) \tag{5.16}$$

where F_o and ω are the amplitude and angular frequency of the driving force. Q, ω_o, and k are the quality factor, angular resonance frequency, and the force constant of the free cantilever, respectively. F_{ts} contains the tip–surface interaction forces. In the absence of tip–surface forces, Eq. 5.15 describes the motion of a forced harmonic oscillator with damping (142). The mathematical expressions for F_{ts} in general can be complex and contain terms for long-range attractive interactions, short-range repulsive interactions, adhesion, and capillary forces (142).

ACKNOWLEDGMENT

J.A.R. and J.H. are grateful for the financial support of the U.S. Department of Energy, Division of Chemical Sciences (DE-AC02-98CH10886 contract). M.F.G. and A.M.A. thank the Spanish "Ministerio de Ciencia y Tecnología" and CSIC for financial support.

REFERENCES

(1) Stout, G.; Jensen, L. *X-ray Structure Determination A Practical Guide*; Wiley: New York, 1989.

(2) Fernández-García, M.; Martínez-Arias, A.; Hanson, J.C.; Rodriguez, J.A. *Chem. Rev.* **2004**, *104*, 4063.

(3) *Defect and Microstructure Analysis by Diffraction*; Snyder, R.L.; Fiala, J.; Bunge, H.J. (Editors); Oxford University Press: New York, 1999.

(4) Egami, T.; Billinge, S.J.L. *Underneath the Bragg Peaks Structural Analysis of Complex Materials, Vol. 7*; Elsevier: Oxford, 2003.

(5) Guinier, A. *X-ray Diffraction in Crystals, Imperfect Crystals and Amorphous Bodies;* Dover: New York, 1994.

(6) Deb, B.; Biswas, T.; Sen, D.; Basumallick, A; Mazumder, S. *J. Nanoparticle Res.* **2002**, *4*, 91.

(7) Ohnuma, M.; Hono, K.; Onnodera, H.; Ohnuma, S.; Fujimori, H.; Pedersen, J. *J. Appl. Phys.* **2000**, *87*, 817.

(8) Jenkins, R.; Snyder, R.L. *Introduction to X-ray Powder Diffractometry*; Wiley: New York, 1996.

(9) Azaroff, L.V. *Introduction to Solids*; McGraw-Hill: New York, 1960.

(10) Kim, J.Y.; Rodriguez, J.A.; Hanson, J.C.; Frenkel, A.I.; Lee, P.L. *J. Am. Chem. Soc.* **2003**, *125*, 10684.

(11) Warren, B. *X-ray Scattering*; Dover: New York, 1990.

(12) Toby, B.H.; Egami, T. *Acta Cryst.* **1992**, *A48*, 336.

(13) Proffen, T.; Billinge, S.J.L. *J. Appl. Cryst.* **1999**, *32*, 572.

(14) Mamontov, E.; Egami, T. *J. Phys. Chem. Solids* **2000**, *61*, 1345.

(15) Chupas, P.A; Hanson, J.C.; Lee, P.L.; Rodriguez, J.A.; Iglesias-Juez, A.; Fernandez-Garcia, M in preparation.

(16) Mamontov, E; Brezny, R; Koranne, M.; Egami, T. *J. Phys. Chem. B* **2003**, *107*, 13007.

(17) Petkov, V.; Trikalitis, P.N.; Bozin, E.S.; Billinge, S.J.L.; Vogt, T.; Kanatzidis, M.G. *J. Am. Chem. Soc.* **2002**, *124*, 10157.

(18) Petkov, V.; Zavalij, P.J.; Lutta, S.; Whittingham, M.S.; Parvanov, V.; Shastri, S. *Phys. Rev. B* **2004**, 085410.

(19) Bateshki, M.; Hwang, S.-J.; Park, D.H.; Ren, Y.; Petkov, V. *J. Phys. Chem. B* **2004**, *108*, 14956.

(20) Norby, P.; Hanson, J. *Catal. Today* **1998**, *39*, 301; Rodriguez, J.H.; and Hanson, J.C.; Ciencia, **2006**, 14, 177.

(21) Grunwaldt, J.-D.; Clausen, B.S. *Topics Catal.* **2002**, *18*, 37.

(22) Walton, R.I.; O'Hare, D. *Chem. Commun.* **2000**, 2283.

(23) Rodriguez, J.A.; Hanson, J.C.; Kim, J.-Y.; Liu, G.; Iglesias-Juez, A.; Fernandez-Garcia, M. *J. Phys. Chem. B* **2003**, *107*, 3535.

(24) Chupas, P.J.; Qiu, X.; Hanson, J.C.; Lee, P.L.; Grey, C.P.; Billinge, S.J.L. *J. Applied Cryst.* **2003**, *36*, 1342.

(25) Fuggle, J.C.; Inglesfield, J.E. *Unoccupied Electronic States*. Springer-Verlag: Berlin, 1992.

(26) *X-ray Absorption*; Konnigsberger, D.C.; Prins, R. (Editors); Wiley: New York, 1987.

(27) See articles in *Chem. Rev*. **2001**, *101*.

(28) Rehr, J.J.; Albers, J.C. *Rev. Modern Phys.* **2000**, *72*, 621.

(29) Egelhoff, W.G. *Surf. Sci. Rep.* **1987**, *6*, 253.

(30) Lytle, F.W. *Synchrotron Rad.* **1999**, *6*, 123.

(31) Levine, I.N. *Molecular Spectroscopy*; Wiley: New York, 1975.

(32) Aberg, T. *Phys. Rev.* **1967**, *16*, 14201.

(33) Muller, J.E.; Jepen, O.; Andersen, O.K.; Wilkins, J.W. *Phys. Rev. Lett.* **1978**, *40*, 720.

(34) Natoli, C.R.; Benfatto, M.; Doniach, S. *Phys. Rev. B* **1986**, *34*, 4682.

(35) Rehr, J.J.; Albers, R.C. *Phys. Rev. B* **1996**, *41*, 8139.

(36) Sayers, D.E.; Stern, E.A.; Lytle, F.W. *Phys. Rev. Lett.* **1971**, *21*, 1204.

(37) Poiarkova, A.V.; Rehr, J.J. *Phys. Rev. B* **1999**, *59*, 948.

(38) Bunker, G. *Nucl. Instrum. Methods Phys. Res.* **1983**, *207*, 437.

(39) Chen, L.X.; Rajh, T.; Wang, Z.; Thurnauer, M.C. *J. Phys. Chem. B* **1997**, *101*, 10688.

(40) Yeung, K.L.; Maira, A.J.; Stolz, J.; Hung, E.; Hu, N.K.-C.; Wei, A.C.; Soria, J.; Chao, K.-J. *J. Phys. Chem. B* **2002**, *106*, 4608.

(41) Luca, V.; Djajanti, S.; Howe, R.F. *J. Phys. Chem. B* **1998**, *102*, 10650.

(42) Han, S.W.; Hoo, H.J.; An, S.J.; Yoo, J.; Yi, G.C. *Appl. Phys. Lett.* **2005**, *85*, 021917.

(43) Winterer, M.; Nitsche, R.; Hahn, H. *Nanost. Mat.* **1997**, *9*, 397.

(44) Rush, G.E.; Chadwick, A.V.; Kosacki, I.; Anderson, H.U. *J. Phys. Chem. B* **2000**, *104*, 9597.

(45) Nachimuthu, P.; Shih, W.-C.; Liu, R.-S.; Jang, L.-Y.; Chem., J.-M. *J. Sol. State Chem.* **2000**, *149*, 408.

(46) Zhu, Z.; Zhang, X.; Benfield, R.E.; Ding, Y.; Gandjean, D.; Zhang, Z.; Xu, X. *J. Phys. Chem. B* **2002**, *106*, 4569.

(47) Nagai, Y.; Yamamoto, T.; Tanaka, T.; Yoshida, S.; Nonaka, T.; Okamoto, T.; Suda, A.; Sugura, M. *Catal. Today* **2002**, *74*, 225.

(48) Xijuan, Y.; Pingbo, X.; Qingde, S. *Phys. Chem. Chem. Phys.* **2001**, *3*, 5266.

(49) Asuka, K.; Iwasawa, Y. *X-ray Absorption Fine Structure for Catalysts and Surfaces*; Iwasawa, Y. (Editor); World Scientific: Singapore, 1996; Chapter 7.

(50) Wachs, I.E. *Catal. Today* **2005**, *100*, 79.

(51) Lemaux, S.; Bansaddik, A.; van der Eerden, A.M.J.; Bitter, J.H.; Koningsberger, D.C. *J. Phys. Chem. B* **2001**, *105*, 4810.

(52) Oshitake, H.; Suguhara, T.; Tatsumi, T. *Phys. Chem. Chem. Phys.* **2003**, *5*, 767.

(53) Rockenberger, J.; Troger, L.; Kornowski, A.; Vossmeyer, T.; Eychmuller, A.; Feldhaus, J.; Weller, H. *J. Phys. Chem. B* **1997**, *101*, 2691.

(54) Clausen, B.S.; Norskov, J.K. *Topics Catal.* **2000**, *10*, 221.

(55) Jentys, A. *Phys. Chem. Chem. Phys.* **1999**, *1*, 4059; and references therein.

(56) Yang, P.; Cai, X.; Xie, X.; Xie, Y.; Hu, T.; Zhang, J.; Liu, T. *J. Phys. Chem. B* **2003**, *107*, 6511.

(57) Fernández-García, M. *Catal. Rev. Sci. Eng.* **2002**, *44*, 59.

(58) Kizler, P. *Phys. Lett. A* **1992**, *172*, 66.

(59) Loudon, R. *Adv. Phys.* **2001**, *50*, 813.

(60) Ferraro, J.R.; Nakamoto, K. *Introductory Raman Spectroscopy*; Academic Press: San Diego, CA, 1994.

(61) Wachs, I.E. In *Handbook of Raman Spectroscopy*; Lewis, I.R.; Edwards, H.G.M. (Editors); Marcel Dekker: New York, 2001; 799.

(62) Richter, H.; Wang, Z.P.; Ley, L. *Sol. St. Commun.* **1981**, *39*, 625.

(63) Parayantal, P.; Pollak, F.H. *Phys. Rev. Lett.* **1984**, *52*, 1822.

(64) Campbell, I.H.; Fauchet, P.M. *Sol. St. Commun.* **1986**, *58*, 739.

(65) Spanier, J.E.; Robinson, R.D.; Zhang, F.; Chan, S.; Herman, I.P. *Phys. Rev. B* **2001**, *64*, 245407.

(66) Kelly, S.; Pollak, F.H.; Tomkiewicz, M. *J. Phys. Chem. B* **1997**, *101*, 2730.
(67) Fangxin, L.; Jinlong, Y.; Tianpeng, Z. *Phys. Rev. B* **1997**, *55*, 8847.
(68) Rajalakshmi, M.; Arora, A.K.; Bendre, B.S.; Mahamuni, S. *J. Appl. Phys.* **2000**, *87*, 2445.
(69) Zhang, W.F.; He, Y.L.; Zhang, M.S.; Yin, Z.; Chen, Q. *J. Phys. D: Appl. Phys.* **2000**, *33*, 912.
(70) Bersani, D.; Lottici, P.P.; Ding, X. *Appl. Phys. Lett.* **1998**, *72*, 73.
(71) Xu, J.F.; Ji, W.; Shen, Z.X.; Li, W.S.; Tang, S.H.; Ye, X.R.; Jia, D.Z.; Xin, X.Q. *J. Raman. Spect.* **1999**, *30*, 413.
(72) Zuo, J.; Xu, C.; Hou, B.; Wang, C.; Xie, Y.; Qian, Y. *J. Raman. Spect.* **1996**, *27*, 921.
(73) Zuo, J.; Xu, C.; Liu, Y.; Qian, Y. *Nanost. Mater.* **1998**, *10*, 1331.
(74) Kim, Y.K.; Jang, H.M. *Solid. State Commun.* **2003**, *127*, 433.
(75) Zhen, Z.; Tan, S.; Zhang, S.; Wang, J.; Jin, S.; Zhang, Y.; Sekine, H. *Jpn. J. Appl. Phys.* **2000**, *39*, 6293.
(76) Tsunekawa, S.; Ishikawa, K.; Li, Z.-Q.; Kawazoe, Y.; Kasuya, A. *Phys. Rev. Lett.* **2000**, *85*, 3440.
(77) McBride, J.R.; Hass, K.C.; Poindexter, B.D.; Weber, W.H. *J. Appl. Phys.* **1994**, *76*, 2435.
(78) Hungría, A.B.; Martínez-Arias, A.; Fernández-García, M.; Iglesias-Juez, A.; Guerrero-Ruiz, A.; Calvino, J.J.; Conesa, J.C.; Soria, J. *Chem. Mater.* **2003**, *15*, 4309.
(79) Wang, X.; Rodriguez, J.A.; Hanson, J.C.; Gamarra, D.; Martínez-Arias, A.; Fernández-García, M. *J. Phys. Chem. B* **2005**, *109*, 19595.
(80) Weber, W.H.; Hass, K.C.; McBride, J.R. *Phys. Rev. B* **1993**, *48*, 178.
(81) Siu, G.G.; Stokes, M.J.; Liu, Y. *Phys. Rev. B* **1999**, *59*, 3173.
(82) Diéguez, A.; Romano-Rodríguez, A.; Morante, J.R.; Bârsan, N.; Weimar, U.; Göpel, W. *Appl. Phys. Lett.* **1997**, *71*, 1957.
(83) Jouanne, M.; Morhange, J.F.; Kanehisa, M.A.; Haro-Poniatowski, E.; Fuentes, G.A.; Torres, E.; Hernández-Tellez, E. *Phys. Rev. B* **2001**, *64*, 155404.
(84) Hinrichs, K.; Schierhorn, A.; Haier, P.; Esser, N.; Richter, W.; Sahm, J. *Phys. Rev. Lett.* **1997**, *79*, 1094.
(85) Glinka, Y.D.; Jaroniec, M. *J. Phys. Chem. B* **1997**, *101*, 8832.
(86) Glinka, Y.D.; Jaroniec, M. *J. Appl. Phys.* **1997**, *82*, 3499.
(87) Williams, D.B.; Carter, C.B. *Transmission Electron Microscopy*; Plenum Press: New York, 1996.
(88) Reimer, L. *Transmission Electron Microscopy*, 4th ed., Springer Series in Optical Sciences: Berlin, 1997; Vol. 36.
(89) Spence, J.C.H. *Mater. Sci. Eng. R* **1999**, *26*, 1.
(90) Yacamán, M.J.; Díaz, G.; Gómez, A. *Catal. Today* **1995**, *23*, 161.
(91) Wang, Z.L. *J. Phys. Chem. B* **2000**, *104*, 1153.
(92) Thomas, J.M.; Terasaki, O.; Gai, P.L.; Zhou, W.Z.; Gonzalez-Calbet, J. *Acc. Chem. Res.* **2001**, *34*, 583.
(93) Bernal, S.; Baker, R.T.; Burrows, A.; Calvino, J.J.; Kiely, C.J.; López-Cartes, C.; Pérez-Omil, J.A.; Rodríguez-Izquierdo, J.M. *Surf. Interf. Anal.* **2000**, *29*, 411.
(94) Yacamán, M.J.; Ascencio, J.A.; Liu, H.B.; Gardea-Torresdey, J. *J. Vac. Sci. Technol. B* **2001**, *19*, 1091.

(95) Wang, Z.L. *Adv. Mater.* **2003**, *15*, 1497.

(96) Datye, A.K. *J. Catal.* **2003**, *216*, 144.

(97) Martínez-Arias, A.; Fernández-García, M.; Iglesias-Juez, A.; Anderson, J.A.; Conesa, J.C.; Soria, *J. Appl. Catal. B* **2000**, *28*, 29.

(98) Iglesias-Juez, A.; Hungría, A.B.; Martínez-Arias, A.; Fuerte, A.; Fernández-García, M.; Anderson, J.A.; Conesa, J.C.; Soria, J. *J. Catal.* **2003**, *217*, 310.

(99) Yin, J.S.; Wang, Z.L. *Phys. Rev. Lett.* **1997**, *79*, 2570.

(100) O'Keefe, M.A.; Buseck, P.R.; Iijima, S. *Nature* **1978**, *274*, 322.

(101) Bernal, S.; Botana, F.J.; Calvino, J.J.; López-Cartes, C.; Pérez-Omil, J.A.; Rodríguez-Izquierdo, J.M. *Ultramicroscopy* **1998**, *72*, 135.

(102) Klenov, D.; Kryukova, G.N.; Plyasova, L.M. *J. Mater. Chem.* **1998**, *8*, 1665.

(103) Sun, K.; Liu, J.; Browning, N.D. *J. Catal.* **2002**, *205*, 266.

(104) Wu, L.; Wiesmann, H.J.; Moodenbaugh, A.R.; Klie, R.F.; Zhu, Y.; Welch, D.O.; Suenaga, M. *Phys. Rev. B* **2004**, *69*, 125415.

(105) Zhang, F.; Wang, P.; Koberstein, J.; Khalid, S.; Chan, S.-W. *Surf. Sci.* **2004**, *563*, 74.

(106) Hansen, P.L.; Wagner, J.B.; Helveg, S.; Rostrup-Nielsen, J.R.; Clausen, B.S.; Topsoe, H. *Science* **2002**, *295*, 2053.

(107) Gai, P.L. *Topics Catal.* **2002**, *21*, 161.

(108) Young, R.; Ward, J.; Scire, F. *Rev. Scientific Inst.* **1972**, *43*, 999.

(109) Young, R.; Ward, J.; Scire, F. *Phys. Rev. Lett.* **1971**, *27*, 922.

(110) Binnig, G.; Rohrer, H. *Rev. Modern Phys.* **1987**, *59*, 615.

(111) Rohrer, H. *Ultramicroscopy* **1992**, *42-44*, 1.

(112) Hamers, R.J. *J. Phys. Chem.* **1996**, *100*, 13103.

(113) Magonov, S.N.; Whangbo, M.-H. *Surface Analysis with STM and AFM*; VCH: New York, 1996.

(114) Tersoff, J.; Hamann, D.R. *Phys. Rev. B* **1985**, *31*, 805.

(115) Hamers, R.J. *J. Vac. Sci. Technol. B* **1988**, *6*, 1462.

(116) Feenstra, R.M. *Phys. Rev. B* **1991**, *44*, 13791.

(117) Kubby, J.A.; Boland, J.J. *Surf. Sci. Reports* **1996**, *26*, 61.

(118) Sautet, P. *Chem. Rev.* **1997**, *97*, 1097.

(119) Carlisle, C.I.; King, D.; Bocquet, M.-L.; Cerdá, J.; Sautet, P. *Phys. Rev. Lett.* **2000**, *84*, 3899.

(120) Golloway, H.; Sautet, P.; Salmeron, M. *Phys. Rev. B* **1996**, *54*, R11 145.

(121) Tersoff, J.; Hamann, D.R. *Phys. Rev. Lett.* **1983**, *50*, 1998.

(122) Tersoff, J. *Phys. Rev. B* **1989**, *40*, 11990.

(123) Chen, C.J. *J. Vac. Sci. Technol. A* **1991**, *9*, 44.

(124) Tsukada, M.; Kobayashi, K.; Isshiki, N.; Kageshima, H.; Uchiyama, T.; Watanabe, S.; Schimizu, T. *Surf. Sci. Rep.* **1991**, *13*, 265.

(125) Tsukada, M.; Kageshima, H.; Isshiki, N.; Kobayashi, K. *Surf. Sci.* **1992**, *266*, 253.

(126) Sacks, W.; Noguera, C. *Ultramicroscopy* **1992**, *42–44*, 140.

(127) Tekman, E.; Ciraci, S. *Phys. Rev. B* **1989**, *40*, 10286.

(128) Sacks, W.; Noguera, C. *J. Vac. Sci. Technol. B* **1991**, *9*, 488.

(129) Sacks, W.; Noguera, C. *Phys. Rev. B* **1991**, *43*, 11612.

(130) Ciraci, S.; Baratoff, A.; Batra, I.P. *Phys. Rev. B* **1990**, *42*, 7618.

(131) Kubby, J.A.; Wang, Y.R.; Greene, W.J. *Phys. Rev. Lett.* **1991**, *65*, 2165.

(132) Diebold, U. *Surface Sci. Reports* **2003**, *48*, 53.

(133) Li, M.; Hebenstreit, W.; Gross, L.; Diebold, U.; Henderson, M.A.; Jennison, D.R.; Schultz, P.A.; Sears, M.P. *Surf. Sci.* **1999**, *437*, 173.

(134) Surnev, S.; Kresse, G.; Sock, M.; Ramsey, M.G.; Netzer, F.P. *Surf. Sci.* **2001**, *495*, 91.

(135) Surnev, S.; Kresse, G.; Ramsey, M.G.; Netzer, F.P. *Phys. Rev. Lett.* **2001**, *87*, 086102.

(136) Norenberg, H.; Briggs, G.A.D. *Phys. Rev. Lett.* **1997**, *79*, 4222.

(137) Batzill, M.; Beck, D.E.; Koel, B.E. *Phys. Rev. B* **2001**, *64*, 245402.

(138) Song, Z.; Cai, T.; Chang, Z.; Liu, G.; Rodriguez, J.A.; Hrbek, J. *J. Am. Chem. Soc.* **2003**, *125*, 8060.

(139) Valden, M.; Lai, X.; Goodman, D.W. *Science* **1998**, *281*, 1647.

(140) Hamers, R.J.; Tromp, R.M.; Demuth, J.E. *Phys. Rev. Lett.* **1986**, *56*, 1972.

(141) Hamers, R.J. *Annu. Rev. Phys.* **1989**, *40*, 531.

(142) García, R.; Pérez, R. *Surf. Sci. Reports* **2002**, *47*, 197.

(143) Martin, Y.; Williams, C.C.; Wickramasinghe, H.K. *J. Appl. Phys.* **1987**, *61*, 4723.

(144) Anselmetti, D.; Lüthi, R.; Meyer, E.; Richmond, T.; Dreier, M.; Frommer, J.E.; Güntherodt, H.-J. *Nanotechnology* **1994**, *5*, 87.

(145) Albrecht, T.R.; Grütter, P.; Horne, D.; Rugar, D. *J. Appl. Phys.* **1991**, *69*, 668.

(146) Giessibl, F.J. *Science* **1995**, *267*, 68.

(147) Sugawara, Y.; Othna, M.; Ueyama, H.; Morita, S. *Science* **1995**, *270*, 1646.

(148) Kitamura, S.; Iwatsuki, M. *Jpn. J. Appl. Phys.* **1995**, *35*, L145.

(149) Smith, R.L.; Rohrer, G.S. *J. Catal.* **1996**, *163*, 12.

(150) Smith, R.L.; Rohner, G.S. *MRS Symp. Proc.*, **1997**, *497*, 53.

(151) Chen, J.; Workman, R.K.; Sarid, D.; Hoper, R. *Nanotechnology* **1994**, *5*, 199.

(152) García, R.; San Paulo, A. *Phys. Rev. B* **1999**, *60*, 4961.

(153) Sahin, O.; Atalar, A. *Appl. Phys. Lett.* **2001**, *78*, 2973.

CHAPTER 6

Techniques for the Study of the Electronic Properties

MARCOS FERNÁNDEZ-GARCÍA

Instituto de Catálisis y Petroleoquímica, CSIC, Campus Cantoblanco, 28049 Madrid, Spain

JOSÉ A. RODRIGUEZ

Chemistry Department, Brookhaven National Laboratory, Upton, NY 11973, USA

6.1. INTRODUCTION

The electronic structure of a solid is affected by size and altered from the continuous electronic levels forming a band, characteristic of bulk or microsized solids, to discrete-like or quantized levels. This is drastically observed when the particle size goes down to the nanometer range and is the origin of the so-called "quantum confinement" terminology referring to this phenomenon. From a solid state point of view, electronic states of confined materials can be considered as being a superposition of bulk-like states with a concomitant increase of the oscillator strength (1). The valence/conduction bandwidth and position observables of a solid oxide are functions of the crystal potential, and this, in turn, is perturbed by the effect of the size in two ways: a short-range effect induced by the presence of ions with a different coordination number and bond distance, and a large-range one, induced by changes in the Madelung potential of the oxide. Theoretical analyses for oxides show a redistribution of charge when going from large periodic structures to small clusters, which is roughly considered small for ionic solids and significantly important for covalent ones (2,3). Chapter 1 describes the most recent theoretical frameworks employed to deal with these physical phenomena, whereas here we will describe their influence in physico-chemical observables obtained by spectroscopical techniques.

In brief, natural consequences of these size-induced complex phenomena are band narrowing and change of band energy (1,4). The thermodynamic limit (at 0 K) for a chemical step/reaction that can be carried out with charged particles (electron or holes

Synthesis, Properties, and Applications of Oxide Nanomaterials, Edited by José A. Rodríguez and Marcos Fernández-Garcia

located in electronic levels) is given by the position of the band edges, the so-called flat band potential. Both size-induced band-related changes modulate the onset energy where absorption occurs, e.g., the band gap energy, changing also the Fermi level of the system and the chemical potential of charge carrier species hosted by these electronic levels. An additional consequence of nanostructure when speaking of semiconductor or insulator oxides is the presence of new electronic states located in the bulk band gap. As is well known, perfect crystals do not display such electronic features and are thus directly or indirectly related to a finite size in the nanometer range. Such mid-gap states are sometimes called "surface states" as the main structural sources of their origin are the presence of a different local symmetry at the surface and/or the presence of additional defects, among which the most typical in nano-oxides are of a punctual nature (oxygen vacancies or interstitial ions) located at/near the surface (3,5,6). The energy distribution of vacancy-related mid-gap energy levels has been experimentally obtained (7–9), allowing the comparison between different samples. Roughly speaking, donor levels are energetically near the conduction band, whereas acceptor ones are located near the valence band and create Gaussian-like distributions with a width proportional to the two-thirds power of the number of defects (10). Such electronic levels display distributions ranging from 0 to a few tens (ca. 0.7) of electronvolts from the corresponding flat band potential with maxima (or depths) located sometimes at 0–100 meV (7,11) and others well above such energies (3,10,12). Both types are phenomenologically differentiated in shallow and deep (mid-gap) electronic levels and display distinctive features when charged particles (electrons or holes) interact with them (13). One last general point to mention here is the absence in nano-particulate materials of band bending under contact with the external media in a heterogeneous chemical reaction (14). This infers an important property to all nanostructured oxides under light excitation, as it facilitates the presence of both charge carrier species, electron and holes, at the surface of the particle, ready to be involved in subsequent chemical steps. This contrasts with the situation of bulk-like semiconductors where one of the charge carriers is typically depleted from the surface area.

These electronic implications of size are mainly studied by using three different techniques based on the interaction between photons and the solid phase. The first one uses (soft and hard) X-ray photons to excite internal or core electrons and to analyze the absorption coefficient as a function of the incident photon energy; this is the so-called X-ray absorption near-edge structure (XANES), which allows the scanning of the (locally projected, see below) density of unoccupied electronic states. A second type of technique is based on optical absorption of visible/ultraviolet photons with an energy that allows to study intra-atomic (typically d–d transitions in transition elements) and charge transfer (near or band gap energies) excitations; they are frequently used to study (optical) band gap energies and unoccupied localized states at the band gap of nanometer solids. A third type of technique, photo-emission, is used to analyze the density of occupied states of solids and may give information about the valence band as well as more inner levels. Several emission optical techniques may also provide information about the electronic properties of solid oxides, but usually they do not add additional input to the main electronic characteristics as studied by the three techniques mentioned above. A main problem when dealing with nanoparticulated solid oxides, which is present in all results obtained with these spectroscopical techniques, comes from the broad size distribution typically obtained during the synthesis of the material, significantly larger than those characteristic of other types of materials. This obscures the detailed investigation of the electronic properties in many occasions.

6.2. XANES

As discussed in Chapter 5 for XAS techniques, XANES spectroscopy probes the electronic state of an absorbent atom of the X-ray radiation. The two XAS techniques are sensible to local order, making them specially suitable for analyzing nanostructured materials, but they differ in that XANES is essentially not affected by thermal effects; in the absence of vibronic or specific (nonfrequent) cases of spin-orbit coupling, the recording temperature does not affect the XANES shape (15–18). Recent work on XANES concerning nanostructured oxides has been published for Al (19), Mg (20), Ti (21–25), V (26), Fe (27–29), Ni (30), Zn (31), Sn-In (32) and Ce (33–37) containing systems, whereas older work is summarized in Refs. 18 and 38. These studies showed that bulk XANES shape is typically reached by clusters of around 80–100 metal atoms, and thus, solids with a typical dimension below 4 nm show a characteristic, distinctive XANES shape. This happens because, in terms of the scattering theory, the different continuum resonances (CRs) are dominated by contributions corresponding to different coordination shells and may not be present or display variations in energy/intensity with respect to the bulk material in the corresponding XANES spectrum of a size-limited material (19,21,28–30). At the moment, not precise information is available concerning differences between surface species present in nanostructured and two-dimensional (2D)-infinite surfaces. A general feature shared by all nanostructured oxide materials with size typically below 10–15 nm is that they display broader CRs with respect to well-crystallized references (19,21,24–26,31,33,39). This can be observed in the inset of Figure 6.1; the derivative spectra give evidence of a larger overlap between CRs. The reason for this can be understood in terms of the f-rule (40). This rule implies that the total absorption over all edges and energies is a constant independent of the final state nature; therefore, disorder, inherent to nanostructured materials, broadens the spectrum without altering the overall intensity (22). The large overlap of CRs usually limits information concerning electronic effects of nanostructure, but some details are still available/resolved when analyzed low-energy (below 1 KeV) absorption edges, even for nearly amorphous materials (41,42).

A strong influence of size can be also found in the $1s \rightarrow d$ pre-edge transition present at K absorption edges ($1s$ absorption spectra; see Figure 6.1a) of transition metal cations inserted in an oxide matrix. Other elements do not present this pre-edge as they have a fully occupied d band. Both a direct $1s$ to nd quadrupole transition and dipole transitions to states with $(n+1)p$ character hybridized with the nd sub-band are possible (18). Typically, the matrix element of a dipole transition (see Eq. 5.8 in Chapter 5) is about two orders of magnitude larger than quadrupole ones, but on the

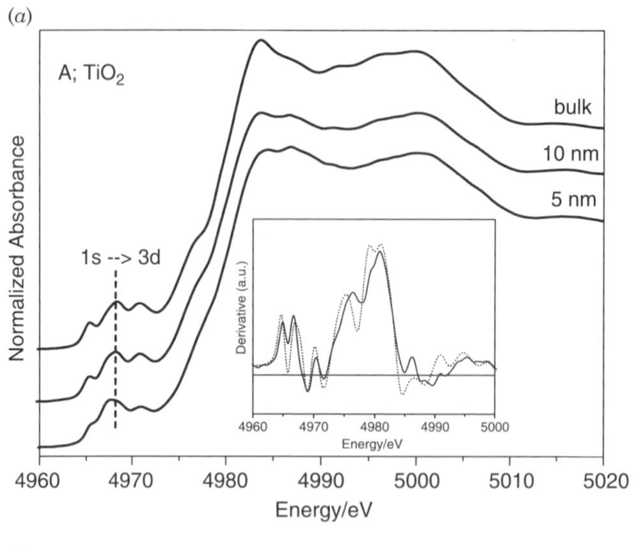

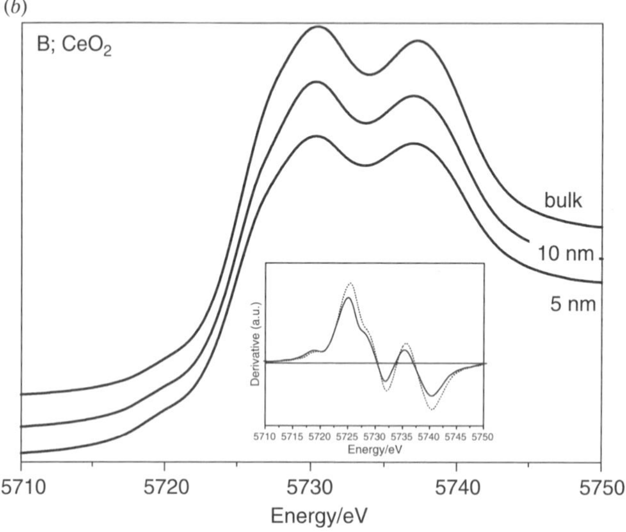

Figure 6.1. (*a*) Ti K-edge XANES spectra of bulk and nanostructured TiO_2 samples. (*b*) Ce L_{III}-edge XANES spectra of bulk and nanostructured CeO_2 samples. Inset, derivative spectra of bulk (dot line) and 5-nm size (full line) specimens.

other hand, the amount of *d* character at or immediately below of the Fermi region is by far larger than the *p* one in transition metal ions. Depending on the particular system we are dealing with, the contributions from dipole and quadrupole can be equivalent or one can be neglected. In particular, for centro-symmetric systems, like cubic systems, only a very small admixture of *p* states is allowed into the *nd* sub-band, whereas in non-centro-symmetric systems, p/d mixing is symmetry allowed. It is obvious that corners, edges, and in general, surface positions are non-centro-symmetric and can be therefore differentiated in the case of many oxides.

The $1s \rightarrow d$ transition can be composed by several peaks, depending on the symmetry and oxidation state of the metal ion constituting the oxide. Reviews of the literature can be found in Refs. 18 and 43. The electronic fine details enclosed in the pre-edge structure are, however, masked by resolution problems when working with most of the ions of chemical interest (transition or lanthanides), but the recent setting-up of resonant inelastic X-ray scattering may soon eliminate such a problem while working with hard X-rays, which allow us to characterize the system in essentially any experimental condition (e.g., in situ) (44). In the first transition row, with current setups, three/two peaks can be observed from Ti to Fe cations in centro-symmetric positions due to the $(s)p$-d mixing of the absorbent/central cation orbitals as well as to the presence of quadrupole transitions to d-like states of neighboring atoms. For some low oxidation states, low spin configurations, and non-centro-symmetric positions of the above-mentioned cations as well as from Co- to Cu-containing oxides, a single (most often dominant but not unique as is accompanied by shoulders) peak is obtained. As an example, Figure 6.1*a* displays the presence of a characteristic peak at the left of the vertical line, indicative of five-fold-coordinated surface Ti ions, not present in the bulk of the material, which therefore, allows us to investigate the characteristics as well as the quantity of surface species in Titania. Data on the second and third transition row are more scarce and usually less well defined due to poor energy resolution coming from the large core-hole lifetime inherent to the high photon energy used (above 17 KeV) (18). As mentioned, this may change in the near future with the removal of the lifetime broadening in XANES spectra by standarizing detection schemes using low-medium energy X-ray photoemission, emission decay, or Auger channels (44,45). Of course, this will be able to give larger resolution in the complete XANES spectrum, enhancing the electronic information extracted by using this technique but at the expense of a strong decrease in signal intensity. On the other hand, a detection scheme based in X-ray emission, like, for example, using the $1snp$ fluorescence channel, would open a window to site-selective XANES and the corresponding speciation of species by chemical, surface/bulk, or other interesting properties (44).

XANES can be used for the characterization of oxide nanoparticles under reaction conditions, varying pressure or temperature as a function of time (3,18). This is illustrated by the Ce L_{III}-edge XANES data shown in Figure 6.2. This figure compares spectra from a $Ce_{0.8}Cu_{0.2}O_2$ nanocatalyst under the water gas shift (WGS) reaction ($CO + H_2O \rightarrow CO_2 + H_2$) at different temperatures (37). As mentioned, the Ce L_{III}-edge is frequently used as a "fingerprint" to characterize the electronic properties of ceria-based materials (18). The electronic transitions behind these XANES features are complex. In pure stoichiometric CeO_2, the Ce L_{III}-edge exhibits two clear peaks frequently labeled B1 and C (18). A third peak, with lower photon energy than B1, can be obtained by curve fitting (18). Included in Figure 6.2 is the spectrum for a $Ce(NO_3)_3 \cdot 6H_2O$ reference, in which the cerium atoms are trivalent (Ce^{3+}). The two main peaks in the spectrum of the $Ce_{0.8}Cu_{0.2}O_2$ sample at room temperature are separated by approximately 7 eV, in agreement with results for pure

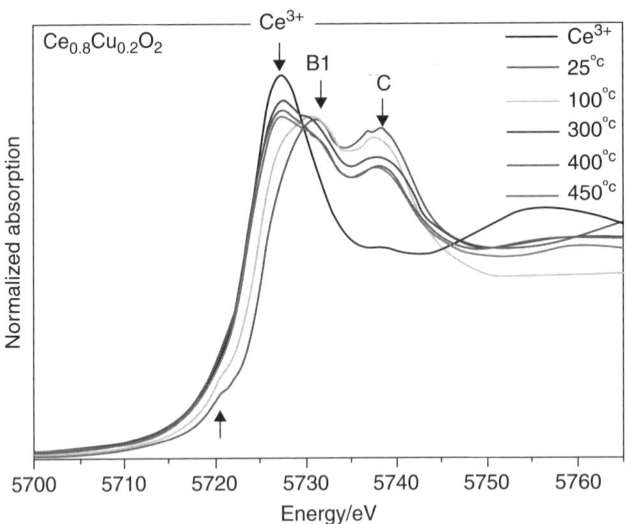

Figure 6.2. Ce L_{III}-edge XANES spectra collected for $Ce_{0.8}Cu_{0.2}O_2$ during the WGS reaction at different temperatures. Included for comparison is the spectrum for a $Ce(NO_3)_3 \cdot 6H_2O$ reference, which contains only Ce^{3+} cations (taken from Ref. 37). See color insert.

CeO_2 nanoparticles (37). Based on a comparison of the intensities of the spectra near the Ce^{3+} position, one can confirm that oxygen vacancies and Ce^{3+} cations are formed during the WGS reaction. The amount of oxygen vacancies and Ce^{3+} is seen to increase with the raise of the reaction temperature up to 300 °C, but decreased at higher temperatures, especially above 400 °C. This type of information is essential for establishing the mechanism of the chemical reaction and studying the behavior of the $Ce_{0.8}Cu_{0.2}O_2$ nanoparticles as catalysts (37). In many technological applications (catalysts, sorbents, sensors, fuel cells, etc.), the electronic properties of oxide nanoparticles undergo important changes during operation and XANES can be a very useful technique to follow these changes in situ (15,18,37,41).

6.3. OPTICAL ABSORPTION

Optical techniques can be devoted to the analysis of metal-oxide electronic structures. As a major part of oxides are semiconductors or insulators, we will first detail such cases. The absorbance $\alpha(\lambda)$ of a solid semiconductor can be calculated by using the relation:

$$T = \frac{(1 - F)^2 \exp^{-\alpha(\lambda)\delta}}{1 - F^2 \exp^{-2\alpha(\lambda)\delta}} \qquad (6.1)$$

where T is the measured transmittance, F is the reflectance, and δ is the optical path length (46). Metal oxides can be direct or indirect semiconductors depending on whether the valence to conduction band electronic transition is dipole allowed or forbidden. In the latter case, the electronic transition is vibrationally allowed through vibronic coupling and phonon assisted. The intensity of the corresponding electronic transition is several orders of magnitude lower for indirect with respect to direct band gap semiconductors (47). Indirect band gap semiconductors thus display a step structure at the absorption onset originated from the momentum-conserving optical phonon absorption and emission and a general stronger dependence of the absorption coefficient with energy near the edge (46,47). The absorption coefficient dependence of energy is stated in Eq. 6.2, where n equals one half or 2 for, respectively, direct and indirect semiconductors:

$$\alpha(\lambda) = B(\hbar\nu - E_g)^n(\hbar\nu)^{-1} \tag{6.2}$$

where B is a constant and E_g is the band gap energy.

As mentioned in Section 6.1, the nanostructure has a strong impact on the electronic structure of oxides producing the so-called quantum size of confinement effects. From a solid-state point of view, nanostructured oxides have discrete, atom-like electronic levels that can be considered as being a superposition of bulk-like states with a concomitant increase in oscillator strength (4,48). Electronic effects related to quantum confinement potentially visible by optical absorption are the shift in band gap energy as a function of primary particle size as well as the discretization of the absorption, presenting a well-defined structure instead of the featureless profile typical of bulk solids (4,48). In metal oxides, however, the influence of nanostructure in optical spectroscopy data is mostly confined to the absorption onset energy as inhomogeneous broadening resulting from the particle size distribution induces broadening of the spectrum and limits the study of the shape of the optical absorption spectrum.

The onset of the optical absorption spectrum occurs at the so-called first exciton (electron-hole pair created after light absorption) or optical band gap energy. The most widespread, simple theoretical framework to study the influence of confinement effects and, particularly, the primary particle size influence in exciton energy, is the so-called effective-mass approximation (EMA) (4,48). Other theories as the free-exciton collision model (FECM) (49) have been implemented and although are much less used may become alternatives in the near future for massive use. The EMA theory assumes parabolic energy bands, infinitely confining potential at the interface of the spherical semiconductor particle, and limits main energy terms to electron-hole interaction energy (Coulomb term) and the confinement energy of electron and hole (kinetic term). For experimental work, it is customary to identify three different energy regions as a function of the average crystalline radius (R) of the semiconductor particle:

(1) $R > a_B$, where a_B is the exciton Bohr radius of the extended/bulk semiconductor, defined as $a_B = a_e + a_h$, where a_e and a_h are the electron and hole Bohr radii, respectively. This is the regime of weak confinement, and the dominant energy is the Coulomb term; a size quantization of the motion of the

exciton occurs (electron-hole pair is treated as a quasi-particle). The energy of the lowest excited state is a function of R as

$$E(R) = E_g + \frac{\hbar^2\pi^2}{2MR^2} \tag{6.3}$$

where E_g is the band gap energy of the extended/bulk semiconductor and M is the mass of the exciton ($M = m_e^* + m_h^*; m^*$ being the effective mass of electron and hole). This indicates a blue shift of $E(R)$ as a function of R^{-2}.

(2) $R < a_B$ or strong confinement region, where the Coulomb term can be treated as a perturbation and electron and holes are treated as confined independent particles, so the exciton is not formed and separate size quantization of electron and hole is the dominant factor. In this case, $E(R)$ for the lowest excited state of a spherical cluster is given by

$$E(R) = E_g + \frac{\hbar^2\pi^2}{2R^2}\left(\frac{1}{m_e^*} + \frac{1}{m_h^*}\right) - \frac{1.786e^2}{\varepsilon_2 R} + 0.248E_R \tag{6.4}$$

where the first corrective term to E_g is the confinement term, the second is the Coulomb correction with ε_2 the dielectric constant of the bulk semiconductor, and the third is the spatial correlation energy correction, being E_R the exciton Rydberg energy (4,48,50). As is well known, this formula overestimates the $E(R)$ energy and is being corrected by using an effective bond-order model (EBOM) to construct the hole Hamiltonian, include spin-orbit coupling, and finite confining potential (51), or by introducing the effect of shape distortion from a spherical particle model (52). Alternative methods including these novel points in the framework of the tight-binding (53) or pseudopotential (54,55) theories have been also published. An important additional challenge to these theories is to develop consistent theoretical estimates of the ε_2 dependence with R (48).

(3) The region between weak and strong confinement has not been thoroughly investigated but has deserved some attempts for specific semiconductors (56). The intermediate regime is precisely defined as the region where $R_e > R_B > R_h$, and here the hole is quasi-localized and the absorption spectrum comes from the oscillation movement of the hole around the center of the nano-crystal, suffering an average potential corresponding to the much faster electron movement (48).

From the above discussion, it can be thus concluded that metal-oxide semiconductors would present, as a first rough approximation, an optical band gap energy with an inverse squared dependence of the primary particle size if quantum confinement dominates the energy behavior of the band gap. Figure 6.3 shows that this happens to be the case for (direct band gap) Fe_2O_3 (57) or (indirect band gap) CdO (58) but not for Cu_2O (59), CeO_2 (36,60), ZnO (61,62), and TiO_2 (63,64). Limited deviations

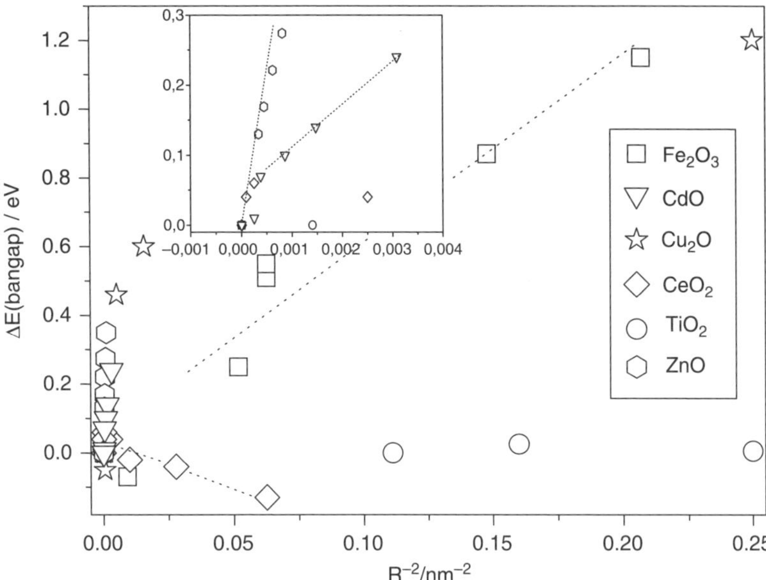

Figure 6.3. Optical band gap energy as a function of the inverse-squared primary particle size for several metal oxides.

from the R^{-2} behavior, as observed for ZnO in Figure 6.3 or SnO_2 (65) can be based on the known fact that the theory (Eqs. 6.3 and 6.4) overestimates the blue shift and can be justified with a proper calculation of electronic states by using simple quantum mechanical methods, whereas marked deviations are usually based on several chemical/physical phenomena not accounted for in the previous discussion. In the case of Cu_2O (59) or CeO_2 (36,60,66), it appears to be directly related to the presence of Cu^{2+} (remarkably for very low particle size) and Ce^{3+} ions at the surface of the nanostructured materials. At the moment, it is not clear whether the presence of these oxidation states are intrinsic to the nanostructure or result from the specific procedure of preparation. The case of WO_3 also shares some difficulties pointed out above. Kubo et al. were able to show that the band gap of this oxide decrease with size from ca. 3.0 to 2.8 eV as a function of R^{-1} (67), but the presence of a variable number of oxygen defects, reduced W redox states, and mid-gap electronic states with size makes this an open question (68). Besides electronic modifications, alien ions, like Cu^{2+} and Ce^{3+} above, induce strain effects and concomitant structural differences in atomic positions with respect to bulk positions. The influence of strain in the absorption spectrum has been nicely demonstrated in the work of Ong et al. for ZnO (69), showing the splitting of the first exciton peak for large values of compressive strain. Strain effects are inherent to nanostructured materials (3,70) and may comprise general, ambiguous term of "surface" effects usually claimed to account for significant deviations from Eqs. 6.3 and 6.4. Surface effects related to the preparation method and particularly important for very low particle size are sometimes observed for certain oxides, as SnO_2 (71) or ZrO_2 (72).

TiO_2 is the other example included in Figure 6.3 having a band gap energy behavior with marked differences from that expected from Eqs. 6.3 and 6.4. Although bulk TiO_2 is an indirect semiconductor, nanostructured TiO_2 materials are likely direct ones (63,73). This may be a general result. As discussed in Ref. 74, the confinement of charge carriers in a limited space causes their wave functions to spread out in momentum space, in turn increasing the likehood of radiative transitions for bulk indirect semiconductors. This may also be the case of NiO (75). The indirect nature of the absorption onset would complicate the analysis of the optical band gap energy due to the above-mentioned step structure of the absorption onset (which includes phonon-related absorption/emission features) (76). Despite this complication, the steady behavior shown in Figure 6.3 cannot be accounted for by small variations in the absorption onset and should be grounded in other physical phenomena. For small primary particle sizes, below 3–4 nm, this has been claimed to be based on a fortuitous cancellation effect between the kinetic/confinement and Coulomb terms in Eq. 6.4, but electronic grounds for this phenomenon still need to be understood. For indirect band gap semiconductors, recent theoretical results (54) suggest the presence of a red shift instead of a blue one with a primary particle size decrease, but recent careful experimental data corroborate the existence of the blue shift qualitatively described by Eq. 6.4 (76).

The nanostructure not only affects the band gap energy, but at the long wavelength of the charge transfer or main absorption band, a weak exponential absorbance decay can be observed. At a given temperature, this region can be modeled following the Urbach empirical formula, yielding an energy parameter describing the width of this exponential absorption edge (77). In certain oxides, the width of Urbach tails has been correlated with the degree of crystallinity, giving thus a quantitative tool to compare this property on structurally similar samples (78).

As happens with the absorption, the emission spectrum followed by photo-luminiscence or fluorescence from nanostructured samples has specific characteristics with respect to bulk systems (4,48). However, they do not usually add further insights into the main electronic structure of metal oxides and will not be treated here. A last word in this section can be said concerning semi-metallic or metallic metal oxides (as VO_2 at $T > 373$ K). Although a metal-to-insulator transition is expected as a function of size (1,4), this has not been analyzed in detail. Therefore, as a rough picture, we may expect the discretization of electronic levels and presence of well-defined transitions, which however, would be typically broadened by the "size distribution" effect.

6.4. VALENCE- AND CORE-LEVEL PHOTO-EMISSION

The roots of core- and valence-level photo-emission can be traced back to the famous article of Einstein explaining the photo-electric effect (79). For a solid, the relationship between the kinetic energy of the photoemitted electron (E_k) and the energy of the incident radiation ($h\omega$) is given by the equation:

$$E_k = h\omega - E_B - \mathrm{WF} \tag{6.5}$$

where E_B is the binding energy of the electron in the solid and WF is the system "work function" (a catch-all term whose precise value depends on both the sample and the spectrometer) (80). Equation 6.5 assumes that the photo-emission process is an elastic process between matter and the external electric field. The photo-emission process is the dominant de-excitation channel for atoms with atomic number Z lower than $30/50$ for K/L edges, whereas Auger electrons dominate de-excitation for $Z > 30/50$. Thus, each characteristic excitation energy ($h\omega$) will give rise to a series of photo-electron peaks that reflect the discrete binding energies of the electrons (with energy lower or equal than the excitation one) present in the system (80). The intensity of the peak is essentially an atomic property and thus allows the quantification of the absorber atom at the surface and near the surface layer (80).

In the period of 1950–1970, pioneer work by the groups of Siegbahn (81), Turner (82), and Spicer (83) showed that spectra with well-resolved ionization peaks could be obtained when using high-resolution electron energy spectrometers combined with X-ray or ultraviolet radiation for photo-excitation. The typical experiments of X-ray photoelectron spectroscopy (XPS) use Mg $K\alpha$ (1253.6 eV) or Al $K\alpha$ (1486.6 eV) radiation to excite the electrons (80). He I (21.21 eV) and He II (40.82 eV) are photon energies frequently used in ultraviolet photoelectron spectroscopy (84). Synchrotrons are a very valuable tool for performing photo-emission experiments (85–87). Electrons orbiting in a synchrotron emit radiation that spans a wide spectral region and is highly collimated and polarized. Using the facilities available at synchrotrons, one can get photo-emission spectra that have simultaneously a very high resolution and excellent signal-to-noise ratio (77,78). The acquisition of spectra can be very fast, and chemical transformations can be followed as a function of time (88–90).

6.4.1. Valence Photo-Emission

Depending on their energy, the levels occupied by the electrons in a nanoparticle can be labeled as core (>50 eV), semi-core, or valence (<20 eV) (80). In the valence region, the electrons occupy de-localized or bonding orbitals. The spectrum in this region consist of many closely spaced levels giving rise to a band structure (91–94). For large oxide nanoparticles, valence bands similar to those of the bulk materials are expected. As the size of the nanoparticles decreases, unique features could appear in the valence region (95,96), and eventually the valence bands could break down into discrete levels in the case of small oxide clusters (97). In their bulk states, some oxides can have metallic character, but most frequently these systems behave as semiconductors or insulators and have a significant gap between the valence and the conduction bands. Due to size effects, a nanoparticle could have a band gap energy differing from that of its bulk state (see Section 6.3 for a detailed analysis), and this could be detected in valence photo-emission (96–98). This technique is also very useful to verify the presence of O vacancies in an oxide nanoparticle. Electronic states associated with O vacancies usually appear above the valence band of the oxide in the band gap (99,100).

6.4.2. Core-level Photoemission

The discovery, during the early days of XPS (80,81), that nonequivalent atoms, either by symmetry or oxidation state, of the same element in a solid gave rise to core-level peaks with different binding energies had a stimulating effect on the development of the field (80). Core-level photo-emission can be very sensitive to changes in the oxidation state of an element in an oxide nanoparticle (3). Thus, different oxides of the same metal element in many cases have substantially different core-level binding energies (101). Figure 6.4 illustrates the utility of core-level photo-emission for the characterization of oxide nanoparticles. These XPS data were acquired in experiments similar to those that produced the STM images of Figure 5.8 in Chapter 5 (102). MoO_3 and Mo_2O_5 were prepared by the oxidation of Mo nanoparticles present on a Au(111) substrate. Initially, there is intermixing of Mo and Au. The formation of the MoO_x nanoparticles induces migration of Mo from inside the Au substrate to the surface, and thus, there is a big increase in the intensity of the signal for the Mo 3*d* core levels (102,103). A Mo → MoO_3 transformation is accompanied by a shift of ~4.3 eV in the Mo 3*d* levels. At small concentrations of molybdenum (~0.05 monolayer), the

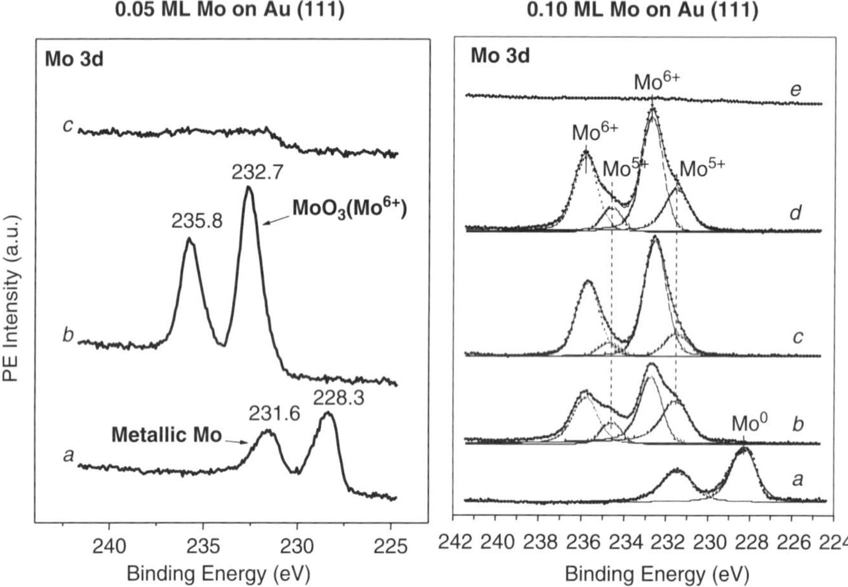

Figure 6.4. Left-side panel: Mo 3*d* photo-emission spectra taken from 0.05 monolayer (ML) of Mo deposited on a Au(111) substrate: (*a*) before oxidation of the Mo nanoparticles, (*b*) after oxidation at 500 K, and (*c*) upon heating the sample to 700 K. Right-side panel: Mo 3*d* spectra taken for a Au(111) substrate covered by 0.10 ML of molybdenum: (*a*) before oxidation, (*b*) after a first oxidation at 500 K, (*c*) after a second oxidation at 500 K, (*d*) upon heating the oxidized sample to 700 K, and (*e*) after final heating to 800 K (taken from Ref. 102).

Mo nanoparticles can be fully oxidized (left-side panel in Figure 6.4, spectrum *b*). On the other hand, after oxidizing a larger concentration of molybdenum (~0.10 monolayer), photo-emission shows a mixture of MoO_3 and Mo_2O_5 (right-side panel in Figure 6.4, spectra *b*, *c*, and *d*) (102).

Core-level binding-energy shifts constitute one of the most widely used diagnostic tools for routine chemical analysis in industrial laboratories (104). A lot of effort has been focused on the interpretation of these shifts (80,104). A core-level binding energy is fundamentally a difference in total energy between the ground and the unbound continuum excited states of a system. This difference can have initial- and final-state contributions (104). To induce a core-level shift with respect to a well-defined reference, one can modify the initial or final state. Thus, a change in the oxidation state of an element can lead to a shift in its core-level binding energies (initial state effect), but a core-level shift also can be produced by a variation in the screening of the hole left by the emitted electron (final state effect) (104). A priori, it is not easy to evaluate final-state contributions to core-level binding-energy shifts (80,104).

Several models, based on physical simplifications of the photo-emission process, have been proposed for the analysis of core-level shifts in oxide compounds (104–111). The physical basis for a core-level shift can be illustrated by a simple charge potential model (80,81):

$$E_i = E_i^R + kq_i + \Sigma 3q_i/r_{ij} \tag{6.6}$$

where E_i is the binding energy of a particular core level on atom *i*, $E_i{}^R$ is an energy reference, q_i is the charge on atom *i*, and the final term of Eq. 6.6 sums the potential at atom *i* due to "point charges" on surroundings atoms *j*. The last term can be expressed as $V_i = \Sigma 3q_i/r_{ij}$. Then, the shift in binding energy for a given core level of atom *i* in two different chemical environments is (80)

$$E_i^{(1)} - E_i^{(2)} = k(q_i^{(1)} - q_i^{(2)}) + (V_i^{(1)} - V_i^{(2)}) \tag{6.7}$$

The first term, $k\Delta q_i$, indicates that an increase in binding energy accompanies a decrease in the valence electron density on atom *i*. The second term depends on the charge distribution of the whole system, is related to the Madelung potential of the oxide, and has an opposite sign to Δq_i in Eq. 6.7 (80,81). For highly ionic oxides, the effects of the Madelung potential can be quite strong and dominate the overall direction of the core-level shifts (80,104). Thus, Hartree–Fock self-consistent field calculations for BaO clusters show that the removal of two *s* electrons leads to an increase of the core-level binding energies in Ba^{2+} over neutral Ba (as expected), but this increase is offset by the Madelung potential and the XPS measurements show a net negative core-level shift. The dominance of these two electrostatic effects in the core-level shifts of all alkaline-earth oxides was also demonstrated (106). From Eq. 6.7, it is clear that one can have a core-level shift in an oxide nanoparticle by changing the oxidation state of atom i, by modifying the Madelung potential, by the creation of O vacancies, or by just moving the atom i to a nonequivalent position (like surface ones) within the structure of the system. It is important to take into consideration that the Madelung potential of a nanoparticle or nanostructure of an oxide can be different from that of the bulk material and affect not only the position but also the intensity of the peaks in an XPS spectrum (112,113).

Equation (6.7) is a first approximation to the problem of core-level shifts and in general should be used only in qualitative terms (80,104). A major simplification of the charge potential model is that it neglects relaxation effects; e.g., it does not take into account the polarization effect of the core hole on the surrounding electrons, both intra-atomic (on atom i) and extra-atomic (on atoms j) (80,104). A precise interpretation of core-level binding energy shifts frequently requires the use of quantum-mechanical calculations (105,106,111). This is the best way to separate contributions from initial and final state effects (114–116). Initial state effects are frequently calculated by the Koopman's theorem, which essentially depends on the "sudden approximation," explained in Section 5.3, to describe the photo-emission process as being realized by one electron, whereas the remaining $N - 1$ electrons of the system are "frozen." The use of the full electron (final-state) wave functions yields the core-hole energy. Note that several final-state wave functions corresponding to a localized/delocalized nature of the hole and the existence of hole-induced charge transfer ($M^{(n+1)+} - L^{-1}$) are usually possible. The latter typically produces the presence of secondary peaks, close to the main but with lower intensity, for transition metals oxides having d holes in the final state. The difference between these two quantities, e.g., the initial state and the core-hole energies, is the relaxation energy (80,104). The relaxation energy may vary from one oxide to another and with the size of the particle, because it is affected by the Madelung potential and this changes with size (104,113).

Charging is a phenomenon that can occur when the conductivity of the oxide nanoparticles or nanostructures is too low to replenish the photo-emitted electron (80,104). This means that a significant net positive charge accumulates in the system and eventually produces a shift in the binding energy position of the core and valence levels. Unfortunately, a fully satisfactory solution to charging problems is not yet available. Several approaches have been proposed (80,104). A flood gun is one of the easiest and most common solutions to charging problems. It produces a huge increase in the concentration of low-energy electrons. One must verify that these "extra" electrons do not induce any structural or chemical transformation in the sample. Whenever possible, it is highly advantageous to deposit the oxide nanoparticles on a conducting support (117,118). During the photo-emission process, electrons spill from the support and neutralize the charging effects. The data in Figure 6.4 show how effective this approach can be (102).

The low-pressure conditions in which XPS spectra are taken ($<10^{-7}$ Torr) can induce the loss of O_2 gas and formation of O vacancies in oxide nanoparticles (112). This can become a serious problem if the oxide system is easy to reduce. For example, measurements of XPS for ceria nanoparticles give a concentration of Ce^{3+} cations that is much larger than that detected with XANES (112).

ACKNOWLEDGMENT

J.A.R. is grateful for the financial support of the U.S. Department of Energy, Division of Chemical Sciences (DE-AC02-98CH10886 contract). M.F.G. thanks the Spanish "Ministerio de Ciencia y Tecnología" and CSIC for financial support.

REFERENCES

(1) Moriarty, P. *Rep. Prog. Phys.* **2001**, *64*, 297.
(2) Noguera, C. *Physics and Chemistry at Oxide Surfaces*; Cambridge University Press: Cambridge, UK, 1996.
(3) Fernandez-García, M.; Martínez-Arias, A.; Hanson, J.C.; Rodriguez, J.A. *Chem. Rev.* **2004**, *104*, 4063.
(4) Yofffre, A. *Adv. Phys.* **1993**, *42*, 173.
(5) Al-Abadleh, H.A.; Grassian, V.H. *Surf. Sci. Report* **2003**, *52*, 63.
(6) Dietbold, U. *Surf. Sci. Report.* **2003**, *48*, 53.
(7) Ikeda, S.; Sugiyama, N.; Murakami, S.; Kominami, H.; Kera, Y.; Noguchi, H.; Uosaki, K.; Torimoto, T.; Ohtani, B. *Phys. Chem. Chem. Phys.* **2003**, *5*, 778.
(8) Boschloo, G.; Fitsmaurice, D. *J. Phys. Chem. B* **1999**, *103*, 228.
(9) Durrant, J.R.; *J. Photchem. Photobiol. A* **2002**, *148*, 5.
(10) Cohen, M.M. *Introduction to the Quantum Theory of Semiconductors*; Gordon: Amsterdam, 1972.
(11) Durrant, J.R. *J. Photchem. Photobiol. A* **2002**, *148*, 5.
(12) Boschloo, G.; Fitsmaurice, D. *J. Phys. Chem. B* **1999**, *103*, 228.
(13) Mora-Seró, I.; Busquet, J. *Nanoletters* **2003**, *3*, 945.
(14) Grätzel, M. *Heterogeneous Photochemical Electron Tranfer*; CRC: Boca Raton, FL, 1989; 106–108.
(15) *X-ray Absorption*; Konnigsberger, D.C.; Prins, R. (Editors); Wiley: New York, 1987.
(16) See articles in *Chem. Rev.* **2001**, *101*.
(17) Rehr, J.J.; Albers, J.C. *Rev. Modern Phys.* **2000**, *72*, 621.
(18) Fernández-García, M. *Catal. Rev. Sci. Eng.* **2002**, *44*, 59.
(19) Cabaret, D.; Sainterit, P.; Idelfonse, P.; Flank, A.M. *J. Phys. Condens. Matter* **1996**, *8*, 3691.
(20) Mozoguchi, T.; Tanaka, I.; Yoshida, M.; Oba, F.; Ogasawara, K.; Adachi, H. *Phys. Rev. B* **2000**, *61*, 2180.
(21) Wu, Z.Y.; Ouward, G.; Gressier, P.; Natoli, C.R. *Phys. Rev. B* **1997**, *5*, 10383.
(22) Farges, F; Brown, G.E., Jr.; Rehr, J.J. *Phys. Rev. B* **1997**, *56*, 1809.
(23) Chen, L.X.; Rajh, T.; Wang, Z.; Thurnauer, M.C. *J. Phys. Chem. B* **1997**, *101*, 10688.
(24) Yeung, K.L.; Maira, A.J.; Stolz, J.; Hung, E.; Hu, N.K.-C.; Wei, A.C.; Soria, J.; Chao, K.-J. *J. Phys. Chem. B* **2002**, *106*, 4608.
(25) McCornick, J.R.; Zhao, B.; Rykov, S.; Wang, H.; Chen, J.G. *J. Phys. Chem. B* **2004**, *108*, 17398.

(26) Sipr, O.; Simunek, A.; Bocharov, S.; Kirchner, Th.; Drager, G. *Phys. Rev. B* **1999**, *60*, 14115.

(27) Wu, Z.Y.; Gota, S.; Jollet, F.; Pollak, M.; Gautier- Soyer, M.; Natoli, C.R. *Phys. Rev. B* **1997**, *55*, 2570.

(28) Chen, L.X.; Liu, T.; Thurnauer, M.; Csencsits, R.; Rajh, T. *J. Phys. Chem. B* **2002**, *106*, 8539.

(29) Wu, Z.; Xian, D.C.; Natoli, C.R.; Marcelli, A.; Paris, E.; Mollans, A. *Appl. Phys. Lett.* **2001**, *79*, 1918.

(30) Van der Laan, G.V.; Zaanen, J.; Sawatzky, G.A.; Karnatak, R.; Estern, J.M. *Phys. Rev. B* **1986**, *33*, 4253.

(31) Chiou, J.W.; et al. *Appl. Phys. Lett.* **2004**, *85*, 3220.

(32) Subramanian, V.; Gnanasekar, K.I.; Rambabu, B. *Solid State Ionics* **2004**, *175*, 181.

(33) Nachimuthu, P.; Shih, W.-C.; Liu, R.-S.; Jang, L.-Y.; Chem, J.-M. *J. Sol. State Chem.* **2000**, *149*, 408.

(34) Nagai, Y.; Yamamoto, T.; Tanaka, T.; Yoshida, S.; Nonaka, T.; Okamoto, T.; Suda, A.; Sugura, M. *Catal. Today* **2002**, *74*, 225.

(35) Soldatov, A.V.; Ivanchenko, T.S.; Della Longa, S.; Kotani, A.; Iwamoto, Y.; Bianconi, A. *Phys. Rev. B* **1994**, *50*, 5074.

(36) Hernández-Alonso, M.D.; Hungría, A.B.; Coronado, J.M.; Martínez-Arias, A.; Conesa, J.C.; Soria, J.; Fernández-García, M. *Phys. Chem. Chem. Phys.* **2004**, *6*, 3524.

(37) Wang, X.; Rodriguez, J.A.; Hanson, J.C.; Gamarra, D.; Martínez-Arias, A.; Fernández-García, M. *J. Phys. Chem. B* **2006**, *110*, 428.

(38) Kizler, P. *Phys. Lett. A* **1992**, *172*, 66.

(39) Asuka, K.; Iwasawa, Y. In *X-ray Absorption Fine Structure for Catalysts and Surfaces*; Iwasawa, Y. (Editor); World Scientific: Singapore, 1996.

(40) Altearelli, M.; Dexter, D.L.; Nussenzveig, H.M. *Phys. Rev. B* **1972**, *6*, 4502.

(41) Chen, J.G. *Surf. Sci. Rep.* **1997**, *30*, 1.

(42) Sokucheyer, G.; et al. *Phys. Rev. B* **2004**, *69*, 245102.

(43) De Groot, F.M.F. *Chem. Rev.* **2001**, *101*, 1779.

(44) De Groot, FMF.; et al. *J. Phys. Chem. B* **2005**, *109*, 20751.

(45) Hamalainen, K.; Siddons, P.; Hastingsm, J.B.; Berman, L.E. *Phys. Rev. Lett.* **1991**, *67*, 2850.

(46) Pannkov, J.I. *Optical Processes in Semiconductors*; Dover: New York, 1971.

(47) Alivisatos, A.P. *J. Phys. Chem.* **1996**, *100*, 13226.

(48) Yoffe, A.D. *Adv. Phys.* **2001**, *50*, 1.

(49) Glinka, Y.D.; Lin, S.H.; Hwang, L.P.; Chen, Y.T.; Tolk, N.H. *Phys. Rev. B* **2001**, *64*, 085421.

(50) Bawendi, M.G.; Wilson, W.L.; Rothberg, L.; Caroll, P.J.; Jedji, T.M.; Steigerward, M.L.; Brus, L.E. *Phys. Rev. Lett.* **1990**, *65*, 1623.

(51) Laheld, V.E.H.; Einevoll, G.T. *Phys. Rev. B* **1997**, *55*, 5184.

(52) Efros, A.L.; Rosen, M.; Kuno, M.; Normal, M.; Norris, D.J.; Bawendi, M. *Phys. Rev. B* **1996**, *54*, 4843.

(53) Von Grumber, H.H. *Phys. Rev. B* **1997**, *55*, 2293.

(54) Tomasulo, A.; Ramakrisna, M.V. *J. Chem. Phys.* **1996**, *105*, 3612.

(55) Wang, L.-W.; Kim, J.; Zunger, A. *Phys. Rev. B* **1999**, *59*, 5678.

(56) Uzomi, T.; Kayanuma, Y.; Yamanaka, K.; Edamatsu, K.; Itoh, T. *Phys. Rev. B* **1999**, *59*, 9826.

(57) Iwamoto, M.; Abe, T.; Tachibana, Y. *J. Mol. Catal. A* **2000**, *55*, 143.

(58) Vigil, O.; Cruz, F.; Morales-Acabedo, A.; Contreras-Puente, G.; Vaillant, L.; Santana, G. *Mat. Chem. Phys.* **2001**, *68*, 249.

(59) Borgohain, K.; Morase, N.; Mahumani, S. *J. Appl. Phys.* **2002**, *92*, 1292.

(60) Suzuki, T.; Kosacki, I.; Petrovsky, V.; Anderson, H.U. *J. Appl. Phys.* **2001**, *91*, 2308.

(61) Viswanaha, R.; Sapra, S.; Satyani, B.; Der, B.N.; Sarma, D.D. *J. Mat. Sci.* **2004**, *14*, 661.

(62) Li, L.; Qui, X.; Li, G. *J. Appl. Phys.* **2005**, *87*, 124101.

(63) Serpone, N.; Lawless, D.; Khairutdinov, R. *J. Phys. Chem.* **1995**, *98*, 16646.

(64) Monticone, S.; Tufeu, R.; Kanaev, A.V.; Scolan, E.; Sánchez, C. *Appl. Surf. Sci.* **2000**, *162–163*, 565.

(65) Deng, H.; Hossenlopp, J.M. *J. Phys. Chem. B* **2005**, *109*, 66.

(66) Tsunekawa, S.; Fukuda, T.; Kasuya, A. *J. Appl. Phys.* **2000**, *87*, 1318.

(67) Kubo, T.; Nishikitani, Y. *J. Electrochem. Soc.* **1998**, *145*, 1729.

(68) Karazhanov, S.Zh.; Zhang, Y.; Wang, L.W.; Masacarenas, A.; Deb, S. *Phys. Rev. B* **2003**, *68*, 233204.

(69) Ong, H.C.; Zhu, A.X.E.; Du, G.T. *Appl. Phys. Lett.* **2002**, *80*, 941.

(70) Fernández-García, M.; Wang, X.; Belver, C.; Iglesias-Juez, A.; Hanson, J.C.; Rodriguez, J.A. *Chem. Mater.* **2005**, *17*, 4181.

(71) Kang, J.; Tsunekawa, S.; Kasuya, A. *Appl. Surf. Sci.* **2001**, *174*, 306.

(72) Kosacki, I.; Petrovsky, V.; Anderson, H.U. *Appl. Phys. Lett.* **1999**, *74*, 341.

(73) Reddy, K.M.; Manorama, S.V.; Reddy, A.R. *Mater. Chem. Phys.* **2002**, *78*, 239.

(74) Iyer, S.S.; Xie, Y.-H. *Science* **1993**, *260*, 40.

(75) Mahmond, S.A.; Akl, A.A.; Kand, H.; Adbel-Had, K. *Physica B* **2002**, *111*, 366.

(76) Matsumoto, T.; Suzuki, J.-I.; Ohmura, M.; Kanemitsu, Y.; Masumoto Y. *Phys. Rev. B* **2001**, *63*, 195322.

(77) Urbach, F. *Phys. Rev.* **1953**, *92*, 1324.

(78) Melsheiner, J.; Ziegler, D. *Thin Solid Films* **1985**, *129*, 35.

(79) Einstein, A. *Ann. Phys. Leipzig* **1905**, *17*, 132.

(80) Briggs, D.; Seah, M.P. *Practical Surface Analysis by Auger and X-ray Photoelectron Spectroscopy*; Wiley: New York, 1983.

(81) Siegbahn, K.; Nordling, C.; Fahlman, A.; Nordberg, H.; Hamrin, K.; Hedman, J.; Johansson, G.; Bergmark, T.; Karlsson, S.E.; Lindgren, J.; Lindberg, B. *Electron Spectroscopy for Chemical Analysis*; Almquist and Wiksells: Stockholm, 1967.

(82) Turner, D.W.; Baker, C.; Baker, A.D.; Brundle, C.R. *Molecular Photoelectron Spectroscopy*; Wiley: London, 1970.

(83) Spicer, W.E. *Phys. Rev.* **1962**, *125*, 1297.

(84) Rabalais, J.W. *Principles of Ultraviolet Photoelectron Spectroscopy*; Wiley: New York, 1977.

(85) Hrbek, J. In *Chemical Applications of Synchrotron Radiation*; Sham, T.K. (Editor); World Scientific: Singapore, 2002.

(86) *Synchrotron Radiation, Techniques and Applications*; Kunz, C. (Editor); Springer-Verlag: Berlin, 1979.

(87) Ferrer, S.; Petroff, Y. *Surf. Sci.* **2002**, *500*, 605.

(88) Polcik, M.; Wilde, L.; Haase, J.; Brenna, B.; Cocco, D.; Comelli, G.; Paolucci, G. *Phys. Rev. B* **1996**, *53*, 13720.

(89) Hrbek, J.; Yang, Y.W.; Rodriguez, J.A. *Surf. Sci.* **1993**, *296*, 164.

(90) Lee, A.F.; Wilson, K.; Middleton, R.L.; Baraldi, A.; Goldoni, A.; Paolucci, G.; Lambert, R.M. *J. Am. Chem. Soc.* **1999**, *121*, 7969.

(91) Soriano, L.; Abbate, M.; Fernández, A.; González-Elipe, A.R.; Sirotti, F.; Rossi, G.; Sanz, J.M. *Chem. Phys. Lett.* **1997**, *266*, 184.

(92) Smith, K.E.; Mackay, J.L.; Henrich, V.E. *Phys. Rev. B* **1987**, *35*, 5822.

(93) Rodriguez, J.A.; Pérez, M.; Jirsak, T.; González, L.; Maiti, A.; Larese, J.Z. *J. Phys. Chem. B* **2001**, *105*, 5497.

(94) Soriano, L.; Abbate, M.; Fernández, A.; González-Elipe, A.R.; Sirotti, F.; Sanz, J.M. *J. Phys. Chem. B* **1999**, *103*, 6676.

(95) Soriano, L.; Abbate, M.; Fernández, A.; González-Elipe, A.R.; Sirotti, F.; Rossi, G.; Sanz, J.M. *Chem. Phys. Lett.* **1997**, *266*, 184.

(96) Soriano, L.; Abbate, M.; Fernández, A.; González-Elipe, A.R.; Sirotti, F.; Sanz, J.M. *J. Phys. Chem. B* **1999**, *103*, 6676.

(97) Rodriguez, J.A.; Campbell, C.T. *J. Phys. Chem. B* **1987**, *91*, 6648.

(98) Chaturvedi, S.; Rodriguez, J.A.; Hrbek, J. *J. Phys. Chem. B* **1997**, *101*, 10860.

(99) Liu, G.; Rodriguez, J.A.; Hrbek, J.; Dvorak, J.; Peden, C.H.F. *J. Phys. Chem. B* **2001**, *105*, 7762.

(100) Rodriguez, J.A.; Azad, S.; Wang, L.-Q.; García, J.; Etxeberria, A.; González, L. *J. Chem. Phys.* **2003**, *118*, 6562.

(101) Wagner, C.D.; Riggs, W.M.; Davis, L.E.; Moulder, J.F.; Muilenberg, G.E. *Handbook of X-ray Photoelectron Spectroscopy*; Perkin-Elmer: Eden Prairie, MN, 1978.

(102) Song, Z.; Cai, T.; Chang, Z.; Liu, G.; Rodriguez, J.A.; Hrbek, J. *J. Am. Chem. Soc.* **2003**, *125*, 8060.

(103) Liu, P.; Rodriguez, J.A.; Muckerman, J.T.; Hrbek, J. *Phys. Rev. B* **2003**, *67*, 155416.

(104) Egelhoff, W.G. *Surf. Sci. Rep.* **1987**, *6*, 253.

(105) Yeh, J.-J.; Lindau, I.; Sun, J.-Z.; Char, K.; Missert, N.; Kapitulnik, A.; Geballe, T.H.; Beasley, M.R. *Phys. Rev. B* **1990**, *42*, 8044.

(106) Bagus, P.S.; Pacchioni, G.; Sousa, C.; Minerva, T.; Parmigiani, F. *Chem. Phys. Lett.* **1992**, *196*, 641.

(107) Wertheim, G.K. *J. Electron. Spectry.* **1984**, *34*, 309.

(108) Hill, D.M.; Meyer, H.M.; Weaver, J.H. *Surf. Sci.* **1984**, *225*, 63.

(109) Vazquez, R.P. *J. Electron. Spectry.* **1991**, *56*, 217.

(110) Sangaletti, L.; Depero, L.E.; Bagus, P.S.; Parmigiani, F. *Chem. Phys. Lett.* **1995**, *245*, 463.

(111) Bagus, P.S.; Freund, H.J.; Minerva, T.; Pacchioni, G.; Parmigiani, F. *Chem. Phys. Lett.* **1996**, *251*, 90.

(112) Zhang, F.; Wang, P.; Koberstein, J.; Khalid, S. *Surf. Sci.* **2004**, *563*, 74.

(113) Perebeinos, V.; Chan, S.-W.; Zhang, F. *Solid State Común.* **2002**, *123*, 295.

(114) Bagus, P.S.; Pacchioni, G.; Parmigiani, F. *Chem. Phys. Lett.* **1993**, *207*, 569.

(115) Hennig, D.; Ganduglia-Pirovano, M.V.; Scheffler, M. *Phys. Rev. B* **1996**, *53*, 10344.

(116) Wu, R.; Freeman, A.J. *Phys. Rev. B* **1995**, *52*, 12419.

(117) Wu, M.-C.; Truong, C.M.; Goodman, D.W. *Phys. Rev. B* **1992**, *46*, 12688.

(118) Borgohain, K.; Singh, J.B.; Rao, M.V.M.; Shripathi, T.; Mahamuni, S. *Phys. Rev. B* **2000**, *61*, 11093.

CHAPTER 7

Post Hartree-Fock and Density Functional Theory Formalisms

FRANCESC ILLAS

Departament de Química Física and Centre especial de Recerca en Química Teòrica, Universitat de Barcelona and Parc Científic de Barcelona, C/ Martí i Franquès 1, E-08028 Barcelona, Spain

GIANFRANCO PACCHIONI

Dipartimento di Scienza dei Materiali, Università di Milano-Bicocca, via R. Cozzi, 53-I-20125 Milano, Italy

7.1. INTRODUCTION

This chapter provides a unified point of view of the different theoretical methods used in the study of electronic structure and a survey of recent representative application on oxide particles. Depending on the computer power, these methods can be applied to a wide number of systems, although applications to oxide nanoparticles are still scarce. This chapter focuses on two families of methods broadly used in computational material science. Both families are based on the principles of quantum mechanics. On the one hand, one has the so-called *ab initio* methods of electronic structures aimed to obtain accurate electronic wave functions and, on the other hand, the methods based on the modern density functional theory are especially well suited to describe very large systems. The general "first principles" term is used to refer to these methods, although the "*ab initio*" term is usually reserved for wave function methods. In the forthcoming discussion, the focus lies mainly on the mathematical foundation and physical significance of these methods rather than on the technical aspects of computer implementation. Details of the methods outlined in this section can be found in specialized references, monographs (1), and textbooks (2,3).

Finally, we would like to point out that one of the main goals of the computational methods described below is to explore potential energy surfaces and, thus, determine geometries of stable molecules, clusters, or nanoparticles but also of intermediates,

Synthesis, Properties, and Applications of Oxide Nanomaterials, Edited by José A. Rodríguez and Marcos Fernández-Garcia

transition state structures, energy barriers and thermochemical properties. This means that often one does not only need to compute accurate energies but also energy gradients and second derivatives with respect to nuclear displacements. Energy derivatives are not trivial, and some methods offer special technical advantages when gradients or higher order derivatives have to be computed. In addition, the proper interpretation of electronic spectra requires simultaneously handling of several electronic states and not only the ground state. The choice of a particular computational method largely depends on the problem to be solved and the computer power at hand. In any case, there is always a compromise between accuracy and feasibility.

7.2. QUANTUM MECHANICS AND WAVE FUNCTIONS

We start this description of methods of electronic structure widely used in either quantum chemistry or solid-state physics by recalling the fundamental aspects of quantum mechanics that are relevant to this purpose; for a deeper insight, see Refs. 4–6. The postulates of quantum mechanics establish that, at a given time t, the state of every physical system is completely described by a state or ket vector, $|\Psi_i\rangle$, belonging to a well-defined vector space, i.e., the Hilbert space denoted by $\boldsymbol{H}$. A complete description of the Hilbert space is beyond the scope of this work; we will just mention that $\boldsymbol{H}$ is a vector space defined on the complex numbers set; it has an inner or scalar product defined, infinite dimension and is complete. This last property is especially important and means that, for any ket vector $|\Psi_i\rangle \in \boldsymbol{H}$, a family of, in principle, infinite and complex numbers, $\{c_{ki}\}$, and an infinite family of ket vectors, $\{|\Phi_k\rangle\}$, exist such that

$$|\Psi_i\rangle = \sum c_{ki}|\Phi_k\rangle \tag{7.1}$$

This state is also called a pure state. As the dot or inner product is well defined in the Hilbert space, the $\{|\Phi_k\rangle\}$ set can be chosen as an orthonormal set

$$\langle\Phi_k|\Phi_l\rangle = \delta_{kl} \quad (1 \text{ for } k = l, \text{ and } 0 \text{ otherwise}) \tag{7.2}$$

without any lack of generality. Now, notice that a simple relationship for c_{ki} can be found by multiplying Eq. 7.1 by $|\Phi_k\rangle$ on the left and assuming that the basis forms an orthonormal set:

$$\langle\Phi_k|\Psi_i\rangle = c_{ki} \tag{7.3}$$

We notice that Eq. 7.1 defines the well-known superposition principle of quantum mechanics. Strictly, the sum on Eq. 7.1 must extend up to infinite because the resolution of the identity in the Hilbert space, within an orthonormal basis, requires

$$\sum |\Psi_i\rangle\langle\Psi_i| = \sum |\Phi_i\rangle\langle\Phi_i| = \hat{1} \tag{7.4}$$

Later on, we will discuss the practical way in which computational methods of the electronic structure, either in quantum chemistry or in solid-state physics, overcome the infinite summation. Now, let us just recall that the goal of any such computational method is to obtain approximate solutions to the eigenvalue problem defined by the Schrödinger time-independent equation

$$\hat{\mathrm{H}}|\Psi_i\rangle = E_i|\Psi_i\rangle \tag{7.5}$$

where $\hat{\mathrm{H}} = \hat{\mathrm{T}} + \hat{\mathrm{V}}_{\mathrm{Ne}} + \hat{\mathrm{V}}_{\mathrm{ee}}$ is the electronic Hamiltonian that is a hermitic operator defined in $\boldsymbol{H}$. In the framework of density functional theory (*vide infra*), the second term is referred to as the external potential. In this way, the first and third term of the electronic Hamiltonian only depend on electronic coordinates. Another important remark is that the electronic Hamiltonian only contains one- and two-body operators, whereas $|\Psi_i\rangle$ contains information about the N electrons in the system. Therefore, one can obtain the total energy by making use of the one- and two-particle density matrices only, which are obtained by integrating over all coordinates of $N - 1$ and $N - 2$ electrons, respectively (7,8). The one- and two-particle density matrices can be obtained from the N-electronic wave function. However, density functional theory provides a wonderful and practical alternative in which only the one-particle density matrix is needed. Both wave function- and density functional-based methods will be described in the forthcoming subsections.

Clearly, solving Eq. 7.5 for $|\Psi_i\rangle$ is equivalent to finding the family of—in principle infinite—$\{c_{ki}\}$ coefficients in Eq. 7.1, and this in turn requires the knowledge of the N-electron basis. Recalling that the eigenstates of a system of N noninteracting electrons supply a basis of the same Hilbert space to which the eigenstates of a system of N interacting electrons belong provides a way to the construction of the N-electron basis. Rigorously speaking, the Hilbert space of a two-electron system, interacting or not, can be constructed as a tensor product of the Hilbert spaces of the individual electrons

$$\boldsymbol{H}(1,2) = \boldsymbol{H}(1) \otimes \boldsymbol{H}(2) \tag{7.6}$$

and the particular vectors of $\boldsymbol{H}(1,2)$, which can be written as a tensor product of eigenstates of $\boldsymbol{H}(1)$ and $\boldsymbol{H}(2)$

$$|\Phi_{ij}(1,2)\rangle = |\Phi_i(1)\rangle \otimes |\Phi_j(2)\rangle = |\Phi_i(1)\Phi_j(2)\rangle = |\Phi_i\Phi_j\rangle \tag{7.7}$$

are eigenstates of the Hamiltonian for two noninteracting electrons

$$\hat{\mathrm{H}}(1,2) = \hat{\mathrm{H}}(1) + \hat{\mathrm{H}}(2) \tag{7.8}$$

with eigenvalues

$$E_{ij}(1,2) = E_i(1) + E_j(2) \tag{7.9}$$

as expected for noninteracting particles. The one-electron states are generally known because the Schrödinger equation, Eq. 7.5, for one-electron systems can be analytically solved. In the case of one-electron atoms or ions, the resulting one-electron states are the well-known atomic orbitals. Now, one realizes that the set of vector states described in Ref. 8 are also a basis set for the states of the two interacting electrons. Therefore, following the superposition principle, Eq. 7.1 may written as

$$|\Psi_k\rangle = \sum c_{ij,k}|\Phi_i\Phi_j\rangle \tag{7.10}$$

which clearly shows that the general form of a two-electron state is a superposition of all possible vector states that can be constructed by choosing a particular one-electron state for each electron in the system. This is generally known as a configuration interaction state because every particular choice of one-electron state for each electron identifies an electronic configuration. The generalization to N-electron states is trivial.

However, the discussion so far has ignored the fact that electrons are fermions, identical indistinguishable particles with one-half spin. To ensure that expectation values that are always the predicted results for a measure are independent of the labels given to identical particles, the vector states must fulfill the Pauli principle. Hence, vector states describing an N-electron system must be antisymmetric with respect to the permutation of the individual states of any pair of electrons. This is

$$\hat{\mathrm{P}}_{ij}|\Psi_k\rangle = -|\Psi_k\rangle \quad \forall i, j \tag{7.11}$$

a condition easily achieved by using a basis of antisymmetric states constructed as

$$\det|\Phi_i\Phi_j\rangle = |\Phi_i(1)\Phi_j(2)\rangle - |\Phi_j(1)\Phi_i(2)\rangle \tag{7.12}$$

which defines the well-known Slater determinants. Finally, we are led to the very important configuration interaction, CI, which forms a general two-electron vector state. Again, the generalization to N-electron states is straight and the CI resulting ket is a linear combination of all Slater determinants that can be constructed by distributing the N-electrons in the one-electron states, included in the, in principle infinite, one-electron basis, in all possible ways:

$$|\Psi_k\rangle = \sum c_{ij,k} \det|\Phi_i\Phi_j\rangle \tag{7.13}$$

The only remaining point concerns the representation of the one-electron states in a given basis. Without entering into the details of the theory of representations of quantum vector states, we just mention that the usual approach consists in using the so-called position representation. This choice leads to the usual definition of the wave function in terms of electron coordinates, $\bar{\mathrm{x}}$; thus finding the key one-to-one correspondence between quantum vector states and quantum wave functions:

$$|\Psi_k\rangle \longleftrightarrow \Psi_k(\bar{\mathrm{x}}) \tag{7.14}$$

and between the Hilbert vector space and the space of quadratically integrable wave functions; both spaces are therefore isomorphous. This one-to-one correspondence extends to the eigenvalue equations

$$\hat{\mathrm{H}}|\Psi_k\rangle = E_k|\Psi_k\rangle \longleftrightarrow \hat{\mathrm{H}}\Psi_k(\bar{\mathrm{x}}) = E_k\Psi_k(\bar{\mathrm{x}}) \tag{7.15}$$

which, as commented above, can be exactly solved for one-electron problems, providing a simple way to obtain one-electron basis sets. From this one-electron basis set, the N-electron basis set can be constructed through Refs. 8 and 13.

Here, we must add that one-electron wave functions that provide adequate basis sets are not limited to those arising from Ref. 16 applied to a given one-electron problem. Any one-electron wave function belonging to the $\boldsymbol{H}$ space of quadratically integrable wave functions can be included in a basis set to construct the N-electron basis, i.e., to construct the set of Slater determinants. Here, we quote that, in nonrelativistic quantum mechanics, one-electron wave functions are always written as (tensor) product of space $\bar{\mathrm{r}}$ and spin ω parts. For the space part, different choices are possible (9–11), use of Hydrogen-like orbitals, Slater type orbitals (STOs), Gaussian type orbitals (GTOs), plane waves (PWs), or even different numerical representations. In practice, a usual but not unique strategy consists in defining a set of contracted GTOs (CGTOs) as

$$\chi_i(\bar{\mathrm{r}}) = \sum d_{pi} g_p(\bar{\mathrm{r}}) \tag{7.16}$$

where the d_{pi} coefficients are obtained from atomic calculations. The next step is to write atomic/molecular orbitals in terms of CGTOs

$$\phi_k(\bar{\mathrm{r}}) = \sum a_{rk}\chi_r(\bar{\mathrm{r}}) \tag{7.17}$$

and, as stated above, spin orbitals, or one-electron functions, are written as a (tensor) product of a spatial orbital and a spin, $\alpha(\omega)$ or $\beta(\omega)$, wave function

$$\Phi_k(r,\omega) = \phi_k(\bar{\mathrm{r}})\alpha(\omega) \quad \text{or} \quad \Phi_k(r,\omega) = \phi_k(\bar{\mathrm{r}})\beta(\omega) \tag{7.18}$$

Now, Slater determinants can be constructed from the spinorbital set and the two-electron CI wave function given by Eq. 7.13 is fully determined. The extension to N-electron CI wave functions is now straightforward.

7.2.1. From Quantum Mechanics to the Computational Method of Electronic Structure

To solve the eigenvalue problem outlined by Eqs. 7.5 or 7.13 requires dealing with complicated differential equations. However, Eq. 7.5 can be transformed to a simpler matrix eigenvalue problem by just multiplying Eq. 7.5 on the left by any $|\Phi_k\rangle$ and using Eq. 7.3

$$\sum \langle \Phi_k|\hat{\mathrm{H}}|\Phi_j\rangle c_{ji} = E_i\, c_{ki} \quad \forall k \tag{7.19}$$

or defining $\langle \Phi_k|\hat{\mathrm{H}}|\Phi_l\rangle = H_{kl}$, one has

$$\sum H_{kj} c_{ji} = E_i\, c_{ki} \quad \forall k \tag{7.20}$$

or

$$\mathbf{HC} = \mathbf{CE} \tag{7.21}$$

which is the general matrix equation to be solved in any electronic structure problem provided an orthonormal set is used.

Transforming the Schrödinger equation to the matrix form given by Eqs. 7.20 or 7.21 reduces the problem of solving differential equations to a simpler eigenvalue problem. However, finding a solution for this matrix equation is impossible in practice for N-electron problems because the matrices involved in the secular problem in Ref. 21 are of an infinite dimension! To convert this equation into a practical working equation two important problems have to be solved. First, one needs to avoid the infinite dimension of the matrices in Eq. 7.21. This problem can be circumvented by projecting this equation into a subspace $\mathbf{S}$ of finite dimension defined by the appropriate projection operator

$$\hat{\mathrm{P}}_S = \sum_{k \in S} |\Phi_k\rangle\langle\Phi_k| \tag{7.22}$$

This permits us to define the restriction of the Hamiltonian to $\mathbf{S}$ as

$$\hat{\mathrm{H}}_S = \hat{\mathrm{P}}_S \hat{\mathrm{H}} \hat{\mathrm{P}}_S \tag{7.23}$$

and, consequently, to project the matrix Eq. 7.21 on $\mathbf{S}$, namely

$$\mathbf{H}_S \mathbf{C}_S = \mathbf{C}_S \mathbf{E}_S \tag{7.24}$$

which provides the matrix equation that needs to be solved in any practical application. We must point out that solving Eq. 7.24 exactly is equivalent to the linear variational method using a trial function belonging to $\mathbf{S}$ and, hence, the eigenvalues of Eq. 7.24 are upper bound to the exact eigenvalues in Ref. 6. As the subspace $\mathbf{S}$ can be systematically improved (not necessarily by simple increase of the dimension of that subspace), the procedure we have just outlined permits us to systematically improve the accuracy of a computational result up to the required accuracy, or to the computational limit. This approach is the basis of the *ab initio* CI methods and, more precisely, of the full configuration interaction (FCI) approach because all possible configurations within the **S** subspace are explicitly included in the FCI wave function. The FCI leads to the exact solution on a given subspace. All *ab initio* wave function-based methods of quantum chemistry intend to approximate the FCI solution because, in practical computations, the FCI matrix can only be diagonalized for rather small systems and using special numerical techniques as commented in the next subsection. Notice that the dimension of the FCI matrix in Ref. 24 grows as $N!$ This fact is a natural consequence of the procedure used to construct an appropriate basis, $\{|\Phi_k\rangle\}$, of N-electron states and the reason for its limited practical use.

7.2.1.1. Wave Function-Based Computational Methods

The preceding discussion made it clear that the best attainable approximation to the wave function and energy of a system of N-electrons is given by the FCI approach. In fact, this is the exact solution in a given finite subspace, and it is independent on the N-electron basis used provided the different N-electron basis considered expands the same subspace. This is the case if the different N-electron basis sets are all constructed from a one-electron basis that are in turn built up from the same primitive set as, for instance, in Eq. 7.16. The invariance of the FCI energy with respect to the N-electron basis is a very important property because it means that the total energy and the final wave function are independent of whether atomic or molecular orbitals are chosen as a one-electron basis to construct the Slater determinants. However, use of atomic orbitals leads to valence bond (VB) theory (8), whereas use of molecular orbitals leads, of course, to the molecular orbital configuration interaction (MO-CI) theory (2). The VB wave functions are relatively easy to interpret (12–14) because the different Slater determinants can be represented as resonant forms. Usually only the valence space is considered within a minimal basis description. The reason for the use of such a limited space is that these VB wave functions are difficult to compute because of the use of a nonorthogonal basis set (15). On the other hand, the MOs are usually taken as orthonormal, a choice that permits us to carry very large CI expansions and to use extended basis sets (16–18). Notice that once a finite set of "$2m$" atomic/molecular spinorbitals is given, the dimension of the N-electron Hilbert subspace is also given. This is because from N electrons and "m" spin orbitals, $m \geq N$, the number of Slater determinants that can be constructed is

$$\dim \text{FCI} = \binom{2m}{N} \text{ or, more precisely } \dim \text{FCI} = \binom{m}{N_\alpha}\binom{m}{N_\beta} \tag{7.25}$$

if the system contains N_α and N_β electrons with alpha and beta spin, respectively. The dimension of the FCI problem, dim FCI, grows so fast that practical computations can be carried out for systems with a small number of electrons. Therefore, the FCI method is often used to calibrate more approximate methods (19).

The simplest N-electron wave function that one can imagine is a single Slater determinant. In this case, there is no eigenvalue problem and the energy is computed as an expectation value. Of course, constraining the wave function to just one Slater determinant largely reduces the variational degrees of freedom of the wave function. In fact, the energy is uniquely defined by the one-electron basis used to construct this particular Slater determinant. The one-electron basis set can be chosen in such a way that the energy expectation value is the lowest possible, i.e., variationally. In practice, one usually chooses

$$|\Psi_0\rangle = \frac{1}{\sqrt{N!}} \det |\Phi_i \Phi_j, \ldots, \Phi_N\rangle \tag{7.26}$$

with spin orbitals defined by Ref. 18 and orbitals defined by Ref. 19, which is the well-known MO-LCAO method originally designed by Roothaan (20). The MO-LCAO scheme permits us to work in a finite subspace from the very beginning. The variational problem reduces to finding the a_{ri} coefficients with the constraint that the spin orbitals remain orthonormal. Mathematically this is equivalent to finding the extreme of a function of the $\{a_{ri}\}$ set with a constraint. This leads to a set of Euler equations that in turn lead to the Hartree–Fock (HF) equations, finally giving the ϕ_k set, although in an iterative way because the HF equations depend on the orbitals themselves. This dependency arises from the fact that the HF equations are effective one-electron eigenvalue equations

$$\hat{\mathrm{f}}\phi_i = \varepsilon_i \phi_i \tag{7.27}$$

where $\hat{\mathrm{f}}$ is the well-known one-electron Fock operator, sum of the kinetic energy, nuclear attraction energy, and the Coulomb and Exchange effective potential operators. These effective potentials average interaction with the rest of electrons, which, of course, is given by the orbitals themselves. The final optimum orbitals are, therefore, those for which the effective average potential used to construct the Fock operator is *exactly* the same as that will be obtained using the solutions of Eq. 7.27 and the effective potential is self-consistent. The optimum orbitals are then named self-consistent, and HF is synonymous of the self consistent field. Solving Eq. 7.27 is not simple, especially for molecules, and Eq. 7.27 is transformed to a matrix form by expanding the orbitals as in Eq. 7.17 and using the same procedure to convert Eq. 7.5 into Eq. 7.21; this leads to

$$\mathbf{FA} = \mathbf{SA}\boldsymbol{\varepsilon} \tag{7.28}$$

where $\boldsymbol{\varepsilon}$ is a diagonal matrix containing the so-called orbital energies, $\mathbf{A}$ is the matrix grouping the coefficients in Eq. 7.17, and $\mathbf{S}$ is the overlap matrix now appearing because the orbitals in Eq. 7.17 are centered in different nuclei and, hence, are not orthogonal. The matrix Eq. 7.28 is also solved iteratively, and the whole procedure is termed the HF-SCF-LCAO method. An important remark here is that, because a variational approach is used, the HF scheme is aimed to approximate the ground state (of a given symmetry) wave function only. Once the SCF process converges $|\Psi_{\mathrm{HF}}\rangle = |\Psi_0\rangle$ and the HF energy is obtained as an expectation value; i.e., $E_{\mathrm{HF}} = \langle\Psi_{\mathrm{HF}}|\hat{\mathrm{H}}|\Psi_{\mathrm{HF}}\rangle$.

The difference between E_{HF} and the FCI energies is known as the electronic correlation energy. In principle, one should extrapolate the electronic correlation to the infinite basis set limit.

The resolution of the HF equations leads to "$2m$" spin orbitals but only N, i.e., the occupied orbitals, are needed to construct the HF determinant in Eq. 7.26. The rest of spin orbitals, unoccupied in the HF wave function, also termed virtual orbitals, can be used to construct additional Slater determinants. A systematic way to do it is substituting one, two, . . . , N, occupied spin orbitals by virtual orbitals leading to Slater determinants with one-, two-, N-substitutions with respect to the HF determinant. The determinants thus constructed are usually referred to as single-, double-, . . . , N-excitations and including all possible excitations leads to the FCI wave function. Clearly, the FCI wave function is invariant with respect to the orbital set chosen to construct the Slater determinants. However, using the HF orbitals has technical advantages because, at least for the ground-state wave function, the HF determinant contribution to the FCI wave function as in Eq. 7.13 is by far the dominant term. The fact that the electronic Hamiltonian includes up to two-electron interactions suggest that double excitations would carry the most important weight in the FCI wave function; this is indeed found to be the case. Therefore, one may design an approximate wave function in which only the reference HF determinant plus all double excited determinants are included. The result is called the doubly excited CI (DCI) method and is routinely used in *ab initio* calculations. The practical computational details involved in DCI are not simple and will not be described here. Adding single excitations is important to describe some properties such as the dipole moment of CO (2), this leads to the SDCI method. Extensions of SDCI by adding triple- or quadruple-excitations (SDTQCI) are also currently used, although the dimension of the problem grows very rapidly.

The truncated CI methods described above are variational, and finding the energy expectation values requires the diagonalization of very large matrices. An alternative approach is to estimate the contribution of the excited determinants by using the Rayleigh–Schrödinger perturbation theory up to a given order. This is the basis of the widely used MP2, MP3, MP4, . . . , methods, which use a particular partition, the Möller–Plesset one, of the electronic Hamiltonian and an HF wave function as a zero-order starting point (21,22). A disadvantage of perturbation theory is that the perturbation series may converge very slowly or even diverge. However, the MP(n) methods have a special advantage over the truncated CI expansions. In the DCI and related truncated CI methods, the relative weight of the different excitations differs from the one in the exact FCI wave function because of the normalization of the truncated wave function. This normalization effect introduces spurious terms, and as a result, the energy of N-interacting molecules does not grow as N. This is the so-called size consistency problem and is inherent to all truncated CIs (2). On the other hand, it can be shown tht the MP series is size-consistent order by order. Successful attempts to render truncated CI expansions size consistent have been reported in the past few years (23). However, the resulting methods are strongly related to a family of methods based on the cluster expansion of the wave function (24). The coupled cluster (CC) form of the wave function can be derived from the FCI one as in the case of the DCI method, although here the terms included are not selected by the degree of excitation with respect to the HF determinant only. The additional condition is that the different terms fulfill the so-called linked cluster theorem (24). The resulting system of equations is complicated and is not usually solved by diagonalization but by means of nonlinear techniques, and the results are not variational (25–27).

The truncated CI and CC (and QCI, which may be viewed as an approximate CC) methods perform well when used to approximate the ground-state wave function, because the HF determinant provides an adequate zero-order approach. However, this is not necessarily the case, especially when several excited states are to be studied. The logical extension of the truncated CI expansion is the so-called multi reference CI (MRCI) approach, where excitations, usually single and double, for a set of reference determinants are explicitly considered (28,29); the method is referred to as the MR(SD)CI method. Energies and MRCI wave functions are obtained by solving the secular Eq. 7.24. Again the concept is quite simple, but solving the eigenvalue problem is not a simple task, especially for excited states (30), and the different computational approaches involve very smart ideas and specialized codes coupled to vector, parallel, or vector-parallel processors (31). For problems of chemical interest, the dimension of the MRCI problem is so large that often a small block of Eq. 7.24 is diagonalized and the effect of the rest is taken up to second order by means of perturbation theory in different partitions. The reference space can be constructed either by selecting important determinants or important orbitals. The first idea is used in the CIPSI (32–34) method, whereas the second one is the basis of CASMP2 (35) and CASPT2 (36–38) methods where CAS stands for complete active space, the active space being defined once a subset of orbitals is chosen and it is complete because an FCI is performed within this orbital space. A general efficient code combining these strategies has been reported recently by Neese (39).

Except for the simplest HF approach, the logic of the methods that we have discussed is based on solving the secular problem in a finite subspace defined by the one-electron, orbital, basis chosen or in finding suitable approximations. In all cases, the orbital set is fixed and, usually, obtained form a previous HF calculation. Then, the contribution of the different Slater determinants in Eq. 7.24 or the cluster amplitudes in the CC methods are obtained either variationally, i.e., the different CI methods, through perturbation theory, i.e., the MP series and related methods, or by mixed approaches, i.e., the CIPSI method. Nothing prevents one to use the variational method to optimize the orbital set and the configuration contribution at the same time. This is the basis of the multiconfigurational self-consistent field (MCSCF) methods, which are the logical extension of HF-SCF to a trial wave function made as a linear combination of Slater determinants (40). The mathematical problem is conceptually very similar to that of the HF-SCF approach, namely finding an extreme of a function (the energy expectation value) with some constraints (orbital orthonormality). The technical problems encountered in MCSCF calculations were much more difficult to solve than those of the single determinant particular case. One problem faced by the earlier MCSCF methods was the poor convergence of the numerical process and the criteria to select the Slater determinants entering into the MCSCF wave function. The first problem was solved by introducing quadratically convergent methods

(41,42) and the second one by substituting the determinant selection by an orbital selection and constructing the MCSCF wave function using the resulting CAS. The resulting MCSCF approach is known as CASSCF and turned to be a highly efficient method (43–45). The CASSCF wave function is always precisely the zero-order wave function in CASMP2 and CASPT2 methods and in CIPSI if desired. The CASSCF wave function has some special features that are worth mentioning. It is invariant with respect to rotations (linear combinations) among active orbitals. When the CAS contains all valence orbitals, the CASSCF wave function is equivalent to the wave function obtained by the spin-coupled VB method (46–48) when all resonant forms involving valence orbitals are included. Before closing this section, we would like to mention that, in practice, configuration state functions (CSFs) are commonly used instead of Slater determinants. A CSF is *simply* a linear combination of determinants with coefficients fixed so as to have an eigenfunction of $\hat{S}^2$, the total square spin operator. The fixed coefficients are often obtained with the assistance of group theory (49). This choice ensures that truncated CIs be spin eigenfunctions and reduces the dimension of the secular problem.

We end this short review on *ab initio* wave function-based methods by noting that all of them can be applied to molecules and nanoparticles, whereas only HF can be extended to account for periodic symmetry (50).

7.2.1.2. Density Functional Theory-Based Computational Methods

The Schrödinger equation—Eqs. 7.5 or 7.15—provides a way to obtain the N-electron wave function of the system and the approximate methods described in the previous section permit reasonable approaches to these wave functions. From the approximate wave function, the total energy can be obtained as an expectation value and the different density matrices, in particular, the one-particle density matrix, can be obtained in a straightforward way as

$$\rho_0(\vec{x}) = N \int \Psi_0^*(\vec{x}_1, \vec{x}_2, \ldots, \vec{x}_N)\Psi_0(\vec{x}_1, \vec{x}_2, \ldots, \vec{x}_N) d_{\vec{x}_2}, \ldots, d_{\vec{x}_N} \tag{7.29}$$

where the integration is carried out for the spin and space coordinates of all electrons but one.

In 1964, Hohenberg and Kohn proved a theorem that states that the inverse of this proposal also holds (51). They proved that for a nondegenerate ground state, the one-electron density determines (up to an arbitrary constant) the external potential and hence the electronic Hamiltonian, the ground-state energy, and the electronic wave function. Here a caveat is necessary because one usually forgets that in nonrelativistic quantum mechanics, the spin coordinates and spin operators are added ad hoc to fulfill the Pauli principle (52). Therefore, stating that the electronic Hamiltonian defines the electronic wave function is not rigorously correct, except for the exact solution in the Hilbert space. This is because the exact Hamiltonian commutes with all spin operators and, hence, a complete set exists of wave functions that are simultaneously eigenfunctions of the Hamiltonian and of the spin operators. However, this does not hold in general for the approximate wave functions. Nevertheless, from the Hohenberg and Kohn theorem, it follows that a universal functional exists so that the ground-state total energy is a functional of the density; i.e., given a density, there is a mathematical rule that permits one to obtain the exact ground-state energy. The resulting theoretical framework is nowadays referred to as density functional theory or simply DFT. As the functional is unknown, one may think that DFT is useless. However, Hohenberg and Kohn also have proven a variational theorem stating that the ground-state energy is an extreme (a minimum) for the exact density, and later, Kohn and Sham proposed a general framework that opened the way for the practical use of DFT.

The Kohn–Sham formalism assumes that there is a fictitious system on N noninteracting electrons experiencing the real external potential and that has exactly the same density as the real system. This reference system permits us to treat the N-electron system as the superposition of N one-electron systems and the corresponding N-electron wave function of the reference system will be a Slater determinant Eq. 7.12. This is important because in this way, DFT can handle discrete and periodic systems. Once the reference system is defined, it is possible to obtain a trial density and all one needs to do is to compute the energy of the real system. Here it is where a model for the unknown functional is needed. To this end, the total energy is written as a combination of terms, all of them depending on the one-electron density only:

$$E[\rho] = T_s[\rho] + V_{\text{ext}}[r] + V_{\text{coulomb}}[\rho] + V_{\text{XC}}[\rho] \tag{7.30}$$

the first term is the kinetic energy of the noninteracting electrons, the second term accounts for the contribution of the external potential, the third one corresponds to the classic Coulomb interaction of noninteracting electrons, and finally, the fourth term accounts for all remaining effects. These last term must take into account the contribution to the kinetic energy because electrons are in fact interacting, the exchange part is due to the Fermi character of the electrons, and the correlation contribution because electron densities are correlated. Obviously, the success of DFT is strongly related to the ability to approximate E_{XC} in a sufficiently accurate way. Now, Eq. 7.30 plus the Hohenberg–Kohn variational theorem permits one to vary the density by varying the orbitals that are now referred to as Kohn–Sham orbitals. In addition, the Kohn–Sham orbitals can be expressed in a given basis set as in Eq. 7.17; when a CGTO basis is used, one has the LCGTO-DF methods (53,54). The orbital variation must preserve the orthonormality of the orbitals in the Kohn–Sham reference system to maintain constant the number of electrons. The overall procedure is then very similar to HF and the orbitals defining the density minimizing the energy in Eq. 7.30 while preserving orthonormality are those satisfying a one-electron eigenvalue problem similar to that expressed as in Eq. 7.27, but here the one-electron operator also explicitly contains the exchange and correlation effective potentials and indeed as local one-electron operators. Mathematically the exchange and correlation potentials entering into the Kohn–Sham one-electron equations are the functional derivative of the corresponding energy contributions in Eq. 7.30. Once $E_{\text{XC}}[\rho]$ (or more precisely $E_{\text{X}}[\rho]$ and $E_{\text{C}}[\rho]$) is known, the effective potentials are also known and solving the Kohn–Sham equations is similar to solving the HF equation with the important difference that here one may find the exact solution if the $E_{XC}[\rho]$ is the exact one. Notice that there is no guarantee that the final electron density thus obtained arises from a proper wave function of the corresponding Hilbert space through Eq. 7.29. This is the famous representability problem, it does not affect the practical use of DFT and will not be further discussed here, and the interested reader is addressed to the more specialized literature (7).

Several approaches to $E_{XC}[\rho]$ have been proposed in the last years with increasing accuracy and predictive power. However, in the primitive version of DFT, the correlation functional was ignored and the exchange part approximated following Slater's $\rho^{1/3}$ proposal; the method was known as SCF-Xα. In 1980, Vosko et al. (55) succeeded in solving the electron correlation for a homogeneous electron gas and in establishing the corresponding correlation potential. The resulting method including also the exchange part is nowadays known as the local density approximation (LDA) to DFT and has been successful in the description of bulk metals and in describing metal surfaces. However, LDA has experienced enormous difficulties in the description of molecules and ionic systems; for instance, LDA incorrectly predicts NiO to be a metal (56,57). This is a general problem of DFT; it is critical for the LDA functional but also appears in the GGA approaches, and only hybrid approaches including a part of nonlocal Fock exchange seem to be able to reproduce the main features of the electronic structure of this kind of system (58). The Kohn–Sham equations were initially proposed for systems with a closed shell electronic structure and hence suitable to study closed-shell singlet electronic states. It has been proposed to apply DFT to open-shell systems by using a spin unrestricted formalism. In this case, different spatial orbitals are used for alpha- and beta- spin orbitals in Eq. 7.18. In the case of LDA, the resulting formalism is known as the local spin density approximation (LSDA) or simply local spin density (LSD). This is similar to the well-known unrestricted Hartree–Fock (UHF) formalism and suffers from the same drawbacks when dealing with open-shell systems (2). This is especially important when attempting to study magnetic systems with many open shells, the unrestricted Kohn–Sham determinant corresponding necessarily to a mixture of the different possible multiplets compatible with the electronic configuration in the electronic configuration represented by the Kohn–Sham determinant. Spin-resctricted formulations of DFT for open shells have been proposed recently (59,60), although for a few simple cases only. Indeed these new approaches have shown inconsistencies in the currently used exchange-correlation potentials (52).

Despite the inherent simplifications, LDA (and LSDA) predictions on molecular geometries and vibrational frequencies are surprisingly good. However, bonding energies are much less accurate and require going beyond this level of theory. This is also the case when dealing with more difficult systems such as biradicals or more delicate properties such as magnetic coupling in binuclear complexes (61) or ionic solids (58,62,63). The DF methods that go beyond the LDA can be grossly classified in gradient corrected (GC) and hybrid methods. In the first set, the explicit calculation of the $E_{XC}[\rho]$ contributions involves not only the density ρ but also its gradient $\nabla\rho$. The number of GC is steadily increasing, but among the ones widely used, we quote the Becke exchange functional (64) (B) and the Perdew–Wang (65,66)

(PW) exchange correlation functional, the latter is usually referred to as the generalized gradient approximation (GGA) and is particularly used in condensed matter and material science. Another popular gradient-corrected correlation functional is the one proposed by the Lee–Yang–Parr (LYP) correlation functional (67) based on the work of Colle and Salvetti on the correlation factor (68). In some cases, one uses B for the exchange and PW for the correlation part—this is usually referred to as BP—or B for the exchange and LYP for the correlation part giving rise to the BLYP method. Some of the more recent revisions of the GC methods aimed to increase the accuracy of predicted chemisorption energies are PBE (69), rPBE (70), and RPBE (71). The term "hybrid functionals" is used to denote a family of methods based on an idea of Becke (72). This approach mixes DF and Fock exchange and local and GC correlation functionals in a proportion that is obtained from a fit to experimental heats of adsorption for a wide set of molecules. The most popular hybrid method is B3LYP, where the number indicates that three parameters are fit to experiment. For a more detailed description about GC and hybrid methods, the reader is referred to general textbooks (73) or to a more specialized literature (7,74).

Aside from the great advances in the development of new exchange-correlation functionals, the question of which functional provides the best chemical accuracy is still under discussion. In the wave function-based methods, it is possible to check the accuracy of a given level of theory by systematic improvement on the basis set used to expand the one-electron functions and on the level of treatment of electron correlation. Unfortunately, this is not completely feasible in the framework of DFT. One can improve the basis set, but there is no way to systematically improve the $E_{XC}[\rho]$. Therefore, one needs to establish the accuracy of the chosen approach by comparing several choices for $E_{XC}[\rho]$. For many systems, this choice is not critical and the use of several functionals permits us to add error bars to the computed quantities. However, in other cases, such as the description of the transition metal-oxide interface, the choice of the functional has been shown (75,76) to be crucial. An extreme case is that of Cu adsorption on MgO: The reported adsorption energies range from a practically unbound Cu atom at the HF level to a moderate adsorption, 0.35–0.90 eV, at the gradient-corrected DF level, to strong adsorption, about 1.5 eV, using the LDA. Ranney et al. (77) have extrapolated the adhesion energy of a single copper atom on MgO from their microcalometric measures of the heat of adsorption and have found a value of 0.7 eV. The comparison of this value with the computed adsorption energies of Ref. 75 indicates that, among the currently used approximation of the exchange-correlation functional, the pure DF ones seem to provide the best answer, whereas the hybrid functionals slightly underestimate the adsorption energy.

7.2.1.3. Quantum Monte Carlo-Based Computational Methods The wave function and DFT methods described in the previous subsections are by far the most widely used in practical applications, including the study of metal-oxide clusters discussed in the next section. However, it is also important to keep in mind that other methods that are currently being developed may become competitive, especially with the ever increasing computer power. A family of methods that deserve a few words is that described by the general Quantum Monte Carlo term. The name comes from the fact that numerical Monte Carlo techniques are used to evaluate integrals involving quantum wave functions. In these methods, one starts from a given reliable trial quantum wave function (78), and in general, the associated energy is obtained by direct integration as a Hamiltonian expectation value. A more detailed discussion can be found in the recent literature (79).

7.3. METAL-OXIDE CLUSTERS: REVIEW OF FIRST PRINCIPLES CALCULATIONS

The theoretical study of oxide nanoparticles and clusters is more recent than that of corresponding clusters of metal elements. This is, in part, due to the intrinsic complexity of oxide electronic structures where the metal oxidation state, the average coordination, the particle size and morphology, the presence of defects and localized electrons (in particular in the *d* shells of transition metal atoms), and so on, are all elements that make the theoretical description of these systems much more complicated than that of simple elemental nanoparticles and clusters. Just to give an idea of the complexity of the problem, consider the question of the stoichiometry of a small cluster or nanoparticle. If one is dealing with a metal, the only parameters to take into account in the calculation are the number of atoms and the particles shape. Then, the main question becomes that to determine the number of low-lying isomers and their relative stabilities. If one deals with a cluster compound of comparable size, the preferred structures do not necessarily have the same stoichiometry of the corresponding bulk phases. Furthermore, the position of the cations and anions in a nanoparticle can be completely different from that assumed in an extended crystal. The number of possibilities is enormous, and a structural analysis solely based on a computational effort is simply not possible, except for very small aggregates containing few atoms. Even in this case, the number of structures is very large, making an exploration of the potential energy surface with first principle methods difficult. For this reason, the number of theoretical studies dedicated to the structure and properties of oxide nanoparticles and clusters is comparatively small with respect, for instance, to that of computational studies dedicated to oxide surfaces and films. In the following, we provide a brief review of theoretical studies on oxide nanoclusters, with the aim to show which kind of properties can be deduced from these calculations, the methods that are most used and the problems that can be encountered. The reader will realize that most of these studies are performed at the DFT level and are restricted to cluster compounds containing at most 20 atoms.

The list of papers reported is probably far from being exhaustive, but it provides a general overview of where the field is at the moment. For convenience, we have grouped the studies according to oxide type, starting from the simple element oxides with closed *d* shells, like alkaline–earth oxides and silica, to move then to transition metal oxides.

7.3.1. Simple Metal Oxides

7.3.1.1. Alkaline–Earth Oxides Not surprisingly, alkaline–earth-oxide clusters have received more attention than other oxides. This is due mainly to the relatively simple electronic structure of these oxides compared with transition metal oxides.

The structures and relative stabilities of neutral stoichiometric $(MgO)_n$ and $(CaO)_n$ ($n = 3, 6, 9, 12, 15, 18$) clusters and of doubly charged non-stoichiometric $(CaO)_n$ ($n = 1$–29) cluster ions have been studied by the *ab initio* perturbed ion plus polarization calculations (80); this is essentially a HF method. The results have been compared with alkaline metal halide clusters, finding substantial structural differences between alkali metal halide and alkaline–earth oxide cluster ions, contrary to what is suggested by the similarities in the experimental mass spectra. The structures of the doubly charged clusters are similar for the two materials, MgO and CaO, whereas the study of the neutrals has revealed interesting structural differences between MgO and CaO. This shows how nanoparticles of these materials can differ from the corresponding bulk counterparts.

The behaviors of bond distances and binding energies for small alkaline–earth-oxide clusters, have been studied by Bawa and Panas employing the DFT-B3LYP approach (81). Here a validation of the applicability of DFT was made on selected clusters by also performing MP4 calculations. It was found that for CaO, it is necessary to include at least two d-polarization functions for a proper description of the cluster binding energy.

The electronic structures of a series of $(MgO)_n$ ($n = 2$–16) clusters cut out from MgO solid have been calculated by means of DFT with the B3LYP hybrid functional. The convergence of the electronic properties and the adsorption properties of (MgO) clusters containing up to 16 MgO units has been investigated by Lu et al. (82) (Figure 7.1). The results demonstrated a good correlation of topologic parameters with the stability of clusters. This is important to set up a good model of a given cluster size without paying the high cost of preliminary calculations.

The structure of various gas-phase isomers of MgO clusters have been studied by *ab initio* perturbed ion calculations (83). For the MgO isomers containing up to 7 MgO units, a full geometrical relaxation was considered. Correlation corrections were included for all cluster sizes using the Coulomb–Hartree–Fock model proposed by Clementi (84), and it was concluded that the inclusion of correlation is crucial to achieve a good description of these systems. The results were used to interpret the experimentally observed magic numbers. Variations in the cluster ionization potential with the cluster size were also studied and related to the structural isomer properties.

The same problem of structure and stability of $(MgO)_n$ clusters with $n = 13$ was addressed in an *ab initio* molecular orbital calculations by Recio et al. (85). Here cube-like clusters were considered and the relative stabilities were studied by analyzing binding energies and dissociation energies involved in the evaporation of a neutral MgO monomer. The calculated magic numbers were found to be in very good agreement with the abundance maxima observed in the mass spectra. The study also analyzed the size-dependence of structural, energetic, and electronic properties of the clusters, finding different rates to reach the corresponding bulk limit. The same group also studied singly ionized stoichiometric clusters of MgO containing up to

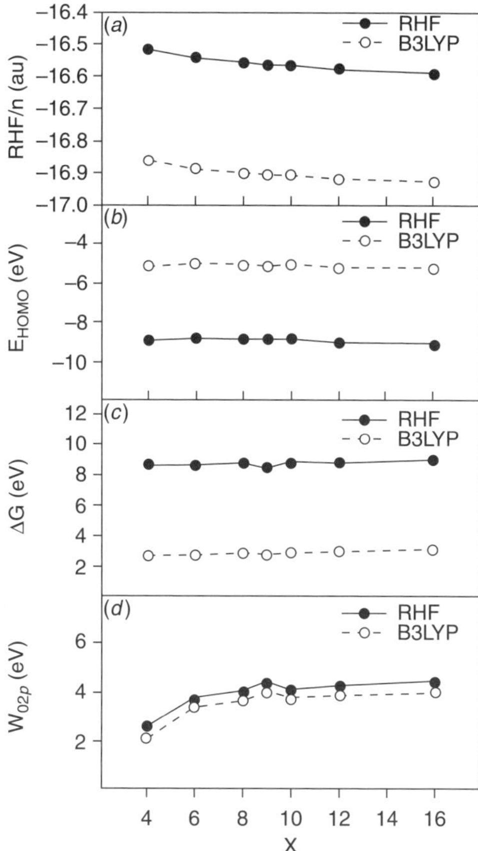

Figure 7.1. Size dependence of the calculated properties of $(MgO)_x$ stoichiometric clusters ($x = 4d, 6c, 8c, 9b, 10c, 12b, 16a$). (*a*) RHF; (*b*) energy of the HOMO level; (*c*) energy gap ΔG; (*d*) width of the O_{2p} band. Reprinted from Ref. 82 with permission.

26 atoms (86). Geometrical parameters of the neutral clusters were optimized at the Hartree–Fock level, whereas for the ionized clusters, the vertical approximation was used. Correlation corrections in the clusters with 2–12 atoms have been included by means of MP2 calculations. The structures based on the $(MgO)_3$ subunit were found to be preferred in comparison with cube-like configurations, although the energy difference decreases with increasing cluster size. The relative stability of neutral and single-ionized clusters was then investigated, and the calculated "magic numbers" for the charged clusters, $(MgO)_n^+$, were found to be also in agreement with the abundance maxima observed in the mass spectra.

In the context of charged MgO clusters, we mention the combined experimental–theoretical study of Gutowski et al. (87) (Figure 7.2) on MgO cluster anions. Photoelectron spectra of $(MgO)_n^-$ ($n = 1$–5) reveal a surprising trend, namely the electron binding energy decreases from $n = 1$ to 4, and then increases from 4 to 5.

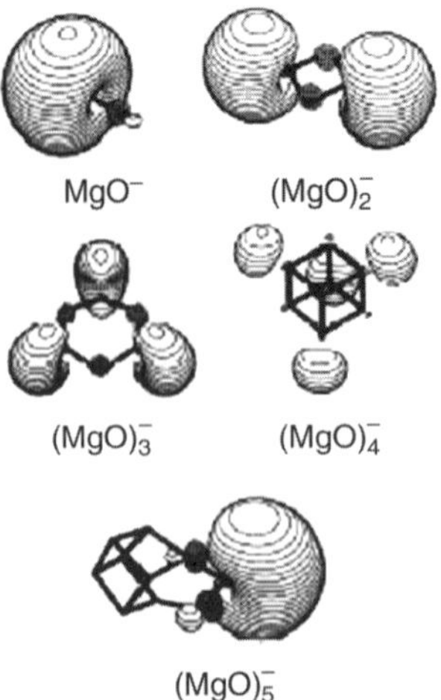

Figure 7.2. Structures of the minimum energy of the negatively charged $(MgO)_x$ ($x = 1$–5) clusters. The single occupied molecular orbitals are also plotted with a 0.02 contour spacing. Reprinted from Ref. 87 with permission.

DFT-B3LYP and MP2 calculations suggested that this pattern is related to the electrostatic interaction between the extra electron and the charge distribution of the neutral cluster. This study shows the important role of the cluster dimensionality in determining important properties like electron trapping energies.

Very few studies have addressed the properties of heavier alkaline–earth oxides. A DFT–LDA calculation has shown that the electronic properties of a BaO cluster are strongly correlated with its equilibrium structure (88). The ground-state structures of Ba_nO_m clusters have been be classified into four categories: (*1*) compact, (*2*) dangling state, (*3*) F-center, and (*4*) stoichiometric. The compact cluster is metallic, as almost no energy gap exists between the highest occupied and the lowest unoccupied molecular orbitals. The energy gap for the dangling state cluster is larger than that for the F-center cluster, whereas the stoichiometric cluster has the largest energy gap. Here, a caveat is also necessary because LDA band gaps are consistently underestimated to the point that insulators such as NiO are erroneously predicted by the LDA to be metals (56,57). Despite this flaw of the LDA, it is likely that the trend described above for the Ba_nO_m clusters is qualitatively correct.

Also the lighter member of the series, beryllium oxide, has received little attention. The atomic structure, stability, and electronic structure of, and interatomic interaction in, the $(BeO)_4$, $(BeO)_{12}$, and $(BeO)_{16}$ nanoclusters have been investigated by *ab initio* Hartree–Fock and DFT calculations using the discrete variational method (89). Their properties have been analyzed as a function of cluster geometry and compared with those of wurtzite-like BeO. Among the beryllium oxide clusters, the polyhedral (fullerene-like) clusters are most stable.

The last study considered here is not based on *ab initio* quantum-mechanics, but it is still interesting in the context of exploring relatively large nanoparticles. This is based on the application of a genetic algorithm for optimizing the geometries of stoichiometric and nonstoichiometric MgO clusters, bound by a simple Coulomb-plus-Born–Mayer potential (90). The genetic algorithm is shown to be efficient and reliable for finding the global minima for these clusters. The variation of the structures of MgO clusters was investigated as a function of the formal charges q on the ions, ranging from $q = 1$ to $q = 2$. Lower charges are found to favor compact, rocksalt-like cuboidal clusters, whereas the higher formal charges favor hollow pseudo-spherical structures. Hexagonal stacks are also found to be stable for small (MgO) clusters.

7.3.1.2. Aluminium-Oxide Clusters Compared with studies on alkaline–earth-oxide clusters, the studies on alumina clusters are scarce, despite the importance of this oxide as a support in catalysis. Fernandez et al. (91) have studied the geometrical and electronic structure of $(Al_2O_3)_n$ clusters with $n \leq 15$ by means of DFT calculations, norm-conserving pseudopotentials, and numerical atomic basis sets. They determined equilibrium geometries and discussed trends obtained for the atomic arrangements (structural isomers, coordination numbers, "amorphous" versus "ordered" structures, etc.) as well as a number of electronic properties (binding energies, HOMO–LUMO gap, density of levels, dipole moments, and Mulliken population charges).

Two other studies have been dedicated to the interaction of alumina clusters with adsorbed molecules. In the first one (92), HF, MP2, QCISD(T), and DFT-B3LYP calculations were used to investigate properties of complexes formed from the association of CH_4 with Al_2O_3, Al_4O_6, and Al_8O_{12} alumina clusters. Methane attaches to a surface Al atom of the cluster, but the rotational motion for methane in these complexes is highly fluxional. Extrapolated interaction energies (using the so-called G2MP2 for the various methane–alumina cluster complexes) are between 14 and 21 kcal/mol, respectively. The electronic structure of the three alumina clusters are very similar, with a charge on the surface Al atom of $+2.2$ to $+2.3$. The results of this study suggest that the adsorption energy for alkane molecules binding to alumina materials depends very strongly on the structure of the binding site. In the second study, using previous results for the equilibrium geometries of stoichiometric $(Al_2O_3)_n$ clusters, Fernandez et al. (93) obtained from a DFT approach the corresponding relaxed structures of the complexes with water for $(Al_2O_3)_n$ with $n \leq 7$. Depending on the initial position of the water molecule relative to the cluster site, the complex evolves to different equilibrium structures, with or without dissociation of H_2O, which energetic, bond lengths, and charge transfer trends depend on the morphology and size of the initial cluster. Dissociation of H_2O with OH^- bound on top of an Al atom and the proton H^+ bound to the second nearest neighbor O, is the dominant process for the reaction $(Al_2O_3)_n + H_2O$.

7.3.1.3. Silica Clusters Silica is probably the most extensively studied oxide either in the form of small clusters or nanoparticles by first principle electronic structure calculations. This is in part related to the importance of amorphous silica in various applications; from optical fibers to microelectronics. Small silica particles may have common features with amorphous silica more than with crystalline forms of this material.

Pereira et al. (94) performed DFT–LDA calculations on the energies and conformations of complex silica-based clusters. They reported calculated structures, charge distributions, and energies of the noncyclic four- and five-silicon chains, the branched trimer and tetramer rings, the double trimer rings, the tetramer plus trimer rings, and the five- and six-silicon rings. The total condensation energy (from the monomer) to form a silica cluster in the gas phase depends essentially on its structure and size. The results show that the stability of the noncyclic clusters decreases with the degree of branching, as observed experimentally.

Magic number silica clusters $[(SiO_2)_nO_2H_3]^-$ with $n = 4$ and 8 have been observed in the XeCl excimer laser (308 nm) ablation of various porous siliceous materials. The structural origin of the magic number clusters has been studied by DFT calculation at the B3LYP level (95,96), with a genetic algorithm as a supplementary tool for global structure searching. The theoretical calculations predict that the first magic number cluster, $(SiO_2)_4O_2H_4$, and its anion will most probably take pseudotetrahedral cage-like structures. The DFT calculations predict the second magic number cluster, $(SiO_2)_8O_2H_4$, and its anion are most probably a mixture of cubic cage-like structural isomers with an O atom inside the cage.

Via another global optimization lead study, the ground-state clusters of silica clusters $(SiO_2)_N$ $N = 6-12$ have been determined by DFT at the B3LYP level (97). This study predicted an extremely energetically favorable structure of the $(SiO_2)_8$ ground-state cluster and a particularly unusual and very symmetric $(SiO_2)_{10}$ non-ground-state cluster [also reported in (Ref. 98)]. These structures are predicted to have highly localized frontier orbitals at the nonbridging oxygen atoms, unique IR spectra, and a propensity to assemble into a variety of higher dimension structures. These structures, if synthesized, are expected to provide not only new porous materials, but also to have potential applications, particularly in surface technology.

Using DFT–GGA theory, the equilibrium geometries, binding energies, ionization potentials, and vertical and adiabatic electron affinities of Si_nO_m clusters ($n \leq 6$, $m \leq 12$) have been calculated (99). It was found that the bonding in Si_nO_m clusters is characterized by a significant charge transfer between the Si and O atoms and is stronger than in conventional semiconductor clusters. The bond distances are much less sensitive to cluster size than seen for metallic clusters. Similarly, calculated energy gaps between the highest occupied and the lowest unoccupied molecular orbital (HOMO–LUMO) of $(SiO_2)_n$ clusters increase with size, as it should be expected for an insulator. The HOMO–LUMO gap decreases as the oxygen content of a Si_nO_m cluster is lowered, eventually approaching the visible range. This study shows that the photoluminescence and strong size dependence of optical properties of small silica clusters could thus be attributed to oxygen defects.

Silica nanoclusters based on two- and three-membered-ring units have also been studied through DFT calculations (100,101). These nanoclusters may present well-defined stoichiometric chain, ring, porous (tubular), and cage-like structures. It was shown that the rings are more favorable than the chains for $(SiO_2)_N$ $N \geq 11$. The calculations also demonstrate that cage-like structures are energetically

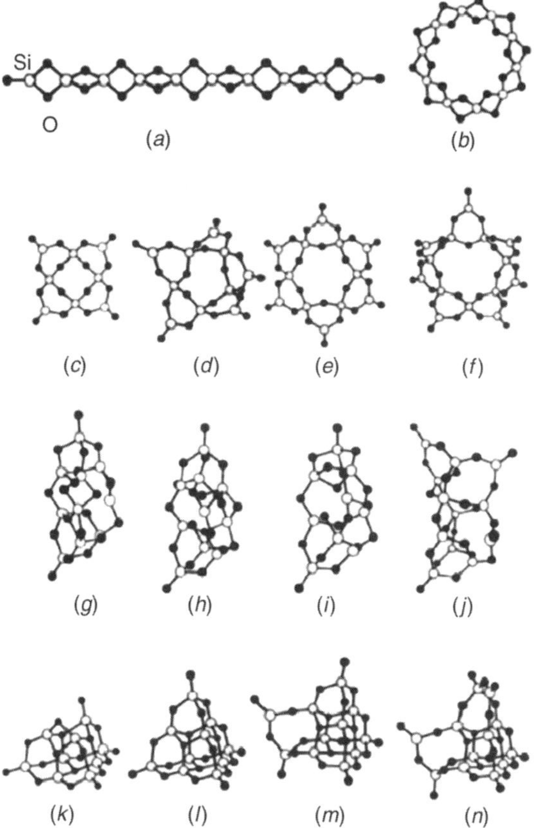

Figure 7.3. Geometries of $(SiO_2)_n$ clusters as predicted by the B3LYP/6-31G* level of theory: (*a*) 2MR-based chain for $n = 10$; (*b*) molecular ring based on 2MRs for $n = 10$; (*c*)–(*f*) 3MR-based molecular rings for $n = 8, 10, 12$, and 14; (*g*)–(*j*) elongated structures for $n = 9$, 10, 11, and 12, respectively; and (*k*)–(*n*) tetrahedral-like structures for $n = 9$, 10, 11, and 12. Reprinted from Ref. 102 with permission.

favorable structural models of larger silica nanoclusters. They possess distinctive IR spectra, characterized by a sharp peak at the vicinity of $1050\,cm^{-1}$ for the clusters with nonbridging oxygen (NBO) defects and by two strongest peaks at the $950–980\,cm^{-1}$ and $1030–1050\,cm^{-1}$ regions, respectively, for the fully coordinated structures.

The energetically favorable configurations of silicon monoxide clusters $(SiO)_n$ for $n \geq 5$ have been studied again by DFT methods (102) (Figure 7.3). These clusters facilitate the nucleation and growth of silicon nanostructures as the clusters contain sp^3 silicon cores surrounded by silicon-oxide layers. The frontier orbitals of $(SiO)_n$ clusters are localized to a significant degree on the silicon atoms on the surface, providing high reactivity for further stacking with other clusters. It was found that the oxygen atoms in the larger clusters prefer to migrate from the center to the exterior surfaces, leading to the growth of sp^3 silicon cores. Atomically well-defined discrete fully coordinated spherical and elongated nanocages constructed from SiO_2 were shown, via DFT calculations, to be structurally stable (103). Generally the spherical nanocages are energetically preferred over the elongated nanocages. With increasing size, both types of nanostructures have been shown to be increasingly thermodynamically stable with respect to known terminated silica nanostructures and bulk quartz. A structural model of silica clusters $(SiO_2)_n$ based on linked three-membered rings has been proposed based on first-principle DFT calculations (104). The three-member ring structures contain cage units of larger sizes and have relatively smaller electronic energy gaps, highly localized frontier orbitals at the exterior atoms, a high thermal stability, a low intrinsic internal strain, and a good extendibility to three-dimensional networks. Using Si_2O_2 rhombuses and Si_3O_3 rings as building blocks, Lu et al. (105) have investigated the structures of medium-sized silicon-oxide clusters by performing DFT calculations on a series of Si_8O_n and $Si_{12}O_n$ clusters with different growth motifs. They proposed a new form for medium-sized silicon-oxide clusters in a form of wheel-like structure composed of Si_3O_3 rings. Clearly, the possibility to study with realistic first principle methods large clusters of oxide materials is limited by the size of the computations involved. The problem has been addressed by Zhang et al. (106) who employed a simple basis set to determine reliable atomic and electronic structures of the silicon-oxide system. The calculations were performed at the HF, MP2, and B3LYP levels. They found that binding energies determined with a single-point energy calculation using a high-level basis set on the geometric structures optimized with the economic basis set provide reasonable agreement with other high-level calculations. The approach has enabled the computation of silicon-oxide clusters as large as containing 16 atoms with considerabe accuracy and using an acceptable level of computation resources.

7.3.1.4. Antimony Oxides The formation of positively charged antimony-oxide clusters has been investigated using time-of-flight mass spectrometry and MP2 calculations as a function of oxygen partial pressure (107) (Figure 7.4). Oxygen-rich clusters gain in stability with increasing cluster size and increasing oxygen partial pressure. To obtain information about structures and general building principles of these clusters, *ab initio* quantum chemistry calculations for the series of neutral and charged $(Sb_2O_3)_n$, clusters have also been performed. Except from a defect center in the cationic series, antimony atoms are trivalent and oxygen atoms divalent. Open structures with terminal oxygen atoms or with Sb–Sb bonds were found to be very high in energy, expressing a principle of the formation of a maximum number of Sb–O bonds. It was also found that characteristic building blocks in the neutral and in the cationic series are eight-membered rings, which are also found in the bulk antimony(III)oxide modification senarmonite, as well as Sb_4O_5 units bridged by oxygen atoms.

7.3.1.5. Zinc Oxides First principle calculations on gas-phase ZnO clusters or nanoparticles do not seem to exist. However, a few studies have been performed

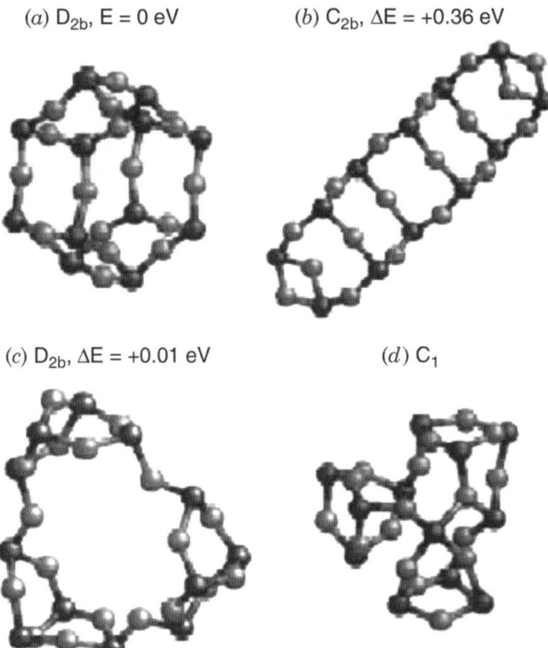

Figure 7.4. The $n = 6$ $Sb_{12}O_{18}$ structures. Structures (*a*) and (*b*) correspond to minima, structure (*c*) corresponds to a second-order saddle point on the potential energy surface. Elongation along the coordinates defined by one of the normal modes with imaginary frequency and subsequent geometry optimization gives the folded-up ring isomer (*d*), which is a real minimum on the potential energy surface (see Ref. 107). Reprinted from Ref. 107 with permission.

with simplified Hamiltonians or even classic potentials. AM1 and MNDO semiempirical electronic calculations, which are simplified forms of the SCF method, have been carried out for $(ZnO)_n$ (n = 11–44) clusters (108). In the second study molecular dynamics simulations of ZnO clusters have been performed (109). At certain magic numbers of atoms, the clusters spontaneously form spheroids reminiscent of fullerenes. Molecular orbital calculations confirm that the spheroids are stable.

7.3.2. Transition Metal Oxides

7.3.2.1. Vanadium Oxides We consider now transition metal oxides with partially filled *d* shells. These systems are notoriously difficult to describe because of the problem of electron and spin localization in this class of materials. For instance, it has already been commented that NiO, a magnetic insulator with an anti-ferromagnetic ground-state ordering, is predicted to be a metal in the DFT–LDA approach (56–58). This has limited the investigation of oxide clusters of transition metals to elements at the left of the Periodic Table, where the *d* orbitals are more diffuse and the problem of localization less severe. Among these systems, vanadium oxides have attracted a lot of interest, also in connection with the use of this material in oxidation catalysis.

Reactivities and collision-induced dissociation of vanadium-oxide cluster cations have been investigated experimentally using a triple quadrupole mass spectrometer coupled with a laser vaporization source and theoretically with HF calculations (110). VO_2, VO_3, and V_2O_5 units are the main building blocks for most of these clusters. The reaction pathways observed for these vanadium-oxide clusters include molecular association, cracking, dehydration, and oxygenation of the neutral hydrocarbons with the reactivities of specific clusters differing from species to species. A more extended analysis of the selectivity of $V_xO_y^+$ clusters in reactions toward ethylene due to the charge and different oxidation states of vanadium for different cluster sizes was shown in a subsequent paper (111). DFT calculations were performed on the reactions between $V_xO_y^+$ and ethylene, allowing us to identify the structure-reactivity relationship. The oxygen transfer reaction pathway was determined to be the most energetically favorable one available to $V_2O_5^+$ and $V_4O_{10}^+$ via a radical-cation mechanism. Investigation of reactions involving gas-phase cationic vanadium-oxide clusters with small hydrocarbons are suitable for the identification of reactive centers responsible for selectivity in heterogeneous catalysis.

The formation of novel vanadium-oxide cluster molecules by oxidative two-dimensional evaporation from vanadium-oxide nanostructures on a Rh(111) metal surface has been studied experimentally and theoretically (112). The structure and stability of the planar V_6O_{12} clusters and the physical origin of their 2D evaporation process have been investigated by high-resolution scanning tunneling microscopy (STM) and DFT calculations. The surface diffusion of the clusters has been followed in elevated-temperature STM experiments, and the diffusion parameters have been extracted, indicating diffusion by hopping of the entire surface-stabilized cluster units.

7.3.2.2. Titanium Oxides Surprisingly, titania nanoclusters have not been studied by many groups, different from the corresponding rutile and anatase surfaces for which several theoretical investigations exist. The electronic and structural properties of TiO_2 species of various sizes, charges, and stoichiometries, ranging from $Ti_nO_m^q$ clusters ($n = 1–3, m - n = 0, 1, q = -1, 0, +1$) to bulk rutile and its (110) surface, have been obtained by total energy calculation based on the DFT in the LSDA (113). The results have been interpreted based on a Bader-type analysis of the total electronic density. Attention was focused on the electron distribution to better understand how the iono-covalent character of the Ti–O bonding and the screening properties vary as a function of the size of the system, the atomic coordination, and the surface orientation.

First-principles quantum chemical calculations at the HF, MP2, and B3LYP levels were used to explore the potential energy hypersurfaces of ionic titanium oxide clusters of molecular dimensions (114). The energetics of all topomers corresponding to global or local minima and saddle points in the potential energy hypersurfaces were computed at the more sophisticated QCISD(T) level. This is one of the few examples of oxide clusters treated with such a sophisticated computational method.

The computed structural, energetic, and spectroscopic properties of the clusters were found to be in line with available experimental results.

In a joint experimental–theoretical study, the niobium–titanium-oxide cluster compound $Ti_2Nb_6O_{12}$ was synthesized from NbO, NbO_2 and TiO_2 (115). It was found that this system crystallizes in an original structure type based on octahedral cluster units (Nb_6O_{12}). The cluster connectivity pattern in $Ti_2Nb_6O_{12}$ and intercluster metal-metal distances are similar to those found in the "Chevrel phases," and band structure calculations indicated that the lowest unoccupied band in $Ti_2Nb_6O_{12}$ resembles the conduction band in the superconducting "Chevrel phases."

Finally, it is worth commenting that a very recent study by Hamad et al. (116) studied the global minima in the potential energy surface of Ti_nO_{2n} clusters with $n = 1-15$. The search was carried out using a novel combination of simulated annealing and Monte Carlo basin hopping simulations together with genetic algorithm techniques, with the energy calculated by means of an interatomic potential. The minina structures are then refined by using DFT. The agreement between the interatomic potentials and DFT is found to be good for large particles ($n = 9-15$). The calculated global minima consist of compact structures in which the Ti atoms tend to reach high coordination with increasing n. In particular, for $n \geq 11$, the particles have at least a central octahedron surrounded by a shell of surface tetrahedra, trigonal bipyramids, and square base pyramids. Hamad et al. (116) point out that the use of these methods opens new possibilities for the study of larger TiO_2 particles and of other oxides.

7.3.2.3. Manganese, Iron, Cobalt, and Nickel Oxides

We consider here nanoclusters of magnetic oxides, like MnO, FeO, CoO, and NiO. They are all characterized by a complex distribution of the d electrons and various possible ferro-, anti-ferro, or ferri-magnetic ordering of the spins—*ab initio* calculations based on DFT calculations carried out at the GGA level have revealed many unusual features of stoichiometric $(MnO)_n$ ($n < 9$) clusters that contrast with their bulk behavior (117). The clusters are ferromagnetic and carry atomic-like magnetic moments ranging from 4 μ_B to 5 μ_B per MnO unit, and the moments are localized at the Mn sites. The $(MnO)_8$ cluster, in particular, exhibits nearly degenerate ferromagnetic and atypical antiferromagnetic solutions with the ferromagnetic structure carrying a net moment of 40 μ_B. It was also found that $(MnO)_2$ and $(MnO)_3$ clusters are unusually stable and form the basis for further growth.

The magic iron-oxide cluster $Fe_{13}O_8$ was discovered by using a reactive laser vaporized cluster source (118), and the equilibrium geometry has been determined from DFT calculations—carried out with the GGA functional and a plane-wave basis set—to be of D_{4h} symmetry. Comparing with $Fe_{13}O_6$, $Fe_{12}O_9$, and $Fe_{14}O_7$, $Fe_{13}O_8$ has a closed electron configuration, a large HOMO–LUMO gap, and a large binding energy, which suggest that it is most stable, in agreement with the experimental observation (Figure 7.5). The main interaction between oxygen and iron atoms in the cluster is the hybridization between O-2p and Fe-3d orbitals. The electronic structure and magnetic properties for this magic iron oxide cluster are discussed in detail in Ref. 118.

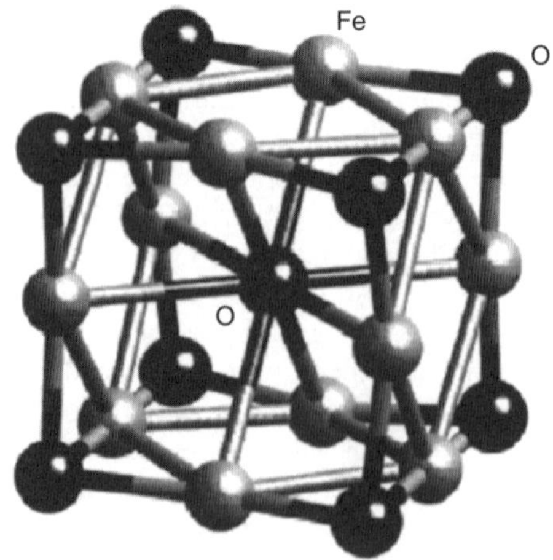

Figure 7.5. Optimized structure of $Fe_{12}O_9$. Reprinted from Ref. 118 with permission.

The M_9O_6 (M = Fe, Co, Ni) oxide clusters have been produced by using reactive laser vaporized cluster source; their structure has been investigated by Sun et al. (119) using also DFT calculations within a plane-wave basis set. The possible equilibrium geometries for these three oxide clusters have a skeleton composed of nine metal atoms with the O atoms capping the triangular faces. In Fe_9O_6 and Co_9O_6 clusters, the oxygen atoms are antiferromagnetically polarized, whereas they are ferromagnetically polarized in Ni_9O_6, which is similar to the case of O atoms adsorbed on the Ni(110) surface.

7.3.2.4. Molybdenum and Tungsten Oxides Photoelectron spectroscopy (PES) and *ab initio* calculations have been combined (120) to investigate the electronic structure of negatively charged MO_n^- clusters (M = W, Mo; n = 3–5). Similar PES spectra were observed between the W and Mo species. A large energy gap between the first and the second PES bands was observed for MO_3^- and correlated with a stable closed-shell neutral cluster. The experimental results have been compared with extensive DFT and *ab initio* CCSD(T) calculations, which were performed to elucidate the electronic and structural evolution for the tungsten oxide clusters. WO_3 is found to be a closed-shell, nonplanar molecule with C_{3v} symmetry. WO_4 is shown to have a triplet ground-state with D_{2d} symmetry, whereas WO_5 is found to be an unusual charge-transfer complex.

Two doubly charged polyoxoanions, $Mo_6O_{19}^{2-}$ and $W_6O_{19}^{2-}$, have been observed in the gas phase using electrospray ionization. Their electronic structures have been investigated using photoelectron spectroscopy and DFT calculations with relativistic effective core potentials (121). Each dianion was found to be highly stable despite the presence of strong intramolecular Coulomb repulsion, which was estimated to be about 2 eV for each system. The valence detachment features originate from electronic excitations involving oxygen lone-pair-type orbitals, and their observed energies have been found to be in excellent agreement with the theoretical vertical detachment energies calculated using time-dependent density functional theory.

7.4. CONCLUSION

In this chapter, we have provided a brief description of the foundations and key features of the methods of electronic structure. These methods are commonly used to approach a variety of problems in molecular chemistry and solid-state physics but are also starting to be used to study oxide nanoparticles. The main difficulty in the application of these sophisticated methods lies in the high computational cost and on the very large number of structures to be explored. In this sense, the combined use of genetic algorithms and *ab initio* calculations seems to open a promising way.

From the examples discussed in the previous section, one can readily see that the field is in its infancy with many references being published just in the last five years. Nevertheless, some general conclusions are already apparent. First, one must realize that the structure of oxide nanoparticles can be completely different from that of the common crystal polymorphs with concomitant new sites for chemistry. Second, magic numbers appear as in the case of metal clusters discovered long ago. However, the interplay between size and stoichiometry leads to a largely increased complexity. Finally, some oxide nanoparticles described above seem to have very special properties that make them good candidates for new technological application. The systematic search for such kind of nanoparticles faces important problems because of their structural and chemical complexity. A successful research will obviously require a strong collaboration between theory and experiment, and in each case, many different strategies will have to be used simultaneously.

ACKNOWLEDGMENT

Financial support from the Italian *Ministero dell'Istruzione, Università e Ricerca*, through a COFIN project, from the Spanish *Ministerio de Ciencia y Tecnologia* project BQU2002-04029-C02-01 and from the *Generalitat de Catalunya* (projects 2001SGR-00043 and *Distinció de la Generalitat de Catalunya per a la Promoció de la Recerca Universitària* awarded to F.I.) is acknowledged.

REFERENCES

(1) *Encyclopedia of Computational Chemistry*; Schleyer, P.V.; Allinger, N.L.; Clark, T.; Gasteiger, J.; Kollman, P.A.; Schaefer, H.F., III; Schreiner, P.R. (Editors); Wiley: Chichester, UK, 1998.

(2) Szabo, A.; Ostlund, N.S. *Modern Quantum Chemistry: Introduction to Advanced Electronic Structure Theory*; Macmillan Publishing: New York, 1982.

(3) Jensen, F. *Introduction to Computational Chemistry*; Wiley: Chistester, UK, 1999.

(4) Dirac, P.A.M. *The Principles of Quantum Mechanics*; Oxford University Press: London, 1958.

(5) von Neumann, J. *Mathematical Foundations of Quantum Mechanics*; Princeton University Press: Princeton, NJ, 1955.

(6) Cohen-Tannoudji, C.; Diu, B.; Laloë, F. *Quantum Mechanics*; Wiley: New York, 1977.

(7) Parr, R.G.; Yang, W. *Density Functional Theory of Atoms and Molecules*; Oxford Science, Oxford University Press: London, 1989.

(8) McWeeny, R. *Methods of Molecular Quantum Mechanics*; Academic Press: London, 1992.

(9) Shavitt, I. *Israel J. Chem.* **1993**, *33*, 357.

(10) Bauschlicher, C.W., Jr.; Langhoff, S.R.; Taylor, P.R. *Adv. Chem. Phys.* **1990**, *77*, 103.

(11) Gaussian Basis Sets; Available at: www.emsl.pnl.gov:2080/forms/basisform.html.

(12) McWeeny, R. *Old and New Approaches in Valence Bond Theory and Chemical Structure*; Klein, D.J.; Trinajstic, N. (Editors); Elsevier: Amsterdam, 1990.

(13) Ohanessian, G.; Hiberty, P.C. *Chem. Phys. Lett.* **1987**, *137*, 437.

(14) Sini, G.; Ohanessian, G.; Hiberty, P.C.; Shaik, S.S. *J. Amer. Chem. Soc.* **1990**, *112*, 1407.

(15) Broer, R.; Nieuwpoort, W.C. *Theor. Chim. Acta* **1988**, *73*, 405.

(16) Knowles, P.J.; Handy, N.C. *J. Chem. Phys.* **1989**, *91*, 2396.

(17) Werner, H.-J.; Knowles, P.J. *J. Chem. Phys.* **1988**, *89*, 5803.

(18) Knowles, P.J.; Werner, H.-J. *Theor. Chim. Acta* **1992**, *84*, 95.

(19) Taylor, P.R. Accurate calculations and calibration. In *Lecture Notes in Quantum Chemistry, Vol. 58*; Roos, B.O. (Editor); Springer-Verlag: Berlin, 1992; 325.

(20) Roothaan, C.C.J. *Rev. Mod. Phys.* **1951**, *23*, 69.

(21) Pople, J.A.; Binkley, J.S.; Seeger, R. *Int. J. Quantum Chem. Symp.* **1976**, *10*, 1.

(22) Raghavachari, K.; Pople, J.A.; Replogle, E.S.; Head-Gordon, M. *J. Phys. Chem.* **1990**, *94*, 5579.

(23) Malrieu, J.P.; Heully, J.L.; Zaitsevskii, A. *Theor. Chim. Acta*, **1995**, *90*, 167.

(24) Paldus, J. Coupled cluster theory in methods of computational molecular physics. In *NATO ASI Series, Series B: Methods in Computational Molecular Physics, Vol. 293*; Wilson, S.; Diercksen, G.H.F. (Editors); Plenum Press: New York, 1992; 99.

(25) Peris, G.; Planelles, J.; Paldus, J. *Int. J. Quantum Chem.* **1997**, *62*, 137.

(26) Bartlett, R.J.; Purvis, G.D. *Int. J. Quantum Chem.* **1978**, *14*, 516.

(27) Purvis, G.D.; Bartlett, R.J. *J. Chem. Phys.* **1982**, *76*, 1910.

(28) Buenker, R.J.; Peyerimhoff, S.D.; Butscher, W. *Mol. Phys.* **1978**, *35*, 771.

(29) Peyerimhoff, S.D.; Buenker, R.J. *Chem. Phys. Lett.* **1972**, *16*, 235.

(30) Bofill, J.M.; Moreira, I. de P.R.; Anglada, J.M.; Illas, F. *J. Comput. Chem.* **2000**, *21*, 1375.

(31) MOLPRO is a package of ab initio programs written by Werner, H.-J.; Knowles, P.J.; Schütz, M.; Lindh, R.; Celani, P.; Korona, T.; Rauhut, G.; Manby, F.R.; Amos, R.D.; Bernhardsson, A.; Berning, A.; Cooper, D.L.; Deegan, M.J.O.; Dobbyn, A.J.; Eckert, F.; Hampel, C.; Hetzer, G.; Lloyd, A.W.; McNicholas, S.J.; Meyer, W.; Mura, M.E.; Nicklaß, A.; Palmieri, P.; Pitzer, R.; Schumann, U.; Stoll, H.; Stone, A.J.; Tarroni, R.; Thorsteinsson, T. Available at: http://www.molpro.net/.

(32) Huron, B.; Rancurel, P.; Malrieu, J.P. *J. Chem. Phys.* **1973**, *75*, 5745.

(33) Evangelisti, S.; Daudey, J.P.; Malrieu, J.P. *Chem. Phys.* **1983**, *75*, 91.

(34) Illas, F.; Rubio, J.; Ricart, J.M.; Bagus, P.S. *J. Chem. Phys.* **1991**, *95*, 1877.

(35) Langhoff, S.R.; Davidson, E.R. *Int. J. Quantum Chem.* **1993**, *7*, 999.

(36) Andersson, K.; Malmqvist, P.-Å.; Roos, B.O.; Sadlej, A.J.; Wolinski, K.J. *Phys. Chem.* **1990**, *94*, 5483.

(37) Andersson, K.; Malmqvist, P.-Å.; Roos, B.O. *J. Chem. Phys.* **1992**, *96*, 1218.

(38) Serrano-Andrès, L.; Merchàn, M.; Nebot-Gil, I.; Roos, B.O.; Fülscher, M. *J. Chem. Phys.* **1993**, *98*, 3151.

(39) Neese, F. *J. Chem. Phys.* **2003**, *119*, 9428.

(40) Hinze, J. *J. Chem. Phys.* **1973**, *59*, 6424.

(41) Sheppard, R.; Shavitt, I.; Simmons, J. *J. Chem. Phys.* **1982**, *76* 543.

(42) Dalgaard, E.; Jorgensen, P. *J. Chem. Phys.* **1978**, *69*, 3833.

(43) Roos, B.O.; Taylor, P.; Siegbahn, P. *Chem. Phys.* **1980**, *48*, 157.

(44) Siegbahn, P.; Heiberg, A.; Roos, B.O.; Levy, B. *Phys. Sci.* **1980**, *21*, 323.

(45) *MOLCAS* Version 5.4, Andersson, K.; Barysz, M.; Bernhardsson, A.; Blomberg, M.R.A.; Cooper, D.L.; Fülscher, M.P.; de Graaf, C.; Hess, B.A.; Karlström, G.; Lindh, R.; Malmqvist, P.-Å.; Nakajima, T.; Neogràdy, P.; Olsen, J.; Roos, B.O.; Schimmelpfennig, B.; Schütz, M.; Seijo, L.; Serrano-Andrès, L.; Siegbahn, P.E.M.; Stålring, J.; Thorsteinsson, T.; Veryazov, V.; Widmark, P.-O. Lund University: Sweden, 2002, Available at: http://www.teokem.lu.se/molcas/.

(46) Gerratt, J. *Adv. At. Mol. Phys.* **1971**, *7*, 141.

(47) Cooper, D.L.; Gerratt, J.; Raimondi, M. *Int. Rev. Phys. Chem.* **1988**, *7*, 59.

(48) Cooper, D.L.; Gerratt, J.; Raimondi, M. *Chem. Rev.* **1991**, *91*, 929.

(49) Pauncz, R. *Spin Eigenfunctions*; Plenum Press: New York, 1979.

(50) Pisani, C.; Dovesi, R.; Roetti, C. In *Hartree-Fock Ab-initio Treatment of Crystalline Systems, Lecture Notes in Chemistry, Vol. 48*; Springer: Heidelberg, 1988.

(51) Hohenberg, P.; Kohn, W. *Phys. Rev. B* **1964**, *136*, 864.

(52) Illas, F.; Moreira, I. de P.R.; Bofill, J.M.; Filatov, M. *Phys. Rev. B* **2004**, *70*, 132414.

(53) Dunlap, B.I.; Rösch, N. *Adv. Quantum Chem.* **1990**, *21*, 317.

(54) Rösch, N.; Krüger, S.; Mayer, M.; Nasluzov, V.A. In *Recent Developments and Applications of Modern Density Functional Theory. Theoretical and Computational Chemistry, Vol. 4*; Seminario, J.M. (Editor); Elsevier: Amsterdam, 1996; 497.

(55) Vosko, S.H.; Wilk, L.; Nusair, M. *Can J. Phys.* **1980**, *58*, 1200.

(56) Terakura, K.; Oguchi, T.; Williams, A.R.; Klüber, J. *Phys. Rev. B* **1984**, *30*, 4734.

(57) Shen, Z.-X.; List, R.S.; Dessau, D.S.; Wells, B.O.; Jepsen, O.; Arko, A.J.; Barttlet, R.; Shih, C.K.; Parmigiani, F.; Huang, J.C.; Lindberg, P.A.P. *Phys. Rev. B* **1991**, *44*, 3604.

(58) Moreira, I. de P.R.; Illas, F.; Martin, R.L. *Phys. Rev. B* **2002**, *65*, 155102.

(59) Filatov, M.; Shaik, S. *Chem. Phys. Lett.* **1998**, *288*, 689.

(60) Filatov, M.; Shaik, S. *Chem. Phys. Lett.* **1999**, *304*, 429.

(61) Illas, F.; Moreira, I. de P.R.; de Graaf, C.; Barone, V. *Theoret. Chem. Acc.* **2000**, *104*, 265.

(62) Martin, R.L.; Illas, F. *Phys. Rev. Lett.* **1997**, *79*, 1539.

(63) de Graaf, C.; Illas, F. *Phys. Rev. B* **2001**, *63*, 014404.

(64) Becke, A.D. *Phys. Rev. A* **1988**, *38*, 3098.

(65) Perdew, J.P.; Wang, Y. *Phys. Rev. B* **1992**, *45*, 1324.

(66) Perdew, J.P.; Chevary, J.A.; Vosko, S.H.; Jackson, K.A.; Pederson, M.R.; Singh, D.J.; Fiolhais, C. *Phys. Rev. B* **1992**, *46*, 6671.

(67) Lee, C.; Yang, W.; Parr, R.G. *Phys. Rev. B* **1988**, *37*, 785.

(68) Colle, R.; Salvetti, O. *J. Chem. Phys.* **1983**, *79*, 1404.

(69) Perdew, J.P.; Burke, K.; Ernzerhof, M. *Phys. Rev. Lett.* **1996**, *77*, 3865.

(70) Zhang, Y.; Yang, W. *Phys. Rev. Lett.* **1998**, *80*, 890.

(71) Hammer, B.; Hansen, L.B.; Nørskov, J.K. *Phys. Rev. B* **1999**, *59*, 7413.

(72) Becke, A.D. *J. Chem. Phys.* **1993**, *98*, 5648.

(73) Levine, I.N. *Quantum Chemistry, 5th ed.*; Prentice-Hall: England Cliffs, NJ, 2000.

(74) Koch, W.; Holthausen, M.C. *A Chemist's Guide to Density Functional Theory*; Wiley-VCH Verlag GmbH: Weinheim, 2000.

(75) Lopez, N.; Illas, F.; Rösch, N.; Pacchioni, G. *J. Chem. Phys.* **1999**, *110*, 4873.

(76) Lopez, N.; Illas, F. *J. Phys. Chem. B* **1998**, *102*, 1430.

(77) Ranney, J.T.; Starr, D.E.; Musgrove, J.E.; Bald, D.J.; Campbell, C.T. *Faraday Discuss.* **1999**, *114*, 195.

(78) Reynolds, P.J.; Ceperley, D.M.; Alder, B.J.; Lester, W.A. *J. Chem. Phys.* **1982**, *77*, 5593.

(79) *Quantum Monte Carlo Methods in Chemistry and Physics, NATO Advanced Institute Series C: Mathematical and Physical Sciences, Vol. 535*; Kluwer Academic: Dordretch, the Netherlands, 1999 and references therein.

(80) Aguado, A.; Lopez, J.M. *J. Phys. Chem. B* **2000**, *104*, 8398.

(81) Bawa, F.; Panas, I. *Phys. Chem. Chem. Phys.* **2001**, *3*, 3042.

(82) Lu, X.; Xu, X.; Wang, N.Q.; Zhang, Q.N. *Int. J. Quant. Chem.* **1999**, *73*, 377.

(83) de la Puente, E.; Aguado, A.; Ayuela, A.; Lopez, J.M. *Phys. Rev. B* **1997**, *56*, 7607.

(84) Clementi, E. *IBM J. Res. Devel.* **1965**, *9*, 2.

(85) Recio, J.M.; Ayuela, A.; Pandey, R.; Kunz, A.B. *Zeit. Physik D* **1993**, *26(Suppl)*, S237.

(86) Recio, J.M.; Ayuela, A.; Pandey, R.; Kunz, A.B. *J. Chem. Phys.* **1993**, *98*, 4783.

(87) Gutowski, M.; Skurski, P.; Li, X.; Wang, L.S. *Phys. Rev. Lett.* **2000**, *85*, 3145.

(88) Chen, G.; Liu, Z.F.; Giong, X.G. *J. Chem. Phys.* **2004**, *120*, 8020.

(89) Enyashin, A.N.; Makurin, Y.N.; Sofronov, A.A.; Kiiko, V.S.; Ivanovskaya, V.V.; Ivanovskii, A.L. *Russ. J. Inorg. Chem.* **2004**, *49*, 893.

(90) Roberts, C.; Johnston, R.L. *Phys. Chem. Chem. Phys.* **2001**, *3*, 5024.

(91) Fernandez, E.M.; Balbas, L.C.; Borstel, G.; Soler, J.M. *Thin Solid Films* **2003**, *428*, 206.

(92) Sawilowsky, E.F.; Meroueh, O.; Schlegel, H.B.; Hase, W.L. *J. Phys. Chem. A* **2000**, *64*, 4920.

(93) Fernandez, E.M.; Eglitis, R.; Borstel, G.; Balbas, L.C. *Phys. Stat. Sol. B* **2005**, *242*, 807.

(94) Pereira, J.C.G.; Catlow, C.R.A.; Price, G.D. *J. Phys. Chem. A* **1999**, *103*, 3268.

(95) Kong, Q.Y.; Zhao, L.; Wang, W.N.; Wang, C.; Xu, C.; Zhang, W.H.; Liu, L.; Fan, K.N.; Li, Y.F.; Zhuang, J. *J. Comput. Chem.* **2005**, *26*, 584.

(96) Bromley, S.T.; Flikkema, E. *J. Chem. Phys.* **2005**, *122*, 114303.

(97) Flikkema, E.; Bromley, S.T. *J. Phys. Chem. B* **2004**, *108*, 9638.

(98) Zhao, M.W.; Zhang, R.Q.; Lee, S.T. *Phys. Rev. B* **2004**, *70*, 205404.

(99) Nayak, S.K.; Rao, B.K.; Khanna, S.N.; Jena, P. *J. Chem. Phys.* **1998**, *109*, 1245.

(100) Bromley, S.T.; Zwijnenburg, M.A.; Maschmeyer, Th. *Phys. Rev. Lett.* **2003**, *90*, 035502.

(101) Zhang, D.J.; Zhao, M.W.; Zhang, R.Q. *J. Phys. Chem. B* **2004**, *108*, 18451.

(102) Zhang, R.Q.; Zhao, M.W.; Lee, S.T. *Phys. Rev. Lett.* **2004**, *93*, 095503.

(103) Bromley, S.T. *Nano Lett.* **2004**, *8*, 1427.

(104) Zhao, M.W.; Zhang, R.Q.; Lee, S.T. *Phys. Rev. B* **2004**, *69*, 153403.

(105) Lu, W.C.; Wang, C.Z.; Ho, K.M. *Chem. Phys. Lett.* **2003**, *378*, 225.

(106) Zhang, R.Q.; Chu, T.S.; Lee, S.T. *J. Chem. Phys.* **2001**, *114*, 5531.

(107) Kaiser, B.; Bernhardt, T.M.; Kinne, M.; Rademann, K.; Heidenreich, A. *J. Chem. Phys.* **1999**, *110*, 1437.

(108) Martins, B.L.; Longo, E.; Andres, J. *Int. J. Quantum Chem.* **1993**, *Suppl. 27*, 643.

(109) Behrman, E.C.; Foehrweiser, R.K.; Myers, J.R.; French, B.R.; Zandler, M.E. *Phys. Rev. A* **1994**, *49*, R1543.

(110) Bell, R.C.; Zemski, K.A.; Kerns, K.P.; Deng, H.T.; Castleman, A.W. *J. Phys. Chem. A* **1998**, *102*, 1733.

(111) Justes, D.R.; Mitric, R.; Moore, N.A.; Bonacic-Koutecky, V.; Castleman, A.W. *J. Am. Chem. Soc.* **2003**, *125*, 6289.

(112) Schoiswohl, J.; Kresse, G.; Surnev, S.; Sock, M.; Ramsey, M.G.; Netzer, F.P. *Phys. Rev. Lett.* **2004**, *92*, 206103.

(113) Albaret, T.; Finocchi, F.; Noguera, C. *Faraday Discuss.* **1999**, *114*, 285.

(114) Tsipis, A.C.; Tsipis, C.A. *Phys. Chem. Chem. Phys.* **1999**, *1*, 4453.

(115) Anokhina, E.V.; Essig, M.W.; Day, C.S.; Lachgar, A. *J. Am. Chem. Soc.* **1999**, *121*, 6827.

(116) Hamad, S.; Catlow, C.R.A.; Woodley, M.; Lago, S.; Mejias, J.A. *J. Phys. Chem. B* **2005**, *109*, 15741.

(117) Nayak, S.K.; Jena, P. *Phys. Rev. Lett.* **1998**, *81*, 2970.

(118) Wang, Q.; Sun, Q.; Sakurai, M.; Yu, J.Z.; Gu, B.L.; Sumiyama, K.; Kawazoe, Y. *Phys. Rev. B* **1999**, *59*, 12672.

(119) Sun, Q.; Sakurai, M.; Wang, Q.; Yu, J.Z.; Wang, G.H.; Sumiyama, K.; Kawazoe, Y. *Phys. Rev. B* **2000**, *62*, 8500.

(120) Zhai, H.J.; Kiran, B.; Cui, L.F.; Li, X.; Dixon, D.A.; Wang, L.S. *J. Am. Chem. Soc.* **2004**, *126*, 16134.

(121) Yang, X.; Waters, T.; Wang, X.B.; O'Hair, R.A.J.; Wedd, A.G.; Li, J.; Dixon, D.A.; Wang, L.S. *J. Phys. Chem. A* **2004**, *108*, 10089.

CHAPTER 8

Parametric Quantum Methods in Modeling Metal Oxide Nanoclusters and Surfaces

F. RUETTE and M. SÁNCHEZ

Laboratorio. de Química Computacional, IVIC, Apartado. 21827, Caracas 1020-A, Venezuela.

8.1. INTRODUCTION

The field of nanoscale particles in science semiconductor physics and catalysis has a rapid development without precedent during the beginning of this century. Small feature sizes result in an increased functionality, faster speed, lower costs in microelectronic processing power, dimension of dynamic memories, and catalytic selective activity. These subjects are of paramount importance because technological achievements of the present and future times are in the research on nanoscopic dimensions.

Techniques for the development of nanodevices in the electronic industry and catalysis involve several experimental preparation procedures (physical and chemical methods; Chapters 3 and 4) and the possibility to carry out theoretical modeling to predict properties such as magnetism, structure, reactivity, conductivity, hardness, chemical selectivity, nucleation, and growth processes. Systematic studies regarding structure-performance relationships are difficult to carry out experimentally, and therefore, here is a spot where theoretical modeling may play an important role. Several recent reviews about preparation and characterization of nanoparticles (1–3) indicate the complexity and importance of this very active new area of scientific research.

Many theoretical methods have been employed to model metal-oxide clusters depending on the desired accuracy and the size of the system. Normally modeling nanoparticles requires a considerable amount of atoms and involves very expensive computational tasks, depending on the level of theory employed. In a recent work (4),

Synthesis, Properties, and Applications of Oxide Nanomaterials, Edited by José A. Rodríguez and Marcos Fernández-Garcia

the complexity of catalytic processes, together with a survey of the most recent and relevant computational *ab initio* works, showed that parametric quantum methods (PQMs) are an alternative for modeling these systems. It is expected that in the same way as quantum modeling has had a great impact on drug-development technologies, its use in the field of nanoscience should have a great future in the prediction of properties for nanometer-sized structures of metal oxides.

The word "semiempirical" means *partly from experiment*. In practice, this denomination has a limitation because different approaches for parameterization are totally based on data obtained from more sophisticated theoretical methods. For example, the extension of PQMs to bimetallic systems is almost impossible due to the lack of molecular experimental data. This is also true for metallic nanoclusters in which their geometry (bond distance, angles) and energetic binding energy are unknown. Therefore, *ab initio* calculations are fundamental to obtain the molecular data to perform parameterization. In this sense, the term "parametric quantum method" (PQM) (based on parameters) is a more universal manner to design semiempirical approaches.

The challenge of modeling nanoparticles is to establish predictability for the new chemical and physical properties of these novel materials that are in the frontier between solid-state physics and molecular chemistry. Microscopic modeling of metal-oxide particles properties is a challenge of great importance in microelectronics, catalysis, magnetic materials, corrosion, semiconductor devices, environmental chemistry, bioengineering, combustion, lasers, solar cells, sensors, and so on. In this sense, a recent review by Fernández-García et al. (1) presents a selection of experimental techniques and theoretical methods to examine the characterization and behavior of nanostructured metal-oxide particles in chemistry.

The reactivity of nanoscale systems for catalytic applications depends on the adsorption energy and dissociation barrier of gas molecules on their surfaces. This is mainly dominated by quantum-size effects that govern the electronic spectra of clusters (active orbitals in reactivity), by the geometrical dynamical fluxionality (structural transitions) and by composition, such as impurity-doping effects (the atomic coordination at the surface depends on their chemical neighborhood).

Qualitative information is very important from a practical point of view. As mentioned by Reynolds (5), "Many industrial problems are simply too large for timely solutions using large-scale *ab initio* calculation. Industrial chemists are often more interested in trends than absolute numbers." The efficiency of PQMs was evaluated by Bredow and Jug (6) by comparison of a benchmark test of the PQM MSINDO-CCM (cyclic cluster model, CCM) and *ab initio* CRYSTAL-SCM (supercell model) using the PWGGA exchange correlation potential and HF method showed that the PQM is 3000 time faster than the *ab initio* ones. In a similar way, Clark (7) talking about the *quo vadis* of semiempirical MO-theory pointed out that PQM is far from dead, if one considers a factor of 10^3–10^4 faster than DFT calculations.

Several developments of great importance have been recently extended to PQMs to give them greater flexibility, potentiality, and the inclusion of an extensive set of tools to aid in the study of molecular structure, electronic properties, and chemical reactions in complex systems: (*1*) Corrections in PQMs for describing H-bonded systems (8). (*2*) Combination of quantum mechanics *ab initio* and DFT with PQM (QM/SE) using different approaches (9). (*3*) Extension to very large systems using PQM/molecular mechanics and applications of QM/MM to complex systems (10–13). (*4*) The use of molecular similarity to build more precise semiempirical structure theories, which have been proposed by Janesko and Yaron (14). (*5*) Parallel programs for evaluation of large systems (NANOPACK) using PQMs, which have been implemented by Berzigiyarov et al. (15). (*6*) Detecting anomalies in reported enthalpies of formation of organic compounds using PQMs, which has been reported by Stewart (16). (*7*) A multi-reference configuration interaction algorithm (MRCI) in an effective valence-shell Hamiltonian, which has been given by Strodel and Tavan (17). (*8*) Transferability of parameters for a sequential derivation of molecular mechanics using strictly local geminal wave functions, which has been proposed Tokmachev and Tchougréeff (18).

In Section 8.2, we present the theoretical foundations of PQMs, indicating the prominent future of these techniques based on the unexplored field of elementary parametric functionals (EPFs). In addition, a review of PQM applications, over the last six years (2000–2005), is carried out to evaluate properties of metal-oxide nanoparticles. Applications of PQMs to metal oxides are less frequent than to organic systems, because of the complexity of the former and the fact that few methods have been parameterized for metallic elements. Different methods are considered here: MSINDO, tight-binding, CATIVIC, INDO/S, ZINDO, MNDO, AM1, and PM3. Finally, some comments and conclusions of this survey are presented in the last section.

8.2. GENERAL FOUNDATIONS OF PARAMETRIC QUANTUM METHODS

Although this chapter is related to applications of PQMs, in this section, a highlight of their fundamental aspects are presented to show the prominent future of these methods in molecular and materials modeling. We do not emphasize the analytical structure of the EPFs that represents basic interactions, because this topic deserves a special chapter due to the multiple variety of these functionals reported in the literature for different methods (6,19,20). Instead, a presentation of the general fundamental formalism is presented to establish a solid PQM theory that supports the systematic improvement of these methods.

Functional analysis can be used to establish the basis of *ab initio* quantum mechanics (21–25) and PQM approaches (26,27). Minimax and variational principles (21) can be set up in terms of accurate simulated analytical energy functionals E_{exa} by parametric energy functionals E_{pa}. The optimal E_{pa} can be expressed in terms of a set of EPFs $\{f_{\mathrm{pa}}\}$ that belong to a family set of parametric functionals ($\boldsymbol{F}_{pa}$) that are associated with intra- and inter-molecular interactions. The minimax principle for the stable state ($E^I, I = 0$) corresponds to the variational principle (27):

$$E^0_{pa}[\boldsymbol{H}_{pa}(R), S] = \min_{\substack{f_{pa}(R)\in \boldsymbol{F}_{pa} \\ \left\{D^{0,a}_{\mu\nu}\right\}}} \overline{E_{pa}}\left(\{D^{0,a}_{\mu\nu}\}, \{f_{pa}\}, R\right) \tag{8.1}$$

where $D^{0,a}_{\mu\nu} = d^a_\mu d^{*a}_\nu$; d^a_μ is the coefficient of an element μ of a basis set $\{\phi\}$, in the expansion of the orthonormal ath molecular orbital (MO) expressed as $\chi^a = \sum_\mu d^a_\mu \phi_\mu$, and R is the set of molecular coordinates. In Eq. 8.1, the parametric total energy functional (E_{pa}) is minimized with respect to sets of coefficients $\{D^{0,a}_{\mu\nu}\}$ and EPFs $\{f_{\text{pa}}\}$. The basis set $\{\phi\}$ belongs to a subspace M of the Hilbert space $\boldsymbol{H}$. $\boldsymbol{F}_{\text{pa}}$ is a family of parametric functionals that must obey the same algebraic properties of a set of analytical functionals $\{\boldsymbol{f}\}(\{\boldsymbol{f}\} = \{(\phi_\mu, h_i(R)\phi_\nu), (\phi_\mu\phi_\nu, G_{ij}(R)\phi_\lambda\phi_\sigma)\})$. The dependence with respect to the basis set is considered by a family $\boldsymbol{F}$ of functionals generated by a set of one- and two-center operators $\Omega = \{h_i(R), G_{ij}(R)\}$ from the exact Hamiltonian operator $\boldsymbol{H}(R)$ applied on the set $\{\phi\}, \phi \in M \in \boldsymbol{H}$. It is important to emphasize that the set $\{f_{\text{pa}}\}$ must come from simulation of a set $\{f_{\text{opt}}\}$ obtained from an optimal basis set $\{\phi\}_{\text{opt}}$. This basis set is unknown, and it is denoted with the acronym OTMBS (optimal transformed minimum basis set) (26,28) that belongs to a subspace $M_{\text{opt}} \in \boldsymbol{H}$. The OTMBS is associated with inter- and intra-molecular interactions over the set of intervals $\{[R_X\text{–}R_Y]\}$ for X and Y atoms. Several orthogonality and localization conditions related to this basis set have been established (26). The OTMBS concept is rationalized by considering different auxiliary basis sets for dissimilar local environments in which electronic interactions are embodied. Note that functionals $\{f_{\text{opt}}(R)\}$ and $\{f_{\text{pa}}(R)\}$ depend on $\{\phi\}_{\text{opt}}$ and R. It means that once a functional is selected, there is implicitly a $\{\phi\}_{\text{opt}}$ for all R in the set of intervals $\{[R_X\text{–}R_Y]\}$.

The set of EPSs associated with nuclear–nuclear, electron–nuclear, and electronic–electronic interactions is evaluated by using simulation techniques of effective Hamiltonians, as has been early proposed by Durand and Malrieu (29) and later by Primera et al. (30). For example, a molecular system $M = W_l X_m Y_n$, with N atoms ($N = l + m + n$) requires the best representation of X, Y, and W atoms and X–X, W–W, Y–Y, X–Y, W–X, and W–Y energy component interactions.

In practice, PQMs are based on parametric expressions related to bond energy functionals ($\boldsymbol{BE}_{\text{pa}}$), such as dissociation energy or heat of formation, that are also defined by means of simulation techniques:

$$\min_{BE^{M_I}_{\text{pa}} \in \boldsymbol{BE}} \left(\sum_M \sum_I \left| BE^{M_I}_{\text{exa}} - BE^{M_I}_{\text{pa}} \right|^2 \right)^{1/2} \tag{8.2}$$

where $BE^{M_I}_{\text{exa}}$ and $BE^{M_I}_{\text{pa}}$ are the exact and parametric bonding energies of the M molecule in the Ith molecular electronic state, respectively. $BE^{M_I}_{\text{exa}}$ values are obtained experimentally or theoretically as accurate as possible, for example, configuration interaction (28,31) or other less-accurate methods, such as DFT. Parameters of $BE^{M_I}_{\text{pa}}$ are calculated not only to reproduce $BE^{M_I}_{\text{exa}}$ but also geometrical, electronic, spectroscopic, and mechanical properties, depending on the research of interest (6,19,20). $\boldsymbol{BE}$ is a family set of parametric binding energy functionals. $BE^{M_I}_t$ ($t = \text{exa}, \text{pa}$) is defined as follows:

$$BE^{M_I}_t = E^{M_I}_t - \sum_X^{N_M} E^X_t \tag{8.3}$$

where E^{M_I} and E^X are total energies of the M molecule in state I and of the atom X of the M molecule, respectively.

For PQMs, the total energy for a molecular system can be easily divided in diatomic (E_{XY}) and monoatomic (E_X) energy terms:

$$E^M = \sum_X E_X + \sum_{\substack{X>Y \\ X,Y\in M}} E_{XY} \tag{8.4}$$

Monoatomic contributions to the total energy depends on EPFs that belong to the X atom, such as $h_{X\mu X\mu}$ and $\gamma_{X\mu X\nu}$, which correspond to core Hamiltonian and electron–electron interactions (Coulomb and exchange), respectively. The diatomic energy can also be expressed in terms of EPFs that depend explicitly on R_{X-Y} distance, such as

$$E_{XY} = f(\{h_{X\mu Y\nu}\}, \{\gamma_{X\mu Y\nu}\}, \{V_{X\mu Y\nu}\}, \{E^C_{XY}\}) \tag{8.5}$$

where $h_{X\mu Y\nu}, \gamma_{X\mu Y\nu}, V_{X\mu Y\nu}$, and E^C_{XY} correspond to resonance, electron–electron repulsion, electron–nucleus attraction, and core–core repulsion functionals, respectively. Differences between PQMs depend on the way E_{XY} and E_X functionals have been defined in terms of EPFs and in the manner that parameterization is performed.

Parametric functionals for molecular and atomic systems may be optimized by nonlinear techniques to find optimal parameters that correspond to a global minimum of an expression that include other properties, such as bond distance and angles. Simulated annealing (SA) (32), evolutionary algorithms (EAs) (33), tabu search (TS) (34), neural network (NN) (35), and ant colonies (AC) (36) techniques can be used to locate a global minimum in nonlinear complicated multivariable functions by minimizing the differences between theoretical multiparametric functionals and experimental data for several properties. The potentiality of PQM methods depends on the feasibility of finding well-defined EPFs with their corresponding parameters. Very few applications of nonlinear optimization techniques to find optimal parameters in PQM methods have been reported in the literature (37–39).

The following considerations for parametric methods may be ascertained: (*1*) Integral calculations are little time consuming because the number of them decreases with the application of neglect differential overlap (NDO) approach, and most of them are evaluated from EPFs; therefore, these methods are faster than the rest of the QC methods. (*2*) The main drawback is the troublesome parameterization process and a correct selection of parametric functionals to ensure accuracy. (*3*) Programs for optimization of parameters and functionals are not available to users, contrarily to standard *ab initio* methods where the optimizations of basis sets, active space, and configuration selection are customary tasks. (*4*) Small molecules can be calculated with relatively high accuracy and short time by *ab initio* methods, and on other hand, PQM can be employed to model large complex systems. For this reason, we assert that parametric and *ab initio* methods may be considered complementary.

8.3. APPLICATIONS OF PQMS METHODS TO METAL-OXIDE SYSTEMS

In this section, we analyze the potentiality of several PQMs with applications to model surface reactions and properties of nanoparticles and solids. We select the most used methods for applications to metal-oxide systems: MSINDO, tight binding, MNDO, AM1, PM3, INDO/S, and CATIVIC.

8.4. MSINDO METHOD

Karl Jug's group has created perhaps the most successful PQM for the first to the fourth row of elements in the Periodic Table with applications to metal-oxide surface processes. Two versions of the symmetrical orthogonalized INDO method (SINDO1) (40–43) and more recently an improved version of SINDO1, modified SINDO (MSINDO) (44–47), have been developed. Several general features characterize the last method: (*1*) Orthogonalization transformation corrections (48) into the Hartree–Fock (HF) equations to improve the simulation of H_{core} matrix elements; (*2*) different set of atomic orbitals for the inter- and intra-atomic interactions; (*3*) extended basis set for hypervalence compounds of third (Al–Cl) and four (Ga–Br) periods (46); (*4*) Zener's pseudopotential is incorporated for including the effect of inner orbitals; (*5*) zero-point energy corrections for the evaluation of heat of formation; and (*6*) an extension to the first row transition elements (Sc–Zn) (47). This information is presented in more detail in a recent review of parametric methods given by Bredow and Jug (6).

Numerous applications of MSINDO to the surface properties of metal oxides have been carried out in recent years by Jug et al. (49–65). Several works related to adsorption and properties of MgO surfaces (49–51), for example, H_2O molecules' adsorption on a defective MgO(100) surface modeled by clusters of $Mg_{162}O_{162}$, $Mg_{144}O_{144}$, $Mg_{200}O_{198}$, and $Mg_{200}O_{200}$ (49). Results of adsorption energy show that MSINDO values are close to experimental findings and reveal that the dissociative adsorption of water mostly occurs on surface-defect sites. They found a very high-energy formation of defects that is compensated with H_2 formation. On the other hand, Tikhomirov and Jug (50) studied several water molecules adsorbed on MgO(100) to simulate a monolayer coverage. They reported that the most stable arrangement is a (2×2) overlayer structure with water molecules tilted on the surface. Formation of strip and bands occur with Mg–OH_2 and lateral O–H interactions between different H_2O molecules. Comparison of the MSINDO adsorption energy (−13 kcal/mol) with the temperature programmed desorption (TPD) experimental value (−16 kcal/mol) shows very close agreement.

Formation of Cu_n clusters ($n = 2$–6, 16, 20, 32, 36, and 52 atoms) on a MgO surface modeled by a $(8 \times 8 \times 3)$ $Mg_{96}O_{96}$ cluster were performed by Geudtner et al. (51) Comparison with DFT calculations is quite reasonable and shows that this PQM can describe correctly interactions of Cu clusters on the modeled MgO(100) surface.

The results predict the formation of a single double layer and then the construction of islands, in agreement with experimental results.

Janetzko et al. (52) investigated the formation of single oxygen vacancies in Al_2O_3 by using the cyclic cluster model (CCM) (53) incorporated to MSINDO and by the supercell model (SCM) employing DFT and HF methods. MSINDO results for cohesive energy of the corundum crystal are in better concordance with experimental data than HF and DFT results. The relaxation process is explained by repulsion between nearest neighbor atoms that form the vacancy.

Some studies for adsorption of small molecules (NO, NH_3, and H_2O) on vanadium- and titanium-oxide surfaces were completed by Homann et al. (54,55). The TiO_2 anatase (100) surface was modeled with $(TiO)_n$ $(H_2O)_m$ $(n = 33–132, m = 17–48)$ clusters using OH and H as saturation groups. In a similar manner, $V_2O_5(001)$ was simulated with $(V_2O_5)_n(H_2O)_m$ $(n = 24–54, m = 20–40)$ clusters. Comparison of experimental adsorption energies (E_{ads}) of NO (−21 kcal/mol) and NH_3 (−31, 36 kcal/mol) on anatase(100) with theoretical values of (−18 to −20) and (−37 to −38 kcal/mol), respectively, are very close. The order found was $|E_{\text{ads}}(NH_3)| > |E_{\text{ads}}(HO_2)| > |E_{\text{ads}}(NO)|$. In the case of a $V_2O_5(001)$ surface, results are in good agreement with DFT and experimental data; i.e., the extrapolated bulk binding energy 925 kcal/mol compares very well with the experimental value of 909 kcal/mol, and adsorption energies are in reasonable agreement with DFT calculations. All of these features suggest that the MSINDO method can be a qualitative and, in some cases, semiquantitative tool to model surface reactions in nanoclusters.

Other multiple applications of MSINDO are molecular dynamics simulations to surface and catalysis by Nair et al. (56) in the interaction of vanadia clusters on a titania (anatase surface) catalyst. They also study the interaction with small molecules $V_xO_yH_z$ $(x = 1, 2; y = 4, 7; z = 3, 4)$ on a TiO_2 support (57). The stability of different structures of $V_2O_7H_4$ and VO_4H_3 was studied for adsorption on (101), (001), and (100) clean surfaces represented by different nanoclusters of TiO_2 ($Ti_{36}O_{72}$). The results show different structures of $V_2O_7H_4$ and VO_4H_3 depending on the crystalline exposed surface.

The high potentiality of PQMs is reflected in calculations by Steveson et al. (58) for the chemical modeling of the structure of aluminum doped anatase and rutile titanium dioxide. They used 29 models of $Ti_xAl_yO_z$ $(x = 53–56, 86–90; y = 1, 2, 4; z = 111, 112, 180)$ obtained from the clusters shown in Figure 8.1, by the inclusion of Al or the substitution of Ti by Al. Results suggested that the replacement of Ti–O–Ti by two Al generates the most stable defect structure in anatase. The following most stable defect corresponds to a single replacement of Ti by Al. An inverse order is obtained for the rutile structure. One important fact is that the relaxation process is very important in the stability of the defects. The Al-doping of anatase clusters is thermodynamically favored as compared with rutile. From these results, they concluded that Al doping favors the rutile–anatase transition.

Investigations for metallic compounds with metal oxides have been carried by Janetzko and Jug (59) with MSINDO, for example, the miscibility of ZnO and ZnS using the $Zn_{48}O_{48}$ cluster as a model. Results show a good agreement between theoretical findings and experimental data. In a similar way, structural and electronic

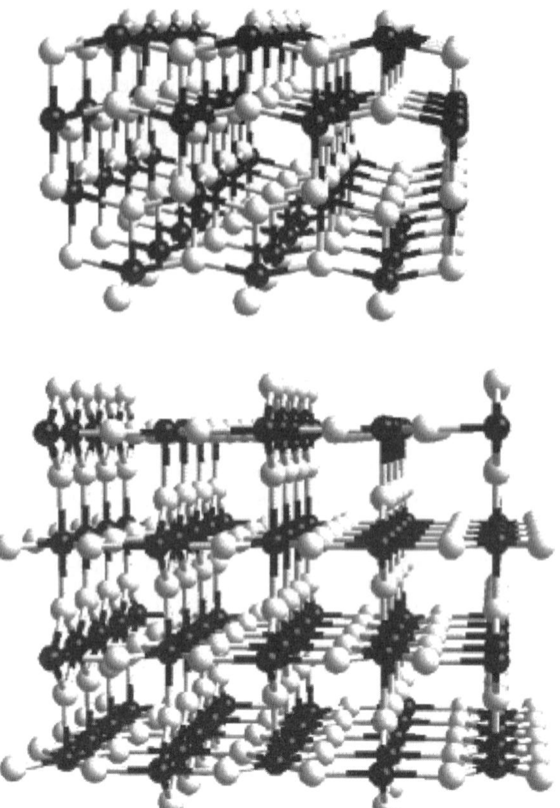

Figure 8.1. Cluster models $Ti_{56}O_{112}$, anatase (top), and $Ti_{90}O_{180}$, rutile (bottom), used for the MSINDO calculations. The clusters were embedded in arrays of 1992 and 1134 pseudo-atoms, respectively (not shown). Titanium atoms are shown in black and oxygen atoms in white. From Ref. 58.

properties of $Li_2B_4O_7$ were modeled with a cyclic cluster of $Li_{16}B_{32}O_{56}$ by Islam et al. (60). The authors concluded that the choice of DFT approaches or MSINDO depends on the complexity of the system and the property that is studied. For example, MSINDO for $Li_2B_4O_7$ gives a better band gap (9.7 eV) than several DFT methods (17.68 (HF+PW), 6.76 (PWGGA), 6.3 (PWGGA-US), 6.2 eV (PWGGA-PAW)) as compared with the experimental estimated value of 9.0 eV. However, PW1PW and B3LYP also give values close to the experimental value (8.88 and 8.85 eV, respectively).

Finally, other applications of MSINDO to surface processes correspond to adsorption of NO, NH_3, and H_2O on V_2O_5/TiO_2 (61); molecular dynamics investigation of oxygen vacancy diffusion in rutile (62); nucleation of CuGa phases on the MgO(100) (63); the mechanism of acidic dissolution at the MgO(100) (64); and extra-framework gallium species in Ga/SAPO-11 catalyst and n-butane transformation (65).

8.5. TIGHT-BINDING METHOD

Tight binding (TB) theory is a heavily parametrized extended Hückel theory with the explicit introduction of pairwise atomic repulsion (66–70). Tight-binding methods are broadly employed to study the electronic structure of transition metal oxides, conductivity, magnetism, fuel cells, elastic constants, surface vacancies, doping, fonon frequencies, band gap, optical absorption, and so on (71–101). Although the main application is mainly on extended solids, it can be used for nanoparticles. Here, we will show a sample of the most recent and varied applications of this technique to metal oxides.

Miyamoto and Kubo's group has applied the TB molecular dynamics method (TBMD) to several systems (71–75), for example, to calculate the electronic structure of In_2O_3 and indium tin oxide (71). They reported band gap values that compare well with the experimental results and underlined that this software is 5000 times faster than the conventional first-principles quantum methods. Furthermore, this group studied the structural and electronic properties of $Li_xMn_2O_4$ cathode materials for a lithium secondary battery with TBMD (72). They found an average potential for the reaction $Li + Mn_2O_4 \rightarrow LiMn_2O_4$ and lattice constants for $LiMn_2O_4$ and λ-MnO_2 in good concordance with the experimental values. In the area of batteries technology, the density of states for Ni/CeO_2 and Cu/Ceo_2 systems used as an anode in solid-oxide fuel cells was evaluated by the same group (73). Their results using the TBMD program explain the electron transfer from metal to oxide and the contributions of Ni and Cu orbitals to the density of states close to the Fermi level in each cermet. They conclude that the effectiveness of the TB approach to the investigation of complex cerment systems was proved. Other work of solid electrolyte fuel cell was carried out by the same authors (74) to study the adsorption and dissociation of O_2 on $La_{1-x}Sr_xMnO_3 - \sigma$(LSM) surfaces. This TB approach has also been successfully applied to investigate adsorption on ultra-fine metal particles on metal-oxide supports (75). The effect of γ-Al_2O_3(100) support on M_{13} nanoclusters (M = Ru, Rh, Pd, Ag, Os, Ir, Pt, Au) was analyzed by the adsorption of NO. They found that NO activation is correlated with the electron transfer from the metal cluster to NO. In particular, the NO adsorption on Pt_{13} cluster is significantly stabilized by the γ-Al_2O_3(100) support.

To compare the structure of free and supported metallic nanoclusters, Mottet et al. (76) used molecular dynamics TB semiempirical potentials to describe metal–metal interactions. Clusters of Ni, Pd, Pt, Cu, Ag, and Au with noncrystalline structures supported in MgO(100) surface were considered. Substrate-induced modifications in morphology and atomic structure were analyzed as a function of the cluster size of Pd clusters and the dynamics of growth and melting of Ag clusters depended on the epitaxial relation with the support.

The stability and changes in the bonding due to a vacancy in SnO_2 crystalline grains have been extensively studied by Mazzone et al. (77–82) using big clusters, varying from 20 to 900 atoms; see, for example, Figure 8.2. They found that grain stability depends on grain composition, shape, and the presence of vacancies, rather than on its size. The studied grains are size comparable with the experimental ones,

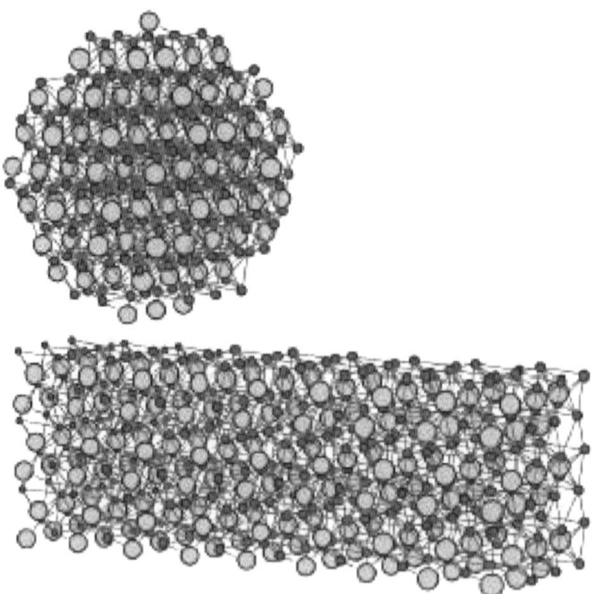

Figure 8.2. Three-dimensional view of a spherical and columnar nanoparticles of SnO_2. Dark and gray grains correspond to O and Sn atoms. From Ref. 81. See color insert.

and it was found that squared structures are favored over filamentary ones and that the exploitation of nanotubes seems hampered by stability problems. Oscillatory variation of fragmentation energy with the number of atoms is found for spherical grains but not for squared and T-shaped grains.

The electronic structure and structural properties of $CuMO_2$ oxides (M = Al, Ga, In) has been studied by Jayalakshmi et al. (83). The calculated equilibrium lattice parameters and bulk modulus are in good agreement with the experimental values. In another context, details of the electronic structure of UO_2 were calculated by Lim et al. (84), who evaluated local and total density of states for three different types of oxygen defects. The local d-states were calculated from the spectral representation of Green's functions. On the other hand, Zainullina et al. (85) studied the monoclinic and cubic phases of Li titanate (Li_2TiO_3) and its protonated analogs ($Li_{1.75}H_{0.25}TiO_3$) and H_2TiO_3, in order to understand the effect of protons on the electronic spectrum and bond strength. The same group also calculated the electron structure of α-Bi_2O_3, β-Bi_2O_3, γ-Bi_2O_3, and γ-Bi_2O_3 phases (86) with the purpose of studying the chemical bonding and oxygen ion migration.

Modeling properties related with spin and magnetism has been extensively carried out by using TB methods. The specific properties of $TiSr_2CoO_5$ and $SrCoO_3$ were interpreted by Pouchard et al. (87) to compare the trivalent and tetravalent Co ions. They found that spin transitions are coupled to metal-isolator transitions where the spin equilibrium may lead to nonintegral oxidation state for mixed valence oxides. In the same way, Jo (88) studied the effect of the 3d spin–orbit interaction on the magnitude of the orbital magnetic moment in vanadium perovskites. He proposed an explanation for the magnetic properties based on spin and orbital coupling and Jahn–Teller distortion in $LaVO_3$ and YVO_3 vanadates. Doublet and Lepetit (89), determined the structure of the dominant transfer and magnetic interactions of β-AV_6O_{15} (A = Sr, Ca, Na) by using extended Hückel TB calculations. Their calculations explain the optical condition and Raman spectra for NaV_2O_5. Magnetic and electronic properties of the Mn site-doped perovskite oxides $LaMn_{1-x}Cr_xO_3$ were studied by Yang et al. (90). The behavior of the total magnetic moment is explained with an amount of Mn in the perovskite and the distortion of its crystal structure. Ohsawa et al. (91) calculated the interlayer coupling constant of perovskite-type ferromagnetic trilayers $LaBaMnO_3/LaNO_3/LaBaMnO_3$, reproducing experimental results, and the tunnel magneto-resistance in magnetic junctions (92). In this field, Mathon (93) evaluated the tunneling magneto-resistance for two Co electrodes separated by a vacuum gap and for an Fe/MgO/Fe(001) junction.

Tunneling resonance spectra have been investigated by Narvaez and Kirzenow (94,95) of a nanoscopic-oxide-coated aluminum island. The electronic properties of vanadium pentoxide (V_2O_5) nanotubes were studied by Enyashin et al. (96,97) to compare the cylindrical zig-zag and the armchair structures. The effect of doping In_2O_3 by lithium using TB methods was studied for Sasaki et al. (98). They found that Li atoms occupied interstitial sites of the crystal and explained the increase in the lattice constants and the decrease of the optical absorption energy. The doping effect was also studied by Gavrichkov et al. (99) in the evolution of the band structure of CuO_2 quasi-particles. Singh et al. (100) studied similar systems to understand the role of CuO-chains on the supercurrent density in bilayered cuprate superconductors, such as $YBa_2Cu_3O_7$. Ivanovskii and Okatov (101) studied ionic conductors in solid solutions of ZrO_2–In_2O_3, ZrO_2–In_2O_3–Tl_2O_3, and ZrO_2–In_2O_3–CaO systems. The injection time of the electron transfer from an excited dye molecule to a metal-oxide–molecule (TiO_2) interface was calculated by Peterson et al. (102), and the results are corroborated by recent experimental studies.

These applications of TB indicate the high versatility of this approach to simulate complex systems related with problems in technology of materials and electronic devices.

8.6. AM1, MNDO, AND PM3 METHODS

MNDO (103) and AM1 (104) are extensively used in theoretical modeling of the molecular structure. The main difference between them is the way core–core repulsions are evaluated. This set of methods has its origin in the Dewar's group and was later improved by Stewart given PM3 (105) and by Thiel's group with different versions of MNDO (106–111). Several reviews of these methods have been presented in the literature (6,20,112). Application of these PQMs to metal-oxide systems is not as much numerous as other methods, because these are more oriented to analyzing organic and biological compound properties.

The PM3 method has been used to model surface processes, such as the Almeida et al. work (113) to study water adsorption on MgO(001) surfaces with different types of vacancies (V, F, and P that correspond to sites where O, Mg, and Mg and O are removed from the surface, respectively). They investigated the interaction of nH_2O molecules on a $(MgO)_{16}$ cluster and found that F and P vacancies promote the water coverage. There is also a decrease of HOMO, LUMO, and gap energy with the increase of the adsorbed water molecules on vacancies. Calculations were cross-checked with DFT calculations. The authors argued that the use of the PM3 method is very advantageous for big systems with a large number of conformations, because it supplies the required physical insight rather than accurate numerical results.

An interesting PM3 application is reported by Cai and Sohlberg (114) for the modeling of methanol, ethanol, propanol, and isopropanol adsorption on a γ-alumina surface (1 1 0 C) by using an $Al_{48}O_{64}$ cluster. They found exothermic alcohols dehydrogenation on vacancy sites. A surface coverage study was performed with two C_2H_5OH molecules, as shown in Figure 8.3. The repulsion between these molecules is an obstacle for a second adsorption with the critical configuration that leads to dehydrogenation. Alkoxide production by removal of the O–H proton is favored over alkoxide formation by the C–OH scission. The alcohol is acting as a Brønsted acid, whereas the surface O_s atoms adjacent to vacancies act like Brønsted bases.

The use of PM3 for elucidating potential intermediates in the Tishchenko reaction mechanism was carried out by Seki et al. (115). They proposed o-$MOCH_2C_6H_4CHO$ (M = Mg, Al) as the active species based on theoretical calculations for the transformation of o-phthalaldehyde to phthalide reaction on MgO using a $(MgO)_{16}$ cluster.

Several applications of AM1 to zeolites have been reported. Lam et al. (116) studied the adsorption of the metronidazole drug in the lattice structure. They found that the nature of the interaction drug-zeolite depends on the size model. The drug adsorption on the zeolite matrix could have practical importance in medicine, because it can decrease the side effects of this drug. Kojima et al. (117,118) used AM1 to study photochemical reactions in zeolite nanocavities. They irradiated *cis*- and *trans*-azobenzenes in NaY, Na-mordenite, and Na-ZSM-5 zeolite to study the ratio change between isomers. The results explain the Na^+ interaction with the lone pair of one of the N atoms and not with the N=N double bond. Kojima et al. also studied the photooxygenation of aromatic alkenes in NaY zeolite (118). They explained the formation of oxygenated products through alkene cation radicals and superoxide anions generated by the excitation of contact charge-transfer complexes stabilized by the strong electrostatic field in the nanocavity.

Other research of hydrocarbon reactions in zeolites was carried out by Moreau et al. (119) using AM1, PM3, and molecular mechanics methods. They evaluated the naphthalene alkylation with ter-butanol by using HY and H-beta zeolites. The selectivity of di(tertbutyl)naphtalenes (DTBN) is analyzed in terms of energy calculations and the kinetic diameters. These two facts explain the high selectivity of 2,6-DTBN with respect to 2,7-DTBN. Large cluster models, such as $(ZnO)_{60}$, were used by Martins et al. (120) to study lateral interaction effects of CO and H_2 adsorbed on the ZnO surface employing the AM1 method. They found that CO interacts mainly with the edge sites between (0001) and (1010) surfaces and that the CO binding energy increases with the number of CO adsorbed molecules on the ZnO surface.

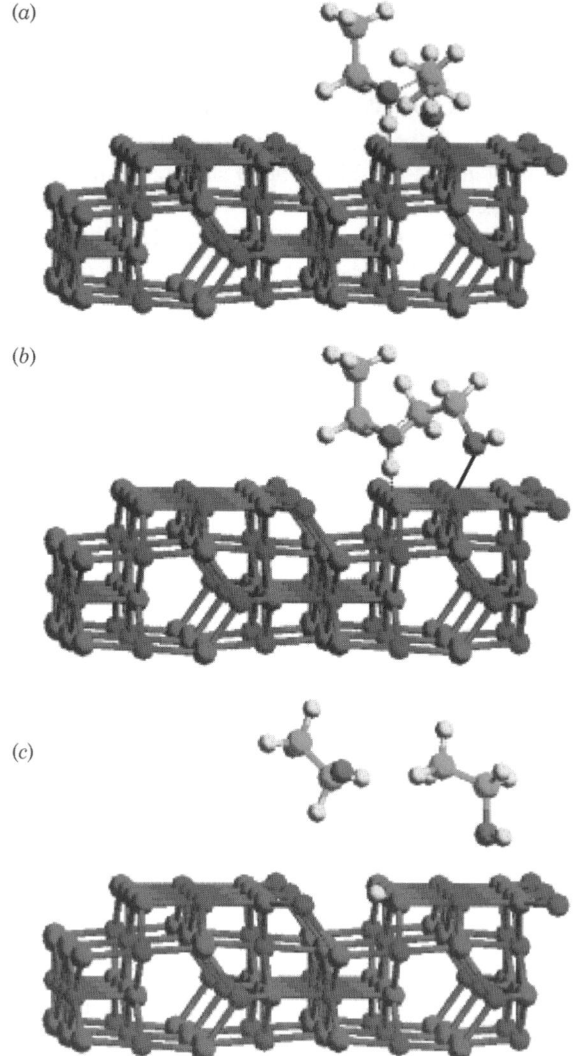

Figure 8.3. Adsorption of (*a*) two C_2H_5OH molecules on $Al_{48}O_{64}$; (*b*) two different configurations and (*c*) optimized configuration. From Ref. 113. See color insert.

Application of a new PM3 modification, named PM5 (121), to study the precipitation of calcium carbonate (vaterite) on gallium oxide was performed by Malkaj et al. (122) using the Ga_4O_6 cluster, $CaCO_3$, and complexes of $Ga_4O_6 \cdot (OH)_x \cdot YCaH_zO_3 \cdot RH_2O(x = 0, 1), R = (0, 1, 2, 3), (Y = 0, 1, 2)$. Several calculations of heat of formation, bond order, bond length, and partial charge on O and Ga atoms were evaluated. Formation of two $CaCO_3$ molecules on Ga_4O_6 is favored, and it initializes the crystallization process.

Several modifications for getting better performance of these methods have been reported. Improvement of PM3 and MNDO has consistently been reached by Repasky et al. (123) using pair distance-directed Gaussian modification (PDDG) that gives rise to PDDG/PM3 and PDDG/MNDO methods applied to compounds with H, C, N, O atoms. On the other hand, new parameterizations for transition metals complexes have been done by Cundary et al. (37,124). Ground-state prediction for transition metal complexes has been successful in comparison with other levels of theory MP2 and DFT, because geometry optimization and frequency calculations typically take from 5 hours to 10 days compared with less than 10 minutes employed with PM3(tm).

8.7. INDO METHOD

The INDO method (intermediate neglect differential overlap) was proposed initially by Pople et al. (125). Several modifications to this approach are given by Zerner's group (126–130) (INDO/S) to study the spectroscopic properties of transition metal and lanthanide species presented in the ZINDO program package.

An interesting application of INDO/S to nanoparticles is reported by Hush et al. (131–133) to analyze electrode conduction. They conclude that the INDO/S method was capable of giving a qualitatively description of all basic electrostatic, resonance, spin, and electronic state effects necessary to properly describe electrode–molecule–electrode conduction; i.e., this PQM is an efficient computational scheme for the study of a wide range of nanoparticle electronic properties.

Calculations for geometry structure and spin and charge densities of Fe(III) oxide nanoclusters with 4 Fe and 6 O atoms, coated with acid surfactants, were carried out by Pomfret et al. (134) using the ZINDO package. The results indicate that the nanoparticle spin density is depending of the surfactant number and charge. They also reported that nanoclusters are transformed from ferri-magnetic to ferro-magnetic when the number of surfactants increases.

Another motivating application of the ZINDO/1 program is reported by Clifford et al. (135) in the recombination dynamics of ruthenium bipyridyl, porphyrin, and phthalocyanine dyes on nanocrystalline TiO_2 films. This system is of particular interest in technological applications in dye-sensitized solar cells and electrochromic windows. They found that the most important factor is limited by the interfacial electron-transfer reaction that depends on the spatial separation of the dye cation HOMO from the electrode surface calculated by the PQM.

Cheng et al. (136) used INDO with single and double CI to study the third-order nonlinear optical responses of spin chain systems A_2CuO_3 (A = Ca, Sr). They found that nonlinear susceptibilities are from the Cu–O chain direction and the two-photon allowed excited states, especially the electron-hole couple excited state within the valence band.

Applications to surface chemistry were performed by Rodríguez et al. (137) to study the adsorption of NO_2 on periodic slabs of ZnO within the DFT approach but in order to investigate localized charges or electronic properties of individual atoms. They performed, however, INDO/S calculations for NO_2 on ZnO clusters, because this method can give qualitative information about the chemisorption process (charge transfers, atoms involved in the bond, bond strength, energy-level spectrum of the adsorption complex) that cannot be easily performed with the DFT method.

A large number of applications with an INDO method, especially modified for ionic/partially covalent solids (138,139), have been carried out by Eglitis et al. (140–157). They studied a series of ABO_3 perovskites (A = Ba, K), (B = Nb, Ta, Ti) and $KNb_x\ Ta_{1-x}O_3$ compounds. Several properties have been analyzed, such as point defects, optical absorption energies, activation energies, electronic structure, polaronic and vibronic excitons, luminescence, effects of impurities, charge transfer, and phase transition. The results explain the basic physical mechanisms and phenomena that occur in complex systems, such as ABO_3 perovskites, in a qualitatively manner.

8.8. CATIVIC METHOD

The PQM named CATIVIC (158) has been developed recently to model catalytic and molecular systems based on simulation techniques (26–28). In the first version, he total energy functional is obtained from basic parametric functionals (26–28,30) similar to MINDO/SR (159), i.e., from diatomic molecules. Details of atomic and molecular parameterization processes used in this method are presented elsewhere (38,39,158). In addition, several theoretical tools have been implemented for evaluating the bond strengths and analyze bond interaction: diatomic energies (DEs), diatomic binding energies (DBEs) (160), critical points of the Laplacian density, and orbital correlation diagrams (158,161).

The philosophy of this PQM is based on a correct representation of the most elementary molecules [diatomic molecule (DM)] and then the extension to more complex systems. The transferability to polyatomic molecules is a crucial task, because it requires a comprehensive parameterization with a wide range of functionals and a selected set of molecules that represent different types of bond coordination.

Several applications of CATIVIC to zeolite ZSM-5 are reported by Ruette et al. (4). For example, the model size effect on zeolite-FeO site for NO adsorption is analyzed using different clusters: $Si_2O_4H_8AlOFe$-NO (A), $Si_9O_{12}H_{20}AlOFe$-NO (B), $Si_{31}O_{43}H_{44}AlOFe$-NO (C), and $Si_{95}O_{157}H_{70}AlOFe$-NO (D) of 19, 46, 123, and 327 atoms, respectively. Validation of parameters is obtained by comparison between CATIVIC and DFT calculations using a $Si_2AlO_4H_8Fe$ cluster (162). It was found that the NO molecule is bonded through the N atom; however, adsorption via the O atom is also feasible. Qualitative information can be obtained from the charge changes in different atoms of the substrate and adsorbate. The charge differences among the A, B, C, and D models with NO adsorbed and the model A without NO indicate that in all cases the Fe atom gains electrons with NO adsorption and that the charge change increases with model size. These results show a significant increase of the electronic density at the adsorption site, Fe atom. The authors conclude that the variation of charge dissipation with cluster size has a strong significance in the reactivity of the catalytic active site. In addition, more realistic cluster size models are convenient because, in many cases, reaction products may be controlled by the size of the zeolite microcavities.

Another use of CATIVIC was on the location of Al and Fe adsorption sites in zeolite frameworks (4). It is well known that Al and Si distribution in zeolites has a strong influence on their catalytic properties. A model cluster of 339 atoms ($AlSi_{79}O_{193}H_{66}$) was employed to model the sinusoidal channel of a natural FMI zeolite known as mutinaite. Total energy data for all 12 different T sites reveal that the most stable tetrahedral sites for Al sitting corresponds to T6 and T2. These two sites have been reported by other theoretical and experimental methods with controversial results.

With respect to the study of stability of Fe atoms in mutinaite on the sinuosoidal channel, calculations were carried out for the ($FeSi_{79}O_{193}H_{67}$) cluster of 340 atoms with a charge-compensating cation (H^+) located at the most stable place of the four oxygen atoms bonded to Fe at each T site. A detailed study of stability of Fe atoms in mutinaite on the sinuosoidal channel was also evaluated for a similar industrial zeolite. Full optimization of all atoms, except edge H atoms, was carried out for the model cluster. Results show that the H^+ location is at those sites that point toward the 10-ring channel. Calculated total energy values show that the most stable sites correspond to T11 and T5. One of these sites has been reported by calculations of DFT, using small model clusters.

A novel modeling of the extra-framework aluminium (EFAL) process in zeolites was also carried out by CATIVIC (4). The formation of EFAL occurs by hydrothermal treatment of the zeolite; therefore, the interaction of H_2O molecules on the Al site was considered. Two clusters were employed: one of 25 atoms ($AlSi_3O_{12}H_9$) that represents a 4-ring in a ZSM5 zeolite and the other of 265 atoms ($AlSi_{63}O_{152}H_{49}$). The results show the formation of a $AlSi_3O_{12}H_9$-H_2O and $AlSi_{63}O_{152}H_{49}$-H_2O penta-coordinate. Successive water molecules (6 H_2O) attack the Al-O bonds in the $AlSi_3O_{12}H_9$ cluster and lead to the formation of different intermediate species:

$$Si_3O_8H_5(AlO_4H) \xrightarrow{H_2O} Si_3O_8H_5(AlO_5H_3) \tag{8.1}$$

$$Si_3O_8H_5(AlO_5H_3) \xrightarrow{H_2O} Si_3O_9H_6(AlO_5H_4) \tag{8.2}$$

$$Si_3O_9H_6(AlO_5H_4) \xrightarrow{H_2O} Si_3O_10H_7(AlO_5H_5) \tag{8.3}$$

$$Si_3O_{10}H_7(AlO_5H_5) \xrightarrow{H_2O} Si_3O_{11}H_8(AlO_5H_6) \tag{8.4}$$

$$Si_3O_{11}H_8(AlO_5H_6) \xrightarrow{H_2O} Si_3O_{11}H_8(AlO_6H_8) \tag{8.5}$$

$$Si_3O_{11}H_8(AlO_6H_8) \xrightarrow{H_2O} Si_3O_{12}H_9 + Al(OH)_3 + 3H_2O \tag{8.6}$$

Reaction **8.1** forms a penta-coordinate complex with an Al–OH_2 bond. Reactions **8.2–8.4** also produce penta-coordinate complexes, but there is the scission of Al–OSi bonds and the formation of Al–OH_2 and SiO–H bonds. In Reaction **8.5**, an hexa-coordinate complex is produced, and finally, the sixth water molecule will form $Al(OH)_3$ as EFAL.

CATIVIC has also been extended to employ an embedded approach for low symmetry systems, such as alumina (163). It is necessary to have an alumina cluster of enough size to avoid artifacts produced by an unsuitable selection of the catalyst model. The growth of a Ni_5 system, supported on an alumina cluster embedded into a large group of charges, was studied. The optimal model cluster was chosen after analyzing the variation of atomic charges and DBEs from several alumina clusters, as shown Figure 8.4. The selected model ($Al_{30}O_{45}Ni$) was embedded into a set of point charges ($Al_{330}O_{643}$) that correspond to a cube in the (110) plane. The embedded

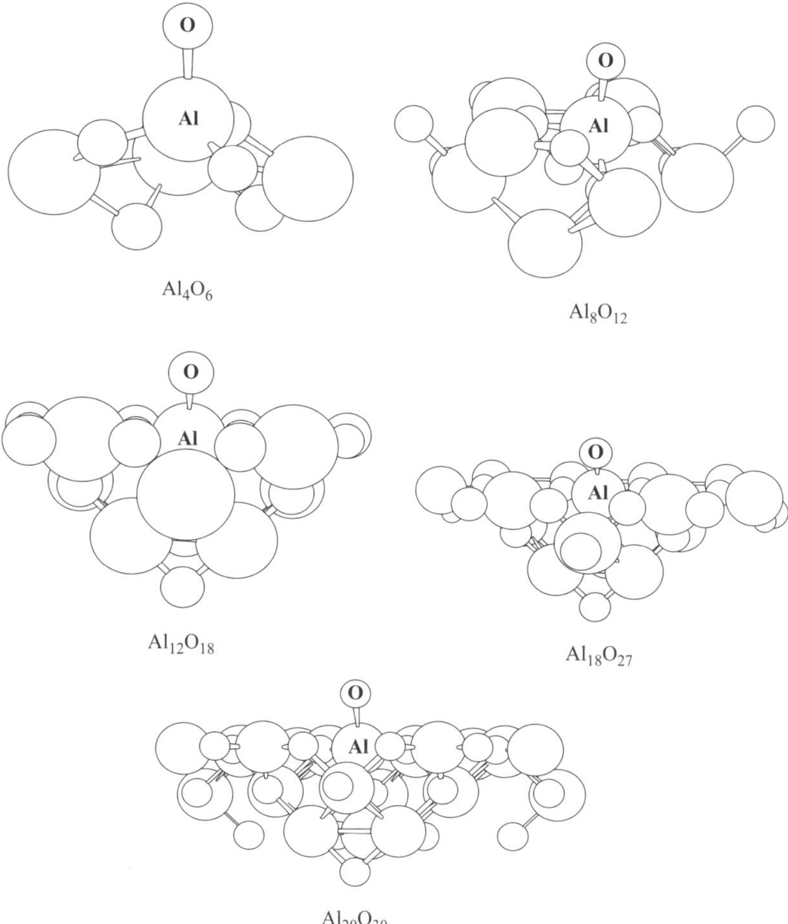

Figure 8.4. Different nanoparticles of AlO_3 used to study property changes with size. From Ref. 163.

approximation improves the modeling of complex systems, such as the effect of the support on surface-active sites. Short- and long-range interactions employed in parametrical methods can be use in the embedded approach. The minimum embedded cluster of alumina in alumina resulted in an $Al_{20}O_{30}$ system whose property variations, such as charge and DBEs, tend to converge to constant values. The effect of the embedding on active site properties produces a decrease of Ni–O and Ni–Ni bond energies and an increase of the positive charge of the most exposed Ni atoms of the Ni_5 cluster that tends to grow in a laminar form; see Figure 8.5.

Calculations for a complex Mo catalytic system $Mo_2O_5(NCS)_2(N_2C_{18}H_{24})_2$ (**1**) in a *trans*-configuration and the reduced species $Mo_2O_4(NCS)_2(N_2C_{18}H_{24})_2$ (**2**) in a *cis* one were evaluated by Griffe et al. (164) to study the mechanism of hydrocarbon oxidation. The (**1**) and (**2**) catalyst structures and a proposed intermediate state (**3**) [*cis*-configuration of complex (**1**)] were analyzed considering bonding properties and total energy. The results indicate that bipyridine ligands show a relatively

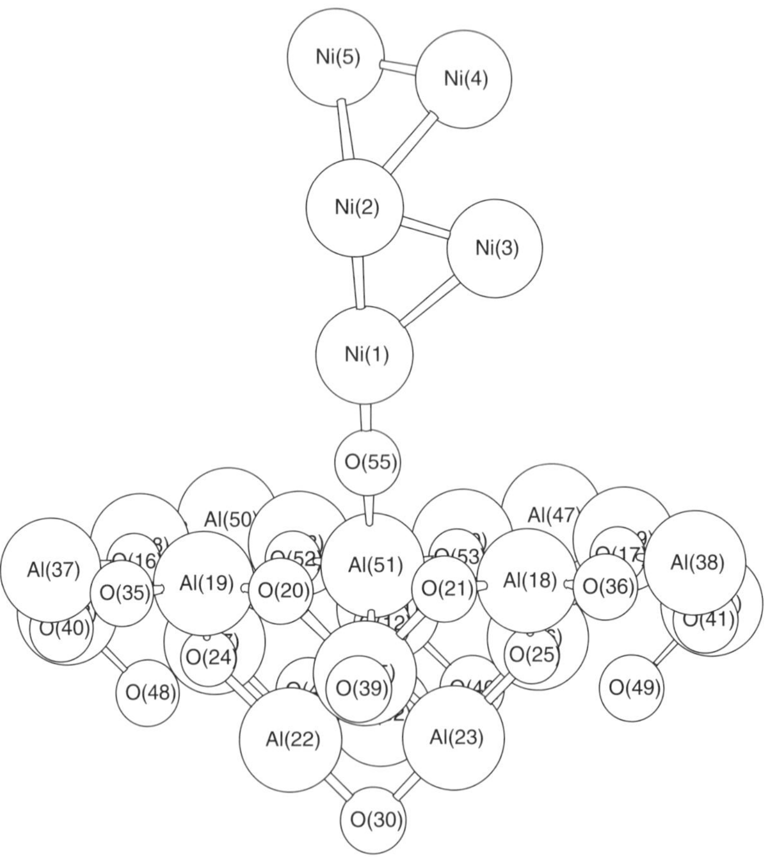

Figure 8.5. Formation of Ni_5 on a $Al_{330}O_{643}$ cluster embedded in charges. From Ref. 163.

weak Mo–N bond (−23 kcal/mol), indicating a high lability of these ligands. Complex (**2**) stabilization is caused by stronger M=O bonds than in the (**1**) system and the formation of a bonding Mo–Mo interaction. The calculated energy barrier for the rotation from (**1**) *trans* to the proposed intermediate (**3**) *cis* is evaluated to be about 28 kcal/mol. Electronic properties indicate that complex (**3**) has more unoccupied orbitals very close to the LUMO than complex (**1**) at Mo and O atoms. It suggests that the former complex should be more reactive for accepting electrons and reducing the Mo site than complex (**1**); therefore, complex (**3**) is a convenient intermediate in the hydrocarbon oxidation reaction. Recent calculations (165) of $H_2C{=}CH_2$ interaction with complex (**3**) confirmed that this intermediate leads to the olefin oxidation with a very small reaction barrier. Interaction of O_2 on the vacancy site created by the hydrocarbon oxidation leads to a species with end-bonded oxygen in the *cis*-configuration.

8.9. CONCLUSIONS AND COMMENTS

The development of new computer hardwares, innovative optimization techniques, and novel softwares have helped to increase the use of sophisticated methods to study transition metal oxides, particularly DFT and HF approaches. Nevertheless, the future of PQMs is very promising, because they are highly competitive for calculations on complex and amorphous systems. Multifaceted processes that require an extensive number of steps are intractable with *ab initio* approaches. The main advantage of PQMs is the small computer time employed in calculations. It is well established that PQMs are about 10^3–10^4 faster than the standard DFT or HF methods.

From the theoretical point of view, the fundamental background of PQMs has, until now, not been clearly explored, in particular, the characterization of EPF spaces. In addition, nonlinear techniques for the search of atomic and molecular parameters have not yet been completely investigated to improve the efficiency and accuracy of PQMs.

It was found in this review that applications of tight-binding methods are very numerous and successful for complex metal-oxide systems, particularly, for the understanding of magnetic, optical, and electric properties. On the other hand, in the last years, the MSINDO method seems to be the most promising PQM software for studying nanoparticles, surfaces, and solids. Otherwise, the INDO approach has been extensively applied to characterize the spectroscopic properties of materials. With respect to MNDO, AM1, and PM3, relatively few applications to clusters and metal oxides have been reported, maybe because these methods are more oriented to organic and biological systems.

Despite the large number of PQM applications, a lot of work is required to consistently improve their use for transition metal-oxide systems. Note also that very few PQMs have been parameterized for the second and third transition metal elements, for instance, Rh, Ru, and Au atoms. In this sense, the CATIVIC method, used for the study of nanoparticles in catalysis, has a great potentiality in the near future, by considering a better definition of EPFs and parameterization for transition metals.

ACKNOWLEDGMENT

Part of the results shown here were performed by the CATIVIC package and computer workstations that have been sponsored by FONACIT, Venezuela, within the S1-2001000907 and G-9700667 contracts.

REFERENCES

(1) Fernández-García, M.; Martínez-Arias, A.; Hanson, J.C.; Rodríguez, J.A. Nanostructured oxides in chemistry: characterization and properties. *Chem. Rev.* **2004**, *104*, 4063–4104.

(2) Capek, I. Preparation of metal nanoparticles in mater-in-oil (w/o) microemulsions. *Adv. Coll. Interf. Sci.* **2004**, *110*, 49–74.

(3) Barth, J.V.; Costantini, G.; Kern, K. Engineering atomic and molecular nanostructures at surfaces. *Nature* **2005**, *437*, 671–679.

(4) Ruette, F.; Sánchez, M.; Sierraalta, A.; Mendoza, C.; Añez, R.; Rodríguez, L.S.; Lisboa, O.; Daza, J.; Manrique, P.; Perdomo, Z.; Rosa-Brussin, M. Application of computational methods to catalytic systems. *J. Mol. Catal. A* **2005**, *228*, 211–225.

(5) Reynolds, C.H. Semiempirical MO methods: the middle ground in molecular modeling. *J. Mol. Struct. (Theochem).* **1997**, *40*, 267–277.

(6) Bredow, T.; Jug, K. Theory and range of modern semiempirical molecular orbital methods. *Theor. Chem. Acc.* **2005**, *113*, 1–14.

(7) Clark, T. Quo vadis semiempirical MO-theory. *J. Mol. Struct. (Theochem).* **2000**, *530*, 1–10.

(8) Bernal-Uruchurtu, M.I.; Ruiz-López, M.F. Basic idea for the correction of semiempirical methods describing H-bonded systems. *Chem. Phys. Lett.* **2000**, *330*, 118–124.

(9) Cui, Q.; Guo, H.; Karplus, M. Combining *ab initio* and density functional theories with semiempirical methods. *J. Chem. Phys.* **2002**, *117*, 5617–5631.

(10) Gogonea, V.; Westerhoff, L.M.; Merz, K.M. Quantum mechanical/quantum mechanical methods. I. A divide and conquer strategy for solving the Schrödinger equation for large molecular systems using a composite density functional semiempirical Hamiltonian. *J. Chem. Phys.* **2000**, *113*, 5604–5613.

(11) Luque, F.J.; Reuter, N.; Cartier, A.; Ruiz-López, M.F. Calibration of the Quantum/Classical Hamiltonian in semiepirical QM/MM AM1 and PM3 methods. *J. Phys. Chem. A* **2000**, *104*, 10923–10931.

(12) Devi-Kesavan, L.S.; Garcia-Viloca, M.; Gao, J. Semiempirical QM/MM potential with simple valence bond (SVB) for enzyme reactions. Application to the nucleophilic addition reaction in haloalkane dehalogenase. *Theor. Chem. Acc.* **2003**, *109*, 133–139.

(13) Pu, J.; Gao, J.; Truhlar, D.G. Combining self-consistent-charge-functional tight-binding (SCC-DFTB) with molecular mechanics by generalized hybrid orbital (GH)) method. *J. Phys. Chem. A* **2004**, *108*, 5454–5463.

(14) Janesko, B.G.; Yaron, D. Using molecular similarity to construct accurate semiempirical electronic structure theories. *J. Chem. Phys.* **2004**, *121*, 5635–5645.

(15) Berzigiyarov, P.K.; Zayets, V.A.; Ginzburg, I.Y.; Razumov, V.F.; Sheka, E.F. Nanopack: Parallel codes for semiempirical quantum chemical calculations of large systems in the sp- and spd-basis. *Int. J. Quantum Chem.* **2002**, *88*, 449–462.

(16) Stewart, J.J.P. Use of semiempirical methods for detecting anomalies in reported enthalpies of formation of organic compounds. *J. Phys. Chem. Ref. Data* **2004**, *33*, 713–724.

(17) Strodel, P.; Tavan, P. A revised MRCI-algorithm coupled to an effective valence-shell Hamiltonian. II. Application to the valence excitations of butadiene. *J. Chem. Phys.* **2002**, *117*, 4677–4683.

(18) Tokmachev, A.M.; Tchougréeff, A.L. Transferability of parameters of strictly local geminals' wave function and possibility of sequential derivation of molecular mechanics. *J. Comp. Chem.* **2005**, *26*, 491–505.

(19) Zerner, M.C. Semiempirical molecular orbital methods. In *Reviews in Computational Chemistry, Vol. 2*; Lipkowitz, K.B.; Boyd, D.B. (Editors); VCH Publishers: New York, 1990; Chapter 8, 313–395.

(20) Thiel, W. Perspectives on semiempirical molecular orbital theory. *Adv. Chem. Phys.* **1996**, *XCIII*, 703–757.

(21) Kato, T. Perturbation theory for linear operators. Springer-Verlag: Berlin, 1966.

(22) Lengyel, B.A. Functional analysis for quantum theorists. *Adv. Quantum Chem.* **1968**, *14*, 1–82.

(23) Roman, P. Some modern mathematics for physicists and other outsiders. Permagon Press: New York, 1978.

(24) Riesz, F.; Sz.-Nagy, B. Functional analysis. Dover: New York, 1990.

(25) Zeidler, E. *Applied Functional Analysis. Applications to Mathematical Physics*; Springer-Verlag: New York; 1991.

(26) Ruette, F.; González, C.; Octavio, A. Fundamental properties of parametric functionals in quantum chemistry. *J. Mol. Struct. (Theochem).* **2001**, *537*, 17–25.

(27) Ruette, F.; Marcantognini, A.M.; Karasiev, V. Functional sets in quantum chemistry. Basic foundations for parametric methods. *J. Mol. Struct. (Theochem).* **2003**, *636*, 15–21.

(28) Romero, M.; Sánchez, M.; Sierralta, A.; Rincón, L.; Ruette, F. Simulation techniques in parametric Hamiltonians. *J. Chem. Infor. Comp. Sci.* **1999**, *39*, 543–549.

(29) Durand, P.; Malrieu, J.P. Effective Hamiltonians and pseudo-operators as tools for rigorous modeling. In *Ab Initio Methods in Quantum Chemistry I*; Lawley, F.P. (Editor); *Adv. Chem. Phys.* **1987**, *68*, 322–412.

(30) Primera, J.R.; Romero, M.; Sánchez, M.; Sierralta, A.; Ruette, F. Analysis of parametric functionals in semiempirical approaches using simulation techniques. *J. Mol. Struct. (Theochem).* **1999**, *469*, 177–190.

(31) Szabo, A.; Ostlund, N.S. *Modern Quantum Chemistry: Introduction to Advanced Electronic Structure Theory, Vol. 70*; MacMillan: New York, 1982; 238.

(32) Kirpatrick, S.; Gellat, C.D.; Vecchi, M.P. *Science* **1983**, *220*, 671–680.

(33) Cai, W.; Shao, X. A fast annealing evolutionary algorithm for global optimization. *J. Comp. Chem.* **2002**, *23*, 427–435.

(34) Fournier, R.; Cheng, J.B.Y.; Wong, A. Theoretical study of the structure of lithium clusters. *J. Chem. Phys.* **2003**, *119*, 9444–9454.

(35) Li, G.-Z.; Yang, J.; Song, H.-F.; Yang, S.-S.; Lu, W.-C.; Chen, N.-Y. Semiempirical quantum chemical method and artificial neural networks applied for λ_{max} computation of some Azo dyes. *J. Chem. Inform. Comp. Sci.* **2004**, *44*, 2047–2050.

(36) Shen, Q.; Jiang, J.-H.; Tao, J.-C.; Shen, G.-L.; Yu, R.-Q. Modified ant colony optimization algorithm for variable selection in QSAR modeling: QSAR studies of cyclooxygenase inhibitors. *J. Chem. Inform. Model.* **2005**, *45*, 1024–1029.

(37) Cundari, T.R.; Jun, D.; Fu, W. PM3(tm) parameterization using genetic algorithms. *Int. J. Quantum Chem.* **2000**, *77*, 421–423.

(38) Ruette, F.; Sánchez, M.; Mendoza, C.; Sieraalta, A.; Martorell, G.; González, C. Calculation of one-center integrals in parametric methods using simulated annealing and Simplex methods. *Int. J. Quantum Chem.* **2004**, *96*, 303–311.

(39) Sánchez, M.; Rodríguez, L.S.; Larrazabal, G.; Galeán, L.; Bello, N.; Ruette, F. Molecular parameter optimization using simulated annealing and evolutionary algorithm techniques in a quantum parametric method (CATIVIC). *Mol. Simul.* in press.

(40) Nanda, D.N.; Jug, K. SINDO1. A semiempirical SCF MO method for molecular binding energy and geometry I. Aproximations and parameterization. *Theor. Chim. Acta* **1980**, *57*, 95–106.

(41) Jug, K.; Nanda, D.N. SINDO1 II. Application to ground states of molecules containing carbon, nitrogen and oxygen atoms. *Theor. Chim. Acta* **1980**, *57*, 107–130.

(42) Jug, K.; Iffert, R.; Schulz, J. Development of SINDO1 for second-row elements. *Int. J. Quantum Chem.* **1987**, *32*, 265–277.

(43) Li, J.; Correa de Mello, P.; Jug, K. Extension of SINDO1 to transition metal compounds. *J. Comp. Chem.* **1992**, *13*, 85–92.

(44) Ahlswede, B.; Jug, K. Consistent modifications of SINDO1: I. Approximations and parameters. *J. Comp. Chem.* **1999**, *20*, 563–571.

(45) Ahlswede, B.; Jug, K. Consistent modifications of SINDO1: II. Applications for first- and second-row elements. *J. Comp. Chem.* **1999**, *20*, 572–578.

(46) Jug, K.; Geudtner, G.; Homman, T. MSINDO parameterization for third-row main group elements. *Int. J. Quantum Chem.* **2000**, *21*, 974–987.

(47) Bredow, T.; Geudtner, G.; Jug, K. MSINDO parameterization for third-row transition metals. *J. Comp. Chem.* **2001**, *22*, 861–887.

(48) Coffey, P.; Jug, K. Semiempirical molecular orbital calculations and molecular energies. A new formula for the β parameter. *J. Am. Chem. Soc.* **1973**, *95*, 7575–7580.

(49) Ahlswede, B.; Homann, T.; Jug, K. MSINDO study of the adsorption of water molecules at defective MgO(100) surfaces. *Surf. Sci.* **2000**, *445*, 49–59.

(50) Tikhomirov, V.A.; Jug, K. MSINDO study of water adsorption molecules on the clean MgO(100) surface. *J. Phys. Chem. B.* **2000**, *104*, 7619–7622.

(51) Geudtner, G.; Jug, K.; Koster, A.M. Cu adsorption on the MgO(100) surface. *Surf. Sci.* **2000**, *467*, 98–106.

(52) Janetzko, F.; Evarestov, R.A.; Bredow, T.; Jug, K. First principles periodic and semi-empirical cyclic cluster calculations for single oxygen vacancies in crystalline Al_2O_3. *Phys. Stat. Sol. B* **2004**, *24*, 1032–1040.

(53) Bredow, T.; Geudtner, G.; Jug, K. Development of the cyclic cluster approach for ionic systems. *J. Comp. Chem.* **2001**, *22*, 89–101.

(54) Homann, T.; Bredow, T.; Jug, K. Adsorption of small molecules on the $V_2O_5(001)$ surface. *Surf. Sci.* **2002**, *515*, 205–218.

(55) Homann, T.; Bredow, T.; Jug, K. Adsorption of small molecules on the anatase(100) surface. *Surf. Sci.* **2004**, *555*, 135–144.

(56) Nair, N.N.; Bredow, T.; Jug, K. Molecular dynamics implementation in MSINDO: Study in silicon clusters. *J. Comp. Chem.* **2004**, *25*, 1255–1263.

(57) Nair, N.N.; Bredow, T.; Jug, K. Toward an understanding of the formation of vanadia-titania catalyst. *J. Phys. Chem. B* **2005**, *109*, 12115–12123.

(58) Steveson, M.; Bredow, T.; Gerson, A.R. MSINDO quantum chemical modelling study of the structure of aluminium-doped anatase and rutile titanium dioxide. *Phys. Chem. Chem. Phys.* **2002**, *4*, 358–365.

(59) Janetzko, F.; Jug, K. Miscibility of zinc chalcogenides. *J. Phys. Chem.* **2004**, *108*, 5449–5453.

(60) Islam, M.M.; Maslyuk, V.V.; Bredow, T.; Minot, C. Structural and electronic properties of $Li_2B_4O_7$. *J. Phys. Chem. B* **2005**, *109*, 13597–13604.

(61) Bredow, T.; Homann, T.; Jug, K. Adsorption of NO, NH_3 and H_2O on V_2O_5/TiO_2 catalyst. *Res. Chem. Intermed.* **2004**, *30*, 65–73.

(62) Jug, K.; Nair, N.N.; Bredow, T. Molecular dynamics investigation of oxygen vacancy diffusion in rutile. *Phys. Chem. Chem. Phys.* **2005**, *7*, 2616–2621.

(63) Geudtner, G.; Jug, K. Nucleation of CuGa phases on the MgO(100) surface. *Z. Anorg. Allg. Chem.* **2003**, *629*, 1731.

(64) Simpson, D.J.; Bredow, T.; Roger, R.S.T.; Gerson, A.R. Mechanism of acidic dissolution at the MgO(100) surface. *Surf. Sci.* **2002**, *516*, 134–146.

(65) Sierralta, A.; Guillén, Y.; López, C.M.; Martínez, R.; Ruette, F.; Machado, F.; Rosa-Brussin, M., Soscún, H. Theoretical study of the Ga /SAPO-11 catalyst. Extra-framework gallium species and n-butane transformation. *J. Mol. Catal.* **2005**, *242*, 233–240.

(66) Harrison, W.A. Tight-binding theory of molecules and solids. *Pure App. Chem.* **1989**, *61*, 2161–2169.

(67) Cohen, R.E.; Mehl, M.J.; Papaconstantopoulos, D.A. Tight-binding total-energy method for transition and noble metals. *Phys. Rev.* **1994**, *50*, 14964–14967.

(68) Papaconstantopoulos, D.A.; Lach-hab, M.; Mehl, M.J. Tight-binding Hamiltonians for realistic electronic structure calculations. *J. Phys. B: Cond. Matt.* **2001**, *296*, 129–137.

(69) Elanany, M.; Selvam, P.; Yokosuka, T.; Takami, S.; Kubo, M.; Imamura, A.; Miyamoto, A. *J. Phys. Chem. B* **2003**, *107*, 1518–1524.

(70) Sasata K.; Yokosuka, T.; Kurokawa, H.; Takami, S.; Kubo, M.; Imamura, A.; Shinmura, T.; Kanoh, M.; Selvam, P.; Miyamoto, A. *Jpn. J. Appl. Phys.* **2003**, *42*, 1859–1864.

(71) Lu, C.; Wang, X.; Govindasamy, A.; Tsuboi, H.; Koyama, M.; Kubo, M.; Ookawa, H.; Miyamoto, A. Tight-binding quantum chemical calculations of electronic structures of indium tin oxide. *Jpn. J. Appl Phys.* **2005**, *44(Pf. 1)*, 2806–2809.

(72) Suzuki, K.; Kuroiwa, Y.; Takami, S.; Kubo, M.; Miyamoto, A.; Inamura, A. Tight-binding quantum chemical molecular dynamics study of cathode materials for lithium secondary battery. *Sol. Stat. Ion.* **2002**, *152–153*, 273–277.

(73) Koyama, M.; Kubo, M.; Miyamoto, A. Large scale calculations of solid oxide fuel cell cermet anode by tight-binding quantum chemistry method. *Appl. Surf. Sci.* **2005**, 244, 598–602.

(74) Ito, Y.; Okazaki, T.; Endou, A.; Kubo, M.; Imamura, A.; Yamamura, Y.; Miyamoto, A. Tight-binding quantum chemical study of oxygen adsorption and dissociation on strontium-doped lanthanum manganites. *Trans. Mat. Res. Soc. Jap.* **2004**, *29*, 3743–3746.

(75) Shibayama, H.; Jung, C.; Kusagaya, T.; Miura, R.; Endou, A.; Kubo, M.; Inamura, A.; Miyamoto, A. Effect of metal oxide supports on catalytic performance of supported ultrafine metal catalysts: density functional and tight-binding quantum chemical study. *Trans. Mat. Res. Soc. Jap.* **2004**, *29*, 3731–3734.

(76) Mottet, C.; Goniakowski, J.; Baletto, F.; Ferrando, R.; Treglia, G. Modeling free and supported metallic nanoclusters: structure and dynamics. *Transitions* **2004**, *77*, 101–113.

(77) Mazzone, A.M. A simple tight-binding model of vacancies in SnO_2 crystalline grains. *Phil. Mag. Lett.* **2004**, *84*, 275–282.

(78) Mazzone, A.M.; Morandi, V. A tight binding study of defects in nanocrystalline SnO_2. *Comp. Mat. Sci.* **2005**, *33*, 346–350.

(79) Mazzone, A.M.; Morandi, V. Defects in nanocrystalline SnO_2 study by tight binding. *Eur. Phys. J. B: Cond. Matt. Phys.* **2004**, *42*, 435–440.

(80) Mazzone, A.M. Stability of SnO_2 nanocrystalline grain: A study at semiempirical level. *Appl. Surf. Sci.* **2004**, *226*, 83–87.

(81) Mazzone, A.M. Electronic charge of SnO_2: A quantum mechanical study effects and adsorption capabilities. *Mat. Sci. Eng.* **2003**, *B103*, 62–70.

(82) Mazzone, A.M. Quantum mechanical study of the shape and stability of SnO_2 nanocrystalline grains. *Phys. Rev.* **2003**, *B68*, 045412–1–7.

(83) Jayalakshmi, V.; Murugan, R.; Palanivel, B. Electronic and structural properties of $CuMO_2$ ($M = Al, Ga, In$). *J. All. Comp.* **2005**, *388*, 19–22.

(84) Lim, H.; Yun, Y.; Park, K.; Song, K.-W. Electronic structure of uranium dioxide. *J. Nuc. Sci. Tech.* **2002**, *(Suppl.3)*, 110–113.

(85) Zainullina, V.M.; Zhukov, V.P.; Denisova, T.A.; Maksimova, L.G. Electronic structure and chemical bonding in monoclinic and cubic $Li_{2-x}H_xTiO_3$ ($0 < x < 2$). *J. Struct. Chem.* **2003**, *44*, 180–186.

(86) Zhukov, V.P.; Zhukovskii, V.M.; Zainullina, V.M.; Medvedeva, N.I. Electronic structure and chemical bonding in bismuth sesquioxide polymorphs. *J. Struct. Chem.* **1999**, *40*, 831–837.

(87) Pouchard, M.; Villesuzanne, A.; Doumerc, J.P. Spin state behavior in some cobaltites (III) and (IV) with perovskite or related structure. *J. Sol. Stat.* **2001**, *162*, 282–292.

(88) Jo, T. On the orbital magnetic moment of V ion in vanadium perovskites. *J. Phys. Soc. Jap.* **2003**, *72*, 155–159.

(89) Doublet, M.-L.; Lepetit, M.-B. Leading interactions in the β-SrV_6O_{15} compound. *Preprint Archive Cond. Matter.* **2005**, 1–7.

(90) Yang, Z.; Ye, L.; Xie, X. Density-functional studies of magnetic and electronic structures for the perovskite oxides La $Mn_{1-x}Cr_xO_3$. *J. Phys: Cond. Matt.* **2000**, *12*, 2737–2747.

(91) Ohsawa, T.; Kubota, S.; Itoh, H.; Inoue, J. Interlayer exchange coupling in perovskite-type magnetic trilayers. *Phys. Rev. B: Cond. Mat. Mat. Phys.* **2005**, *71*, 212407–1–4.

(92) Kimura, G.; Ohsawa, T.; Itoh, H.; Inoue, J. Tunnel magnetoresistance of junctions with twisted magnetization. *Trans. Mag. Soc. Jap.* **2005**, *5*, 13–17.

(93) Mathon, J. Theory of spin-dependent tunneling in magnetic junctions. *J. Phys. D: Appl. Phys.* **2002**, *35*, 2437–2442.

(94) Narvaez, G.A.; Kirzenow, G. Understanding tunneling experiments of metallic nanoparticles: Single-particle versus many-body phenomena. *Phys. Stat. Sol. B: Basic Res.* **2002**, *230*, 457–461.

(95) Narvaez, G.A.; Kirzenow, G. Electronic structure and tunneling resonance spectra of nanoscopic aluminium islands. *Phys. Rev. B. Cond. Matt. Mat. Phys.* **2002**, *65*, 121403/1–121403/4.

(96) Enyashin, A.N.; Ivanovskaya, V.V.; Makurin, Y.N.; Ivanovskii, A.L. Electronic band structure of scroll-like divanadium pentoxide nanotubes. *Phys. Lett. A.* **2004**, *326*, 152–156.

(97) Enyashin, A.N.; Ivanovskaya, V.V.; Makurin, Y.N.; Ivanovskii, A.L. Computer modeling of electronic properties of scroll-like V_2O_5-based nanotubes. Los Alamos National Laboratory. *Preprint Arch. Cond, Matt.* **2003**, 1–6

(98) Sasaki, M.; Kiyoshima, R.; Kohiki, S.; Matsushima, S.; Oku, M.; Shishido, T. Preparation and characterization of lithium doped indium sesqui-oxide. *J. All. Comp.* **2001**, 322, 220–225.

(99) Gavrichkov, V.A.; Ovchinnikov, S.G.; Borisov, A.A.; Goryachev, E.G. Evolution of the band structure of quasiparticles with doping in copper oxides on the basis of a generalized tight-binding method. *J. Exp. Theor. Phys.* **2000**, *91*, 369–383.

(100) Singh, M.P.; Ajay, G.; Gupta, B.R.K. Role of CuO chains on the supercurrent density in bilayer cuprate superconductors. *Ind. J. Phys. A* **2003**, *77A*, 441–447.

(101) Ivanovskii, A.L.; Otakov, S.V. Electronic structure of fluorite-like ionic conductors: Solid solutions in ZrO_2-In_2O_3, ZrO_2-In_2O_3-Tl_2O_3, and ZrO_2-In_2O_3-CaO systems. *Zhur. Fis. Khim.* **2000**, *74*, 1433–1437.

(102) Peterson, A.; Ratner, M.; Karlsson, H.O. Injection time in the metal oxide-molecule interface calculated within the tight-binding model. *J. Phys. Chem. B* **2000**, *104*, 8498–8502.

(103) Dewar, M.J.S.; Thiel, W. Ground states of molecules, 38. The MNDO method. *J. Am. Chem. Soc.* **1977**, *99*, 4899–4907.

(104) Dewar, M.J.S.; Zoebisch, E.G.; Healy, E.F.; Stewart, J.J.P. AM1: A new general purpose quantum mechanical model. *J. Am. Chem. Soc.* **1985**, *107*, 3902–3909.

(105) Stewart, J.J.P. Optimization of parameters for semiempirical methods I and II. *J. Comp. Chem.* **1989**, *10*, 209–264.

(106) Kolb, M.; Thiel, W. Beyond the MNDO model: Methodical considerations and numerical results. *J. Comp. Chem.* **1993**, *14*, 775–789.

(107) Thiel, W.; Voityuk, A.A. Extension of MNDO to d orbitals: Parameters and results for the second-row elements and for the zinc group. *J. Phys. Chem.* **1996**, *100*, 616–626.

(108) Patchkovskii, S.; Thiel, W. NMR chemical shifts in MNDO approximation: Parameters and results for H, C, N, and O. *J. Comp. Chem.* **1999**, *20*, 1220–1245.

(109) Weber, W.; Thiel, W. Orthogonalization corrections for semiempirical methods. *Theor. Chem. Acc.* **2000**, *103*, 495–506.

(110) Koslowski, A.; Beck, M.E.; Thiel, W. Implementation of a general multireference configuration interaction procedure with analytic gradients in a semiempirical context using the graphical unitary group approach. *J. Comp. Chem.* **2003**, *24*, 714–726.

(111) Khandogin, J.; Gregersen, B.A.; Thiel, W.; York, D.M. Smooth solvation method for d-orbital semiempirical calculations of biological reactions. 1. Implementation. *J. Phys. Chem. B* **2005**, *109*, 9799–9809.

(112) Thiel, W. Semiempirical methods. In *Modern Methods and Algorithms of Quantum Chemistry, Proceedings*, NIC series, *Vol. 3*, Grotenderst, J. (Editor); 2000; 261–283.

(113) Almeida, A.L.; Martins, J.B.L.; Longo, E.; Taft, C.A.; Murgich, J.; Jalbout, A.F. Theoretical analysis of water coverage on MgO(001) surfaces with defects and without F, V, P, type vacancies. *J. Mol. Struct. (Theochem)*. **2003**, *664–665*, 111–124.

(114) Cai, S.; Sohlberg, K. Adsorption of alcohols on γ-alumina (110C). *J. Mol. Catal. A* **2003**, *193*, 157–164.

(115) Seki, T.; Tachikawa, H.; Yamada, T.; Hattori, H. Synthesis of phthalide-skeleton using selective intramolecular Tishchenko reaction over solid base catalyst. *J. Catal.* **2003**, *217*, 117–126.

(116) Lam, A.; Rivera, A.; Rodríguez-Fuentes, G. Theoretical study of metronidazole adsorption on clinoptilolite. *Micropor. Macropor. Mat.* **2001**, *49*, 157–162.

(117) Kojima, M.; Takagi, T.; Goshima, T. Photoisomerization of azobencene in zeolite cavities. *Mol. Crys. Sci. Tech. A: Mol. Crys. Liq. Crys.* **2000**, *344*, 179–184.

(118) Kojima, M.; Nakajoh, M.; Matsubara, C.; Hashimoto, S. Photooxygenation of aromatic alkenes in zeolite nanocavities. *J. Chem. Soc. Perkin Trans.* **2002**, *2*, 1894–1901.

(119) Moreau, P.; Liu, Z.; Fajula, F.; Joffre, J. Selectivity of large pore zeolite in the alkylation of naphthalene with *tert*-butyl alcohol: analysis of experimental results by computational modeling. *Catal. Today* **2000**, *60*, 235–242.

(120) Martins, J.B.L.; Taft, C.A.; Lie, S.K.; Longo, E. Lateral interaction of CO and H_2 molecules on ZnO surfaces. An AM1 study. *J. Mol. Struct. (Theochem)*. **2000**, *528*, 161–170.

(121) Stewart, J.J.P. *MOPAC 2002*; Tokyo, Japan: Fujitsu Limited, 1999.

(122) Malkaj, P.; Chrissanthopoulos, A.; Dalas, E. Understanding nucleation of calcium carbonate on gallium oxide using computer simulation. *J. of Crystal Grow.* **2004**, *264*, 430–437.

(123) Repasky, M.P.; Chandrasekhar, J.; Jorgensen, W.L. PDDG/PM3 and PDDG/MNDO: Improved semiempirical methods. *J. Comp. Chem.* **2002**, *23*, 1601–1622.

(124) Buda, C.; Flores, A.; Cunadri, T.R. *De novo* prediction of the ground state structure of transition metal complexes using semiempirical and *ab initio* quantum mechanics. Coordination isomerism. *J. Coord. Chem.* **2005**, *58*, 575–585.

(125) Pople, J.A.; Beveridge, D.L.; Dobosh, P.A. Approximate self-consistent molecular orbital theory. V. Intermediate neglect of differential overlap. *J. Chem. Phys.* **1967**, *47*, 2026–2033.

(126) Ridley, J.E.; Zerner, M.C. An intermediate neglect of differential overlap technique for spectroscopy: Pyrrole and for azines. *Theor. Chim. Acta.* **1973**, *32*, 111–134.

(127) Bacon, A.D.; Zerner, M.C. An intermediate neglect of differential overlap technique for spectroscopy of transition metal complexes: Fe, Co and Cu chlorides. *Theor. Chim. Acta* **1979**, *53*, 21–54.

(128) Culberson, J.C.; Knappe, P.; Rösch, N.; Zerner, M.C. An intermediate neglect of differential overlap (INDO) technique for lanthanide complexes: Studies of lanthanide halides. *Theor. Chim. Acta* **1987**, *71*, 21–39.

(129) Anderson, W.P.; Cundari, T.R.; Drago, R.S.; Zerner, M.C. Utility of the semiempirical INDO-1 method for the calculation of the geometries of 2nd-row transition-metal species. *Inorg. Chem.* **1990**, *29*, 1–3.

(130) Neto, J.D.D.; Zerner, M.C. New parameterization scheme for the resonance integrals ($H_{\mu\nu}$) within the INDO/1 approximation. Main group elements. *Int. J. Quantum Chem.* **2001**, *81*, 187–201.

(131) Reimers, J.R.; Hush, N.S. The need for quantum mechanical treatment of capacitance and related properties of nanoelectrodes. *J. Phys. Chem. B* **2001**, *105*, 8979–8988.

(132) Jeffrey, R.R.; Hush, N.S. The need for quantum-mechanical treatment of capacitance and related properties of nanoelectrodes. *J. Phys. Chem. B* **2001**, *105*, 8979–8988.

(133) Reimers, J.R.; Shapley, W.A.; Lambropoulos, N.; Hush, N.S. An atomistic approach to conduction between nanoelectrodes through a single molecule. *Ann. N.Y. Acad. Sci.* **2002**, *960*, 110–130.

(134) Pomfret, M.B.; Shew, C.-H.; Yang, N.-L.; Ulman, A. Self-consistent calculation of geometry and spin density of an iron oxide nanoparticle in acid surfactants. *Polymer Preprints* **2002**, *43*, 461–462.

(135) Clifford, J.N.; Palomares, E.; Nazeeruddin, M.K.; Grätzel, M.; Nelson, J.; Li, X.; Long, N.J.; Durrant, J.R. Molecular control recombination dynamics in dye-sensitized nanocrystalline TiO2 films: Free energy vs distance dependence. *J. Am. Chem. Soc.* **2004**, *126*, 5225–5233.

(136) Cheng, W.-D.; Zhang, H.; Wu, D.-S.; Chen, D.-G.; Wang, H.-X. Electronic structure and frequency-dependent third-order nonlinear optical properties of spin-1/2 chain systems A_2CuO_3 (A = Ca, Sr). *Phys. Rev. B: Cond. Matt. Mat. Sci.* **2003**, *68*, 045118/1–045118/6.

(137) Rodríguez, J.A.; Jirsak, T.; Dvorak, J.; Sambasivan, S.; Fischer, D. Reaction of NO_2 with Zn and ZnO: Photoemission, XANES, and density functional studies on the formation of NO_3. *J. Phys. Chem. B* **2000**, *104*, 319–328.

(138) Shluger, A.; Stefanovich, E. Models of self-trapped exciton and nearest-neighbor defect pair in SiO_2. *Phys. Rev. B* **1990**, *42*, 9664–9673.

(139) Stefanovich, E.; Shidlovskya, E.; Shluger, A.; Zakharov, M. Modification of the INDO calculation scheme and parameterization for ionic crystals. *Phys. Stat. Sol. B* **1990**, *160*, 529–540.

(140) Eglitis, R.I.; Kotomin, E.A.; Borstel, G. Large scale computer modeling of point defects in ABO_3 perovskites. *Phys. Status Sol. C* **2005**, *2*, 113–119.

(141) Vikhnin, V.S.; Eglitis, R.I.; Borstel, G. Polaronic excitons in ferromagnetic oxides: Phenomenological and microscopical theory and manifestation of polaronic exciton phase. *Ferroelectics* **2004**, *299*, 21–33.

(142) Vikhnin, V.S.; Kapphan, S.; Eglitis, R.I. Localized polaronic exciton and active impurity problems in incipient and relaxor ferroelectrics. *Ferroelectics.* **2004**, *299*, 11–20.

(143) Eglitis, R.I.; Fuks, D.; Dorfman, S.; Kotomin, E.A.; Borstel, G.; Trepakov, V.A. Large scale quantum chemical modeling of the phase transitions in KTN solid solutions. *AIP Conf. Proc.* **2003**, *677*, 231–240.

(144) Eglitis, R.I.; Trepakov, V.A.; Kapphan, S.E.; Borstel, G. Quantum chemical modeling of "green" luminescence in self active perovskite-type oxides. *Sol. Stat. Comm.* **2003**, *126*, 301–304.

(145) Eglitis, R.I.; Kotomin, E.A.; Borstel, G.; Vikhnin, V.S. Computer modeling of point defects in perovskite crystals. *Ferroelectrics* **2002**, *268*, 479–484.

(146) Eglitis, R.I.; Trepakov, V.A.; Kapphan, S.E.; Borstel, G. Large scale computer modelling of Li impurities in $KTaO_3$ and $K_{1-x}Li_xTa_{1-y}Nb_yO_3$ perovskite solid solutions. *Crys. Eng.* **2002**, *5*, 227–233.

(147) Eglitis, R.I.; Fuks, D.; Dorfman, S.; Kotomin, E.A.; Borstel, G. Large scale modelling of of the phase transitions in $KTa_{1-x}Nb_xO_3$ perovskite solid solutions. *Mat. Sci. Semicond. Proc.* **2003**, *5*, 153–157.

(148) Eglitis, R.I.; Trepakov, V.A.; Kapphan, S.E.; Borstel, G. Quantum chemical modelling of electron polarons and "green" luminescence in $PbTiO_3$ perovskite crystals. *J. Phys: Cond. Matt.* **2002**, *14*, L647–L653.

(149) Eglitis, R.I.; Kotomin, E.A.; Trepakov, V.A.; Kapphan, S.E.; Borstel, G. Quantum chemical modelling of electron polarons and "green" luminescence in PbTiO3 perovskite crystals. *J. Phys: Cond. Matt.* **2002**, *14*, L647–L653.

(150) Eglitis, R.I.; Vikhnin, V.S.; Trepakov, V.A.; Kotomin, E.A.; Kapphan S.E., Borstel, G. Theoretical prediction and experimental confirmationof charge transfer vibronic excitons and their phase in ABO_3 perovskite crystals. *Mat. Res. Soc. Symp. Proc.* **2002**, *718*, 317–322.

(151) Eglitis, R.I.; Kotomin, E.A.; Borstel, G. Quantum chemical modelling of "green" luminescence in ABO_3 perovskites. *Eur. Phys: J.: Cond. Matt. Phys.* **2002**, *27*, 483–486.

(152) Eglitis, R.I.; Kotomin, E.A.; Borstel, G. Calculations of the electronic and atomic structure and diffusion of point defects in $KNbO_3$ perovskite crustal and relevant KTN solid solutions. *Mat. Res. Soc. Symp. Proc.* **2002**, *718*, 83–88.

(153) Vikhnin, V.S.; Eglitis, R.I.; Kapphan, S.E.; Kotomin, E.A.; Borstel, G. A new phase in ferroelectric oxides: The phase of charge transfer vibronic excitons. *EuroPhys. Lett.* **2001**, *56*, 702–708.

(154) Vikhnin, V.S.; Eglitis, R.I.; Kapphan, S.E.; Borstel, G.; Kotomin E.A. Polaronic-type excitons in ferroelectric oxides: Microscopic calculations and experimental manifestation. *Phys. Rev. B: Cond. Matt. Mat. Phys.* **2002**, *65*, 104304/104304/11.

(155) Vikhnin, V.S.; Eglitis, R.I.; Kotomin, E.A.; Kapphan, S.E.; Borstel, G. New polaronic-type excitons in ferroelectric oxides: Nature and experimental manisfestation. *AIP Conf. Proc.* **2001**, *582*, 228–238.

(156) Eglitis, R.I.; Borstel, G. Quantum chemical modelling of polarons and perovskite solid solutions. *Comp. Mat. Sci.* **2001**, *21*, 530–534.

(157) Devreese, J.T.; Fomin, V.M.; Pokatilov, E.P.; Kotomin, E.A.; Eglitis, R.I.; Zhukovskii, Y.F. *Phys. Rev. B: Cond. Matt. Mat. Phys.* **2001**, *63*, 184304/1–184304/6.

(158) Ruette, F.; Sánchez, M.; Martorell, G.; González, C.; Añez, R.; Sierraalta, A.; Rincón, L.; Mendoza, C. CATIVIC: A parametric quantum chemistry package for catalytic reactions. *I. Int. J. Quantum Chem.* **2004**, *96*, 321–332.

(159) Blyholder, G.; Head, J.; Ruette, F. Semiempirical calculation method for transition metals. *Theor. Chim. Acta (Berlin)* **1982**, *60*, 429–444.

(160) Ruette, F.; Poveda, F.M.; Sierraalta, A.; Rivero, J. A theoretical tool for the analysis of adsorbate surface bond interactions. *Surf. Sci.* **1996**, *349*, 241–247.

(161) Sierraalta, A.; Ruette, F.; Machado, E. Topology of densities from parametric methods: A predictive tool. *Int. J. Quantum Chem.* **1998**, *70*, 113–123.

(162) Sánchez, M.; Añez, R.; Sierraalta, A.; Rosa-Brussin, M.; Soscum, H.; Ruette, F. Model size effects on calculated electronic and bonding properties of adsorption sites in zeolites. *Ciencias* **2005**, *13*, 94–102.

(163) Poveda, F.; Fernández-Sanz, J.; Ruette, F. A parametric embedding method for catalytic modeling. *J. Mol. Catal.* **2003**, *191*, 101–112.

(164) Griffe, B.; Agrifoglio, G.; Brito, J.L.; Ruette, F. Theoretical study of dimeric dioxo-μ-oxo and oxo-bis (μ-oxo) of molybdenum complexes used in catalytic oxidations reactions. *Catal. Today* **2005**, *107–108*, 388–396.

(165) Griffe, B.; Agrifoglio, G.; Brito, J.L.; Ruette, F. Theoretical study of the mechanism of olefine oxidation on dioxo-μ-oxo molybdenum complex. *J. Mol. Catal.* in submission.

CHAPTER 9

Atomistic Models and Molecular Dynamics

D.C. SAYLE and T.X.T. SAYLE

DEOS, Cranfield University, Defense Academy of the United Kingdom Shrivenham, Swindon, UK

9.1. INTRODUCTION

A milestone for atomistic computer simulation will be the ability to model "real" systems, which comprise, for example, 10^{23} atoms. This is unlikely to be realized in the near future, even with the almost exponential increase in computational power coupled with grid computing (for example, see Ref. 1); currently, "only" trillions of atoms are possible. Accordingly, the simulator must resort to a variety of mechanisms such as imposing periodic boundary conditions or continuum descriptions of the material away from the site of interest rather than treat every ion explicitly. Perhaps the most important scientific field associated with the 21st century is that of nanomaterials, which is agreeable to the atomistic simulator in that it is now possible to model every single ion comprising the system explicitly, at least in one dimension.

In this chapter, we first describe briefly the theoretical methods, which include the interionic potentials used to describe the interactions between the oxygen and metal ions, together with the simulation strategies used to generate realistic atomistic structures. We have taken the liberty of ensuring the mathematics in this chapter is straightforward rather than rigorous. This is to help the nonexpert gain a rudimentary understanding of atomistic simulation to help gauge its accuracy, scope, and limitations without the pain of assimilating a forest of equations. We have, where necessary, provided references that are more rigorous for the interested reader.

Equipped with these theoretical methods, we show how they can be used to generate realistic atomistic models of oxide nanomaterials. We start with isolated nanoparticles of MnO_2 and explore how microstructural features (such as dislocations, grain-boundaries, microtwinning, vacancies, substitutionals interstitials,

Synthesis, Properties, and Applications of Oxide Nanomaterials, Edited by José A. Rodríguez and Marcos Fernández-Garcia

intergrowths, and morphology) can be introduced into the model. We next look at CeO_2 nanoparticles and investigate how the morphology of the crystal changes as one traverses to the nanoscale. Here, the atomistic models provide insights into the possible structures of edges and corners, which comprise a high proportion of the surface area of the nanoparticle and are difficult to characterize experimentally. Edge and corner sites are also likely to be more reactive compared with ions accommodating plateau regions. At this point we take a short diversion to structural characterization and use the atomistic models to calculate important properties. In particular, the model is used to determine how easy it is to extract an oxygen ion from corner, edge, or plateau sites of the nanoparticle compared with oxygen extraction from the surface of the parent (bulk) material.

Isolated nanoparticles have three dimensions at the nanoscale, whereas supported thin films have only one dimension—thickness—at the nanoscale. We show how simulation can be used to generate models for oxide-supported oxide thin films. These models can then be used to predict the considerable influence that the substrate has in directing the structure of the thin film deposited thereon. Indeed, one can tune the properties of the thin film, by careful choice of substrate. We illustrate this by showing how the $CeO_2(110)$ surface can be exposed in preference to the more stable $CeO_2(111)$ by depositing a thin film of ceria on an yttrium stabilized zirconia substrate. This is important because the $CeO_2(110)$ surface is catalytically more active compared with the $CeO_2(111)$.

In the final section, we investigate the structure of nanoparticles that is supported on a substrate and consider the structural implications of encapsulating a nanoparticle into a host oxide material.

9.2. THEORETICAL METHODS

In this section, we describe potential models that can be used to represent the interactions between the ions comprising the nanomaterial, the computational codes used to perform the simulations, and finally, the strategies for generating atomistic models (atom coordinates) for the oxide nanomaterials.

9.2.1. Interatomic Potentials

Many oxides are highly ionic, and therefore, we can describe them accurately using the Born model of the ionic solid (2). For nanomaterials, where a high proportion of the ions are located at a surface or interface, the interatomic potential must be reliable at describing ions in sites with low coordination. Clearly, the simplest way to address this is to use the same interatomic potential models to describe surface and interface ions as bulk ions. Indeed, this approach has enjoyed considerable success, for example, see Refs. 3–5. Conversely, as (inevitably) systems become more complex and the need for higher levels of accuracy arise, surface-specific potential models may play a more significant role in the future.

In the Born model of the ionic solid, all ions are all assigned a charge. For a highly ionic material such as MgO, formal charges (i.e., −2 for oxygen and +2 for magnesium) have been shown to work well. Conversely, for oxides that are less ionic, partial charges are sometimes more appropriate. The (attractive) interaction energy between the charge on the oxygen and magnesium ion is given by

$$E_{\text{attractive}} = \frac{q_{\text{Mg}} q_{\text{O}}}{r_{\text{Mg-O}}} \tag{9.1}$$

where q_{Mg} and q_{O} are the charges of the magnesium and oxygen ions, respectively, and $r_{\text{Mg-O}}$ is the distance between the two ions. Equation 9.1 suggests that the ions would prefer to move closer and closer to each other to achieve a more stable (lower energy) configuration. However, as the ions move closer to one another, the "electron clouds" of the two ions start to repel each other. This short-range repulsion can be described using various equations, one of which is given below:

$$E_{\text{repulsive}} = A\exp\left(\frac{-r_{\text{Mg-O}}}{\rho}\right) \tag{9.2}$$

Two variable parameters are associated with Eq. 9.2: A and ρ, which can be fitted to the experimental properties. For example, if one increased the value of parameter A, the repulsive interaction between the two ions increases and the ions would move further apart. Thus, the A parameter used to describe MgO ($a_{\text{MgO}} = 4.2$ Å) may be smaller compared with the A parameter used to describe BaO, which has a higher lattice parameter ($a_{\text{BaO}} = 5.5$).

Setting a particular value for A might give an accurate bond distance. However, ions are rarely at their equilibrium distance, and therefore, one needs to describe the repulsive interaction between the two ions over a range of distances. Accordingly, the parameters A and ρ are fitted *together* and to a range of experimental data. This may include, for example, the lattice parameter, elastic constants, and dielectric constants. This enables the potential model to accurately describe the oxide over a range of interatomic distances and is particularly important when simulating the system at a particular temperature, other than 0 K, where the ions are vibrating or indeed diffusing. Gale and Rohl provide a comprehensive treatment of atomistic potentials and how they can be derived (6).

The total energy of the system is the sum of all interactions between all *pairs* of ions comprising the oxide nanoparticle:

$$E_{\text{total}} = \sum (E_{\text{attractive}} + E_{\text{repulsive}}) \tag{9.3}$$

The energy is shown graphically for a diatomic molecule as a function of interionic separation in Figure 9.1. The minimum of the graph corresponds to the equilibrium bond distance where the attractive Coulombic term balances exactly the short-range

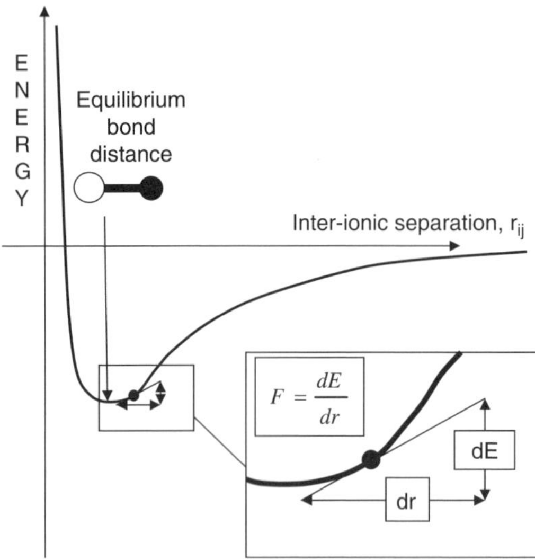

Figure 9.1. Energy of a diatomic molecule, calculated as a function of interionic separation (r). The force acting upon the ions at any particular interatomic separation can be determined by calculating the gradient (illustrated in the inset).

repulsion between the electron clouds. Starting from the equilibrium distance, as the ions move closer together, there is greater repulsion because of the electron clouds and the energy becomes more positive (less stable). Similarly, as the ions move further away from the equilibrium position, there is less repulsion from the electron clouds but also the Coulombic interaction is reduced and again the energy becomes more positive. The force F on the ions at each point can be calculated by measuring or calculating the gradient at a particular bond distance. At distances less than the equilibrium bond distance, the gradient (force) will be negative, indicating that the force between the two ions tries to move them apart. Conversely, at distances greater than the equilibrium bond distance, the gradient will be positive and the force acts to move the ions closer together. At the minimum energy position, the force is zero.

In nature, the ions comprising the oxide nanoparticle will exist in a low- or minimum-energy configuration. This is not necessarily the lowest energy possible [global minimum energy (6)] as most systems comprise, for example, defects; rather, the ion positions are such that the system exists in a "local" energy minimum position. For a diatomic molecule, it is easy to move the two ions into the lowest energy position (Figure 9.1). However, for a system comprising, for example, 10,000 or more atoms, the problem is more challenging. Indeed, there are many mechanisms for exploring low-energy structures and five important methods are described briefly in the following section.

9.2.2. Simulation Strategies

9.2.2.1. Energy Minimization (6) Here, the ions are moved (iteratively) from a particular starting configuration to a lower energy configuration. This can be illustrated by considering the energy minimization of a diatomic molecule:

- Start with a (best guess) bond distance (Figure 9.1).
- Calculate the gradient.
- If the gradient is positive, reduce the bond distance; if it is negative, increase the bond distance.
- This process is repeated by recalculating the gradient until the gradient becomes zero, which corresponds to a minimum energy position.

In reality, the problem is never this simple because a nanomaterial will have $3N$-6 coordinates, where N is number of atoms, and therefore, we have to minimize in $3N$-6-dimensional space. However, the basic procedure is that outlined above although there are many variations.

9.2.2.2. Monte Carlo—MC (7–9) As the name implies, the ions are moved in a random fashion—determined by a roll of the dice—to generate a low-energy configuration. For example, a starting configuration is generated and then all ions are moved in a random direction (determined by a roll of the dice). If the new configuration is lower in energy, the configuration is accepted. Conversely, if the configuration is higher in energy, the configuration is rejected. However, central to this method is that it allows the system to go uphill in energy as well as down and therefore escape from local energy minima that do not relate to realistic structures. In particular, instead of rejecting *all* structures that are higher in energy than the previous one, a further roll of the dice is performed. It is this second roll of the dice that decides whether a higher energy structure is indeed rejected.

9.2.2.3. Genetic Algorithms—GA (10) Here, the atomistic structures of successive generations are constructed based on attributes from the parent that are deemed desirable. Desirability in this sense means, for example, low energy or realistic bond distances or chemically sound. Highly desirable configurations are assigned a high breeding probability and will likely generate lots of "daughter structures," whereas undesirable structures are unlikely to "procreate." It is probably worth noting that daughter structures may have more than two parents!

9.2.2.4. Evolutionary Simulation (11) Here, the atomistic model is evolved in a similar fashion to experimental synthesis. For example, molecular beam epitaxy involves the deposition of molecules onto a substrate, which then adhere to its surface to form a coating or thin film. To generate the analogous model, instead of placing the whole of the thin film on top of the substrate in one go, the molecular deposition process is simulated. In this way, the atomistic model evolves during the simulation and in so doing captures (hopefully) some of the structural features observed experimentally. Variations on this theme, and there are many, may include simulating the nucleation and growth of a material from solution or crystallization from a melt.

9.2.2.5. Molecular Graphics This important, powerful, and sometimes overlooked approach provides a wealth of visualization tools to move and manipulate atom coordinates in three dimensions (3D). It is used as a graphical user interface (GUI) to construct chemically/physically or intuitively correct starting configurations prior to the simulation, to animate and thus "observe" the simulation, to analyze final structures, and to help one comprehend complex (crystal) structures from experimental data (i.e., X-ray diffraction) that are difficult to understand in two dimensions (2D).

9.2.2.6. Molecular Dynamics—MD (12) This simulation strategy enables one to simulate the movement of the ions comprising the system at a particular temperature and as a function of time. The method uses Newton's laws of motion to calculate and change the positions of all ions as they move, vibrate, or collide with one another. For example, the force F acting upon a particular ion i at time t is given by

$$F_i(t) = m_i a_i(t) \tag{9.4}$$

where m_i is the mass of the ion and a_i its acceleration. At the start of the simulation, the positions $r_i(x, y, z)$ of all ions, comprising the nanoparticle, are defined. The forces acting on each ion can then be calculated by adding up all pair-wise interactions between all ions comprising the oxide nanoparticle using an interatomic potential such as that described in Eq. 9.3. The force, which is the gradient of the energy graph, Figure 9.1, is given by

$$F = \frac{dE}{dr} \tag{9.5}$$

First, the accelerations of the ions can be calculated by rearranging Eq. 9.4 and dividing by the mass of the ion:

$$a_i(t) = \frac{F_i(t)}{m_i} \tag{9.6}$$

Once the accelerations have been calculated, the velocities can be updated after a particular time interval δt:

$$v_i(t + \delta t) = v(t) + a_i(t)(\delta t) \tag{9.7}$$

Finally, the coordinates of the atom positions can be changed based on the atom velocities:

$$r_i(t + \delta t) = r_i(t) + v_i(t)(\delta t) \tag{9.8}$$

At this point, all required information has been calculated and the cycle repeats; i.e., the force, Eq. 9.5, is recalculated with the new atomic coordinates.

The time interval δt used for a particular simulation has to be chosen carefully and must be smaller than the time required for a bond to vibrate. This is to ensure that the energy and forces are updated *before* the ions have a chance to move too near to one another. Unfortunately, a bond vibration is of the order of 10^{-15} seconds, and therefore, the biggest limitation of MD simulation is the short period of time one can run the simulation. For example, to simulate just one second in real time would require the calculation of over 10^{15} cycles, which is, at present, well beyond even the fastest computer processors. Moreover, parallelization of the problem does not really help as the equations have to be solved sequentially and therefore introducing more parallel processors only enables the number of ions in the system to be increased, rather than facilitating any increase in the simulation time. However, as one might have anticipated, various innovative and imaginative methods have been introduced to help counter this somewhat debilitating limitation.

Many computational codes have been written to allow the user to perform MD simulations (for example, collaborative computational projects; www.ccp.ac.uk). For most of the work described in this chapter, the DL_POLY code was used to perform the dynamical simulations (13). This (parallel) code was developed by Daresbury Laboratory in the mid-1990s for the molecular simulation community in the United Kingdom. After more than 20 years since its inception, this code enjoys a worldwide user base. Indeed, over 600 licenses have been issued. A philosophy, central to the DL_POLY MD code, is that its source code is freely available (12), and therefore, the users can modify, add, and share new subroutines to the code to suit their particular simulation. As such, the code has been continually evolving since its inception.

In practice, it is rare that any one simulation technique is used in isolation; rather, the methods are combined (especially with molecular graphics), which results in a more powerful technique.

In this section, we have provided a brief and necessarily simple overview of some atomistic simulation methods and procedures. For a more in-depth and authoritative treatment, we suggest that the interested reader refer to *Molecular Modeling, Principles, and Applications* by Andrew R. Leach (14), which provides an excellent and detailed reference source. In the following section, we describe how the simulation techniques are applied to the study of oxide nanomaterials.

9.3. NANOPARTICLES

In this section, we describe the evolutionary simulation procedure, used to generate the atomistic models described in this chapter, in more detail and illustrate the procedure using MnO_2 as an example. Next, we explore the structure, morphology, and catalytic activity of CeO_2 nanoparticles and compare it with the parent material.

9.3.1. Amorphization and Recrystallization (A&R) of MnO_2

Many atomistic studies, used to investigate the structure and energetics of a system, proceed by defining the basic atomistic structure ("by hand"), which is then simulated using static or dynamical methods. However, for systems that comprise complex 3D microstructures, such as nanoparticles, the starting structural models can prove challenging to generate. For example, the simulator needs to include within the atomistic model the morphological structure, including the particular low-energy surfaces that are exposed. And although many simulation codes are available to generate individual surfaces (such as Ref. 15), for a nanoparticle, one also needs to consider the atomistic structure of edges, where a pair of surfaces meet, and vertices, where three or more surfaces meet, together with the implications of dipolar surfaces (16). One also requires a representation of the defect chemistry, which may include point defects that exist within the bulk or surface regions of the nanoparticle, including both intrinsic and extrinsic defects such as vacancies, interstitials, and substitutionals together with clustering or segregation of these defects. For larger nanoparticles, defects such as dislocations and/or grain boundaries are likely to be present. And although the introduction and simulation of a specific and isolated grain boundary or dislocation is now routine within the perfect "bulk" material (5,17), it remains challenging to introduce these into, for example, a 50,000-atom nanoparticle.

These structural and microstructural features are likely to influence, or indeed govern, the properties of the oxide nanomaterial. Accordingly, if the simulator is to offer predictions of a nanoparticle that are both accurate and can be used reliably by the experimentalist, the atomistic models must include the microstructural features alluded to above. To introduce these features "by hand" is certainly a daunting, if not intractable, prospect, and therefore, an alternative simulation strategy must be sought to generate models with this complexity.

Materials synthesis inevitably involves some kind of "crystallization" process. Indeed, it is the crystallization process that controls the (micro)structure and hence the properties of the material. Moreover, by modifying the crystallization process (whether crystallization from solution, vapor deposition, molecular beam epitaxy, ball milling, etc.), one can exact some control over the microstructure and hence the properties of the material. Clearly, the ideal way of capturing, within a single atomistic model, the microstructural features observed experimentally, is to simulate the crystallization process. Indeed, many theoretical studies explore the crystallization process. For example, Piana and Gale performed some highly detailed MD simulations on the growth and dissolution of urea crystals (18). Similarly, Hamad et al. used MD to explore the embryonic stages of ZnS nanobubbles (19). These outstanding and elegant dynamical atomistic simulations capture much of the important features associated with the nucleation, growth, and dissolution processes. Indeed, these approaches would prove ideal in generating models for nanoparticles. Unfortunately, they would also prove far too computationally expensive (at present) to use them to generate models for oxide nanoparticles comprising, for example, 50,000 atoms. An alternative approach to this problem, and one that can routinely accommodate 100,000+ atoms with the computational facilities available today, is Amorphization and Recrystallization (A&R) (20–22). We illustrate this evolutionary technique in the following section by considering the example of MnO_2 nanoparticles.

9.3.1.1. Microstructure MnO_2 is a widely studied electrochemical material that accommodates over 14 polymorphs (23) and comprises an internal interconnecting tunnel structure with channel sizes commensurate with that of cations spanning H^+ to Cs^+. Accordingly, they can act as host lattices for the insertion, storage, and extraction of cations and, hence, charge, which is central to their ability to store energy. Among the several varieties of MnO_2, the form known as γ-MnO_2 exhibits the best electrochemical activity. This material comprises a wealth of microstructural features. Indeed, it has been proposed that this rich microstructure facilitates the exemplary properties of this material. The microstructural features include: de Wolff disorder, which is the intergrowth of domains comprising single and double channels, microtwinning, and a variety of point defects such as cation vacancies. Clearly, the ability of A&R to evolve, from an amorphous precursor, the crystalline MnO_2 structure together with microstructural features observed experimentally is a stringent validation of the A&R strategy and provides a useful demonstration of the technique. In the following section, we describe briefly the generation and structure of MnO_2 nanoparticles about 10 nm in diameter. Further details have been published elsewhere (22).

9.3.1.2. Evolution of the Atomistic Model A “cube” of MnO_2, comprising about 25,000 atoms, was constructed and is shown in Figure 9.2*a*. The coordinates of all ions comprising this cube were then changed to *increase* the lattice parameter by 36%, and MD was then performed on this system at 2000 K. The MnO_2 can perhaps be compared with an elastic band, which is tensioned and then released. As soon as MD is applied to this tensioned configuration, the Mn and O ions “implode.” And as they accelerate and move toward the center of the cube, they hit other ions and bounce off. The culmination of all accelerations and collisions results in the amorphization of the MnO_2, which is shown in Figure 9.2*b*. Essentially, the energy introduced into the system to tension the nanoparticle results in its amorphization. During prolonged MD simulation, the MnO_2 nanoparticle starts to recrystallize. Snapshots of the recrystallization are shown in Figure 9.3*a*–*f* together with the energy of the system in Figure 9.3*g*: From 0–1000 ps, the MnO_2 is amorphous. At about 1000 ps, a crystalline seed evolves within the amorphous MnO_2 and the Mn and O ions start to condense onto the surface of this seed increasing its size, Figure 9.3*a*, *b*. The crystallization front propagates from this seed to encompass the whole nanoparticle; recrystallization is essentially complete after about 3000 ps. The energy of the system goes down during the recrystallization with the difference in energy between the amorphous structure and crystalline structure reflecting the latent heat of crystallization. Finally, the nanoparticle is cooled, under MD, to 0 K. The complete simulation is computationally intensive and requires about 5 days using 96 processors of a SunFire F15K Galaxy-class configuration supercomputer, Cambridge University, United Kingdom.

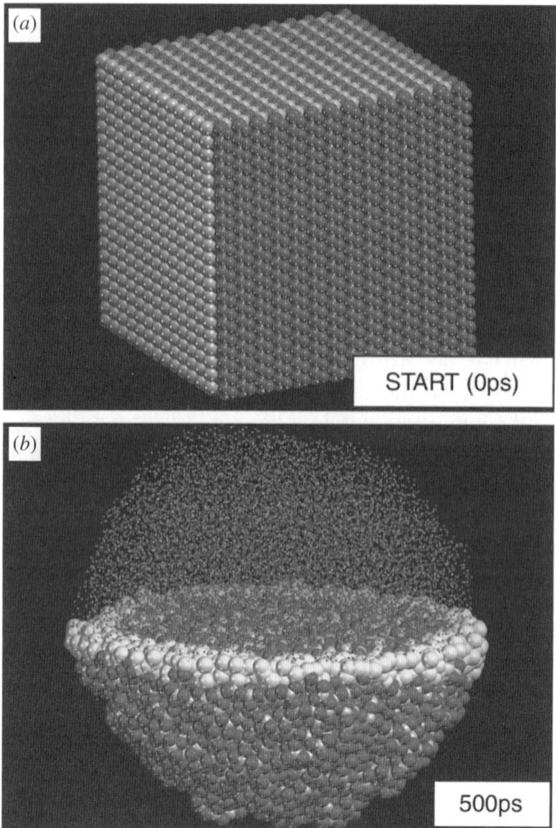

Figure 9.2. Atom positions corresponding to (*a*) the starting structure and (*b*) the amorphous configuration (after 500 ps of MD simulation) of the MnO_2 nanoparticle. A sphere model representation of the atoms position has been used. In (*b*), the atoms comprising the top half of the sphere are represented using a smaller sphere radius to show more clearly the amorphicity of the nanoparticle. Manganese ions are light gray, and oxygen is dark gray. Some ions are colored white to show more clearly that the nanoparticle is spherical. See color insert.

9.3.1.3. Final Structure Inspection of the final, 0 K structure, shown in Figure 9.4, reveals a crystalline nanoparticle, which accommodates predominantly domains that conform to the pyrolusite structure (23)—a low-energy polymorph of MnO_2. Figure 9.4*a* shows a representation of the Mn and O atom positions comprising the nanoparticle. The oxygen sublattice forms a close packed array and gives rise to octahedral sites, 50% of which are filled by Mn to give the MnO_2 stoichiometry. In particular, it is the filling of these octahedral sites that gives rise to the particular polymorph. In Figure 9.4*a*, a rectangle has been drawn that encompasses two oxygen planes and an atomic plane of Mn sandwiched between. These three planes are shown (plan view) in Figure 9.4*b*. Inspection of the filling by Mn reveals straight

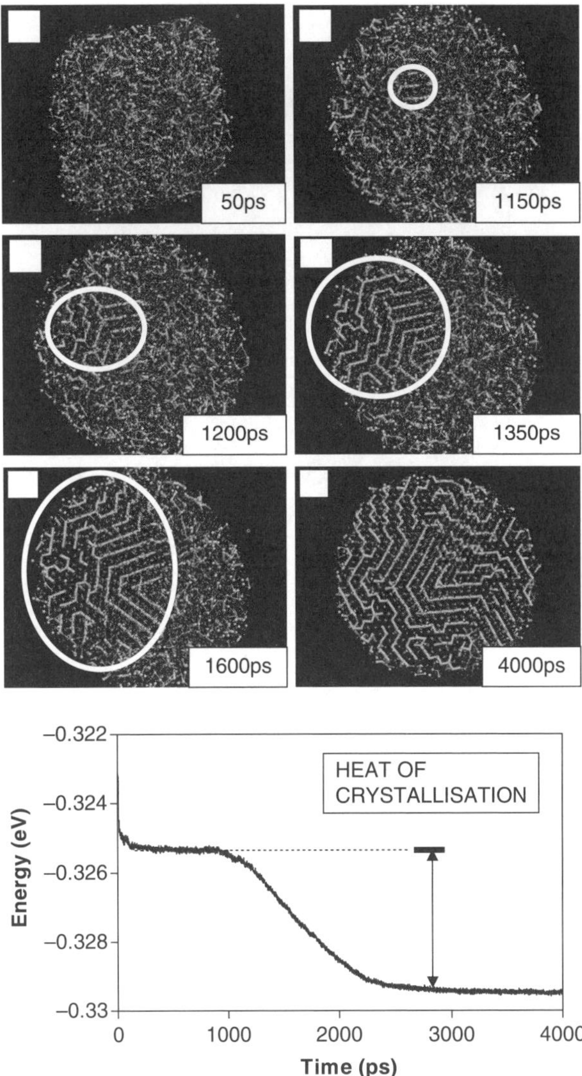

Figure 9.3. Snapshots, taken at various time intervals, of a slice cut through the MnO_2 nanoparticle depicting the recrystallization: (*a*) after 50 ps of MD simulation; (*b*) after 1150 ps; (*c*) after 1200 ps; (*d*) after 1350 ps; (*e*) after 1600 ps; and (*f*) after 4000 ps. The energy of the system (10^6 eV), calculated as a function of time, is shown in (*g*) and can be usefully correlated with the structures (*a*)–(*f*).

single parallel lines of Mn, which correspond to the pyrolusite polymorph. These lines also change direction, which corresponds to microtwinned domains as observed experimentally (24); an enlarged segment is depicted in Figure 9.4*c*. Detailed analysis of the final structure, using graphical techniques, also reveals the presence of vacancies, which have also been observed experimentally.

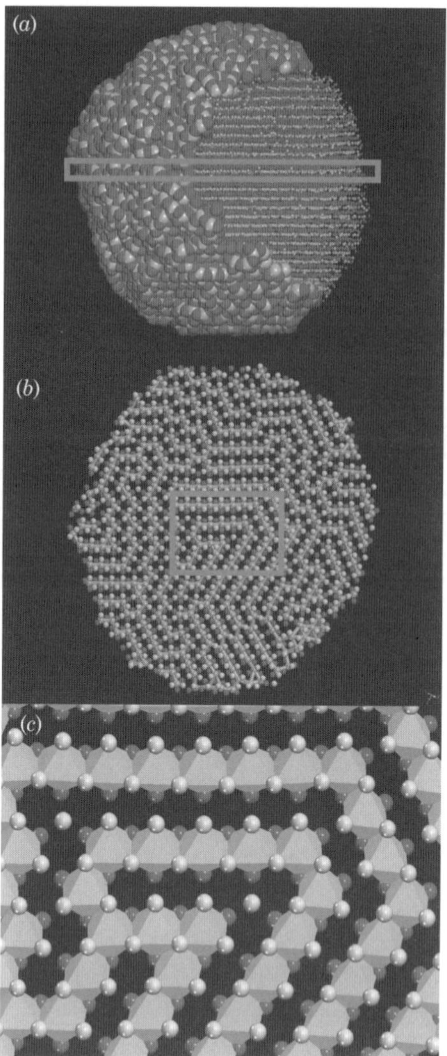

Figure 9.4. Structure of the fully recrystallized MnO_2 nanoparticle: (*a*) sphere model representation of the atom positions. Two different sphere radii have been used for the atoms to aid understanding of the figure; a slice cut through this figure (blue rectangle) is shown in plan view in (*b*), which depicts the microtwinning of the MnO_2. An enlarged segment is shown in (*c*). Notation: (*a*) Manganese atoms are colored gray, and oxygen is red; (*b*) Manganese are colored gray, O atoms above the plane of the paper are red, and O ions below the plane of the paper are yellow. The manganese atoms are shown as ball-and-stick representation to show more clearly the microtwinning; (*c*) manganese are shown as gray polyhedra indicating the octahedral sites they occupy, oxygen atoms above the plane of the paper are red, and oxygen ions below the plane of the paper are yellow. See color insert.

In summary, A&R has been used to generate models for MnO_2 nanoparticles that are realistic in that they reflect the low-energy polymorphic configuration (pyrolusite) and include complex microstructural features that have been observed experimentally. Accordingly, we suggest therefore that the A&R strategy is an appropriate simulation tool one can use to generate realistic models of oxide nanoparticles.

9.3.2. CeO_2

The second isolated nanoparticle that we consider is ceria, CeO_2. This material has enjoyed much attention recently because its high oxygen conductivity, at moderate temperatures, makes it a potential candidate as a component of a fuel cell (25). Here, we explore its catalytic properties.

In the early 1980s, three-way catalysts (TWCs) were found to have the ability to oxidize CO and hydrocarbons, while simultaneously reducing NO_x to form less toxic products such as CO_2, H_2O, and N_2. These conversions are attained in a narrow window of the air-to-fuel ratio (26). CeO_2 is a common promoting component in TWC, which has been attributed to its low Ce^{4+}/Ce^{3+} redox potential and high oxygen-defect mobility, but what are the implications for this material when one traverses down to the nanoscale?

9.3.2.1. Morphology The atomistic structure of CeO_2 nanoparticles has been well characterized by several groups using transmission electron microscopy (27,28). In particular, the shape of CeO_2 nanoparticles, about 10 nm in diameter, can be described as conforming to truncated octahedra. Specifically, the nanoparticles were observed to expose {111} surfaces, truncated by {100}—within a transmission electron microscopy (TEM) micrograph, these polyhedra project into hexagons. TEM images of these nanoparticles are presented (with permission) in Figure 9.5*b* together with a schematic illustrating the proposed morphological shape, based on the TEM micrographs.

Atomistic simulation can be used to predict the crystal morphology by calculating the energy of low index surfaces, and this method has been shown to be remarkably accurate. Intuitively, the more stable the surface, the more that particular surface is exposed in the morphology. Indeed, this strategy has been shown to be a powerful simulation tool and has enjoyed much success (29), not the least because of its inherent simplicity. The calculated energies of the low index surfaces of CeO_2 are $(111) > (110) > (100)$, and therefore, the simulation would, in accord with the experiment, predict {111} to dominate (30). Surface energy calculations also predict that the morphology of CeO_2 should comprise more {110} than {100}. However, this is not supported experimentally; rather the only other face observed is the {100}, which is surprising because this surface is dipolar (16).

It is inevitable that a simplistic simulation approach, such as predicting crystal morphologies based solely on surface energy calculations, will fail in certain circumstances. This is because crystal morphologies can be preparation dependent (31). Accordingly, simulation strategies, used to predict crystal morphologies, must

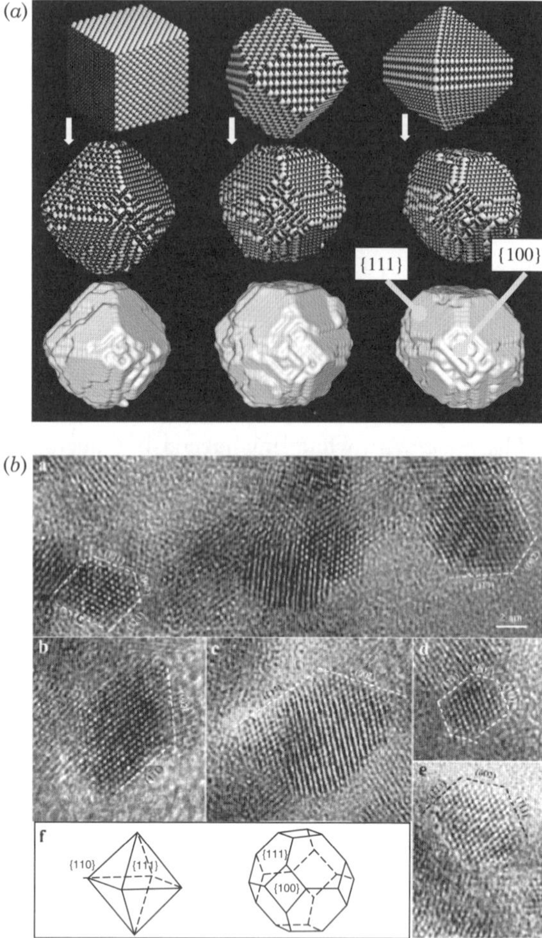

Figure 9.5. (*a*) Molecular graphics representations of the theoretical models, (*b*) high-resolution transmission electron micrographs for CeO_2 nanoparticles. (*a*) Graphical representation of the atom positions comprising each nanoparticle. Top: starting configurations; middle: final configurations; bottom: final configurations with surface rendering rather than sphere model representation. Cerium ions are represented by light gray spheres and oxygen the dark gray spheres. (*b*) HRTEM reproduced with permission from Ref. 27.

include features pertaining to nucleation and growth (32). This added complexity may facilitate more accurate predictions but at a cost of losing simplicity of approach.

In the following section we describe how A&R can be used to predict the crystal morphology of CeO_2 nanoparticles. This should also prove a useful validation of the A&R method because detailed TEM images are available to compare. It is worth bearing in mind that the A&R method has to accommodate two unexpected features associated with the morphology of CeO_2. First the nonexistence of (relatively low-energy) $CeO_2\{110\}$ surfaces, and second the ability for A&R to evolve {100} surfaces and necessarily quench the dipole, which is associated with the (100) surface.

9.3.2.2. Evolution of the Atomistic Model A&R was used to predict the morphological shape and atomistic configuration of CeO_2 nanoparticles about 8 nm in diameter. Full details regarding the simulation can be found elsewhere (33) and follow a very similar procedure as that described above for MnO_2 nanoparticles. The morphologies of the CeO_2 nanoparticles, predicted using A&R, are shown in Figure 9.5*a*; TEM images are also shown in Figure 9.5*b*, as a comparison. The theoretical models do indeed exhibit truncated octahedral morphologies in accord with the experiment, and in contrast to surface energy predictions, the predicted morphologies do not (at first sight) expose {110} surfaces. However, upon closer inspection of the atomistic models, Figure 9.5*a*, we note that the edges, formed where two {111} surfaces intersect, could perhaps be described as {110}. Also, steps on the {111} also appear commensurate with {110}. TEM images are primarily 2D projections, and therefore, structural features that are "3D" are difficult to identify experimentally. Accordingly, we suggest that the atomistic models, which can be observed and manipulated in 3D using graphical techniques, are valuable in that they can be used to help interpret TEM images and therefore aid the experiment in characterizing the full 3D atomistic structure of CeO_2 nanoparticles.

9.3.2.3. Dipolar Surfaces Fluorite-structured {100} surfaces are dipolar and therefore inherently unstable (34). Surface energy calculations on dipolar surfaces have necessitated quenching the dipole prior to simulating the system with static or dynamical methods, which is normally achieved by physically rearranging the ions "by hand" (35). Conversely, A&R is a simulation method in which the structure/atom positions evolve during the simulation. Accordingly, this approach does not need (or indeed allow!) the simulator to manually move ions with an aim of quenching the dipole. However, close inspection of the $CeO_2\{100\}$ at the end of the simulation (Figure 9.6) revealed a 50% reduction in the number of oxygen ions at the {100} surfaces. Specifically, the simulation has evolved a structure in which the dipoles associated with the {100} surfaces have been quenched. Moreover, because these surfaces have evolved, we suggest that they are likely to be more realistic than dipolar surfaces that have had their dipoles quenched manually (based on chemical intuition) because the simulated crystallization (A&R) reflects (in part) crystallization that occurred during the experimental synthesis. However, this assumption must be tested on a system where the real structure of the (quenched) dipolar surface is unambiguous—down to the individual positions of the surface ions, which is, at this time, a contentious issue (34).

9.3.2.4. Oxidation of CO to CO_2 Using CeO_2 Nanoparticles Once a realistic model of the CeO_2 nanoparticle has been generated, this model can be used to predict its catalytic activity. In particular, atomistic simulation can be used to predict whether the nanoparticle would be catalytically more or less active compared with

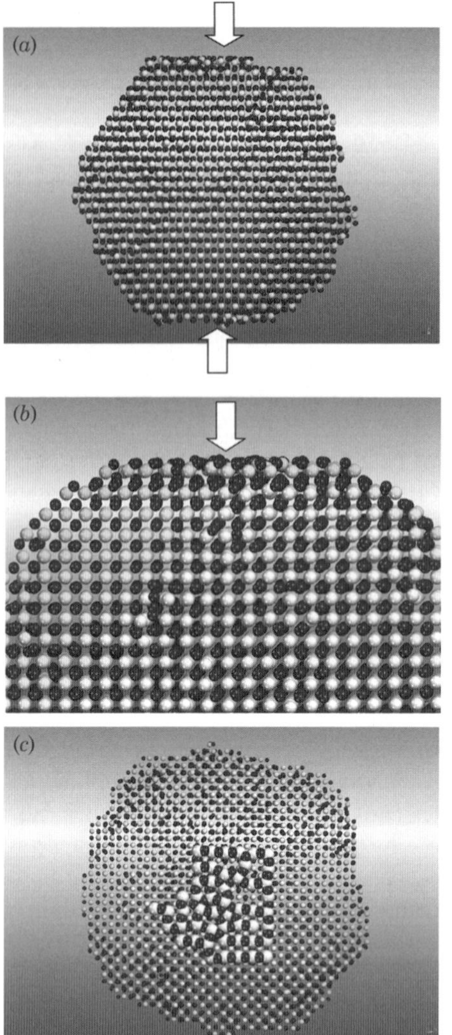

Figure 9.6. (*a*) Sphere model representations of the atom positions comprising a slice cut through a CeO_2 nanoparticle revealing the atomistic structure of dipolar {100} surfaces. Notice the similarity between this model structure and the experimental TEM in Figure 9.5*b*. The arrows indicate the {100} surfaces. (*b*) Enlarged view of the full nanoparticle, showing more clearly the dipolar surface. (*c*) Plan view showing the surface layer as large spheres and the rest of the crystal as smaller spheres and shows clearly the 50% filling by oxygen of the surface atomic layer, which is necessary to quench the dipole.

the parent material. For example, the oxidation of CO to CO_2 involves extracting oxygen from the surface of CeO_2. Clearly, if one could make the surface oxygens more facile to extraction, one can, potentially, fabricate a more active catalyst. It has been shown previously that extracting oxygen from the step site of CeO_2 is easier than from the plateau (36,37). This seems intuitive in that an oxygen ion accommodating a step site is lower coordinated compared with an oxygen ion at a plateau site and therefore less strongly bound to the surface. One can then argue that if more step sites can be generated at the surface of the catalyst, then theoretically, one should be able to generate a more active catalyst. One way of achieving this (discussed later) is to deposit a thin film of CeO_2 on a substrate, which acts as a template in directing the CeO_2 to expose a reactive surface. The other approach is to traverse to the nanoscale.

At the risk of stating the obvious, the dictionary definition of a polyhedron is *a solid figure consisting of more than four plane faces. A pair of faces will meet along an edge and three or more edges meet at a vertex.* Accordingly, as a particle decreases in size, the proportion of ions accommodating edge and vertex sites, compared with those occupying plateau regions, increases. In addition, the number of surface-to-bulk atoms also increases. Eventually, at the nanoscale, the number of edge sites becomes commensurate with plateau sites and surface ions commensurate with bulk ions. Indeed, one can perform a simple count of the ions in the CeO_2 nanoparticle (Figure 9.5*a*) to confirm this. Accordingly, the CeO_2 nanoparticle should comprise a high concentration of labile (low-coordinated) surface oxygen species.

To explore the catalysis of the CeO_2 nanoparticle, we need to determine how difficult it is to extract an oxygen ion from its surface. The simulation procedure that can be used to calculate this is relatively straightforward. In particular, we choose an oxygen ion to remove; we then extract this ion out of the surface and replace two Ce^{4+} with Ce^{3+} (to accommodate for the charge imbalance). The energy of this system is then calculated and compared with the energy of the same system but with an oxygen ion removed from a different position. Using this procedure, one can generate a table of oxygen location versus vacancy formation energy. However, when one inspects the atomistic model (Figure 9.5), a daunting number of different surface oxygen species is available for extraction; the nanoparticle comprises 15,972 ions, many of which are surface or near-surface species. Accordingly, an oxygen vacancy formation energy "averaged" over the whole nanoparticle was calculated. Specifically, 266 oxygens were removed from the surface of the starting (prior to amorphization) configuration together with 532 Ce^{4+} species replaced by Ce^{3+}. The rationale underlying introducing the defects prior to the amorphization is that the A&R will direct the vacancies and Ce^{3+} species into a range of low-energy configurations, thereby providing a more realistic average reduction energy for the nanoparticle. The nanoparticle comprises, in total, 15,972 ions, and therefore, removal of 266 oxygens results in a reduced nanoparticle with composition $CeO_{1.95}$. This system was amorphized and then recrystallized; further details pertaining to the simulation can be found elsewhere (38). The final structure is shown in Figure 9.7.

Inspection of the reduced nanoparticle (Figure 9.7*a–c*) reveals an octahedral morphology comprising {111}, truncated by {100}, which is similar to the unreduced nanoparticle (Figure 9.5). The Ce^{3+} ions can be seen to decorate step and corner sites in addition to positions on {111} terraces. A simple count revealed that 20% of the total number of Ce^{3+} species occupy positions on {111} terraces, 26% decorate {111} step (17%) or corner (9%) positions, and 11% occupy positions on {100}. The remaining

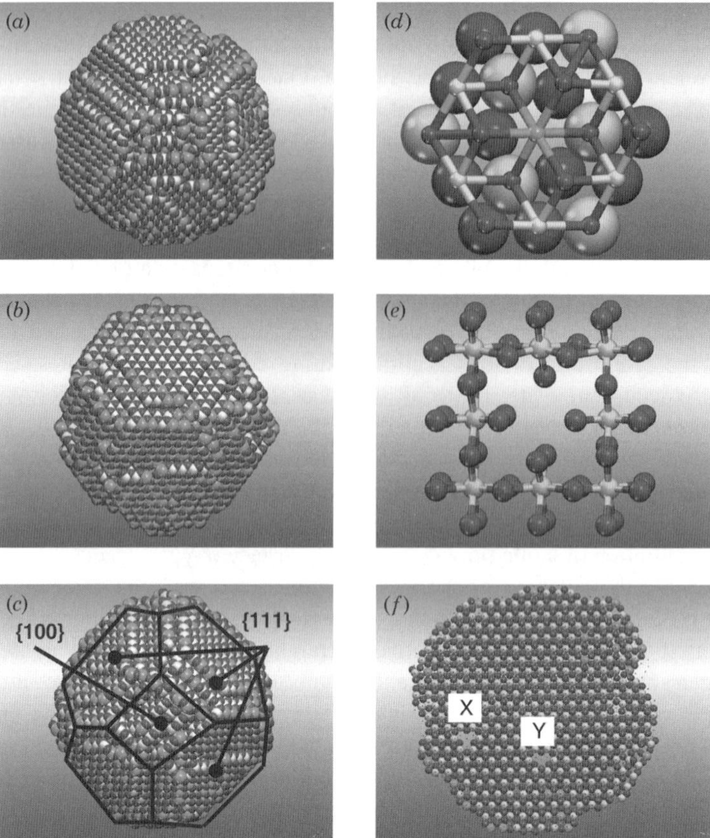

Figure 9.7. Sphere model representation of the reduced ceria ($CeO_{1.95}$) nanoparticle. (*a*)–(*c*) Depict three different views of the same nanoparticle showing the truncated octahedral morphology, similar to that of the unreduced nanoparticle (Figure 9.5). The Ce^{3+} ions are clearly seen to decorate surface steps and edges in addition to plateau regions on {111} and {100}. In (*c*), the atomistic structure is annotated with lines to show more clearly the surfaces. (*d*) Segment cut from the reduced nanoparticle depicting an oxygen vacancy bound to an adjacent Ce^{3+}. (*e*) Structure of a Ce^{4+} vacancy together with two adjacent oxygen vacancies, which have evolved to quench the charge imbalance associated with the vacancy. (*f*) A slice, cut through the nanoparticle, showing two complex defect clusters labeled X and Y. X comprises, using the Kroger–Vink notation, $[V_{Ce}^{''''}, 3V_{O}^{\bullet\bullet}]^{2-}$ and Y is $[V_{Ce}^{''''}, 2V_{O}^{\bullet\bullet}]^{0}$. Ce^{4+} is colored white, Ce^{3+} is green, and oxygen is red. See color insert.

43% of the Ce^{3+} species occupy positions within the bulk of the nanoparticle. We note, however, that most of these "bulk" Ce^{3+} species occupy positions one or two atomic planes below the surface rather than deep inside and near the center of the nanoparticle. Locating oxygen vacancies is much more difficult compared with locating Ce^{3+} species because a vacancy is not a physical entity within an atomistic model, and the relaxation of the lattice surrounding an O^{2-} vacancy or Ce^{3+} is sometimes

TABLE 9.1. Calculated Energies Associated with Oxidizing CO to CO_2 Using Lattice Oxygen From CeO_2. Energies are Calculated in Electronvolts. For the Nanoparticle, the Energy is an Average (Reaction 9.1) Calculated Over all Defects.

Bulk*	(111)*	(110)*	(310)*	Nanoparticle
3.1	−0.6	−2.8	−1.9	−1.4

*Taken from Ref. 37.

considerable. Consequently, a simple count of the vacancies, together with their location, was not possible, although oxygen vacancies were identified to occupy step and corner sites in addition to {111} terrace sites. In addition, oxygen vacancies were observed to lie one oxygen subplane below the surface oxygen ions.

The energy required to extract an oxygen ion from the bulk and from the "perfect" (111), (110), and (310) surfaces of CeO_2 was calculated by Sayle et al. (36,37). From these values, the energy associated with oxidizing carbon monoxide to carbon dioxide, using an oxygen ion extracted from a ceria surface, was calculated as follows:

$$CO(g) + CeO_2 \longrightarrow CO_2(g) + V_O^{\bullet\bullet} + 2Ce'_{Ce} \quad \textbf{(9.1)}$$

These values are presented, together with the average energy required to extract an oxygen ion from the surface of a CeO_2 nanoparticle to oxidize the carbon monoxide, in Table 9.1. We note that the more negative the energy, the more labile the oxygen, and therefore, the easier it is to remove from the lattice to promote the oxidation of CO.

The simulations suggest that it is easier to extract oxygen from the surface of a CeO_2 nanoparticle compared with the bulk parent material. In particular, the average energy required to extract an oxygen ion from the surface of the ceria nanoparticle is lower compared with the (most stable) $CeO_2(111)$ surface associated with the parent material. Moreover, the average energy required to extract oxygen from the nanoparticle is calculated to lie between those energies calculated previously for extracting an oxygen from the bulk, (111), (110), or (310) surfaces. This we suggest is a realistic value because Figure 9.7 reveals that O^{2-} vacancies and Ce^{3+} ions populate all conceivable positions in the nanoparticle, including bulk, terrace, and step sites and comprise a variety of configurations spanning simple isolated defects to complex multidefect clusters. For illustration, the atom positions showing the structure of three defect clusters are shown in Figure 9.7*d–f*.

In conclusion, the simulations predict that ceria nanoparticles can offer more reactive surfaces compared with the parent material, because of the higher proportion of step/corner sites and, therefore, facile oxygen species. Accordingly, ceria nanoparticles may help promote the oxidation of CO to CO_2, which is central to the performance of ceria-based TWCs.

9.3.2.5. Reducibility of CeO_2 Nanoparticles Using Monte Carlo In 1996, Cordatos et al. generated atomistic models for CeO_2 nanoparticles Ce_nO_{2n} (n = 2–20 and $n = 50$) using a Monte Carlo approach coupled with simulated annealing (39). The authors found that the $Ce_{50}O_{100}$ nanoparticle exhibited the fluorite structure and exposed the (111) surface. The reducibility of the clusters was determined from the difference in lattice energies of Ce_nO_{2n} and Ce_nO_{2n-1}, where charge neutrality was maintained by changing two Ce^{4+} ions to Ce^{3+}. The authors concluded that the reduction of ceria nanoparticles is structurally sensitive and that larger crystals are more difficult to reduce.

9.3.2.6. Aggregated Nanoparticles and Grain Boundaries Experimentally it is well known that during the synthesis of oxide nanoparticles, the nanoparticles are not always monodispersed; rather, they aggregate together—as shown by HRTEM—for example, see Ref. 27. The interfacial structure between the two nanoparticles can (for example) be coherent (if perfectly aligned), can form a twin-boundary (if they are aligned but misoriented by a particular angle) or form a more

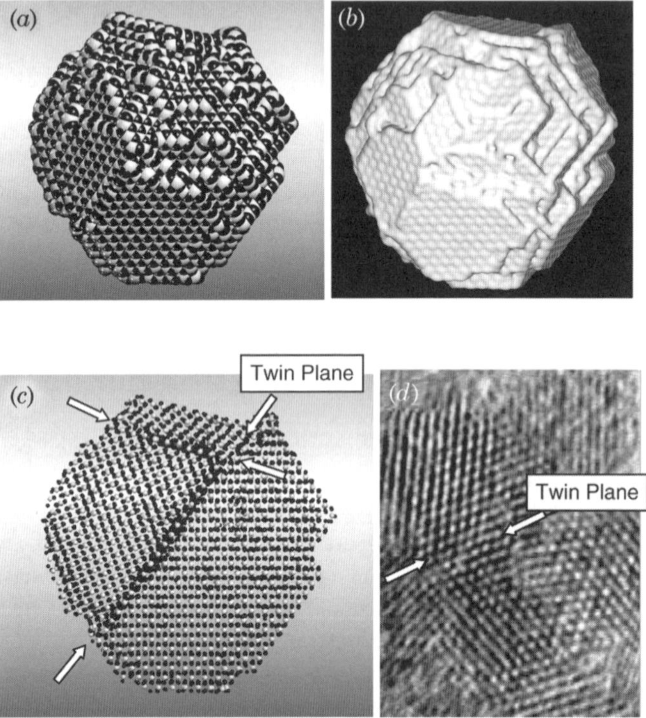

Figure 9.8. Representations of a CeO_2 nanoparticle that comprises two twin grain boundaries. (*a*) Sphere model representation. (*b*) Surface-rendered representation. (*c*) Sphere model representation with the sphere diameters reduced to show more clearly the twin grain boundaries that run through the nanoparticle. Arrows are included to highlight the boundary planes. (*d*) HRTEM of a CeO_2 nanoparticle showing the twin grain-boundary plane to compare with the theoretical models, reproduced with permission from Ref. 27.

general and complex grain boundary structure. Examples (HRTEM) of all such grain-boundary structures can be found in Refs. 27 and 28. The properties of a grain boundary have been shown to differ profoundly compared with the perfect parent material (17,40,41). Clearly, for aggregated nanoparticles, the number of grain boundaries will be considerable. Accordingly, if one is to make predictions pertaining to the properties of nanoparticles, then models for nanoparticles that comprise grain boundaries must be generated. And using atomistic computer simulation, it is possible to generate such structures. For example, Figure 9.8 shows a model for a CeO_2 nanoparticle that comprises two twin-boundaries. This nanoparticle, which comprises about 21,000 atoms, was generated using an analogous procedure as that used to generate MnO_2 or CeO_2 nanoparticles, described above.

9.4. SUPPORTED THIN FILMS

Thus far, we have explored, using atomistic simulation, the structure of "isolated" nanoparticles and find the structure and chemistry to be quite different compared with their bulk counterparts. In the next section, we consider oxide thin films supported on an oxide substrate. Here, in contrast to a nanoparticle, which has all three dimensions at the nanoscale, the thin film exists at the nanoscale in only one dimension—its thickness.

9.4.1. Ion Deposition

Atomistic models for oxide thin films with nanoscale thicknesses, supported on an oxide substrate, can be generated by depositing those ions comprising the oxide thin film, onto the surface of the substrate using MD. This type of approach can be classed as evolutionary and reflects or simulates the basic operation of experimental deposition techniques such as molecular beam epitaxy. We illustrate the technique using CaO/MgO(001) as a model system.

9.4.1.1. CaO/MgO(100) Snapshots (taken during MD simulation) depicting the growth of CaO deposited on MgO(100) are shown in Figure 9.9. More details pertaining to this study can be found in Ref. 20. Initially, ions are fired from a source to the surface of the MgO substrate, and as the ions hit the surface, they move, under MD, into low-energy configurations. For example, cations and anions diffuse across the surface via a hopping mechanism, and if they come in close contact, they are likely to adhere to form a diatomic. Further ions "diffusing" across the surface will add progressively to this diatomic to form a larger cluster. The structure of this CaO cluster is clearly influenced profoundly by the interactions from the underlying substrate. Eventually, with continued deposition, a second layer will start to evolve followed by a third and so forth. It is interesting to note that the Ca and O ions swap positions with Mg and O ions of the underlying MgO(100) substrate. This is because

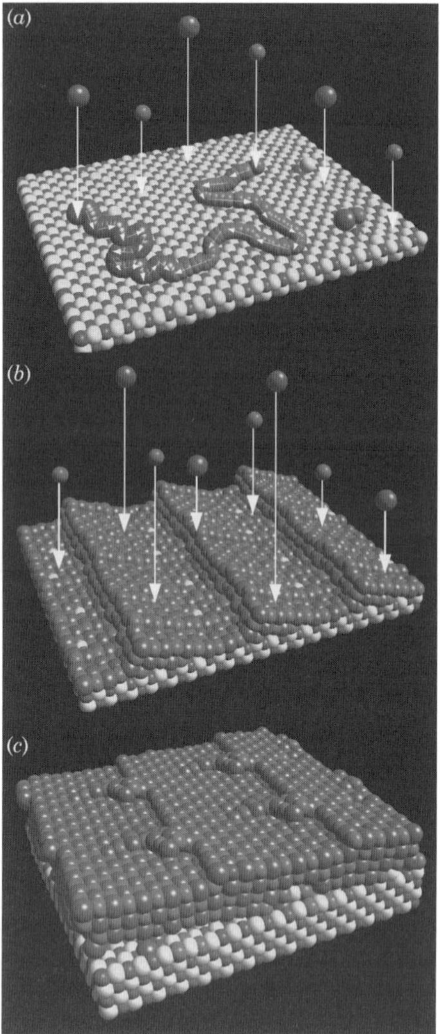

Figure 9.9. Sphere model representation of the atom positions during the simulated ion deposition of CaO onto MgO(001): (*a*) at the start of the simulation; (*b*) after just over one monolayer has been deposited and (*c*) after the deposition of just over three monolayers of CaO onto the MgO substrate. Notice the structural changes that occur as the effective number of monolayers deposited increases.

an energetically facile surface diffusion mechanism involves an exchange process between the deposited ions and the ions occupying the surface atomic layer of the substrate. This suggests that ionic materials may not be grown on a substrate with a similar structure without significant intermixing across the interfacial regions. This phenomenon was explored in more depth by Harris et al. (9).

The nature of the substrate, on which the ions are deposited, also has a significant influence on the structure of the oxide thin film deposited thereon. For example, substrates that comprise a high concentration of surface steps provide energetically favorable nucleation sites for ions deposited thereon. This is because of the increased coordination of an ion that adheres at a step, compared with a plateau position. The surface(s) exposed by the substrate will also influence profoundly the thin film because the substrate can act as a template in directing the structure of the overlying thin film (42). This is especially relevant for films with nanothicknesses, because the interfacial interactions will comprise a significant component of the total energy.

9.4.1.2. Growth Rates A major limitation with MD, in the context of ion deposition, is the timescale generally accessible, which using (present day) computational resources is still only of the order of nanoseconds. If the whole of the thin-film is deposited within this timescale, this corresponds to thin-film growth rates of at least meters per second; experimentally, depositions are typically of the order of "atomic layers per second." One solution to reduce the simulated deposition speed is to increase the temperature of the simulation. This increases the diffusivity of the ions in the simulation to levels commensurate with experiment (at a much lower and more realistic a temperature). However, one must be cautious in using this approach because at high temperatures, the diffusion mechanisms may change, which may result in artificial structures. Methods for circumventing timescale issues, pertaining to MD simulation, are receiving considerable attention at present, and various innovative strategies are being devised—see, for example, Refs. 8 and 43.

9.4.1.3. Incommensuration Another important issue associated with simulating oxide thin films with nanoscale thicknesses is the incommensurate relationship that exists between the lattice parameter of the substrate and the thin film deposited thereon. For example, consider the CaO/MgO system in which the lattice parameters of the component materials are $a_{CaO} = 4.8$ Å and $a_{MgO} = 4.2$ Å. To create a model for a CaO thin film supported directly on top of the MgO substrate with all ions in alignment (coherent interface—see Ref. 44) would require the CaO to be compressed by about 13%. This is easily achievable for a single atomic layer because the energetically favorable cation–anion interactions across the interface will compensate for the energy required to strain a single atomic layer of CaO to bring it into alignment with the underlying MgO substrate. However, for each additional layer of CaO added, an extra strain energy term would also need to be added. Eventually, at a particular *critical thickness* (45,46), the strain energy would outweigh the energy associated with the interactions across the interface between the CaO and the MgO (47). Above the critical thickness, the structure and configuration of the overlying thin film may change with respect to the underlying substrate to reduce the strain. (Micro)structural changes include, for example, the formation of superlattices, misfit dislocations, low interfacial densities, and grain boundaries (48). Clearly, atomistic simulation must be able to reproduce these microstructural features if reliable predictions are to be made using these models. And in the following section, we describe how such microstructural features can be introduced into the atomistic models.

9.4.2. Near-Coincidence Site Lattice (NCSL) Theory

One way of addressing the difference (incommensurate relationship) between the lattice parameter of the thin film and that of the substrate is to use a near-coincidence site lattice theory to predict epitaxial relationships with low associated misfits. In the following sections, we describe how this approach is used, in conjunction with atomistic simulation, to generate models of thin films.

9.4.2.1. BaO/MgO Sayle et al. (49) used an NCSL theory, coupled to atomistic simulation, to predict superlattices for BaO ($a_{\mathrm{BaO}} = 5.5$ Å) thin films supported on MgO ($a_{\mathrm{MgO}} = 4.2$ Å). For example, for the BaO(100)/MgO(100) system, 10 BaO unit cells (55.0 Å) of the thin film are nearly lattice matched with 13 MgO unit cells (54.6 Å) of the substrate. Clearly, 54.6 Å and 55.0 Å are in near coincidence, and the BaO overlayer only needs to be expanded by a small amount, from 54.6 to 55.0 Å, to facilitate coincidence. This particular NCSL configuration is associated with a residual misfit F calculated, to be about +0.7%, as follows:

$$F = \frac{n a_{\mathrm{BaO}} - m a_{\mathrm{MgO}}}{\frac{1}{2}(n a_{\mathrm{BaO}} + m a_{\mathrm{MgO}})} \tag{9.9}$$

where n is the number of BaO unit cells (equal to 10) and m is the number of MgO unit cells (equal to 13). Clearly, the energy required to accommodate a 0.7% misfit is much lower than that required to accommodate the full 27% misfit associated with the fully coherent system (i.e., $n = m = 1$).

In addition to coincidences found by matching integer unit cells of the thin film with that of the substrate, coincidence can also be found by rotating one material with respect to the other. For the simplest case—where the thin film is the same material as the substrate—a twist grain boundary will result (17). For the BaO/MgO system, a configuration associated with a low misfit and favorable interfacial interactions involves a 45° rotation of the BaO overlayer with respect to the MgO substrate and was observed experimentally by Cotter et al. (50); atomistic models for this structure have also been generated (51).

9.4.2.2. CeO_2/Al_2O_3 When a thin film is supported on a substrate, the supported thin film will respond structurally to the substrate to facilitate a low-energy configuration. This can include, for example, dislocation evolution and will influence the defect chemistry of the system. In this respect, Sayle et al. explored how easy it was to extract an oxygen ion from the surface of a CeO_2 thin film, when supported on an Al_2O_3 substrate, compared with oxygen extraction from the parent material (52). The calculations suggested that supporting the CeO_2 thin film on top of Al_2O_3 might promote the migration of oxygen from the interfacial region to the surface, a process of catalytic importance.

9.4.2.3. NiO/ZrO$_2$ Fisher and Matsubara extended the NCSL/atomistic simulation approach to generate heterophase interfaces between materials with rectangular surface symmetries, and used the method to simulate eight heterophase interfaces between NiO and cubic ZrO_2 (53). They observed that many interfaces were found to comprise disordered, open structures with correspondingly high energies as a consequence of the strong repulsive forces between like-charged ions in close proximity. The exception was the NiO(111)/ZrO_2(100), which was observed to exhibit good coherency across the interphase boundary, where the NiO and ZrO_2 shared an oxygen plane. The rationale underlying this study was to demonstrate the ability of simulation-based methods for identifying interface configurations associated with strong cohesive forces holding the materials together for optimizing the properties of composites of technologically important materials.

A limitation with using a "NCSL strategy" is that the simulator has to decide which particular NCSL to construct and then use atomistic simulation (i.e., energy minimization and/or MD) to determine whether the configuration is of low energy and therefore likely to be structurally realistic. An alternative is to use an evolutionary method to guide the generation of the interfacial configurations and structures. These methods, which includes simulated ion deposition in Section 9.4.1, do not require (or even allow!) the simulator to prespecify a particular interfacial configuration.

9.4.3. Amorphization and Recrystallization

Here, the oxide thin film is placed on top of an oxide substrate and then amorphized. MD simulation, applied to this system (amorphous thin film on top of crystalline substrate) for long duration, results in the recrystallization of the thin film (20,21). In contrast to the recrystallization process observed for "isolated" nanoparticles in Section 9.3, the underlying (crystalline) substrate is central in directing the structure of the overlaying (amorphous) thin film. In particular, the structure of the substrate facilitates the crystallization of the thin films by acting as a pseudo-"nucleating seed." In addition, the simulator does not have to wait for a crystalline seed to spontaneously evolve from the amorphous sea of ions, and therefore, the simulation time, and hence computational cost, is reduced.

9.4.3.1. SrO/MgO(100) Snapshots illustrating the A&R process as applied to an SrO thin film, about 1 nm thick, supported on an MgO(001) substrate, are presented in Figure 9.10. Inspection of the final crystalline structure (Figure 9.10*d*), using molecular graphics, revealed that the SrO thin film generated conformed to 8 lattice spacings for the overlying SrO thin film, lattice matched with 10 lattice spacings of the underlying MgO substrate (NCSL configuration). This supercell corresponds to a lattice misfit of about -2%, based on the lattice parameters of the component materials: $a_{SrO} = 2.57$ Å and $a_{MgO} = 2.10$ Å, which is much smaller than the "bulk" misfit (+20%). In addition, mixed screw-edge dislocations, shown in Figure 9.11, were observed to have evolved within the SrO thin film, which help both to accommodate

Figure 9.10. Sphere model representations of the atom positions comprising the SrO/MgO(001) system simulated using A&R: (*a*) start of the simulation; (*b*) after 0.25 ps of MD, the SrO overlayers start to buckle under the considerable strain imposed; (*c*) after 0.75 ps of MD, here the SrO overlayer is completely amorphous; (*d*) final, 0 K, structure after the complete recrystallization of the SrO overlayer. Oxygen is represented by the dark spheres strontium the light spheres, and magnesium the smaller spheres. Only two layers of the MgO substrate are shown to improve clarity of the figure.

the misfit associated between the thin film and underlying support and to remove deleterious interfacial interactions such as cations in close proximity (54,55).

9.4.3.2. MgO/BaO(100) Figure 9.12 shows the structure of a thin film of MgO supported on a BaO(100) substrate. Here the recrystallization resulted in the evolution of a nanopolycrystalline MgO film. The nanocrystallites range from about 200 to 2000 $Å^2$ in size and are rotated, with respect to the underlying BaO surface normal by various angles. The structure of the grain boundaries that are formed between neighboring grains are consistent with grain boundaries observed experimentally (56,57). An STEM image of an MgO grain-boundary is included in the figure for comparison and has been reproduced with permission (57). Further details of this study can be found in Ref. 55.

9.4.3.3. CeO_2/YSZ In Section 9.3.2, on CeO_2 nanoparticles, it was suggested that by traversing down to the nanoscale, one is able to increase the proportion of reactive to unreactive CeO_2 surfaces simply by virtue of the fact that if one goes small enough, then the number of oxygen ions accommodating reactive step/corner

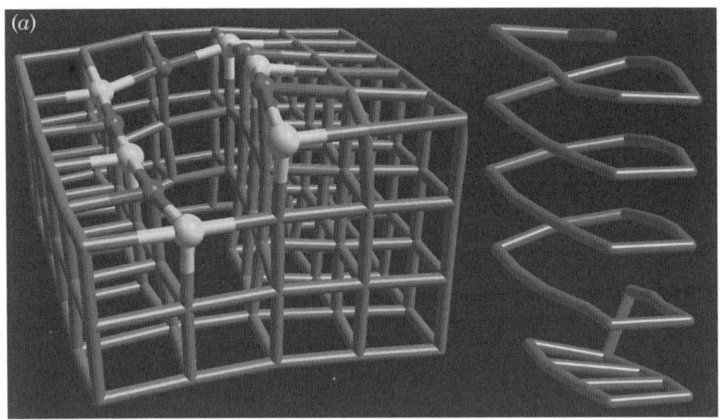

Figure 9.11. Stick model representations of two screw-edge dislocations that evolved within the SrO thin film. (*a*) Left: part of the SrO lattice (gray) that surrounds the dislocation core, right: dislocation core structure. (*b*) Another dislocation, similar to that in (*a*), but here the core traverses the SrO thin film in a clockwise fashion [as opposed to anticlockwise in (*a*)]. (*c*) Core structure, shown in red, illustrating how the dislocation core "lies" within the surrounding lattice. Oxygen is colored red, and strontium is yellow. See color insert.

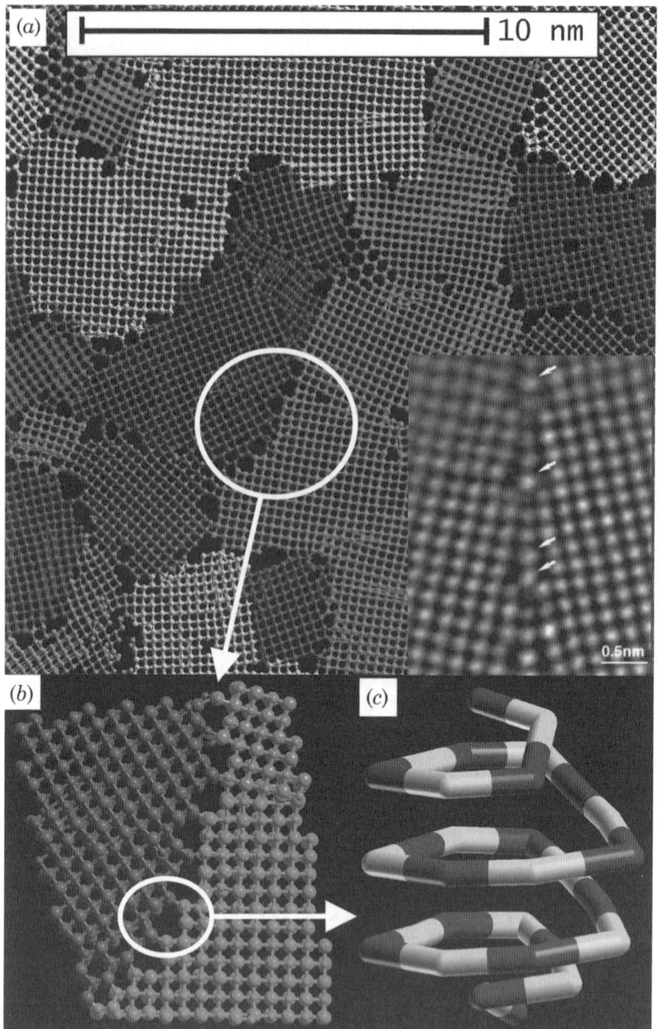

Figure 9.12. Atom positions comprising a nanopolycrystalline MgO thin film supported on a BaO substrate. (*a*) Plan view of the system showing the various nanocrystallites comprising the MgO, which are colored to highlight the individual nanocrystallites. The MgO substrate is not shown to maintain clarity of the figure. The inset shows a STEM image to compare and has been reproduced, with permission, from Ref. 57. (*b*) Enlarged segment of (*a*) showing more clearly the grain-boundary structure. (*c*) An enlarged segment of (*b*) depicting the core structure of the grain boundary and revealing its "screw" in addition to "edge" character. See color insert.

sites will be commensurate to oxygen ions comprising more stable positions and therefore unreactive surfaces. An alternative strategy is to fabricate thin films with nanoscale thicknesses. In particular, if a material, which is deposited onto a substrate, is sufficiently thin, then it is likely that the substrate will act as a template in directing the structure of the thin-film deposited thereon, including the particular surface it exposes.

For example, the (111) surface is energetically the most stable ceria surface. It is therefore exposed preferentially compared with less-stable surfaces (58). However, the energy required to create oxygen vacancies on the (111) surface, which is directly linked to its catalytic activity (36,37), is higher compared with the (110) surface. One therefore desires a mechanism for fabricating CeO_2(110) in preference to CeO_2(111). One way is to traverse to the nanoscale, as described, the other is to deposit an ultra-thin film of CeO_2 on top of a substrate material. The rationale underlying the latter is that the substrate may act as a template (42) in directing the structure of the overlying thin CeO_2 film. A simulation study was therefore performed to explore this possibility.

A CeO_2 film, about 2 nm thick, was deposited on top of an YSZ(110) substrate (YSZ: yttrium stabilized zirconia). The CeO_2 was first amorphized and then recrystallized. Full details of the simulation can be found in Ref. 59. It was anticipated that the YSZ(110) exposed at the surface would direct the CeO_2 thin film to expose the (catalytically active) (110) surface also.

Figure 9.13 shows graphically the atom positions comprising the CeO_2/YSZ(110) system. Inspection of these figures revealed that the YSZ(110) substrate did indeed influence the structure of the CeO_2 thin film. However, surprisingly, although the CeO_2 exposed the (110) surface as anticipated, it exposed also the (111) and dipolar (100). In particular, the CeO_2 thin-film overlayer was observed to be nanopolycrystalline with each of the nanocrystalline domains exposing a particular face at the surface—either (110), (100), or (111). It was proposed that the lattice misfit between the YSZ substrate and CeO_2 thin film deposited thereon was in part responsible for the nanopolycrystalline structure of the CeO_2 thin film.

9.4.4. Melting and Seeded-Crystallization

Similar to A&R, this evolutionary strategy involves crystallization from a melt. However, to reduce the computational time required for the system to evolve "naturally" a crystalline seed, seeds were manually introduced into the liquid.

9.4.4.1. FeO(Nanocrystalline)/FeO A study by Phillpot et al. describe a simulation strategy for generating models for nanopolycrystalline FeO with grain sizes of about 5 nm using MD simulation (60). In particular, the authors deposited an ultra-thin, molten film of FeO on top of an FeO substrate. Within the molten film, they introduced 16 randomly oriented crystalline "seeds." The system was then cooled gradually under MD. During the crystallization, the (molten) Fe and O ions condensed onto the seeds propagating their structure, and because the seeds were misoriented, the crystallization resulted in the formation of a nanopolycrystalline film with general grain boundaries as propagated by the orientation of the seeds. Inspection of the grain boundaries revealed them to be structurally realistic—similar to those observed experimentally (44).

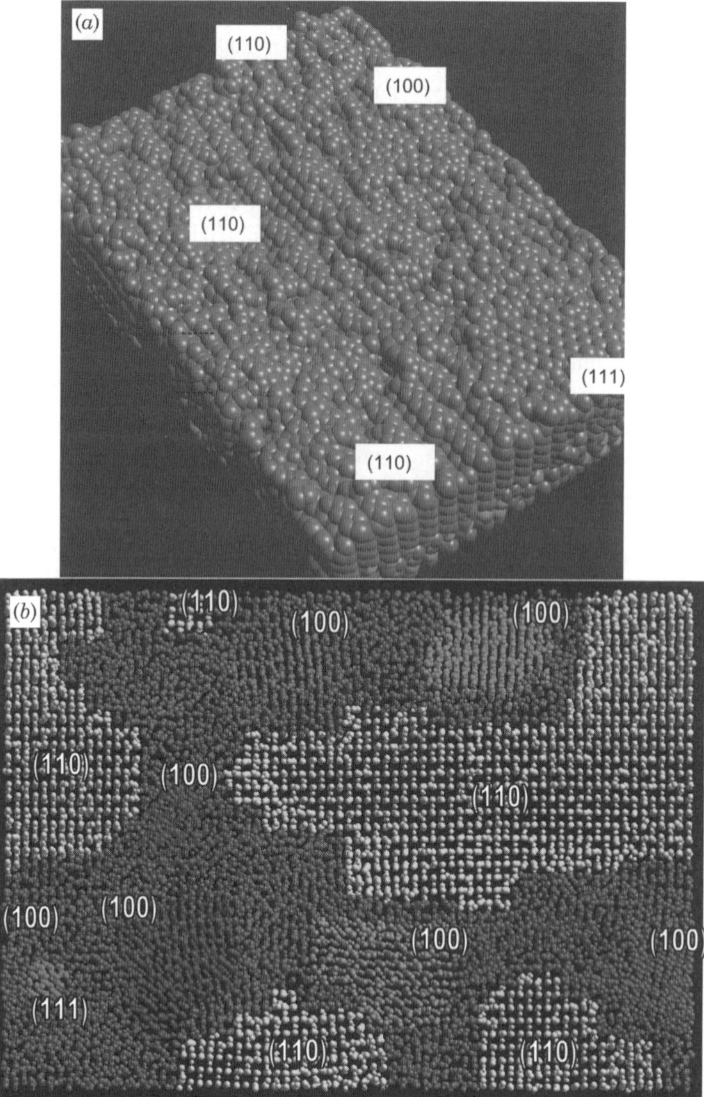

Figure 9.13. Sphere model representation of the atom positions comprising the CeO_2/YSZ system. (*a*) Perspective view and (*b*) plan view in which the various nanocrystallites are colored. The figures are annotated to reveal which particular planes—(111), (110), or (100)—are exposed at the surface.

9.4.5. Temperature-Assisted Dynamics

Another strategy for overcoming the debilitatingly short timescales accessible using MD simulations is temperature-assisted dynamics (TAD). This strategy is derived from a variety of methods labeled "hyperdynamics," which were developed by Voter et al. (43). The strategy involves the use of simulations, performed at high temperatures, to gauge the evolution of a system at a lower (realistic) temperature. For example, the energy required for an ion to move out of the surface of a material is so high that the probability of it happening at room temperature—and within timescales accessible to MD—is low. Conversely, if we increase the temperature, the probability of it occurring within the same timescale increases. Accordingly, the simulation is run at sufficiently high a temperature to ensure the process occurs with the timescale accessible. Once the transition is detected (i.e., the ion moves out of the surface), the attributes of this process are recorded and characterized. This process, together with all other related important processes, are then transferred back to a simulation performed at a more realistic (low) temperature. For a more comprehensive treatment of the method, see Ref. 43.

9.4.5.1. BaO/SrO Harris et al. used TAD to explore the surface diffusion and heteroepitaxial growth of BaO on an SrO substrate. Crucially, it was found that the surface diffusion proceeds most favorably by an exchange mechanism involving the surface layer. For example, as a barium ion diffuses across the SrO surface, it was found that an energetically favorable pathway, in addition to a simple hopping across the surface, is for the barium ion to swap places with a strontium ion located within the surface atomic layer of the underlying SrO substrate (8,9). This has severe implications for growing sharp interfaces between different oxide materials using, for example, molecular beam epitaxy. Mixing can also lead to surface contamination and subsequent changes in surface properties. It is likely that this phenomenon will also occur at the surface of nanomaterials—nanoparticles and thin films, which will present problems to applications using nanodevices.

9.5. SUPPORTED NANOPARTICLES

Thus far we have seen that changing one or more of the dimensions of a system to the nanoscale can influence, sometimes profoundly, its microstructure and hence properties. Clearly it is important to be able to exercise control over the structure to fine-tune or indeed optimize desirable properties (40). To this end, one can envisage supporting oxide nanoparticles on an oxide substrate: Indeed, it is well known that small metal particles, supported on an oxide substrate, can be catalytically more active compared with the parent materials (61).

9.5.1. MO/BaO (M = Sr, Ca, Mg)

Models for SrO, CaO, and MgO nanoparticles were generated by placing a rectangular 25,000-atom block of each oxide onto the surface of a BaO(100) substrate. The oxide nanoparticles were then amorphized and recrystallized. Further details can be found in Ref. 62. During the recrystallization, the interactions between the nanoparticle and the underlying BaO substrate were found to have a profound influence on the structure of the nanoparticle. Indeed, these interactions facilitate a wealth of microstructural features that evolve during the recrystallization.

For the SrO/BaO(100) system ("low" −7% misfit; Eq. 9.9), the SrO nanoparticle was found to lie coherent with the underlying BaO substrate. For the CaO/BaO(100) system ("medium" −15% misfit), only small regions of coherence were observed between the CaO and the BaO with dislocations evolving within CaO regions that were misaligned with respect to the underlying BaO. For the MgO/BaO system ("high" −31% misfit), the misfit is so high that no regions of coherence between the MgO nanoparticle and BaO substrate could be identified. This study also proposed the existence of a critical *area* for dislocation; the concept of a critical thickness is well known (62). If confirmed experimentally, this prediction may have important implications with respect to the field of microelectronic circuits.

In Figure 9.14, the structure of the MgO/BaO(100) system is shown. Close inspection of the nanoparticle reveals a variety of microstructural features, including:

1. *Morphology*: The nanoparticle is not a rectangular block; rather, it comprises four crystallites interconnecting at various angles. The central region, colored red, exposes the MgO(100) at the surface and at the interface, i.e., MgO(100)/BaO(100). The missoriented crystallites, colored yellow, blue, and gray, emanating from this central region, appear to exhibit triangular pyramidal morphologies. The nanoparticle exhibits a high density of surface steps.
2. *Grain Boundaries*: The four interconnecting crystallites are misaligned, which gives rise to various grain-boundary structures. Inspection of these structures in Figure 9.14*a* reveals that they are complex (general) and therefore difficult to assign to a particular coincidence site lattice (CSL) description. Moreover, bending of the lattice planes compounds this difficulty.
3. *Epitaxial Configurations*: The crystallites expose the MgO(100) and MgO(111) surfaces at the interface (we note that rocksalt {111} surfaces are dipolar). No epitaxial configuration could be identified. This is attributed to the high misfit associated with this system.
4. *Point Defects*: The interfacial region (Figure 9.14*b*), is highly defective and includes voids, vacancies, and cation intermixing across the MgO/BaO boundary.
5. *Decorations*: Inspection of Figure 9.14*a* reveals that the surface of the MgO nanoparticle is decorated with many Ba ions that have migrated from the BaO substrate.
6. *Dislocations*: Dislocations were observed to have evolved within the MgO nanoparticle; the core structure of a mixed screw-edge dislocation is shown in Figure 9.14*c*.

This study provides an illustration that supporting a oxide nanoparticle, on an oxide substrate, can influence profoundly the structure and microstructure of the nanoparticle and is likely to modify its properties.

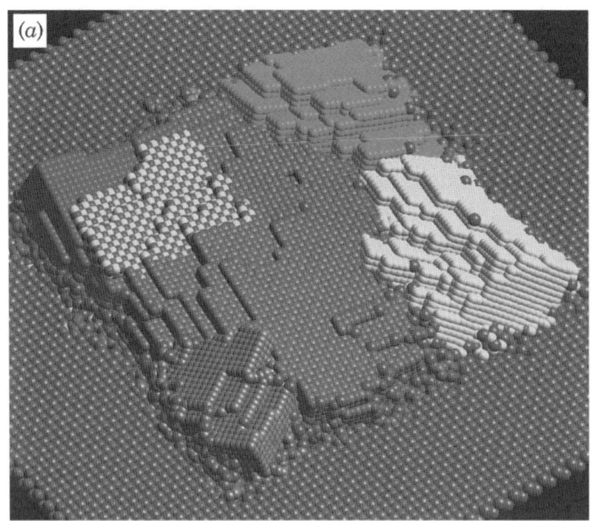

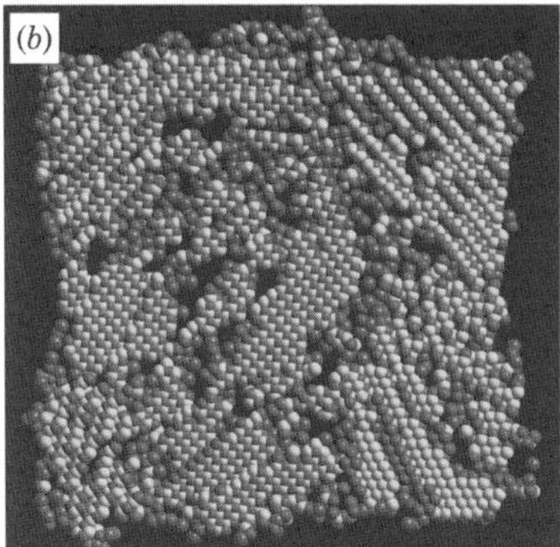

Figure 9.14. (*a*) Perspective view, with sphere model representations, of the atom positions comprising a 25,000-atom MgO nanoparticle supported on a BaO(001) substrate. The various misoriented crystallites are colored. (*b*) Interfacial region illustrating the considerable defective nature of the interfacial region. (*c*) Mixed screw-edge dislocation that was identified to have evolved within the MgO nanoparticle. See color insert.

9.6. ENCAPSULATED NANOPARTICLES

An important area within nanoscience is the embedding of nanoparticles within a host lattice. For example, various materials have been encapsulated within microporous materials such as zeolites (63). More recently, materials have been encapsulated within single and multiwall carbon nanotubes (64), leading to the fabrication of one-dimensional materials. The intense scientific interest in these systems is hardly surprising because with an appropriate choice of encapsulating lattice, one can potentially exact control over the structure and properties of the encapsulated material. This has inspired many researchers to explore materials that might act as the host lattice and materials that might prove suitable to be encapsulated. By mixing and matching, an almost limitless number of possible systems can be envisaged. One can also envisage encapsulation of a nanoparticle within a fully dense host lattice. Indeed, Jeng and Shen have explored the structure of NiO nanoparticles encapsulated within CaO by sintering and annealing NiO and CaO powders (65). Atomistic simulation can also be used to predict the structure and morphology of encapsulated nanoparticles. One such study by Sayle and Parker explored the influence of introducing an oxide nanoparticle into the near-surface region of an oxide host (66). In particular, models for oxide encapsulated oxide nanoparticles were generated by constructing a block of a rocksalt-structured oxide. The surface region of this block, comprising about 25,000 atoms, was then amorphized and, a spherical nanoparticle, about 700 atoms in size, was introduced. The whole system was then recrystallized and the system analyzed structurally. The procedure is illustrated, for CaO encapsulation within an MgO host, in Figure 9.15.

The introduction of the amorphous CaO nanoparticle into the amorphous MgO substrate is shown in Figure 9.15*a*. Figure 9.15*b* shows the system after recrystallization. In this figure, the CaO nanoparticle has been extracted out of the host lattice to show more clearly the morphology of the nanoparticle and the void, within the MgO host, it occupies. Figure 9.15*c* shows the system with a segment of the simulation cell cutaway. This enables one to see more clearly the nanoparticle within the host lattice. In Figure 9.15*d*, a slice cut through the system is shown to reveal that the CaO nanoparticle rotates slightly with respect to the MgO host. The driving force for this to occur was attributed to reducing the lattice misfit associated with the configuration, while maximizing favorable cation–anion interactions across the (curved) interfacial region. Careful analysis, using graphical techniques, reveals that the CaO nanoparticle exposes {100}, {110}, and {111} facets to the MgO host at the interfacial regions. In addition, this procedure was performed for (BaO and CaO) nanoparticles encapsulated in MgO and (SrO and MgO) nanoparticles encapsulated in BaO. The structure of MgO, BaO, CaO, and SrO nanoparticles, which have been extracted out of their respective host lattices to reveal more clearly the morphology they accommodate, are shown in Figure 9.16*a–d*, respectively.

The study raised several issues pertaining to the encapsulation of nanoparticles. In particular, the nanoparticles were identified to be highly strained. For example, the SrO nanoparticle, encapsulated within BaO, revealed that the SrO maintained full alignment of counter-ions across the interfacial regions. The lattice misfit associated with the SrO/BaO system is about -7%, and therefore, the SrO is tensioned to maintain this aligned configuration. It was proposed that the energetically favorable cation–anion interactions across the interfacial region outweighed the energetically deleterious effect of tensioning the lattice. Clearly, if the size of the nanoparticle was increased, then the energy required to tension the lattice, which operates effectively in three-dimensions, would increase faster than the increase in the favorable interfacial interactions, which is a surface phenomenon and inherently 2D. This suggests that there is a "critical" size associated with encapsulated nanoparticles, above which misfit

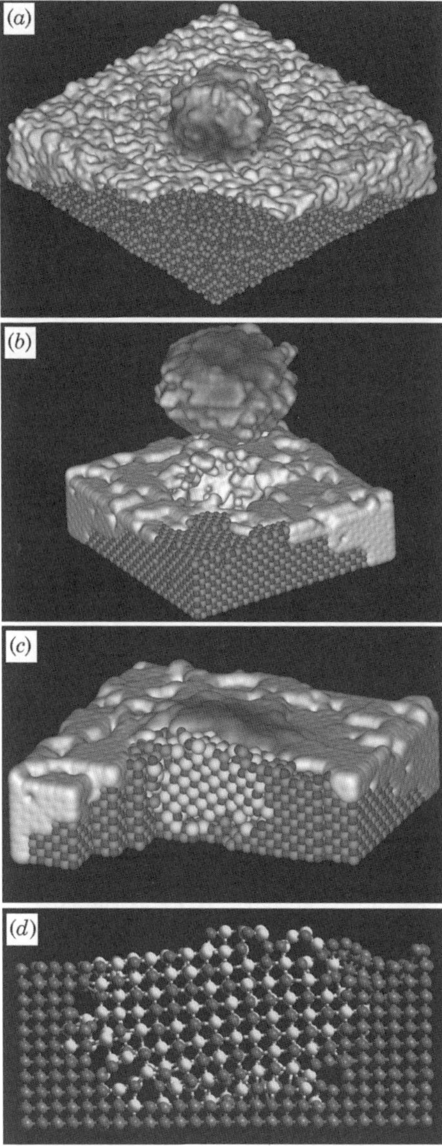

Figure 9.15. Representation of the atom position illustrating the encapsulation of a CaO nanoparticle within an MgO host lattice. (*a*) Amorphous starting structure showing the CaO nanoparticle being introduced into the MgO host. (*b*) Final recrystallized structure showing the structure of the CaO nanoparticle and the void in the MgO that it occupies (the CaO nanoparticle has been physically extracted to reveal the structure). (*c*) Perspective view with part of the simulation cell cut-away to reveal the CaO nanoparticle encapsulated by its MgO host. (*d*) Slice cut through the simulation cell to reveal more clearly how the CaO nanoparticle lies with respect to the host. Calcium is colored light gray, magnesium is medium gray, and oxygen is dark gray. Surface rendering has also been used to aid interpretation of the structures. Light gray rendering is the MgO host and light/dark rendering, the CaO nanoparticle. See color insert.

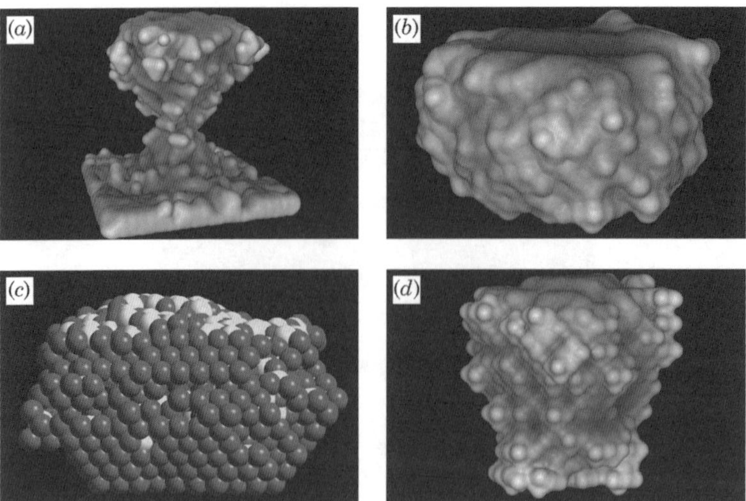

Figure 9.16. Representations of the atom positions of oxide nanoparticles (shown) encapsulated within an oxide host lattice (not shown). (*a*) MgO nanoparticle (in BaO host), (*b*) BaO nanoparticle (in MgO host), (*c*) CaO nanoparticle (in MgO host), and (*d*) SrO nanoparticle (in BaO host). (*a*), (*b*), and (*d*) are shown with surface rendering and (*c*) using a sphere model representation of the atom positions.

dislocations evolve. Indeed, dislocations/commensurate structures were observed for the other three systems, which are all associated with a higher lattice misfit. The critical size was predicted to depend on a combination of both the nanoparticle and the host within which it is encapsulated.

In summary, the encapsulated nanoparticles exhibited a range of morphologies, expose a variety of facets at the nanoparticle/host lattice interface, and are observed to rotate within the cavity in which they occupy. The structure and nature of the nanoparticles reflect the lattice misfit between the nanoparticle and the host lattice. Further details can be found in Ref. 66.

9.7. CONCLUSIONS

In this chapter, we have explored primarily how models of oxide nanomaterials, with full atomistic detail, can be generated using computer simulation. These include isolated nanoparticles, ultra-thin films supported on a substrate, supported nanoparticles and encapsulated nanoparticles. Once a model is available, a wealth of electronic, chemical, and mechanical properties can then be calculated and the results used predictively to aid experiment: Properties amenable to calculation include, for example, diffusion coefficients, ionic conductivity (41), catalytic activity (included in this chapter), elastic constants, and bulk and shear moduli, Youngs moduli, piezoelectric constants, phonons, dielectric constants, and surface energies (15). A well-established and popular computer code for calculating a variety of properties using such atomistic models is the General Utility Lattice Program (GULP), which is freely available from Ref. 6.

The atomistic models derived are also useful as starting structures for quantum-mechanical (QM) simulations. This is because it is less expensive computationally to generate models using atomistic simulation compared with QM.

It is tempting to think, especially when viewing the animations of MnO_2 crystallizing (Figure 9.3), and the resulting nanoparticle structures, that one is observing "real" crystallization at the atomistic level. However, one must exercise caution in that the simulation is bound by (artificial) simulation constraints implicit in the methodology, not the least by the potential models describing the interactions between the ions. Conversely, it is very surprising that such highly complex microstructural features, observed experimentally, such as the dislocations, grain boundaries, morphologies, surface structures, and isolated and associated point defects, have simply evolved in a purely artificial way within the simulations described above. Accordingly, we suggest that the simulated A&R we have discussed in this chapter must, at least in part, reflect crystallization that occurs in nature, and that the models generated using this approach are realistic.

If used with a degree of caution, models generated using atomistic simulation can help experiment unravel the rich structural complexity of nanomaterials and aid in the fabrication of nanomaterials with improved, tunable, or indeed new properties.

REFERENCES

(1) Schissel, D.P. Grid computing and collaboration technology in support of fusion energy sciences. *Phys. Plasmas* **2005**, *12(5)*, 058104.

(2) Stoneham, A.M. Interatomic potentials for condensed matter. *Phys. A, B, C* **1985**, *131*, 69–73.

(3) Lewis, G.V.; Catlow, C.R.A. Potential models for ionic oxides. *J. Phys. C: Solid State Phys.* **1985**, *18*, 1149–1161.

(4) Duffy, D.M.; Harding, J.H. Growth of polar crystal surfaces on ionized organic substrates. *Langmuir* **2004**, *20*, 7637–7642.

(5) Walker, A.M.; Slater, B.; Gale, J.D.; Wright, K. Predicting the structure of screw dislocations in nanoporous materials. *Nat. Mater.* **2004**, *3*, 715–720.

(6) Gale, J.D.; Rohl, A.L. The General Utility Lattice Program (GULP). *Molec. Sim.* **2003**, *29*, 291–341.

(7) Mohn, C.E.; Lavrentiev, M.Y.; Allan, N.L.; Bakken, E.; Stolen, S. Size mismatch effects in oxide solid solutions using Monte Carlo and configurational averaging. *Phys. Chem. Chem. Phys.* **2005**, *7*, 1127–1135.

(8) Harris, D.J.; Lavrentiev, M.Y.; Harding, J.H.; Allan, N.L.; Purton, J.A. Novel exchange mechanisms in the surface diffusion of oxides. *J. Phys. Cond. Matt.* **2004**, *16*, L187–L192.

(9) Harris, D.J.; Farrow, T.S.; Harding, J.H.; Lavrentiev, M.Y.; Allan, N.L.; Smith, W.; Purton, J.A. Surface diffusion and surface growth in nanofilms of mixed rocksalt oxides. *Phys. Chem. Chem. Phys.* **2005**, *7*, 1839–1844.

(10) Johnston, R.L. Evolving better nanoparticles: Genetic algorithms for optimising cluster geometries. *Dalton Trans.* **2003**, *22*, 4193–4207.

(11) Sayle, D.C.; Johnston, R.L. Evolutionary techniques in atomistic simulation: Thin films and nanoparticles. *Curr. Opin. Solid State Mater. Sci.* **2003**, *7*, 3–12.

(12) Smith, W.; Forester, T.R. *DL_POLY*. Council for the Central Laboratory of the Research Councils, Daresbury Laboratory: Daresbury, Warrington, UK, 1996; available at: http://www.dl.ac.uk/TCSC/Software/DLPOLY.

(13) Smith, W.; Yong, C.W.; Rodger, P.M. DL_POLY: Application to molecular simulation. *Molec. Sim.* **2002**, *28*, 385–471.

(14) Leach, A.R. *Molecular Modelling, Principles and Applications, 2nd ed.*; Pearson Education: Essex, UK, 2001.

(15) Gale, J.D.; Rohl, A.L. MARVIN—A new computer code for studying surfaces and interfaces and its application to calculating the crystal morphologies of Corundum and Zirconia. *J. Chem. Soc. Faraday Trans.* **1995**, *91*, 925–936.

(16) Harding, J.H. The simulation of general polar boundaries. *Surf. Sci.* **1999**, *422*, 87–94.

(17) Duffy, D.M. Grain boundaries in ionic crystals. *J. Phys. C: Solid State Phys.* **1986**, *19*, 4393–4412.

(18) Piana, S.; Gale, J.D. Understanding the barriers to crystal growth: Dynamical simulation of the dissolution and growth of urea from aqueous solution. *J. Amer. Chem. Soc.* **2005**, *127*, 1975–1982.

(19) Hamad, S.; Cristol, S.; Catlow, C.R.A. Simulation of the embryonic stage of ZnS formation from aqueous solution. *J. Amer. Chem. Soc.* **2005**, *127(8)*, 2580–2590.

(20) Sayle, D.C.; Maicaneanu, S.A.; Slater, B.; Catlow, C.R.A. Exercising control over the influence of the lattice misfit on the structure of oxide-oxide thin film interfaces. *J. Mater. Chem.* **1999**, *9*, 2779–2787.

(21) Sayle, D.C. The predicted 3-D atomistic structure of an interfacial screw-edge dislocation. *J. Mater. Chem.* **1999**, *9*, 2961–2964.

(22) Sayle, T.X.T.; Catlow, C.R.A.; Maphanga, R.R.; Ngoepe, P.E.; Sayle, D.C. Generating MnO_2 nanoparticles using simulated amorphization and recrystallisation. *J. Amer. Chem. Soc.* **2005**, *127(37)*, 12828–12837; DOI: 10.1021/ja0434073.

(23) Chabre, Y.; Pannetier, J. Structural and electrochemical properties of the proton gamma-MNO2 system. *Progr. Solid State Chem.* **1995**, *23*, 1–130.

(24) Hill, M.R.; Freeman, C.M.; Rossouw, M.H. Understanding gamma-MnO2 by molecular modeling. *J. Solid State Chem.* **2004**, *177*, 165–175.

(25) Yu, H.B.; Kim, J.H.; Lee, H.I.; Scibioh, M.A.; Lee, J.; Han, J.; Yoon, S.P.; Ha, H.Y. Development of nanophase CeO2-Pt/C cathode catalyst for direct methanol fuel cell. *J. Power Sources* **2005**, *140*, 59–65.

(26) Di Monte, R.; Kaspar, J. Nanostructured CeO2-ZrO2 mixed oxides. *J. Mater. Chem.* **2005**, *15*, 633–648.

(27) Wang, Z.L.; Feng, X.D. Polyhedral shapes of CeO2 nanoparticles. *J. Phys. Chem. B* **2003**, *107*, 13563–13566.

(28) Zhang, F.; Jin, Q.; Chan, S.W. Ceria nanoparticles: Size, size distribution, and shape. *J. Appl. Phys.* **2004**, *95*, 4319–4326.

(29) Titiloye, J.O.; Parker, S.C.; Osguthorpe, D.J.; Mann, S. Predicting the influence of growth additives on morphology of ionic crystals. *J. Chem. Soc. Chem. Commun.* **1991**, *20*, 1494–1496.

(30) Vyas, S.; Grimes, R.W.; Gay, D.H.; Rohl, A.L. Structure, stability, and morphology of stoichiometric ceria crystallites. *J. Chem. Soc. Faraday Trans.* **1998**, *94*, 427–434.

(31) Chen, H.-I.; Chang, H.-Y. Synthesis and characterization of nanocrystalline cerium oxide powders by two-stage non-isothermal precipitation. *Solid State Commun.* **2005**, *133*, 593–598.

(32) Rohl, A.L. Computer prediction of crystal morphology. *Curr. Opin. Solid State Mater. Sci.* **2003**, *7*, 21–26.

(33) Sayle, T.X.T.; Parker, S.C.; Sayle, D.C. Shape of CeO2 nanoparticles using simulated amorphisation and recrystallisation. *J. Chem. Soc. Chem. Commun.* **2004**, *21*, 2438–2439.

(34) Norenberg, H.; Harding, J.H. The surface structure of CeO2(001) single crystals studied by elevated temperature STM. *Surf. Sci.* **2001**, *477*, 17–24.

(35) Stanek, C.R.; Bradford, M.R.; Grimes, R.W. Segregation of Ba2+, Sr2+, Ce4+ and Zr4+ to UO2 surfaces. *J. Phys.-Cond. Matt.* **2004**, *16*, S2699–S2714.

(36) Sayle, T.X.T.; Parker, S.C.; Catlow, C.R.A. The role of oxygen vacancies on Ceria surfaces in the oxidation of carbon monoxide. *Surf. Sci.* **1994**, *316*, 329–336.

(37) Sayle, T.X.T.; Parker, S.C.; Catlow, C.R.A. Surface oxygen vacancy formation on CeO2 and its role in the oxidation of carbon monoxide. *Chem. Soc. Chem. Commun.* **1992**, *14*, 977–978.

(38) Sayle, T.X.T.; Parker, S.C.; Sayle, D.C. Oxidising CO to CO_2 using ceria nanoparticles. *Phys. Chem. Chem. Phys.* **2005**, *7*, 2936–2941.

(39) Cordatos, H.; Ford, D.; Gorte, R.J. Simulated annealing study of the structure and reducibility in ceria clusters. *J. Phys. Chem.* **1996**, *100*, 18128–18132.

(40) Sata, N.; Eberman, K.; Eberl, K.; Maier, J. Mesoscopic fast ion conduction in nanometre-scale planar heterostructures. *Nature* **2000**, *408*, 946–949.

(41) Sayle, D.C.; Doig, J.A.; Parker, S.C.; Watson, G.W.; Sayle, T.X.T. Computer aided design of nano-structured materials with tailored ionic conductivities. *Phys. Chem. Chem. Phys.* **2005**, *7*, 16–18.

(42) Sayle, D.C.; Watson, G.W. Inducing polycrystallinity within supported oxide thin films using template buffer layers. *J. Phys. Chem. B* **2002**, *106*, 3778–3787.

(43) Voter, A.F.; Montalenti, F.; Germann, T.C. Extending the time scale in atomistic simulation of materials. *Annu. Rev. Mater. Res.* **2002**, *32*, 321–346.

(44) Sutton, A.P.; Baluffi, R.W. *Interface in Crystalline Materials, Monographs on the Physics and Chemistry of Materials, Vol. 51*; Oxford University Press: New York, 1995.

(45) Dong, L.; Schnitker, J.; Smith, R.W.; Srolovitz, D.J. Stress relaxation and misfit dislocation nucleation in the growth of misfitting films: A molecular dynamics simulation study. *J. Appl. Phys.* **1998**, *83*, 217–227.

(46) Lu, Y.F.; Przybylski, M.; Trushin, O.; Wang, W.H.; Barthel, J.; Granato, E.; Ying, S.C.; Ala-Nissila, T. Strain relief in Cu–Pd heteroepitaxy. *Phys. Rev. Lett.* **2005**, *94*, 146105.

(47) Schnitker, J.; Srolovitz, D.J. Misfit effects in adhesion calculations. *Model. Sim. Mater. Sci. Eng.* **1998**, *6*, 153–164.

(48) Chambers, S.A. Epitaxial growth and properties of thin film oxides. *Surf. Sci. Rep.* **2000**, *39*, 105–180.

(49) Sayle, T.X.T.; Catlow, C.R.A.; Sayle, D.C.; Parker, S.C.; Harding, J.H. Computer—Simulation of thin-film heteroepitaxial ceramic interfaces using a near-coincidence-site lattice theory. Physics of condensed matter structure defects and mechanical properties. *Philosoph. Mag. A* **1993**, *68*, 565–573.

(50) Cotter, M.; Campbell, S.; Egdell, R.G. Growth of ordered BaO overlayers on MgO(001). *Surf. Sci.* **1988**, *197*, 208–224.

(51) Sayle, D.C.; Catlow, C.R.A.; Dulamita, N.; Healy, M.J.F.; Maicaneanu, S.A.; Slater, B.; Watson, G.W. Modelling oxide thin-films. *Mole. Sim.* **2002**, *28 (6–7)*, 683–725.

(52) Sayle, D.C.; Sayle, T.X.T.; Parker, S.C.; Harding, J.H.; Catlow, C.R.A. The stability of defects in the ceramic interfaces, MgO/MgO and CeO_2/Al_2O_3. *Surf. Sci.* **1995**, *334*, 170–178.

(53) Fisher, C.A.J.; Matsubara, H. Molecular dynamics simulations of interfaces between NiO and cubic ZrO2. *Philosoph. Mag.* **2005**, *85*, 1067–1088.

(54) Sayle, D.C.; Watson, G.W. Dislocations, lattice slip, defects, and rotated domains: The effect of a lattice misfit on supported thin-film metal oxides. *Phys. Chem. Chem. Phys.* **2000**, *2*, 5491–5499.

(55) Sayle, D.C.; Watson, G.W. Simulated amorphisation and recrystallisation: Nanocrystallites within meso-scale supported oxides. *J. Mater. Chem.* **2000**, *10*, 2241–2243.

(56) Yan, Y.; Chisholm, M.F.; Duscher, G.; Maiti, A.; Pennycook, S.J.; Pantelides, S.T. Impurity-induced structural transformation of a MgO grain boundary. *Phys. Rev. Lett.* **1998**, *17*, 3675–3678.

(57) Pennycook, S.J.; Dickey, E.C.; Nellist, P.D.; Chisholm, M.F.; Yan, Y.; Pantelides, S.T. A combined experimental and theoretical approach to atomic structure and segregation at ceramic interfaces. *J. Eur. Ceram. Soc.* **1999**, *19*, 2211–2216.

(58) Gritschneder, S.; Namai, Y.; Iwasawa, Y.; Reichling, M. Structural features of CeO2(111) revealed by dynamic SFM. *Nanotechnology* **2005**, *16*, S41–S48.

(59) Sayle, D.C.; Maicaneanu, S.A.; Watson, G.W. Atomistic models for CeO_2(111), (110), and (100) nanoparticles, supported on yttrium-stabilized zirconia. *J. Amer. Chem. Soc.* **2002**, *124*, 11429–11439.

(60) Phillpot, S.R.; Keblinski, P.; Wolf, D.; Cleri, F. Synthesis and characterization of a polycrystalline ionic thin film by large-scale molecular-dynamics simulation. *Interface Sci.* **1999**, *7*, 15–31.

(61) Diebold, U. The surface science of titanium dioxide. *Surf. Sci. Rep.* **2003**, *48*, 53–229.

(62) Sayle, D.C.; Doig, J.A.; Maicaneanu, S.A.; Watson, G.W. Atomistic structure of oxide nanoparticles supported on an oxide substrate. *Phys. Rev. B* **2002**, *65*, 245414.

(63) Viswanadham, N.; Shido, T.; Iwasawa, Y. Performances of rhenium oxide-encapsulated ZSM-5 catalysts in propene selective oxidation/ammoxidation. *Appl. Catal. A Gen.* **2001**, *219*, 223–233.

(64) Monthioux, M. Filling single-wall carbon nanotubes. *CARBON* **2002**, *40*, 1809–1823.

(65) Jeng, M.L.; Shen, P.Y. Thermally activated rotation of Ni1-xO particles within CaO grains. Materials Science and Engineering—A structural materials properties. *Microstructure Processing* **2000**, *287*, 1–9.

(66) Sayle, D.C.; Parker, S.C. Encapsulated oxide nanoparticles: The influence of the microstructure on associated impurities within a material. *J. Amer. Chem. Soc.* **2003**, *125*, 8581–8588.

PART IV

PHYSICOCHEMICAL PROPERTIES OF OXIDE NANOMATERIALS

The focus of Chapters 5–9 was on the study and characterization of the structural and electronic properties of oxide nanomaterials using experimental and theoretical methods. The special structural and electronic properties of oxide nanostructures affect, or determine, in many cases the physical and chemical properties of these systems. In Chapter 1, quantum theory was used to analyze some of the links that exist among the structural, electronic, and chemical properties of oxide nanomaterials. A decrease in the average size of an oxide particle leads to quantum confinement that may induce new electronic states and change the magnitude of the band gap with a strong influence in the chemical reactivity. In this part of the book, the focus is on the physico-chemical properties of oxide nanomaterials. Many molecules interact with oxide surfaces through acid–base reactions (Chapter 10). An oxide nanoparticle can expose one or more types of basic anion (formally O^{2-}) sites and acidic metal cation (M^{n+}) sites. A Lewis acid–base bond arises from the asymmetric sharing of an electron pair between a donor molecule (the base) and an acceptor (the acid): $B: + A \rightarrow B:A$. The relative "strength" of the acid and base sites in a nanomaterial depends on the local surface structure, the nature of the metal counter-ion, and the type of the adsorbate–adsorbate interactions (Chapter 10). The use of probe molecules to extract information about the chemical properties of oxide nanostructures is a well-established and widely practiced approach and involves the combination of simple adsorbates (CO, NO, NO_2, SO_2, C_2H_2, C_2H_4, etc.) and spectroscopic techniques (FTIR, Raman, EPR, NMR, X-Ray absorption spectroscopy, photoemission, etc.), which are sensitive to the detection of the molecule in its adsorbed form (Chapter 11). A series of systematic studies has investigated the adsorption of molecules on oxide nanostructures from gas- and liquid-phase environments (Chapters 11 and 12). The enhanced chemical reactivity of oxide nanomaterials can have a strong impact on environmental catalysis and environmental remediation (Chapter 12). In many physico-chemical processes, the transport properties and oxygen-handling capabilities of an oxide nanostructure are an important issue (Chapter 13). Thus, the chemical reactivity may depend on the existence of surface *and* lattice defects. The effects of the nanoscale in oxides with

Synthesis, Properties, and Applications of Oxide Nanomaterials, Edited by José A. Rodríguez and Marcos Fernández-Garcia

ionic and mixed ionic/electronic conductivity have given rise to the fledgling science of "nanoionics," encompassing applications from fuel cells and lithium-ion batteries to sensors and catalysts. Chapter 13 presents recent experimental and theoretical findings for the transport and oxygen-handling properties of nanoscaled oxides.

CHAPTER 10

Theoretical Aspects of Oxide Particle Stability and Chemical Reactivity

YE XU and WILLIAM A. SHELTON

Computer Science and Mathematics Division, Oak Ridge National Laboratory, Oak Ridge, TN 37831

WILLIAM F. SCHNEIDER

Department of Chemical and Biomolecular Engineering, and Department of Chemistry and Biochemistry, University of Notre Dame, Notre Dame, IN 46556

10.1. INTRODUCTION

Molecular-level simulation is increasingly a key complement to experiment in the exploration of metal-oxide nanoparticle structure, composition, and reactivity. Unlike classic, purely ionic models of oxide bonding, "first-principles" methods treat the quantum-mechanical interactions within and between atoms explicitly, allowing the structure and reactivity of particles and their surfaces to be simulated without recourse to empirical parameterization and at a resolution that is often difficult to access experimentally. Density functional theory (DFT) (1,2) in particular has emerged as the leading tool for these studies because of its balance of theoretical rigor and computational tractability. When properly applied, DFT simulation provides reliable qualitative and often quantitative descriptions of molecular and surface reactivity as well as of bulk material properties. Several other chapters in this book address the various theoretical details of first-principles simulation applied to oxides. In this chapter, we focus on the use of these tools to describe and explain chemical reactivity at oxide surfaces, from simple gas adsorption to redox reactivity to the phase stability of oxide nanophases. Excellent complementary references include the books by Henrich and Cox (3) and by Noguera (4).

Synthesis, Properties, and Applications of Oxide Nanomaterials, Edited by José A. Rodríguez and Marcos Fernández-Garcia

10.2. ADSORPTION ON BASE METAL OXIDES

The simplest class of oxide reactivity is surface gas adsorption, typically of Lewis acidic or basic adsorbates. A Lewis acid–base bond arises from the asymmetric sharing of an electron pair between a donor molecule (the base) and an acceptor (the acid):

$$\text{B:} + \text{A} \longrightarrow \text{B:A}$$

A metal-oxide surface can expose one or more types of basic oxide anion (formally O^{2-}) and acidic metal cation (M^{n+}) sites. The relative "strength" of these acid and base sites depends on the local surface structure, the nature of the metal counter-ion, and the nature of the adsorbate. We take as one an example the adsorption of Lewis acids on the oxide MgO, which has been extensively studied both experimentally (5–11) and theoretically (7,12–16). Simple cubic MgO preferentially cleaves along the (100) surface, exposing both Lewis acid Mg^{2+} and base O^{2-} sites. Figure 10.1 shows a typical periodic boundary condition model (or "supercell" model) (17) of an oxide surface: Here a three-layer-thick slab of ions is included within a simulation cell that is periodically replicated in three dimensions. This representation allows size effects to be probed in one dimension (by varying the slab thickness), whereas the other two dimensions are periodic and unbounded. In this particular example, the surface bonding is highly localized and not sensitive to slab thicknesses greater than two layers. Within this periodic description, the electronic wave functions are conveniently described in terms of plane waves; this combined with a "pseudopotential" description

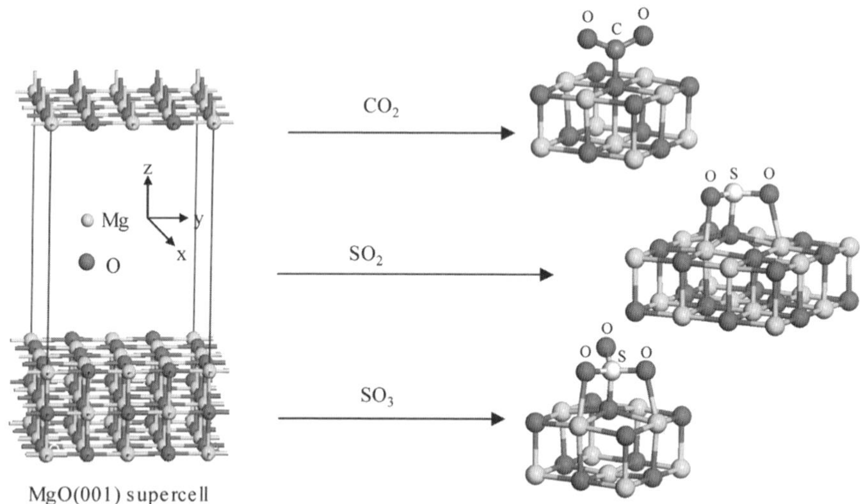

Figure 10.1. (Left) Periodic supercell model of the MgO(001) surface. (Right) Detail views of the adsorbate DFT-calculated structures of SO_2 adsorbed on MgO in both a Lewis base (left) and acid (right) conformation (15). See color insert.

of the chemically inert core electrons produces a highly efficient computational model within which the adsorption of gas molecules can be probed.

Plane-wave, pseudopotential DFT calculations performed within the PW91 generalized gradient approximation (GGA) show that the Lewis acids CO_2, SO_2, and SO_3 all preferentially bond at MgO basic sites to form surface species that can be qualitatively described as carbonates (CO_3^{2-}), sulfites (SO_3^{2-}), and sulfates (SO_4^{2-}) (15,16). As shown in Figure 10.1, the adsorbate structures all adopt the geometries expected for these ions in the bulk (e.g., approximately trigonal planar for carbonate, approximately tetrahedral for sulfate). The relative strength of adsorption increases with the Lewis acidity of the adsorbates and down the Periodic Table with increasing basicity of the metal oxides (13,16), as shown in Figure 10.2.

Several factors can influence the details of this acid–base interaction. For instance, step defects expose ions with reduced coordination number that have enhanced basicity. DFT calculations show that SO_2 adsorption at MgO step edges is increased by approximately 20 kcal mol^{-1} over that at (100) terraces (15). The qualitative acid–base nature of adsorption is unchanged, but the stronger interaction causes more pronounced distortions of the adsorbate molecule. One convenient way to probe these distortions computationally and to make a direct connection with the experiment is through comparison of calculated and observed spectroscopic properties, in particular, vibrational spectra. Distortions of adsorbates are reflected in shifts in characteristic vibrational modes; in this example, adsorbate bonding at the surface diminishes the internal S–O bonds and causes a red shift in their vibrational modes. The calculated vibrational spectra of terrace- and step-bound SO_2 provides a satisfactory description of the observed vibrational spectrum of SO_2-exposed MgO (15).

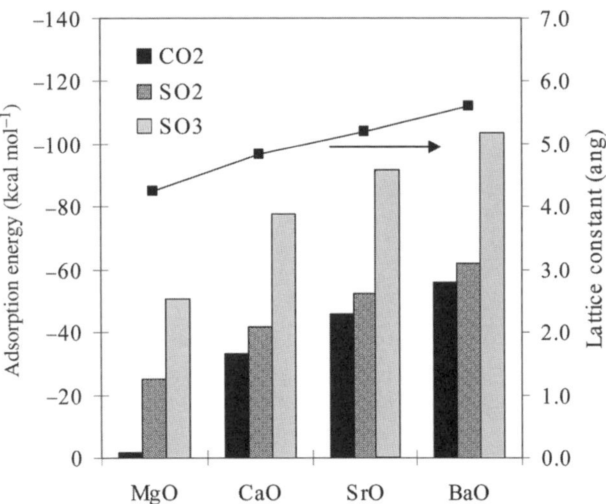

Figure 10.2. PW91 DFT-calculated adsorption energies of Lewis acids on the (001) surfaces of the alkaline-earth oxides. Reproduced with permission from Ref. 16.

Surface reactivity can also depend sensitively on the presence and concentration of other species on that surface. One can imagine a surface to present an array of reactive sites to a gas, with each site either vacant or occupied by an adsorbate. In the simplest realization, the occupation of one site has no influence on the occupation of another, and thus, the adsorption energy is the same for the first molecule to react with the surface as it is for the last. This is essentially the Langmuir adsorption model, which describes the physisorption and chemisorption of many gases on surfaces well (18). Often adsorbates will interfere with one another to some extent either sterically or electronically, causing the adsorption energy to decrease only slightly as the surface coverage increases. Such appears to be the case for CO_2 and SO_x adsorption on MgO, where the computed properties of isolated adsorbates agree well with the observed properties of adsorbates at higher coverage. Of course, at very high exposures, oxide surfaces evolve toward the corresponding metal salt structures, which have their own unique reactivity patterns.

The nitrogen oxides NO, NO_2, and NO_3 provide one interesting counterexample of this relatively simple Lewis acid–base model with weak coverage dependence (14,16,19–24). These "NO_x" molecules exhibit strong "cooperative" interactions between neighbors that dramatically enhance their adsorption strength (16,23–26). This cooperative bonding arises from spontaneous electron transfer between adsorbates and can only be captured in quantum-mechanical models of surface reactivity. As with SO_2, DFT calculations identify two binding modes for a single NO_2 on MgO(001), one in which NO_2 acts as a donor (base) toward Lewis acid surface sites:

Basic NO_2: $$\text{—Mg}^{2+}\text{—O}^{2-}\text{—Mg}^{2+}\text{—} + NO_2 \longrightarrow \text{—Mg}^{2+}\text{—O}^{2-}\text{—Mg}^{2+}\text{—} \quad (\text{O–N–O bound via both O atoms to the two Mg}^{2+}) \qquad \textbf{(10.1)}$$

and another in which NO_2 acts as an acceptor (acid) toward Lewis base surface sites:

Acidic NO_2: $$\text{—Mg}^{2+}\text{—O}^{2-}\text{—Mg}^{2+}\text{—} + NO_2 \longrightarrow \text{—Mg}^{2+}\text{—}\overset{NO_2\uparrow}{\text{O}^{2-}}\text{—Mg}^{2+}\text{—} \qquad \textbf{(10.2)}$$

The left and right sides of Figure 10.3 show the lowest-energy configurations obtained from DFT simulations for these two adsorption modes (16,23). The Mg^{2+} ions are weak Lewis acids, and both isolated SO_2 and the isostructural NO_2 only weakly physisorb to these sites. Unlike SO_2, however, NO_2 binds no more strongly to the basic O^{2-} sites of MgO(001). Neither the acidic nor basic configurations are consistent with the experimentally observed formation of chemisorbed nitrite (NO_2^-) and nitrate (NO_3^-) upon exposure of single-crystal MgO to NO_2 (19,20).

Qualitative molecular orbital/electron counting arguments can be used to rationalize the weak binding of an individual NO_2 on MgO(001) (23,25). MgO is an ionic and insulating main group oxide with a substantial band (or HOMO–LUMO) gap. NO_2 is odd electron and thus a much better *one*-electron donor and acceptor than it is a Lewis acid or base, as reflected by its large electron affinity (52 kcal mol^{-1}) and

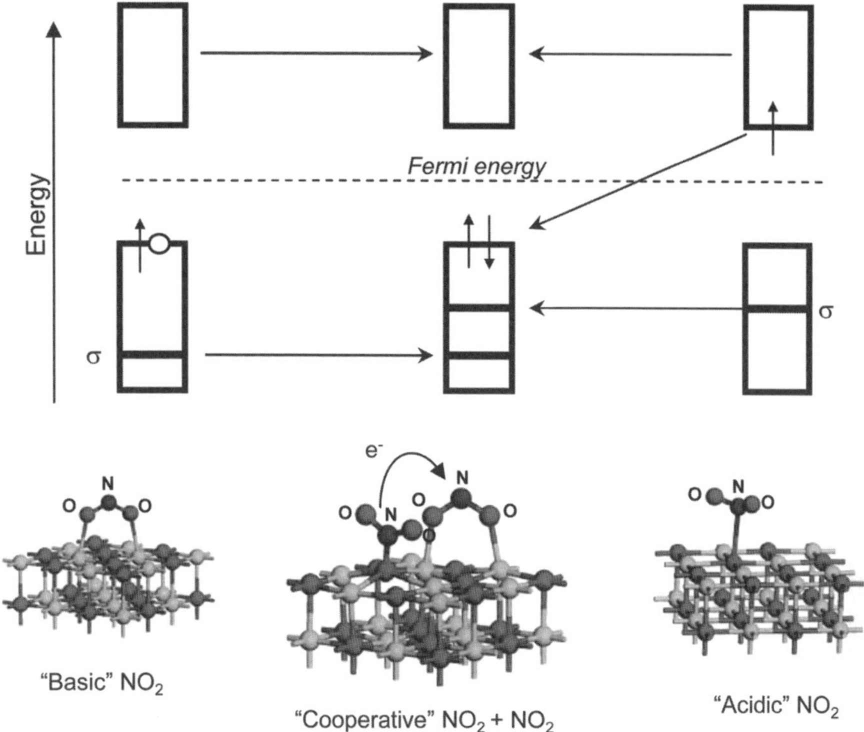

Figure 10.3. PW91 DFT-calculated adsorption geometries and qualitative molecular orbital diagrams for isolated and adsorption of NO_2 on MgO(001). Formation of the cooperative pair from isolated adsorbates is calculated to be exothermic by 16 kcal mol^{-1} (16, 23). See color insert.

moderate ionization potential (221 kcal mol^{-1}). As shown on the left and right sides of Figure 10.3, NO_2 bound in either the basic or acidic configuration is deficient or in excess one electron, respectively, which in electronic terms produces either a partially filled highest occupied molecular orbital (HOMO) or lowest unoccupied molecular orbital (LUMO) level (25). These unsatisfied valences are energetically costly, and as a result, NO_2 is weakly bound in either configuration.

As indicated in the center of Figure 10.3, the excess electron and hole can potentially be satisfied by bring the two NO_2 into proximity on the surface, allowing electron transfer to occur between the two rather than just with the surface:

$$\text{“Cooperative” } NO_2: \quad \underset{-Mg^{2+}-}{\overset{NO_x}{|}} + \underset{-O^{2-}-}{\overset{NO_x}{|}} \longrightarrow \underset{-Mg^{2+}-O^{2-}-}{\overset{\overset{e^-}{\curvearrowleft}}{NO_x^{-}\;NO_x^{+}}} \qquad (10.3)$$

Supercell DFT calculations confirm this hypothesis. The center of Figure 10.3 shows the substantial geometric relaxations that accompany NO_2 surface pairing: The basic

NO_2 relaxes toward the surface as a nitrite (NO_2^-), and the acidic NO_2 flattens and closely approaches a surface O to form a nitrate (NO_3^-, where one of the O comes from the MgO surface). Analysis of the distribution of electron charge around the adsorbates is consistent with a one-electron transfer between the two NO_2. The total bonding energy of a pair of NO_2 to the MgO surface is calculated to be twice that of the two NO_2 in isolation. Both energetically and structurally, then, the paired NO_2 is consistent with the experimentally observed chemisorption on NO_2 on MgO as a mixed nitrite and nitrate (19,20). NO_2 chemisorption on MgO(001) can thus be described as arising from an unusual "self-cooperative" interaction.

This cooperative interaction is present not only in these extended surface models, but also in fact, the same phenomenon is predicted to persist in MgO clusters containing 50 atoms or fewer (25,26). These small clusters provide an attractive alternative to supercell models not only for exploring the consequences of finite particle size, but also for the flexibility they offer in treating the electronic structure of the material. For instance, unlike supercell models, DFT-based cluster models can treat neutral and charged adsorbate states with equal ease. This flexibility is particularly useful for analyzing cooperative bonding. The DFT cluster results in Figure 10.4 show that the NO_2^+ cation combines much more readily with a base site on MgO than does the NO_2 neutral; similarly, the NO_2^- anion is a better base and associates more strongly with acidic MgO sites than does the neutral (adsorption energies of the ions are increased by 95 and 23 kcal mol^{-1}, respectively, over neutral NO_2) (25). These charged states are unlikely to persist in isolation, but they can be paired up to form a net neutral aggregate equivalent to a cooperative pair. A thermodynamic cycle can be used to relate the energies of the neutral, charged, and paired adsorbates (24); the results show that the ease with which an electron is transferred between two NO_2, combined with the considerable increase in adsorption energy of the ions over the neutrals, accounts for the cooperative chemisorption effect on MgO.

Similar cooperative effects are observed in models of the heavier members of the alkaline-earth oxide family, although the relative contribution of the effect diminishes with increasing oxide basicity (16,27,28). Numerous experimental studies provide

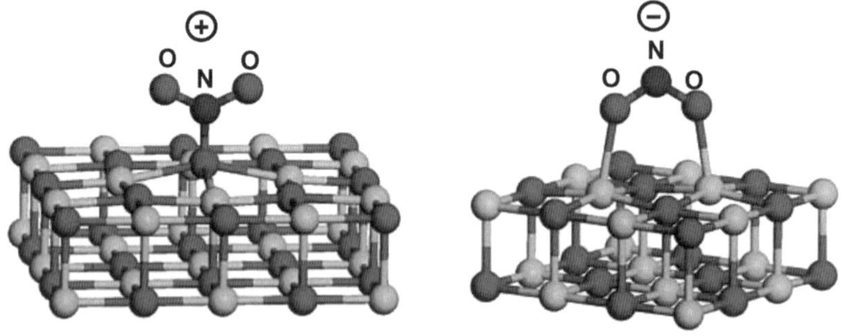

Figure 10.4. PW91 DFT-calculated geometries of (left) NO_2^+ and (right) NO_2^- adsorbed on $(MgO)_n$ clusters (25).

qualitative support for the cooperative model beyond MgO, including the observation of mixtures of NO_x surface species on BaO thin films (29) and mixed BaO/MgO powders (30). In fact, complicated NO_x surface species were noted in very early work on CaO powder as an NO adsorbant (31). Outside the alkaline-earth family, NO_2 has been found to form mixtures of nitrite and nitrate species on Al_2O_3, TiO_2, and Fe_2O_3 powders (32). Perhaps most compelling, experimentally derived kinetic models of NO_x uptake on supported baria (33) bear the clear signature of a cooperative adsorption effect that agrees nearly quantitatively with that predicted (16).

Thus, even the relatively simple alkaline-earth oxides can produce complicated adsorption behavior. Cooperative adsorption is not expected to be restricted solely to NO_x; the thermodynamic cycle analysis suggests that any combination of readily reducible and readily oxidizable adsorbates may exhibit the same effect (Figure 10.5). The nitrogen oxides have very low ionization potentials and high electron affinities that make them particularly suitable candidates for cooperative adsorption. However, many other odd-electron molecules, in particular, the halogen oxides (XO_x) and hydrogen oxides (HO_x), share these characteristics and likely participate in some form of cooperative chemisorption on oxides.

As one additional example of these concepts and models, a similar approach has been used to determine the chemistry and structural effects of water on the low-energy, hexagonal (0001) α-alumina surface (34,35). Figure 10.6 shows a top-view of this surface, which like the MgO(001) surface, exposes both basic and acidic centers. Water molecularly adsorbs at the acidic Al^{3+} sites. This molecular water is only metastable, however, and is calculated to dissociate with negligible activation barrier via at least two pathways to produce surface hydroxyls of two types:

$$O^{2-}\text{—}Al^{3+}\text{—}O^{2-} \xrightarrow{H_2O} O^{2-}\text{—}Al^{3+}(\leftarrow OH_2)\text{—}O^{2-} \longrightarrow (H^+)O^{2-}\text{—}Al^{3+}(\leftarrow [OH]^-)\text{—}O^{2-} \qquad \textbf{(10.4)}$$

Like the NO_2 pairs discussed above, the stability of this dissociated state can be considered to arise from a cooperative interaction between H and OH to make

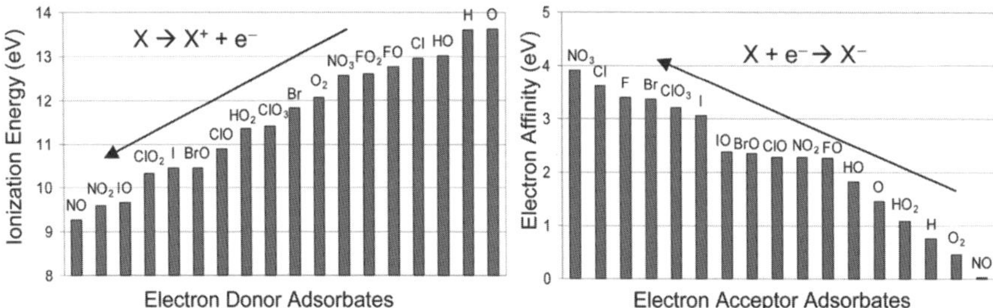

Figure 10.5. Experimental (left) ionization potentials and (right) electron affinities of selected odd-electron small molecules. Low ionization potentials and high electron affinities favor cooperative chemisorption.

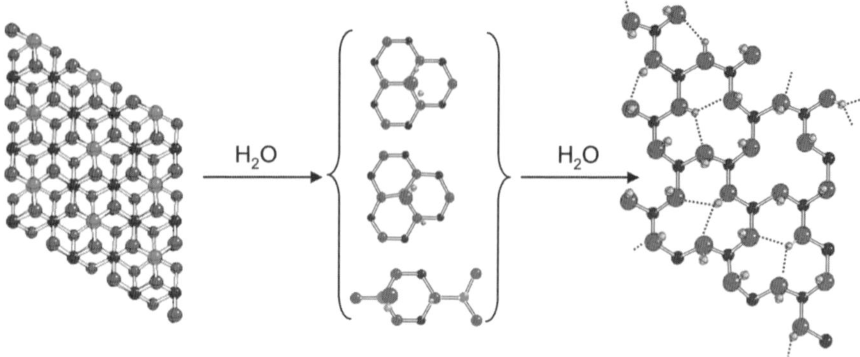

Figure 10.6. DFT models of the (left) pristine (0001) α-Al_2O_3 surface, molecularly and dissociatively adsorbed H_2O at low coverage (center), and the (right) disordered, fully hydroxylated surface (34,35). Reprinted with permission from Ref. 35.

H^+/OH^- pairs. One way to explore the coverage dependence of this chemistry is through a direct dynamical simulation of a monolayer of water. The Car–Parrinello molecular dynamics (36) (CPMD) approach is an efficient scheme for studying the dynamics of a system using DFT. CPMD simulations of a water monolayer confirm the ready dissociation of water and reveal additional cooperative water interactions that further enhance the rate of water dissociation (34,35).

These alumina results are consistent with the experimentally observed coverage-dependent adsorption energy of water on alumina (37–39). Furthermore, they lead to the conclusion that the most stable α-alumina surface in the presence of water is in fact fully hydroxylated, with the surface Al ions replaced with three surface H that are both disordered and dynamic (Figure 10.6). Synchrotron experiments (40) and subsequent calculations (41–44) confirm this picture of the water-modified alumina surface. Recent calculations on θ-, (45) κ-, and γ-aluminas (46,47) demonstrate that the high surface areas of these phases result directly from surface OH groups similar to those on the α-phase. Thus, the binding of metal and especially metal-oxide particles to alumina surfaces are most likely mediated by hydroxyls, and the hydroxylated α-alumina surface is a good model for these interactions.

As with MgO, the interaction of H_2O with finite-sized Al_2O_3 clusters has also been explored, including the influence of edge effects (35). Undercoordinated edge ions again are found to be associated with enhanced reactivity for water adsorption and dissociation, and this behavior is expected to persist to nanoscale particles.

10.3. OXIDATION–REDUCTION REACTIONS AT TRANSITION METAL-OXIDE SURFACES

Unlike the base metal oxides, the transition metal oxides often exhibit a variety of phases associated with different metal oxidation states, and because of the accessibility of these oxidation states, their surface reactivity is not limited to the adsorption reactions described above. A particularly exciting area of research today concerns the catalytic activity of these transition metal oxides, particularly of the noble to near-noble metals that form oxides more reluctantly than do base metals. A good example is oxidations catalyzed by Ru; the Ru metal shows low activity for CO oxidation under the conditions of a typical low-pressure surface science experiment, but when exposed to more realistic oxygen pressures, the Ru surface becomes both oxidized and activated for oxidation chemistry (48–54). Similarly, the activity of Ag catalysts for ethylene epoxidation has been attributed to formation of a surface-oxide phase (55), and surface oxides have been implicated in enhanced activity of Pt toward CO oxidation (56). Thus, many metals are found to form surface oxides in situ under conditions of catalytic importance.

Just as Pt is the prototypical model for catalytic oxidation on metals, RuO_2 has emerged as a preferred model for these "environmental" oxides: It is a stable oxide phase over a wide range of temperature and O_2 partial pressures, and it has a well-characterized (110) cleavage plane that is convenient for both surface science experiments and molecular simulations (49–52,57–61). As an example of the redox reactivity possible on transition metal-oxide surfaces, we briefly consider here the interaction of an oxygen atmosphere with RuO_2(110). The inset to Figure 10.7 shows a supercell model of the stoichiometric (110) surface of rutile-like RuO_2, which exposes both under-coordinated Ru cations as well as two types of O anions. Formal electron counting indicates that the exposed Ru ions are reduced relative to the bulk, whereas the Ru ions lying between bridging O are oxidized. This unequal charge distribution suggests that the (110) surface might be susceptible both to oxidation, e.g., by addition of oxygen to the stoichiometric surface, and to reduction, e.g., by removal of oxygen to create a sub-stoichiometric surface. DFT calculations can be used to explore the entire range of possible stoichiometries, and by normalizing these to the surface area, a phase diagram of surface structures can be constructed. Reuter et al. have discussed these surface phase diagrams in detail (53,62).

Figure 10.7 shows the results of supercell DFT calculations of the surface energy of RuO_2(110) as a function of the coverage of surface oxygen, calculated using the plane-wave DFT code VASP with PAW core potentials (63,64). The cusp in the plot near 1 J m^{-2} corresponds to the stoichiometric surface termination. Points to the left correspond to removal of bridging O from the surface, i.e., to surface reduction. The surface energy is seen to increase linearly with the decrease in occupation of these bridging oxygen sites, corresponding to a vacancy formation energy of approximately 60 kcal mol^{-1} with respect to molecular O_2. By way of contrast, an estimate of the MgO(001) vacancy formation energy using comparable methods is 140 kcal mol^{-1}, illustrating the more ready reducibility of the transition metal-oxide surface. Although surface reduction is endothermic, oxidation beyond stoichiometry is found to be exothermic. Points further to the right of the cusp correspond to the addition of extra O atoms at under-coordinated Ru sites. These excess O are bound by approximately 20 kcal mol^{-1}, again nearly independent of the surface coverage, and reflect a "super-stoichiometric" surface-oxide state. The existence of these excess

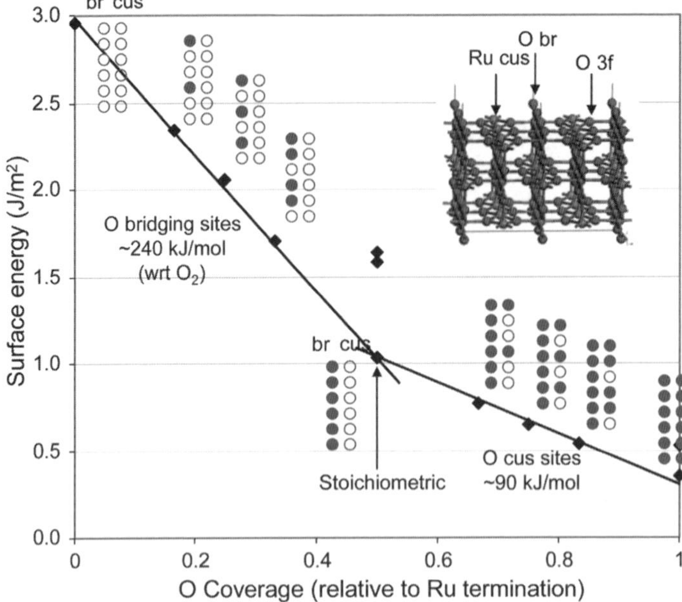

Figure 10.7. DFT-calculated coverage-dependent oxygen binding energies on $RuO_2(110)$. Open and filled circles indicate unoccupied and occupied O sites, respectively. Two distinct binding regimes are found, corresponding to occupation of bridging and coordinately unsaturated Ru sites.

O have been observed experimentally (65) in TPD experiment and have been proposed to compete with bridging O in the catalytic CO oxidation activity of RuO_2 (49,50,53,57).

As with water dissociation on α-Al_2O_3, a key question in this examination of oxygen reactivity at $RuO_2(001)$ is the facility of the O_2 dissociation process. The surface energies in Figure 10.7 correspond to dissociated O, but these are likely accessible only through some O_2 dissociation mechanism. DFT calculations combined with transition state theory can be used to explore the kinetics of these dissociations. Figure 10.8 shows the results of "nudged elastic band" (NEB) (66) calculations of O_2 dissociation across two coordinatively unsaturated Ru and between one of the Ru and an O bridge site. These NEB calculations work by stretching a virtual "elastic band" between reactant and product states; minimizing the energy of that elastic band produces a minimum energy pathway from reactants to products. Although the absolute magnitude of the derived activation energies calculated within DFT must be viewed even more cautiously than absolute adsorption energies, the indications from these calculations (Figure 10.8), consistent with the experiment, are that O_2 dissociation from metastable adsorbed molecular O_2 is facile across both sites studied, with activation energies that scale with the binding energy of the products.

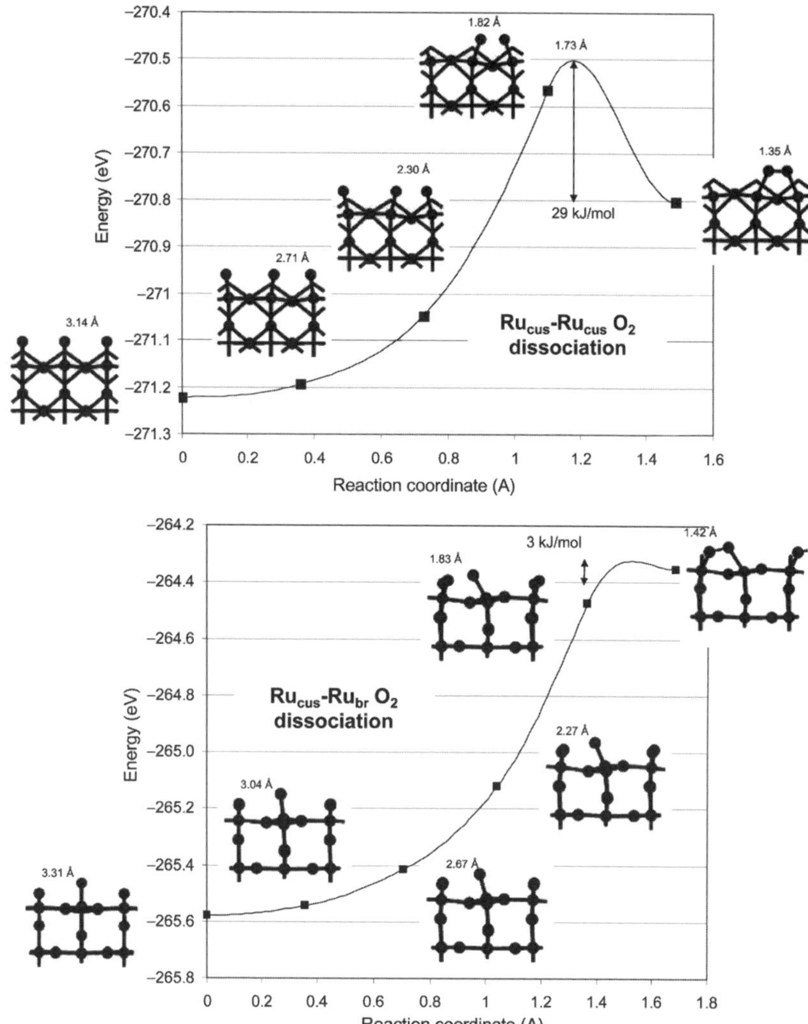

Figure 10.8. Nudged elastic band calculations of the reaction paths for dissociation of O_2 across (top) two coordinatively unsaturated Ru and between a coordinatively unsaturated Ru and (bottom) a vacant bridge site.

10.4. THE METAL-TO-METAL-OXIDE TRANSITION AT THE ATOMIC SCALE

The alumina/H_2O and ruthenia/O_2 results above illustrate the interplay between the metal-oxide surface and the surrounding environment that determines surface structure, composition, and reactivity. Although complicated, to a large extent, these interactions can be understood from basic concepts of acid–base and redox chemistry. An important question is how these phenomena translate down to the smallest oxide particle sizes, where size effects may qualitatively alter reactivity. As an illustrative example, we consider here the oxidation behavior of atomic-scale particles of Pt, a metal that resists oxidation in the bulk, by examining the equilibrium between metal and oxide:

$$Pt_x + y/2O_2 \leftrightarrows Pt_xO_y \qquad \textbf{(10.5)}$$

as a function of x and y.

The power and versatility of DFT simulation makes it an ideal method for investigating metal-oxide nanoparticles. The DFT results of this section (67,68), like most of those above, were obtained using the Vienna Ab-initio Simulation Package (VASP) (64) with a well-established and robust set of parameters, including the PW91 exchange-correlation functional (69). A particular challenge in modeling the thermodynamics of finite clusters is the determination of energy-optimal structures. DFT calculations work by locating energy minima in the vicinity of some starting structural guess; although the appropriate structures for bulk oxides are often well established, those structures are often not applicable to small particles. Even for a particle containing just a handful of atoms, the range of possible structures to search over is already quite large, and the situation becomes more complex with each additional atom.

For the Pt_xO_y clusters, a combination of manual searches over configuration space combined with DFT-based finite-temperature molecular dynamics (70) (MD) can be used to generate and screen structures. As an example, we consider the discrete Pt monoxide (Pt_xO_x) and dioxide (Pt_xO_{2x}) homologs to the most stable bulk Pt oxide phases. The formation energies of the $x = 1$–5 and $x = 10$ structures are plotted in Figure 10.9 against the number of Pt atoms. The size dependency of the formation energy is systematic and dramatic. Smaller Pt oxide clusters have substantially greater formation energies than the larger Pt oxide clusters or the respective bulk oxides, indicating that smaller Pt clusters have greater affinity for oxygen than larger clusters or bulk Pt metal. In fact, the oxidation of a Pt atom or Pt dimer is over 1 eV/Pt more exothermic than the formation of either bulk PtO or bulk PtO_2. As the size of the cluster (x) increases, oxidation generally becomes less exothermic, although the variations are not always monotonic: Whereas the Pt_xO_{2x} formation energy rises steadily toward that of bulk PtO_2, that of the Pt_xO_x series reaches a minimum around $x = 4$ before rising toward bulk PtO. This is an example of a quantum size effect, indicating in this case that monoxides in the size range (Pt_4O_4–Pt_6O_6) are more resistant to conversion to the dioxides than either smaller or larger clusters.

Oxidation substantially alters the molecular and electronic structures of the Pt clusters, and the effect is markedly size-dependent. The close-packed structures of the pure Pt_x clusters give way to open or one- or two-dimensional shapes. The smallest Pt_xO_x clusters are predicted to prefer closed triangle, quadrangle, pentagon, and so on, structures, with each Pt atom coordinated to two O atoms (forming O–Pt–O subunits) and each O atom occupying a Pt–Pt bridge position. The Pt_xO_{2x} series, on the other hand, forms various open, curved chains, in which the middle Pt atoms are each

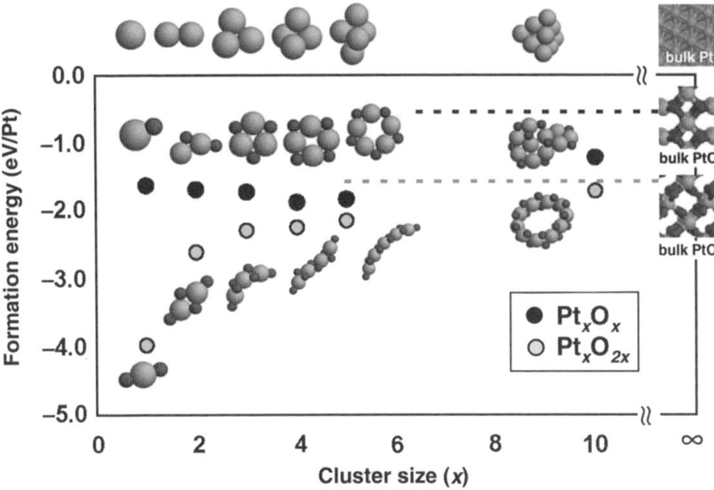

Figure 10.9. DFT-calculated formation energies (with respect to Pt_x and O_2 gas) of Pt_xO_x (filled circles) and Pt_xO_{2x} (open circles) clusters. Black and gray dashed lines indicate the formation energies of bulk PtO (−0.54 eV) and β-PtO_2 (−1.58 eV), respectively. In the structural representations, small and large spheres represent O and Pt atoms, respectively. The pure Pt clusters are included for comparison. Reproduced with permission from Ref. 67.

coordinated to four O atoms (forming flat Pt–4O subunits) and each terminal Pt atom coordinated to three. As the length grows, the chain structure eventually becomes flexible enough to form a closed loop so that the terminal Pt atoms are also fully coordinated. The Pt–Pt distances in these small Pt monoxide and dioxide clusters are shorter than or comparable with the GGA-PW91 calculated Pt–Pt bond length in the bulk Pt metal, 2.83 Å (the experimental value is 2.77 Å). As the cluster size increases, particularly in the Pt_xO_{2x} series, the Pt–Pt distance begins to approach the bulk Pt oxides over 3 Å. Of particular note is that clusters prepared from fragments extracted from the bulk oxide lattice are substantially less stable than these open structures.

The electronic structures of small Pt oxide clusters are also very different from large oxide clusters or bulk oxides of the same stoichiometry, as indicated by a decomposition of the electron charge into individual atomic contributions (71,72). The results for the two series of small Pt oxide clusters are listed in Tables 10.1 and 10.2. In the Pt_xO_x series, the amount of charge lost by each Pt atom is always balanced by the charge gained by each O atom. However, the extent of electron transfer from Pt to O in molecular PtO is only about half of that in bulk PtO. The extent of electron transfer in molecular PtO_2 is also appreciably less than that in bulk PtO_2. Here the positive charge on each Pt atom in a Pt_xO_{2x} cluster is twice the amount of negative charge on each O atom. If the extent of charge transfer between the Pt and the O atoms in the bulk oxides is taken to be the ionic limit, then the Pt–O bond in molecular PtO and PtO_2 and the other small Pt oxide clusters is appropriately described as possessing significant covalent characters. The level of ionicity depends on cluster size: As the cluster size increases, it converges toward the bulk level.

TABLE 10.1. Net Spin and Bader Atomic Charges of the Pt_xO_x Clusters.

Cluster	Spin	Atom	Charge (e)
PtO	2	**Pt**	**+0.48**
		O	−0.48
Pt_2O_2	0	**Pt**	**+0.99, +0.36**
		O	−0.70, −0.65
Pt_3O_3	2	**Pt**	**+0.77**
		O	−0.77
Pt_4O_4	0	**Pt**	**+0.77**
		O	−0.77
Pt_5O_5	2	**Pt**	**+0.76**
		O	−0.76
$Pt_{10}O_{10}$	0	**Pt**	**+0.76**
		O	−0.76
bulk PtO	0	**Pt**	**+0.96**
		O	−0.96

Although only two stoichiometries are considered in Figure 10.9, a transition metal particle of a given size x in a given environment could in principle adopt any stoichiometry y consistent with the equilibrium (5). The thermodynamics of equation (5) for any value of x and y can in principle be determined directly from DFT total energy calculations of the minimum energy reactants and products. As an example, Figure 10.10 shows the calculated minimum energy structures along the Pt_2O_y

TABLE 10.2. Net Spin and Average Bader Atomic Charges of the Pt_xO_{2x} Clusters.

Cluster	Spin	Atom	Charge (e)
PtO_2	0	**Pt**	**+1.27**
		O	−0.63
Pt_2O_4	0	**Pt**	**+1.30**
		O	−0.65
Pt_3O_6	0	**Pt**	**+1.29**
		O	−0.65
Pt_4O_8	2	**Pt**	**+1.33**
		O	−0.67
Pt_5O_{10}	0	**Pt**	+1.39
		O	−0.69
$Pt_{10}O_{20}$	0	**Pt**	**+1.41**
		O	−0.71
bulk PtO_2	0	**Pt**	**+1.70**
		O	−0.85

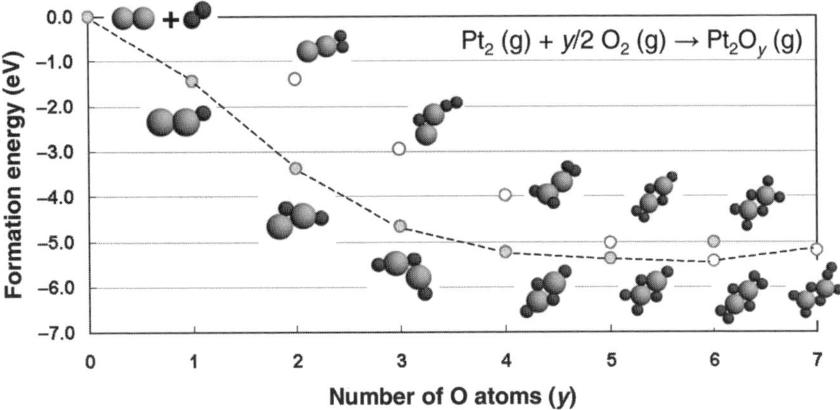

Figure 10.10. DFT-calculated minimum energy structures and 0 K formation energies of the Pt_2 oxide clusters, including dissociated O_2 (filled circles) and partially associated O_2 (open circles) (68) .

series, $y = 0$–7, derived as above by systematically exploring a variety of structures in search of the lowest energy configuration (68). As y increases, the O atoms are observed to alternate between terminal and bridging positions, suggesting a balance between the preference for fully coordinating the Pt and O and the need to avoid excessive repulsion between the O atoms. As Pt_2 takes on more O atoms, the Pt–Pt distance increases from 2.3 to 3.1 Å, and at the largest values of y, the O atoms pair up as O_2 molecules. The DFT results indicate a very shallow energy minimum around Pt_2O_5, which is slightly more oxidized than the Pt_2O_4 described above.

To translate these 0 K DFT energies to finite temperature and pressure, further details about the thermochemical system must be specified. One interesting and relevant limit is to imagine Pt particles to be dispersed and immobile on a two-dimensional support and in equilibrium with an ideal gaseous reservoir of O_2 of temperature T and pressure p_{O_2}. Based on this description, a thermodynamic potential $\omega_{Pt_xO_y}$ can be defined that is minimized when equilibrium is established between these immobile Pt_xO_y particles and the O_2 reservoir. All necessary temperature-dependent quantities can ultimately be related to the heat capacity of the particles, which can be calculated from the DFT-derived structures and vibrational spectrum of the particles using standard statistical mechanical models (68,73). $\omega_{Pt_xO_y}$ is evaluated for each composition at whatever T and p_{O_2} that is of interest, and the lowest ω indicates the equilibrium composition.

The results of application of this algorithm to the Pt_2O_y series are shown in the form of a T-p_{O_2} phase diagram in Figure 10.11. The ranges of T (from 200 to 1000 K) and p_{O_2} (from 10^{-10} to 10 bar) encompass virtually all conditions of practical interest. A cursory inspection shows that equilibrium shifts toward lower oxygen content as temperature increases or pressure decreases, in line with basic physical insights. Under

Figure 10.11. T-p_{O_2} phase diagram for Pt_2 clusters immobilized on a support. The O_2 pressure is plotted on a logarithmic scale. See color insert.

ambient conditions (room temperature and ca. 0.2 bar for p_{O_2}; cf. the orange circle in Figure 10.11), the stable composition is the dioxide. At higher temperatures and lower pressures typical, for instance, of catalytic oxidations (cf. the red ellipse in Figure 10.11), the equilibrium O:Pt ratio decreases to Pt_2O_3. Only at the very highest temperatures and lowest pressures does the fully reduced Pt_2 become thermodynamically preferred. A similar tendency toward oxidation is found for Pt_3 particles, although as indicated in Figure 10.9, the energy "gap" between monoxide and dioxide phases is decreased and the lower oxide phases become more stable over a wider range of T and p_{O_2}.

The results are primarily meant to illustrate the nonintuitive compositional behavior that becomes possible as metal-oxide particles decrease to the nanoscale. The Pt_2, Pt_3, or Pt_{10} particles discussed here are clearly smaller than are typically encountered in practical applications of nanooxides, and one computational challenge is to extend the results more fully into the nanoscale. Nonetheless, the results are still of value in interpreting experiment. For instance, Alexeev et al. recently reported on an EXAFS analysis of Pt–Pt and Pt–O bonding and coordination in Pt particles supported on γ-Al_2O_3 under oxidizing and reducing conditions (74). They found that the smallest (~11 Å) Pt particles in their experiment are particularly susceptible to structural modification by the gas environment, including the replacement of Pt–Pt coordination by Pt–O bonds upon oxidation and recovery of Pt–Pt bonding upon reduction. Although these particles contain on the order of 25 Pt atoms and thus are much larger than those considered here, the qualitative behavior is consistent with facile oxidation and reduction expected from the smaller particles.

10.5. CONCLUSIONS

As we have seen, metal oxides can exhibit a wide range of reactivity patterns, and first-principles simulation based on density functional theory provides a powerful tool for exploring and explaining those patterns. Clearly neither all aspects of oxide reactivity nor all available simulation tools and methods can be covered in the limited space available here. Rather, we hope to have demonstrated several relevant examples of how these tools can be productively applied to topical problems in oxide reactivity: to nonlinear coverage effects and cooperativity at metal-oxide surfaces, as seen in the adsorption chemistry of NO_x; to the thermodynamics and kinetics of surface reactions and their impact on surface composition, as seen in the surface hydration of α-alumina; to the variable surface compositions possible at redox active transition metal-oxide surfaces, as illustrated by the reactions of oxygen with the ruthenium oxide surface; and finally to the dramatic influence of finite size effects on the stability of oxide particle phases, as illustrated for atomic-scale platinum particles. In each case, we have seen how the molecular level details accessible through simulation—molecular and electronic structures, thermodynamics, kinetics, and spectroscopies—can be used to inform the experiment and to enrich our understanding of metal-oxide chemistry. As theoretical methods and computational capabilities continue to advance, simulation will only grow in importance as a tool in metal-oxide research.

REFERENCES

(1) Parr, R.G.; Yang, W. *Density-Functional Theory of Atoms and Molecules*; Oxford University Press: New York, 1999.

(2) Kohn, W.; Becke, A.D.; Parr, R.G. Density functional theory of electronic structure. *J. Phys. Chem.* **1996**, *100*, 12974–12980.

(3) Henrich, V.E.; Cox, P.A. *The Surface Science of Metal Oxides*; Cambridge University Press: Cambridge, 1994.

(4) Noguera, C. *Physics and Chemistry at Oxide Surfaces*; Cambridge University Press: Cambridge, 1996.

(5) Goodsel, A.J.; Low, M.D.J.; Takezawa, N. Reactions of gaseous pollutants with solids. II. Infrared study of sorption of SO_2 on MgO. *Environ. Sci. Technol.* **1972**, *6*, 268.

(6) Schoonheydt, R.A.; Lunsford, J.H. Infrared spectroscopic investigation of the adsorption and reactions of SO_2 on MgO. *J. Catal.* **1972**, *26*, 261–271.

(7) Rodriguez, J.A.; Jirsak, T.; Freitag, A.; Larese, J.Z.; Maiti, A. Interaction of SO_2 with MgO(100) and Cu/MgO(100): Decomposition reactions the formation of SO_3 and SO_4. *J. Phys. Chem. B* **2000**, *104*, 7439.

(8) Babaeva, M.A.; Tsyganenko, A.A.; Filimonov, V.N. IR spectra of adsorbed SO_2. *Kinet. Katal.* **1984**, *25*, 921.

(9) Benistel, M.; Waqif, M.; Saur, O.; Lavalley, J.C. The structure of sulfate species on magnesium oxide. *J. Phys. Chem.* **1989**, *93*, 6581–6582.

(10) Waqif, M.; Saur, O.; Lavalley, J.C.; Perathoner, S.; Centi, G. Nature and mechanism of formation of sulfate species on copper/alumina sorbent catalysts for SO_2 removal. *J. Phys. Chem.* **1991**, *95*, 4051.

(11) Stark, J.V.; Park, D.G.; Lagadic, I.; Klabunde, K.J. Nanoscale metal oxide particles/clusters as chemical reagents. Unique surface chemistry on magnesium oxide as shown by enhanced adsorption of acid gases (sulfur dioxide and carbon dioxide) and pressure dependence. *Chem. Mater.* **1996**, *8*, 1904.

(12) Pacchioni, G.; Clotet, A.; Ricart, J.M. A theoretical study of the adsorption and reaction of SO_2 at surface and step sites of the MgO(100) surface. *Surf. Sci.* **1994**, *315*, 337–350.

(13) Pacchioni, G.; Ricart, J.M.; Illas, F. Ab initio cluster model calculations on the chemisorption of CO_2 and SO_2 proble molecules on MgO and CaO(100) surfaces. A theoretical measure of oxide basicity. *J. Am. Chem. Soc.* **1994**, *116*, 10152–10158.

(14) Karlsen, E.J.; Nygren, M.A.; Pettersson, L.G.M. Comparative study on structures and energetics of NO_x, SO_x, and CO_x adsorption on alkaline-earth-metal oxides. *J. Phys. Chem. B* **2003**, *107*, 7795–7802.

(15) Schneider, W.F.; Li, J.; Hass, K.C. Combined computational and experimental investigation of SO_x adsorption on MgO. *J. Phys. Chem. B* **2001**, *105*, 6972–6979.

(16) Schneider, W.F. Qualitative differences in the adsorption chemistry of acidic (CO_2, SO_x) and amphiphilic (NO_x) species on the alkaline earth oxides. *J. Phys. Chem. B* **2004**, *108*, 273–282.

(17) Martin, R.M. *Electronic Structure: Basic Theory and Practical Methods*; Cambridge University Press: Cambridge, 2004.

(18) Masel, R.I. *Principles of Adsorption and Reaction on Solid Surfaces*; Wiley: New York, 1996.

(19) Rodriguez, J.A.; Jirsak, T.; Sambasivan, S.; Fischer, D.; Maiti, A. Chemistry of NO_2 on CeO_2 and MgO: Experimental and theoretical studies on the formation of NO_3. *J. Chem. Phys.* **2000**, *112*, 9929–9939.

(20) Rodriguez, J.A.; Jirsak, T.; Kim, J.-Y.; Larese, J.Z.; Maiti, A. Interaction of NO and NO_2 with MgO (100): Photoemission and density-functional studies. *Chem. Phys. Lett.* **2000**, *330*, 475–483.

(21) Di Valentin, C.; Figini, A.; Pacchioni, G. Adsorption of NO and NO_2 on terrace and step sites and on oxygen vacancies of the CaO(100) surface. *Surf. Sci.* **2004**, *556*, 145–158.

(22) Branda, M.M.; Di Valentin, C.; Pacchioni, G. NO and NO_2 adsorption on terrace, step, and corner sites of the BaO surface from DFT calculations. *J. Phys. Chem. B* **2004**, *108*, 4752–4758.

(23) Schneider, W.F. Fundamental concepts in molecular simulation of NO_x catalysis. In *Environmental Catalysis*; Grassian, V.H. (Editor); CRC Press: New York, 2005.

(24) Schneider, W.F.; Hass, K.C.; Miletic, M.; Gland, J.L. Dramatic cooperative effects in adsorption of NO_x on MgO(001). *J. Phys. Chem. B* **2002**, *106*, 7405–7413.

(25) Miletic, M.; Gland, J.L.; Schneider, W.F.; Hass, K.C. First-principles characterization of NO_x adsorption on MgO. *J. Phys. Chem. B* **2003**, *107*, 157–163.

(26) Miletic, M.; Gland, J.L.; Hass, K.C.; Schneider, W.F. Characterization of adsorption trends of NO_2, nitrite, and nitrate on MgO terraces. *Surf. Sci.* **2003**, *546*, 75–86.

(27) Broqvist, P.; Gronbeck, H.; Fridell, E.; Panas, I. Characterization of NO_x species adsorbed on BaO: Experiment and theory. *J. Phys. Chem. B* **2004**, *108*, 3523–3530.

(28) Broqvist, P.; Panas, I.; Fridell, E.; Persson, H. NO_x storage on BaO(100) surface from first principles: A two channel scenario. *J. Phys. Chem. B* **2002**, *106*, 137–145.

(29) Schmitz, P.J.; Baird, R. *J. Phys. Chem. B* **2002**, *106*, 4172–4180.

(30) Hess, C.; Lunsford, J.H. Mechanism for NO_2 storage in barium oxide supported on magnesium oxide studied by in situ Raman spectroscopy. *J. Phys. Chem. B* **2002**, *106*, 6358.

(31) Low, M.D.J.; Yang, R.T. Reactions of gaseous pollutants with solids. V. Infrared study of the sorption of NO on CaO. *J. Catal.* **1974**, *34*, 479–489.

(32) Underwood, G.M.; Miller, T.M.; Grassian, V.H. Transmission FT-IR and Knudsen cell study of the heterogeneous reactivity of gaseous nitrogen dioxide on mineral oxide particles. *J. Phys. Chem. A* **1999**, *103*, 6184–6190.

(33) Olsson, L.; Persson, H.; Fridell, E.; Skoglundh, M.; Andersson, B. A kinetic study of NO oxidation and NO_x storage on Pt/Al_2O_3 and $Pt/BaO/Al_2O_3$. *J. Phys. Chem. B* **2001**, *105*, 6895–6906.

(34) Hass, K.C.; Schneider, W.F.; Curioni, A.; Andreoni, W. The chemistry of water on alumina surfaces: Reaction dynamics from first principles. *Science* **1998**, *282*, 265–268.

(35) Hass, K.C.; Schneider, W.F.; Curioni, A.; Andreoni, W. First-principles molecular dynamics simulations of H_2O on α-Al_2O_3(0001). *J. Phys. Chem. B* **2000**, *104*, 5527–5540.

(36) Remler, D.K.; Madden, P.A. Molecular dynamics without effective potentials via the Car-Parrinello approach. *Mol. Phys.* **1990**, *70*, 921–966.

(37) Elam, J.W.; Nelson, C.E.; Cameron, M.A.; Tolbert, M.A.; George, S.M. Adsorption of H_2O on a single-crystal α-Al_2O_3(0001) surface. *J. Phys. Chem. B* **1998**, *102*, 7008–7015.

(38) McHale, J.M.; Auroux, A.; Perrotta, A.J.; Navrotsky, A. Surface energies and thermodynamic stability in nanocrystalline alumina. *Science* **1997**, *277*, 788.

(39) McHale, J.M.; Navrotsky, A.; Perrotta, A.J. Effects of increased surface area and chemisorbed H_2O on the relative stability of nanocrystalline γ-Al_2O_3 and α-Al_2O_3. *J. Phys. Chem. B* **1997**, *101*, 603–613.

(40) Eng, P.J.; Trainor, T.P.; Brown Jr., G.E.; Waychunas, G.A.; Newville, M.; Sutton, S.R.; Rivers, M.L. Structure of the hydrated α-Al_2O_3(0001) surface. *Science* **2000**, *288*, 1029.

(41) Di Felice, R.; Northrup, J.D. Theory of the clean and hydrogenated Al_2O_3(0001)–(1×1) surfaces. *Phys. Rev. B* **1999**, *60*, R16287.

(42) Wang, X.-G.; Chaka, A.; Scheffler, M. The effect of the environment on α-Al_2O_3(0001) surface structures. *Phys. Rev. Lett.* **2000**, *84*, 3650–3653.

(43) Lodziana, Z.; Nørskov, J.K.; Stoltze, P. The stability of the hydroxylated(0001) surface of α-Al_2O_3. *J. Chem. Phys.* **2003**, *118*, 11179–11188.

(44) Tepesch, P.D.; Quong, A.A. First-principles calculations of α-alumina(0001) surface energies with and without hydrogen. *Phys. Stat. Sol.* **2000**, *217*, 377.

(45) Lodziana, Z.; Topsoe, N.Y.; Nørskov, J.K. A negative surface energy for alumina. *Nat. Mater.* **2004**, *3*, 289–293.

(46) Digne, M.; Sautet, P.; Raybaud, P.; Euzen, P.; Toulhoat, H. Use of DFT to achieve a rational understanding of acid-basic properties of [gamma]-alumina surfaces. *J. Catal.* **2004**, *226*, 54–68.

(47) Digne, M.; Sautet, P.; Raybaud, P.; Toulhoat, H.; Artacho, E. Structure and stability of aluminum hydroxides: A theoretical study. *J. Phys. Chem. B* **2002**, *106*, 5155.

(48) Peden, C.H.F. Carbon monoxide oxidation on model single-crystal catalysts. In *Surface Science of Catalysis*; Dwyer, D.J.; Hoffmann, F.M. (Editors); ACS: New York, 1992; 143–159.

(49) Knapp, M.; Seitsonen, A.P.; Kim, Y.D.; Over, H. Catalytic activity of the $RuO_2(100)$ surface in the oxidation of CO. *J. Phys. Chem. B* **2004**, *108*, 14392–14397.

(50) Wendt, S.; Seitsonen, A.P.; Over, H. Catalytic activity of $RuO_2(110)$ in the oxidation of CO. *Catal. Today* **2003**, *85*, 167–175.

(51) Over, H.; Kim, Y.D.; Seitsonen, A.P.; Wendt, S.; Lundgren, E.; Schmid, M.; Varga, P.; Morgante, A.; Ertl, G. Atomic-scale structure and catalytic reactivity of the $RuO_2(110)$ surface. *Science* **2000**, *287*, 1474.

(52) Over, H.; Muhler, M. Catalytic CO oxidation over ruthenium—bridging the pressure gap. *Prog. Surf. Sci.* **2003**, *72*, 3–17.

(53) Reuter, K.; Scheffler, M. First-principles atomistic thermodynamics for oxidation catalysis: Surface phase diagrams and catalytically interesting regions. *Phys. Rev. Lett.* **2003**, *90*, 046103.

(54) Reuter, K.; Scheffler, M. Composition, structure, and stability of $RuO_2(110)$ as a function of oxygen pressure. *Phys. Rev. B* **2001**, *65*, 035406-035401-035406-035411.

(55) Li, W.-X.; Stampfl, C.; Scheffler, M. Why is a noble metal catalytically active? The role of the O-Ag interaction in the function of silver as an oxidation catalyst. *Phys. Rev. Lett.* **2003**, *90*, 256102.

(56) Hendriksen, B.L.M.; Frenken, J.W.M. CO oxidation on Pt(110): Scanning tunneling microscopy inside a high-pressure flow reactor. *Phys. Rev. Lett.* **2002**, *89*, 0461011.

(57) Reuter, K.; Frenkel, D.; Scheffler, M. The steady state of heterogeneous catalysis, studied by first-principles statistical mechanics. *Phys. Rev. Lett.* **2004**, *93*, 116104–116105.

(58) Kim, S.H.; Wintterlin, J. Atomic scale investigation of the oxidation of CO on $RuO_2(110)$ by scanning tunneling microscopy. *J. Phys. Chem. B* **2004**, *108*, 14565–14569.

(59) Assmann, J.; Narkhede, V.; Khodeir, L.; Löffler, E.; Hinrichsen, O.; Birkner, A.; Over, H.; Muhler, M. On the nature of the active state of supported ruthenium catalysts used for the oxidation of carbon monoxide: Steady-state and transient kinetics combined with in situ infrared spectroscopy. *J. Phys. Chem. B* **2004**, *108*, 14634–14642.

(60) Wang, J.; Fan, C.Y.; Jacobi, K.; Ertl, G. The kinetics of CO oxidation on $RuO_2(110)$: Bridging the pressure gap. *J. Phys. Chem. B* **2002**, *106*, 3422–3427.

(61) Liu, Z.-H.; Hu, P.; Alavi, A. Mechanism for the high reactivity of CO oxidation on a ruthenium—oxide. *J. Chem. Phys.* **2001**, *114*, 5956–5957.

(62) Reuter, K.; Stampfl, C.; Ganduglia-Pirovano, M.V.; Scheffler, M. Atomistic description of oxide formation on metal surfaces: The example of ruthenium. *Chem. Phys. Lett.* **2002**, *352*, 311.

(63) Kresse, G.; Joubert, J. From ultrasoft pseudopotentials to the projector augmented-wave method. *Phys. Rev. B* **1999**, *59*, 1758–1775.

(64) Kresse, G.; Furthmüller, J. Efficiency of ab-initio total energy calculations for metals and semiconductors using a plane-wave basis set. *Comp. Mat. Sci.* **1996**, *6*, 15–50.

(65) Kim, Y.D.; Seitsonen, A.P.; Wendt, S.; Wang, J.; Fan, C.; Jacobi, K.; Over, H.; Ertl, G. Characterization of various oxygen species on an oxide surface: $RuO_2(110)$. *J. Phys. Chem. B* **2001**, *105*, 3752–3758.

(66) Jónsson, H.; Mills, G.; Jacobsen, K.W. Nudged elastic band method for finding minimum energy paths of transitions. In *Classical and Quantum Dynamics in Condensed Phase Simulations*; Berne, B.J.; Ciccotti, G.; Coker, D.F. (Editors); World Scientific: Singapore 1998; 389–405.

(67) Xu, Y.; Shelton, W.A.; Schneider, W.F. Effect of particle size on the oxidizability of Pt clusters. *J. Phys. Chem. A* **2006**, *110*, 5839–5846.

(68) Xu, Y.; Shelton, W.A.; Schneider, W.F. The thermodynamic equilibrium compositions, structures, and reactivities of Pt_xO_y ($x = 1$–3) clusters predicted from first principles. *J. Phys. Chem. B* **2006**, *110*, 16591–16599.

(69) Perdew, J.P.; Chevary, J.A.; Vosko, S.H.; Jackson, K.A.; Pederson, M.R.; Singh, D.J.; Fiolhais, C. Atoms, molecules, solids, and surfaces: Applications of the generalized gradient approximation for exchange and correlation. *Phys. Rev. B* **1992**, *46*, 6671–6687.

(70) Nosé, S. A unified formulation of the constant temperature molecular dynamics methods. *J. Chem. Phys.* **1984**, *81*, 511.

(71) Henkelman, G.; Arnaldsson, A.; Jónsson, H. A fast and robust algorithm for Bader decomposition of charge density. *Comp. Mat. Sci.* **2005**, in press.

(72) Bader, R.F.W. *Atoms in Molecules—A Quantum Theory*; Oxford University Press: Oxford, 1990.

(73) McQuarrie, D.A. *Statistical Mechanics*; Harper & Row: New York, 1976.

(74) Alexeev, O.S.; Li, F.; Amiridis, M.D.; Gates, B.C. Effects of adsorbates on supported platinum and iridium clusters: Characterization in reactive atmospheres and during alkene hydrogenation catalysis by x-ray absorption spectroscopy. *J. Phys. Chem. B* **2005**, *109*, 2338–2349.

CHAPTER 11

Adsorption of Probe Molecules on Nanostructured Oxides

JAMES A. ANDERSON and RUSSELL F. HOWE

Department of Chemistry, King's College, University of Aberdeen, Aberdeen AB24 3UE

11.1. GENERAL INTRODUCTION

The use of probe molecules to extract information about a solid surface is a well-established and widely practiced approach and involves the combination of a simple, usually gaseous adsorptive and a spectroscopic technique, which is sensitive to the detection of the molecule in its adsorbed form. Electronic and geometric characteristics of the adsorption site are reflected in particular modifications to the adsorbed molecule that are then detected by the application of an appropriate spectroscopic technique that is sensitive to these modifications, which may take the form of changes in bond angles, bond distances, polarisability, dipole moment, distribution of electronic charge and so on. By far the most commonly applied technique used in conjunction with probe molecules is Fourier transform infrared (FTIR), although the use of this procedure employing Raman, electron paramagnetic resonance (EPR), nuclear magnetic resonance (NMR), and X-ray techniques is also commonplace, and these will all be reviewed in this chapter. As the particular probes employed are, to a great extent, technique specific, this chapter will look at each of the common techniques and their associated probes in turn. Numerous existing reviews have dealt with this particular subject matter, and we will draw the reader's attention to these at relevant points throughout the chapter. However, few, if any, to the best of our knowledge, focus on the use of this approach to extract fundamental information specific to the peculiarities of nanostructured oxide materials, and this will be the main objective of this chapter.

Synthesis, Properties, and Applications of Oxide Nanomaterials, Edited by José A. Rodríguez and Marcos Fernández-Garcia

11.2. CHEMISORPTION, ADSORPTION MECHANISM, AND SURFACE SITES

Unlike the situation with probe molecule chemisorption on supported metal surfaces (1), adsorption on oxide surfaces is dominated by the ionic nature of the solid that leads to the predominance of interactions that are acid/base and donor/acceptor in nature (2). Adsorption may involve exposed cations that act as Lewis acid centers and interact with probe molecules such as water or ammonia that act as donors of electrons, whereas exposed oxide anions may exhibit basic character and accept protons after heterolytic dissociation of the adsorbate. Oxide anions may result in oxidation of the adsorbate with a subsequent reduction of the adsorbent. The oxidation/reduction process may be viewed in terms of change in oxidation state of the adsorbate with release or capture of electrons or in terms of a transfer of oxygen such as occurs when the adsorptive is an organic molecule such as an alcohol or an aldehyde (2). Loss of oxygen ions, which become transferred to an adsorbate as a neutral oxygen atom, imply a localized decrease in the oxidation state of the cations in the surface. The ability of an oxide to release oxygen is a function of the degree of coordinative unsaturation that influences the strength to which the lattice holds an ion. Similarly, the acidic and basic properties of the oxide are likely to be heavily modified as the extent of coordinative unsaturation increases. This coordinative unsaturation is a key feature to small nanocrystallites and thus is likely to play an important part in the extent to which a probe molecule might distinguish between these types of structures and the sites on a stoichiometric low-index extended surface. The following sections outline the sensitivity of various adsorbates to these distinguishing features of nanostructured oxides, which we will review on a technique basis.

11.3. FTIR

As outlined, this is the most widely employed technique and is probably the procedure that involves the greatest number of potential probe molecules (3). However, in addition to the use of probe molecules to extract information indirectly regarding particular characteristics of nanostructured oxide, the technique may also be used directly to obtain information pertinent to nanostructure, particularly in the far-IR region. Spectra of these structures are often dominated by surface mode absorptions, whereas transverse optical modes corresponding to bulk absorptions are often indiscernible. For example, in a study of NiO of particle size range 3–16 nm, weak absorption bands in the 60–100-cm^{-1} region were ascribed to surface-induced transverse acoustic modes (ϖTA), which became infrared active due to the breakdown of translational symmetry in the nanocrystallites (4). The use of basic probe molecules to determine information regarding surface acid sites on oxide materials is well established following the pioneering papers involving ammonia (5) and pyridine (6). Similarly, extensive studies have involved the use of CO (7) and NO (8) and small organic molecules such as methanol to determine the nature of exposed Lewis acid and general redox properties of oxide surfaces.

11.3.1. Carbon Monoxide

Although this molecule has found widespread use in the characterization of supported metals surfaces when used in conjunction with FTIR (1), it has also been employed for oxide surfaces as an effective probe to characterize cationic surface sites in terms of their nature, oxidation state, coordination environment and coordinative unsaturation, and location at faces, edges, or corners of microcrystallites (7). Although CO contains a lone pair of electrons in the 5σ orbital of C, it is only a weak electron donor molecule. The bonding of CO to transition metal-oxide surfaces is significantly enhanced by back-donation of *d* electrons and adsorption strength is increased in cases where CO can act as both a donor and an acceptor of electrons (2). Adsorption will thereby be favored by low oxidation states in the surface of the adsorbent nanoparticulate oxide. When applied to oxide surfaces terminated by surface hydroxyl groups, adsorbed CO undergoes hydrogen bonding from which information may be determined using the IR frequencies regarding the proton acid strength (7). CO may also be a reactive probe. Although it is seldom dissociated on metal-oxide surfaces, it often reacts with adsorbed or lattice oxygen to form CO_2 gas or carbonate anions. In this respect, observations from studies using powdered oxides often differ significantly from those involving surfaces of defect-limited oxides with long-range ordering such as single crystals. For example, studies of CO adsorption on single-crystal alumina surfaces by infrared absorption reflection spectroscopy did not detect evidence for CO dissociation or chemisorptive reaction (9). On the other hand, the formation of formate, carbonate and bicarbonate species on powdered aluminas is well documented (10–13). As the formation of carbonates involves a nucleophilic attack of adsorbed CO molecules by oxygen anions, the relative basicity of an oxide may be gauged by the detection of infrared bands due to these species. Strongly basic oxides such as baria readily form carbonates irrespective of whether the oxide is presented as a single crystal (2) as a complex oxide containing barium such as $YBa_2Cu_3O_{7-x}$ (2) or when presented as nanoparticulates supported on other oxides such as silica or alumina (14,15). For transition metals forming less basic oxides, the formation and detection of carbonate by FTIR following CO adsorption offers a route to the study of the evolution of basicity as a function of oxide nanoparticle size. The interaction between CO and oxide anions of MgO is described in detail later in this chapter. On other oxides, CO_2 rather than carbonate may be formed after adsorption of CO. The desorption energies of CO_2 may then be used to distinguish between release from extended, defect-free oxide surfaces and surfaces that have a high density of defect sites. For example, on ZnO (0001), CO_2 has a binding energy of 34.4 on the former and 43.6 kJ mol^{-1} on the latter (16). These may be more readily distinguished in a TPD where the former is released at ca. 120 K, where the latter requires an additional 30 K to achieve the same objective (17).

When CO adsorption is performed at room temperature, uptake can be low and limited to only the strongest of Lewis acid sites. Under these conditions, the frequency shift with respect to the equivalent vibrational mode in the gas phase (2143 cm^{-1}) may often be correlated with the strength of the surface Lewis acid sites (18,19). Such an analysis allows one to compare acid strengths of exposed cations on, for example, zirconia/sulphated zirconia and δ-alumina, where the latter were found to be greater (20). More importantly, in the context of the current review, it offers the possibility to relate nanoparticulate oxide size with trends in surface Lewis acidic properties. Adsorption at low temperatures (e.g., 77 K) reveals a series of bands in the 2250–2150-cm^{-1} range, some of which can be attributed to adsorption on sites that are unique to nanoparticulate oxide materials. For example, adsorption of CO at 77 K on three samples of tetragonal zirconia with average particle sizes of 5, 10, and 15–30 nm (21,22) gave bands at ca. 2155, 2175, and 2190 cm^{-1}, the relative intensities of which varied according to annealing temperature (and thus particle size). The band at 2155 cm^{-1} was assigned to adsorption on bridging OH groups that were thought to be localized at crystallographically defective sites such as steps, corners, and small patches of high index crystal planes. These were diminished in abundance as the crystallite size increased. Bands at 2175 and 2190 cm^{-1} were assigned to adsorption on coordinatively unsaturated (cus) surface cationic Zr^{4+} centers that acted as Lewis acid sites. The contrasting evolution of the relative abundance of the two species as a function of firing temperature/particle size led the authors to assign the former to CO adsorbed on σ-coordinated to cus Zr^{4+} centers localized in extended patches of regular, low-index crystal planes constituting the "top" termination of the flat, scale-like crystallites of t-ZrO_2 (21). The 2190-cm^{-1} band was ascribed to similar centers but where the sites were located at defective configurations that constituted the "side" termination of the crystallites. This feature, representing Lewis acid sites of greater strength, was more abundant on the sample with the smallest (5 nm diameter) particles (21), and its greater relative strength also allowed detection at ambient (room) temperature (23).

On some oxides such as MgO (24), CO may interact with both the cationic sites (Mg^{2+}) as well as the anionic surface sites (O^{2-}). FTIR spectra of CO on MgO show a degree of complexity that increases significantly in going from the perfect Mg(100) surface to powdered MgO with increasing surface area. The peculiar surface properties of high-surface-area MgO have been attributed to the presence of not only (100) terraces, edges and steps but also to inverse step and corner sites where three-, four-, and five-coordinated Mg cations are found. As the surface area increases, the relative proportions of three-coordinated/four-coordinated and four-coordinated/five-coordinated Mg^{2+} ions increases, and these are responsible for the presence of different IR bands between 2200 and 2100 cm^{-1} due to different CO adducts. The formation of $(C_nO_{n+1})^{2-}$ species that lead to the high complexity of CO absorption spectra in the region 1650–1150 cm^{-1} originate from a common CO_2^{2-} precursor species formed by attack of CO on three and four coordinated O^{2-} ions present at defects sites such as kinks and edges and corners (9). The presence of three and four coordinated Mg cations in nanoparticulate MgO (4–6 nm) leads to a CO desorption temperature of 190 K compared with 60 K from a (100) single-crystal face cleaved in vacuum (24). Nanoparticles of MgO or CaO may be prepared by sol-gel methods that create nanoparticles of the oxides with polyhedral or hexagonal shapes that produce structures with more defects than the standard cubic structure (25), thus influencing the interactions the solid is likely to undergo with probes such as CO (26).

Much of the interest regarding interactions between nanostructured oxides and CO is related to the need for appropriate gas-sensing devices and adequate air pollution monitoring in addition to a desire to produce materials that can catalytically convert CO to CO_2 at low temperatures for space shuttle and other applications. In terms of gas-sensing devices, the adsorption of CO on an activated titania surface resulted in a decrease in the transmitted IR energy, suggesting an increase in electrical conductivity (27). This adsorption may result in reduction of exposed titanium cations to Ti^{3+} centers, although this reducing effect is suppressed in the presence of water (27).

The interaction of CO with cobalt oxides has been studied owing to the high activity of these oxides for CO oxidation as well as other reactions. Both CoO and CoO–MgO solid solutions have received attention in terms of the activity of edges, steps, and extended faces with CO (28–31). As observed for TiO_2 (27), adsorption of CO on Co_3O_4 is strongly influenced by the presence of moisture (32). An IR band at 2060 cm^{-1} on the dry sample due to adsorbed CO at Lewis centers was not consistent with expectations for adsorption on five-coordinate Co(II). The low frequency was attributed to back-donation of electronic density from surface states. This feature was absent when coadsorbed with water, suggesting a similar adsorption center for both species. Defect and step sites were thought to stabilize various oxygen species, including CoO_2 (985 cm^{-1}), O–Co–O (974 cm^{-1}), and O–Co–O–Co–O (934, 944, 955 cm^{-1}) (32). The role of isolated and interacting oxygen vacancy defects is significant here given that the CoO(100) plane is effectively inert to oxygen adsorption (33). In the case of CoO–MgO solid solutions (29,30), these oxygen species may be stabilized by surface Mg^{2+} ions. Given the FTIR evidence for weak adsorption of CO, yet high CO oxidation activity over Co_3O_4 (32), it might be argued that the presence of coordinatively unsaturated Mg^{2+} ions in a CoO–MgO solid solution would be detrimental for the reaction given that considerably greater adsorption strength of three- and four-coordinated Mg cations compared with those in a (100) crystal surface (24).

11.3.2. Nitric Oxide and Other Oxides of Nitrogen

Nitrogen oxides adsorb in an even more diverse range of binding modes and central atom oxidation states on oxide surfaces than CO. Adsorption of NO on oxide surfaces may create the nitrosonium ion, NO^+, or dissociate to yield N_2O plus O. Nitrite formation is also possible. NO_2 may yield nitrate, which may coordinate to the oxide surface in a range of structures. N_2O is only very weakly adsorbed on oxide surfaces. An understanding of the diversity of adsorbed NOx species and the interpretation of their IR spectra may be gained through the reviews of Hadjiivanov (8) and Kung and Kung (34). The use of NOx as an adsorption probe of nanostructured oxides using FTIR is of considerable relevance given the use of such oxide materials in catalysts for NOx reduction (35), N_2O decomposition (36), and NOx detection for indoor sensor applications (37). The nanoparticulate Ga–Al–Zn complex oxide is effective in the simultaneous removal of *N*-containing polycyclic aromatic compounds (NPACs) present in diesel-engine emissions and lean NOx (35). The reduction of nitrates from solution is known to be effectively catalysed by nanoparticulate Fe/Fe_2O_3 (38). In the case of NOx reduction using hydrocarbons, Chen et al. (39) have found that small Fe_2O_3 particles, electron-deficient ferric oxide nanoclusters, isolated iron ions, and oxygen-bridged iron binuclear complexes all coexist in Fe/MFI catalysts. These Fe species could be largely distinguished by their FTIR spectra of adsorbed NO. Small Fe_2O_3 particles and ferric oxide nanoclusters could not strongly adsorb NO or NO_2. Binuclear oxo-ions were thought to be the active centers for the oxidation of NO to NO and for the formation of nitro and NO_3^- groups. The observation that small particulates of Fe_2O_3 did not strongly chemisorb either NO or NO_2 is consistent with conclusions of earlier IR studies (40,41). Joyner and Stockenhuber (42) used EXAFS to examine Fe/MFI catalysts prepared by various ion exchange methods and found isolated iron ions, iron oxo-nanoclusters, and microscopic iron-oxide particles coexisting in their samples. On the basis of their quantitative extended X-ray absorption fine structure analysis, they interpreted results in terms of iron oxo nanoclusters of an average composition of Fe_4O_4. These clusters have short Fe–Fe distances, and these sites were proposed as the active sites for the SCR of NO.

11.3.3. Water

Water is potentially a useful probe of oxide surfaces because it has a large dipole moment and possesses a lone pair of electrons through which it may interact with exposed metal ions (2). Strong adsorption at steps and defects may involve dissociation unlike adsorption on defect-free low-index surface planes where adsorption may be weak (2). The sensitivity of dissociation to defect sites may offer opportunities to determine characteristics unique to nanoparticulate oxide materials. Water is generally found to be more strongly adsorbed at Lewis acid sites than CO (32). A study of titania nanoparticles of dimensions 2.5–5 nm (43) showed that such small particles consisted of a Ti_2O_3 overlayer covering a rutile phase core. FTIR of the OH stretching region showed bands at 3635, 3645, and 3680 cm^{-1} due to tetrahedral coordinated vacancies and a band at 3400 cm^{-1} due to associated hydroxyl groups. The unique features were two strong absorbances at 3750 and 3840 cm^{-1}, which were attributed to octahedral vacancies and are absent in investigations of bulk rutile (44). The authors suggest that these features are indicative of defect attributes that develop as the nanoparticle size reduces and related these to a surface relief structure in the form of the surface Ti_2O_3 overlayer. Other studies of the same oxide (21-nm particles) suggest that the interactions with water are sensitive to pretreatment conditions (27). The surface affinity of titania nanoparticles is important in terms of the latter's gas sensing properties and FTIR of hydroxyl groups is often used to follow surface modifications, e.g., following grafting using hexamethyldisilazane (45). The FTIR studies of the adsorption of water on alumina nanoparticles revealed the presence of coordinatively unsaturated cations and anions, which acted as Lewis acid and Lewis base sites, respectively (46). FTIR studies of nanocrystalline mixed oxides such as $Co_{0.2}Zn_{0.8}Fe_2O_4$ (5–65 nm) appear to show that small particles retain water until higher than expected temperatures (47).

11.3.4. *N*-Containing Bases

Nitrogen-containing bases such as ammonia and pyridine are widely used to distinguish between Lewis and Brønsted acidity on oxide and supported oxide surfaces

(5,6,34). Pyridine is often preferred as an IR probe molecule of finely divided metal-oxide surfaces, because it is (*1*) more selective and stable than NH_3, (*2*) more strongly adsorbed than CO and CH_3CN, and (*3*) relatively more sensitive to Lewis acid site strength than NO (48). Unlike other bases, its molar absorption coefficients are often determined, thus allowing quantitative analysis of site densities (49,50). Lewis acid sites are detected by coordinated molecules, which give rise to IR-absorption(s) due to the ν8a mode of νCCN vibrations at 1630–1600 cm^{-1}, where the higher the frequency, the stronger the assumed acidity of the site. Formation of protonated species are identified by (*1*) an ν8a-absorption at ca. 1630 cm^{-1}, (*2*) a ν19b-absorption at 1550–1530 cm^{-1}, as well as (*3*) νN^+-H and δN^+-H absorptions occurring at ca. 2450 and 1580 cm^{-1}, respectively, indicative of Brønsted acidity (48). Additionally, pyridine may undergo surface reactions at temperature above 298 K that may be used to determine the characteristics of surface base sites (OH^- and O^{2-}) that may be involved in acid–base pair site formation. In the gas phase, from which pyridine and ammonia are generally administered, the basicity of pyridine is less than that of ammonia (13), although this is reversed in the liquid phase where solvation effects are involved. Another fundamental difference between the two adsorbates lies in the molecular dimensions, those being 0.127 and 0.313 nm^2 for ammonia and pyridine, respectively (34). Infrared bands at ca. 1450 (δ as NH_4^+) and ca. 1610 cm^{-1} (δ as NH_3) are generally used to indicate the detection of Brønsted and Lewis acid sites, respectively.

The generation of acid sites on mixed metal oxides is described by the model of Tanabe (51) and involves charge imbalance at M_1–O–M_2 sites, where M_1 and M_2 refer to the component metal species. As the electronic properties of nanoparticulate oxides are expected to differ from their bulk counterparts, this charge imbalance might also be expected from single-component oxides. The relationship between particle size and surface acidity for titania was studied by Nishiwaki et al. (52) for particle sizes, which ranged from ca. 5 to 150 nm. They used a series of pH indicators with pK_a values ranging from 4.0 (1-phenylazo-2-naphthylamine) to −5.6 (Benzalacetophenone) that detected both Brønsted and Lewis acid sites. Results suggest that the acid strength increased with decreasing particle size, suggesting the creation of new, strong acid sites on the smallest oxide particles (5 nm). The creation of these sites was attributed to the presence of many oxygen vacancies that generated large numbers of dangling bonds with energy levels localized in the band-gap region between the valence and conduction bands. The local charge imbalances discussed above were related to electrons trapped in these levels. Divergence from the Tanabe model (51) for mixed oxides in terms of the predicted relative proportions of Brønsted and Lewis acid sites (50) may find some relevance in such a study.

11.3.5. Other Probes

Another feature of nanoparticulate oxides results from their small size and, consequentially, high surface area. This gives rise to large adsorption capacities, normally only found for carbons and other ultra-high-surface-area materials. However, Kahleel et al. (53) attributed the high adsorption capacity of MgO not to a surface area effect

(SA = ca. 350 m^2g^{-1}), but also due to enhanced surface reactivities that are due to the unusual crystal shapes (polyhedra) with a high ratio of edge and corner surface sites that are inherently more reactive toward incoming adsorbates. FTIR was used to study acetaldehyde adsorption on MgO, and results were interpreted in terms of multilayer dissociative adsorption, which took place as a result of surface interaction with the carbonyl bond followed by aldehydic hydrogen dissociation. Samples showed higher adsorptive capacity than carbon or commercial magnesia. Other organic compounds, including propionaldehyde, trimethylacetaldehyde, benzaldehyde, acetone, dimethylamine, *N*-nitrodimethylamine, and methanol, were also investigated in a similar fashion and found to give similar results (53).

11.4. RAMAN

In terms of application in the area of oxide nanomaterials, Raman spectroscopy is most widely employed in studies where bulk analysis is performed rather than through its use with probe molecules and the study of surface-specific characteristics. This is a consequence of the technique's high sensitivity to the phonon characteristics with respect to the crystalline nature of the solid. The phonon confinement model is most widely used to describe the spectral features, with the phonon momentum being of the order of $1/d$, where d is the particle diameter for nanocrystalline materials. This leads to shifts to lower frequencies in the transverse optic (TO) vibrational features as the size of the nanocrystals are decreased. For example, Ge crystallites of between 2.3 and 5 nm embedded in ZnO give a Raman active feature at 300 cm^{-1} attributed to a Ge–Ge TO mode, which is red-shifted as the particle size is decreased (54). Further details are given in a recent review that encompasses this subject (24). The use of selective nanovolume Raman spectroscopy that allows spectral selection of nanoparticles would appear to be a promising approach for the study of such materials (55).

In terms of adsorption studies, titania has been one of the more widely studied nanoparticulate oxides with reporting of both rutile and anatase as adsorbents (56–58). Thin layers of anatase particles deposited on gold substrates previously roughened through electrochemical oxidation and reduction cycles can give rise to surface-enhanced Raman scattering (SERS) effects. The electromagnetic contribution to the enhancement extends spatially toward the first layer(s) of the TiO_2 nanoparticles, resulting in enhanced intensity, not only of the bands corresponding to bulk anatase, but also of those corresponding to vibrations of molecules adsorbed on the semiconductor nanoparticles (58). Even though the authors found enhanced intensities of features due to adsorbed phthalic acid on their anatase nanoparticles of 30 nm diameter, they conclude that further benefits would be obtained by a reduction of particle size as a consequence of the relationship between the SERS enhancement factor and the distance from the gold surface (58). The presence of uncoordinated titanium atoms at the surface of nanocrystalline titania facilitate the binding of adsorbates such as *tert*-butylpyridine (57) through the nitrogen atom, although in the presence of water, acetonitrile, or methanol, competition for such sites made adsorption of glycine unfavorable (56). Carboxylic acids in the deprotonated form were thought to bridge two such uncoordinated Ti atoms (56).

11.5. EPR SPECTROSCOPY

EPR spectroscopy is an extremely high-sensitivity spectroscopic technique for observing species that are paramagnetic. In the context of oxide nanomaterials, these include transition metal ions in bulk oxide phases or as dopants substituted into diamagnetic hosts, bulk and surface defects (trapped electrons, anion vacancies, interstitial transition metal ions), supported transition metal ions, and adsorbed paramagnetic molecules or ions. In this section, attention is focused on adsorbed species that are either paramagnetic or interact with paramagnetic surface sites.

For further background on EPR spectroscopy, the reader is referred to several textbooks and review chapters (59–62). All published studies of species adsorbed on oxide nanomaterials to date have used conventional field-swept CW–EPR spectroscopy. Pulsed EPR techniques (63) are, however, now commercially available, and they have been successfully applied, for example, to study adsorption at transition metal centers in zeolites (64). The double resonance technique of ENDOR can also be extremely powerful in probing adsorbate metal ion interactions, but it has been most successful when applied to well-defined zeolite systems (65). This section will consider further only the more routinely applicable CW–EPR technique.

In the conventional CW–EPR experiment, the two pieces of spectroscopic information derived from observed spectra are the g-tensor and the hyperfine coupling tensor. As experiments with oxide nanomaterials are inevitably performed on polycrystalline powders rather than single crystals, the spectra observed represent the sum of spectra from individual crystallites, and the resulting powder patterns are never as well resolved as single-crystal spectra. Reliable extraction of all spectroscopic parameters with high accuracy can therefore be difficult, although computer simulation of the observed spectrum as a check on the reliability of the parameters extracted is usually helpful. The g-tensor contains information about the extent of coupling between the electron spin angular momentum and the orbital angular momentum, and hence, it depends on the nature of the orbital containing the unpaired electron. The hyperfine coupling tensor describes interactions between the electron spin and nearby nuclear spins. The g- and hyperfine tensors together should identify uniquely the paramagnetic species being observed, and they provide information about its interaction with its surroundings. One particular strength of EPR as a surface technique is its sensitivity: Surface species can be detected down to concentrations of 10^{14} g^{-1} or less. The method can also be applied in principle in an in situ manner, at least for processes occurring at temperatures below about 573 K. However, maximum sensitivity is achieved by recording spectra at as low a temperature as possible, and there may be reasons why spectra cannot be detected at all above 77 K.

To illustrate how an observed EPR spectrum can be related to the nature of the species concerned, we consider the superoxide ion O_2^-. This species has been observed on many different oxide surfaces, and a comprehensive review of the older literature

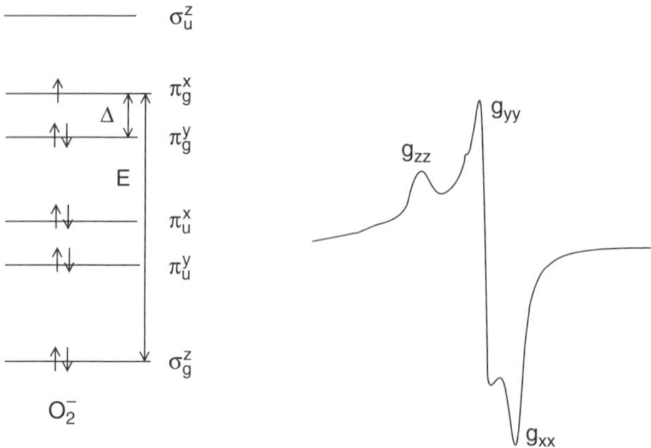

Figure 11.1. Energy level diagram and EPR spectrum of the superoxide ion O_2^-.

has been given by Che and Tench (66). In a simple ionic model of O_2^-, the unpaired electron occupies one of two degenerate π anti-bonding molecular orbitals. When the O_2^- is adsorbed on a surface, the degeneracy of the π orbitals is lifted, as shown in Figure 11.1. Along the direction of the orbital containing the unpaired electron (taken as the x-axis in Figure 11.1) there is to first order no coupling between the electron and the orbital angular momentum, so that the position of resonance for those radicals who have their x-axis aligned with the magnetic field is at the free-electron value:

$$H = h\nu/g_e\beta \quad \text{or} \quad g_{xx} = g_e$$

Along the y- and z-directions however, spin orbit coupling causes the value of g to increase. In a simple first-order treatment $g_{yy} = g_e + 2\lambda/E$ and $g_{zz} = g_e + 2\lambda/\Delta$, where λ is the spin orbit coupling constant and Δ and E are the energy level separations shown. Note in particular the sensitivity of the g_{zz} tensor component to the splitting of the π orbitals. Thus, radicals with their y or z axes aligned along the magnetic field will give signals at lower magnetic field than the free electron value. In a polycrystalline sample, all possible orientations of the radical are present, but the principal components of the g-tensor g_{xx}, g_{yy}, and g_{zz} can still be determined from the turning points (maxima, minima, and points of inflection) in the first derivative of the absorption spectrum.

If the unpaired electron interacts with a nuclear spin I, the EPR spectrum is expected to show hyperfine splitting into 2I + 1 lines. In the case of powder spectra, there will in general be a set of 2I + 1 lines associated with each of the three components of the g-tensor, corresponding to the three principal components of the hyperfine coupling tensor (strictly speaking, this is true only if the g- and hyperfine tensors have coincident principal axes, which is a commonly made assumption). In the case of O_2^-, the naturally abundant isotope ^{16}O has I = 0, so that no hyperfine splitting is seen. The radical can, however, be generated with ^{17}O-enriched oxygen; in this case, I = 5/2, and the spectrum of adsorbed O_2^- should show 3 sets of 11 lines if both oxygens are equivalent (n equivalent nuclei of spin I will give $2n\text{I} + 1$ lines), or 3 sets of 36 lines (6 lines each split into a further 6) if the oxygens are not equivalent. In practice, the spectrum will be further complicated by the presence of singly labeled and unlabeled oxygen species if the level of enrichment is less than 100% (as is always the case). On the other hand, the ^{17}O hyperfine coupling tensor for O_2^- is highly anisotropic, with large coupling only along the x-direction, so that only one set of hyperfine lines for each species is normally resolved, centered on g_{xx}. Figure 11.2 shows a typical spectrum of O_2^- obtained with ^{17}O enriched oxygen, obtained in this case by adsorption of oxygen on a silica supported molybdena catalyst (67). This shows the superposition of three signals from doubly, singly, and unlabeled O_2^-, respectively. The hyperfine patterns obtained show clearly that in this case, both oxygen atoms in the O_2^- are magnetically equivalent and interacting equally with the oxide surface. [On other surfaces, inequivalence is attributed to a partial covalent bonding to the surface (66)].

The EPR spectra of adsorbed radicals may show additional hyperfine coupling resulting from interaction of the electron spin with a nuclear spin or spins on surface ions on which the radical is adsorbed. For example, Figure 11.3 shows the spectrum of $^{16}O_2^-$ adsorbed on a ^{95}Mo enriched MoO_3–SiO_2 catalyst. In this case, the surface

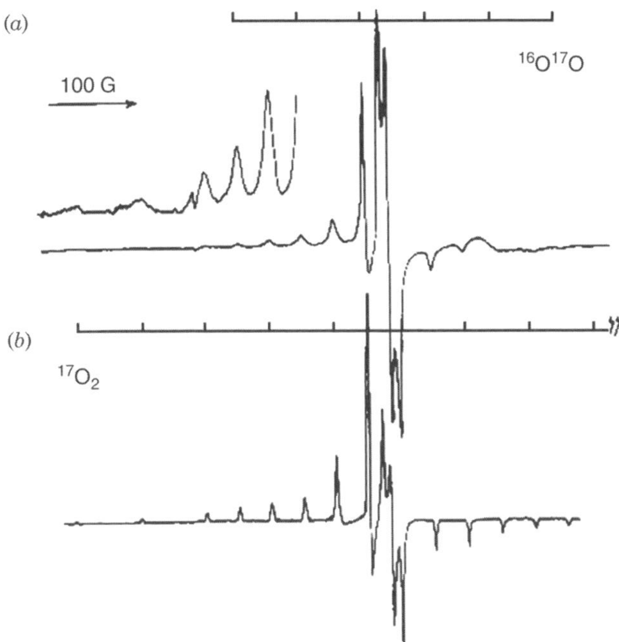

Figure 11.2. EPR spectrum of O_2^- generated with ^{17}O enriched oxygen adsorbed on a molybdena-silica catalyst. (a) observed; (b) simulated. Reprinted from Ref. 67 with permission of the Royal Society of Chemistry.

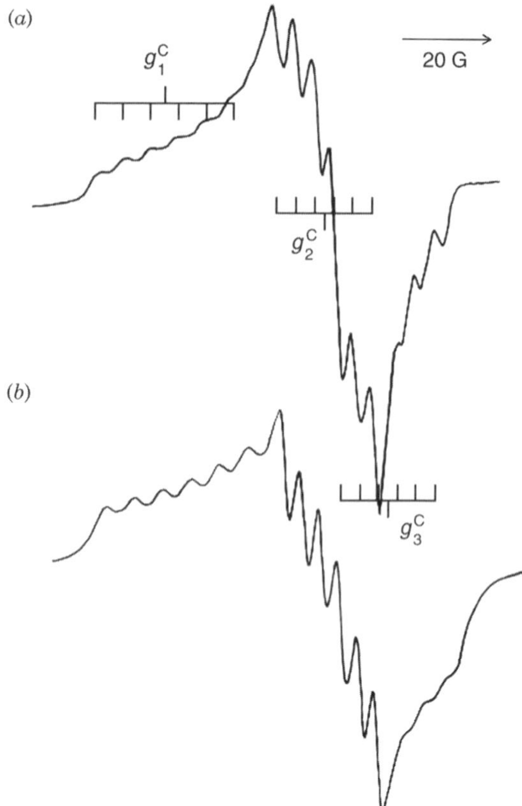

Figure 11.3. EPR spectrum of O_2^- generated by adsorption of oxygen on a ^{95}Mo enriched molybdena-silica catalyst. (*a*) observed; (*b*) simulated. Reprinted from Ref. 67 with permission of the Royal Society of Chemistry.

site on which O_2^- is adsorbed is Mo^{6+}. ^{95}Mo has $I = 5/2$, so each of the three *g*-tensor features in the O_2^- spectrum is split into 6 line, confirming the nature of the adsorption site.

Analysis of the *g*- and hyperfine tensor components from spectra such as those shown here not only identifies the particular radical formed, but also provides detailed information about the electronic structure and the nature of any interaction with surface sites. Signal intensities, when properly integrated and calibrated, can be used to monitor the extent of electron transfer or other radical forming processes as a function of experimental parameters.

11.5.1. Probes Used with Nanocrystalline Magnesium Oxide

EPR spectroscopy has been used by several groups to study surface defects and adsorbed species on nanocrystalline magnesium oxide. Magnesium oxide has a simple cubic structure and is highly ionic, allowing in principle an easier understanding of its surface structure and reactivity than might be the case for more complex mixed or partially covalent oxides. Many studies have shown that the reactivity of oxide surfaces in general is strongly influenced by the presence of extended and point defects, including low coordination sites, impurity atoms, and oxide ion vacancies. EPR studies of MgO, coupled with other spectroscopic probes and theoretical calculations, have allowed a detailed picture of the surface to be established. The reader is referred to a comprehensive review article by Spoto et al. (68) for a full account of the past 30 years of research on the surface properties of MgO.

The number of point defects such as low coordination sites and oxide ion vacancies is expected to increase dramatically as the crystallite size of a metal oxide is reduced. Evidence for low coordination sites on highly dispersed MgO crystallites comes from ultraviolet (UV) spectroscopy and photoluminescence measurements (68). EPR spectroscopy of adsorbed NO has been used to probe these sites. NO is paramagnetic, so that weakly physically adsorbed NO is expected to give an EPR spectrum (69–71). Most adsorbed NO is dimerized, however, and not EPR visible, but low coordination metal ion sites may interact sufficiently strongly with NO to allow monomeric physically adsorbed species to be seen. The electrostatic field due to a surface cation will lift the degeneracy of the pi anti-bonding orbitals of NO in a similar manner to that described above for the superoxide ion. In the case of NO, it is the high-field g-tensor component (g_{zz}), which is most sensitive to the electrostatic field, and this shifts to lower field as the orbital degeneracy is further lifted. Martino et al. resolved three distinct forms of physisorbed NO on a nanocrystalline MgO ($300\,m^2g^{-1}$) at 77 K, which they attribute to three different exposed cation sites with different electrostatic field strengths (71). (All three have similar values of $g_{xx} \sim 1.995$, $g_{yy} \sim 1.997$, but g_{zz} varying between 1.89 and 1.96.)

At higher temperatures, a paramagnetic form of chemisorbed NO is seen, identified as a NO_2^{2-} radical formed by reaction of NO with low-coordination, highly reactive O^{2-} surface sites. This species is distinguished from the physisorbed NO by having g-tensor components greater than the free electron value, and it has been shown that this is consistent with theoretical calculations for a bent NO_2^{2-} species having two nearly equivalent N–O bond lengths (70).

Dissociative adsorption of hydrogen on MgO is known from FTIR studies to involve heterolytic H–H bond cleavage to form Mg–H and O–H species (71). When UV light and adsorbed hydrogen are both present, at least four different paramagnetic species have been described by Sterrer et al. (72): a so-called F_S^+(H) species, which is a trapped electron in close proximity to a surface hydroxyl group, showing ^{1}H hyperfine coupling, and three different F_S^+ species corresponding to electrons trapped on different sites not coupled to adjacent protons.

The trapped electron species present on pre-irradiated and/or reduced MgO are highly reactive. For example, exposure to oxygen results in formation of the superoxide ion O_2^- through electron transfer. Chiesa et al. (73) have recently reported an in-depth study of this species formed on a highly ordered nanocrystalline sample of MgO ($300\,m^2g^{-1}$), using ^{17}O isotopic labeling to assist in the complete characterization. By working at low oxygen coverages, these workers could obtain significantly better quality spectra than had been possible previously. Figure 11.4 shows, for

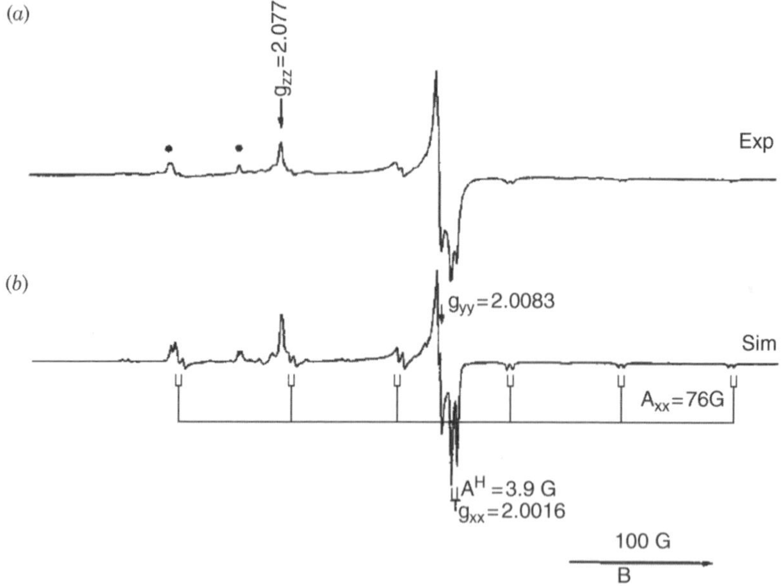

Figure 11.4. EPR spectrum of O_2^- formed by adsorption of ^{17}O enriched oxygen on MgO. Reprinted from Ref. 73 with permission of the American Institute of Physics.

example, the spectrum obtained when 28% ^{17}O enriched oxygen was employed. This shows the 6-line hyperfine splitting centered on g_{xx} expected for a singly labeled O_2^- (at this level of enrichment, the doubly labeled species is not seen). Each ^{17}O hyperfine component is further split into a doublet due to a superhyperfine coupling to nearby protons . This was confirmed by showing that the doublet splitting disappeared when the MgO surface was pretreated with D_2 rather than H_2 (the nuclear moment for deuterium is approximately six times smaller than that of the proton).

Paramagnetic adsorbed radicals are also produced when trapped electrons on nanocrystalline MgO are exposed to other gases. For example, the trapped electron color centers are bleached by exposure to N_2O through the reaction;

$$(e^-)_{trapped} + N_2O \longrightarrow O^- + N_2$$

The O^- radical anion formed gives a characteristic EPR signal whose g-tensor components (particularly $g_{perpendicular}$) are strongly sensitive to the local crystal field environment. Di Valentin et al. (74) identified up to five different O^- species on a high-surface-area nanocrystalline MgO, which were attributed to quenching of different trapped electron sites on the MgO surface. Theoretical modeling of possible surface sites using an embedded cluster model is described. The presence of the O^- species at the surface of the MgO and not in the bulk was confirmed by their instantaneous reaction with gas phase O_2 to form the paramagnetic O_3^- radical anion.

A more unusual surface species is formed when trapped electron centers on MgO are exposed to dinitrogen. In this case, a reversible electron transfer occurs at low temperature to form the N_2^- radical anion (75). A single well-resolved signal is observed, which the authors take to mean that only the most energetic of the trapped electron sites are capable of forming this species. As well as the expected ^{14}N or ^{15}N hyperfine coupling from two equivalent nitrogen atoms, the N_2^- signal also shows ^1H superfine splitting, confirming that the adsorbed species is located in close proximity to the F_S^+(H) sites. A similar N_2^- species could also be generated on high-surface-area nanocrystalline CaO surfaces.

Several studies have reported the formation of paramagnetic species when CO_2 adsorbed on MgO is UV-irradiated. Teramura et al. (76) recently described the identification of CO_2^- and CO_3^- radical anions when CO_2 adsorbed on a commercial (100 m^2g^{-1}) MgO was irradiated. The scheme shown in Figure 11.5 was presented by these authors to account for the formation of these species and their rapid reaction with hydrogen or methane. They suggest that both radicals are generated from adsorbed bidentate carbonate species (detected by FTIR spectroscopy), but that it is the CO_2^- species that is active in the photocatalytic reduction of CO_2.

Nanocrystalline oxides are often prepared via a sol-gel route via alkoxide precursors. Richards et al. have shown that when nanocrystalline MgO is prepared from a magnesium methoxide precursor, EPR spectroscopy provides evidence for the presence of residual methoxy groups in the oxide (77). In particular, the nanocrystalline MgO shows two EPR signals after calcination: a singlet assigned to F^+

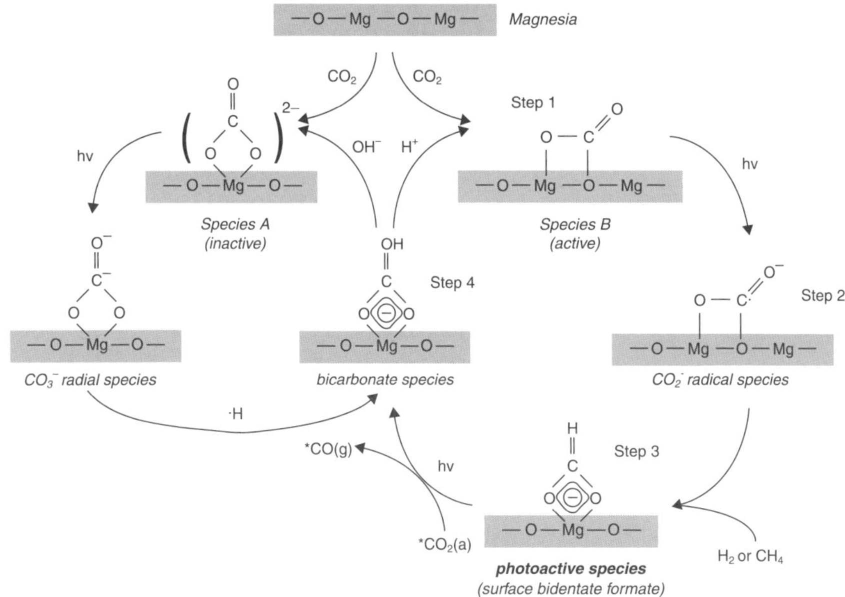

Figure 11.5. Reaction scheme for the photocatalytic reduction of CO_2. Reprinted from Ref. 76 with permission.

trapped electron centers, and a triplet (3 lines with an intensity ratio of 1:2:1 and a splitting of 18 gauss), which is assigned to O–$CH_2^{\bullet}$ radicals formed from residual methoxide groups in the sample. Both signals were observed with similar intensities in the presence and absence of oxygen. The fact that they were not broadened in the presence of paramagnetic gas phase oxygen suggests, however, that both species are located within the bulk of the nanocrystals and not at the surface.

11.5.2. Probes Used with Nanocrystalline Titanium Dioxide

The widespread use of titanium dioxide as a photocatalyst has led to numerous EPR studies of paramagnetic species formed on the titania surface, particularly under UV irradiation. UV irradiation of TiO_2 in vacuo at low temperatures produces EPR signals characteristic of Ti^{3+}, corresponding to trapped conduction band electrons, and O^-, corresponding to trapped valence band holes (78). In the presence of oxygen, however, electrons are scavenged by adsorbed oxygen to form the superoxide ion. Oxygen may also react with trapped holes (O^-) to form the ozonide ion, O_3^-. Coronado et al. have reported a comparison of adsorbed oxygen radicals formed in this way on a range of nanostructured titania samples of different average crystallite sizes, ranging from 6 to 100 nm (79). The highest concentrations of oxygen radicals were observed on samples having the smallest crystallite sizes. Five different signals were observed: two different forms of trapped holes (described as Ti^{4+}–$O^{2-}$$Ti^{4+}$–$O^-$ and Ti^{4+}–O^-–Ti^{4+}–OH^- respectively), the superoxide ion O_2^-, the ozonide ion O_3^-, and a signal assigned to the hydroperoxyl radical HO_2. This latter signal did not appear to show, however, the expected ^1H hyperfine coupling for HO_2, so that its identification remains tentative. The HO_2 radical (with ^1H hyperfine splitting about all three g-tensor components) has been observed on a TS-1 titanosilicate surface exposed to hydrogen peroxide (80), whereas on TiO_2, only two different superoxide species were detected. Decomposition of H_2O_2 is presumed to occur via hydroxyl radicals and HO_2, which on TiO_2 surfaces is deprotonated to form O_2^-:

$$H_2O_2 \longrightarrow 2HO^{\bullet}$$

$$HO^{\bullet} + H_2O_2 \longrightarrow H_2O + HO_2^{\bullet}$$

$$HO_2^{\bullet} \longrightarrow H^+ + O_2^-$$

Adsorbed oxygen radicals are widely assumed to be important in the photocatalytic oxidation of organic molecules over titania surfaces, and EPR evidence for this has recently been reported by Attwood et al. (81). These authors carried out UV irradiation of a fully dehydroxylated titania surface in the presence of acetone:oxygen mixtures and obtained the spectrum shown below in Figure 11.6. The high field signals ($g = 1.989$ and $g = 1.972$) are due to Ti^{3+} signals resulting from trapped conduction band electrons. The low field signal has an orthorhombic g-tensor, each component of which is split into a triplet due to coupling of the unpaired electron to two equivalent protons. The authors also reported the corresponding spectrum obtained with ^{17}O

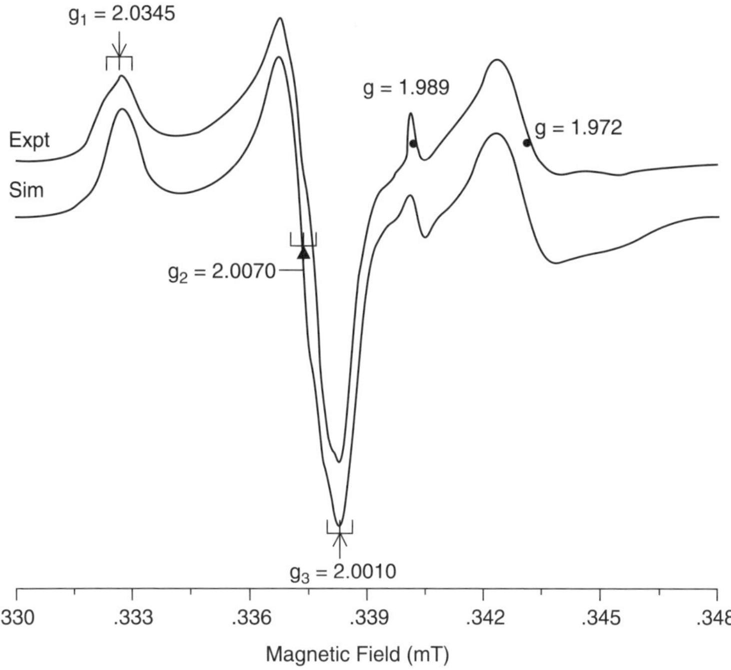

Figure 11.6. EPR spectrum of the $CH_3OCH_2OO^{\bullet}$ radical formed on irradiation of TiO_2 in the presence of oxygen and acetone. Reprinted from Ref. 81 with permission.

enriched O_2, which showed ^{17}O hyperfine coupling from two inequivalent oxygens (unlike the two equivalent oxygens typically seen for the superoxide ion).

The spectroscopic evidence together with plausible mechanistic arguments was used to assign the signal observed to the alkyl peroxy radical $CH_3COCH_2O_2^{\bullet}$, generated by hole transfer to adsorbed acetone and subsequent reaction with oxygen:

$$CH_3COCH_3 + O^- \longrightarrow CH_3COCH_2^{\bullet} + OH^-$$
$$CH_3COCH_2^{\bullet} + O_2 \longrightarrow CH_3COCH_2OO^{\bullet}$$

The peroxy radical was stable only at temperatures below 150 K. Its observation here was the first direct detection by EPR of an adsorbed reaction intermediate in the photocatalytic oxidation of a hydrocarbon.

11.5.3. Probes Used with Nanocrystalline Ceria

Cerium dioxide is an inexpensive oxide frequently incorporated into oxidation catalysts to promote combustion of hydrocarbons and oxidation of CO. As a semiconductor, it has a band gap of 2.94 eV (slightly less than that of anatase), and its potential as a photocatalyst alternative to titania has been discussed (82). Hernandez-Alonso et al. have described recently an EPR study of radicals formed on nanocrystalline CeO_2 during the catalytic photo-oxidation of toluene (83). A reductive pretreatment of the CeO_2 generates surface defects (Ce^{3+} cations coordinated to oxide ion vacancies), which subsequently adsorb oxygen molecules as superoxide species. At least four different superoxide species were observed, the relative concentrations of which varied according to the preparation conditions used for the nanocrystalline ceria. Spin trapping experiments were also carried out with aqueous suspensions of the ceria to show that, under these conditions, hydroxyl radicals are formed (the spin trapping experiment involves reaction of short-lived unstable radicals with a spin trap molecule, in this case DMPO, to form a stable-free radical that can be identified from its EPR characteristics.) However, when the catalytic performance of different ceria preparations for the gas-phase photo-oxidation of toluene was compared with the yields of trapped hydroxyl radicals from the same catalysts in aqueous suspension, no correlation was observed, leading the authors to conclude that hydroxyl radicals do not play a role in the gas-phase photocatalytic oxidation.

11.6. X-RAY ABSORPTION SPECTROSCOPY

The techniques of extended X-ray absorption fine structure (EXAFS) and X-ray absorption near-edge spectroscopy (XANES) are normally regarded as bulk characterization methods, giving, respectively, local structural and electronic information about the centers being probed. In the case of oxide nanomaterials, the fraction of metal ion centers present at the surfaces of the nanocrystallites may be sufficient, however, to allow surface coordination and reactivity to be explored. XANES spectroscopy measures electronic transitions of core-level electrons to high lying excited states; these occur as features immediately below the X-ray absorption edge of the element concerned, which corresponds to photoionization of the core-level electron. EXAFS, on the other hand, measures backscattering of photoionized electrons from neighboring atoms, which manifests itself as oscillations on the high energy side of the X-ray absorption edge. EXAFS thus provides information about the distances, numbers, and types of surrounding nearest neighbor atoms. References 84 and 85 should be consulted for further background on these two closely related techniques. Fernandez-Garcia has recently written a comprehensive review of the use of in situ XANES spectroscopy for analysis of catalyst systems under reaction conditions (86).

XAS evidence for low coordination surface atom sites in nanocrystalline titania has been reported by several groups. Bulk (well crystalline) anatase, in which Ti^{4+} is octahedrally coordinated, shows a characteristically low-intensity three-component XANES spectrum below the Ti K-edge, which has been theoretically explained in terms of dipole forbidden 1s to 3d transitions. Luca et al. showed that for a series of nanocrystalline anatase samples prepared by sol-gel methods, a fourth band characteristic of five-coordinate titanium appeared and increased in relative intensity as the average crystallite size decreased (87). Rajh et al. subsequently confirmed the assignment of this band to low-coordination surface titanium sites by showing that the band was removed and the original three-band XANES spectrum characteristic of octahedral Ti^{4+} was restored when nanocrystalline titania surfaces were exposed to ascorbic acid, which coordinates to Ti^{4+} (88).

The presence of low-coordination surface Ti^{4+} in nanocrystalline titania can also be seen in the Ti K-edge EXAFS. The difficulty here is that the observed EXAFS is a superposition of the EXAFS spectra from bulk and surface atoms, so that extraction of quantitative data for coordination numbers and bond lengths is not easy. Figure 11.7 shows a comparison of the Ti K-edge XANES and EXAFS for a 50-nm anatase sample, a 1.9-nm nanocrystalline titania sample, and the 1.9-nm sample after adsorption of ascorbic acid (taken from Ref. 88). The reduction in intensity of the XANES peak due to five-coordinate Ti^{4+} on adsorption of ascorbic acid and the corresponding increase in intensity of the first peak in the EXAFS due to the first coordination shell oxygen atoms are clearly visible. It can be seen that the Ti–O nearest neighbor distance in the nanoparticles is significantly shorter than in anatase, and this is unchanged on adsorption of ascorbic acid. The intensities of the next nearest neighbor and more remote Ti–O peaks in the EXAFS are greatly reduced in the nanocrystalline sample, and this difference is also unaffected by ascorbic acid adsorption.

There are few examples of the application of XANES and EXAFS spectroscopy to the direct observation of adsorbed molecules. The X-ray absorption edges for light elements lie in the soft X-ray region below 4 keV, where UHV techniques must be employed, and working with nanocrystalline powders is difficult. One recent successful application is the study by Rodriguez et al. of mixed ceria-zirconia

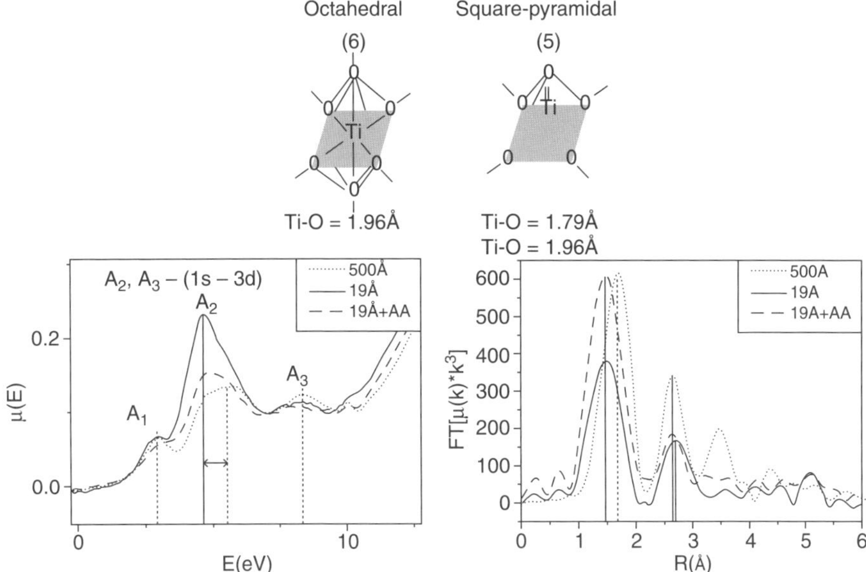

Figure 11.7. Ti K-edge XANES and EXAFS spectra of nanocrystalline anatase before and after adsorption of ascorbic acid. Reprinted from Ref. 88 with permission.

nanoparticles (89). Figure 11.8 compares sulphur K-edge XANES spectra obtained by adsorption of SO_2 onto three different nanocrystalline ceria-zirconia oxides and on to single-crystal surfaces. On the single-crystal surfaces, SO_2 adsorbs as sulphite and sulphate species through reaction with surface oxide ions. On the nanocrystalline surfaces, on the other hand, there are oxide ion vacancies and metal ions exposed at edge and corner sites that can fully dissociate SO_2 to form metal sulphide species.

The increasing availability of soft X-ray beam lines with capability for handling high-surface-area polycrystalline samples means that many more applications of this technique for studying adsorbed probe molecules may be anticipated. A conventional hard X-ray beam line has been used by Jing et al. to study the adsorption of organic arsenic species on nanocrystalline titania (90). The arsenic K-edge occurs at around 11.8 keV, which means that UHV methods are not required to observe the adsorbed species. Both XANES and EXAFS data were measured to show that monomethylarsenic acid forms bidentate surface complexes on the titania surface, whereas dimethylarsenic acid forms a monodentate complex. The authors used these

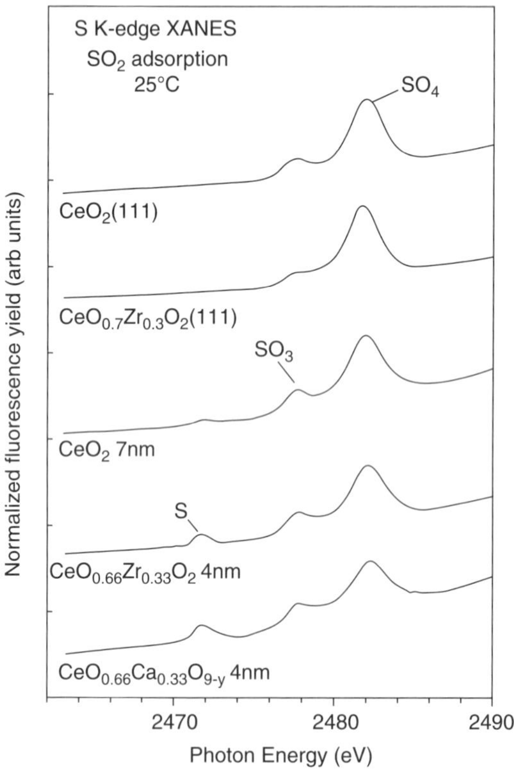

Figure 11.8. Sulphur K-edge XANES spectra of SO_2 adsorbed on nanocrystalline ceria-zirconia and on single-crystal surfaces. Reprinted from Ref. 89 with permission from Elsevier.

results to build a model for multisite complexation of arsenic species present in contaminated ground waters onto titania as a possible remediation pathway.

A recent advance in this area is the development of a combined X-ray absorption and high-resolution X-ray fluorescence spectroscopy for the study of multicomponent oxide catalysts by the group of Iwasawa (91). This technique is capable of observing XANES and EXAFS with high sensitivity from elements present at low concentrations in the presence of high-concentration elements with similar atomic numbers (e.g., V and Ti). Reference (91) describes the use of this technique to characterize vanadium coordination and its modification by adsorbed species in titania-supported vanadium catalysts containing as little as 0.6wt% vanadium.

REFERENCES

(1) Anderson, J.A.; Fernandez-García, M. *Supported Metals in Catalysis*; ICP: London, 2005.

(2) Henrich, V.E.; Cox, P.A. *The Surface Science of Metal Oxides*; Cambridge University Press: Cambridge, 1994.

(3) Ryczkowski, J. *Catal. Today*. **2001**, *68*, 263.

(4) Biju, V.; Abdul Khadar, M. *Spectrochimica Acta Part A* **2003**, *59*, 121.

(5) Peri, J.B. *J. Phys. Chem.* **1965**, *69*, 231.

(6) Parry, E.P. *J. Catal.* **1963**, *2*, 371.

(7) Hadjiivanov, K.I.; Vayssilov, G.N. *Advanc. Catal.* **2002**, *47*, 307.

(8) Hadjiivanov, K.I. *Catal. Rev. Sci. Eng.* **2000**, *42*, 71.

(9) Hsiao, G.S.; Erley, W.; Ibach, H. *Surf. Sci.* **1988**, *405*, 465.

(10) Parkyns, N.D. *J. Chem. Soc.* **1965**, 1910.

(11) Gopal, P.G.; Schneider, R.L.; Watters, K.L. *J. Catal.* **1987**, *105*, 366.

(12) Busca, G.; Lorenzelli, V. *Mater. Chem.* **1982**, *7*, 89.

(13) Anderson, J.A.; Rochester, C.H. In *Encyclopedia of Surface and Colloid Science*; Marcel Dekker: New York, 2002; 2528.

(14) Coronado, J.M.; Anderson, J.A. *J. Molec. Catal.* **1999**, *138*, 83.

(15) Paterson, A.J.; Fernandez-Garcia, M.; Anderson, J.A. *Stud. Surf. Sci Catal.* **2000**, *130*, 1331–1336.

(16) Wang, J.; Hokkanen, B.; Burhaus, U. *Surf. Sci.* **2005**, *577*, 158.

(17) Wang, J.; Funk, S.; Burghaus, U. *Catal. Letts.* **2005**, *103*, 219.

(18) Zaki, M.I.; Knözinger, H. *Spectrochim Acta A* **1987**, *43*, 1455.

(19) Della Gatta, G.; Fubini, B.; Ghiotti, G.; Mortearra, C. *J. Catal.* **1976**, *43*, 90.

(20) Morterra, C.; Cerrato, G.; Bolis, V.; Di Ciero, S.; Signoretto, M. *J. Chem. Soc. Faraday Trans.* **1997**, *93*, 1179.

(21) Morterra, C.; Cerrato, G.; Di Ciero, S. *Appl. Surf. Sci.* **1998**, *126*, 107.

(22) Morterra, C.; Cerrato, G.; Pinna, F. *Spectrochimica Acta A* **1999**, *55*, 95.

(23) Signoretto, M.; Pinna, F.; Strukul, G.; Cerrato, G.; Morterra, C. *Catal. Letts.* **1996**, *36*, 129.

(24) Fernandez-García, M.; Martínez Arias, A.; Hanson, J.C.; Rodriguez, J.A. *Chem. Rev.* **2004**, *104*, 4063.

(25) Klabunde, K.J.; Stark, J.; Koper, O.; Mobs, C.; Park, D.G.; Decker, S.; Jiang, Y.; Lagadic, I.; Zhang, D. *J. Phys. Chem.* **2004**, *100*, 12142.

(26) Spoto, G.; Gribov, E.N.; Ricchiardi, G.; Damin, A.; Scarano, D.; Bordiga, S.; Lamberti, C.; Zecchina, A. *Prog. Surf. Sci.* **2004**, *76*, 71.

(27) Baraton, M.I.; Melhari, L. *Nanostruct. Mater.* **1998**, *10*, 699.

(28) Gugglielminotti, E.; Coluccia, S.; Garrone, E.; Cerruti, L; Zecchina, A. *J. Chem. Soc. Faraday Trans.* **1979**, *75*, 97.

(29) Zecchina, A.; Spoto, G.; Borello, E.; Giarnello, E. *J. Phys. Chem.* **1984**, *88*, 2582.

(30) Zecchina, A; Spoto, G.; Garrone, E.; Bossi, A. *J. Phys. Chem.* **1984**, *88*, 2587.

(31) Zecchina, A; Scarano, D.; Bordiga, S.; Lamberti, C. *Adv. Catal.* **2001**, *46*, 265.

(32) Grillo, F.; Natille, M.M.; Glisenti, A. *Appl. Catal. B. Environmental* **2004**, *48*, 267.

(33) Mackay, J.L.; Henrich, V.E. *Phys. Rev. B.* **1989**, *39*, 6156.

(34) Kung, M.C.; Kung, H. *Catal. Rev. Sci. Eng.* **1985**, *27*, 425.

(35) Shen, S.C.; Hidajat, K.; Yu, L.E; Kawi, S. *Catal. Today* **2004**, *98*, 387.

(36) Centi, G.; Arena, G.E.; Perathoner, S. *J. Catal.* **2003**, *216*, 443.

(37) Francioso, L.; Forleo, A.; Capone, S.; Epifani, M.; Taurino, A.M.; Siciliano, P. *Sens. Actuators B* **2006**, *114*, 646.

(38) Chen, S.-S.; Hsu, H.-D.; Li, C.-W. *J. Nanopart. Res.* **2004**, *6*, 639.

(39) Chen, H.-Y.; El-Malki, El.-M.; Wang, X.; van Santen, R.A.; Sachtler, W.M.H. *J. Molec. Catal. A: Chem.* **2000**, *162*, 159.

(40) Yuen, S.; Chen, Y.; Kubsh, J.E.; Dumesic, J.A; Topsøe, N.; Topsøe, H. *J. Phys. Chem.* **1982**, *86*, 3022.

(41) Rochester, C.H.; Topham, S.A. *J. Chem. Soc. Faraday Trans. 1* **1979**, *75*, 1259.

(42) Joyner, R.; Stockenhuber, M. *J. Phys. Chem. B* **1999**, *103*, 5963.

(43) Madhu Kumar, P.; Badrinarayanan, S.; Sastry, M. *Thin Solid Films* **2000**, *358*, 122.

(44) Suda, Y.; Morimoto, T. *Langmuir.* **1987**, *3*, 786.

(45) Baraton, M.-I.; Merhari, L. *J. Eur. Ceram. Soc.* **2004**, *24*, 1399.

(46) Dokhale, P.A.; Sali, N.D.; Madhu, P. *Mater. Sci. Eng. B* **1997**, *49*, 18.

(47) Dey, S.; Ghose, J. *Mate. Res. Bull.* **2003**, *38*, 1653.

(48) Zaki, M.I.; Hasan, M.A.; Al-Sagheer, F.A; Pasupulety, L. *Coll. Surf. A* **2001**, *190* 261.

(49) Rosenberg, D.J.; Coloma, F; Anderson, J.A. *J. Catal.* **2002**, *210*, 218.

(50) Anderson, J.A.; Fergusson, C.; Rodriguez-Ramos, I.; Guerrero-Ruiz, A. *J. Catal.* **2000**, *192*, 344.

(51) Tanabe, K. In *Solid Acids and Bases*; Academic Press: New York, 1970.

(52) Nishiwaki, K.; Kakuta, N.; Ueno, A.; Nakabayashi, H. *J. Catal.* **1989**, *118*, 498.

(53) Kahleel, A.; Li, W.; Klabunde, K.J. *Nanostruct. Mater.* **1999**, *12*, 463.

(54) Pal, U.; García Serrano, J. *Appl. Surf. Sci.* **2005**, *246*, 23.

(55) Athalin, H.; Lefrant, S. *J. Nanopart. Res.* **2005**, *7*, 89.

(56) Ojamäe, L.; Aulin, C.; Pedersen, H.; Käll, P.L. *J. Coll. Inter. Sci.* **2006**, *296*, 71.

(57) Shi, C.; Dai, S.; Wang, K.; Pan, X.; Kong, F.; Hu, L. *Vibra. Spectr.* **2005**, *39*, 99.

(58) Lana-Villarreal, T.; Pérez, J.M.; Gómez, R. *Surf. Sci.* **2004**, *572*, 329.

(59) Weil, J.A; Bolton, J.R.; Wertz, J.E. *Electron Paramagnetic Resonance: Elementary Theory and Practical Applications*; Wiley Interscience: New York, 1994.

(60) Lunsford, J.H. EPR methods in heterogeneous catalysis. In *Catalysis Science and Technology*; Anderson, J.R.; Boudart, M. (Editors); Springer-Verlag: New York, 1987; 227–256.

(61) Dyrek, K.; Che, M. *Chem. Rev.* **1997**, *97*, 305–331.

(62) Chiesa, M.; Giamello, E. In *Encyclopaedia of Analytical Chemistry*; Wiley: New York, 2000.

(63) Schweiger, A. *Angewandte Chemie (Int. Ed.)* **1991**, *30*, 265–292.

(64) Hartmann, M.; Kevan, L. *Chem. Rev.* **1999**, 635–660.

(65) Biglino, D.; Li, H.; Erickson, R.; Lund, A.; Yahiro, H.; Shiotani, M. *Phys. Chem. Chem. Phys.* **1999**, *1*, 2887–2896.

(66) Che, M.; Tench, A.J. *Advanc. Catal.* **1982**, *31*, 87.

(67) Seyedmonir, S.; Howe, R.F. *J. Chem. Soc. Faraday Trans. I.* **1984**, *80*, 2269–2285.

(68) Spoto, G.; Gribov, E.N.; Ricchiardi, G.; Damin, A.; Scarano, D., Bordiga, S.; Lamberti, C.; Zecchina, A. *Progr. Surf. Sci.* **2004**, *76*, 71–146.

(69) Lunsford, J.H. *J. Chem. Phys.* **1967**, *46*, 4347.

(70) Di Valentin, C.; Pacchioni, G.; Chiesa, M.; Giammelo, E.; Abbet, S.; Heiz, U. *J. Phys. Chem. B* **2002**, *106*, 1637.

(71) Martino, P.; Chiesa, M.; Paginini, M.C.; Giamello, E. *Surf. Sci.* **2003**, *527*, 80–88.

(72) Sterrer, M.; Berger, T.; Stankic, S.; Diwald, O.; Knoezinger, E. *Chem. Phys. Chem.* **2004**, *5*, 1695–1703.

(73) Chiesa, M.; Giamello, E.; Paganini, M.C.; Sojka, Z.; Murphy, D. *J. Chem. Phys.* **2002**, *116*, 4266–4274.

(74) Di Valentin, C.; Ricci, D.; Pacchioni, G.; Chiesa, M.; Paganini, M.; Giamello, E. *Surf. Sci.* **2002**, *521*, 104–116.

(75) Chiesa, M.; Giamello, E.; Murphy, D.M.; Pacchioni, G.; Paganini, C.; Soave, R.; Sojka, Z. *J. Phys. Chem. B* **2001**, *105*, 497–505.

(76) Teramura, K.; Tanaka, T.; Ishikawa, H.; Kohno, Y.; Funabiki, T. *J. Phys. Chem. B* **2004**, *108*, 346–354.

(77) Richards, R.M.; Volodin, A.M.; Bedilo, A.F.; Klabunde, K. *Phys. Chem. Chem. Phys.* **2003**, *5*, 4299–4305.

(78) Howe, R.F.; Graetzel, M. *J. Phys. Chem.* **1987**, *91*, 3906–3910.

(79) Coronado, J.M.; Maira, A.J.; Conesa, J.C.; Leung, K.Y.; Augugliaro, V.; Soria, J. *Langmuir* **2001**, *17*, 5368–5374.

(80) Antcliff, K.L.; Murphy, D.M.; Griffiths, E.; Giamello, E. *Phys. Chem. Chem. Phys.* **2003**, *5*, 4306–4316.

(81) Attwood, A.; Edwards, J.; Rowlands, C.; Murphy, D.M. *J. Phys. Chem. A* **2003**, *107*, 1779–1782.

(82) Coronado, J.M.; Maira, A.J.; Martinez-Arias, A.; Conesa, J.C.; Soria, J. *J. Photochem. Photobiol. A* **2002**, *150*, 213–221.

(83) Hernandez-Alonso, M.; Hungria, A.B.; Martinez-Arias, A.; Fernandez-Garcia, M.; Coronado, J.M.; Conesa, J.C.; Soria, J. *Appl. Catal. B* **2004**, *50*, 167–175.

(84) Konigsberger, D.C.; Prins, R. *X-ray Absorption*; Wiley: New York, 1988.

(85) Schulze, D.G. *Synchrotron X-ray Methods in Clay Science*; Clay Minerals Society: Chantilly, VA, 1999.

(86) Fernandez-Garcia, M. *Catal. Rev.* **2002**, *44*, 59–121.

(87) Luca, V.; Djajanti, S.; Howe, R.F. *J. Phys. Chem. B* **1998**, *102*, 10650–10657.

(88) Rajh, T.; Nedeljkovic, J.; Chen, L.; Poluektov, O.; Thurnauer, M. *J. Phys. Chem. B* **1999**, *103*, 3515–3519.

(89) Rodriguez, J.; Wang, X.; Liu, G.; Hanson, J.; Hrbek, J.; Peden, C.; Iglesias-Juez, A.; Fernandez-Garcia, M. *J. Molec. Catal.* **2005**, *228*, 11–19.

(90) Jing, C.; Meng, X.; Liu, S.; Baidas, S.; Patraju, R.; Christodoulatos, C.; Korfiatis, G. *J. Coll. Interface Sci.* **2005**, *290*, 14–21.

(91) Izumi, Y.; Kiyotaki, F.; Yagi, N.; Vlaicu, A.; Nisawa, A.; Fukushima, S.; Yoshitake, H.; Iwasawa, Y. *J. Phys. Chem. B* **2005**, *109*, 14884–14891.

CHAPTER 12

Chemical Properties of Oxide Nanoparticles: Surface Adsorption Studies from Gas- and Liquid-Phase Environments

JOHN M. PETTIBONE, JONAS BALTRUSAITIS, and VICKI H. GRASSIAN

Departments of Chemistry and Chemical and Biochemical Engineering, University of Iowa, Iowa City IA 52242

12.1. OXIDE NANOPARTICLES: INTRODUCTION

Single-crystal surfaces provide model systems in which fundamental aspects of the surface chemistry of oxides can be unraveled. However, oxide particles, often in the nanometer size range, are abundant in natural and engineered environmental systems. As reviewed by Zhang and Banfield, sulfides, oxalates, phosphates, and oxide minerals are produced through biological pathways where microbially induced precipitates form nanoscale minerals as a result of metabolic activity of microorganisms (1). Nanoparticles formed in the environment by inorganic pathways through precipitation or restructuring include amorphous silica, hydrous aluminosilicates, oxides, and oxyhydroxides (1). Metal oxides because of enhanced chemical properties on the nanoscale could be important materials for environmental remediation (2). Nanomaterials as sorbents, catalysts, or enhanced filters are possible uses for these materials, with TiO_2 nanoparticles often having the most desirable properties (3). Because of the presence of oxide nanoparticles in the natural environment and the use of oxide nanoparticles in environmental remediation, there is a great deal of interest in better understanding oxide nanoparticles and the unique properties found in oxide particles of this size range.

Because the focus of nanoscience and nanotechnology is on size effects, Figure 12.1 shows a schematic of the relative sizes of nanoparticles of MgO and

Synthesis, Properties, and Applications of Oxide Nanomaterials, Edited by José A. Rodríguez and Marcos Fernández-Garcia

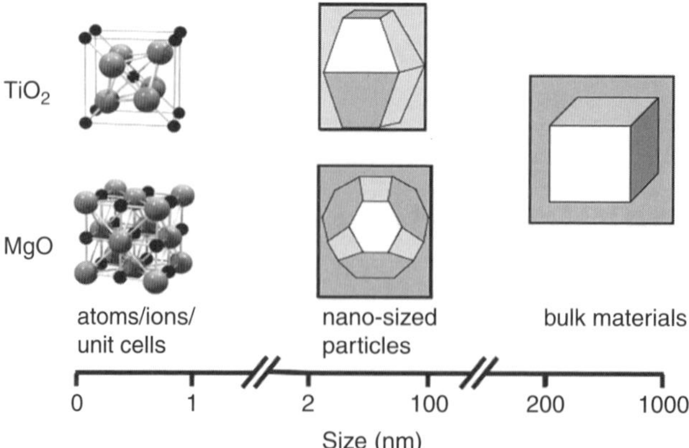

Figure 12.1. Schematic of the relative sizes of nanoparticles of MgO and TiO_2 to its unit cells and bulk structures. Reprinted from Ref. 4 with permission from Elsevier.

TiO_2 compared with their unit cells and bulk materials (4). As discussed in several previous book chapters and in several papers in the literature (5–8), the properties of the oxides are found to change as a function of particle size. For example, there is an increase in the band gap for semiconductors as the particle size decreases. These quantum size effects will alter the reactivity and optical properties of these particles. In addition, the melting point, magnetic moment, specific heat, morphology, and particle shape have all found to be size-dependent properties. Most interestingly from the perspective of this chapter, is the effect of size on the surface chemistry of nanoparticles. Enhanced surface reactivity is found for nanoparticles. This enhanced reactivity is not only due to an increase in surface area with reduced size, but is also due to inherent chemical properties of nanoparticles when compared with their micron-sized counterparts (2,5,8).

For oxide single-crystal surfaces, there is a low number density for step edges, kinks, and other defect sites. These sites are typically the most reactive sites where adsorption and catalysis can occur. In the case of oxide nanoparticles, there is a significantly greater defect site density at the surface because of the increased number of edge and step sites and thus potentially enhanced reactivity. It is this enhanced reactivity that makes the chemical properties of oxide nanoparticles of great interest.

12.2. OXIDE NANOPARTICLES: STRUCTURE AND COMPOSITION

Only within the last 20 years has the atomic structure of individual nanoparticles been unraveled with the development and availability of high-resolution transmission electron microscopy (HRTEM) and scanning tunnel microscopy (STM). Two factors with considerable importance in determining the shape and stability of oxide nanoparticles are the balance in energy between the surface and the bulk as well as maintaining the stoichiometry in a nano-sized particle. Particles created in the nano-sized range may contain several hundred to thousands of ions, many of which will be surface ions (9), upwards of 40% for MgO (10), with incomplete coordination spheres (1,7). The increase in the number of "exposed" surface ions relative to "unexposed" near-surface ions and bulk ions will affect the energy of these very small particles.

Nanoparticles may undergo phase transformation to minimize their total free energy. The dependence of phase stability on crystal size has been previously discussed in some detail (1,8). For instance, the γ-phase of alumina is more stable than the α-phase in the nanocrystalline size range with surface areas greater than $125\ m^2\ g^{-1}$ (10,11). Furthermore, the thermodynamically most stable phase for TiO_2 is anatase for crystal size less than 11 nm brookite for sizes greater than 11 and less than 35 nm and rutile for sizes greater than 35 nm (12).

Another factor in stability oxide nanoparticle is maintaining the chemical stoichiometry. There is a chemical gradient in ion concentration from the surface to the core of the nanoparticle. The structure of oxide nanoparticles varies continually from the center to the surface of the particle (6), but these differences have been difficult to discern. It is only under ultra-high vacuum (UHV) conditions and high annealing temperatures that surfaces of metal oxides are terminated with the metal cations. Under non-UHV conditions, metal-oxide single crystals and nanoparticles are typically oxygen terminated and metal cations tend to be fully coordinated in the bulk (7). Such high concentrations of negative oxide ions on the surface cause surface atoms to relax and/or react with ambient gases, e.g., water-forming hydroxyl groups (5). Under ambient conditions, particles tend to minimize the surface energy by hydration, protonation, and surface reconstruction. Residual hydgroxyl groups on the surface of nanoparticles from reaction with ambient water vapor are found to persist even after relatively high thermal treatment (5).

Under conditions that eliminate interactions with surrounding molecules, such as vacuum and heating treatments, oxide nanoparticles are found to display crystallographic low and high index faces so as to increase the coordination of surface ions (6). Because of the different faces found on nanoparticles, "average" surface energies are usually measured making it difficult to determine the dominant surface face (9). Such faceting in nanoparticles increases the edge/corner density and provides an increase in the number of reactive surface sites (5). Although surface reconstruction is somewhat well understood for single-crystal surfaces, it is poorly understood for nanoparticles.

In a step toward understanding surface reconstruction on the nanoscale, Gilbert et al. have recently combined pair distribution function and extended X-ray absorption fine structure to better understand the structure of ZnS nanoparticles (13). They have shown that there is strain and structural disorder throughout the particle. Electronic properties will be affected by the disorder in the particle. This type of structural information is important to determine as theoretical studies of nanoparticles are currently underway. Additional information about the inherent structural nature of nanoparticles and the stability of these particles is needed if the potentially unique chemical properties associated with these materials can be truly exploited to its maximum potential in nanotechnology.

12.3. OXIDE NANOPARTICLES: PREPARATION

There have been several articles that summarize recent advances in the preparation methods and characterization techniques of oxide nanoparticles (2,4,14–17). The phase of the nanoparticle can depend on the method used in its preparation. Some preparation methods produce amorphous materials, whereas other methods produce nanocrystalline materials. Increasing the stability and enhancing the dispersion of a narrow size range of the nanoparticles are primary goals in various synthetic methods (14).

The many industrial and technological uses of magnetic nanoparticles have resulted in the development of new synthetic routes for the production of well-crystallized iron-oxide particles (18). In general, iron-oxide nanoparticles are stabilized by a surfactant or polymer coating (15). In addition, different oxidation states of iron can result depending on the nature of the synthesis and the precursor molecules used. The oxidation of Fe_3O_4 to γ-Fe_2O_3 nanoparticles has been observed to occur as samples age with time (15).

12.4. OXIDE NANOPARTICLE SURFACE ADSORPTION: THERMODYNAMICS, SIZE EFFECTS, AND DEFECT SITES

Metal-oxide nanoparticles are often considered unique chemical stoichiometric reagents for the adsorption of pollutants (5). A thermodynamic model to predict and better understand the behavior of nanoparticle adsorption was first presented in detail by Zhang et al. (8). The model is based on first-principles of thermodynamics for the equilibrium constant and Gibbs free energy:

$$K_{\text{ads}} = \exp\left(-\frac{\Delta G^{\circ}_{\text{ads}}}{RT}\right) \tag{12.1}$$

where K_{ads} is the adsorption constant, $\Delta G^{\circ}_{\text{ads}}$ is the standard state Gibbs free energy, R is the gas constant, and T is the temperature. The free energy of the system is the spontaneous driving force for adsorption onto the solid. It can also be represented by the interfacial tension (surface tension) γ_{o}, which is taken to be independent of curvature. Integration over the surface area A, between the solution and the particle results in

$$\Delta G^{\circ}_{\text{ads}} = \int_{r=\infty}^{r} \gamma_{\circ} dA. \tag{12.2}$$

By approximating the particle as a sphere and distinguishing the interfacial tension of a site with an adsorbed species $\gamma_{\text{o}(S-I)}$, and an empty site, $\gamma_{\text{o}(S)}$, a ratio of adsorption constants for two particles A and B, where the radius of A, r_A, is less than the radius of B, r_B, yields the following;

$$\frac{K_{ads}(A)}{K_{ads}(B)} = \exp\left[\frac{-3\,V_m(\gamma_{o(S-l)} - \gamma_{o(S)})}{RT} \cdot \left(\frac{1}{r_A} - \frac{1}{r_B}\right)\right] \tag{12.3}$$

where V_m is the molar volume of the sphere. Because the interfacial tension is always lower when there is an adsorbed species, and the radius of A is less than B, the ratio of the adsorption constants is always greater than one. The equation correlates the size of the particle with change in surface tension. Experimental results have suggested that surface tension changes with size of particle; however, this consideration is not included in the above analysis.

A similar but somewhat more detailed thermodynamic approach was developed by Lu et al., in which the γ_o size dependence was taken into account (19). The comparison between the theoretically and experimentally determined adsorption constants determined by Lu et al. and Zhang et al., respectively, are presented in Table 12.1. The table shows that the ratio of K_{ads} (6 nm)/K_{ads} (16 nm) is quite good for acetic, valeric, and oxalic acids but not as good for adipic acid. However, the theoretical calculations do qualitatively predict the trend in the enhancement of these mono- and di-carboxylic acids.

Using a different approach to understanding nanoparticle surface adsorption, Kakkar et al. performed density functional calculations to better understand the adsorption of formaldehyde on MgO nanoparticle surfaces (10). The focus of this theoretical study was to understand on a molecular level the bonding of formaldehyde to the different sites available on the MgO nanoparticle surface. The calculations showed that surface reactivity is highly dependent on the nature of the surface site (steps, kinks, corners, edges, etc.).

TABLE 12.1. Enhanced Adsorption Constants K_{ads} for a Series of Carboxylic Acids on TiO_2 Nanoparticles Less Than 10 nm: Theory Compared with Experiment.

	K_{ads} (6 nm)/K_{ads} (16 nm)	
Carboxylic Acid	Theory	Experiment
Acetic acid	3.4	3.4
Valeric acid	5.9	6.8
Oxalic acid	20.9	23.5
Adipic acid	46.1	69.9

12.5. ENHANCED CHEMICAL PROPERTIES OF OXIDE NANOPARTICLE SURFACES IN ENVIRONMENTAL CATALYSIS AND ENVIRONMENTAL REMEDIATION

As discussed in Section 12.4, the current understanding of the influence of size on the chemical properties is centered on both the nature of surface sites and the thermodynamic driving force for adsorption. The size-dependence of the surface-free energy explains the higher adsorptive properties of nano- versus micron-sized materials. In addition, changes in crystal shape will result in increasing the density of defect sites such as edges, corners, and point defects. The unique chemical properties of several different types of oxide nanoparticles, including Mg, Ca, and Fe, have been shown. These oxides are better catalysts and adsorbents for acidic gases (2).

The most well-studied nanocrystalline oxide thus far in terms of chemical reactions is MgO. Klabunde et al. have shown that nanocrystalline MgO has enhanced reactivity toward the adsorption of several acidic gases at room temperature, including SO_2, CO_2, HCl, HBr, SO_3, and H_2S (20,21). Changes in surface bonding as a function of particle size have been shown for specifically SO_2 adsorption. The adsorption mode for SO_2 is monodendate on small crystals and bidendate on larger ones (20). At the same gas pressure, the capacity of MgO nanocrystallites to chemisorb SO_2 is higher by a factor of three at room temperature compared with larger particle sizes. Similar results were observed for other acidic gases.

Destructive adsorption of chlorinated hydrocarbons has been studied on the surface of both CaO and MgO nanoparticles by Klabunde et al. (20,22). Noncatalytic adsorption/decomposition of CCl_4, $CHCl_3$, and C_2Cl_4 on CaO surfaces have been studied in the temperature range from 473 to 773 K (22). Reaction products and optimum reaction temperatures are shown as follows:

$$\mathrm{CaO + CCl_4 \xrightarrow{673\,K} CaCl_2 + Cl_2CO} \tag{12.1}$$

$$\mathrm{CaO + Cl_2CO \xrightarrow{673} CaCl_2 + CO_2} \tag{12.2}$$

$$\mathrm{3CaO + 2CHCl_3 \xrightarrow{573-673\,K} 3CaCl_2 + 2CO + H_2O} \tag{12.3}$$

$$\mathrm{3CaO + C_2Cl_4 \xrightarrow{773\,K} 2CaCl_2 + CaCO_3 + C} \tag{12.4}$$

It can be seen from the reaction scheme above that the formation of $CaCl_2$, $CaCO_3$, and carbon (graphite) inhibits the reaction as these solid products form coatings on the nanoparticle surface. However, the above reactions can be catalyzed by the presence of a small amount of a transition metal oxide such as Fe_2O_3. For example, the catalyzed reaction of CCl_4 destruction can be written as (20,23)

$$\mathrm{2[Fe_2O_3]CaO + CCl_4 \xrightarrow{673\,K} 2[Fe_2O_3]CaCl_2 + CO_2} \tag{12.5}$$

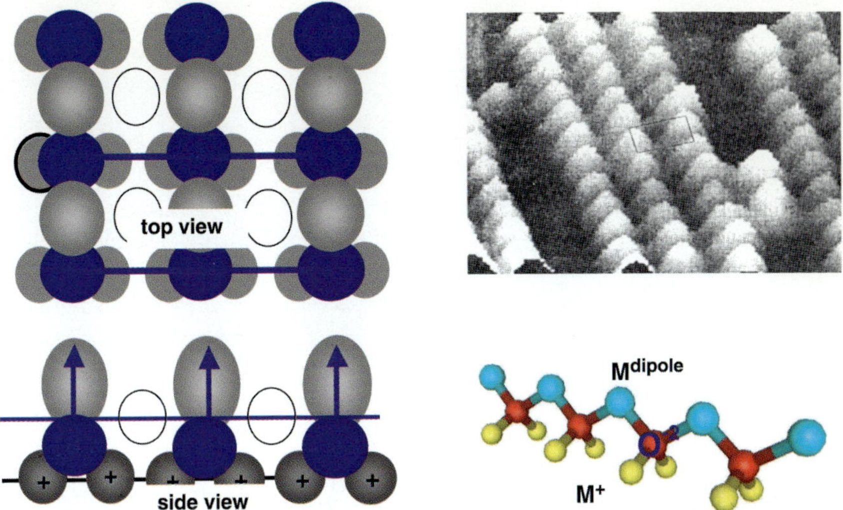

Figure 1.9. STM image (98) and the corresponding models (1) for the surface atomic valencies of Cu(110)–(2 × 1)–O^{2-}. The STM gray scale is 0.85 Å, much higher than that of metallic Cu on a clean (110) surface (0.15 Å). The single "O^{2-}: Cu^{dipole}: O^{2-}" chain zigzagged by the nonbonding lone pairs and composed of the tetrahedron.

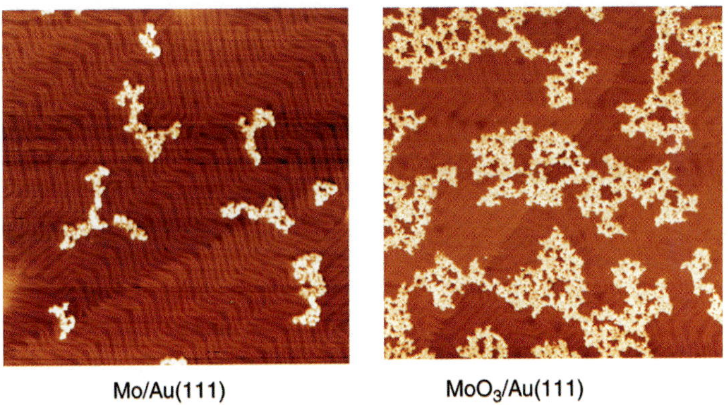

Figure 5.8. STM images recorded before and after oxidizing Mo nanoparticles on a Au(111) template. Image size: 190 × 190 nm^2 (taken from Ref. 138).

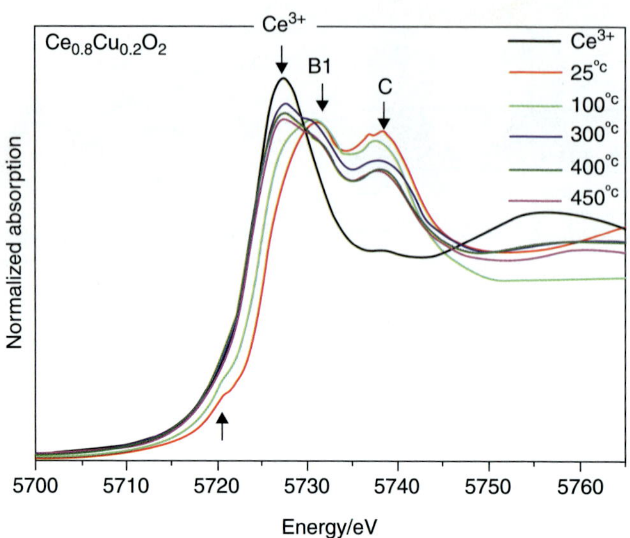

Figure 6.2. Ce L_{III}-edge XANES spectra collected for $Ce_{0.8}Cu_{0.2}O_2$ during the WGS reaction at different temperatures. Included for comparison is the spectrum for a $Ce(NO_3)_3 \cdot 6H_2O$ reference, which contains only Ce^{3+} cations (taken from Ref. 37).

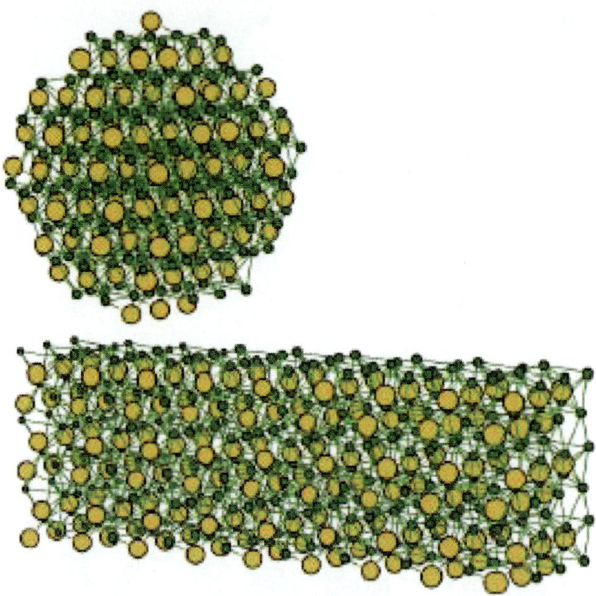

Figure 8.2. Three-dimensional view of a spherical and columnar nanoparticles of SnO_2. Dark and gray grains correspond to O and Sn atoms. From Ref. 81.

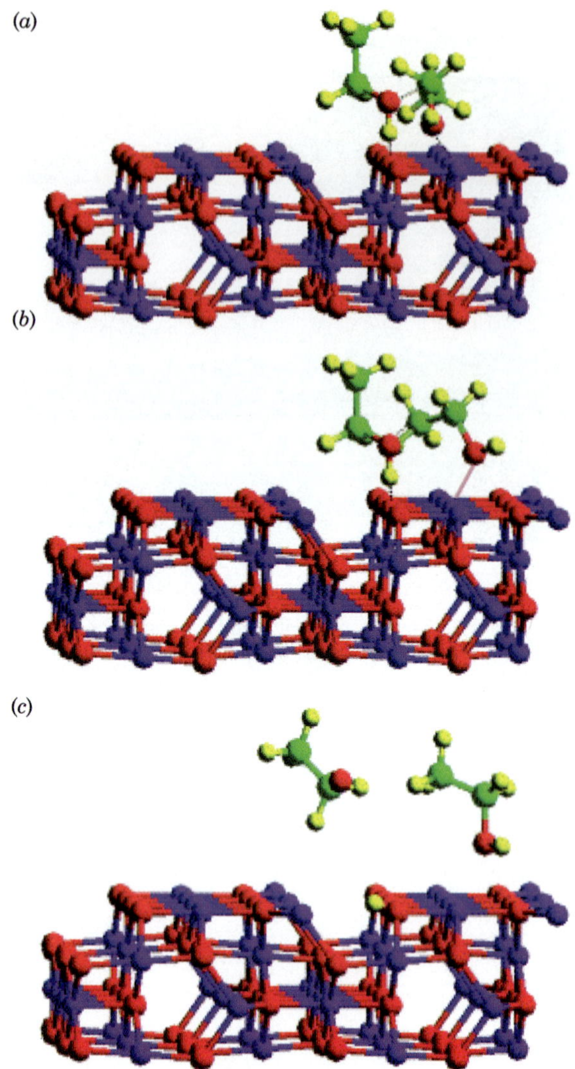

Figure 8.3. Adsorption of (*a*) two C_2H_5OH molecules on $Al_{48}O_{64}$; (*b*) two different configurations and (*c*) optimized configuration. From Ref. 113.

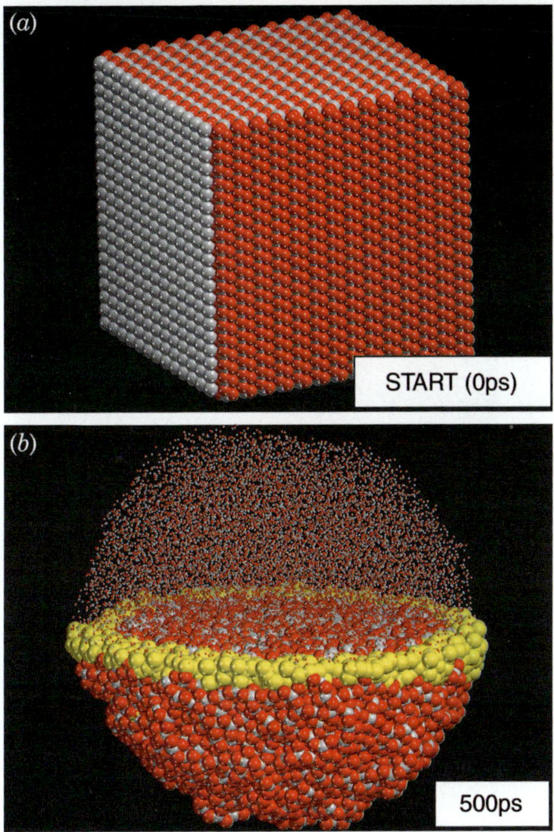

Figure 9.2. Atom positions corresponding to (*a*) the starting structure and (*b*) the amorphous configuration (after 500 ps of MD simulation) of the MnO_2 nanoparticle. A sphere model representation of the atoms position has been used. In (*b*), the atoms comprising the top half of the sphere are represented using a smaller sphere radius to show more clearly the amorphicity of the nanoparticle. Manganese ions are light gray, and oxygen is dark gray. Some ions are colored white to show more clearly that the nanoparticle is spherical.

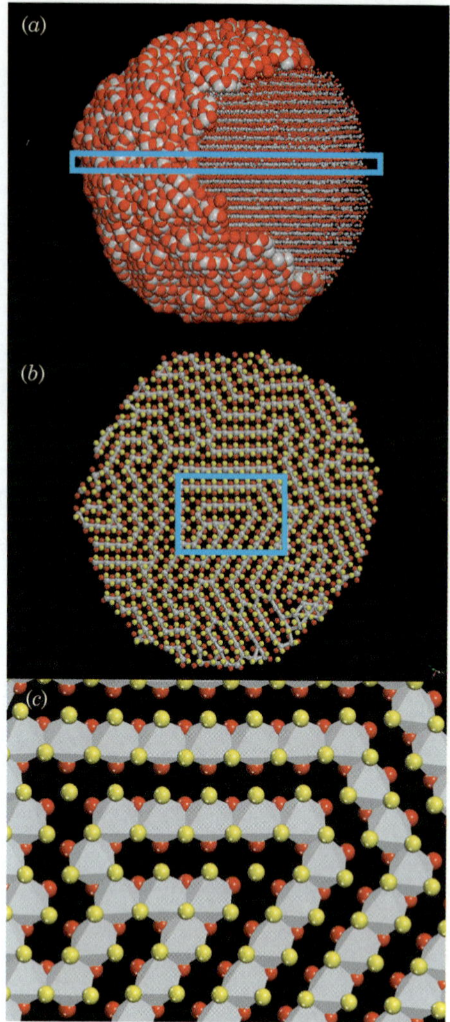

Figure 9.4. Structure of the fully recrystallized MnO_2 nanoparticle: (*a*) sphere model representation of the atom positions. Two different sphere radii have been used for the atoms to aid understanding of the figure; a slice cut through this figure (blue rectangle) is shown in plan view in (*b*), which depicts the microtwinning of the MnO_2. An enlarged segment is shown in (*c*). Notation: (*a*) Manganese atoms are colored gray, and oxygen is red; (*b*) Manganese are colored gray, O atoms above the plane of the paper are red, and O ions below the plane of the paper are yellow. The manganese atoms are shown as ball-and-stick representation to show more clearly the microtwinning; (*c*) manganese are shown as gray polyhedra indicating the octahedral sites they occupy, oxygen atoms above the plane of the paper are red, and oxygen ions below the plane of the paper are yellow.

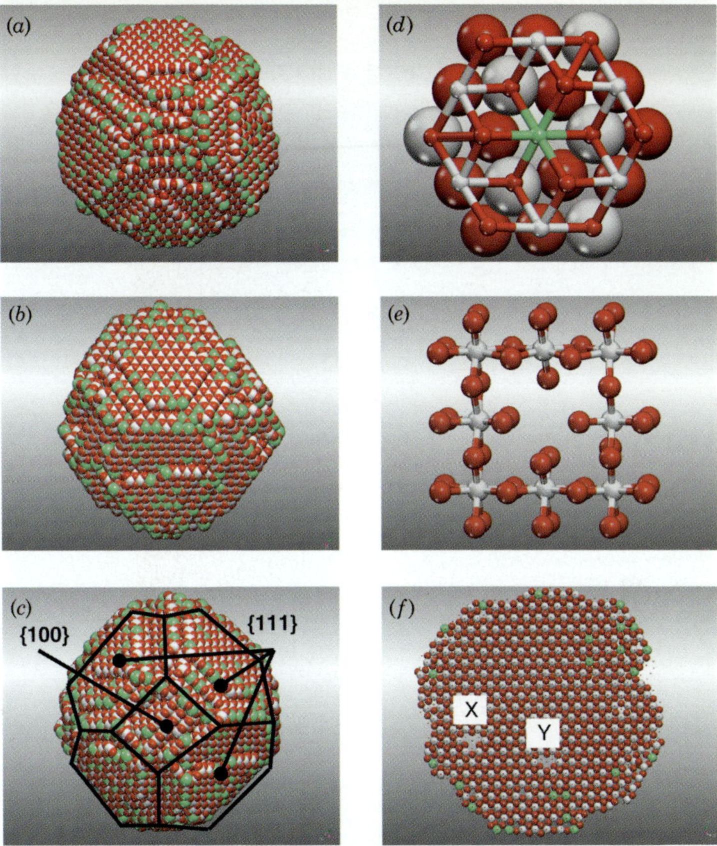

Figure 9.7. Sphere model representation of the reduced ceria ($CeO_{1.95}$) nanoparticle. (*a*)–(*c*) Depict three different views of the same nanoparticle showing the truncated octahedral morphology, similar to that of the unreduced nanoparticle (Figure 9.5). The Ce^{3+} ions are clearly seen to decorate surface steps and edges in addition to plateau regions on {111} and {100}. In (*c*), the atomistic structure is annotated with lines to show more clearly the surfaces. (*d*) Segment cut from the reduced nanoparticle depicting an oxygen vacancy bound to an adjacent Ce^{3+}. (*e*) Structure of a Ce^{4+} vacancy together with two adjacent oxygen vacancies, which have evolved to quench the charge imbalance associated with the vacancy. (*f*) A slice, cut through the nanoparticle, showing two complex defect clusters labeled X and Y. X comprises, using the Kroger–Vink notation, $[V_{Ce}^{''''}, 3V_{O}^{\bullet\bullet}]^{2-}$ and Y is $[V_{Ce}^{''''}, 2V_{O}^{\bullet\bullet}]^{0}$. Ce^{4+} is colored white, Ce^{3+} is green, and oxygen is red.

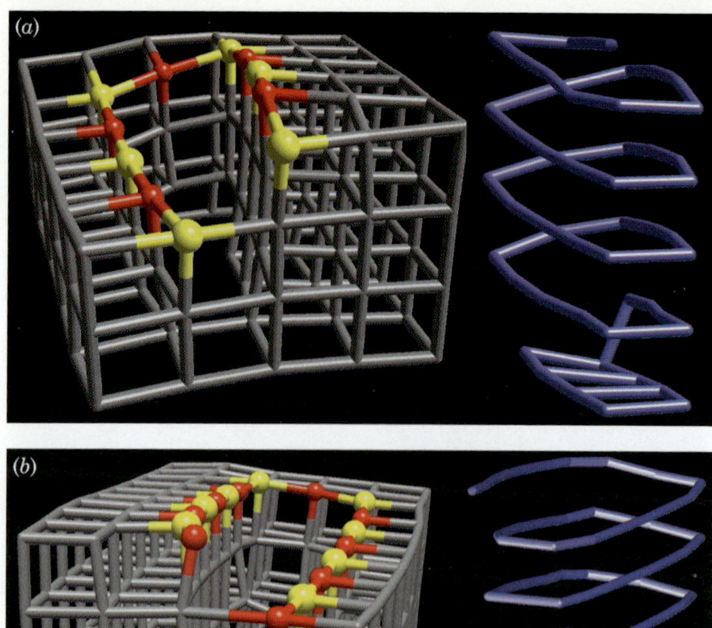

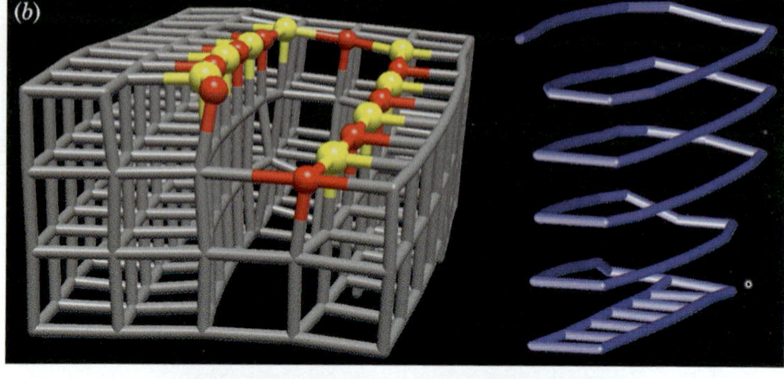

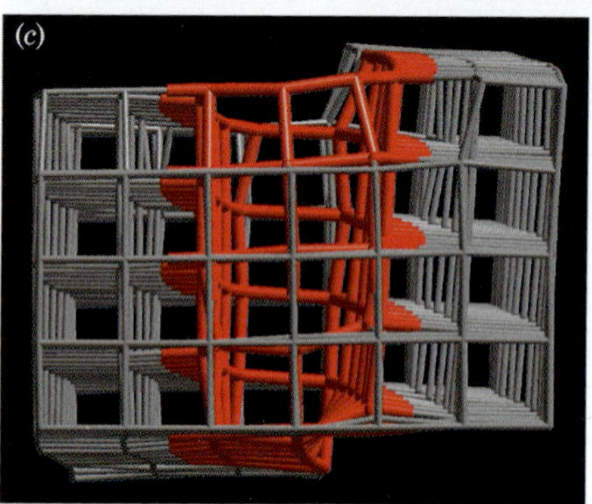

Figure 9.11. Stick model representations of two screw-edge dislocations that evolved within the SrO thin film. (*a*) Left: part of the SrO lattice (gray) that surrounds the dislocation core, right: dislocation core structure. (*b*) Another dislocation, similar to that in (*a*), but here the core traverses the SrO thin film in a clockwise fashion [as opposed to anticlockwise in (*a*)]. (*c*) Core structure, shown in red, illustrating how the dislocation core "lies" within the surrounding lattice. Oxygen is colored red, and strontium is yellow.

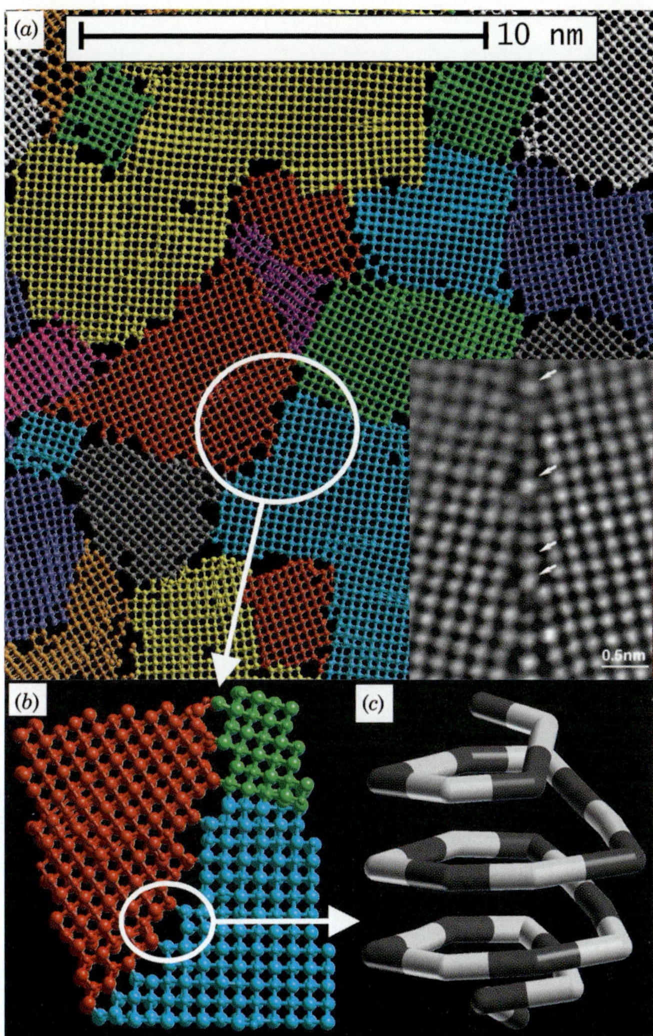

Figure 9.12. Atom positions comprising a nanopolycrystalline MgO thin film supported on a BaO substrate. (*a*) Plan view of the system showing the various nanocrystallites comprising the MgO, which are colored to highlight the individual nanocrystallites. The MgO substrate is not shown to maintain clarity of the figure. The inset shows a STEM image to compare and has been reproduced, with permission, from Ref. 57. (*b*) Enlarged segment of (*a*) showing more clearly the grain-boundary structure. (*c*) An enlarged segment of (*b*) depicting the core structure of the grain boundary and revealing its "screw" in addition to "edge" character.

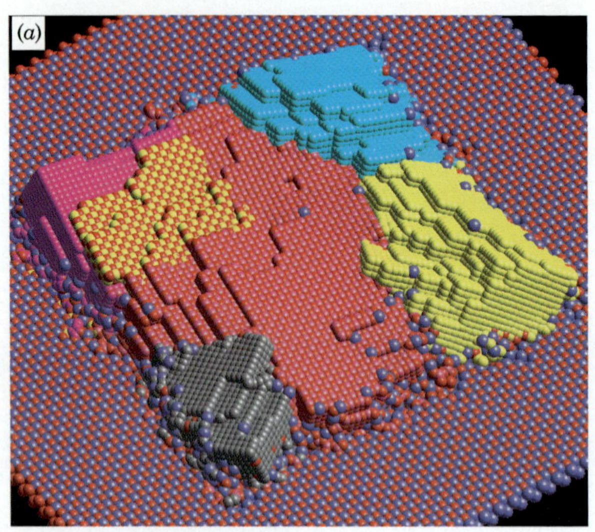

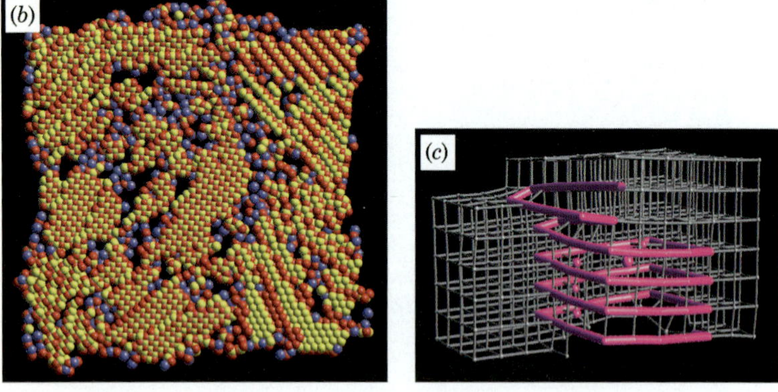

Figure 9.14. (*a*) Perspective view, with sphere model representations, of the atom positions comprising a 25,000-atom MgO nanoparticle supported on a BaO(001) substrate. The various misoriented crystallites are colored. (*b*) Interfacial region illustrating the considerable defective nature of the interfacial region. (*c*) Mixed screw-edge dislocation that was identified to have evolved within the MgO nanoparticle.

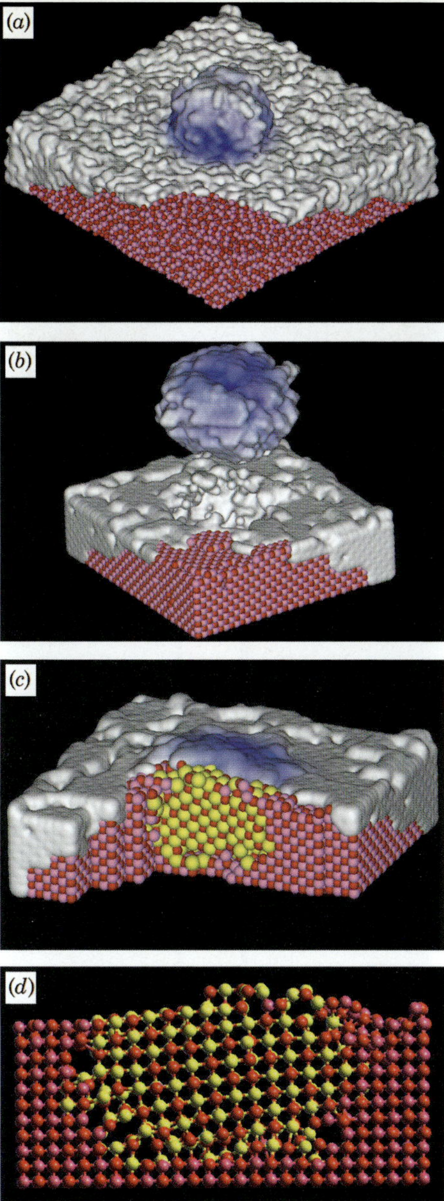

Figure 9.15. Representation of the atom position illustrating the encapsulation of a CaO nanoparticle within an MgO host lattice. (*a*) Amorphous starting structure showing the CaO nanoparticle being introduced into the MgO host. (*b*) Final recrystallized structure showing the structure of the CaO nanoparticle and the void in the MgO that it occupies (the CaO nanoparticle has been physically extracted to reveal the structure). (*c*) Perspective view with part of the simulation cell cut-away to reveal the CaO nanoparticle encapsulated by its MgO host. (*d*) Slice cut through the simulation cell to reveal more clearly how the CaO nanoparticle lies with respect to the host. Calcium is colored light gray, magnesium is medium gray, and oxygen is dark gray. Surface rendering has also been used to aid interpretation of the structures. Light gray rendering is the MgO host and light/dark rendering, the CaO nanoparticle.

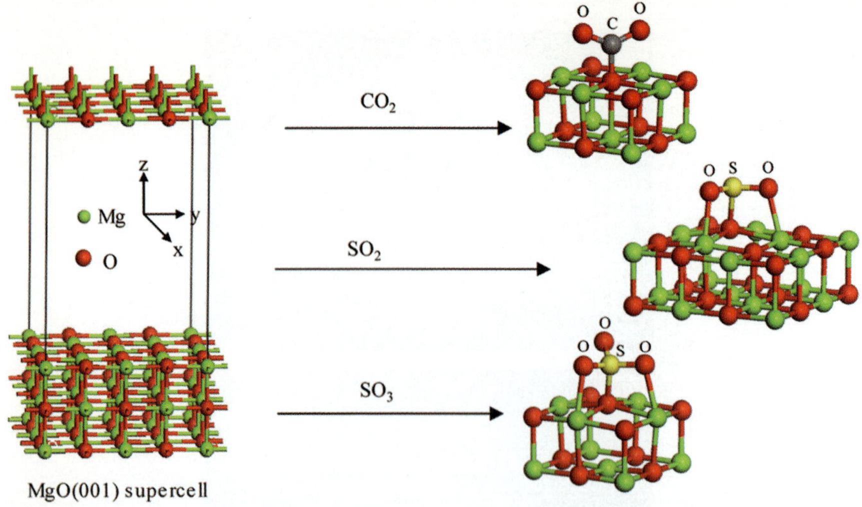

Figure 10.1. (Left) Periodic supercell model of the MgO(001) surface. (Right) Detail views of the adsorbate DFT-calculated structures of SO_2 adsorbed on MgO in both a Lewis base (left) and acid (right) conformation (15).

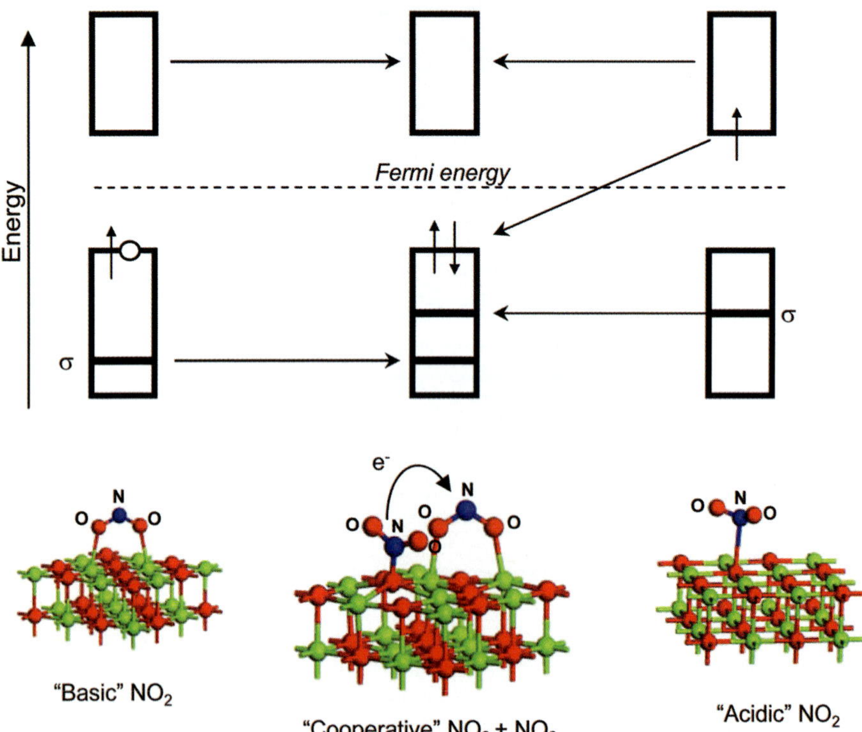

Figure 10.3. PW91 DFT-calculated adsorption geometries and qualitative molecular orbital diagrams for isolated and adsorption of NO_2 on MgO(001). Formation of the cooperative pair from isolated adsorbates is calculated to be exothermic by 16 kcal mol^{-1} (16, 23).

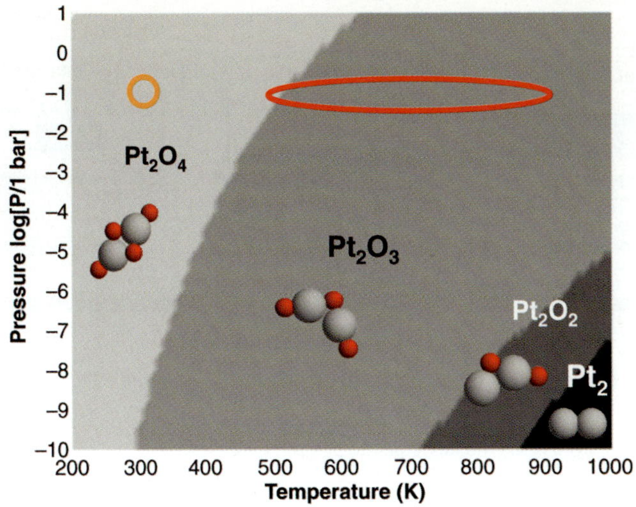

Figure 10.11. T-p_{O_2} phase diagram for Pt_2 clusters immobilized on a support. The O_2 pressure is plotted on a logarithmic scale.

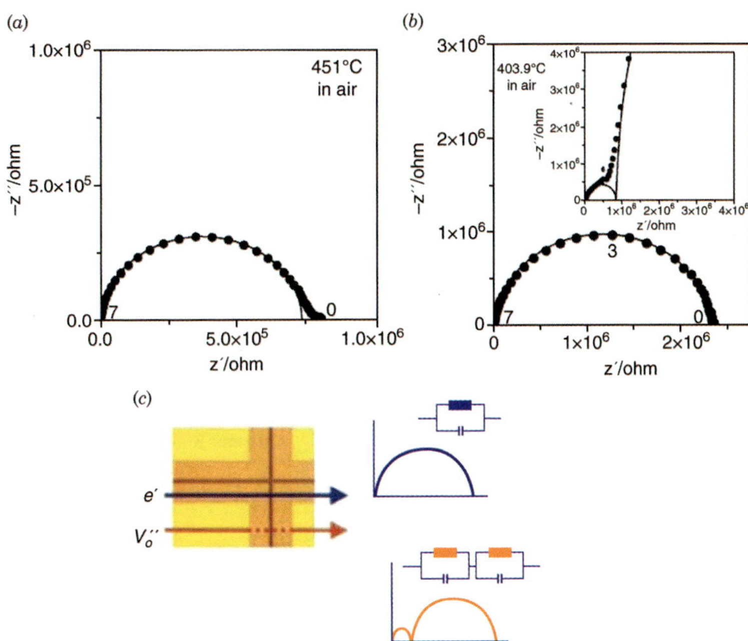

Figure 13.9. Impedance spectra of (*a*) pure nanocrystalline ceria and (*b*) 0.15 mol% Gd-doped nanocrystalline ceria. (Numbers adjacent to data points refer to the logarithm of the measurement frequency.) (*c*) Space-charge model. Reproduced from Ref. 47 with permission from The Electrochemical Society Inc.

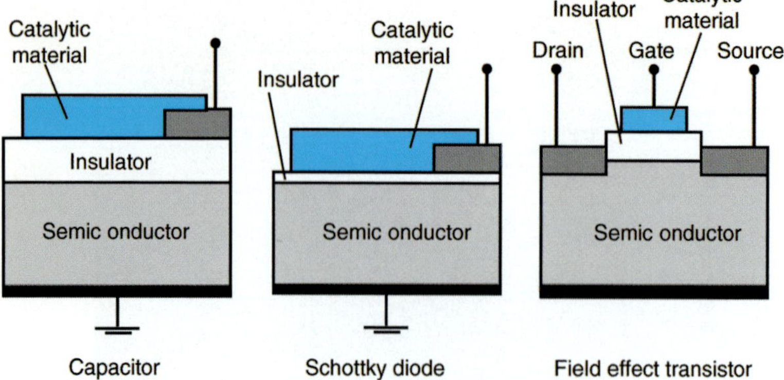

Figure 15.6. Schematic picture of FET devices, capacitor, Schottky diode, and transistor. The catalytic gate material turns the devices into gas sensors.

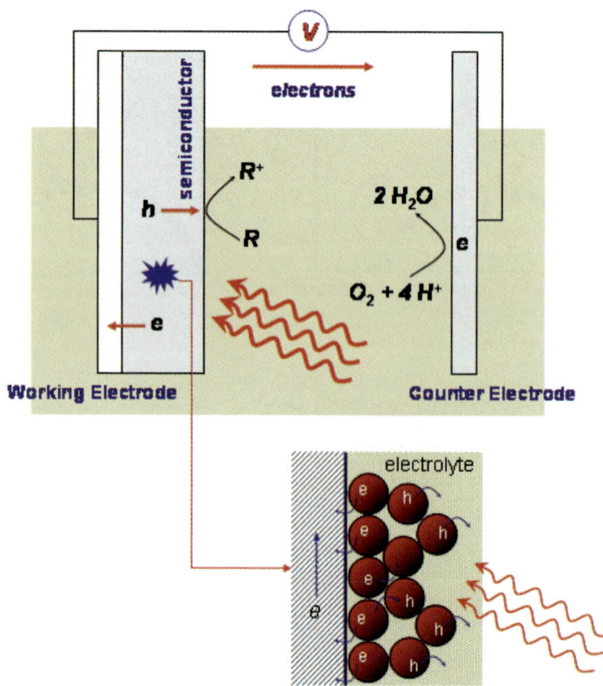

Figure 17.3. Reaction scheme of a typical potential-assisted photo-catalytic process.

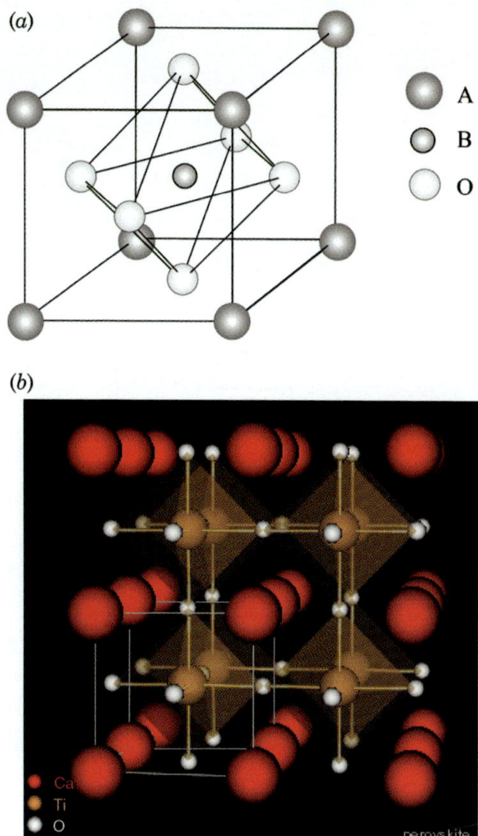

Figure 18.5. Example of perovskite-like structure: (*a*) unit cell (76) and (*b*) example of $CaTiO_3$ tri-dimensional array (77).

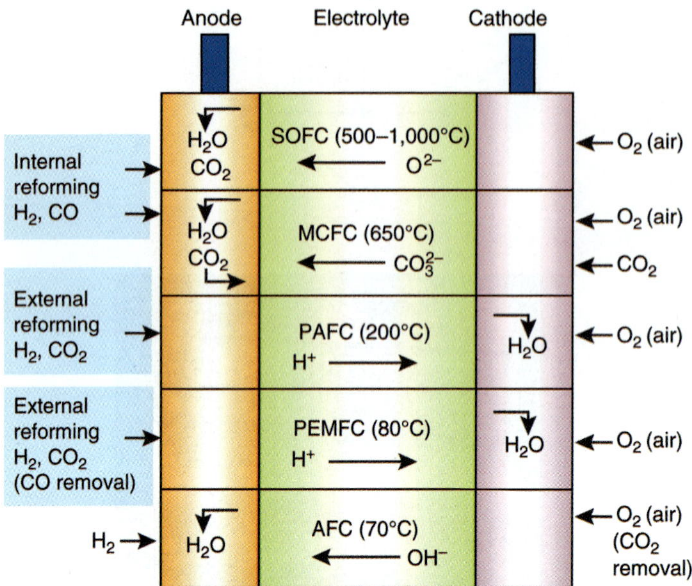

Figure 21.2. Different types of fuel cells (taken from Ref. 12).

Similar reactivity has been observed with oxides of vanadium and manganese (23). The proposed basic steps of these reactions can be summarized as follows. First, CCl_4 attacks the V_2O_5 center on the nanoparticle forming VCl_x. Second, a solid ion–ion (Cl^-/O^{2-}) exchange occurs between the core basic oxide and the VCl_x species, most likely at defect sites and enhanced by the reaction high temperature and low melting points of the transition metal chlorides. This step is followed by surface migration to defective sites where reaction with CCl_4 and ion–ion exchange can further repeat itself. The same catalysts described above are found to catalyze destruction of other environmental toxins such as COS, CS_2, SO_2, and $(CH_3O)_2P(O)CH_3$ (23–25).

In quantitative batch reactor studies, Anschutz and Penn have shown that crystalline iron oxyhydroxides exhibit a size-dependent reactivity (26). In particular, the reductive dissolution of iron oxyhydroxides was found to be 20 times higher for particles smaller than 5 nm in diameter. As the reaction chemistry of iron oxides is driven by redox processes, these results suggest that the chemical properties of iron oxides will be significantly different on the nanoscale.

Other recent studies have discussed the potential for oxide nanoparticles to be important in the adsorption of heavy metals in the environment (27,28). In one study, it was shown that TiO_2 nanoparticles exhibited a smaller K_{ads} for Cd^{2+} compared with larger particles, in contrast to the carboxylic acid studies. It was proposed that there were enhanced repulsive electrostatic interactions between cadmium ions adsorbed on the surface of the nanoparticles that caused the lower measured K_{ads}.

12.6. ATR–FTIR SPECTROSCOPY—AN EXPERIMENTAL TOOL TO INVESTIGATE LIQUID- AND GAS-PHASE SURFACE AND CHEMICAL ADSORPTION PROPERTIES OF OXIDE NANOPARTICLES

As we begin to further our understanding of the inherent unique chemical properties and reactivity of oxide nanoparticles, a variety of analytical tools will be needed. To better understand some of the surface adsorption studies discussed above, spectroscopic probes become increasingly important. For surface adsorption studies, there are few techniques that work well in both gas- and liquid-phase environments. Attenuated total reflectance–Fourier transform infrared (ATR–FTIR) spectroscopy is a useful technique for in situ investigations of liquid/solid and gas/solid interfaces (29–37). In general, multiple internal reflection (MIR) techniques differ from traditional transmission IR techniques in that very short effective path lengths on the order of tens of microns can be achieved. This makes MIR techniques powerful tools for measuring IR spectra of liquid solutions with strong IR absorbing solvents such as water. The internal reflection element (IRE) is usually a high refractive index crystal such as Ge and ZnSe. When the IR beam enters the crystals, it is internally reflected creating a probe radiation at the reflection point that is an exponentially decaying wave (evanescent wave) on the order of a micron or less. Because of the limited penetration depth, the reflectivity is a measure of the interaction of the evanescent wave with the sample. The resulting spectrum becomes characteristic of the sample (29). The penetration depth of the evanescent wave d_p is defined as the distance required for the amplitude of the electric field to fall to e^{-1} of its value at the surface. It depends on the refractive index of the IRE n_1, the sample n_2, angle of incidence θ, and the wavelength of the incident light λ, as follows (29,30):

$$d_p = \frac{\lambda}{2\pi n_1 \sqrt{\sin^2\theta - (n_2/n_1)^2}} \tag{12.4}$$

The absorbance spectra obtained from ATR differ from conventional transmission methods in that correction for varying depth of penetration with wavelength is necessary. With proper subtraction, spectra due to adsorption can be easily obtained.

MIR techniques have been used in several studies investigating important environmental systems at the molecular level. These investigatons include studies of surface complexation and dissolution of minerals (31–37). For example, in a study by Duckworth and Martin, surface complexation and dissolution of hematite by C_1–C_6 dicarboxylic acids was investigated using ATR–FTIR spectroscopy (31). From the ATR–FTIR spectra coupled to batch reactor measurements, surface structure and an equilibrium binding constant were determined (31).

A schematic of a commercially modified ATR–FTIR cell that is currently being used for surface adsorption studies on oxide nanoparticles is shown in Figure 12.2 (38). The cell is designed to investigate the adsorption of molecules from either the gas or liquid phase. For gas-phase studies, the flow reactor shown in Figure 12.3 is coupled to the ATR–FTIR cell (39). Adsorption studies of gases on the oxide nanoparticles as a function of relative humidity can be investigated with this flow reactor system. Examples of the use of ATR–FTIR spectroscopy for surface adsorption studies on oxide nanoparticle surfaces from both the gas and liquid phases are described below.

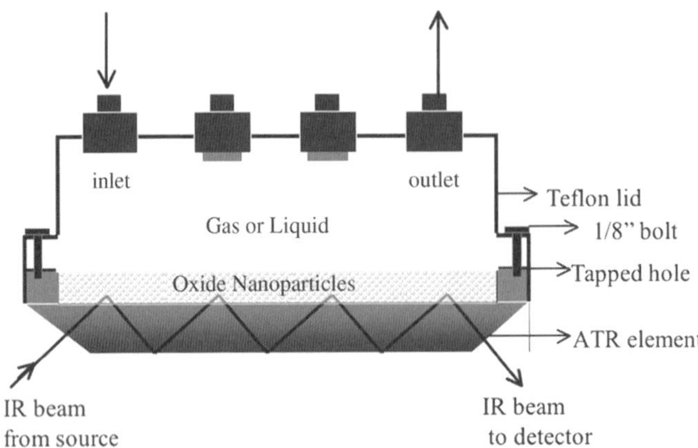

Figure 12.2. Schematic of the ATR–FTIR cell used to investigate surface adsorption on nanoparticles from both gas-phase and liquid-phase environments. Modified from Al-Hosney et al. (38). Reproduced with permission of the PCCP Owner Societies.

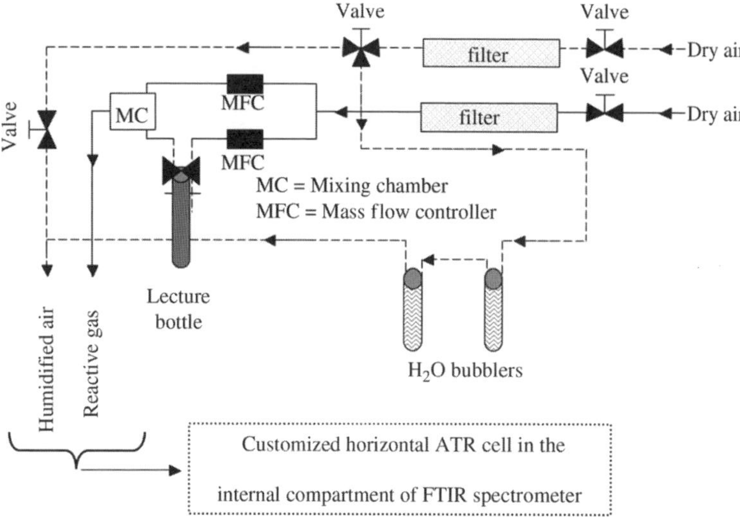

Figure 12.3. Schematic diagram of a flow cell reactor coupled to an ATR–FTIR horizontal cell. The apparatus can be used to measure the uptake of water as well as the uptake of reactive gases in the presence of water. This allows for rate measurements to be made as a function of relative humidity. Modified from Al-Hosney et al. (39). Reproduced with permission of the PCCP Owner Societies.

12.7. ATR–FTIR SPECTROSCOPY—EXAMPLES OF STUDIES OF ADSORPTION ON OXIDE NANOPARTICLES FROM GAS- AND LIQUID-PHASE ENVIRONMENTS

Here we demonstrate the use of ATR–FTIR spectroscopy as a tool to investigate the oxide nanoparticle/gas and oxide nanoparticle/liquid interface. Two different types of oxide nanoparticles have been investigated—nanoparticulate titanium dioxide and iron oxides. Electron microscopy shows the different nanoparticle aggregates and structures for these different samples. Figure 12.4 shows transmission electron microscopy (TEM) images of TiO_2 (anatase) nanoparticles. The particles are typically in an aggregate state, and this can be easily seen in the image shown on the left in Figure 12.4. These aggregates can be dispersed by ultra-sonication of the sample before dispersing them on to the TEM grid. Once dispersed, it is easy to see the 2 to 5-nm-diameter particles (5-nm according to manufacture specifications, Nanostructured and Amorphous Materials, Inc.) Scanning electron microscopy images of iron-oxide nanoparticles (Alfa Aesar) show particle aggregates on the order of 35 nm in diameter (see Figure 12.5). In contrast to the TiO_2 nanoparticles, these particles when viewed at higher magnification show that these 35-nm particles are made up of ca. 5-nm particles that have grown together to form the 35-nm larger interwoven aggregates that cannot be dispersed upon ultra-sonication. The surface areas of these two oxide nanoparticles, TiO_2 and Fe_2O_3, were determined with an automated BET surface area apparatus to be very similar, 219 $m^2\ g^{-1}$ and 229 $m^2\ g^{-1}$, respectively.

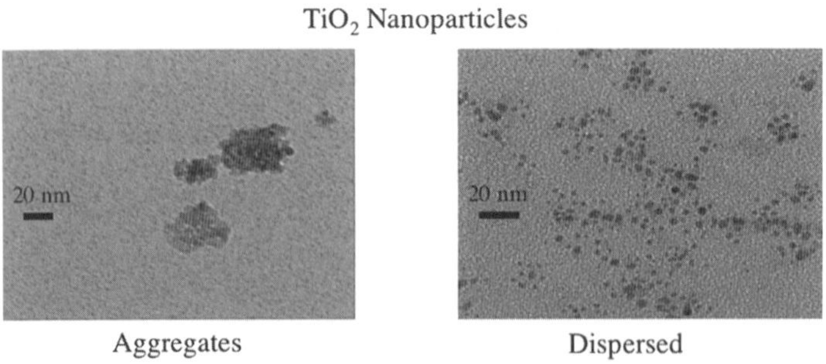

TiO_2 Nanoparticles

20 nm 20 nm

Aggregates Dispersed

Figure 12.4. Transmission electron micrographs (TEM) of TiO_2 (anatase) nanoparticles in the 2–5 nm size range. The nanoparticles tend to aggregate (left image), but upon sonication, they can be dispersed on the TEM grids as isolated particles (right image).

Figure 12.5. Scanning electron micrograph (SEM) of Fe_2O_3 (hematite and ferrihydrite) nanoparticles. The nanoparticles appear around 35 nm in size in the SEM image. A close-up view shows that these 35-nm particles are made up of ca. 5-nm particles that have grown together to form the 35-nm interwoven aggregates.

Several examples of the use of ATR–FTIR spectroscopy with these particles are shown in Figures 12.6–12.9. First, the ATR–FTIR spectra for acetic acid adsorption on 5-nm TiO_2 nanoparticles at pH = 4.5 are presented in Figure 12.6. As discussed, acetic acid is of interest as it has been shown in batch reactor studies that there is enhanced adsorption on particles less than 10 nm in diameter. The spectral range shown is from 1200 to 1900 cm^{-1}. This is the spectral range where absorptions due to acetic acid and adsorbed acetate are found.

To make these measurements, a thin layer of the oxide nanoparticles was coated on the surface of the ATR crystal (ZnSe). This was done by preparing a suspension of the TiO_2 nanoparticles in aqueous solution and allowing the solution to evaporate on the ATR crystal surface. Distilled water was then run through the ATR–FTIR cell, and a

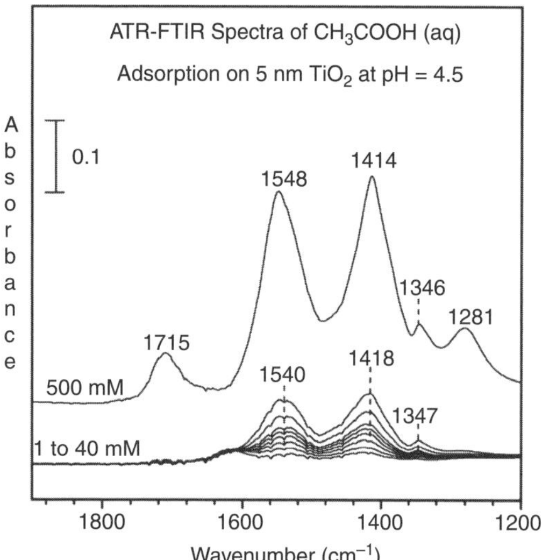

Figure 12.6. ATR–FTIR spectra of acetic acid uptake on 5-nm TiO_2 anatase particles at pH = 4.5 and T = 298 K as a function of acetic acid concentration. The bottom spectra were recorded for concentrations ranging from 1 to 40 mM. At these lower concentrations, absorptions due to adsorbed acetate dominate the spectra. At 500-mM acetic acid concentrations, absorptions due to solution phase acetic acid begin to dominate the spectrum. See text for further details.

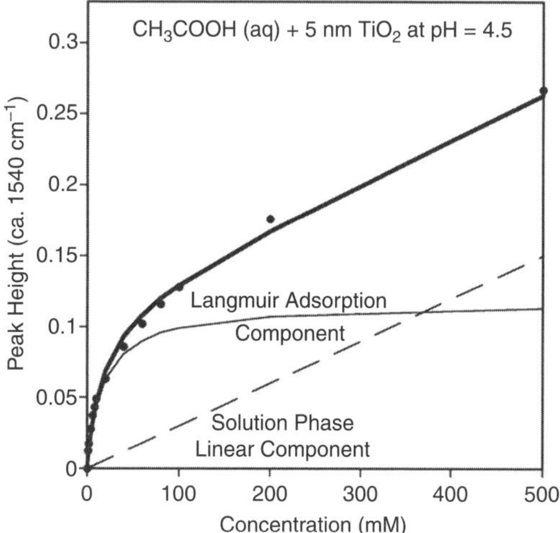

Figure 12.7. Analysis of the ATR–FTIR acetic acid adsorption spectra on 5-nm TiO_2 anatase particles at pH = 4.5. Filled circles represent the peak height of the absorption band at ca. $1540\,cm^{-1}$ plotted as a function of acetic acid concentration. The dark line through the data points represents a fit to the data that consists of two contributions to the peak at ca. $1540\,cm^{-1}$—an adsorption component that is fit with a Langmuir adsorption model and a linear solution phase component. The Langmuir adsorption constant determined for this fit is $58\,M^{-1}$.

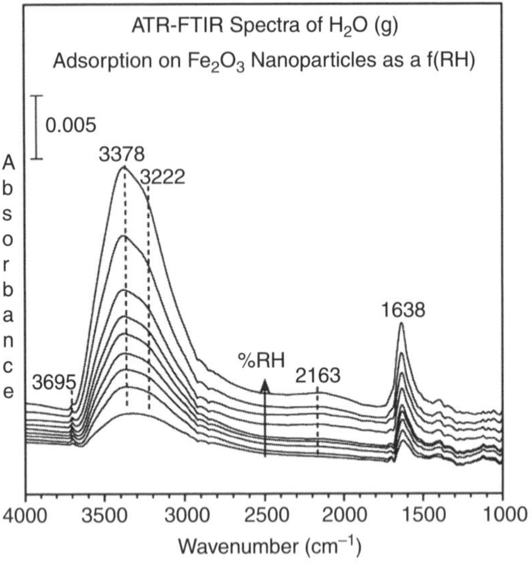

Figure 12.8. Representative ATR–FTIR spectra of water uptake on Fe_2O_3 nanoparticles as a function of relative humidity at T = 298 K (%RH = 10, 20, 30, 40, 50, 60, 70, 81, and 92).

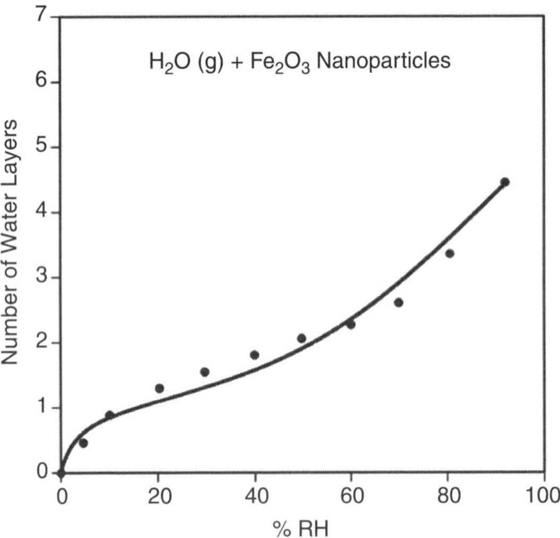

Figure 12.9. Analysis of the ATR–FTIR spectra of water uptake on Fe_2O_3 nanoparticles as a function of relative humidity at T = 298 K. The amount of adsorbed water on the Fe_2O_3 nanoparticle surface as a function of relative humidity. The data were analyzed by fitting the integrated absorbance of the water bending mode at 1638 cm^{-1}, a modified-BET isotherm. See text for further details.

reference spectrum was collected. A solution of known acetic acid concentration and pH was then added to the ATR–FTIR cell; this solution was allowed to equilibrate for 15 minutes. Solution concentrations from 1 to 500 mM were used in these studies. At low concentrations, between 1 and 40 mM, absorptions due to adsorbed acetic acid are evident and dominate the spectrum. These absorption bands can be seen at 1347, 1418, and 1540 cm^{-1}. These bands can be assigned to adsorbed acetate on the surface (40).

At higher concentrations, near 500 mM, the absorptions bands in the spectrum shift to 1346, 1414, and 1548 cm^{-1}. In addition, two more absorption bands are apparent at 1281 and 1715 cm^{-1} at the higher concentrations. The pK_a of acetic acid is 4.67. Therefore, at a pH of 4.5, below the pK_a, the solution phase is mostly composed of CH_3COOH.

The infrared data were then analyzed by assuming the peak height for each of the absorption bands had two contributions to the observed spectra; the first component is attributed to adsorbed acetic acid, and the second component is attributed to solution phase acetic acid. The first component assumed a Langmuir-type adsorption isotherm could successfully model the surface adsorption; i.e., adsorption was confined to a single-layer adsorption with one molecule per binding site, there was no interaction between the adsorbed molecules, and the uptake could be described as

$$\theta = \frac{K_{ads}C}{1 + K_{ads}C} \tag{12.5}$$

where θ is the coverage and C is the solution phase concentration. Using the peak at ca. 1540 cm^{-1}, K_{ads} was calculated by a least-squares fitting method. In the calculation, the molar absorptivity multiplied by the path length was determined from the slope of the high concentration data and Beer's law. The absorbance due to solution phase was then subtracted out to give the Langmuir isotherm and K_{ads} for acetic acid on TiO_2 nanoparticles. The results of this analysis are shown in Figure 12.7. The K_{ads} determined is 58 M^{-1}. Ongoing studies using ATR–FTIR spectroscopy for larger particles is underway. Further analysis of the infrared data will require that peak frequencies between small and larger particles be compared to determine whether the adsorption mode changes as a function of particle size. This analysis is ongoing.

The next example of using ATR–FTIR spectroscopy to study adsorption at the oxide nanoparticle/gas interface is shown in Figure 12.8. In particular, water adsorption as a function of relative humidity on nanoparticulate iron oxide is investigated. For these experiments, the iron-oxide powder was evenly dispersed over the ZnSe crystal. The relative humidity above the sample was controlled and measured. The infrared spectra recorded as a function of increasing relative humidity is shown in Figure 12.8. Water absorption bands in the O–H stretching and bending regions grow as the relative humidity increases.

The infrared data were analyzed by integrating the water bending mode absorption and fitting the data to a modified-BET isotherm. The modified equation has been in

Goodman et al. (41). It has the form of

$$V = \left(\frac{V_m c\left(\frac{P}{P_o}\right)}{1-\left(\frac{P}{P_o}\right)}\right)\left(\frac{1-(n+1)\left(\frac{P}{P_o}\right)^n + n\left(\frac{P}{P_o}\right)^{n+1}}{1+(c-1)\left(\frac{P}{P_o}\right) - c\left(\frac{P}{P_o}\right)^{n+1}}\right) \tag{12.6}$$

where

$$c = \exp - \left[\frac{\Delta H_1^o - \Delta H_2^o}{RT}\right] \tag{12.7}$$

and ΔH_1^o is the standard enthalpy of adsorption of the first layer, ΔH_2^o is the standard enthalpy of adsorption on subsequent layers and is taken as the standard enthalpy of condensation for water, R is the gas constant, T is the temperature, V is the volume of gas adsorbed at equilibrium pressure P, V_m is the volume of gas necessary to cover the surface of the adsorbent with a complete monolayer, P is the equilibrium pressure of the adsorbing gas, and P_o is the saturation vapor pressure of the adsorbing gas at that temperature. The ratio P/P_o is the activity; multiplying the activity by 100 gives the percent relative humidity. In this modified-BET equation, n is the maximum number of layers of the adsorbing gas and is related to the pore size and properties of the adsorbent. The fit to the data and the data points are shown in Figure 12.9. The fitting parameters used for the curve in Figure 12.9 are $n = 9$, $\Delta H_1^o = 52\,\text{kJ/mol}$, $\Delta H_2^o = 44\,\text{kJ/mol}$, (therefore, $c = 28$ at room temperature). Further studies of water adsorption as a function of iron-oxide nanoparticle size may reveal a size dependence on the heat of adsorption of the first layer.

12.8. FUTURE OUTLOOK

Oxide nanoparticles show unique surface reactivity, including enhanced adsorption capacity, larger equilibrium constants, and enhanced redox chemistry that cannot be explained by surface area effects alone. To better understand on a molecular level the reactivity of oxide nanoparticle surfaces, a greater number of spectroscopic studies are needed to complement batch reactor measurements. Ideally a multitechnique approach should be used in studies of oxide nanoparticle reactivity and structure so that structure-property relationships can be discerned. These techniques should include spectroscopy, microscopy, particle analysis, and diffraction methods, some of these most likely being synchrotron-based techniques. A better understanding of surface site distributions and the quantification of these sites is important. A better understanding of the electronic structure and its relationship to chemical reactivity is also needed.

In this chapter, the potential use of ATR–FTIR spectroscopy is discussed. This technique is somewhat unique in that surface adsorption can be investigated at both the nanoparticle oxide/gas and nanoparticle oxide/liquid interfaces. The molecular nature of oxide nanoparticle surfaces will most likely be different in these two phases. The quantification of surface adsorption as a function of size could be done with this technique under these different conditions. The results should be revealing and may provide additional insight into the chemical properties of oxide nanoparticles.

ACKNOWLEDGMENT

This material is based upon work supported by the National Science Foundation under Grant No. EAR0506679. Any opinions, findings, and conclusions or recommendations expressed in this material are those of the author(s) and do not necessarily reflect the views of the National Science Foundation. In addition, although the research described in this article has been funded in part by the Environmental Protection Agency through grant number EPA RD-83171701-0 to VHG it has not been subjected to the Agency's required peer and policy review and therefore does not necessarily reflect the views of the Agency and no official endorsement should be inferred.

Furthermore, the authors would like to acknowledge Dr. David Cwiertny for Mossbauer measurements on the iron oxide particles and Dr. Kevin Knagge for his contributions to the TiO_2 solution phase ATR-FTIR measurements. The authors also acknowledge the staff of the Central Microscopy Research Facility on the University of Iowa campus.

REFERENCES

(1) Banfield, J.F.; Zhang, H. Nanoparticles in the environment. In *Nanoparticles and the Environment, Vol. 44, Reviews in Mineralogy & Geochemistry*; Banfield, J.F.; Navrotsky, A. (Editors); The Mineralogical Society of America: Washington, DC, 2001; 1–58.

(2) Ranjit, K.T.; Medine, G.; Jeevanandam, P.; Martyanov, I.N.; Klabunde, K.J. Nanoparticles in environmental remediation. In *Environmental Catalysis*; Grassian, V.H. (Editor); CRC Press, Taylor & Francis Group: Boca Raton, FL, 2005; 391–420.

(3) Obare, S.; Meyer, G. Nanostructured materials for environmental remediation of organic contaminants in water. *Toxic/Hazardous Subst. Environ. Eng.* **2004**, *A39*, 2549–2582.

(4) Al-Abadleh, H.A.; Grassian, V.H. Oxide surfaces as environmental interfaces. *Surf. Sci. Rep.* **2003**, *52*, 63–161.

(5) Lucas, E.; Decker, S.; Khaleel, A.; Seitz, A.; Fultz, S.; Ponce, A.; Li, W.; Carnes, C.; Klabunde, K.J. Nanocrystalline metal oxides as unique chemical reagents/sorbents. *Chem. Eur. J.* **2001**, *7*, 2505–2510.

(6) Jefferson, D.A.; Tilley, E.E.M. The structural and physical chemistry of nanoparticles. *Partic. Matt.* **1999**, *5*, 63–84.5.

(7) Waychunas, G. Structure, aggregation and characterization of nanoparticles. In *Nanoparticles and the Environment, Vol. 44, Reviews in Mineralogy & Geochemistry*; Banfield, J.F.; Navrotsky, A. (Editors); The Mineralogical Society of America: Washington, DC, 2001; 105–166.

(8) Zhang, H.; Penn, R.L.; Hamers, R.J.; Banfield, J.F. Enhanced adsorption of molecules on surfaces of nanocrystalline particles. *J. Phys. Chem. B* **1999**, *103*, 4656–4662.

(9) Navrotsky, A. Thermochemistry of nanomaterials. In *Nanoparticles and the Environment, Vol. 44, Reviews in Mineralogy & Geochemistry*; Banfield, J.F.; Navrotsky, A. (Editors); The Mineralogical Society of America: Washington, DC, 2001; 73–103.

(10) Kakkar, R.; Napoor, P.; Klabunde, K.J. Theoretical study of the adsorption of formaldehyde on magnesium oxide nanosurfaces: Size effects and the role of low-coordinated and defect sites. *J. Phys. Chem. B* **2004**, *108*, 18140–18148.

(11) McHale, J.M.; Auroux, A.; Perrotta, A.J.; Navrotsky, A. Surface energies and thermodynamic phase stability in nanocrystalline aluminas. *Science* **1997**, *277*, 788–791.

(12) Zhang, H.; Banfield, J.F. Understanding polymorphic phase transformation behavior during growth of nanocrystalline aggregates: Insights from TiO_2. *J. Phys. Chem. B* **2000**, *104*, 3481–3487.

(13) Gilbert, B.; Huang, F.; Zhang, H.; Waychunas, G.A.; Banfield, J.F. Nanoparticles: Strained and stiff. *Science* **2004**, *305*, 651–654.

(14) Schmidt, H. Nanoparticles by chemical synthesis, processing to materials and innovative applications. *Appl. Organometal Chem.* **2001**, *15*, 331–343.

(15) Feltin, N.; Pileni, M.P. New technique for synthesizing iron ferrite magnetic nanosized Particles. *Langmuir* **1997**, *13*, 3927–3933.

(16) Fernandez-Garcia, M.; Martinez-Arias, A.; Hanson, J.C.; Rodriguez, J.A. Nanostructured oxides in chemistry: Characterization and properties. *Chem. Rev.* **2004**, *104*, 4063–4104.

(17) Cushing, B.L.; Kolesnichenko, V.L.; O'Connor, C.J. Recent advances in the liquid-phase syntheses of inorganic nanoparticles. *Chem. Rev.* **2004**, *104*, 3893–3946.

(18) Rancourt, D.G. Magnetism of earth, planetary, and environmental nanomaterials. In *Nanoparticles and the Environment, Vol. 44, Reviews in Mineralogy & Geochemistry*; Banfield, J.F.; Navrotsky, A. (Editors); The Mineralogical Society of America: Washington, DC, 2001; 217–292.

(19) Lu, H.M.; Wen, Z.; Jiang, Q. Size dependent adsorption on nanocrystal surfaces. *Chem. Phy.* **2005**, *309*, 303–307.

(20) Klabunde, K.J.; Mulukutla, R. Chemical and catalytic aspects of nanocrystals. In *Nanoscale Materials in Chemistry*; Klabunde, K.J. (Editor); John Wiley and Sons: New York, 2001; 223.

(21) Carnes, C.L.; Klabunde, K.J. Unique chemical reactivities of nanocrystalline metal oxides toward hydrogen sulfide. *Chem. Mater.* **2002**, *14*, 1806–1811.

(22) Koper, O.; Li, Y.X.; Klabunde, K.J. Destructive adsorption of chlorinated hydrocarbons on ultrafine particles of calcium-oxide. *Chem. Mater.* **1993**, *5*, 500–505.

(23) Jiang, Y.; Decker, S.; Mohs, C.; Klabunde, K.J. Catalytic solid state reactions on the surface of nanoscale metal oxide particles. *J. Catal.* **1998**, *180*, 24–35.

(24) Decker, S.; Klabunde, J.S.; Khaleel, A.; Klabunde, K.J. Catalyzed destructive adsorption of environmental toxins with nanocrystalline metal oxides. Fluoro-, chloro-, bromocarbons, sulfur, and organophosophorus compounds. *Environ. Sci. Technol.* **2002**, *36*, 762–768.

(25) Stark, J.V.; Park, D.G.; Lagadic, I.; Klabunde, K.J. Nanoscale metal oxide particles/clusters as chemical reagents. Unique surface chemistry on magnesium oxide as shown by enhanced adsorption of acid gases (sulfur dioxide and carbon dioxide) and pressure dependence. *Chem. Mater.* **1996**, *8*, 1904–1912.

(26) Anschutz, A.J.; Penn, R.L. Reduction of crystalline iron(III) oxyhydroxides using hydroquinone: Influence of phase and particle size. *Geochem. Trans.* **2005**, *6*, 60–66.

(27) Waychunas, G.A.; Kim, C.F.; Banfield, J.F. Nanoparticulate iron oxide minerals in soils and sediments: Unique properties and contaminant scavenging mechanisms. *J. Nanopart. Res.* **2005**, *7*, 409–433.

(28) Gao, Y.; Wahi, R.; Kan, A.T.; Falkner, J.C.; Colvin, V.L.; Tomson, M.B. Adsorption of cadmium on anatase nanoparticles-effect of crystal size and pH. *Langmuir* **2004**, *20*, 9585–9593.

(29) Hind, A.R.; Bhargava, S.K.; McKinnon, A. At the solid/liquid interface: FTIR/ATR—the tool of choice. *Adv. Coll. Interf. Sci.* **2001**, *93*, 91–114.

(30) Mirabella, F.M. Internal-reflection spectroscopy. *Appl. Spectrosc. Rev.* **1985**, *21*, 45–178.

(31) Duckworth, O.W.; Martin, S.T. Surface complexation and dissolution of hematite by C-1-C-6 dicarboxylic acids at pH = 5.0. *Geochim. Cosmochim. Acta* **2001**, *65*, 4289–4301.

(32) Tejedor-Tejedor, M.I.; Anderson, M.A. In situ attenuated total reflection Fourier-transform infrared studies of the goethite (alpha-FeOOH)-aqueous solution interface. *Langmuir* **1986**, *2*, 203–210.

(33) Degenhardt, J.; McQuillan, A.J. In situ ATR-FTIR spectroscopic study of adsorption of perchlorate, sulfate, and thiosulfate ions onto chromium(III) oxide hydroxide Thin. *Langmuir* **1999**, *15*, 4595–4602.

(34) Mielczarski, J.A.; Cases, J.M.; Barres, O. In situ infrared characterization of surface products of interaction of an aqueous xanthate solution with chalcopyrite, tetrahedrite, and tennantite. *J. Coll. Interf. Sci.* **1996**, *178*, 740–748.

(35) Guan, X.; Liu, Q.; Chen, G.; Shang, C. Surface complexation of condensed phosphate to aluminum hydroxide: An ATR-FTIR spectroscopic investigation. *J. Coll. Interf. Sci.* **2005**, *289*, 319–327.

(36) Hug, S.; Sulzberger, B. In situ Fourier transform infrared spectroscopic evidence for the formation of several different surface complexes of oxalate on TiO_2 in the aqueous phase. *Langmuir* **1994**, *10*, 3587–3597.

(37) Yoon, T.H.; Johnson, S.B.; Brown, G.E. Jr. Adsorption of Suwannee River fulvic acid on aluminum oxyhydroxide surfaces: An in situ ATR-FTIR study. *Langmuir* **2004**, *20*, 5655–5658.

(38) Al-Hosney, H.A.; Grassian, V.H.; Water, sulfur dioxide and nitric acid adsorption on calcium carbonate: A transmission and ATR-FTIR study. *Phys. Chem. Chem. Phy.* **2005**, *7*, 1266–1276.

(39) Al-Hosney, H.A.; Carlos-Cuellar, S.; Baltrusaitis, J.; Grassian, V.H. Heterogeneous uptake and reactivity of formic acid on calcium carbonate particles: A Knudsen cell reactor, FTIR and SEM study. *Phys. Chem. Chem. Phys.* **2005**, *7*, 3587–3595.

(40) Liao, L.; Lien, C.; Lin, J. FTIR study of adsorption and photoreactions of acetic acid on TiO_2. *Phys. Chem. Chem. Phys.* **2001**, *3*, 3831–3837.

(41) Goodman, A.L.; Bernard, E.B.; Grassian, V.H. A spectroscopic study of nitric acid and water adsorption on oxide particles: Enhanced nitric acid uptake kinetics in the presence of adsorbed water. *J. Phys. Chem. A* **2001**, *105*, 6443–6457.

CHAPTER 13

Transport Properties and Oxygen Handling

GLENN C. MATHER

Instituto de Cerámica y Vidrio, CSIC, C/ Kelsen 5, Campus Cantoblanco, 28049 Madrid, Spain

ARTURO MARTÍNEZ-ARIAS

Instituto de Catálisis y Petroleoquímica, CSIC, C/ Marie Curie 2, Campus Cantoblanco, 28049 Madrid, Spain

13.1. INTRODUCTION

The electrical and catalytic properties of oxides may be substantially altered in sintered bodies and particles with nano-sized grains in comparison with coarser grained analogues. The modification of grain size in the nanoregime introduces an additional parameter for controlling a material's characteristics; this has recently enabled materials scientists to tailor properties that are not open to manipulation at larger dimensions. For example, the dominant charge carrier and conductivity of solids may be tuned by engineering the composition and volume of interfaces (grain boundaries) and by accurately controlling the interfacial spacing (grain size). The effects of the nanoscale in oxides with ionic and mixed ionic/electronic conductivity have given rise to the fledgling science of "nanoionics," encompassing applications from fuel cells and lithium-ion batteries to sensors and catalysts (1–5). In this chapter, we explore the developing theory of the transport and oxygen-handling properties of nanoscaled oxides, describe relevant experimental techniques, and highlight the most pertinent results from the rapidly expanding literature in the area.

Synthesis, Properties, and Applications of Oxide Nanomaterials, Edited by José A. Rodríguez and Marcos Fernández-Garcia

13.2. THEORETICAL ASPECTS

13.2.1. Electrical Conduction in Oxides

The total electrical conductivity σ of a solid can be expressed as

$$\sigma = \sum_j \sigma_j \tag{13.1}$$

which is simply the sum of the partial conductivities associated with each type of charge carrier σ_j. Each partial conductivity is defined by

$$\sigma_j = c_j Z_j q \mu_j \tag{13.2}$$

where c_j is the carrier density (per m^3), $Z_j q$ is the effective charge (in coulombs), and μ_j is the mobility (m^{2}V^{-1}s^{-1}) of the *j*th species.

In solids with more than one charge carrier, the contribution to the total conductivity from each charge carrier is

$$t_j = \frac{\sigma_j}{\sigma} \tag{13.3}$$

where t_j is known as the transference number.

13.2.1.1. Electronic Conduction The number of electronic charge carriers in a metal oxide is a function of the band gap (E_g) between bonding (valence) and anti-bonding (conduction) bands. Intrinsic electronic conduction occurs when electrons are excited across the band gap, forming an electron (e$'$) in the anti-bonding band and leaving behind an electron hole (h$^{\bullet}$) in the bonding band. The bond that forms between the metal atom and oxygen is primarily due to electron transfer from the metal to the more electronegative oxygen. In terms of the band structure, the bonding electrons are localized on oxygen such that the valence band is O 2*p* in character, whereas the anti-bonding orbitals are associated with the metal. The bond state is essentially dominated by the Coulomb integrals of the Schrödinger equation, although polarization effects may lead to some back transfer and covalent character (resonance integrals). The magnitude of the band gap is effectively governed by the electropositivity of the metal and the outer-shell configuration of the cation in the lattice. In the absence of *d*-electron effects, electropositive atoms form large band gap oxides that are essentially insulating. The intrinsic densities of electrons (n_i) and holes (p_i) at temperature T conform to Boltzmann statistics:

$$n_i = p_i = (N_c N_v)^{1/2} \exp(-E_g/2kT) \tag{13.4}$$

N_c and N_v are the effective density of states in the conduction and valence bands:

$$N_{c,v} = 2(2\pi m^*_{e,h} kT/h^2)^{3/2} \tag{13.5}$$

(The effective mass of electrons or holes $m^*_{e,h}$ is frequently approximated as $m_{e^-} \approx m^*_{e,h}$.)

In the case of transition metal oxides, bonding states above the O 2*p* level may lead to partially occupied states, a vanishing band gap, and metallic conduction. The extent of orbital overlap and conduction behavior is influenced by the intra-atomic electron–electron correlation energy (U), the anion-to-metal charge-transfer energy (Δ), and the oxygen and metal bandwidths (6).

The excited electron may bind with its hole to form a quasiparticle known as an exciton, whose energy is slightly less than that of the band gap. The distance between electron and hole defines the Bohr radius a_{exc} of the exciton (7):

$$a_{\text{exc}} = \frac{4\pi\varepsilon_o\hbar^2}{m_o e^2}\varepsilon_\infty\left(\frac{1}{m_e^*} + \frac{1}{m_h^*}\right) \tag{13.6}$$

A Frenkel exciton has a radius of the order of the unit-cell parameter and is found in materials with a small dielectric constant. The much larger species that forms in materials with a higher dielectric constant is known as a Mott–Wannier exciton. The probability of the excited electron and hole recombining is limited both by the difficulty of losing the excess energy that stabilizes the exciton and the requirement for the electron and hole wave functions to overlap.

The number of electrons may be enhanced substantially above intrinsic levels on introducing nonstoichiometry either by doping with an impurity of lower valence or by reducing the oxygen content. For example, ceria is readily reduced to $CeO_{1.8}$ at 1000 °C (8) as a result of an unoccupied 4*f* energy level in the gap between Ce 5*d* and O 2*p*. The defect reaction for reduction may be written in Kröger–Vink notation as

$$O_O^\times = V_O^{\bullet\bullet} + 1/2O_2 + 2e' \tag{13.7}$$

The electrons are charge balanced by much less mobile oxygen vacancies; hence, the conductivity of ceria at high temperature or in reducing atmospheres is dominated by electron transport. The electron does not travel alone but is accompanied by the polarization which it induces, thereby trapping itself in a potential well and reducing its mobility. Large polarons are formed when the electronic species move in a broad band (typically O 2*p* orbitals), in which the coupling between electrons and phonons is weak; small polarons form in narrow bands where the coupling is stronger. In the latter case, the electron is effectively trapped on an ion (in our example as Ce^{3+}) and can only migrate by thermal activation to a neighboring ion of higher valence (Ce^{4+}) according to an Arrhenius law:

$$\sigma_e = A\exp\left(\frac{E_a}{RT}\right) \tag{13.8}$$

where A is the pre-exponential factor and E_a is the activation energy for conduction. The electron mobility is significantly reduced with respect to band conduction, and the corresponding activation energy may be as high as 0.5 eV. Analogously, electron holes may form in oxidizing atmosphere or as a result of doping with a higher valence cation. The conduction is referred to as n-type or p-type hopping-type semiconduction depending on whether the principal charge carriers are, respectively, electrons or holes.

13.2.1.2. Ionic Conduction In an analogous manner to hopping-type semiconduction, ionic conduction takes place when ions can hop from site to site within the crystal lattice as a result of thermal activation (9). The movement of matter via ionic conduction under a concentration gradient, termed diffusion, is usually interpreted on the basis of a simplified version of Fick's second law (10):

$$\frac{\partial C}{\partial t} = D\frac{\partial^2 C}{\partial x^2} \tag{13.9}$$

where C is the concentration of the diffusing substance, t is time, and D is an average diffusion coefficient at the coordinate of interest, x. From an analysis of concentration profiles as a function of t and x, different expressions can be derived from Eq. 13.9 for the diffusion coefficient; these are analysed further in Section 13.2 on experimental techniques.

Four types of mechanism have been observed for ion migration, as shown schematically in Figure 13.1.

The interstitial mechanism involves the direct jump of an interstitial defect from one interstitial site to the next (Figure 13.1a). The interstitialcy mechanism (Figure 13.1b)

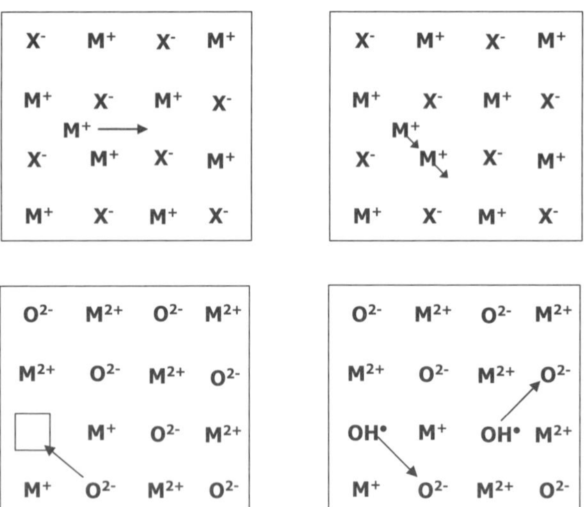

Figure 13.1. Ion migration mechanisms in ionic solids: (a) direct interstitial mechanism, (b) interstitialcy mechanism, (c) vacancy mechanism, and (d) Grotthüs mechanism.

is characterized by an ion pushing a neighboring ion from its lattice site into an interstitial site so that the original interstitial ion can occupy the vacated position, resulting in a correlated motion of interstitial and lattice ions. This type of diffusion is observed in sodium β-alumina $(Na_2O)_{1+x}(Al_2O_3)_{11}$, in which sodium ions move within the conduction planes by knocking each other off their regular sites. The Ag^+ ions in AgCl also diffuse by the interstitialcy mechanism; the much higher mobility of the Ag^+ ions in the cation sublattice with respect to the Cl^- ions leads to a phenomenon sometimes referred to as sublattice melting.

Very commonly, ionic conductivity is induced by aliovalent doping of a phase to achieve a high concentration of mobile vacancies. A material of considerable technological importance that exploits the formation and mobility of oxygen vacancies is yttrium stabilized zirconia (YSZ), a solid electrolyte used in oxygen-gas sensors and solid-oxide fuel cells (SOFCs). Replacement of some of the Zr sites in stoichiometric ZrO_2 with Y results in the formation of oxygen vacancies in order to maintain charge balance:

$$Y_2O_3 \underset{ZrO_2}{\longrightarrow} 2Y'_{Zr} + 3O_O^{\times} + V_O^{\bullet\bullet} \tag{13.1}$$

The oxygen vacancies become sufficiently mobile at high temperature under an electrochemical potential gradient to permit an oxygen-ion flux through the electrolyte (Figure 13.1*c*), which is balanced by the equivalent electron flux through an external circuit. Each oxygen vacancy is positively charged with respect to the electroneutral oxygen sublattice and is charge balanced by two Y on Zr sites. It is conceptually easier to treat the motion of a positively charged vacancy, although it is the negatively charged oxide ions that physically move (in the opposite direction to the vacancy). In contrast to CeO_2 at high temperature or under reducing conditions, the electrons that charge balance the oxygen vacancies in ZrO_2 are associated with Y^{3+} and are not mobile. Analogously, in the family of phases known as NASICONs (Na-SuperIonic Conductor), for example $Na_{1+x}Zr_2P_{3-x}SiO_{12}$, sodium vacancies are formed for $x < 0.3$. Sodium-ion conductivity can then take place in channels throughout the structure via the vacancy mechanism.

The fourth type of migration mechanism that occurs in oxides is unique to high-temperature proton conductors. These materials have the potential for application in several electrochemical devices, including hydrogen gas-separation membranes, hydrogen and humidity sensors, and protonic ceramic fuel cells (PCFCs) (11–14). The most well-known proton-conducting oxides are perovskites (ABO_3) such as $SrCeO_3$ and $BaZrO_3$, with large, basic cations, which are doped with a lower valence cation on the B site to form oxygen vacancies, e.g., $SrCe_{0.9}Yb_{0.1}O_{3-\delta}$. Water is incorporated when OH^- fills an oxygen vacancy and the complementary proton becomes associated with a lattice oxygen according to

$$H_2O + O_O^{\times} + V_O^{\bullet\bullet} \longrightarrow 2OH_O^{\bullet} \tag{13.2}$$

The protons migrate through the lattice via a Grotthüs mechanism (Figure 13.1*d*), in which the proton is transferred from one lattice oxide ion to a neighboring oxide ion in a "pass-the-parcel" fashion.

13.2.1.3. Interfacial Effects Charged species in polycrystalline bodies, principally defects and impurities, tend to segregate to the grain boundaries, thereby lowering the strain and electrostatic potential of the system (3,15). The difference in chemical potential is compensated by a difference in electrical potential such that the electrochemical potential across the boundary is constant. In effect, the grain boundaries bring about charge separation. The compensating counter charge within the grain is distributed adjacent to the interface, giving rise to a space-charge region, in which bulk defects of like charge to the boundary are depleted and defects of opposite charge accumulate. The width of the space-charge around a defect of opposite charge decays monotonically with the Debye length λ. The extent of λ for two defects of equivalent but opposite charge z is inversely proportional to the square root of the defect concentration at an infinite distance from the interface C_b according to

$$\lambda = \left(\frac{\varepsilon RT}{2z^2F^2C_b}\right)^{1/2} \tag{13.10}$$

The defect concentration within the boundary layer is a function of temperature T, the activity of the component, and doping content. In ionic or electronic conductors, the bulk mobile defect may accumulate in the space-charge region, giving rise to an additional contribution to the conductivity parallel to the interface (16):

$$\Delta\sigma^{\parallel} = (2\lambda/L)(\bar{\sigma}_{vL} + \bar{\sigma}_{iL}) \tag{13.11}$$

where $\bar{\sigma}_{vL}$ and $\bar{\sigma}_{iL}$ are the mean space-charge conductivities due to vacancies and interstitials, respectively, and L is the interfacial spacing.

In microstructured ceramics, the fractional cross-sectional area of grain boundaries lying parallel to the current flux is not large enough to ensure high ionic conductivities (3,17). For an idealized brick-layer model (Figure 13.2) with cubic grains of dimension

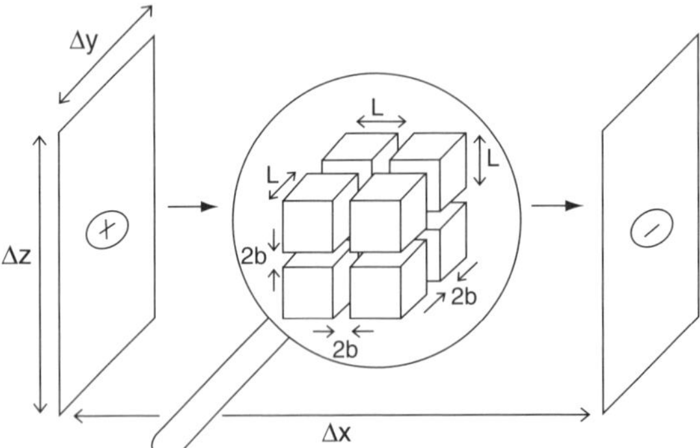

Figure 13.2. Brick-layer model of ceramic microstructure with grains of length L and grain boundaries of width $2b$ (17). Reprinted from Ref. 16 with permission from Elsevier.

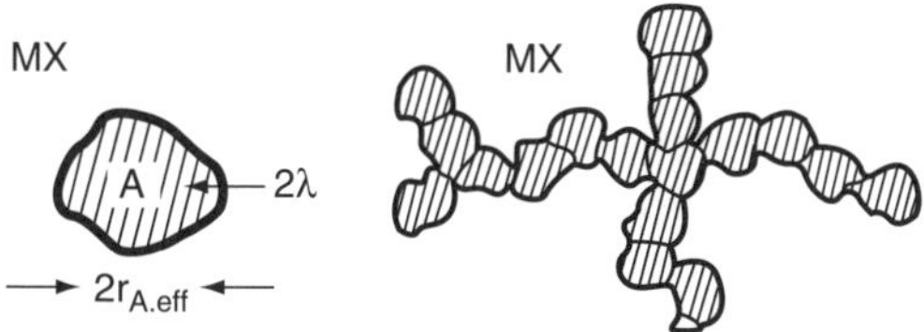

Figure 13.3. (*a*) The conducting boundary layer of an isolated particle A in the MX matrix makes no significant contribution to conductivity. (*b*) If a continuous network of A particles form, then the contribution to the conductivity is significant. Reprinted from Ref. 16 with permission from Elsevier.

L separated by grain boundaries of width $2b$, in which the conductance is equal in the grains and grain boundaries, the grain boundary conductivity σ_{gb} must be greater than the grain conductivity by

$$\sigma_{gb} = \sigma_b \left(\frac{L}{4b} \right) \tag{13.12}$$

In a microstructured ceramic with an average grain size of 1 μm and grain boundary width of 1 nm, a grain-boundary conductivity of $\sigma_{gb} = 500\,\sigma_b$ is required to simply double the total conductivity.

The enhanced conductivity in space-charge regions may be exploited more effectively by decorating the grain boundaries with a second phase (heterogeneous doping) (4). Grain-boundary engineering of this type has been used to transform insulators into conductors, anion conductors into cation conductors, and to enhance the conductivity of mixed conductors. For example, the adsorption of Ag^+ cations to form cation vacancies takes place in AgCl grains adjacent to additions of basic Al_2O_3 particles, transforming the intersticialcy conduction mechanism to a vacancy-controlled one (18). This concept is illustrated in Figure 13.3. A conducting boundary layer that forms around an isolated particle plays a small role in conduction (Figure 13.3*a*). However, if the grain-boundary paths are continuous and percolation is achieved, then a significant contribution to the conductivity may result from the enhanced transport in the space-charge regions (Figure 13.3*b*).

13.2.2. Electrical Conduction in Nanostructured Oxides

On reducing grain size to the nanoscale, the fraction of grain boundaries, and hence surface atoms, within the bulk of the material becomes extremely large. Surface or interfacial properties start to dominate bulk properties, and the conventional distinction between bulk and grain boundaries is no longer always appropriate. Maier has classified the behavioral impact on properties in terms of "trivial size effects" and "true size effects" (4,5). The former term refers to an alteration of the macroscopic behavior resulting from a sufficiently large increase in the volume fraction of interfaces, but the interfaces are essentially noninteracting and transport properties per se do not vary with the interfacial volume fraction. When the interfaces are close enough to interact, for example when space-charge regions overlap, then "true-size effects" are encountered. In a nanostructured ceramic, in which the dimensions of the grains and grain boundaries are comparable, conduction may be altered in several ways.

Impurity phases that block ion transport at grain boundaries, for example siliceous phases in YSZ, are diluted on reducing the grain size due to the much larger surface area of the grains they cover. As a result, grain-boundary resistivities may be substantially reduced. This is a trivial size effect with important technological consequences.

A true size effect results from space-charge effects of neighboring boundaries if the distance between boundaries is so small that the space-charge regions interact (19). If the grain size is less than four times the Debye length, then the center of the grain is not electrically neutral; that is, local charge neutrality is not satisfied anywhere in the sample. Figure 13.4 shows this effect for a symmetrical system (grain or film) of thickness L.

The boundary conditions are finite on both sides of the grain, and the effective boundary conductivity is enhanced compared with a conventional grain boundary (semi-infinite conditions) by a "nanosize factor" g given by

$$g \cong \frac{4\lambda}{L}\left[\frac{C_o - C_b}{C_o}\right]^{1/2} \tag{13.13}$$

where C_o is the concentration of the majority mobile defect in the first layer adjacent to the interface and C_b is the core or background concentration.

In addition to these modifications to the electrical potential, the chemical potential of an ionic or polaronic defect is altered if the region encompassing the polarized neighborhood of the defect impinges on the boundary or is confined by the particle size (20). In general, the interfacial spacing may not be large enough to separate

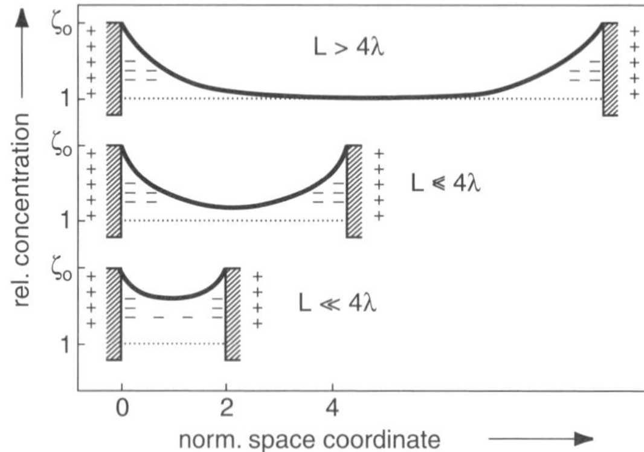

Figure 13.4. Defect profiles in structures with dimension L. The bulk defect concentration is not reached when $L \ll 4\lambda$, where λ is the Debye length (19). Reprinted from Ref. 19 with permission from Elsevier.

defect pairs. A quantum confinement effect results when the Bohr radius of an exciton approaches the grain size, spatially confining the electron-hole pair. When this happens, the energy of the lowest excited state increases and the increased band gap produces a blue shift in the optical spectra. Other defect pairs, such as ionic defect pairs, may also not be able to be separated if the particle size is small enough. The effects of quantum confinement in dimensionally reduced electronic conductors, such as observed in quantum wires and dots, form the rich and burgeoning field of "nanoelectronics."

Other complications arise at the nanoscale due to anisotropies and curvatures. The increased curvature of small particles foments an enhanced internal pressure and chemical potential in comparison with a large single crystal. As a consequence, nanocrystals exhibit reduced melting points (21).

13.2.3. Surface Phenomena

The principal phenomena involved in oxygen handling at the surface of oxides are adsorption and diffusion; multilayer adsorption and associated transport phenomena have been addressed in several reviews (22,23). From a chemical point of view, phenomena related to monolayer adsorption can generally be classified according to the nature of the chemisorption process—bond cleavage, adsorption site, electron transfer, and so on (24,25)—or the acid–base and/or redox characteristics of the oxide that govern these processes. Theoretical treatments of oxygen chemisorption on oxides are addressed in other chapters of this book. It is generally accepted that oxygen incorporation from the gas to the solid phase takes place through electron transfer from the oxide following the redox scheme: $O_2 \leftrightarrow O_2{}^{-} \leftrightarrow O_2{}^{2-} \leftrightarrow 2O^{-} \leftrightarrow 2O^{2-}$. Several experimental techniques like electron paramagnetic resonance (EPR) and Raman and infrared spectroscopy have been used to identify these different species, and are therefore powerful techniques for analyzing oxygen-handling characteristics (26); results in this respect for nanostructured CeO_{2-x} samples are outlined below (27–29).

On a microscopic scale, an atom or a molecule diffuses over a surface by performing a series of jumps between minima imposed by the surface potential. Different theoretical analyses of surface diffusion may be derived from application of a general Hamiltonian:

$$H = p^2/2m + V(r) + H_s \tag{13.14}$$

where the first two terms are the kinetic and potential energies of the diffusing substance and the last term relates to the energies of the substrate (30,31). Expressions for the diffusion matrix as a function of the microscopic surface geometry have been derived by Ala-Nissila and Ying (30). The classic diffusion regime consists of activated jumps between the minima and follows an Arrhenius-type temperature dependence:

$$D = D_0 \exp \frac{E_D}{RT} \tag{13.15}$$

where D_0 is a prefactor and E_D is the activation energy for diffusion. A quantum regime is also observed at low temperature when the diffusion is dominated by tunneling of atoms through the potential barrier, for which a temperature-independent diffusion coefficient is appropriate (31). The surface diffusion coefficient can also be given as a function of microscopic parameters as (23,32)

$$D = \frac{kd^2}{\tau} \tag{13.16}$$

where d is the average jump distance, τ is the correlation time for the motion, and k is a numerical factor that depends on the dimensionality of the system.

Simply from theoretical considerations, the surface of an oxide nanostructure [inherently defective or even amorphous (33)] must induce changes in the diffusion properties of oxygen, although there are few systematic studies of such phenomena.

13.3. EXPERIMENTAL TECHNIQUES

13.3.1. Impedance Spectroscopy

One of the most useful and widely employed techniques for the characterization of electroceramics, not least nanostructured materials, is impedance spectroscopy. The technique involves the measurement of the current (or voltage) response to an alternating voltage (or current) in a sweep of frequencies from above dc to the megahertz range. The separation of various microstructural and electrode contributions to the resistance is then usually possible as a consequence of their different relaxation times to the alternating signal.

The ac response of the test system (ceramic body and electrodes) is measured in the frequency domain by applying a single-frequency voltage:

$$V(\omega) = V_o e^{j\omega t} \quad (j = \surd(-1)) \tag{13.17}$$

resulting in an ac current:

$$I(\omega) = I_o e^{j(\omega t+\phi)} \tag{13.18}$$

of the same excitation frequency (where ω is the angular frequency, $\omega = 2\pi f$) but with phase difference (ϕ) with respect to the applied voltage.

The impedance is defined as

$$Z(\omega) = \frac{V(\omega)}{I(\omega)} \tag{13.19}$$

and

$$Z(\omega) = Z_o e^{-j\phi} = Z_o \cos\phi - jZ_o \sin\phi \tag{13.20}$$

Equation 13.20 may be expressed as the complex impedance Z^*, which is a vector quantity with real and imaginary components:

$$Z^* = Z'(\text{real}) - jZ''(\text{imaginary}) \tag{13.21}$$

For an ideal resistor, there is no phase difference between the applied voltage and the response current ($\phi = 0°$):

$$Z^* = Z' = R \tag{13.22}$$

An ideal capacitor ($\phi = \pi/2$) allows the passage of an ac, but not a dc, current because:

$$I(\omega) = C\left(\frac{dV(\omega)}{dt}\right) = \omega C V_o \sin(\omega t + \pi/2) \tag{13.23}$$

hence,

$$I_o = j\omega C V_o \tag{13.24}$$

It follows that

$$Z^* = -jZ'' \quad \text{and} \quad Z'' = \frac{1}{\omega C} \tag{13.25}$$

The ac current for an ideal capacitor is thus dependent on the magnitude of the capacitance C and on the exciting frequency, whereas the current in an ideal resistor is independent of frequency, essentially exhibiting Ohmic behavior.

The impedance Z^* at each frequency may be represented as a point in a complex plane with magnitude $|Z|$ and direction ϕ, whose magnitude is expressed by the real (Z') and imaginary ($-Z''$) components. An ideal resistor and an ideal capacitor are then represented as points on the Z' and $-Z''$ axes, respectively (Figure 13.5). The frequency dependence of capacitance results in a vertical spike along Z''.

In most ceramic materials, the impedance for each microstructural component (the bulk and grain boundaries) may be represented by an equivalent circuit composed of a parallel combination of resistance and capacitance (Figure 13.6). The impedance

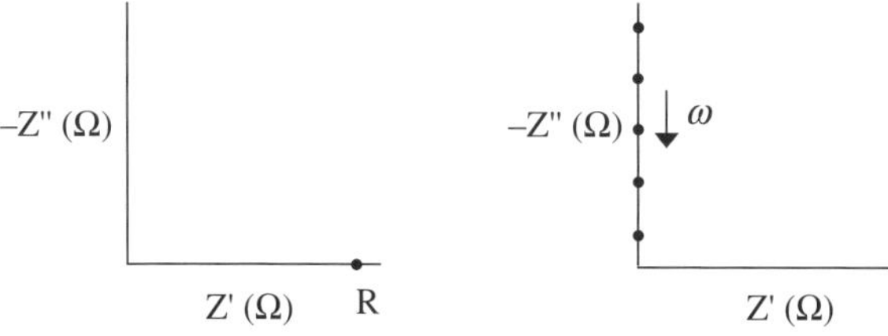

Figure 13.5. Complex plane plots of the impedance of (*a*) an ideal resistor and (*b*) an ideal capacitor.

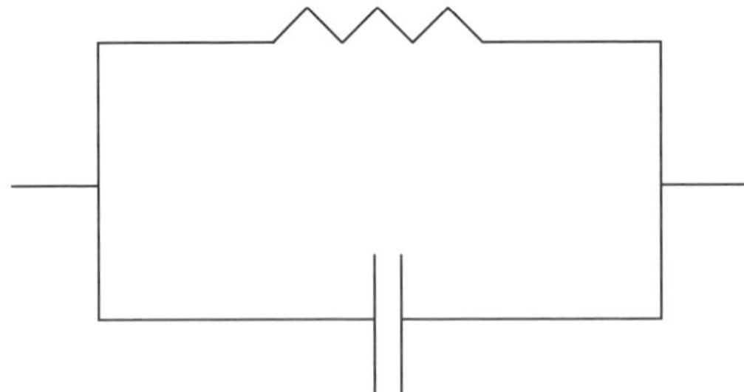

Figure 13.6. Parallel combination of resistance and capacitance representing a microstructural component.

for this case is given by the addition of the resistance and capacitance in parallel ($j\omega C$), in accordance with Kirchoff's rules:

$$\frac{1}{Z^*} = \frac{1}{R} + j\omega C \tag{13.26}$$

Taking reciprocals:

$$Z^* = \frac{R}{1 + (\omega RC)^2} - R\frac{j\omega RC}{1 + (\omega RC)^2} \tag{13.27}$$

Real and imaginary parts are thus given by

$$Z' = \frac{R}{1 + (\omega RC)^2}; \quad Z'' = R\frac{\omega RC}{1 + (\omega RC)^2} \tag{13.28}$$

which may be reformulated as

$$\left(Z' - \frac{R}{2}\right)^2 + (Z'')^2 = \left(\frac{R}{2}\right)^2 \tag{13.29}$$

This is the equation of a circle in the complex Z^* plane with radius $R/2$, centered at $R/2$. The frequency response of the parallel RC element therefore gives rise to a semicircle in the first quadrant of the plane of maximum height $R/2$ with intercepts on the Z' axis in the limits of ω with values of R ($\omega = 0$) and 0 ($\omega = \infty$). The semicircle is at a maximum at frequency $\omega_{\max}$, when Z' and $-Z''$ are equal, such that

$$\omega_{\max} RC = 1 \tag{13.30}$$

The product RC gives the time constant ($1/\omega_{\max}$) of the parallel RC element, which is proportional to the discharge time of the capacitive element.

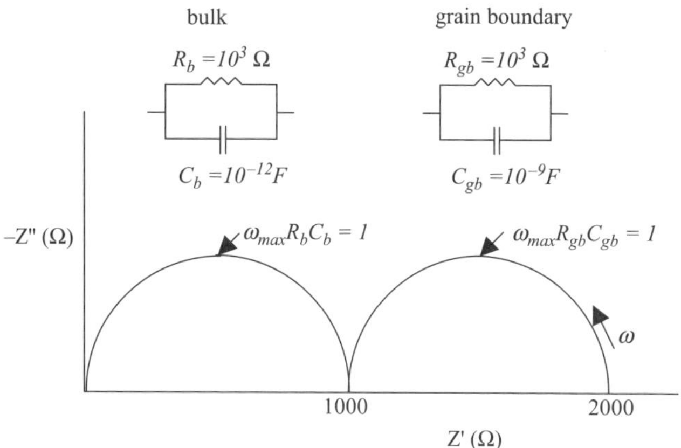

Figure 13.7. Complex Z^* plot for a ceramic with typical "brick-layer" microstructure; parallel RC circuits represent bulk and grain-boundary contributions.

Generally speaking, each microstructural component may be separated or resolved in the frequency domain provided their relaxation frequencies differ by three decades or more. This is normally the case in dense conventional ceramic materials, for which typical grain and grain-boundary capacitances are of the order 10^{-12} F and 10^{-9} F, respectively. Such a microstructure may be simulated with the brick-layer model (Figure 13.2). If the resistance of the grain boundaries is significant, then the path of least resistance for the conducting species is through the grains and across the grain boundaries. The equivalent circuit is two parallel RC elements in series, each of which gives rise to a semicircle in the complex Z^* plane (Figure 13.7).

13.3.2. Tracer Diffusion Methods

Analysis of diffusion by tracer methods provides information on macroscopic diffusion properties. A tracer is a radioactive or stable isotope that can be tracked either by its radioactive emission or by its mass. In the case of oxygen, diffusion properties are typically examined by carrying out oxygen isotopic exchange measurements with ^{18}O-containing oxygen mixtures under isothermal or temperature-programmed conditions. The oxygen exchange activity has been shown to correlate with the catalytic activity of oxide materials for oxidation reactions (34,35). Exchange with other oxygen-containing molecules, commonly isotopically labeled CO_2 (10), has also proved useful in examining diffusion properties of oxide materials. In all cases, the exchange of isotopically enriched molecules with $^{16}O^{2-}$ ions of the oxide leads to changes in the fraction of ^{18}O-containing molecules in the gas phase, which may be monitored with a mass spectrometer. According to Boreskov and Novakova (34,36), three types of exchange take place on oxide surfaces upon interaction with oxygen: (*1*) homoexchange (mechanism *I* or R^0), which occurs with passive promotion but without appreciable participation of the oxide; (*2*) simple heteroexchange (mechanism R^1 or R') with participation of one oxide ion; and (*3*) multiple heteroexchange (mechanism R^2 or R''), which involves two oxide ions from the solid. These types of exchange are described by the following equations:

$$R^0 : {}^{18}O_2(g) + {}^{16}O_2(g) \longrightarrow 2\,{}^{18}O^{16}O(g) \tag{13.3}$$

$$\begin{aligned} R^1 : {}^{18}O_2(g) + {}^{16}O(s) &\longrightarrow {}^{18}O^{16}O(g) + {}^{18}O(s) \text{ or} \\ {}^{18}O^{16}O(g) + {}^{16}O(s) &\longrightarrow {}^{16}O^{16}O(g) + {}^{18}O(s) \end{aligned}$$

$$\begin{aligned} R^2 : {}^{18}O_2(g) + 2\,{}^{16}O(s) &\longrightarrow {}^{16}O_2(g) + 2\,{}^{18}O(s) \text{ or} \\ {}^{18}O^{16}O(g) + 2\,{}^{16}O(s) &\longrightarrow {}^{16}O_2(g) + {}^{16}O(s) + {}^{18}O(s) \text{ or} \\ {}^{18}O^{16}O(g) + 2\,{}^{18}O(s) &\longrightarrow {}^{18}O_2(g) + {}^{16}O(s) + {}^{18}O(s) \end{aligned}$$

where suffixes g and s refer to the gas and solid phases, respectively. The overall rate constant of the exchange process K is usually defined by the sum of the rates for each of the three processes, whereas the exchange rate constant R is the sum of the individual rate constants of mechanisms R^1 and R^2. Identification of the type of exchange mechanism taking place at a certain temperature is usually achieved by following the concentration of the atomic fraction of ^{18}O in the gas phase as a function of the exchange temperature (37). Models have thus been developed to extract surface and bulk diffusion coefficients at a determinate temperature from analysis of the evolution of the different oxygen isotopomers as a function of time (10,37). For this, the experimental conditions are set such that it is possible to identify and separate the prevailing exchange process. Surface exchange necessarily occurs first and may be limited by surface migration, allowing the surface diffusion coefficient to be determined from (37–39)

$$D_S = \pi/4 \left(\frac{S_1}{C_x I_0} \right)^2 \quad \text{with } S_1 = \frac{dN_e^t}{d\sqrt{t}} \tag{13.31}$$

where N_e^t is the number of exchanged oxygen atoms, C_x is the initial concentration of ^{18}O (37), and I_0 is the specific perimeter of metallic particles deposited on the oxide that act as oxygen activators. The inclusion of metallic particles at the sample surface avoids difficulties in dissociatively activating oxygen on the pure oxide surface and allows the separation of the activation step from the other processes involved in the exchange. The latter stages of exchange under isothermal conditions are usually dominated by bulk diffusion (10,37). The oxygen bulk diffusion coefficient D_b can then be extracted from the final part of the exchange curve using (10):

$$\ln(\alpha_g^t - \alpha_s^0) = -\frac{\rho S}{N_g} \sqrt{\frac{4 D_b t}{\pi}} + \ln(\alpha^* - \alpha_s^0) \tag{13.32}$$

where ρ and S are the densities of oxygen atoms in the bulk and surface of the oxide, respectively, N_g is the total amount of oxygen atoms in the gas phase, and α_g^t, α_s^0, and α^* are the atomic fractions of ^{18}O at time t in the gas phase, at time 0 in the oxide and at equilibrium, respectively. Recently, models that account for cases when separation of the various exchange processes is not possible have been proposed and refined on the basis of the different possible species involved in exchange (40,41).

Isotopic exchange depth profiling (IEDP) is a widely employed method for obtaining oxygen diffusion parameters (42). Following sample preparation and exchange with ^{18}O-enriched oxygen, the concentration profiles of ^{18}O are determined by secondary ion mass spectroscopy (SIMS) (32,43). In the SIMS technique, the surface of the specimen is bombarded with a beam of primary ions, fomenting a continuous atomization of the sample. These sputtered secondary ions can then be detected in a mass spectrometer and the isotopic concentration profile measured. The isotope depth profile is fitted on the basis of theoretical profiles in order to obtain the surface exchange coefficient k_s and the bulk oxygen tracer diffusion coefficient D from the equation proposed by Crank (44):

$$C'(z,t) = \operatorname{erfc}\left(\frac{z}{2\sqrt{Dt}}\right) - \left[\exp\left(\frac{k_s z + k_s^2}{D}\right) \times \operatorname{erfc}\left(\frac{z}{2\sqrt{Dt}} + k_s\sqrt{\frac{t}{D}}\right)\right] \tag{13.33}$$

where $C'(z,t)$ is the normalized concentration of ^{18}O as a function of depth z and diffusion time t (45).

13.4. APPLICATION TO NANOSCALED OXIDES

Several groups have studied CeO_{2-x} for its very high catalytic activity and high oxide-ion mobility when doped with a suitable trivalent cation. Tuller et al. have shown that CeO_{2-x} exhibits enhanced n-type conductivity in reducing atmosphere of the order of 10^4 greater than microcrystalline ceria (46). The oxygen partial pressure (pO_2) dependence of the conductivity was used to separate ionic and electronic contributions. Generally speaking, the ionic component in oxides is essentially independent of pO_2, whereas the electronic conductivity shows a power-law dependence. Figure 13.8 shows the conductivity of nano- and microcrystalline CeO_{2-x} as a function of pO_2.

The enhanced electronic component to the conductivity has been attributed to easier vacancy formation at the grain boundaries in the nanostructured phase according to Eq. 13.7, with the formation of a higher concentration of conducting electrons in the bulk. Chiang et al. (46) associated the ease of reduction in the nanocrystalline phase with a reduction in the effective reduction enthalpy. More recently, Kim and Maier (47) attributed the behavior of nanocrystalline ceria to a space-charge effect, in which the positive carriers (oxygen vacancies and holes) are depleted and negative carriers (the conduction electrons) are concentrated in the space-charge regions. The ionic

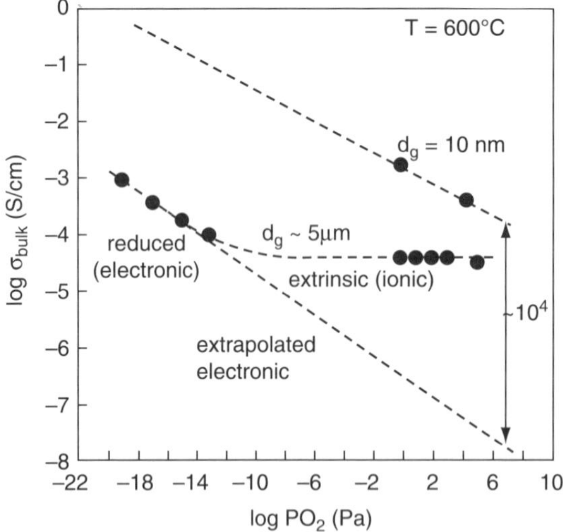

Figure 13.8. Conductivity of nanocrystalline ($L \sim 10\,\text{nm}$) and microcrystalline ($L \sim 5\,\mu\text{m}$) CeO_{2-x} as a function of oxygen partial pressure. The conductivity of the microcrystalline sample exhibits well-defined electronic and ionic regimes, whereas the nanocrystalline sample is predominantly electronic over the entire pO_2 range. From Ref. 46, with kind permission of Springer Science and Business Media.

and electronic contributions were determined in this case using electrochemical polarization with an electronically blocking YSZ electrode (complete cell configuration, Pt/CeO_{2-x}/YSZ/Pt). The space-charge model accounts for the difference between the impedance spectra of Gd-doped and undoped nanocrystalline ceria (Figure 13.9). The undoped material (Figure 13.9*a*) shows one semicircle in the complex impedance plane due to high electron conductivity in the space-charge zones, which bypasses the bulk component. In the Gd-doped phase (Figure 13.9*b*), the concentration of oxygen vacancies is much higher, giving rise to a high frequency bulk ionic response and an additional low-frequency arc, which results from grain-to-grain resistance to ionic transport.

The quantum confinement predictions for band-gap energies break down in thin films of CeO_2; the band gap energy decreases for films with grain sizes below 20 nm (48). This has been associated with the unsuitability of the effective mass approximation in CeO_2 for grain sizes less than 20 nm; instead, the effective mass decreases for smaller grain sizes. In contrast, thin films of $Zr(Y)O_{2-\delta}$ of thickness 0.7 μm show a quantum confinement effect with increasing band gap for grain sizes smaller than 30 nm (49). There is a blue shift in the optical emission spectra of both $Zr(Y)O_{2-\delta}$ and CeO_2 films in the regime where the effective mass approximation is valid. The role of quantum confinement effects on the thermodynamics of electronic and ionic defects in nanoscaled particles where space-charge effects are significant is an issue of considerable interest for the solid-state electrochemical community (1,50).

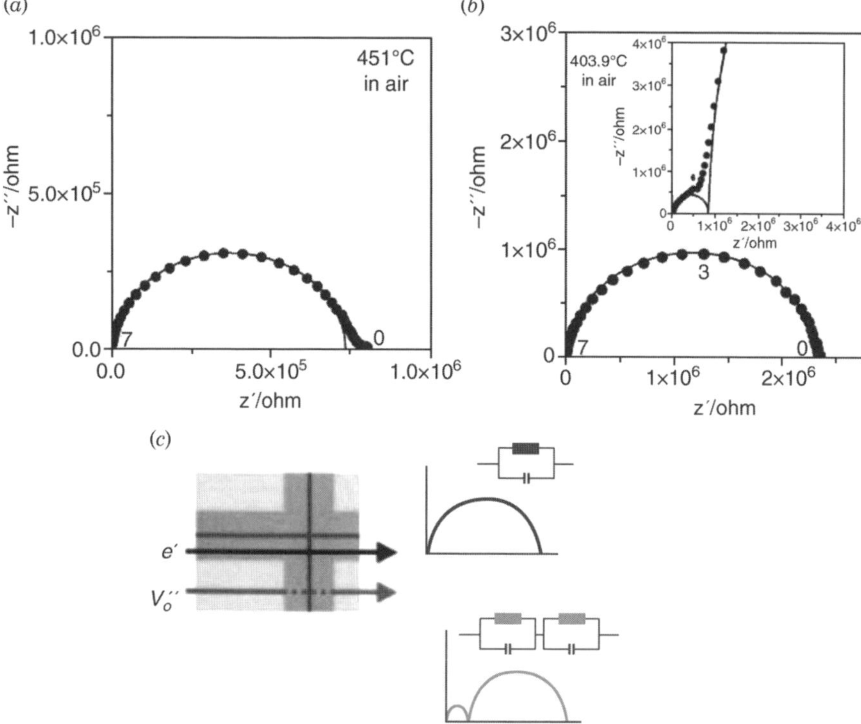

Figure 13.9. Impedance spectra of (*a*) pure nanocrystalline ceria and (*b*) 0.15 mol% Gd-doped nanocrystalline ceria. (Numbers adjacent to data points refer to the logarithm of the measurement frequency.) (*c*) Space-charge model. Reproduced from Ref. 47 with permission from The Electrochemical Society Inc. See color insert.

The reported transport properties of nanocrystalline zirconia do not yet give as consistent a picture as those of nanostructured ceria. Contrary to reports of improved ionic conductivity in the nano-regime, space-charge effects are expected to decrease rather than increase ionic conductivity (49,51). In addition, theoretical molecular dynamic simulations by Fisher and Matsubara (52) suggest that the intrinsic grain-boundary ionic conductivity is lower than that of the bulk in 8 mol% YSZ. Correspondingly, Mondal et al. (53) demonstrated that YSZ with average grain size of 50 nm exhibits grain and grain-boundary arcs in the impedance spectrum with higher conductivity and lower activation energy in the bulk than in the grain boundaries, but the values are similar to those of microcrystalline samples. In this case, the grain size may still be too large to present significant size effects. Perhaps of more practical interest is the considerable drop observed in grain-boundary resistance in stabilized zirconia on decreasing grain size which is associated with a decrease in the relative concentration of silicate phases [e.g., $Si_2Ca_3O_7$ in Ca-doped ZrO_2 (54)], which coat the grains and block ionic transport. Tuller et al. have pointed out that the grain-boundary resistance due to blocking from impurities may be exceptionally low in nanocrystalline electrolytes (3). A caveat to this is that the doped nanocrystalline CeO_2 and ZrO_2 materials for use as electrolytes in SOFCs may offer less ionic conductivity in comparison with microcrystalline ceramics due to interfacial resistance to ionic transport at grain-to-grain contacts, in addition to space-charge regions with depleted vacancy concentrations.

There are few studies of nanocrystalline oxide proton conductors. Nonetheless, one report (55) indicates that the electrical conductivity of thin films of $SrCe_{0.95}Yb_{0.05}O_{3-\delta}$ with nanosized grains is largely determined by grain boundary/interfacial effects, resulting in enhanced proton conductivity and faster kinetics in hydrogen-containing atmospheres. The time characteristics and sensitivity to hydrogen concentration indicate that the nanostructured films have potential as hydrogen sensors. A recent review of solid-state proton conductors, encompassing phosphates, inorganic-organic composites and nanomaterials has been given by Yaroslavstev (56).

Some of the most dramatic and tangible effects of nanostructure on the ionic transport of oxides are in the field of Li^+-ion batteries. The room-temperature operation of these devices does not lead to excessive coarsening of the nanostructure, in contrast to the higher operating temperature of SOFCs, whose nanoscaled components would most probably also require nanocomposite architectures to circumvent stability problems.

Li^+-ion conducting ceramic electrolytes are a major research focus for their potential application in an all solid-state Li^+-ion battery with improved durability, cost, safety, and ease of sealing. An established example of heterogeneous doping at the nanoscale is the improvement of Li^+-ion conductivity in LiI infiltrated into nanoporous Al_2O_3, which leads to an enhanced Li^+-ion vacancy conductivity in the space-charge regions of the LiI component (16,57,58). Conductivities as high as 10^{-3} Scm^{-1} at 60 °C can be achieved, which is within the range where application as a battery electrolyte becomes feasible. Analogously, oxide additions to polymer electrolytes can improve conductivity and stability. "Soggy-sand electrolytes" have recently been developed that consist of the addition of nanosized SiO_2 particles to non-aqueous liquid electrolytes (59).

The diminished transport lengths of the nanoscale may also bring about improved performance and storage capacity in Li^+-ion battery electrodes. The standard positive electrode is the intercalation compound $LiCoO_2$, which can reversibly accommodate Li without significant structural modification. $LiCoO_2$ and similar intercalation oxides (A_xMO_2: A = Li, Na; M = Mn, Fe, Co, Ni) exhibit an increasing open-circuit voltage with decreasing particle size in the range 3–15 nm for $x < 0.2$, which is rationalized in terms of a decrease in the Fermi energy (60). The storage capacity is strictly limited by the storage capacity of the host phase, however. Much higher storage capacity than the state-of-the-art cathodes has recently been demonstrated in nanoscaled CoO, which can reversibly store Li on reduction as a nanocomposite of Co/Li_2O (61).

The mechanism of oxygen exchange in various metal oxides is described in several reports (34,36,62–64). For most oxides (Cr_2O_3, MnO_2, Fe_2O_3, Co_3O_4, NiO, CuO, and ZnO), exchange takes place via two (R^1 and R^2) or three mechanisms simultaneously (Reaction **13.3**), with the relative prominence of each mechanism dependent on temperature. However, the R^2 mechanism appears to be dominant in systems with greater oxygen-vacancy diffusion, such as V_2O_5, δ-$VOPO_4$, and CeO_2, and, under certain conditions, CuO and Ag_2O (64). Recently, it was found that there was no significant difference in the isotopic exchange rate in MgO samples of different particle size (between 10 and 200 nm) and the same crystal termination plane ({1 0 0}). In contrast, the specific activity was highly dependent on the exposed surface, following the order, {1 1 1} or {1 1 0} *mean* surface planes > {1 0 0} > hydroxyl-terminated {1 1 1}, suggesting that clusters of low-coordination Mg^{2+}–O^{2-} ion pairs may play a beneficial role in exchange activity (65).

Redox processes involved in oxygen handling by nanosized CeO_{2-x} samples have been examined by employing different spectroscopic techniques (27–29). Among them, electron paramagnetic resonance (EPR) and Raman have shown to be most suitable to identify and detect the species formed during the redox processes that involve electron transfers from the partially reduced oxide to the oxygen molecules within the scheme $O_2 \leftrightarrow O_2^- \leftrightarrow O_2^{2-} \leftrightarrow 2O^- \leftrightarrow 2O^2$ (27,29); typical examples are shown in Figures 13.10 and 13.11. In addition to the information gained on oxygen-handling properties of this type of system, advantage has been taken of the sensitivity of the spectral features of the O_2-derived chemisorbed species to the chemical environment of the adsorption site for exploring the characteristics of surface defects (27–29). These techniques are thus powerful tools for analyzing oxygen chemisorption sites in this type of material. In this sense, it is remarkable the relatively strong agreement between earlier proposals for the configuration of oxygen vacancy defects in nanoscaled CeO_{2-x} samples, indicated by O_2-probe EPR experiments (27,66), and more recent STM observations of surface defects on CeO_2(1 1 1) samples (67,68).

Oxygen diffusivities of nanosized particles of metal oxides have been measured using isotopic exchange, either by analyzing the exchange rate with the gas phase as a function of time at constant temperature (37,63,64,69,70) or by obtaining the ^{18}O depth profile of the solid by SIMS after a pretreatment with isotopically labeled, oxygen-containing molecules (71). All reports concur that oxygen exchange takes place in a two-step procedure with well-defined rate constants for each step. One process is fast, occurring at or near the surface region, and the other is slow, characteristic of the bulk. Under suitable experimental conditions, surface and bulk oxygen diffusion coefficients can be measured independently (Eqs. 13.31 and 13.32), but it is pertinent to restate that the conceptual difference between bulk and surface can be obscure in nanostructured materials.

As mentioned, surface diffusion coefficients for nano-oxide systems are scarce in the literature (37,72); some comparative data are presented in Table 13.1 (following the approach of Eq. 13.32). Significant differences in surface diffusion among oxides have been attributed to variations in surface basicity and strength of the metal–oxygen bond: greater surface basicity and weaker metal–oxygen bonds are associated with higher rates of oxygen surface diffusion (37). The average primary particle sizes in these cases are not generally reported; hence, no firm conclusions can be drawn with respect to size effects. It is worth mentioning, however, that CeO_2 calcined at different temperatures in the range 773–1173 K exhibited substantial differences in the surface diffusion coefficient, suggesting that particle size can have a profound

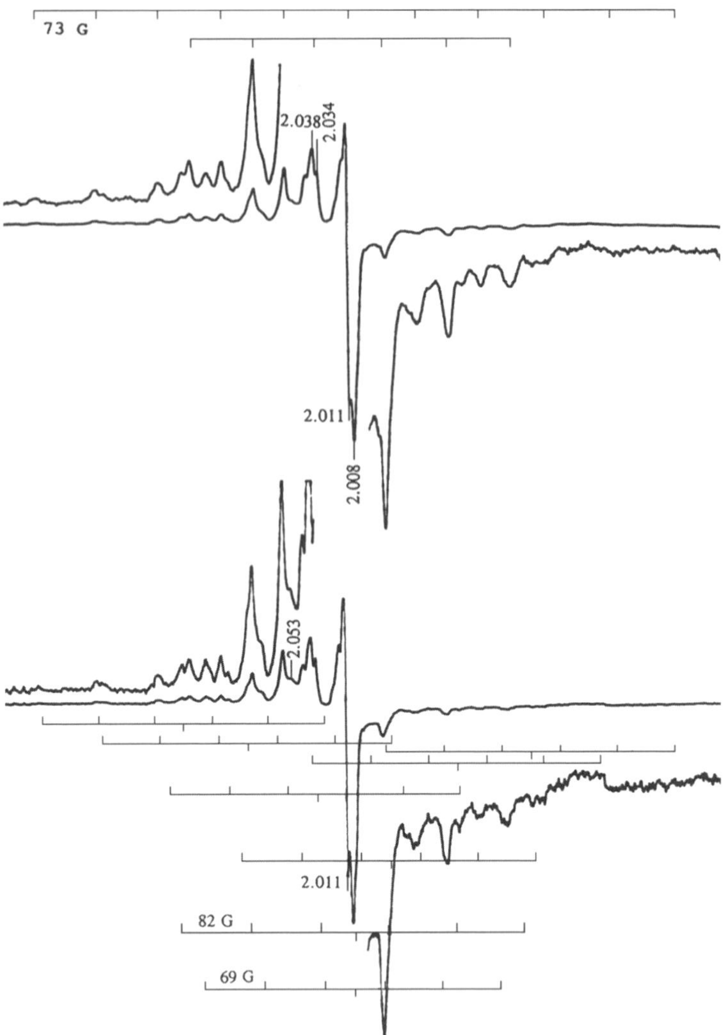

Figure 13.10. EPR spectra following low-temperature adsorption of ^{17}O-enriched oxygen mixtures on CeO_2 outgassed at 773 K. The hyperfine structure allows unambiguous identification of several O_2^- species. See Ref. 27 for details.

impact on surface diffusion (33). Oxygen bulk diffusion coefficients for a series of nanostructured oxides, determined from oxygen-exchange experiments, are also shown in Table 13.1 (37,69,70). Other exchange data (measured with isotopically labeled CO_2) are available for a series of nanostructured Ca-doped and Tb-doped CeO_2, and for Ce–Zr mixed oxides, with primary particle sizes well below 10 nm (73–75). The results of such exchange experiments for $Ce_{1-x}Tb_xO_{2-\delta}$ are shown in Figure 13.12. Analysis of these plots clearly shows, as mentioned, the presence of a

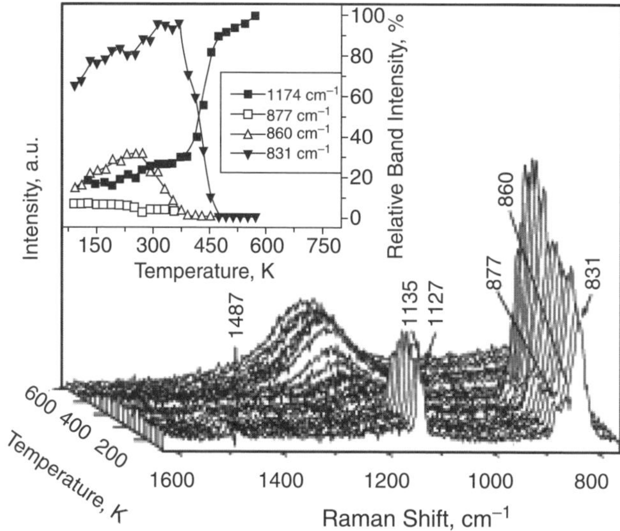

Figure 13.11. Raman spectra of oxygen adsorbed at low temperature on CeO_2 reduced with H_2 at 673 K followed by flushing with He at the indicated temperatures. O_2^- and O_2^{2-} species are clearly identified in the spectra by bands at 1135–1127 and 877–831 cm^{-1}, respectively, whereas a broad band at ca. 1174 cm^{-1} has tentatively been attributed to special oxide anions (29). Reprinted with permission from Ref. 29.

TABLE 13.1. Oxygen Surface and Bulk Diffusion Coefficients (D_s and D_b, respectively) Measured by Tracer Methods for Different Nanoscaled (based on XRD) Metal Oxides. Only Data Done Under Fairly Similar Conditions are Included for Comparative Purpose.

Metal Oxide	D_s (m^2 s^{-1})*	D_b (m^2 s^{-1})*	Reference
SiO_2	3.0×10^{-20}		37
γ-Al_2O_3	2.0×10^{-18}	7.0×10^{-23} (723 K)	37
MgO	1.0×10^{-17}		37
ZrO_2	2–6×10^{-18}	6.0×10^{-22}	37, 72
CeO_2	5.7×10^{-16}	5.0×10^{-22} (623 K)	37
CeO_2/Al_2O_3	3.7×10^{-18}	1.1×10^{-22} (573 K)	37
PdO		1–20×10^{-22} (573–673 K)	70

*Data measured at 673 K unless otherwise indicated.

first fast exchange zone, which corresponds essentially to surface exchange, whereas in the final part of the exchange, bulk diffusion dominates (Eq. 13.32). In this latter region (see inset of Figure 13.12), it can be inferred that the oxygen bulk diffusion coefficient decreases with increasing dopant content, which is also the case on doping with Ca (73–75). A maximum in oxide-ion conductivity is generally observed as a

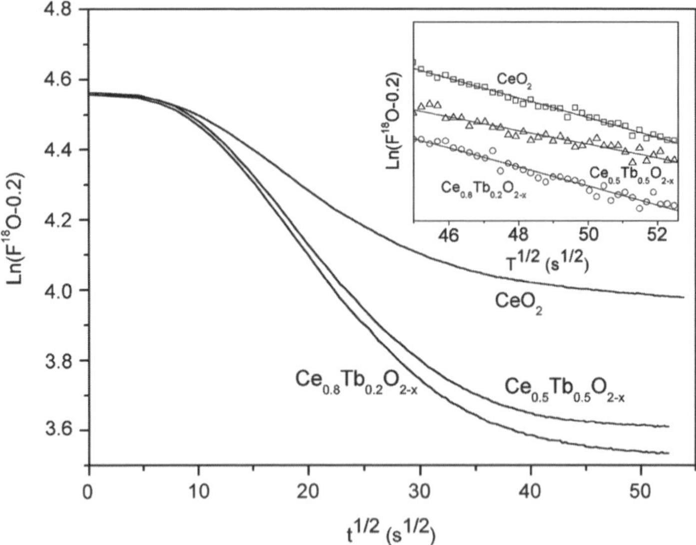

Figure 13.12. Results of isotopically labeled carbon dioxide exchange at 573 K over the indicated samples. The inset shows in detail the linear zone employed for estimation of bulk diffusion coefficients. Reprinted from Ref. 72 with permission.

function of dopant content (8,76). The oxygen vacancies that form on doping with Tb or Ca (74,75) enhance oxygen exchange and mobility at low dopant concentrations, but ordering of dopant cations and vacancies takes place at high concentrations that inhibits vacancy migration (76). For $Ce_{1-x}Tb_xO_{2-\delta}$ samples with large particle size, a maximum in ionic conductivity has been observed for $x = 0.25$ (76). However, for nanosized $Ce_{1-x}Tb_xO_{2-\delta}$, lower oxygen diffusivity is observed for $x = 0.20$ with respect to $x = 0$ (Figure 13.12). This suggests that the maximum oxygen diffusivity in nanosized $Ce_{1-x}Tb_xO_{2-\delta}$ particles occurs at lower Tb content than for extended crystals and that a gradual decrease in oxygen transport occurs with Tb doping, at least up to $x = 0.5$ (75). In the case of Ca-doped samples, the influence of carbonate impurities in the form of $CaCO_3$ may also be of relevance in explaining the significant decrease of the bulk oxygen diffusion coefficient (an order of magnitude) with calcium content (73,74).

13.5. CONCLUSIONS

The principal features of nanostructured ion- and mixed ion/electron-conducting oxides have been discussed. The effects on charge-carrier concentrations and surface-exchange kinetics can be significant due to both the large volume of interfaces and the reduction of the interfacial spacing. The overall properties may thus be governed by interfacial properties, often in dramatic fashion. It is no surprise, then, that the field of nanoionics is a burgeoning discipline that is expected to play an essential role in materials research in the coming years. As has been highlighted in this chapter, there are already developments with significant potential for Li^+-ion batteries, fuel cells, sensors, and catalysts. When applied to both new and existing materials, the powerful additional degree of freedom offered through nanoscaled engineering is likely to lead to many further advances.

ACKNOWLEDGMENTS

The authors thank the financial support by CICyT (Project MAT2003-03925) and Comunidad de Madrid (project ENERCAM S-0505/ENE-304) under which this work has been performed.

REFERENCES

(1) Schoonman, J. *Solid State Ionics* **2000**, *135*, 5.
(2) Schoonman, J. *Solid State Ionics* **2003**, *157*, 319.
(3) Tuller, H.L. *Solid State Ionics* **2000**, *131*, 143.
(4) Maier, J. *Nat. Mater.* **2005**, *4*, 805.
(5) Maier, J. *J. Electroceram.* **2004**, *13*, 593.
(6) Goodenough, J.B.; Zhou, J.-S. In *Localized to Itinerant Electronic Transition in Perovskite Oxides*; Goodenough, J.B. (Editor); Springer-Verlag: Berlin, 2001; 17–113.
(7) Dexter, D.L.; Knox, R.S. *Excitons*; Wiley-Interscience: New York, 1965.
(8) Toft Sørensen, O. (Editor); *Nonstoichiometric Oxides*; Academic Press: New York, 1981.
(9) West, A.R. *Solid State Chemistry and its Applications*; John Wiley & Sons: Chichester, 1984.
(10) Kakioka, H.; Ducarme, V.; Teichner, S.J. *J. Chim. Phys.* **1971**, *68*, 1715.
(11) Coors, W.G. *J. Power Sources* **2003**, *118*, 150.
(12) Norby, T. *Solid State Ionics* **1999**, *125*, 1.
(13) Bonanos, N. *Solid State Ionics* **1992**, *53–56*, 967.
(14) Mather, G.C.; Figueiredo, F.M.; Fagg, D.P.; Norby, T.; Jurado, J.R; Frade, J.R. *Solid State Ion.* **2003**, *158*, 333.
(15) Maier, J. *Physical Chemistry of Ionic Materials (Ions and Electrons in Solids)*; John Wiley & Sons: Chichester, 2004.
(16) Maier, J. *Prog. Solid St. Chem.* **1995**, *23*, 171.
(17) Maier, J. *Ber. Bunsenges. Phys. Chem.* **1986**, *90*, 26.
(18) Maier, J. *J. Phys. Chem. Solids* **1985**, *46*, 309.
(19) Maier, J. *Solid State Ionics* **1987**, *23*, 59.
(20) Maier, J. *Z. Phys. Chem.* **2003**, *217*, 415.
(21) Buffat, P. *J. Borel Phys. Rev. A* **1976**, *13*, 2287.
(22) Choi, J.-G.; Do, D.D.; Do, H.D. *Ind. Eng. Chem. Res.* **2001**, *40*, 4005.

(23) Rigby, S.P. *Langmuir* **2003**, *19*, 364.

(24) Calatayud, M.; Markovits, A.; Menetrey, M.; Mguig, B.; Minot, C. *Catal. Today* **2003**, *85*, 125.

(25) Calatayud, M.; Markovits, A.; Minot, C. *J. Molec. Struct.* **2004**, *709*, 87.

(26) Che, M.; Tench, A.J. *Adv. Catal.* **1983**, *32*, 1.

(27) Soria, J.; Martinez-Arias, A.; Conesa, J.C. *J. Chem. Soc. Faraday Trans.* **1995**, *91*, 1669.

(28) Descorme, C.; Madier, Y.; Duprez, D. *J. Catal.* **2000**, *196*, 167.

(29) Pushkarev, V.V.; Kovalchuk, V.I.; d'Itri, J.L. *J. Phys. Chem. B* **2004**, *108*, 5341.

(30) Ala-Nissila, T.; Ying, S.C. *Phys. Rev. B* **1990**, *42*, 10264.

(31) Chen, L.Y.; Ying, S.C. *Phys. Rev. Lett.* **1994**, *73*, 700.

(32) Heitjans, P.; Indris, S. *J. Phys. Condens. Matt.* **2003**, *15*, 1257 *and references therein.*

(33) Fernández-García, M.; Martínez-Arias, A.; Hanson, J.C.; Rodríguez, J.A. *Chem. Rev.* **2004**, *104*, 4063.

(34) Boreskov, G.K. *Adv. Catal.* **1964**, *15*, 285.

(35) Grootendorst, E.J.; Verbeek, Y.; Ponec, V. *J. Catal.* **1995**, *157*, 706.

(36) Novakova, J. *Catal. Rev.* **1970**, *4*, 77.

(37) Martin, D.; Duprez, D. *J. Phys. Chem.* **1996**, *100*, 9429.

(38) Kramer, R.; Andre, M. *J. Catal.* **1979**, *58*, 287.

(39) Cavanagh, R.R.; Yates, Jr. J.T. *J. Catal.* **1981**, *68*, 22.

(40) Galdikas, A.; Descorme, C.; Duprez, D. *Solid State Ionics* **2004**, *166*, 147.

(41) Galdikas, A.; Duprez, D.; Descorme, C. *Appl. Surf. Sci.* **2004**, *236*, 342.

(42) Manning, P.S.; Sirman, J.D.; Kilner, J.A. *Solid State Ionics* **1997**, *93*, 125.

(43) De Souza, R.A.; Chater, R.J. *Solid State Ionics* **2005**, *176*, 1915.

(44) Crank, J. *Mathematics of Diffusion, 2nd ed.*; Oxford University Press: Oxford, 1975.

(45) Vannier, R.N.; Skinner, S.J.; Chater, R.J.; Kilner, J.A.; Mairesse, G. *Solid State Ion.* **2003**, *160*, 85.

(46) Chiang, Y.-M.; Lavik, E.B.; Kosacki, I.; Tuller, H.L.; Ying, J.Y. *J. Electroceram.* **1997**, *1*, 7.

(47) Kim, S.; Maier, J. *J. Electrochem. Soc.* **2002**, *149*, J73.

(48) Kosacki, I.; Suzuki, T.; Anderson, H.U. *Ceram. Trans.* **2000**, *108*, 275.

(49) Kosacki, I.; Petrosvsky, V.; Anderson, H.U. *Appl. Phys. Lett.* **1999**, *74*, 34152.

(50) Tuller, H.L. *J. Electroceram.* **1997**, *1*, 203.

(51) Guo, X.; Maier, J. *J. Electrochem. Soc.* **2001**, *148*, E121.

(52) Fisher, C.A.J.; Matsubara, H. *J. Eur. Ceram. Soc.* **1999**, *19*, 705.

(53) Mondal, P.; Klein, A.; Jaegermann, W.; Hahn, H. *Solid State Ion.* **1999**, *119*, 331.

(54) Aoki, M.; Chiang, Y.-M.; Kosacki, I.; Lee, J.-R.; Tuller, H.L.; Liu, Y.J. *J. Am. Ceram. Soc.* **1996**, *79*, 1169.

(55) Kosacki, I.; Anderson, H.U. *Solid State Ionics* **1997**, *97*, 429.

(56) Yarolslavstev, A.B. *Solid State Ionics* **2005**, *176*, 2935.

(57) Liang, C.C. *J. Electrochem. Soc.* **1973**, *120*, 1289.

(58) Maekawa, H.; Tanaka, R.; Sato, T.; Fujimaki, Y.; Yamamura, T. *Solid State Ionics* **2004**, *175*, 281.

(59) Bhattacharyya, A.J.; Dollé, M.; Maier, J. *Electrochem. Solid State Lett.* **2004**, *7*, A432–A434.

(60) Han, S.D.; Treuil, N.; Campet, G.; Portier, J.; Delmas, C.; Lassegues, J.C.; Pierre, A. *Mater. Sci. Forum* **1994**, *152–153*, 217.

(61) Poizet, P.; Laruelle, S.; Grugeon, S.; Dupont, L.; Tarascon, J.-M. *Nature* **2000**, *407*, 496.

(62) Winter, E.R.S. *J. Chem. Soc. A* **1968**, 2289.

(63) Doornkamp, C.; Clement, M.; Ponec, V. *J. Catal.* **1999**, *182*, 390.

(64) Doornkamp, C.; Clement, M.; Gao, X.; Deo, G.; Wachs, I.E.; Ponec, V. *J. Catal.* **1999**, *185*, 415.

(65) Mellor, I.M.; Burrows, A.; Coluccia, S.; Hargreaves, J.S.J.; Joyner, R.W.; Kiely, C.J.; Martra, G.; Stockenhuber, M.; Tang, W.M. *J. Catal.* **2005**, *234*, 14.

(66) Martinez-Arias, A.; Soria, J.; Conesa, J.C.; Seoane, X.L.; Arcoya, A.; Cataluña, R. *J. Chem. Soc. Faraday Trans.* **1995**, *91*, 1679.

(67) Nörenberg, H.; Briggs, G.A.D. *Phys. Rev. Lett.* **1997**, *79*, 4222.

(68) Esch, F.; Fabris, S.; Zhou, L.; Montini, T.; Africh, C.; Fornasiero, P.; Comelli, G.; Rosei, R. *Science* **2005**, *309*, 752.

(69) Karasuda, T.; Aika, K.-I. *J. Catal.* **1997**, *171*, 439.

(70) Yeung, J.A.; Bell, A.T.; Iglesia, E. *J. Catal.* **1999**, *185*, 213.

(71) Ishihara, I.; Kilner, J.A.; Honda, M.; Sakai, N.; Yokokawa, H.; Takita, Y. *Solid State Ion.* **1998**, *113–115*, 593.

(72) Duprez, D. *Stud. Surf. Sci. Catal.* **1997**, *112*, 13.

(73) Fernández-García, M.; Martínez-Arias, A.; Guerrero-Ruiz, A.; Conesa, J.C.; Soria, J. *J. Catal.* **2002**, *211*, 326.

(74) Iglesias-Juez, A.; Hungría, A.B.; Gálvez, O.; Fernández-García, M.; Martínez-Arias, A.; Guerrero-Ruiz, A.; Conesa, J.C.; Soria, J. *Stud. Surf. Sci. Catal.* **2001**, *138*, 347.

(75) Hungría, A.B.; Martínez-Arias, A.; Fernández-García, M.; Iglesias-Juez, A.; Guerrero-Ruiz, A.; Calvino, J.J.; Conesa, J.C.; Soria. J. *Chem. Mater.* **2003**, *15*, 4309.

(76) Shuk, P.; Greenblatt, M.; Croft, M. *Chem. Mater.* **1999**, *11*, 473.

PART V

INDUSTRIAL/TECHNOLOGICAL APPLICATIONS OF OXIDE NANOMATERIALS

The physical and chemical properties of oxide nanomaterials described in the previous chapters of this book in many cases are crucial for technological applications. Within the last decade, many areas of the industry have witnessed the advent of nanoscience. The fifth and final part of this book is focused on technological uses of nanostructured oxides as adsorbents, sensors, ceramic materials, photo-devices, and catalysts for reducing environmental pollution, transforming hydrocarbons, and producing H_2. In future energy and environmental applications, technologies based on adsorption will play a major role in air and water purification, desulfurization of fuels, the removal of carbon monoxide from hydrogen for fuel cell applications, and so on. New and better adsorbents are required to meet the challenges. Nanostructured metal oxides are expected to play a prominent role as effective adsorbents for the above-mentioned applications (Chapter 14). Furthermore, the detection of various species in gas mixtures or liquid solutions, in order to evaluate the impact of these species in the quality of a commercial product or their effects on the environment (odors, toxic species etc.), is a challenging task that can be accomplished by employing oxide nanomaterials in sensor manufacturing (Chapter 15). The electron and oxygen-ion conducting capabilities of many oxides improve when going from the bulk to the nanoscale. In recent years, several oxide nanostructures have been successfully used for the fabrication of photovoltaic, photoelectronic, and electrochemical devices (Chapter 16), and they can have a very high activity and good performance as photocatalysts (Chapter 17). The electronic perturbations induced by quantum confinement and the structural defects usually associated with oxide nanoparticles have a strong impact on catalysts used for the combustion of hydrocarbons (Chapter 18), the prevention of environmental pollution by NO_x (Chapter 19) or SO_x species (Chapter 20), and the production of H_2 by hydrocarbon steam reforming or the water-gas shift reaction (Chapter 21). Low-cost oxide-conducting nanoarrays are very promising materials for building electrodes and

Synthesis, Properties, and Applications of Oxide Nanomaterials, Edited by José A. Rodríguez and Marcos Fernández-Garcia

other components in fuel cells (Chapter 21). The last chapter of the book deals with oxide nanomaterials in ceramics. These systems can lead to advanced ceramics that exhibit superior mechanical properties, corrosion/oxidation resistance, and thermal, electrical, optical, or magnetic properties (Chapter 22).

CHAPTER 14

Adsorbents

PETHAIYAN JEEVANANDAM and KENNETH J. KLABUNDE

111, Willard Hall, Department of Chemistry, Kansas State University, Manhattan, KS, 66506, USA

14.1. INTRODUCTION

In future energy and environmental applications, technologies based on adsorption will play a major role, e.g., in air and water purification, desulfurization of fuels, and removal of carbon monoxide from hydrogen for fuel cell applications (1). Currently available adsorbents such as activated alumina, zeolite, activated carbon, and silica gel cannot fulfill the stringent rules by the government regulatory agencies. New and better adsorbents are required to meet the challenges. Nanostructured metal oxides are expected to play a prominent role as effective adsorbents for the above-mentioned applications (2). Nanoscale materials belong to a unique family of compounds (3). They are considered as materials that form a bridge between molecular and condensed matter. Nanoparticles consist of tiny particles with size ranging from about 2 to 10 nm. They fall in a regime where neither quantum chemistry nor the classic theory of physics can be applied. The extensive delocalization of valence electrons in nanoparticles varies with the size leading to a variety of interesting physical and chemical properties (4). Several properties depend on the size of nanoparticles, and the examples include optical, magnetic, and mechanical properties. For example, surfaces can be more reactive for nanoparticles (5), coercive force in magnetic materials can be changed (6), band gaps in semiconductors can be altered (4), melting points can change, and mechanical properties such as hardness and plasticity can be modified by a change in size (7,8). Throughout this chapter, we will come across terms such as nanoparticle, nanocrystal, nanoscale material, nanophase material, and nanostructured material. Let us clarify these terms in detail (2). Nanoparticle is a particle with size varying from 1 to 1000 nm, and it could be amorphous, single crystalline, or even an aggregate of small crystallites. A nanocrystal is a single crystal with a size of few nanometers; a nanoparticle need not necessarily be crystalline. Nanophase,

Synthesis, Properties, and Applications of Oxide Nanomaterials, Edited by José A. Rodríguez and Marcos Fernández-Garcia

nanoscale, or nanostructured material all refer to a solid that has nanometer size in either one (wire) or two (film) or three dimensions (particle).

The dependence of physicochemical properties of nanoparticles on size opens a huge number of possibilities of new materials that can be produced. Generally, nanomaterials can be classified into three types: metals, semiconductors, and insulators. In the case of metals and semiconductors, the changes in their physicochemical properties with size are well documented; e.g., the progression from discrete molecular orbitals to a continuum of energy levels (bands) has been demonstrated. However, in the case of insulators, most of which are metal oxides, the size effects are not clearly understood, but they seem to be mainly based on differences in surface reactivity (9,10). Surface effects in nanoparticles are very important in shaping new advances in technologies for the present and the future. In the next section, we shall discuss the effect of nanosize on the surface reactivity of nanostructured metal oxides. Whenever we mention nanostructured oxides or nanoscale oxides, they mean the same.

14.1.1. Reactivity of Nanostructured Metal Oxides: Effect of Nanosize

The surface effects in nanoparticles depend on the particle size (a nanoparticle is defined here as a loose aggregate of nanocrystals) and crystallite size. When the particle size is small, a larger fraction of atoms would be on the surface of nanoparticles (9,10). Nanocrystalline oxides, in general, possess small crystallites; e.g., MgO nanocrystals prepared by an aerogel method (hereafter referred to as AP–MgO) have a crystallite size of about 4 nm (9,10), AP–CaO has a crystallite size of about 7 nm (9–11) while nanocrystalline CuO (NC–CuO) has a size varying from 7 to 9 nm, and NC–NiO has a crystallite size of about 3–5 nm (12). As the crystallites are small, the surface area of these oxides is usually very high (Table 14.1) (13), ranging from about 400 to 800 m^2/g. The surface area of AP–MgO varies from 250 to 500 m^2/g,

TABLE 14.1. Surface Area, Pore Volume, and Pore Diameter of Nanostructured Metal Oxides Prepared by the MAP Process. Reprinted from Ref. 13 with Permission of The Royal Society of Chemistry.

Sample	Surface Area (m^2/g)	Pore Diameter (nm)	Pore Volume (cm^3/g)
AP–$MgAl_2O_4$	639	18	2.88
AP–$CaAl_2O_4$	517	28	3.56
AP–$SrAl_2O_4$	112	32	0.89
AP–$BaAl_2O_4$	135	31	1.05
AP–MgO	385	21	2.03
AP–CaO	67	13	0.22
AP–SrO	16	13	0.05
AP–Al_2O_3	637	20	3.17

whereas a conventionally prepared MgO sample (referred to as CP–MgO, prepared by boiling commercial MgO in water followed by dehydration) has a surface area varying from 130 to 250 m^2/g. The surface area of a commercial MgO sample (referred to as CM–MgO), however, is only about 10–30 m^2/g. A similar trend is observed for other nanostructured metal oxides such as aluminum oxide and calcium oxide (AP–Al_2O_3 and AP–CaO). AP–Al_2O_3 possesses a surface area of about 800 m^2/g, whereas a commercial Al_2O_3 sample has a surface area of only about 100 m^2/g (14). The surface areas of AP–CaO, CP–CaO, and CM–CaO are about 120–150 m^2/g, 50–100 m^2/g, and about 1–3 m^2/g, respectively.

The increased surface area in nanostructured metal oxides leads to an increased reactivity as they are able to react by a greater extent compared with normal crystals (e.g., CM–MgO). There are more molecules of metal oxide in nanocrystals, available on the surface, for the reaction with the incoming molecules. However, it is also now well established that nanoscale oxides possess *unique* surface chemistry (9,10). In general, metal-oxide nanocrystals (AP–metal oxides) exhibit higher surface chemical reactivities compared with CP (microcrystals) and CM (normal crystals) oxides; the surface chemistry of AP–MgO is different from that of CP–MgO and CM–MgO. For example, AP–MgO adsorbs about six molecules of SO_2/nm^2, whereas CM–MgO adsorbs only about 0.7 molecules of SO_2/nm^2 (15) (Table 14.2). AP–MgO has higher adsorption capacities for acid gases such as HCl, HBr, NO, and SO_3 compared CP–MgO (16). AP–CaO is 30 times more effective in destroying CCl_4 compared with CM–CaO (9,10,17). In the destructive adsorption of dimethyl methyl phosphonate (DMMP), AP–MgO is about 50 times more effective compared with CM–MgO (18,19). One might think that the increased reactivity for nanostructured metal oxides is due to increased surface areas, but it has been proved that it is not the only factor. High surface area is beneficial, but nanocrystals show *intrinsically greater chemical reactivities* compared with microcrystals (CP-oxides) and normal crystals (CM oxides) (9,10,20). The factors responsible for the higher reactivity of nanoscale oxides are believed to be

(*1*) unusual morphology of the crystallites and aggregates (20); in contrast to normal crystals (e.g., CM–MgO), oxide nanocrystals (e.g., AP–MgO) possess

TABLE 14.2. Sulfur Dioxide Adsorption on Nanoscale Metal Oxides. Reprinted from Ref. 14 with Permission.

Sample	Surface Area (m^2/g)	Total Molecules SO_2/nm^2 Adsorbed	Molecules of SO_2/nm^2 Physisorbed	Molecules of SO_2/nm^2 Chemisorbed
AP–Al_2O_3	800	3.5	1.8	1.7
CM–Al_2O_3	100	3.5	3.0	0.45
AP–Al_2O_3/MgO	750	6.8	2.9	3.9
CM–MgO	30	0.68	0.51	0.17
AP–MgO	500	6.0	1.5	4.5

high index crystal faces that are more reactive than the normal (100) face, especially because lower coordination ions such as Mg^{2+}_{3c}, Mg^{2+}_{4c}, O^{2-}_{3c}, and O^{2-}_{4c} occur on the high index faces;

(*2*) a high ratio of edge ions to surface ions;

(*3*) lattice disorder and defects; morphological studies by transverse electromagnetic microscopy (TEM), XRD, and atomic force microscopy (AFM) measurements conclude that nanocrystals possess more defects compared with microcrystals (e.g., CP–MgO); the defects can be vacancies (Scottky-type, e.g., missing Mg^{2+} or O^{2-} ions) or unusual configurations of corners, crystal planes, or edges; and

(*4*) smaller particle size; the diffusion of reagent molecules becomes easy with nanoscale oxides because the particle size is small and diffusing only through top-most layers of the aggregated nanocrystals is enough to reach the core of the particles. In terms of the chemical nature of the surface too, nanostructured oxides are different compared with CP and CM oxides (9,10,21).

It was found, by careful studies on the residual surface hydroxyl groups, AP–MgO nanocrystals possesses only about 5% bridged hydroxyl groups, whereas CP–MgO possesses about 22% of bridged hydroxyl groups. In other words, AP–MgO possesses few geminal hydroxyl groups compared with CP–MgO (22). Also, the hydroxyl groups on nanocrystalline MgO (AP–MgO) are less acidic compared with the hydroxyl groups on CP–MgO. As nanoparticles are usually sensitive to air and water, creative and skillful synthetic procedures are needed to make them in pure form. In the following section, the synthesis of nanostructured metal oxides, by a process called MAP, and studies on their textural properties are described.

14.2. SYNTHESIS OF NANOSCALE METAL OXIDES BY THE MAP (MODIFIED AEROGEL PROCESS) AND THEIR TEXTURAL PROPERTIES

The aerogel method is widely used in synthesizing nanoscale oxides with very high surface area. This method was originally developed by Kistler in 1932 in an attempt to produce high-surface-area silica (23). It involves hydrolyzing an alkoxide precursor in a solvent (e.g., ethanol), and a gel is formed at the end (sol-gel process). The gel, also called an alcogel, is subsequently subjected to a supercritical (also called hypercritical) drying process where the gel is heated in an autoclave above a temperature and pressure corresponding to the supercritical temperature and pressure of the solvent. This way, the pore structure of the gel is not destroyed due to surface tension effects and the porous nature is retained after drying. The gelation process was slow, and to make the gelation process faster, Teichner introduced organic solvents instead of water during the gelation/hydrolysis step which reduced the synthesis time (24). Additional modification of the synthesis was to add large amounts of hydrocarbons such as toluene during the gelation process, and this further enhanced the rate of gelation (25,26). The addition of the solvent with low dielectric constant reduces the surface tension of the solvent mixture (e.g., toluene-ethanol), and it facilitates the removal of solvent during the alcogel-to-aerogel transformation. The product after the supercritical drying is called an aerogel because air replaces the solvent in between the pores of the material. Thermal dehydration of the aerogels (usually hydroxides) in vacuum results in aerogel oxides with a high surface area. If the gel is dried under ambient conditions instead of supercritical drying, the porous network structure is destroyed due to surface tension effects. The product obtained in this case is called a xerogel, and xerogels usually possess lower surface areas compared with the aerogels. Also, oxides prepared by the aerogel method have very low densities; e.g., AP–MgO has a density of only $0.3\,cm^3/g$ (20). The MAP has been used to produce various nanostructured metal-oxide and mixed metal-oxide systems, and multi-gram quantities of metal-oxide nanoparticles can be produced by this method.

14.2.1. Intimately Mixed Nanostructured Metal Oxides

A combination of two metals in an oxide matrix or the combination of two metal oxides at the molecular level can lead to materials with novel structural or electronic properties; e.g., they can give rise to superior catalytic and chemical reactivity compared with constituent metal oxides (27). The MAP process for the production of intimately intermingled mixed metal oxides has been promising to obtain homogeneous and molecular level mixing of two or more metal oxides. Hydrolyzing two metal alkoxides simultaneously during the gelation process, followed by the supercritical drying, yields intimately mixed metal oxides (13,28). The basic synthetic procedure is illustrated are as follows (28):

$$M_a(OR)_x + M_b(OR)_x + H_2O + \text{solvent} \longleftrightarrow M_aM_b(OH)_x + ROH + \text{solvent} \quad \text{(hydrolysis step)} \tag{14.1}$$

$$M_aM_b(OH)_x + ROH + \text{solvent} \longrightarrow [M_aM_b(OH)_x]_y \text{ with high surface area} \quad \text{(gelation step)} \tag{14.2}$$

$$[M_aM_b(OH)_x]_y \longrightarrow M_aM_bO_2 \text{ or } M_a(M_b)_2O_4 \text{ nanocrystalline powder} \quad \text{(thermal dehydration step)} \tag{14.3}$$

In one specific example, an intimately mixed 1:1 molar ratio mixture of MgO and Al_2O_3 was prepared (13,28). The specific surface area of the oxide remained as high for AP–Al_2O_3 itself ($\sim 640\,m^2/g$), indicating that the aluminum-oxide matrix served as a template for MgO moieties without loss in its surface area. The stronger Lewis basicity of the MgO nanocrystals was preserved and in fact enhanced due to the higher surface area. The adsorptive performance of intimately mixed oxides was found to be better than that of constituent metal oxides, e.g., nanostructured Al_2O_3 (AP–Al_2O_3), under a certain set of standard conditions, adsorbs about 3.5 molecules of SO_2/nm^2, whereas AP–MgO adsorbs about 6.0 molecules, but the intimately mixed sample (AP–$Al_2O_3 \cdot MgO$) adsorbs about 6.8 SO_2 molecules, (Table 14.2). It is also interesting to note that one can tune the acid/base sites in the intimately mixed oxide, during the MAP synthetic process, to get better adsorption properties. The enhanced reactivity of AP–$Al_2O_3 \cdot MgO$ has been attributed to the Lewis base nature of the tiny MgO crystals that are well dispersed within the large pore volume and surface area of the aluminum-oxide matrix.

14.2.2. Carbon-Coated Nanoscale Metal Oxides

Nanostructured metal oxides (e.g., AP–MgO) are sometimes partially deactivated due to the adsorption of water either as liquid or as vapor. To avoid the adsorption of water and prevent deactivation, carbon has been deposited on the surface of AP–MgO (29,30). Magnesium-oxide–carbon-core/shell composites with graphitic carbon as the shell material can be prepared. The carbon aerogel was added to the already formed MgO nanocrystals and followed by heat treatment, the core-shell structure could be obtained (29). The carbon-coated MgO nanocrystals can also be prepared by heating AP–MgO under flowing butadiene at high temperatures (500 °C) (30). The carbon-coated oxide nanocomposites offer some unique advantages with regard to adsorption selectivity compared with the uncoated nanocrystalline oxides, e.g., minimization of water adsorption and in favor of better adsorption of polar organic molecules. The simultaneous availability of hydrophilic (oxide phase) as well as hydrophobic phase (carbon phase) on the surface makes these core-shell composite materials very interesting for further adsorption studies.

14.2.3. Morphologies of Nanostructured Metal Oxides

Nanocrystals of metal oxides (mainly those belonging to alkaline-earth metals family) can be prepared with unique shapes, e.g., polyhedrons, hexagonal platelets, and cubes. These metal oxides can be stable, and they are highly ionic with high melting points. In Figure 14.1, TEM images of MgO normal crystals (CM–MgO), microcrystals (CP–MgO), and nanocrystals (AP–MgO synthesized by the MAP process) are shown (9,10,21). In Figure 14.2, the TEM images of normal crystals and nanocrystals of SrO (CM–SrO and AP–SrO) are shown (31). It can be clearly noticed that the morphologies of the nanocrystals are remarkably different compared with the micro- and normal crystals. Let us discuss the morphologies of the oxide materials in detail.

CM–MgO exhibits cube-like morphology, CP–MgO exhibits plate-like morphology, but AP–MgO shows a porous web-like morphology. CM–SrO shows large platelets and cubes (Figure 14.2*a*), whereas AP–SrO, produced by the MAP process followed by the dehydration procedure, possesses SrO strands aggregated into porous brushes (Figure 14.2*b*). Each particle shows a dense body with a tail of crystalline needles, and "flat-fish" type particles of size about 100 nm could be seen (31). It is interesting to compare the low- and high-magnification TEM images of AP–MgO (Figures 14.1*c* and 14.3). The TEM image at lower magnification (Figure 14.1*c*) shows the presence of porous web-like aggregates (size ~1400 nm). Small angle X-ray

Figure 14.1. TEM images of (*a*) normal polycrystalline MgO (CM–MgO), (*b*) CP–MgO, and (*c*) nanoscale MgO (AP–MgO).

scattering (SAXS) studies also provided additional evidence for the presence of these web-like structures (20). The high-resolution TEM (HRTEM) image of nanoscale MgO (AP–MgO; Figure 14.3) shows very interesting facets of its morphology (20). Numerous "cubic" crystallites (size about 4 nm) can be seen, and they aggregate (these aggregates can be seen in the low-resolution TEM image, Figure 14.1*c*) into polyhedral structures with plenty of edge/corner sites. The polyhedral structure of AP–MgO suggests high concentrations of edge and corner sites on the surface. The edges of the nanoparticles are rough, and the embryonic formation of pores between the crystalline structures can be noticed. The number and nature of the surface defects determine the reactivity of nanoscale metal oxides, and it is believed that these

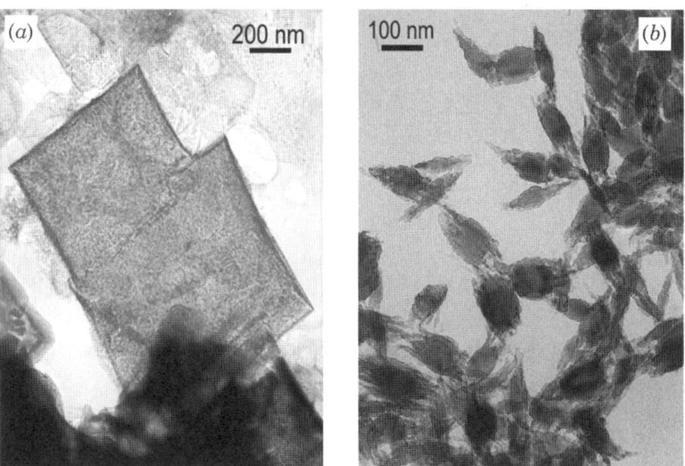

Figure 14.2. TEM images of (*a*) normal polycrystalline SrO and (*b*) nanocrystalline SrO (AP–SrO). Reprinted from Ref. 31 with kind permission of Springer Science and Business Media.

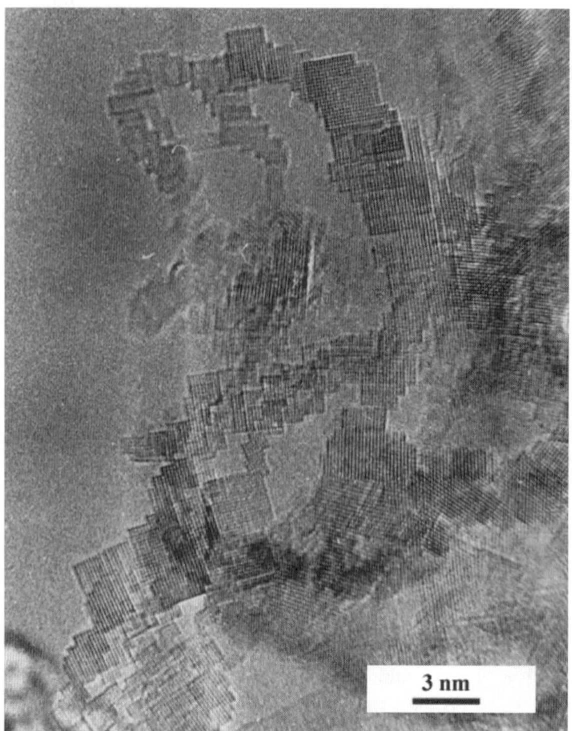

Figure 14.3. HRTEM image of nanoscale MgO (AP–MgO). Reprinted from Ref. 20 with permission.

defects get introduced during the synthesis of oxides. It has also been demonstrated that the acid/base properties are stronger at corner and edge sites in nanocrystalline oxides.

The HRTEM images of intimately mixed oxides also show interesting features. One such HRTEM picture for the mixed metal-oxide AP–$Al_2O_3 \cdot MgO$ is shown in Figure 14.4. Boehmite (AlOOH) planes mixed with MgO nanocrystals can be seen (13,28). The distance between the boehmite planes was found to be about 15 Å, and the increased lattice spacing compared with the pristine boehmite has been attributed to the intercalation of tiny MgO moieties within the boehmite sheets. The powder XRD pattern for AP–$Al_2O_3 \cdot MgO$ indicated only the presence of small MgO crystallites, and no peaks due to Al_2O_3 were observed. In Figure 14.5, a HRTEM image of the core/shell graphitized nanocrystalline MgO–carbon composite is shown (30). The carbon deposits (size as about a few nanometers) at active sites, such as on MgO crystal edges and corners, can be clearly seen. The carbon nano-regimes (hydrophobic) intimately intermingled with the metal-oxide nano-regimes (hydrophilic) are expected to allow selective physisorption of organic molecules. In the next section, surface areas and pore structures in nanostructured metal oxides, prepared by the MAP process, are briefly discussed.

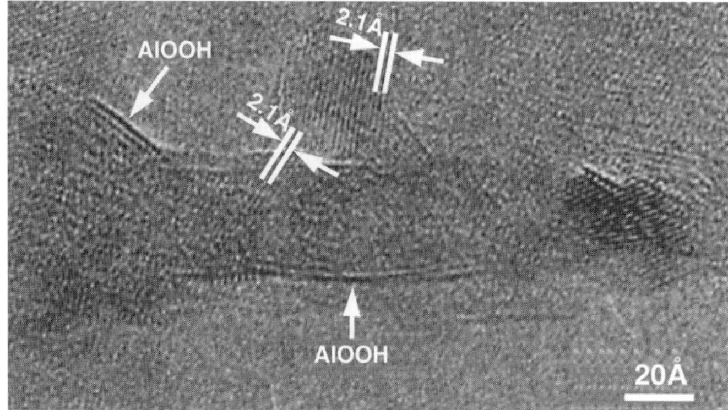

Figure 14.4. HRTEM image of the intimately mixed oxide, AP–$Al_2O_3 \cdot MgO$. Tiny MgO nanocrystals mixed with boehmite (AlOOH) planes can be seen. Reprinted from Ref. 13 with permission of The Royal Society of Chemistry.

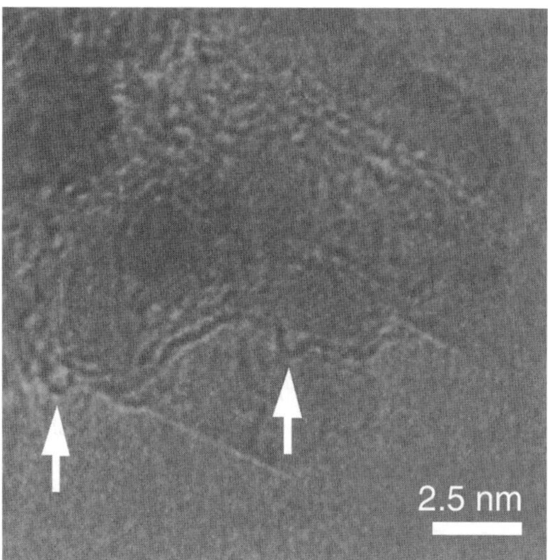

Figure 14.5. HRTEM image of carbon-coated AP–MgO; graphitic thin layers (indicated by arrows) on AP–MgO crystals can be seen. Reprinted in part from Ref. 30 with permission.

14.2.4. Surface Areas and Pore Structures

The specific surface areas of the nanoscale oxides (AP samples) are very large compared with the conventionally prepared (CP) and commercial (CM) oxides. The nanoscale metal oxides show adsorption/desorption isotherms indicative of the presence of bottleneck pores. The pore volume of nanostructured metal oxides are also usually high compared with CP and CM oxides; e.g., CP–MgO does not show pore structure at all, whereas AP–MgO has a pore volume of about 2 cm^3/g. Some metal oxides prepared by the MAP process have lower surface areas compared with other oxides prepared by the same procedure; e.g., AP–SrO has a surface area of about $20\,m^2/g$, which is much lower than that of AP–CaO ($\sim 100\,m^2/g$) and AP–MgO ($\sim 400\,m^2/g$). The surface areas, pore volumes, and pore diameters of the nanostructured metal oxides prepared by the MAP process are given in Table 14.1. The aerogel-prepared oxides were found to be resistant to collapse under the application of pressure, e.g., compaction of fine powder into a pellet (20). Due to their polyhedral nanocrystal shapes and their tendency to form web-like aggregates, these fine powders retain their high surface areas as well as small crystallite sizes on compaction. In the next section, the effect of compaction on the textural properties (pore size, pore volume, and surface area) of nanoscale oxides is discussed.

14.2.5. Densification (Compaction) of Nanostructured Metal Oxides

Nanostructured metal oxides readily form porous aggregates as observed by TEM and SAXS studies. If these porous aggregates of nanocrystals (i.e., nanoparticles) can be pressed into pellets and at the same time if they retain their high surface areas, large pore volumes, and pore sizes, it will be very advantageous for their technological applications. Indeed, studies on compaction of nanoscale metal oxides by pelletization led to very interesting observations (15,20,32). The results are shown in Figure 14.6, and Tables 14.3 and 14.4. The surface areas, pore volumes, and pore sizes did not drop significantly after compaction, and they are in the ranges where they can be useful as adsorbents. The compaction process was found to yield porous pellets that have narrow pore size distributions. A unique feature of nanostructured metal oxides is that they have particle shapes/morphologies that do not allow their densification under moderate pressure. In each case (AP–MgO, AP–Al_2O_3, NC–NiO, and NC–CuO), crystallites of 2 to 5 nm self-aggregate into highly porous web-like structures. The initially formed aggregates are tighter in the case of nanocrystalline CuO and NiO (see Table 14.4) than for AP–MgO. This is evident if we compare the starting surface areas (140 and $351\,m^2/g$) and pore volumes (0.22 and 0.35 cc/g) for these oxides (NC–CuO and NC–NiO). Nanocrystalline NiO was found to be the most susceptible oxide to densification under increasing pressure at room temperature. According to the hysteresis curves for adsorption–desorption for the nanoscale metal oxides, the pores were found to be cylindrical and open at both ends for all pressures except at 20,000 psi where the pores were found to be slit shaped.

14.2.6. A New Family of Porous Inorganic Adsorbents

The MAP procedure leads to the formation of very fine powders with high surface area. High surface area and large pores are desirable for applications based on adsorption, and the morphology required for such properties would be a random fibrous structure. This is indeed observed for many nanoscale metal oxides synthesized by the MAP process, indicating that large pores are accessible in these materials (20)

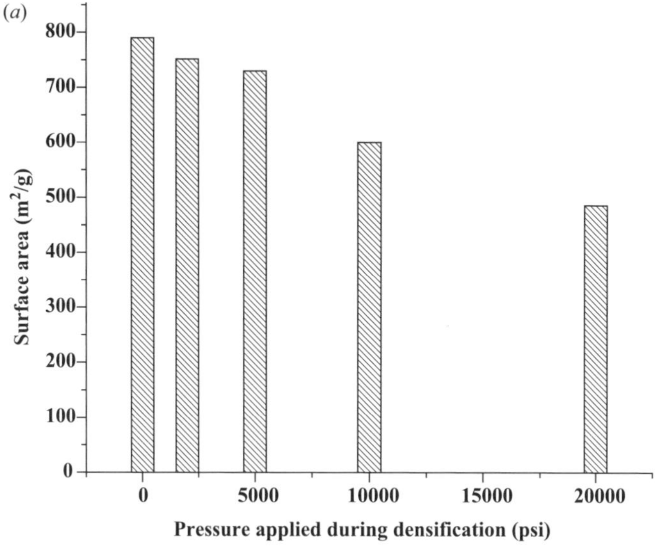

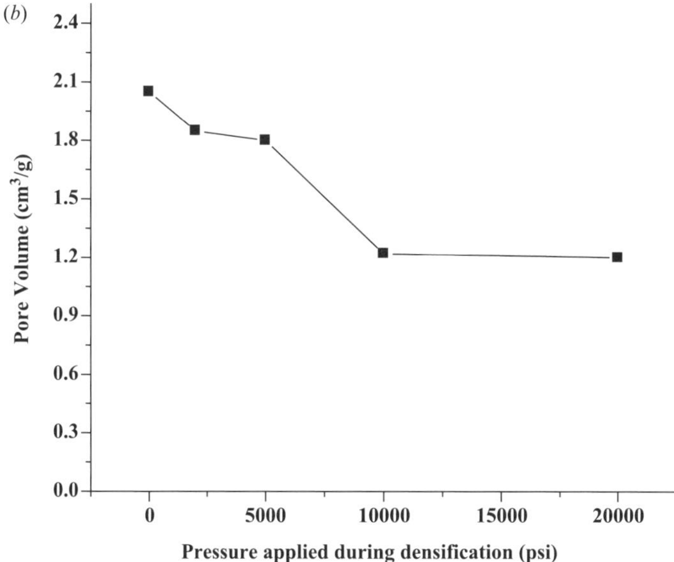

Figure 14.6. Effect of compaction on (*a*) surface area and (*b*) pore volume of nanostructured aluminum oxide (AP–Al_2O_3).

(see TEM Figures 14.1*c* and 14.3). When pellets are made out of these materials, the porous nature must be maintained, and this is indeed observed for the nanostructured metal oxides (see Figure 14.6 and Tables 14.3 and 14.4). Upon consolidation as pellets, their high surface area is retained, and the pore volumes and pore size openings can be decreased in the pellets in a controlled manner with the compaction

TABLE 14.3. Surface Areas, Pore Volumes, Average Pore Size Openings, and Primary Nanocrystal Size of Powders and Pellets of Nanoscale Metal Oxides.

Sample	Surface Areas Powder (pellet)[a,b] (m^2/g)	Pore Volume (cc/g)	Pore Opening (nm)	Crystallite Size (nm)
AP–Al_2O_3	800 (730)	2.0 (1.8)	11 (10)	<2 (<2)
AP–MgO	364 (366)	0.9 (0.6)	9.8 (6.4)	4 (4)
AP–Al_2O_3/MgO	775 (535)	1.8 (0.7)	11 (7.5)	<2 (<2)

[a]Pelletizing pressure = 5000 psi.
[b]Values in parenthesis indicate those of pellets.

pressure. The HRTEM studies indicate that the nanoscale metal oxides have a high concentration of surface edge/corner sites. It has been calculated for a AP–MgO nanocrystal (size ~4 nm), the percentage of edge ions to total surface ions is about 20%, whereas for a microcrystal MgO (CP–MgO), the ratio is only about 0.5%; a higher concentration of edge sites is present in nanocrystals. The chemical properties of nanostructured metal oxides can be altered or manipulated by incorporating other suitable metal oxides. The compacted pellets of nanoscale metal oxides retain their adsorbent properties and capacities for a wide variety of organic and acid gas molecules. The unique adsorption chemistry of these materials have been well studied

TABLE 14.4. Effect of Compaction on the Textural Properties of AP–$Mg(OH)_2$, NC–NiO, and NC–CuO.

Pressure of Compaction (psi)	Surface Area (m^2/g)	Total Pore Volume (cm^3/g)	Av. Pore Diameter (nm)
	AP–$Mg(OH)_2$		
0	364	0.90	9.8
5,000	366	0.59	6.4
10,000	383	0.44	4.6
20,000	342	0.29	3.4
	NC–CuO		
0	140	0.22	14
5,000	98	0.20	10
10,000	96	0.20	8.5
20,000	80	0.18	8
	NC–NiO		
0	351	0.35	4.6
5,000	210	0.12	4.1
10,000	205	0.12	4.0
20,000	200	0.10	3.8

(9,10,21), and examples include sorption properties such as

(*1*) enhanced adsorption of SO_2; microcrystalline MgO (CP–MgO) adsorbs about 2 molecules/nm^2, whereas nanocrystalline MgO (AP–MgO) adsorbs about 6 molecules/nm^2; and

(*2*) nanocrystals of ZnO and MgO adsorb paraoxon, a mimic of the chemical warfare agent, VX, destructively, at room temperature. Hence, the nanoscale metal oxides, including the intimately mixed metal oxides, constitute a new family of porous materials that exhibit unique adsorption properties.

14.2.7. Relationship to Zeolites and Other Mesoporous Materials

Nanostructured metal oxides, produced by the MAP process, constitute a novel family of inorganic porous materials (20). These materials can be made up of a wide array of metal oxides or mixed metal oxides, and they possess high surface areas in addition to large pores and pore volumes. The metal-oxide nanocrystals have unique morphologies and possess high concentrations of reactive edges, corners, and defect sites. Pellets made out of these materials retain their porous nature. These novel features impart these materials a tremendous level of flexibility toward engineering-desired chemical properties, e.g., Lewis acidity or basicity or soft/hard basicity or acidity. There are, of course, other very useful porous inorganic materials such as zeolites, MCM-41 (a mesoporous silicate), and related silicas, and certain types of porous clays, e.g., montmorillonite and kaolinite. The main differences between these porous materials and nanoscale metal oxides have to do with the nature of the pores. Zeolites have much smaller pores that are well ordered, and clays, usually, are with two-dimensional-layered structures with plate-like morphology. Whether these porous materials are superior to the nanocrystalline oxides depends solely on the intended application. The microporosity of zeolites is known to pose diffusional limitations for the reactant and product molecules and on the reaction rate. The mesoporous molecular sieves related to the MCM-41 family do not possess high acidity as that of zeolites. Also, MCM-type mesoporous materials have poor hydrothermal stability and cannot survive harsh conditions such as high temperatures and pressures. Nanoscale metal oxides are increasingly finding applications as better adsorbents compared with conventional adsorbents (9,10,21). In the forthcoming sections, the uses of nanostructured metal oxides for the adsorption of various contaminants are discussed in detail.

14.3. DESTRUCTIVE ADSORPTION BY NANOSTRUCTURED METAL OXIDES

The ability of an adsorbent to adsorb an incoming adsorbate molecule and chemically destroy it is called destructive adsorption; destructive adsorption can also be called as dissociative chemisorption. Nanostructured metal oxides such as AP–MgO, AP–CaO, AP–TiO_2, AP–Al_2O_3, and AP–ZnO behave as effective adsorbents for acid gases, chemical warfare agents, and air pollutants. These metal oxides have brought a new dimension for the adsorption of toxic chemicals. The adsorbate molecules are chemically dismantled/detoxified, and the toxic molecules are converted into nontoxic entities. Destructive adsorbents based on nanocrystalline oxides are increasingly finding use in tackling chemical and biological warfare agents. In the following sections, destructive adsorption on nanoscale metal oxides at ambient and high temperatures is discussed.

14.3.1. Adsorption at Ambient Temperatures

Some organophosphorus compounds and chemical warfare agents can be destroyed by destructive adsorption on nanostructured metal oxides at room temperature. They are discussed below in detail.

14.3.1.1. Organophosphorus Compounds A few organophosphorus compounds are surrogates of chemical warfare agents, and the chemical structure of these compounds is shown in Figure 14.7. Traditionally, adsorbents based on charcoal, thermal decomposition, and catalytic methods have been used to detoxify these hazardous chemicals. These methods have certain disadvantages; e.g., the catalysts become inactive due to pollutants containing heteroatoms. If a material adsorbs the toxic organophosphorus compound, destroys it by adsorption, and immobilizes the products of destructive adsorption on its surface, it will be of tremendous potential and one such example is discussed below.

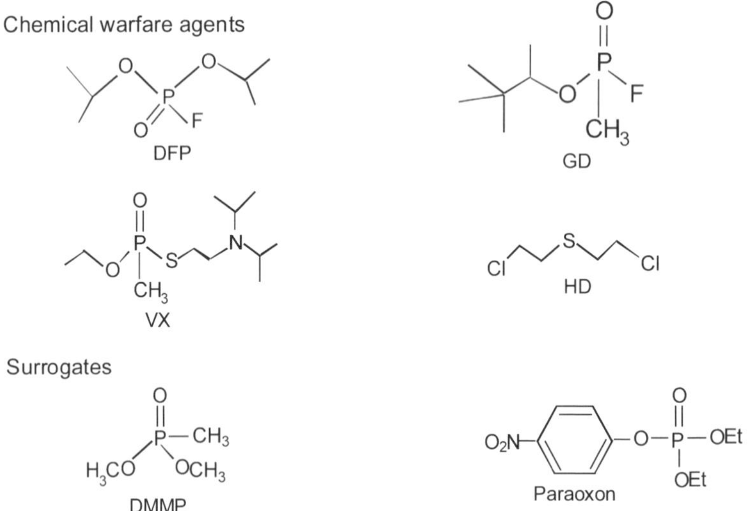

Figure 14.7. Chemical structures of chemical warfare agents and their surrogates.

14.3.1.1.1. Adsorption of Paraoxon Paraoxon (diethyl 4-nitrophenyl phosphate) is a mimic of chemical warfare agent VX (O–ethyl–S–(2-diisopropylamino) ethyl-methyl phosphonothioate). It can be destructively adsorbed on nanostructured metal oxides such as AP–MgO (33), AP–Al_2O_3, or even intimately mixed oxides such as AP–MgO · Al_2O_3 (13,28), prepared by the MAP process. For example, 100 mg of AP–MgO can destroy about 15 μL of paraoxon in a pentane solution at room temperature. Under the same conditions, 100 mg of AP–Al_2O_3 can adsorb about 16 μL of paraoxon. Figure 14.8 shows a comparison of AP–MgO with activated carbon toward paraoxon adsorption in pentane at room temperature (33). It can be seen that the performance of nanoscale AP–MgO (surface area = 400 m^2/g) is much better than that of activated carbon (surface area = 900 m^2/g). Activated carbon adsorbs only about 4.5 μL of paraoxon, and commercial MgO (CM–MgO) does not adsorb any appreciable amount of paraoxon. The destructive adsorption capability of nanoscale metal oxides toward paraoxon adsorption is summarized in Table 14.5 (13,28). Calculations indicate that about one molecule of paraoxon is adsorbed per squared nanometer area of AP–MgO. The mechanism of paraoxon adsorption on nanostructured metal oxides has been studied in detail by Rajagopalan et al. (33) and Carnes et al. (12) by ^{13}C, ^{31}P nuclear magnetic resonance (NMR) and infrared (IR) spectroscopy. It was proved that activated carbon does not destroy the paraoxon molecule on adsorption, and the paraoxon can be easily extracted using a solvent such as toluene. On the other hand, destructive adsorption of paraoxon takes place on the surface of nanoscale metal oxides. First, the P–OAr (Ar = aromatic ring) bond is broken, and this leads to the adsorption of p-nitrophenoxy anions and ethoxy groups on the surface of the oxide. The presence

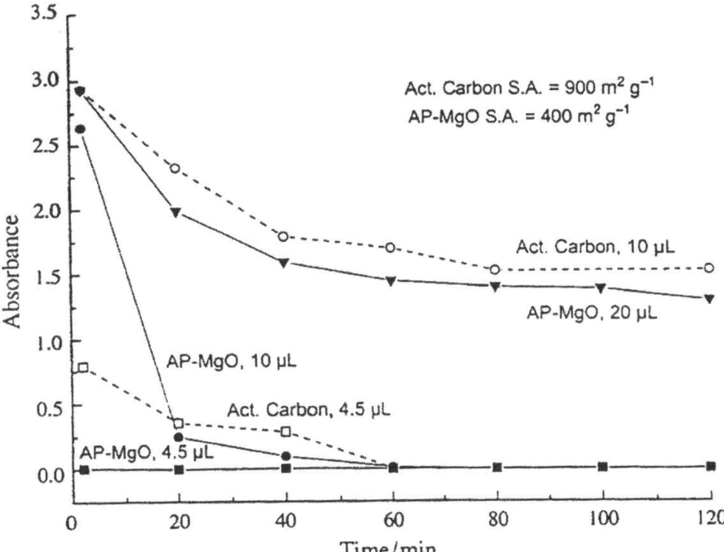

Figure 14.8. Adsorption of paraoxon from pentane at room temperature by AP–MgO as monitored by UV-Vis spectroscopy. Reprinted from Ref. 33 with permission.

TABLE 14.5. Destructive Adsorption of Paraoxon on Nanostructured Metal Oxides. Reprinted from Ref. 13 with Permission of The Royal Society of Chemistry.

Sample	Surface Area (m^2/g)	Pore Volume (cm^3/g)	Amount of Paraoxon Adsorbed (μL)	Molar Ratio (moles of paraoxon to moles of material)
AP–MgO	385	2.03	15	1:36
AP–CaO	67	0.22	4	1:113
AP–SrO	16	0.05	2	1:104
AP–Al_2O_3	637	3.17	16	1:14
AP–$MgAl_2O_4$	639	2.88	16	1:10
AP–$CaAl_2O_4$	517	3.56	16	1:9
AP–$SrAl_2O_4$	112	0.89	6	1:19
AP–$BaAl_2O_4$	135	1.05	6	1:17

of nitrophenoxy groups on the surface imparts a pale yellow color to the metal-oxide nanoparticles. The performance of AP–SrO and AP–CaO toward paraoxon adsorption was found to be poor, and this has been attributed to the lower surface area of these oxides (see Table 14.5).

The intimately mixed metal oxides prepared by the MAP process (e.g., AP–$Al_2O_3 \cdot MgO$, AP–$Al_2O_3 \cdot CaO$) also perform well for the destructive adsorption of paraoxon (13,28). Although the intimately mixed oxides (e.g., AP–$Al_2O_3 \cdot MgO$ and AP–$Al_2O_3 \cdot CaO$) adsorb roughly the same amount of paraoxon (16 μL per 100 mg of the oxide), the kinetics of adsorption was found to be faster for intimately mixed oxides compared with that of individual oxides (AP–MgO or AP–Al_2O_3) or even a physical mixture of AP–MgO and AP–Al_2O_3; AP–$Al_2O_3 \cdot MgO$ destroys about 16 μL paraoxon in only 20 minutes. The reason for the faster kinetics is believed to be that AP–MgO nanocrystals (Lewis base) is intimately mixed/incorporated onto the alumina matrix with high surface area and pore volume. This helps in the rapid destructive adsorption of the incoming paraoxon molecules. The performance of other mixed metal oxides such as AP–SrO · Al_2O_3 and AP–BaO · Al_2O_3 was found to be poor. Both oxides were able to adsorb only about 6-μL paraoxon per 100 mg of the adsorbent. The poor performance in these cases is ascribed due to lower surface area and poor intimate mixing of these oxides, as revealed by TEM studies. It is now clear that nanoscale metal oxides are advantageous over conventional adsorbents such as activated carbon for the destructive adsorption of organophosphorus compounds. This is because activated carbon only physisorbs the surrogate molecules but does not destroy them. On the other hand, nanoscale metal oxides not only adsorb the surrogate molecules but also destroy them into nontoxic moieties. In the next section, destructive adsorption of chemical warfare agents on nanostructured oxides is discussed.

14.3.1.2. Chemical Warfare Agents (CWAs) Chemical warfare agents (CWAs), also called nerve agents, present a great concern for the whole world.

Examples of CWAs include VX, GD (pinacolyl methylphosphono-fluoridate), and HD (bis(2-chloroethyl)sulfide) (see Figure 14.7), and most of these toxic compounds are polar organic liquids based on P(V) organophosphorus esters. CWAs react irreversibly with acetylcholine esterase enzyme and prevent it from controlling the human central nervous system. Disposal of CWAs safely without producing harmful chemicals is very important. Various researchers have described the decontamination procedures for the CWAs. Studies by Yang et al. (34–36) have demonstrated two effective chemical methods for decontaminating the CWAs:

(*1*) nucleophilic substitution reactions, and

(*2*) oxidation reactions.

The nucleophilic substitution reactions are done with basic H_2O_2 solutions. The nucleophile in this case is HO_2^- ion, and it neutralizes VX, Sarin (GB), and Soman (GD) agents. Also, it has been found that HO_2^- ion is a better nucleophile than OH^-. However, it has been pointed out that the temperature of the neutralization reactor must be kept low to prevent the H_2O_2 decomposition. For the disposal of CWAs by oxidation reactions (35), aqueous bleach [NaOCl or $Ca(OCl)_2$], $KMnO_4$, aqueous peroxydisulfate, or gaseous chlorine are used. Some CWAs that are difficult to decontaminate with OH^- could be oxidized with the above oxidizing agents; e.g., VX could be oxidized with aqueous hypochlorite (ClO^-) or hypochlorus acid (HClO). Hydrolysis of HD was also studied as a means of decontamination. It is fast in a very dilute solution, but at higher concentrations (~0.1 M), the hydrolysis is very slow and complicated (37). Possible improvements, such as the addition of organic liquids and the using of different nucleophiles, did not help in decontaminating this agent. Other methods of decontamination of CWAs such as catalytic decomposition (38), electrochemical decomposition (39), and bio-degradation (40) have been reported.

Recently, a convenient method of decontaminating CWAs agents has been demonstrated using nanoscale metal oxides (41–45). It has been found that these oxides (e.g., AP–MgO and AP–CaO) destructively adsorb CWAs such as VX, GD, and HD at room temperature. The destructive adsorption by nanoscale metal oxides was found to be fundamentally different from the adsorption of nerve agents on conventional adsorbents such as activated carbon. Activated carbon only adsorbs the CWAs, but it does not destroy these molecules. On the other hand, AP–MgO and AP–CaO destructively adsorb VX, GD, and HD at room temperature. These nerve agents (VX and GD) are hydrolyzed according to Scheme 14.1. It was found that after hydrolysis, the phosphonate products exist as complexes bound on the surface of oxides and EA-2192, a very toxic chemical, is not produced when nanoscale metal oxides are used for decontamination (EA-2192 is always produced if basic hydrolysis is adapted as a method of decontamination). For the hydrolysis of HD on AP–MgO (Scheme 14.1), it was found that HCl is eliminated and the products are thiodiglycol (TG) and divinyl sulfide (DVS) in equal amounts (50:50). HD on AP–CaO undergoes autocatalytic dehydrohalogenation and produces 80% of divinyl sulfide (DVS), along with thiodiglycerol (TG). ^{13}C NMR studies on the destructive adsorption on nanostructured metal oxides indicate that the decontamination products are present as surface bound alkoxides on

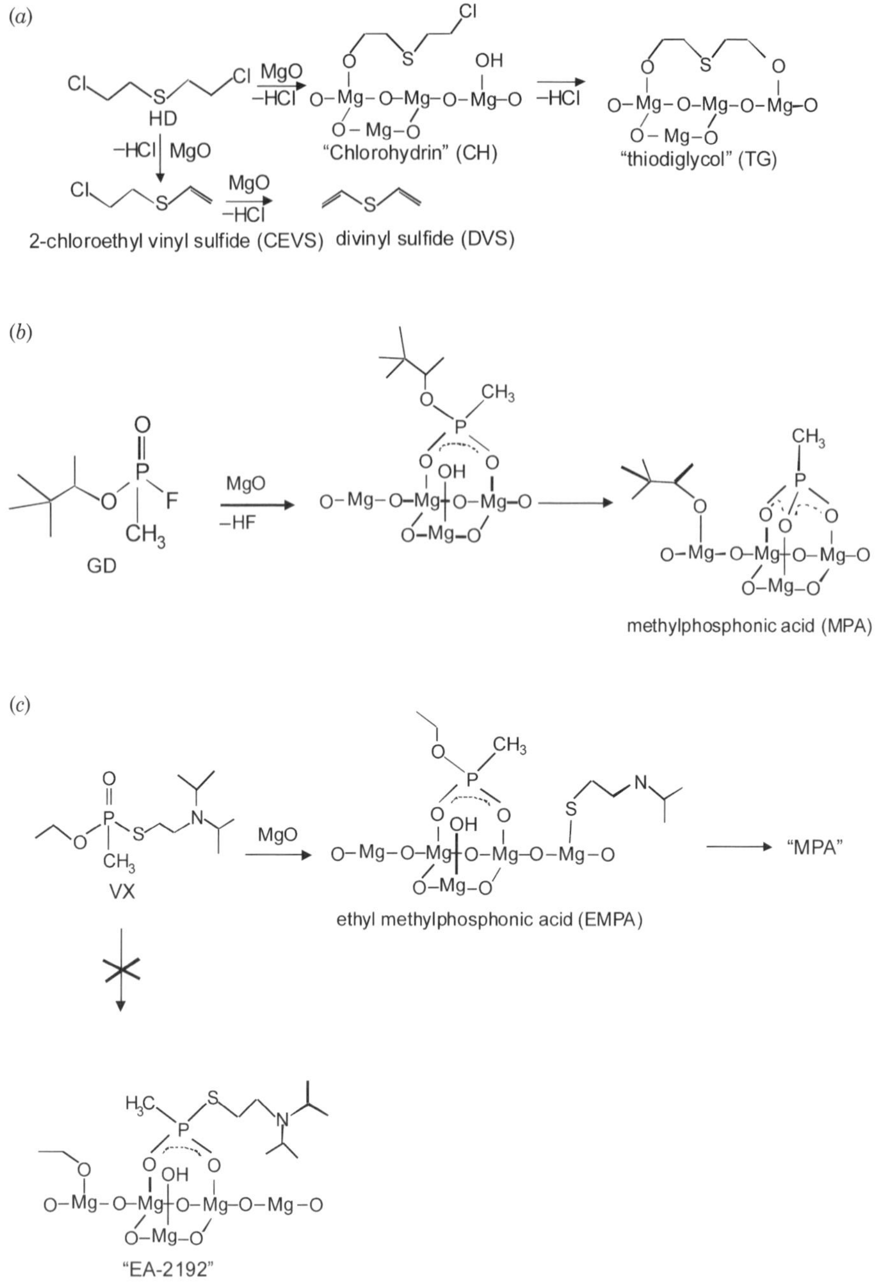

(a)

HD; MgO, −HCl → "Chlorohydrin" (CH); −HCl → "thiodiglycol" (TG)

HD; −HCl, MgO → 2-chloroethyl vinyl sulfide (CEVS); MgO, −HCl → divinyl sulfide (DVS)

(b)

GD; MgO, −HF → ... → methylphosphonic acid (MPA)

(c)

VX; MgO → ethyl methylphosphonic acid (EMPA) → "MPA"

VX ✗→ "EA-2192"

Scheme 14.1. Reaction of HD, GD, and VX with AP–MgO. Reprinted from Ref. 45 with permission.

the surface of oxides. Using nanoscale metal oxides for the decontamination of CWAs seems to be an alternative method for effectively destroying the CWAs. Working with a nontoxic powder such as AP–MgO is convenient because handling and storage will be easy in addition to having no liquid waste.

14.3.2. Adsorption at High Temperatures

Nanoscale metal oxides also destructively adsorb toxic chemicals at high temperatures. In the following section, various destructive adsorption reactions at high temperatures are discussed.

14.3.2.1. Chlorocarbons Because of their beneficial characteristics, a variety of chlorine-containing compounds is widely used in our modern society. Chlorocarbons are present as refrigerating agents, solvents (e.g., CCl_4, $CHCl_3$), heat transfer liquids, lubricants, washing liquids, and so on. Although substitution of these toxic compounds (some are carcinogenic) by nontoxic chemicals has been done in few cases, some chlorinated chemicals are still in common use. Safe disposal of chlorocarbons without producing toxic chemicals as byproducts is of great importance. The reported methods for destroying chlorine-containing compounds such as the catalytic oxidation of chlorocarbons to HCl and CO_2 and incineration at high temperatures have some disadvantages. For example, the incineration method requires very high temperatures (~1500 °C) to prevent the formation of harmful chemicals such as furans and dioxins. A simple noncatalytic method that will only yield environmentally benign products will be very useful. In this respect, nanostructured metal oxides have been effectively used for the destructive adsorption of chlorocarbons (17,46). A brief description of destructive adsorption of chlorocarbons, especially CCl_4, on nanoscale metal oxides is given below.

The reaction between a metal oxide and CCl_4 yields a metal chloride and CO_2 (9)

$$2MgO + CCl_4 \longrightarrow 2MgCl_2 + CO_2 \qquad \Delta H^\circ = -334\,kJ/mol$$

The reactions between alkaline-earth metal oxides and chlorocarbons such as CCl_4 are energetically feasible from a thermodynamic point of view. The exothermicity of the reactions is because the standard heats of formation for the metal chlorides are large compared with that of the oxides. Although the reactions are exothermic, surface effects are known to influence these reactions because these are gas/solid reactions; nanoscale metal oxides can have a tremendous effect on these reactions. Indeed, nanoscale metal oxides have been found to destructively adsorb chlorocarbons at high temperatures (400–500 °C) (see Table 14.6). The chlorinated compounds are destroyed by adsorption, and the products depend on the nature of chlorocarbon and on the ratio of metal oxide to the chlorocarbon. CCl_4 is destroyed by mineralization to a metal chloride and CO_2 gas, and both products are environmentally nontoxic. The reactivity order toward the destructive adsorption of chlorocarbons for the oxides is found as AP–CaO > CP–CaO ≫ CM–CaO. Nanocrystals of calcium

TABLE 14.6. Destructive Adsorption of CCl_4 with Nanostructured Metal Oxides at 500 °C. Reprinted from Ref. 14 with Permission.

Sample	Breakthrough	Saturation	Molar Ratio[a]
AP–Al_2O_3	57	98	1.44 mol of CCl_4:1 mol of Al_2O_3
CM–Al_2O_3	2	17	1 mol of CCl_4:16 mol of Al_2O_3
AP–(1/1) Al_2O_3/MgO	25	88	1.8 mol of CCl_4:1 mol of Al_2O_3/MgO
CM–MgO	1	4	1 mol of CCl_4:32 mol of MgO

[a]Theoretical molar ratio; 1 mol of CCl_4:2 mol of MgO, 1.5 mol of CCl_4:1 mol of Al_2O_3, 2 mol of CCl_4:1 mol of Al_2O_3/MgO.

oxide (AP–CaO) are more effective in decomposing CCl_4 compared with nanocrystalline magnesium oxide (AP–MgO), although it has more surface area ($\sim$400 m^2/g) compared with AP–CaO ($\sim$125 m^2/g). Even for samples having the same surface area (AP–CaO = 130 m^2/g and CP–CaO = 107 m^2/g), AP–CaO is superior in destructively adsorbing CCl_4 (47). This proves that the higher *intrinsic reactivity* of nanoscale oxides (e.g., AP–CaO) is the reason for their superior performance and surface area is not the only factor. It was found that the mobility of oxygen and chlorine ions determine the rate of the destructive adsorption reaction. Mixed metal oxides (e.g., AP–MgO · Al_2O_3) also behave as good adsorbents for CCl_4 at high temperatures, and the adsorptive reaction was found to be stoichiometric (Table 14.6).

14.3.2.1.1. Second-Generation Destructive Adsorbents Although nanoscale metal oxides such as AP–CaO and AP–MgO (they can be referred to as first-generation destructive adsorbents) do destructively adsorb CCl_4 well, they do not react completely with CCl_4 despite that the reactions are exothermic and thermodynamically feasible. The kinetic factors for the migration of chloride and oxygen ions were found to be the rate-limiting steps. To circumvent this problem, chemical modification of the first-generation adsorbents were carried out. The basic idea was to have a transition metal-oxide coating on the surface of the primary adsorbent that will be a catalyst for the reaction and thereby increase the efficiency of nanoparticles of the substrate oxide toward CCl_4 adsorption. It is known that the standard heats of formation favor the formation of alkaline-earth metal chlorides ($MgCl_2$ or $CaCl_2$) compared with the corresponding oxides (MgO or CaO). On the other hand, for the transition metal analogs, the standard heats of formation favor the formation of their oxides instead of the chlorides. In other words, the transition metal oxides are thermodynamically more favored than the transition metal chlorides. The transition metal oxide on the surface can react first with the incoming CCl_4 forming $FeCl_2$ and this can further react with the substrate oxide (AP–CaO) to replenish the iron oxide until all oxide is consumed.

Core-shell type (M_xO_y)MgO and (M_xO_y)CaO (M = Mn, Fe, Co, Ni) adsorbents were explored as second-generation destructive adsorbents (47). When nanoscale metal oxides (e.g., AP–CaO) are coated with transition metal oxides such as Fe_2O_3, CoO, or NiO, the reactivity toward destructive adsorption could be enhanced

TABLE 14.7. Efficiencies for Destructive Adsorption of CCl_4 on Nanoscale Metal Oxides. Reprinted from Ref. 9 with Permission.

Sample	Surface Area (m^2/g) Before	Weight% Transition Metal Loading	Breakthrough Number[a]	Performance Efficiency[b]
CM–MgO	10–30	0	0	0.12
CM–CaO	1.3	0	0	<0.01
CP–MgO	190	0	4	0.16
CP–CaO	100	0	1	0.23
AP–MgO	450	0	4	0.17
AP–CaO	140	0	1	0.31
[Fe_2O_3] CP–MgO	150	0.88	8	0.23
[Fe_2O_3] CP–CaO	90	1.6	2	0.44
[Fe_2O_3] AP–MgO	400	1.4	25	0.36
[Fe_2O_3] AP–CaO	130	1.6	9	0.51
[Mn_2O_3] AP–MgO	290	4.5	40	0.42
[NiO] AP–CaO	130	1.6	3	0.39
[CoO] AP–CaO	80	1.6	10	0.30

[a]Number of 1-μL injections of CCl_4 before breakthrough of any trace of CCl_4.

[b]Number of moles CCl_4 destroyed according to the stoichiometric equation $2MO + CCl_4 \rightarrow 2MCl_2 + CO_2$. A maximum value of 0.50 would be expected.

(Table 14.7). For example, when AP–CaO nanocrystals are coated with Fe_2O_3, NiO, or V_2O_5, they have about 1.5–2.0 times more reactivity toward CCl_4 adsorption compared with AP–CaO alone; the metal-oxide coating was found to be just less than a monolayer. Without the metal-oxide coating, only the surface reaction took place. The mechanism of destructive adsorption of CCl_4 was also studied and it is given in Scheme 14.2. It was found that the coated iron oxide acts as a surface catalytic

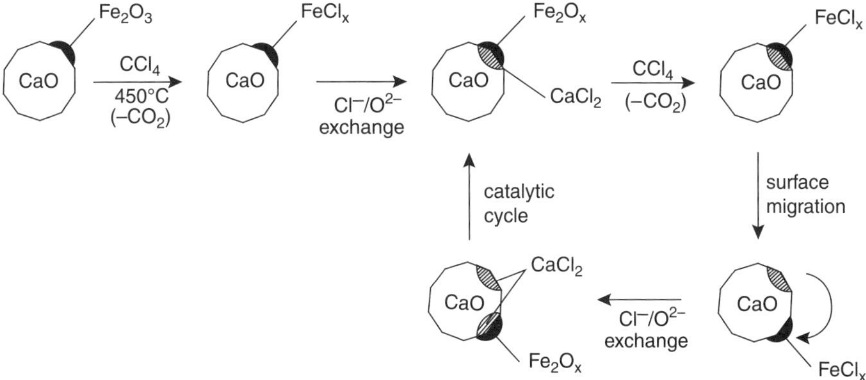

Scheme 14.2. Mechanism of destructive adsorption of CCl_4 on Fe_2O_3 coated AP–CaO (second-generation destructive adsorbent). Reprinted from Ref. 47 with permission.

agent, and it reacts first with CCl_4 to form iron chloride and CO_2. The $FeCl_2$, in turn, reacts with the metal oxide (AP–CaO) to regenerate the iron oxide; i.e., iron chloride exchanges chloride with oxide anion on AP–CaO. This way, the nanostructured oxide surface is continuously replenished, and the destructive adsorption proceeds until all oxide is consumed. Evidence has been found for a structural change that has taken place on coating the oxide nanocrystals with the transition metal oxide, probably due to the lattice mismatch between the metal oxide (AP–CaO) and the transition metal-oxide coating (47). In conclusion, we can say that nanoscale metal oxides behave as stoichiometric agents for the destructive adsorption of chlorocarbons. In the case of conventional metal oxides as adsorbents, only the surface of the oxide reacts and a protective layer of products is formed, which hinders further progress of the reaction. But nanocrystals have more exposed surface area, and they behave remarkably well as stoichiometric reagents for the adsorption.

14.3.2.1.2. Nanoscale Metal Oxides as Adsorbents for Other Chlorine- and Sulfur-Containing Compounds Nanoscale metal oxides also destructively adsorb other chlorine-containing compounds such as chlorinated benzenes. For example, AP–MgO and AP–CaO destructively adsorb mono-, di-, and trichlorobenzenes at temperatures close to 700–900 °C (48). However, in the presence of hydrogen, the reaction temperature can be lowered down to ~500 °C. The reaction products were found to be benzene, CO, CO_2, and H_2O. One significant finding was that the volatile chlorine-containing compounds were never emitted during the destructive adsorption of chlorinated benzenes. AP–MgO is more reactive than AP–CaO, and toxic by-products are formed only with AP–CaO. Decomposition of chloromethanes were also carried out on nanostructured metal oxides, and the studied compounds included CH_3Cl, CH_2Cl_2, and $CHCl_3$ (47). Mixed results were observed for the destructive adsorption on AP–CaO. $CHCl_3$ and CH_3Cl did not adsorb on AP–CaO, whereas CH_2Cl_2 destructively adsorbed on it (activity = 30%). The iron-oxide-coated AP–CaO performed better compared with the uncoated oxide, and the activity in this case was close to 50%.

The reaction of C_2Cl_4 with AP–CaO and Fe_2O_3 coated AP–CaO gave reaction ratios of 0.36 and 0.46, respectively (the expected theoretical reaction ratio is 0.5) (47). The products after the destructive adsorption of C_2Cl_4 on AP–CaO were found to be $Ca(OCl)_2 \cdot Ca(OH)_2$, $Ca(OCl)_2 \cdot 4H_2O$, and $CaCl_2$. The destructive adsorption reaction of 1, 1, 2, 2-tetra chloroethane on AP–CaO was also successful, and the reaction ratio was about 0.4 (theoretical reaction ratio = 0.5) (47). The final product after the destructive adsorption in this case was found to be $CaCl_2 \cdot 2H_2O$. The destructive adsorption results for trichloroethene was very similar to the above examples, and a reaction ratio of 0.58 was obtained for AP–CaO (theoretical ratio = 0.66).

Nanoparticles of CaO (AP–CaO) and Fe_2O_3 coated AP–CaO have been shown to adsorb sulfur dioxide, at about 500 °C, in larger amounts compared with the microcrystalline oxide (CP–CaO) (49). For AP–CaO, the reaction ratio was found to be 0.51 (the theoretical ratio is unity), whereas for Fe_2O_3-coated AP–CaO, the reaction ratio was found to be enhanced to 0.94. The solid products of the destructive adsorption were identified as calcium sulfite ($CaSO_3$), calcium sulfate ($CaSO_4$), and calcium sulfide (CaS). Nanoscale metal oxides also behave as destructive adsorbents for carbon disulfide (CS_2) and carbonyl sulfide (COS). For example, the activity for AP–CaO was close to 50%, and the products of the adsorption were CO_2 and the solid form of CaS. It was also observed that during the destructive adsorption, the surface area of AP–CaO dropped drastically from about 100 to about 15 m^2/g.

14.3.2.2. Organophosphorus Compounds Nanoparticles of metal oxides have also been found to destructively adsorb some organophosphorus compounds at high temperatures. Under this category, the destructive adsorption reactions of dimethyl methyl phosphonate (DMMP) and diisopropyl phosphorofluoridate (DFP) on nanostructured metal oxides are discussed below.

14.3.2.2.1. Dimethyl Methyl Phosphonate (DMMP) DMMP is a good mimic of CWAs. The compound is nontoxic, and it contains the similar types of bonds that are present in some common CWAs. Understanding the destructive adsorption of DMMP should shed light on understanding the chemistry of safe disposal of CWAs. Nanoparticles of MgO (AP–MgO) have been shown to destructively adsorb DMMP at about 500 °C as shown as follows (18,19):

$$3CH_3(OCH_3)_2P{=}O \xrightarrow{6MgO_{(surface)}} CH_3(OCH_3)PO_{(a)} + 2CH_3(OCH_3)P_{(a)} + OCH_{3(a)} + 2HCOOH_{(g)} + 2H_{(a)}$$

It was found that, to decompose one DMMP molecule, two MgO surface molecules are required and high surface area is crucial for this purpose (AP–MgO has a surface area of about 500 m^2/g). The products of destructive adsorption are formic acid, which is volatile, and OCH_3 [$CH_3(CH_3O)P$] and $CH_3(CH_3O)PO$ groups, which are immobilized on the surface of the oxide. Nanoscale calcium oxide (AP–CaO) also destroys DMMP at about 500 °C (decomposition efficiency = 70–78%). The nanostructured metal oxides perform better in terms of capacity and breakthrough number (the number of 1-μL injections of CCl_4 before any trace of CCl_4 starts appearing in the GC-chromatogram) compared with other metal oxides; e.g., AP–MgO shows a capacity of 0.48 mol of DMMP/mol of AP–MgO and a higher breakthrough number of 30 μL. In contrast, CP–MgO shows a capacity of 0.23 mol/mol of CP–MgO and a much lower breakthrough number of 6 μL.

14.3.2.2.2. Diisopropyl Phosphorofluoridate (DFP) It has been shown that AP–MgO can destructively adsorb diisopropyl phosphorofluoridate, which is also the mimic of a chemical warfare agent, GD. The adsorption takes place with the breaking of P–F bond (33). Infrared and NMR studies have indicated that with the cleavage of the P–F bond, a P–O–O bridge is formed, and this is followed by the formation of P–O–C bonds. The adsorptive performance of the other two analogous oxides (CP–MgO and CM–MgO) was poor compared with that of AP–MgO.

14.3.3. Nanostructured Metal Oxides for Air Purification

The presence of volatile organic compounds and acid gases causes great concern for the indoor air quality. The most commonly used adsorbents for adsorbing these toxic chemicals have disadvantages; e.g., activated carbon is not effective for acid gases such as SO_2. Also, activated carbon can adsorb organic molecules, but because the molecules are not destructively adsorbed (only weak van der Waals forces are responsible for adsorption), the adsorbed molecules can desorb over time. This affects the performance of the conventional adsorbents. Nanoscale metal oxides such as AP–MgO and AP–Al_2O_3 have been shown to destructively adsorb indoor air pollutants such as acetaldehyde (50).

The adsorption capacity for various MgO oxides (AP–MgO, CP–MgO, and CM–MgO) and activated carbon, for acetaldehyde adsorption, is given in Figure 14.9. AP–MgO (which consists of nanocrystals) and CP–MgO (which consists of microcrystals) adsorb acetaldehyde in large amounts compared with the commercial MgO; e.g., one mole of AP–MgO adsorbs about one mole of CH_3CHO at room temperature. Activated carbon (both treated and untreated) and commercial MgO could not adsorb appreciable amounts of CH_3CHO. It was found that even on compaction of AP–MgO fine powder into pellets (pressure = 4000 psi), the efficiency of CH_3CHO adsorption was affected only by a small magnitude; more than 85% of adsorption efficiency was retained. Also, on exposure to air and water, the adsorption efficiencies of the metal-oxide nanoparticles were retained. These excellent properties are beneficial for the purpose of their application. Fourier transform IR (FTIR) studies have proven that the adsorbed acetaldehyde is destroyed irreversibly, and a multilayer dissociative adsorption takes place on the surface of AP–MgO and AP–Al_2O_3. It was

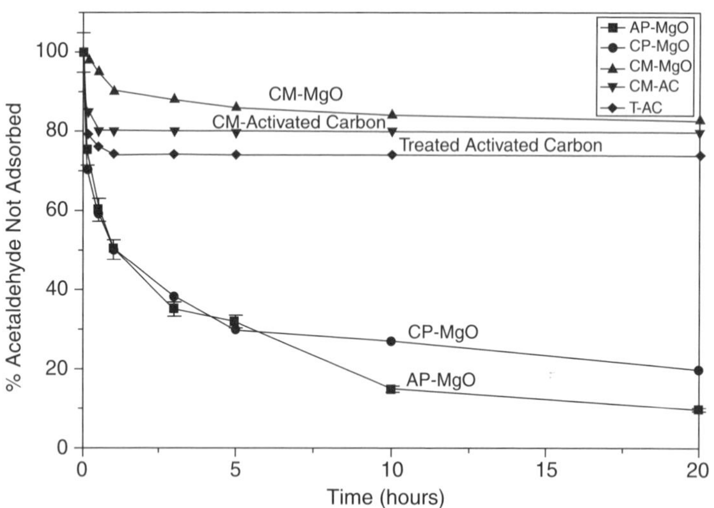

Figure 14.9. Adsorption of acetaldehyde on AP–MgO and other adsorbents. Reprinted from Ref. 50 with permission.

also found that other common organic chemicals, such as acetone, ammonia, and methanol, also adsorb well on metal-oxide nanoparticles (e.g., AP–MgO).

14.3.4. Adsorption of Halogens and Use of the Metal-Oxide–Halogen Adducts

Large amounts of halogens (e.g., Cl_2, Br_2, I_2, etc.) adsorb on nanoparticles of metal oxides (51). Halogens have been used as disinfectants (e.g., as bactericides) for quite a long time. However, in their free form, the use of halogens is limited because they are gases, toxic, and corrosive. Nanoscale metal oxides form stable solid adducts with halogens such as Cl_2, Br_2, I_2, and even inter-halogen compounds such as ICl and IBr (51). The nanoscale oxides also have good capacities for adsorbing halogens; e.g., AP–MgO can adsorb about 7 wt% chlorine, about 18 wt% bromine, and about 20 wt% iodine. The commercial MgO sample, on the other hand, adsorbs only about 1 wt% of these halogens. The commercial oxides cannot hold halogens even for a week, but the metal-oxide nanoparticles can retain the halogens for weeks or even months. By Raman spectroscopic studies, the strength of adsorption between the halogens and the oxide surface has been shown to be stronger in the case of nanoscale-oxide–halogen adducts compared with the corresponding adducts prepared from commercial metal oxides. The X–X (X = halogen) band positions are shifted by a larger magnitude compared with the band positions for free halogens in the case of nanoscale-oxide–halogen adducts (51). The halogen molecules in the adsorbed state are chemically more reactive than in the gas phase; i.e., adsorption makes the X–X bond weaker. The nanoscale-oxide–halogen adducts possess excellent bactericidal properties; e.g., AP–MgO–Cl_2 can kill bacteria, spores, and other toxins (52).

14.4. CHEMICALLY MODIFIED NANOSTRUCTURED METAL OXIDES AS ADSORBENTS

Numerous possibilities exist for the chemical modification of the nanostructured metal oxides, e.g., by incorporating other metal ions during the synthesis (modified aerogel process) or afterward. The modification can be done for a particular application in mind. Recently, it has been demonstrated that silver-incorporated nanoscale aluminum oxide can adsorb thiophene and related molecules from a pentane solution at room temperature (53). Removal of thiophene and other sulfur-containing compounds from commercial fuels is of great recent scientific interest. The modified nanostructured aluminum oxide was prepared by impregnating silver acetyl acetonate on nanoparticles of aluminum oxide (AP–Al_2O_3). Incorporation of soft Lewis acid sites (e.g., silver ion) was found to be necessary for the adsorption of thiophene-related molecules. The silver impregnated nanoscale aluminum oxide (Ag–AP–Al_2O_3) has 10 times more capacity compared with that of activated carbon; Ag–AP–Al_2O_3 has a capacity of ~0.022 mmol/g for thiophene, whereas activated carbon has only about 0.002 mmol/g thiophene, under the same conditions. The silver impregnated commercial aluminum oxide, also prepared under similar conditions, did not perform well toward thiophene adsorption. The modified adsorbent Ag–AP–Al_2O_3 retains its adsorption properties even after compaction, and it can be used after regeneration by heating in air at modest temperatures (170–250 °C). Detailed characterization studies have indicated that the active site in silver impregnated nanostructured Al_2O_3 is the Ag^+ ion coordinated to a carbonate ion. The silver ions exist in the form of β-Ag_2CO_3, which is stabilized on the surface of Al_2O_3, and Ag nanoparticles are not responsible for the adsorption of thiophene molecules.

14.5. MISCELLANEOUS ADSORBENTS BASED ON NANOSTRUCTURED METAL OXIDES

Nanoparticles of MgO (AP–MgO) have also been reported to be a good destructive adsorbent for dehydrochlorination of 1-chlorobutane, and the products of the reaction are butene and HCl (54). MgO is converted into $MgCl_2$ during the process, however, because $MgCl_2$ by itself is a good catalyst, the reaction continues. The effect of pelletization on the destructive adsorption has also been studied. Due to diffusion limitations, the reactant (1-chlorobutane) could not diffuse into the pellets all the way. The initially formed $MgCl_2$ blocks the pores in the pellet, and $MgCl_2$ forms a layer on the outside of the pellet. It was found that the MgO layer inside the pellet does not contribute to the dehydrochlorination.

A composite adsorbent, prepared by the sol-gel method, and based on nanostructured SiO_2–TiO_2 (TiO_2 loading ~13%) has been shown to be effective for the removal of elemental mercury vapor (55). Mercury has a significant environmental and health impact, and its safe removal is of great interest. The composite adsorbs mercury by a synergistic adsorption and photo-catalytic oxidation under ultraviolet (UV) light. The capacity of the adsorbent was found to be about 1500 μg of Hg per gram of the adsorbent. The UV-light photocatalytic oxidation activates the TiO_2 surface, thereby enhancing the absorption capacity, and hence, there was no need for continuous UV light irradiation. The composite adsorbent could be easily regenerated by rinsing with an acid.

Nanocrystalline MgO (crystallite size ~12–23 nm and surface area ~107 m^2/g), prepared by a combustion synthesis using magnesium nitrate and glycine, has been shown to be effective in removing fluoride ions from drinking water (56). The adsorption efficiency was found to be close to about 90%, and the adsorbent can be regenerated by NaOH treatment. Mehrotra et al. found that nanocrystalline alkaline-earth oxides (MgO, CaO, and Al_2O_3) prepared by the sol-gel process behave as better adsorbents for air purification compared with activated carbon (57). The adsorption of aldehydes and ketones was found to proceed by the dissociation of organic molecules in which the carbonyl oxygen is coordinated to the metal cation and the hydrogen from the aldehyde group transfers to oxygen ion on the surface of oxide. Also these authors have found that hydrolyzing a hetero-bimetallic alkoxide [e.g., $Mg(Al(isoprop)_4)_2$] yields a more uniform structure compared with hydrolyzing individual alkoxides. Adsorption of heavy metal ions such as Pb^{2+} on nanocrystalline γ-Fe_2O_3 and γ-Fe_2O_3-thiourea complex composite has been reported by Mallikarjuna et al. (58).

It was found that the efficiency of the former adsorbent toward lead adsorption is about 15%, whereas for the latter, it is about 50%. Nanocomposites based on silica and iron(III) oxide, prepared by the sol-gel process, has been used for the removal of toxic As (V) ions from aqueous solutions (59). It was found that the presence of magnetic field affects the adsorption process.

14.6. CONCLUSIONS

Nanoscale metal oxides offer unique and novel opportunities for applications based on adsorption. They behave as destructive adsorbents for acid gases, polar organophosphorus compounds, chlorocarbons, indoor air pollutants, and even CWAs. The destructive adsorption reactions can be carried out at ambient conditions or at moderately high temperatures. The nanoparticles of metal oxides serve as effective adsorbents due to the following reasons:

(*1*) They possess high surface areas and have a large surface-to-bulk ratio compared with conventional oxides;

(*2*) they have unusual shape and high number of reactive edges, corners, and defect sites that impart a higher surface reactivity;

(*3*) properties such as Lewis acidity and Lewis basicity can be tailored for a specific application; and

(*4*) nanostructured metal oxides can be pelletized while maintaining the high surface areas of the fine powders.

Thus, nanoscale metal oxides represent a new family of porous inorganic sorbents that exhibit unusual adsorption properties. These porous metal oxides can adsorb halogens, and the halogen adducts exhibit bactericidal and sporicidal properties. A great deal of fundamental and applied research is yet to be carried out in this very promising and interesting area, e.g.,

(*1*) desulfurization of commercial fuels (1); the Environmental Protection Agency (EPA) requires that the sulfur level in commercial gasoline and diesel fuels be reduced to about ~15 ppm from current levels of few hundreds of ppm. Gasoline is the most convenient fuel for fuel cell, but it must have less than 1-ppm level of sulfur to avoid platinum catalyst poisoning. Currently available adsorbents cannot meet this goal, and a breakthrough is needed in these cases.

(*2*) There is a need to invent new adsorbents for CO_2 capture and sequestration in the generation of power (60); nanostructured metal oxides can be explored for this purpose.

(*3*) Other applications that remain to be explored are NO_x removal, CO removal from H_2, and CH_4 for storage on-board. Nanostructured metal oxides are novel promising materials, and serious research efforts can bring remarkable breakthroughs in technologies based on adsorption.

REFERENCES

(1) Yang, R.T. *Adsorbents: Fundamentals and Applications*; Wiley-Interscience: New Jersey, 2003.

(2) Klabunde, K.J. (Editor); *Nanoscale Materials in Chemistry*; Wiley-Interscience: New York, 2001.

(3) Klabunde, K.J. *Free Atoms, Clusters, and Nanoscale Particles*; Academic Press: San Diego, CA, 1994.

(4) Steigerwald, M.L.; Brus, L.E. *Acc. Chem. Res.* **1990**, *23*, 183–188.

(5) Koper, O.B.; Lagadic, I.; Volodin, A.; Klabunde, K.J. *Chem. Mater.* **1997**, *9*, 2468–2480.

(6) Morrish, A.H.; Haneda, K.; Zhou, X.Z. In *Nanophase Materials, Synthesis, Properties, Applications*; Hadijipanyis, G.C.; Seigel, R.W. (Editors); Kluwer Academic Publications: London, 1994; 515–536.

(7) Lindsay, D.M.; Wang, Y.; George, T.F. *J. Cluster Sci.* **1990**, *1*, 107–126.

(8) Andres, R.P.; Averback, R.S.; Brown, W.L.; Brus, L.E.; Goodard III, W.A.; Kaldor, A.; Louie, S.G.; Moskovits, M.; Peercy, P.S.; Riley, S.J.; Siegel, R.W.; Spaepen, F.; Wang, Y. *J. Mater. Res.* **1989**, *4*, 704–736.

(9) Klabunde, K.J.; Stark, J.V.; Koper, O.; Mohs, C.; Park, D.G.; Decker, S.; Jiang, Y.; Lagadic, I.; Zhang, D. *J. Phys. Chem.* **1996**, *100*, 12142–12153.

(10) Ranjit, K.T.; Medine, G.; Jeevanandam, P.; Martyanov, I.N.; Klabunde, K.J. In *Environmental Catalysis*; Grassian, V.H. (Editor); CRC Press: Boca Raton, FL, 2005; 391–420.

(11) Koper, O.; Li, Y.X.; Klabunde, K.J. *Chem. Mater.* **1993**, *5*, 500–505.

(12) Carnes, C.L.; Stipp, J.; Klabunde, K.J.; Bonevich, J. *Langmuir* **2002**, *18*, 1352–1359.

(13) Medine, G.M.; Zaikovski, V.; Klabunde, K.J. *J. Mater. Chem.* **2004**, *14*, 757–763.

(14) Carnes, C.L.; Kapoor, P.N.; Klabunde, K.J. *Chem. Mater.* **2002**, *14*, 2922–2929.

(15) Stark, J.V.; Park, D.G.; Lagadic, I.; Klabunde, K.J. *Chem. Mater.* **1996**, *8*, 1904–1912.

(16) Stark, J.V.; Klabunde, K.J. *Chem. Mater.* **1996**, *8*, 1913–1918.

(17) Koper, O.; Lagadic, I.; Klabunde, K.J. *Chem. Mater.* **1997**, 9, 838–848.

(18) Li, Y.X.; Klabunde, K.J. *Langmuir* **1991**, *7*, 1388–1393.

(19) Li, Y.X.; Koper, O.; Atteya, M.; Klabunde, K.J. *Chem. Mater.* **1992**, 4, 323–330.

(20) Richards, R.; Li, W.; Decker, S.; Davidson, C.; Koper, O.; Zaikovski, V.; Volodin, A.; Rieker, T.; Klabunde, K.J. *J. Am. Chem. Soc.* **2000**, *122*, 4921–4925.

(21) Lucas, E.; Decker, S.; Khaleel, A.; Seitz, A.; Fultz, S.; Ponce, A.; Li, W.; Carnes, C.; Klabunde, K.J. *Chem. Eur. J.* **2001**, *7*, 2505–2510.

(22) Itoh, H.; Utamapanya, S.; Stark, J.; Klabunde, K.J.; Schlup, J.R. *Chem. Mater.* **1993**, *5*, 71–77.

(23) Kistner, S.S. *J. Phys. Chem.* **1932**, *36*, 52–64.

(24) Teichner, S.J. *Aerogels*; Springer-Verlag: Berlin, 1985; 225.

(25) Diao, Y.; Walawender, W.P.; Sorenson, C.M.; Klabunde, K.J.; Ricker, T. *Chem. Mater.* **2002**, *14*, 362–368.

(26) Utamapanya, S.; Klabunde, K.J.; Schlup, J.R. *Chem. Mater.* **1991**, *3*, 175–181.

(27) Chi, Y.; Chaung, S.S. *J. Phys. Chem. B* **2000**, *104*, 4673–4683.

(28) Medine, G.M. Ph.D. Thesis, Manhattan, KS: Department of Chemistry, Kansas State University, 2004.

(29) Bedilo, A.F.; Sigel, M.J.; Koper, O.B.; Melgunov, M.; Klabunde, K.J. *J. Mater. Chem.* **2002**, *12*, 3599–3604.

(30) Heroux, D.S.; Volodin, A.M.; Zaikovski, V.I.; Chesnokov, V.V.; Bedilo, A.F.; Klabunde, K.J. *J. Phys. Chem. B* **2004**, *108*, 3140–3144.

(31) Medine, G.M.; Klabunde, K.J.; Zaikovskii, V. *J. Nanopart. Res.* **2002**, *4*, 357–366.

(32) Klabunde, K.J.; Koper, O.; Khaleel, A. *U.S. Patent* 6093236, 2001.

(33) Rajagopalan, S.; Koper, O.; Decker, S.; Klabunde, K.J. *Chem. Eur. J.* **2002**, *8*, 2602–2607.

(34) Yang, Y.C. *Acc. Chem. Res.* **1999**, *32*, 109–115.

(35) Yang, Y.C.; Baker, J.A.; Ward, J.R. *Chem. Rev.* **1992**, *92*, 1729–1743.

(36) Yang, Y.C. *Chem. Ind.* **1995**, *9*, 334–337.

(37) Yang, Y.C.; Szafraniec, L.L.; Beaudry, W.T.; Ward, J.R. *J. Org. Chem.* **1988**, *53*, 3293–3297.

(38) Zhou, X.L.; Sun, Z.J.; White, J.M. *J. Vac. Sci. Technol. A* **1993**, *11*, 2110–2116.

(39) Franklin, T.C.; Nnodimele, R.; Kerimo, J. *J. Electrochem. Soc.* **1993**, *140*, 2145–2150.

(40) Kilbame, II, J.J.; Jackowski, K. *J. Chem. Technol. Biotechnol.* **1996**, *65*, 370–374.

(41) Klabunde, K.J.; Medine, G.; Bedilo, A.; Stoimenov, P.; Heroux, D. *ACS Symp. Ser.* **2005**, *890*, 272–276.

(42) Wagner, G.W.; Procell, L.R.; Koper, O.B.; Klabunde, K.J. *ACS Symp. Ser.* **2005**, *891*, 139–152.

(43) Wagner, G.W.; Bartram, P.W.; Koper, O.; Klabunde, K.J. *J. Phys. Chem. B* **1999**, *103*, 3225–3228.

(44) Wagner, G.W.; Procell, L.R.; O'Connor, R.J.; Munavalli, S.; Carnes, C.L.; Kapoor, P.N.; Klabunde, K.J. *J. Am. Chem. Soc.* **2001**, *123*, 1636–1644.

(45) Wagner, G.W.; Koper, O.; Lucas, E.; Decker, S.; Klabunde, K.J. *J. Phys. Chem. B* **2000**, *104*, 5118–5123.

(46) Koper, O.; Klabunde, K.J. *Chem. Mater.* **1997**, *9*, 2481–2485.

(47) Decker, S.P.; Klabunde, J.S.; Khaleel, A.; Klabunde, K.J. *Environ. Sci. Technol.* **2002**, *36*, 762–768.

(48) Li, Y.X.; Li, H.; Klabunde, K.J. *Environ. Sci. Technol.* **1994**, *28*, 1248–1253.

(49) Decker, S.; Klabunde, K.J. *J. Am. Chem. Soc.* **1996**, *118*, 12465–12466.

(50) Khaleel, A.; Kapoor, P.N.; Klabunde, K.J. *Nanostruct. Mater.* **1999**, *11*, 459–468.

(51) Stoimenov, P.K.; Zaikovski, V.; Klabunde, K.J. *J. Am. Chem. Soc.* **2003**, 125, 12907–12913.

(52) Stoimenov, P.K.; Klinger, R.L.; Marchin, G.L.; Klabunde, K.J. *Langmuir* **2002**, *18*, 6679–6686.

(53) Jeevanandam, P.; Klabunde, K.J.; Tetzler, S.H. *Micropor. Mesopor. Mater.* **2005**, *79*, 101–110.

(54) Gupta, P.P.; Hohn, K.L.; Erickson, L.E.; Klabunde, K.J.; Bedilo, A.F. *AIChE J.* **2004**, *50*, 3195–3205.

(55) Pitoniak, E.; Wu, C.; Londeree, D.; Mazyck, D.; Bonzongo, J.; Powers, K.; Sigmund, W. *J. Nanopart. Res.* **2003**, *5*, 281–292.

(56) Nagappa, B.; Chandrappa, G.T. *Trans. Ind. Ceram. Soc.* **2005**, *64*, 87–93.

(57) Singh, A.; Mehrotra, R.C. *Chemtracts* **2001**, *14*, 659–661.

(58) Mallikarjuna, N.N.; Ventataram, A. *Talanta* **2003**, *60*, 139–147.

(59) Peleanu, I.; Zaharescu, M.; Rau, I.; Crisan, M.; Jitianu, A.; Meghea, A. *Separation Sci. Technol.* **2002**, *37*, 3693–3701.

(60) Noble, R.D.; Agrawal, R. *Ind. Eng. Chem. Res.* **2005**, *44*, 2887–2892.

CHAPTER 15

Gas Sensors

DOINA LUTIC and MEHRI SANATI

School of Technology and Design, Växjo University, 35195 Växjö, Sweden

ANITA LLOYD SPETZ

S-SENCE and Div. of Applied Physics, Linköping University, 58183 Linköping, Sweden

15.1. INTRODUCTION

Detection of various species in gas mixtures, in order to evaluate the impact of these species on the quality of the product or on their effect on the environment (odors, toxic species etc.), is a challenging subject. A practical approach can be achieved by using so-called "chemical sensors." A chemical sensor consists of a system, including a sensitive layer, called a *receptor*, and a device that transforms the atomic scale interaction into a mechanical or electrical response called a *transducer*. The receptor interacts with the sensed molecules by physical adsorption or by weak chemisorption often followed by chemical reactions like combustion. These phenomena occur at the molecular or atomic level. The transducer functioning is especially dependent on the microstructure of the sensor surface (1,2).

The interaction between molecules from a fluid (gas or liquid) phase and the receptor results in a change of the physical properties of the receptor, such as electron density, optical properties, mass, or temperature. The most common sensor systems contains metallic oxides as receptors and deal with the changes that occur during the interaction of a gas with a solid surface at the level of its electronic charge density and charge carrier mobility. The sensing layer is deposited on the surface of an electronic device (most common a resistor or a solid electrolyte, but it may also be a capacitor, diode, transistor or resonator) and the interaction with the gas changes its function (1,2).

One of the main problems involved in the research of chemical sensors is the high number of parameters—hundreds—that interfere during the measurement. The main

Synthesis, Properties, and Applications of Oxide Nanomaterials, Edited by José A. Rodríguez and Marcos Fernández-Garcia

effects that must be considered in evaluating a chemical sensor are as follows (3):

- The *lack of specificity* for a given chemical species.
- The *cross-sensitivity*, meaning that the signal from one species can change if this is in a mixture with other compounds.
- The great *temperature dependence*, because both adsorption and chemical reactions are widely influenced by the temperature.
- The "memory effect" of the sensor, meaning modification of the signal after a certain period of use and hence the need for frequent recalibration.
- Drift problems after long-term use, i.e., rather low stability in time.

The *response* of a sensor to chemical compounds can be expressed in various ways (4):

- The difference between the value of the sensor signal in the absence and presence of the sensed species.
- The ratio between the values of the sensor signal in the absence and presence of the species to be detected.
- The derivative of the initial change of the signal due to a chemical species.
- The integral change of the signal due to a chemical species.

15.2. METAL-OXIDE–GAS INTERACTIONS

15.2.1. The Effect of Metal-Oxide–Gas Interactions

The role of metal oxides as sensing layers in transducers is closely related to the specific interaction between the oxide-based sensor surface and the gas to be measured. Some changes in the electronic structure can also occur in the bulk phase. The parameters that change in most cases in sensor signals are an electric measure reflected into the following (5):

1. The ***changes of the electrical resistance (or conductivity) values*** of the semiconductor layer (5,6), as a function of the oxygen partial pressure at the sensor surface, are used to measure its value when using tin or titanium oxides. A tin or titanium-based sensor usually consists of a thick film deposited on a ceramic tube, heated by a coil inserted inside it. The temperature range for sensing gases is 150–500 °C, because the metallic oxides are insulators rather than semiconductors at low temperatures (7) and the chemisorption is an activated process, which can be assimilated to a chemical reaction. The sensors have to be stabilized before measuring the gases by heating in an oxygen-containing atmosphere for a certain time before sensing. When the sensor detects reducing (combustible) gas, the resistance across the tin oxide decreases in accordance with the measured gas concentration. The chemical reactions involved in the sensing process assume the possibility of SnO_2 reducing to SnO in the presence

of reducing detected species (8,9):

$$SnO_2 + CO \longrightarrow SnO + CO_2 \quad (15.1)$$

$$SnO + 1/2\,O_2 \longrightarrow SnO_2 \quad (15.2)$$

The presence of these reactions is confirmed by slow oxidation of CO on SnO_2 even in the absence of oxygen in the gas phase (8). However, adsorption of oxygen and formation of oxygen ions through uptake of electrons from the surface layer of SnO_2 forming a depletion layer in the surface of the SnO_2 grains also occurs (10). In the case of TiO_2, the exposure to hydrogen and hydrogen-containing gases at elevated temperatures produce oxygen vacancies, which are known as donors and act as oxygen adsorption sites.

2. The ***voltage changes*** when oxygen crosses a semiconducting zirconia layer and can be used to measure oxygen pressure. The oxygen vacancies associated with the presence of the trivalent dopants in the framework of the tetravalent zirconium ion allow the molecular oxygen to be adsorbed as an O^{2-} ion at temperatures above 300 °C, using a pair of electrons of the zirconium atom (5,8):

$$O_{2(\text{gas})} + 4e^- \longrightarrow 2O^{2-}_{(\text{electrolyte})} \quad (15.3)$$

The O^{2-} ions thus formed are mobile and can move in the solid structure, which acts as a solid electrolyte. The O^{2-} ions formed at one face of the zirconia layer circulate to the opposite face, where they are neutralized and desorbed as O_2 molecules.

$$2O^{2-}_{(\text{electrolyte})} \longrightarrow O_{2(\text{gas})} + 4e^- \quad (15.4)$$

The oxygen transfer is accompanied by production of an electromotive voltage U, which can be measured. There is a direct proportionality between the developed voltage and the oxygen partial pressure at the two sides of the zirconia layer. This makes it possible to use the zirconia cell as a potentiometric device (Figure 15.1*a*) (5,11,12).

Zirconium oxide doped with yttrium oxide can be used as an "oxygen pump" in order to measure its concentration in a gas flow. If a current is applied between the two sides of the layer, the circulation of the O^{2-} ions is stimulated and a current intensity can be measured. This current intensity is a measure of the amount of oxygen transported through the layer. This device allows the oxygen concentration to be measured by an amperometric method (8).

Maskel (12) reviewed the zirconia-based sensors, indicating that this material can be used in thick or in thin sensitive layers. A pump-gauge sensor manufactured by two-twin zirconia layers separated by a gold seal and leaving a void volume between the oxide sheets is presented in Figure 15.1*b*.

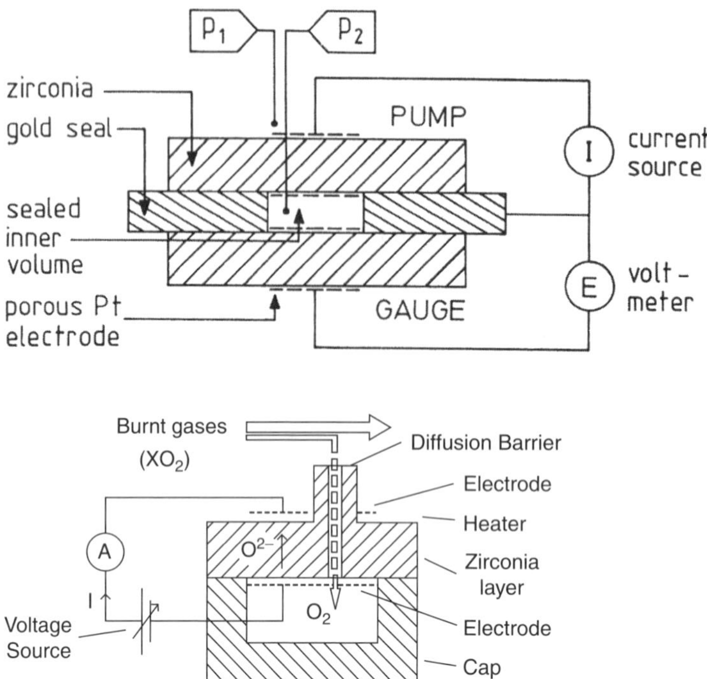

Figure 15.1. (*a*) Linear oxygen sensor layout. (*b*) Linear oxygen sensor layout with a diffusion limiting hole into the cavity. Reprinted from Refs. 5 and 12 with kind permission from Elsevier.

The access of oxygen in the inner volume is possible by a laser-made drill. The device allows simultaneous use of the oxygen-pumping cell and the electromotive force measuring device. The validity of the Nernst Eq. 15.1 giving the correlation between the electromotive force E and the pressures outside (p_1) and inside (p_2) the cell was confirmed by experimental data:

$$E = \frac{RT}{4F} \ln \frac{p_1}{p_2} \tag{15.1}$$

where R is the universal gas constant and F is the Faraday number.

3. The sensors made by materials for which a catalytic interaction occurs with the gas to be dosed involve measuring a certain gas by the ***changes that occur in the surface layer charge, as a consequence of the chemical reaction*** or by correlating the chemical reaction heat with the changes of the electrical resistance values (5).

In addition, similar phenomena as in the case of zirconia occur at the interface between a semiconductor oxide, a metal, and a gas phase. This kind of structure is called "triple-phase boundary." The oxygen adsorption takes place on the sites (*) situated at the interface of metal and oxide (Reaction **15.5**), the metal also being the source for the electrons necessary to transform oxygen into O^{n-} ions ($n = 0.5$, 1 or 2). Several oxygen anions were identified on the solid surfaces, as O_2^-, O_2^{2-}, and O^{2-}. The first two species have an electrophilic character, whereas the latest is nucleophilic. The nature of the adsorbed species is essential in defining the mechanism of interaction with the gas species to be detected (13,14).

The anionic oxygen species then diffuses through the semiconductor oxide layer, using the oxygen vacancies existing in the layer (V) (Reaction **15.6**). If a reducing species (carbon monoxide, hydrocarbon species) also exists in the gas phase, it could react with the oxygen, either with the adsorbed species or with the anion species (Reactions **15.7** and **15.8**). The oxygen adsorption can thus define the baseline for a sensor, whereas the modifications in the surface electrical charge associated with the decrease of the amount of adsorbed oxygen (or the increase of the electron density, due to oxygen chemical transformation into products) may be used for measuring the reducing gases. The chemical reactions involved in this case are as follows:

$$O_{2(gas)} + 2^* \longleftrightarrow 2O_{(ads)} \quad \textbf{(15.5)}$$

$$O_{(ads)} + V + 2e^- \longleftrightarrow O^{2-}_{(electrolyte)} + {}^* \quad \textbf{(15.6)}$$

$$CO_{(gas)} + O_{(ads)} \longleftrightarrow CO_{2(gas)} \quad \textbf{(15.7)}$$

$$CO_{2(gas)} + O^{2-}_{(electrolyte)} \longleftrightarrow CO_{2(ads)} + V + 2e^- \quad \textbf{(15.8)}$$

The reactions (Reaction **15.6**) and (Reaction **15.8**), which occur simultaneously, are cathodic and anodic reactions. The sensor output signal will therefore register a sum between the contribution of oxygen adsorption and of CO oxidation. The sensitivity and selectivity of the sensor are defined by the nature of the metal and oxide joining in the triple phase boundary. Other reducing gases like hydrogen and ammonia interact with the zirconia in a similar way as CO. Decomposition of the gases preferably occurs on dopants like Pt, Pd, or Au followed by spillover of hydrogen atoms to the metal oxide. The role of dopants is further described below (15–17).

To detect an oxidizing species, the mechanism of detection must take into account different reactions and elementary processes. Nitrogen oxides, NO, and NO_2, are often simultaneously present in, e.g., exhaust gases from cars and flue gases from the combustion in boilers. The former acts as reducing agent, whereas the latter is oxidant in a gas mixture containing oxygen. The following reactions have been proposed to describe the detection mechanism:

$$NO + O_2\text{: } O_2 + 4e^- \longrightarrow 2O^{2-} \quad \textbf{(15.9)}$$

$$NO + O^{2-} \longrightarrow NO_2 + 2e^- \quad \textbf{(15.10)}$$

$$NO_2 + O_2\text{: } NO_2 + 2e^- \longrightarrow NO + O^{2-} \quad \textbf{(15.11)}$$

$$2O^{2-} \longrightarrow O_2 + 4e^- \quad \textbf{(15.12)}$$

Another hypothesis considers that NO_2 can easily dissociate at the surface of tin oxide and that its detection mechanism is essentially identical to that of NO (5).

A thin film of In_2O_3–SnO_2 (1:1 molar ratio) prepared by the sol-gel technique (18) deposited by spin coating was used as sensor for NO_2, because the thin film containing both chemical species is more conductive than either of the two species. This is due to the replacement or to the interstitial insertion of an In^{3+} by an Sn^{4+} ion, which delivers an extra electron that improves the conductivity and to the increase of the number and the dispersity of oxygen vacancies in the crystal lattice. The exposure of the sensor samples to NO_2 decreased the conductivity value, due to the capture of the electrons from the conduction band. Exposing the sensor to NO_2 containing gas causes the conductivity to decrease steeply, whereas recovering occurs relatively fast after circulating a flow of air over the sensor surface. The sensor response depends to a large extent on the value of the temperature. The sensitivity, measured in terms of I_{air}/I_{gas}, reaches the maximum value at 250 °C on each of the simple oxides, but also on the mixed oxide. In this case, the response is much higher than with either of the simple oxides; see Figure 15.2 (18).

The narrow temperature range where the sensitivity of the mixed-oxide sensor is very high must be associated with the optimal adsorption conditions for oxygen and the detected species. It can be seen from Figure 15.2 that the mixed sensor follows the same behavior as SnO_2, even if the structure determined by X-ray diffraction is that of In_2O_3 and the sensitivity of indium oxide decreases with temperature.

The detection mechanisms on the two oxide surfaces are different to a certain extent. Both oxides contain on the surface certain amounts of anionic vacancies $V_O^{\bullet}$ and partially reduced cations, In^{2+} and Sn^{2+}. They act as basic adsorption sites for

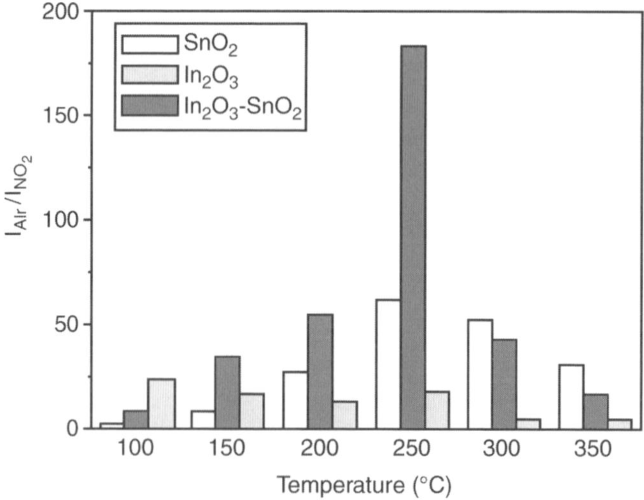

Figure 15.2. The response versus temperature to NO_2 of In_2O_3–SnO_2 as compared with the single materials. Reprinted from Ref. 18 with kind permission from Elsevier.

NO_2, which can react with the sensor surface as follows:

$$NO_{2(gas)} + In^{2+} \longrightarrow (In^{2+} - NO_{2(ads)}) \longrightarrow (In^{3+} - O^-_{ads}) + NO_{(gas)} \quad \mathbf{(15.13)}$$

$$NO_{(gas)} + In^{2+} \longrightarrow (In^{2+} - NO_{(ads)}) \longrightarrow (In^{3+} - O^-_{ads}) + \tfrac{1}{2}N_{2(gas)} \quad \mathbf{(15.14)}$$

$$NO_{2(gas)} + Sn^{2+} \longrightarrow Sn^{3+} - NO^-_{2(ads)} \quad \mathbf{(15.15)}$$

On tin oxide, this reaction has a strong competition by two oxygen atoms adsorbed as $O^-_{2(ads)}$ and its dissociation state, $O^-_{(ads)}$. These species are serious competitors on adsorption sites for NO_2, which could instead be oxidized to NO_3^-:

$$NO_{2(gas)} + O^-_{(ads)} \longrightarrow NO^-_{3(ads)} \quad \mathbf{(15.16)}$$

As the basic character of indium oxide is stronger, the NO_2 adsorption is dominating, whereas on SnO_2, the reaction between NO_2 and $O^-_{(ads)}$ is by far preferred.

At low temperatures, the role of the basic character of the surface is more important, so the sensitivity of In_2O_3 is increased, whereas at higher temperatures, the role of catalytic reaction increases. In the case of mixed oxide, the high activity is supposed to be based on a synergetic effect of the good conductivity due to association of the two metallic species and to the higher global dispersity and high defect density.

4. **The role of the noble metal additives** on the sensor surfaces can be explained in terms of electronic changes occurring in the oxide superficial charge distribution or by the chemical influence (catalytic reactions promoted by the dopants) (15). In the first case, the reaction with the gas to be detected takes place on the metallic cluster and the metal-oxide support has only the role of a transducer. In the second case, the semiconductor acts as a chemical catalyst and the role of the doping metal consists in increasing the reaction rate of the detected gas molecules. The dopants act as adsorption sites for the species to be detected. They then migrate to the oxide surface in a "spillover" process. Usually platinum acts as catalyst, whereas metals as palladium are related to the electronic mechanism. These mechanisms are closely related to the interaction between the gas and the dopants, or the oxide and the dopants. In practice, doping leads to both metallic clusters and noble metals in ionic state, the latter being predominant. The affirmation is sustained by the large increase of the value of the electrical resistance of SnO_2, when synthesized by precipitation of $SnCl_4$ with ammonia and simultaneously doped by soluble sources of Pt, Pd, and Au. It can be explained by the fact that the noble metals are highly dispersed on SnO_2 grains and act as oxygen O^{2-} adsorption sites, capturing thus the superficial electrons responsible for conduction. The sensitivities to CO and CH_4, in terms of $(R_0 - R_{CO})/R_{CO}$ and $(R_0 - R_{CH_4})/R_{CH_4}$, differ as a function of the metallic dopants and dry or humid air. If the influence of the first two parameters on the sensitivity due to peculiar interactions is easily anticipated, the large differences brought by the presence of water vapor in the air are less expected. The presence of water strongly influences the sensor properties of the Pt- and Pd-containing layers and has practically no influence on the

Au-containing sensor. Pt and Pd exist in the metallic state, but also as ions, whereas gold is mostly metallic. The water seems to be involved in activating some adsorption sites and deactivating others, because the temperature for the maximum sensitivity is different for dry or humid air.

The same group (19) investigated in detail the sensor properties of Pd/SnO_2 for methane, propane, and toluene in dry and humid air, by designing an equipment able to monitor the composition of the gas flow that has passed over the sensor. The three hydrocarbons were converted into carbon dioxide and water based on the stoichiometric ratios between hydrocarbons and oxygen in the gas flow. The oxygen demands for the oxidation of the three species were, in molar ratios, 9, 5, and 2. The sensitivity was highest for the toluene and decreased in the order toluene > propane > methane, inversely to that expected if the oxygen adsorption step would be a limiting step. The results in terms of sensitivity indicate that the necessity of oxygen for the total combustion is not a limiting factor. The presence of water vapor in the gas influences to a large extent the sensitivity, most probably because of its competitive adsorption on the same sites as hydrocarbons on the sensor surface.

An ingenious device was designed and tested by Ozawa et al. (16) based on adsorption and catalytic reactions on the sensor surface, following the procurement of a very effective and fast sensor for fire alarm. The main organic compounds that appear when a fire is started, as shown by simulation of a fire and analysis by a GC–MS, were proved to be 5-methyl-2-furaldehyde, furfural, benzene, and toluene. The active surface of the sensor, consisting of Pd-impregnated alumina, was intermittently heated, 0.4 s "ON" and 9.6 s "OFF". During the "OFF" period, the organic compounds adsorb on the sensor surface and are combusted during the "ON" period. The combustion step develops a very sharp and high signal after about 100 ms, and then a flat signal due to burning of the organic material arriving on the surface during the remaining 0.3 s before the heat is switched "OFF". During the "OFF" period, the signal is null. The time chosen for the "OFF" state was determined experimentally, in order to obtain the highest output during the "ON" time. The sensitivity of this device to 5-methyl-2-furaldehyde and furfural is extremely high.

Recently Koziej et al. (20) investigated the detection mechanism involved in sensing propane by thick-layer tin-oxide based sensors. This investigation was performed by simultaneous diffuse reflectance infrared Fourier-transformed spectroscopy (DRIFT) during the catalytic conversion and resistance measurements, carried on in a specially developed setup. They concluded that propane was dissociated into H and $^{\bullet}C_3H_7$ radicals on adjacent acid–base sites Sn^{4+}–O^{2-}, not to other intermediates as propene or acetone, as it could have been assumed. Subsequently, the propyl radicals reacted with the molecular oxygen ions O_2^- to give formate and acetate ions, which were then oxidized to CO_2.

15.2.2. Role of the Layer Structure

The optimization of the sensitive layer can be achieved by the so-called structural *engineering of metal-oxide films* (21). The enhancement of the layer properties refers to parameters such as gas response and speed of response, selectivity, and stability. These challenges can be reached by searching the best technological methods for doping the metallic oxide as well as for fine-tuning the crystallites morphology. As the sensing properties are directly related to the accessibility of the adsorption sites by the species in the measured gas, the optimization of the solid must take into account that the gas molecules must reach the sites as easily as possible. It means that the more easily available adsorption sites the surface offers, the larger the sensing potential of the surface will be. The morphology of the solid is defined by a series of parameters such as film thickness and porosity, crystallite size, shape, and their microscopic structure, the space orientation of the crystallite planes, and the amount, positioning, and bounding nature of the dopants. This fact requires consideration of the following:

- During grain growth, the *geometric shape* changes from spherulites of nanometer range size to different polygonal faceted shapes, at the micrometer range size.
- Most sites participating in adsorption are situated on the *external surface*; this is why it is important to know whether crystallites are arranged in an ordered or disordered manner, whether there are intergrowth of crystals with changes of the growing main direction, or whether there are inter-grain contacts.
- Every *crystallographic plane* has its combination of *surface electron parameters*, including surface electron density, levels of electron energy, availability of the surface for adsorption–desorption processes, the energy state of the surface, and the concentration of adsorption sites.
- The *surface defects or bulk defects* (*dopants*) induce partial control in the growing process as well as in promoting adsorption.

It is well known from the theory of heterogeneous catalysis that the best positions that favor the adsorption process are in places where the local irregularity is high, namely, corners, edges, and plane break offs, sites with abnormal electronic density in comparison with the bulk. The experimental approaches proved that the most efficient sites for gas adsorption are the boundaries between the oxide grains, the so-call "necks" between two neighboring crystallites (22). The sensitivity of SnO_2 was proven to reach a maximum for a particle size close to 6 nm, when prepared by a hydrothermal procedure followed by calcination (23). At this size, on the one hand, the specific surface area is higher than in the case of larger particles, and on the other hand, the area cannot any longer be considered as a flat surface, but the curvature of the surface should be taken into account. This fact requires consideration of the fact that the density of defect sites on the surface increases to a large extent. These defects are usually strong reactive sites and enhance the overall surface reactivity. This behavior is effective in thick films as well as in thin films (24). Korotcenkov et al. (25) consider that a polycrystalline film of particles could be schematically presented as an equivalent circuit, in which the small grains, the inter-grain contacts, and the agglomerated and the inter-agglomerate contacts are resistors alternatively bounded in series or in parallel (Figure 15.3) (21).

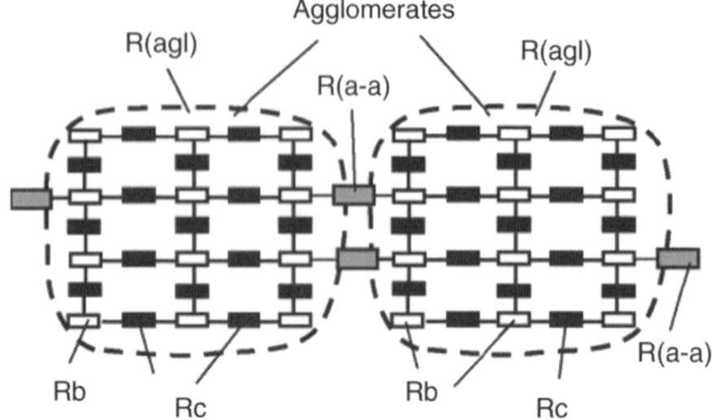

Figure 15.3. Schematic representation of a metal-oxide film as serial and parallel resistors; R_b—resistance of grains; R_{ag}—resistance of agglomerates $R_{a\text{-}a}$—resistance of interagglomerate contacts; and R_c—resistance of intergrain contacts. Reprinted from Ref. 21 with kind permission of Elsevier.

This approach allows putting into evidence that the porosity of the agglomerates and the gas-sensing matrix play an important role in the overall value of the layer resistance. The low gas permeability of agglomerates increases the influence of inter-agglomerate contacts, which do not interact with the gas to be detected. The use of small-sized crystallites in a sensor is a reliable way to achieve maximum gas sensitivity and fast response. At the same time, excessive decreasing of the grain size and the increase of the layer thickness lead to a lower penetration depth for the gas, due to the decrease of the porosity of the layer.

Korotcenkov et al. (26,27) showed that the crystal growth direction of the SnO_2 and In_2O_3 has an important influence on the sensing properties of the layer. In equilibrium conditions, the crystals grow in the direction of the face with the highest growing energy and consequently adjacent faces appear. Comparison between two samples of SnO_2 obtained in equilibrium and nonequilibrium conditions in the detection of CO in wet and dry air outlined different sensitivities as a function of the predominant growing direction of the crystals. Water adsorption is possible on oxygen sites from faces of the less stable crystals, and the sensitivity of the samples prepared in nonequilibrium conditions to CO is much higher. The following reactions can explain the experimentally observed increasing of the conductance due to a release of free electrons:

$$H_2O + (Sn)O_{ads}^{n-} \longrightarrow (SnOH)^- + (O_{ads}H)^{(n-1)-} \quad (n = 1 \text{ or } 2) \qquad \textbf{(15.17)}$$

$$CO_{gas} + O_{ads}^- \longrightarrow CO_2 + e^- \qquad \textbf{(15.18)}$$

$$CO_{gas} + (SnOH)^- + (O_{ads}H)^{(n-1)-} \longrightarrow CO_2 + H_2O + ne^- + (Sn)O_{ads}^{n-} \qquad \textbf{(15.19)}$$

Sasahara et al. (28) prepared sensors for toluene sensing by a screen-printing technique of Pd-impregnated mesoporous silicas as catalytic sensing layer, using the same quality silicas as references. The original silica has sharp and uniform-sized cylindric pores of 2.2 nm diameter and specific surfaces between 666 and 1502 m^2g^{-1}, whereas after impregnation and calcination, the values decreased between 19 and 897 m^2g^{-1}. The highest response of the sensor in terms of $(V_g - V_a)/V_a$ (V_g and V_a being the output voltages) was obtained on the higher surface area material, as expected.

Xu et al. (29) and Sberveglieri (30) showed that not only the chemical composition, but also the preparation methods have an important influence on the sensing properties of the layer. Xu et al. prepared a series of sensors based on tungsten oxide promoted with molybdenum oxide. The layer was prepared by impregnation of tungsten powder with molybdenum salt, painting of the tungsten oxide layer with molybdenum oxide, and vapor deposition of molybdenum oxide on tungsten oxide. The electrical resistance of the layers was measured in ammonia and an ammonia + NO containing mixture in 2% oxygen and nitrogen. The measurements showed that the sensitivity to 5-ppm ammonia in the $O_2 + N_2$ mixture was the same regardless of the preparation method. In contrast, the sample prepared by the evaporation method was more than two-folds more sensitive to ammonia, when 100-ppm NO was added to the tested gas. This behavior is supposed to be due to formation of new sites for ammonia sensing and to the poisoning of the sensing sites for NO by the MoO_3 doping. The hypothesis is supported by the SEM images, showing that the MoO_3 is standing as fine fibers on the WO_3 grains.

Stankova et al. (31) underline the role of the morphology of tungsten oxides on the sensing properties to ammonia, ethanol, and nitrogen dioxide. An increase of the annealing temperature deepens the roughness profile of the surface, as atomic force microscopy investigations showed. There is a close relationship between the surface irregularities and sensitivity potential of the surface to the gases, more obvious in the case of ethanol.

The *n*-type semiconductor oxides were also find to be suitable for the detection of oxidizing species such as O_3, Cl_2, and NO_2 (32). Indium oxide doped with Ce, Sn, Ni, Fe, and MoO_3 give good results in the concentration range from 1 to 10000 ppb, depending to a high extent on the dopants nature. These gases induce an increase of the resistance, due to electron capture, whereas the species are adsorbed. The nanocrystalline films have a better behavior than the thick films. The sensing mechanism is related to the adsorption that can occur with or without dissociation of the O_3 molecule:

$$O_3 + e^- \longrightarrow O_{2ads} + O^-_{ads} \tag{15.20}$$

$$O_3 + e^- \longrightarrow O_3^- \tag{15.21}$$

The XPS investigations indicate the first dissociation mode as more consistent.

15.2.3. Influence of the Chemical Composition of the Layer

As the interaction between the gas and the sensor receptor occurs at a microscopic level, recent development of nanotechnology plays an important role in obtaining gas sensors (33). The nature of the particle distribution in the layer defines the transducer, which is responsible for transforming the interaction into measurable electrical signals. The chemical and structural stability of the layer depends to a big extent on its chemical composition. The maximum response rate will be given by the minimum agglomeration of the particles (21).

The mixed oxides used for sensors can be grouped in three categories, as follows (34):

- *Chemical Well-Defined Compounds*: A distinctive chemical species is formed by reaction between two simple oxides:

$$ZnO + SnO_2 \longrightarrow ZnSnO_3 \text{ or } Zn_2SnO_4 \quad \textbf{(15.22)}$$

$$CdO + SnO_2 \longrightarrow Cd_2SnO_4 \quad \textbf{(15.23)}$$

$$CdO + In_2O_3 \longrightarrow CdIn_2O_4 \quad \textbf{(15.24)}$$

$$SnO_2 + WO_3 \longrightarrow \alpha\text{-}SnWO_4 \quad \textbf{(15.25)}$$

- *Solid Solutions*: TiO_2–SnO_2.
- *Mechanical Mixtures*: TiO_2–WO_3.

Researchers (21,35) indicate that the sensor performances improve significantly when binary oxides or complex multicomponent materials are used instead of simple oxides. Moreover, inclusion of catalytically active elements (noble metals or other transition metals) and even inert impurities can have a favorable effect on sensor performance. The role of the impurities can be extremely diverse, from favoring the formation of and/or stabilizing of a certain phase, the size of the crystallites and their growth direction, acting as catalytic promoters, favoring the occurrence of the proper valence state of the receptor, the electronic state of the surface, the surface potential, and the interparticle geometrical and energy barriers. The effects on sensor properties concern changes of concentration of charge carriers, chemical and physical concentration of metal-oxide matrix, and electronic and physical-chemical properties of the surface etc.

The electronic structure of the multicomponent oxides changes the bulk electronic structure, band gap, Fermi level position, and transport properties especially in the first two cases above, whereas the changes in the surface properties mainly occurs at the boundaries between grains of different chemical composition. These phenomena are directly involved in the sensing mechanism.

The influence of doping on electrical properties of tin oxide was investigated by Szczuko et al. (36). The doping with In, Sn, and Nb was performed by the wet method, and the surface composition was analyzed by XPS. A 5% mole ratio was found to be the maximum ratio of enrichment of the surface in dopants species; over this limit, a second phase appears. Concentrations of dopants of 0.1–10% lead to an important decrease of crystallite size from 220 nm for pure SnO_2 to 50 nm in the case of In and Sb and to 30 nm in the case of the sample doped with 0.1–2%Nb. The effect of the

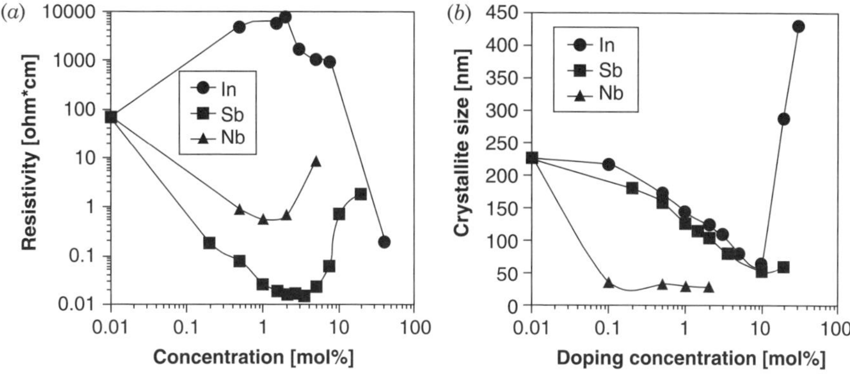

Figure 15.4. The influence of an SnO_2 film by the doping concentration of In, Sb, Nb on (*a*) the resistivity and (*b*) the crystalline size. Reprinted from Ref. 36 with kind permission of Elsevier.

doping on the resistivity was very different from one dopant to another; a decrease in size also reduces the resistivity in the case of Nb and In-doped SnO_2, whereas doping with Sb led to resistivity increase; see Figure 15.4 (36).

Doping the thin (30–40 nm) SnO_2 film with Ag and Pd nanoclusters by successive ionic layer deposition (SILD) leads to an important increase of the response to 0.1–0.5% H_2 and CO sensing using air as a reference gas 5–10 times in comparison with pure SnO_2 film (25,37). The measurements were done in terms of resistance changes, and the results were strongly dependent on the operating temperature, number of deposition cycles, and nature of dopants.

Timmer et al. (38) reviewed information about the ammonia-sensing sensors. The ammonia molecules are chemisorbed, and the electron pair from the nitrogen atom increases the conductance of the layer. Wang et al. (39) investigated the sensing properties of a large number of WO_3 sensors doped with alkali metals, earth metals, and a long series of transitional and f-block metals in the Periodic Table. Some dopants increase to a large extent the sensitivity to NH_3 (Gd, La, Fr, Si, Y, Ti), whereas others decrease and even cancel the sensing properties (Li, Na, Rb, Cs, Pb, Ba).

The selective detection performances of CO in a gas atmosphere containing H_2 was investigated by Yamaura et al. (40) on simple oxide-based sensors, prepared by the hydrolysis and screen-printing method (41,42). From 24 different oxides, the best behavior regarding the CO sensing properties was noticed for In_2O_3, SnO_2, Fe_2O_3, ZnO, Y_2O_3, and TiO_2. To test the cross-sensitivity in the presence of H_2, the indium oxide was doped with alkali metals, earth metals, and a long series of transition metals. In terms of sensitivity to CO, important increases in sensitivity were noticed for Na-, K-, and Rb-doped films, whereas the other species had no effect or even significantly decreased the sensitivity. When a mixture of CO and H_2 was used as a test gas, the sensitivity increased almost nine times for the Rb-doped sensor, whereas Na, K, Ca, Y, Mo, and Sm increased almost four times and the other dopants had negligible effect.

Vaishnav et al. (43,44) investigated the indium–tin-oxide sensors for the detection of inferior alcohols, methanol, and ethanol. The catalytic properties of the indium-tin surface could be improved by deposition of a thin layer (10 μm) of supplementary catalytic material on the sensing layer. The nature of this supplementary layer was Cu, MgO, or CaO in the case of methanol and MgO for ethanol.

The first step in the methanol catalytic conversion routes involved in the detection can occur by its oxidation to formaldehyde by reaction with O^- or to formic acid by reaction with O_2^- species adsorbed on the surface:

$$CH_3OH + O^- \longrightarrow CH_2O + H_2O + e^- \quad \mathbf{(15.26)}$$

$$CH_3OH + O_2^- \longrightarrow HCOOH + H_2O + e^- \quad \mathbf{(15.27)}$$

The oxygen chemisorption, which forms these two ion species, is associated with the presence of copper.

In the case of ethanol, a net difference can occur if the catalytic sites on the surface are acidic or basic (either O^- and O_2^- can participate in the reaction):

$$CH_3{-}CH_2{-}OH_{(gas)} + O^-_{ads} \longrightarrow H_2C{=}CH_2 + H_2O + e^- \text{(on acid sites)} \quad \mathbf{(15.28)}$$

$$CH_3{-}CH_2{-}OH_{(gas)} + O^-_{2ads} \longrightarrow CH_3{-}CH{=}O + H_2O + e^- \text{(on basic sites)} \quad \mathbf{(15.29)}$$

The electrons released in these reactions result in the decrease of the electrical resistance value of the surface. The maximum sensitivity of the sensors occurs at the optimal catalytic reaction temperatures, about 625 K for methanol and 723 K for ethanol, respectively.

Dutta and De Lucia (45) briefly reviewed the role of some additives in the sensing layer on its stabilization. Ten percent lanthanum mixed with the anatase phase of TiO_2 leads to preserving this polymorphic form even up to 800 °C, avoiding the transformation to rutile. This material is suited for preparing sensors for CO detection, by adding CuO or Au. The sensitivity of Au–La_2O_3–TiO_2 was remarkable at temperatures up to 600 °C. The explanation consists in the ability of Au to adsorb CO, whereas TiO_2 acts as an adsorbed oxygen species reservoir.

Setkus et al. (46) investigated the kinetics of the sensor response as a function of the nature of the dopants on the CO and H_2 sensing properties of tin-oxide thin-film sensors modified with metallic impurities. The response time of a sensor is the time duration to reach 90% of the total sensor response. The tin-oxide sensors were doped with Pt, Mo, W, and Ag. All dopants modified both the sensor sensitivity, due to their specific catalytic action, and the transient of the response. Kinetic models were elaborated for the reactions involved in the sensors' response, and the kinetic parameters were determined from experimental data. Important modifications appear in the values of the adsorption coefficients and in the probability of bimolecular interaction between adsorbed oxygen and the detected species. The metallic layer of dopants has very little influence on the adsorption of hydrogen, and it is a clear disadvantage for the adsorption of CO. Regarding the reaction rate, Pt and W greatly increase the probability of CO to participate in a bimolecular interaction, whereas all metals are favorable for a H_2 bimolecular reaction.

15.3. DETECTION PRINCIPLES

The sensing properties are based on changes in measurable electrical properties of the sensor in the presence of certain gas species. The changing of the properties is associated with chemical interactions that occur at atomic level.

A *receptor* is generally a solid layer with specific sites at atomic level on its surface. Changes occur in contact with a gas as a consequence of interactions by physical adsorption or chemisorption. For oxide-based sensors, the interaction changes several physical parameters such as electrical charge, polarizability, heat generation, potential changes, resonating frequency, or optical properties. The depth of the charge layer L_d is defined by the Debye length (7):

$$L_d = (kT\varepsilon/q^2n)^{1/2} \tag{15.2}$$

where k, T, ε, q, and n are the Boltzmann constant, absolute temperature, dielectric constant, electron charge, and charge-carrier concentration, respectively.

The interaction occurring at the molecular level of the receptor is transformed in a measurable output signal by a *transducer*. The transducer refers then to the receptor microstructure, because it has an essential influence on the output signal. The *operation mode* of the output signal can be different, depending on the *output measure* that have been chosen: electrical resistance changes, capacitance changes, FET diode, and transistor voltage signals, thermovoltage, and oscillation frequency change due to mass changes; see Figure 15.5 (10).

There is a tight link among the nature of receptor, the transducer structure, and the operation mode. According to the changes of physical properties, the following types of sensor systems can be developed (1,2,4):

1. Based on changes of electrical resistance or impedance—thick-film metal-oxide sensors, thin-film metal-oxide sensors on interdigitated electrodes, conducting polymers.
2. Changes of electrical current—amperometric and potentiometric sensors.
3. Changes of electrical capacitance—metal-oxide semiconductor capacitors or plate capacitors.
4. Changes of work function or polarization—field effect transistors (FET).
5. Changes of mass—bulk or surface acoustic wave resonators.
6. Changes of temperature—pellistors, thermistors, thermopiles.

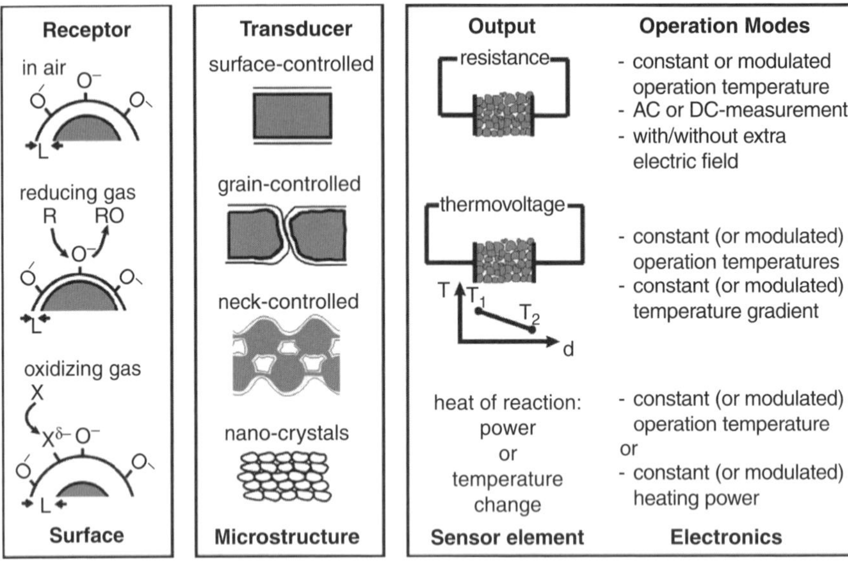

Figure 15.5. Some examples of different principles of metal oxides as sensing material. Reprinted from Ref. 10 with kind permission of Elsevier.

7. Changes of optical adsorption or reflection due to changes in refractive index, optical layer thickness—optical sensors.
8. Changes in incidence angle for the start of a surface plasmon due to changes in refractive index—SPR sensors.

For the oxide-based sensors, the modification of the electrical properties of the layer is the most widely used in evaluation of the interaction with a gas. The main parameters for describing a chemical sensor are as follows: the detection mechanism, the sensitivity profile, its concentration range, the speed of response, the operation temperature, and its lifetime.

15.3.1. Conductivity Measuring Sensors

The type of gas sensors that are by far the most widely used are based on the fact that the resistivity of a conductor is a measure of the mobile charge carriers in the surface region of the device. A wide variety of resistive gas sensors have been designed and tested. Many sensors are already commercially applied, and considerable research work has been focused to tin oxide, which has proved to be a good sensor material.

Lee et al. (47) used for detection of isobutane a sensor having sensing layers on both sides of an alumina plate. On one side, a layer of In_2O_3 doped with tin oxide was deposited, and on the other side, an alumina layer doped with platinum was deposited.

Under each sensitive layer, a conductive platinum layer was sputtered. This platinum layer has a triple role: acts as a heater, as a resistor for a measuring system, and has a catalytic effect in combustible gas sensing at high gas concentrations. The proper conductivity and maximum sensitivity for detection of 1% butane in air was achieved when the In–Sn layer was between 12% and 16% Sn. The Pt-based sensor worked on the principle of a combustible catalytic sensor mechanism. The heat emerged when the gas burned on the sensor surface, and led to a linear increase of output voltage with a rate of 0.153 mV/ppm up to 1.5% butane; then a portion of nonlinear increase follows up to 3% butane, and finally a slight slower linear increase appears up to 10% butane.

The so-called "sensitivity model for the equilibrium state," consisting of an analysis of the transient responses after a steep change between two equilibrium states was used by Setkus (48) to analyze the SnO_2 sensor behavior in CO sensing. In principle, the method considers the elementary steps involved in the catalytic reaction for elaborating a kinetic model, and then it compares the results issued from this model with those from an experimental approach. The working principle of the transducer is the modification of electrical resistance when the catalytic reaction takes place. In oxygen-rich atmosphere, the chemisorbed oxygen acts as trap for the bulk electrons on the surface of the solid. Any change in the oxygen coverage due to oxidation reactions will be reflected in changes of the electrical resistance of the layer. Setkus considers in elaboration of the modeling that the catalytic reaction occurs via a classic Langmuir–Hinshelwood mechanism, meaning that both initial species (oxygen and CO) react from adsorbed state on the solid and that the reaction product is immediately desorbed from the surface, delivering the sites to resume the cycle. The mathematical model gives the expression of a ratio between the actual electrical resistance and the initial resistance. The expression is deduced in the conditions of reaching the catalytic reaction equilibrium, as a function of time; the kinetic constants of the elementary steps; adsorption and desorption coefficients; the height of the potential barrier at the boundaries; and the elementary charge. For an 80–120-nm-thick sensing layer made by 8-nm-sized particles, the experimental results fitted excellently with the mathematical model.

On a porous ceramic body sensor based on semiconducting oxides such as SnO_2, ZnO, and Fe_2O_3, heated between 300 and 700 °C, the adsorption of oxygen seems to take place in the neck region between the grains (33). As electrons are requested for oxygen adsorption, a simultaneous accumulation of negative charge occurs at the surface of the grains, whereas a complementary positive charge called the depletion layer appears in the oxide layer. The negative charge makes the surface attractive for reducing gases as hydrogen or carbon monoxide, which react with the adsorbed oxygen, which follows by an increase of the conductivity of the layer (33).

The resistance changes occurring in semiconductive oxides as a consequence of the interaction with different gases were investigated by Williams (49) on oxides with different character, *n* or *p*-type. The chromia-titania (1–45% Ti) increases its resistance even to traces of reducing gases; and thus, it can be classified as a *p*-type oxide. It responds to CO, volatile organic compounds, H_2S, NH_3. Indium, tin, tungsten, and molybdenum form *n*-type oxides, which increase the resistance in the presence of ozone, chlorine, and nitrogen dioxide. In the case of indium and molybdenum oxides, the sensitivity to reducing gases is explained by oxygen vacancies, which offers a flexible behavior.

15.3.2. Ion Conductor Sensors

Miura et al. (50) made a classification of the types of potential devices based on a solid electrolyte used for sensing redox gases. According to his concept, three main groups of sensors can be defined, namely, equilibrium potential, electrochemical pumping current, and mixed potential.

The first category deals with sensors in which electrochemical reversible reactions are involved and in which the sensing electrode is assisted by a reference electrode. For example, the yttria stabilized zirconia works on the basis of the reaction:

$$O_2 + 4e^- \rightleftharpoons 2O^{2-} \tag{15.30}$$

In the case of its use as an electrochemical pump, the reduction takes place after the diffusion of the ions through the electrolyte layer, by the contribution of an applied potential.

The mixed-potential electrodes are characterized by allowing occurrence of oxidation and reduction reactions at the same time, by the contribution of specific sites. This concept was first introduced to explain the zirconia sensors non-nernstian behavior of an oxygen sensor in the mixture of air and fuel. An electrochemical cell including yttria stabilized zirconia (YSZ) should be defined as air, Pt | YSZ | Pt, and CO + air.

The catalytic effect of platinum makes the oxidation of CO possible, by a purely chemical reaction:

$$CO' + \tfrac{1}{2}O_2' \longrightarrow CO_2' \tag{15.31}$$

O_2', CO', and CO_2' denote oxygen in the neighborhood of a three-phase boundary.

The catodic reaction is

$$O_2' + 4e^- \longrightarrow 2O^{2-} \tag{15.32}$$

and the corresponding electrochemical cell: P'_{O2}, Pt | YSZ | Pt, P_{O2} (P'_{O2} being the partial pressure of O_2 after the reaction). CO can also participate directly in the electrochemical oxidation:

$$CO' + O^{2-} \longrightarrow CO_2' + 2e^- \tag{15.33}$$

and the corresponding electrochemical cell is P'_{CO}, Pt | YSZ | Pt, P_{O2}. The two-electrochemical reactions form a distinct cell. The simultaneous progresses of these reactions determine the non-nernstian response of the YSZ.

The mixed-potential type gas sensors find application in the design of gas sensors for species such as CO, NO_x, H_2, H_2S, NH_3, and hydrocarbons (51).

15.3.3. Field-Effect Sensors

Lundström et al. invented the field-effect sensors about 30 years ago (52). They experienced that metal–insulator–semiconductor field-effect transistors having a palladium gate could be used as hydrogen sensors. The field effect involves devices like capacitors, diodes, and transistors (53). A schematic representation of these devices is presented in Figure 15.6.

The *capacitor* consists of a semiconductor with an ohmic contact on the one side and an insulator layer on the other side. On the insulator material, a catalytic, mostly porous layer, e.g., of Pt, Pd, Rh, and RuO_2 is applied as a gate contact together with a thick metallic contact, which allows the bonding. For example, in a hydrogen or hydrocarbon gas environment, the hydrogen-containing molecules dissociate on the catalytic metal sites and hydrogen atoms spill over to the insulator and adsorb predominantly as OH groups. This was recently investigated by DRIFT spectroscopy using model surfaces of Pt- or Ir-impregnated SiO_2 powder. Clear evidences for OH groups and even eventually OH_2^+ groups were found on the SiO_2 sites during hydrogen and ammonia exposure in air (54,55). The OH groups on the insulator form a polarized layer that will introduce an electric field in the insulator, and for a positively biased n-type semiconductor, the depletion area will increase; the OH groups act as an extra-applied positive voltage. Thus, the polarization of the insulator surface causes a voltage shift of the capacitance versus voltage curve, proportional to the concentration of hydrogen. For the sensing of other gases, e.g., ammonia, the presence of triple-phase boundaries, where gas, metal, and insulator are all in contact, are necessary to obtain dissociation of the molecule in such a way that hydrogen atoms are released to the insulator to form the polarized layer in the same way as for hydrogen (56–61).

For the Schottky diode used as a gas sensor, an interfacial layer of a thin insulator is shown to be advantageous for the gas response (62,63). The thin insulator also is the weak part of the device, easily destroyed because current flows through the insulator

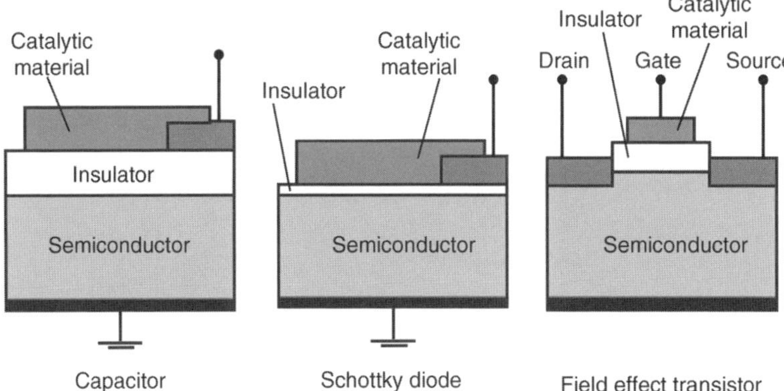

Figure 15.6. Schematic picture of FET devices, capacitor, Schottky diode, and transistor. The catalytic gate material turns the devices into gas sensors. See color insert.

during the operation. The field-effect transistor combines in an advantageous manner the properties of the capacitor, with a thicker and more stable insulator and the simple electric circuitry for operating the diode (53). The working principle is the same for all three devices, hydrogen or hydrogen-containing molecules dissociate on the gate material, and the dissociated hydrogen atoms diffuse to the metal–oxide interface and form a polarized layer as described above. The sensor signal is the voltage at a constant current, and the polarization of the metal insulator interface shows up as a voltage change in the signal (Figure 15.7) (64). This shift is thus a measure of the intensity of the interaction between the gas and the catalytic surface, proportional to the ratio of reducing to oxidizing species (65).

Salomonsson et al. (66–68) investigated the sensing properties of a silicon carbide-based field-effect capacitor prepared by deposition of ruthenium oxide or ruthenium nanoparticles as the gate material. The insulator layer on top of the silicon carbide is a stack about 80 nm thick of silicon dioxide and then silicon nitride that is densified, resulting in a top layer of silicon oxide. The sensing layer was deposited by dropping a RuO_2 or Ru suspension on the substrate surface, followed by annealing. The sheet resistance of the annealed RuO_2 layer was investigated between room temperature up to 500 °C in air; the values are comprised between 7 and 14 Ω/cm^2. The gas sensitivity of the MISiC capacitor was tested for hydrogen, carbon monoxide, nitrogen oxide, propene, and ammonia diluted in air. The voltage shift depends on

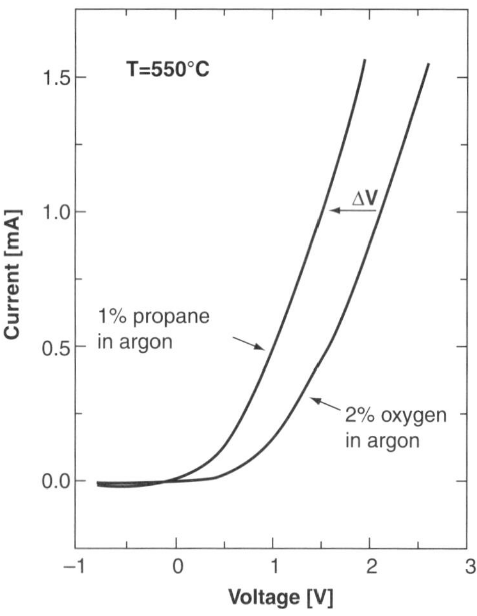

Figure 15.7. I-V characteristics for a MISiC sensor working as a diode. Reprinted from Ref. 64 with the permission of Elsevier.

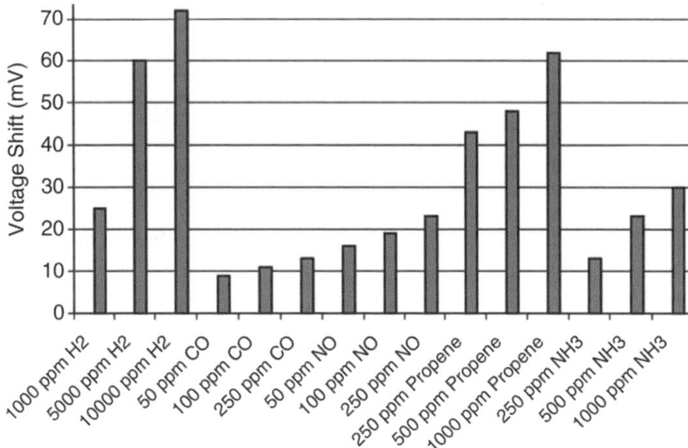

Figure 15.8. Response patterns in terms of voltage shifts of $RuO_2/SiO_x/Si_3N_4/SiO_2$/4H–SiC capacitive sensors at 400 °C to different gases. Reprinted from Ref. 66 with the permission of Elsevier.

the sensing material, the temperature, the nature of the gas, and its concentration (Figure 15.8) (56). It is interesting to consider the fact that in the case with nanoparticles used as a gate material in a FET device, it is the charging of the gate material that will introduce the sensor response. This should be compared with, e.g., the measurements of resistivity changes introduced by the gas interaction. The most information will of course be obtained from a multicomponent device, where both the field effect from the charging of the gases and the resistivity changes are measured simultaneously (61).

Åbom et al. (69) used capacitor-type MIS field-effect sensors with SnO_2, SiO_2, and Al_2O_3 as catalytic material for the detection of hydrogen, ammonia, propene, and acetaldehyde. A platinum layer, 8 nm, was deposited on the insulator layer prior to an oxide deposition. The properties of the layer in terms of chemical composition were determined by AES spectra. The morphology revealed by SEM showed that the oxide layer up to about 56 nm allows a certain fraction of the Pt porous layer on the surface to remain uncovered by the oxide. The catalytic effect of platinum on the gas interaction is essential to achieve detection of the species at hand. For SnO_2 layers thicker than 56 nm, the shift voltage is null, indicating that the access of molecules to the platinum is mandatory and that the reactions involved in sensing occur at metal-oxide grain boundaries. A new series of sensors was prepared, with a thicker, more dense, Pt layer (60 nm) deposited on top of the 8-nm layer before oxide deposition. The SnO_2 and SiO_2 layers, less than 56 nm in thickness, deposited on this thicker platinum layer gave better responses to ammonia than the Pt layer alone. This outlines the role played in the sensing in terms of both the role of the catalytic platinum metal and the role of the oxide in oxygen and hydrogen adsorption. These results were confirmed in a new setup of experiments with thin or thick Pt films, cosputtered Pt and SiO_2, and overlayers of SiO_2 (69).

Inspired by the results above, cosputtered films of SiO_2/Pt or Ir were successfully processed and tested as sensing materials for reducing gases like NH_3, H_2, and propene in SiC-based FET devices (70). It was also shown that a cosputtered film of Ir and SiO_2 increased the long-term stability as tested at 500 °C in hydrogen/oxygen intermittent pulses (71).

15.4. PROCESSING OF METAL OXIDES AS SENSING LAYERS

As the sensing mechanism is controlled at the nanoscale level, the use of nanoparticles in sensor technology can bring benefits in the "four S" requirements: sensitivity, selectivity, stability, and speed of response (72). The wide use of sensors nowadays based on nanoparticles supports the special importance that we pay to their synthesis methods. Nevertheless, a distinctive problem consists in their deposition on the active substrate area, in order to give a sensor device. The adherence of the nanoparticles on the substrate is a key step in preparation of a chemical sensor based on oxides.

15.4.1. Nanoparticle Synthesis

The preparation of oxide nanoparticles for use in sensor manufacturing involves fulfilling some requirements: control of particle size, obtaining a narrow size distribution, even monodispersity if possible, passivation of the surface, and control of particle shape (33). Achievement of these demands must also take into account the possible degradation of the synthesized materials, which should be avoided. For applications in sensors, the nanoparticle grain density should be high enough to display a maximum number of grain boundaries, but not too dense, in order to avoid the effect of slow diffusion of molecules to the interaction sites.

The synthesis of nanoparticles can be performed in gas or liquid phase. The synthesis of solids in wet medium can be accomplished by precipitation, sol-gel transformation, and wet impregnation.

The *precipitation method* is the extremely simple **liquid-phase** method and based on the formation of a solid when the concentration of a species in a solvent exceeds the solubility limit. The quality of the product obtained by precipitation depends on the conditions for precipitation: purity of the initial reagents and nature of impurities, the values of the concentrations of the solutions used as reagents, the pH value, the rate of reagents added, temperature, stirring regime, maturation time, temperature, and so on. The choice of the precipitation conditions involves knowledge about the nucleation and growth process, whose rates will to a large extent have an impact on product quality (compositional homogeneity, uniformity of shape and size, crystallinity, etc.) (73).

A widely used wet synthesis method is the *sol-gel method*. It consists of the controlled transformation of a precursor (usually a salt of metal or metalloid element) in a solid phase, by the intermediate formation of a sol (a colloidal solution), then by its transformation in gel, and finally by obtaining a solid product. The formation of sol is generally the result of the salt hydrolysis, in adequate conditions of pH, salt/solvent ratio, and sometimes use of a catalyst (74,75) resulting in oligomeric species. The sol can be transformed in gel, precipitated, or deposited on solid surfaces as a thin film. Transformation in gel is in fact a condensation of the oligomers, in which the solvent is included in large spaces delimited by polymeric species formed by condensation. The gel is then transformed into a solid by evaporation or extraction of the solvent. In most cases, the condensation begins before the hydrolysis ends. The equilibrium between these steps is tuned by the choice of composition and operating conditions. The gel can be transformed into powders, fibers, coatings, monoliths, or solids with ordered pore structure, depending on the conditions chosen for solvent elimination (heating and/or vacuum treatment, solvent extraction, etc.). The method is widely used for production of metal oxides and ceramic powders (76), by including more than one species in the initial solution. Activation is generally the next step, made by calcination at proper temperatures.

An alternative method for doping the solid can be wet impregnation, which can be performed either on the as-synthesised initial solid or on the calcined form (73).

The complexity of the process and the high number of variables interfering in the synthesis makes control of the chemical purity of the product and the nature of the crystalline phases obtained by wet synthesis a difficult task.

The nanoparticle powder needs then to be deposited in a stable manner on the sensor surface. The simplest technique to do this is the so-called "*screen printing*." It consists in preparing a paste by mixing the nanoparticle powder with a solvent or a polymer solution and by applying it on the surface by the use of a mask. The stabilization of the solid deposited by screen-printing is achieved by thermal treatments (26,39).

Gas-phase synthesis of particles is usually a continuous process. Several groups of methods can be involved in nanoparticle processing. For oxide nanoparticles, the most widely used methods include *physical vapor deposition* (PVD), *chemical vapor deposition* (CVD), and *aerosol processing* (76). The aerosol synthesizing can in turn be achieved by a wide choice of methods, one of which can include the so-called "homogeneous nucleation in the gas phase," achievable by several various heating and evaporation methods involving physical and chemical reactions (furnace flow-type reactors, laser reactors, flame reactors, plasma reactors, and sputtering), the electrospray techniques, and the homogeneous nucleation in aerosol droplets (spray conversion technique) (33).

Some of these techniques are extremely sophisticated and costly, whereas others are extremely simple and easy to perform. Which one should be chosen depends on the nature of the product. The gas-phase processing may give better results in terms of some parameters: The purity of the product can be higher, because the use of solvent is avoided (even in the water, traces of minerals can be found). The aerosol processes can lead to complex chemical structures useful in producing multicomponent materials. The product quality is more easily controlled in terms of particle size, crystallinity, agglomeration, porosity, chemical homogeneity, and stoichiometry, by adjusting the process parameters or by adding extra processing steps as sintering or size fractioning. The aerosol processing is much cheaper than vacuum synthesis and faster in terms of deposition rate. The chemical segregation in the aerosol droplet is minimized, because the formed phases have to stay within the particle.

Comini et al. (72) investigated the sensor qualities of SnO_2 *nanobelts, synthesized in the vapor phase*, by oxidation of Sn vapors. The vapors were obtained by the decomposition of SnO at 800–1000 °C:

$$2SnO \longrightarrow Sn\,(l) + SnO_2 \qquad \textbf{(15.34)}$$

then transported by an argon flow, condensed on a substrate as micron-size droplets, then oxidized by an oxygen flow:

$$Sn + O_2 \longrightarrow SnO_2 \qquad \textbf{(15.35)}$$

The layer was tested as gas sensor for CO, NO_2, and ethanol. The maximum sensitivity was achieved at 300–350 °C, a value that assures a convenient amount of oxygen adsorbed on the surface that can act as an oxidizing agent. The sensor responses, in terms of relative variation of the conductance, are much higher than in the case of SnO_2 layers (200% for 30-ppm CO at 350 °C, 900% for 200-ppb NO_2 at 300 °C, and 2500% for 10-ppm ethanol at 350 °C).

15.4.2. Deposition of Particles by Aerosol Technology

The deposition of the oxide particles on the substrate surface in order to obtain a sensitive layer is sometimes a challenging task. The stable deposition of particles on a substrate is an essential step to realize a high-quality sensor.

If the active layer is formed by nanoparticles obtained by aerosol technology, the deposition techniques are specific for this method. A reference book on the subject (77) indicates that the immobilization of particles on the surface is connected with the concept of adhesion of solid from aerosol at the surface of the substrate. The adhesion can be achieved by several methods such as diffusion, thermal precipitation, inertial impact, and deposition in electric fields.

The adhesion on the surface is governed by van der Waals type, electrostatic, and forces associated with surface tension. These forces depend on the deposited particles (chemical composition, size, shape, "age" etc.), the nature of the surface (chemical composition, roughness, electrical properties), and the operating conditions (temperature, aerosol velocity, humidity, duration of contact, etc.).

The van der Waals forces are long-range acting and relatively weak in intensity. Nevertheless, the asperities present on the substrate surface can create good contact with particles, and through thermal treatments, chemical reactions can occur in solid state, thus fixing strongly the particles on the surface; the step is named annealing. Kennedy et al. (78) emphasized that this step is essential to form the conducting necks between single particles, avoiding the growth of their initial sizes. The extent of structure transformation is controlled by temperature and annealing time.

The deposition by diffusion results in particles that stick to the solid surface, as a result of the collision between the particles and the solid. The diffusion of particles to the wall is due to the particle concentration gradient, taking into account that once stuck on the surface, the particle concentration in the aerosol in the layer next to the surface becomes zero. The deposition by diffusion is the method most used (21,25–27,79) in deposition of sensitive thin layers.

Thermophoresis is another phenomenon that can occur in the aerosol as an effect of temperature. A force associated with the temperature gradient acts on the particle, forcing it to move in order to decrease the temperature. When a cold surface is put in contact with a warm gas, thermophoresis favors the particle deposition on the surface.

The deposition by impact is based on the different behavior of the solid particles and the gas molecules, when the aerosol flow is passed through a nozzle and the jet is directed to a flat plate that deflects the flow to form a 90° bend to the streamlines. The solid particles have higher inertia and collide with the substrate surface and get stuck on it, whereas the gas easily changes direction.

Electric fields can be used in deposition of particles either by spraying ions in the region of particle formation and growth (corona discharge) or by attracting the ionized particles to electrodes. The corona effect also reduces the possible residence time in the high-temperature region in the furnaces, thus better controlling their size uniformity (80–83). The electrically charged particles can be deposited in a very convenient way by applying an electric field to direct them to the substrate surface. Kruis et al. (84) used a neutralizer containing a radioactive source to charge particles smaller than 20 nm. The particles were deposited on the substrate by applying a 5-kV voltage. Chen et al. (85) showed that a rigorous control of a particle move could be achieved by a proper choice of the deposition area design, the flow value, and the applied voltage value. The deposition parameters were defined by Deppert et al. (86): the bias voltage, inlet configuration, distance susceptor-inlet, aerosol flow, particle size, and the substrate temperature. The method was used by Kennedy et al. for the deposition of SnO_2 nanoparticles for sensors (78). Gourari et al. (87) prepared tin-manganese oxide-based sensors for hydrogen sensing by electrostatic deposition, applying a 10-kV voltage between the substrate and the needle used for spraying the precursor solutions. At a substrate temperature of 400 °C and a solution concentration between 10^{-2} and 10^{-1} mol L^{-1}, aggregates of grains were formed in the range of 1–10 μm. The sensor was sensitive for hydrogen concentrations higher than 800 ppm in nitrogen, and the best response was noticed between 350 °C and 400 °C. In terms of chemical composition of the sensing layer, the optimal SnO_2/Mn_2O_3 ratio was 10/1.

15.4.3. More Methods for Synthesizing of Metal-Oxide Layers

Two more methods used especially for processing of catalytic metal-oxide layers for sensors are reviewed here.

The *rheotaxial growth* and thermal oxidation (RGTO), very advantageous to obtain sensor materials, was developed by Sberveglieri (30) and then used by many authors (88–90) to obtain thin-oxide films of one or more oxides.

The technique consists of a two-step process: First, a metallic thin film is deposited on a substrate maintained at a temperature slightly higher than the metal melting point, and then an oxidation step transforms the metal into an oxide. Almost spherical droplets, isolated from each other, compose the metallic layer. The oxidation step performed on this layer preserves the granular structure, but the increase in film thickness with 30–40% creates connections between agglomerates, so-called "necks," associated with the sensing active sites of the material. The structure of the material is spongy, a fact that gives a high surface-to-volume ratio, which is favorable for gas sensors. More than one metal can be used at a time to give a better sensor. Radecka et al. (89) investigated the properties of Sn–Ti mixed oxides (5% Ti), obtained by oxidation at 450–800 °C for 18 hours, for methane and hydrogen sensing at concentrations between 0 and 1000 ppm. The sensitivity was expressed in terms of a relative change in electrical conductance. The layers had a good sensitivity to both reducing species. The crystalline nature of the material was confirmed by X-ray diffraction, and the uniform grain size was confirmed by scanning electron microscopy.

Szuber et al. (90) showed that the droplet size during the RGTO depends on the chemical nature of the substrate and performed a study to determine the optimal substrate temperature in order to obtain a maximum surface coverage; for SnO_2, the value was 265 °C. In what concerns the sensor properties, the sensor response to NO_2 depended severely on the measurement temperature. For concentrations of NO_2 between 50 and 200 ppm in synthetic air, the optimal value was 200 °C. It is interesting to note that the value of the substrate temperature during deposition of the sensing layer also had an important effect on the sensitivity. At lower measurement temperature, the samples prepared at a higher substrate temperature exhibited a higher response.

Comini et al. (88) prepared sensors by two successive RGTO depositions. They deposited a first layer of tungsten and molybdenum (8/2 weight ratio) on the substrate by sputtering and oxidized it in humid air to obtain a porous oxide with a discontinuous structure that is, presumably, with a high resistivity. This layer was used as a template for a SnO_2 sensing layer, obtained by the RGTO technique, using different times for the droplet deposition step, between 30 and 130 minutes. The conductivity of the samples deposited on the WO_3–MoO_3 layer increased by more than one order of magnitude as compared with the pure SnO_2, most probably due to the improvement of the connections between the grains in the film. These sensors were tested for conductivity and for the sensitivity to CO, in dry and humid air. The SnO_2 layer deposited by RGTO on WO_3–MoO_3 was more sensitive to CO than those of pure SnO_2, and the response is better for the layer obtained with a longer droplet duration.

The *spray pyrolysis* was widely used by Korotcenkov et al. (25,76,91–94) to obtain sensors with a different composition of the layer. The parameters that influence the growth rate and the deposition process are as follows: the substrate temperature, the input pressure of the carrier gas, the reactor design and working parameters (temperature, flowing regime, geometry, etc.), the distance between the exit from the nebulizer and the substrate, the precursor nature, the stability and concentration, and the time of spraying. The crystallite sizes are generally in the order of tens of nanometer. The maximum gas sensitivity was obtained for layers made by small crystallite sizes. The external forms of nanocrystals can be pyramids, bipyramids, and prisms. The deposited layer varies from some tens of nanometers to some hundreds of nanometers. Due to the more dense packing in a layer containing more than one statistical monolayer of particles, the resistance values of the layer decrease with thickness. For a SnO_2 film used for sensing hydrogen, the resistance value of a 15–35-nm film decreased with more than one order of magnitude in the presence of 2% H_2 in air as compared with air (79,91). The pyrolysis temperature has a strong influence on the particle morphology and sensing properties; the general behavior indicates that the particle size increases with temperature, and the sensitivity to gas decreases. The measurement temperature is also an important factor that influences the sensitivity. For a constant layer thickness, the response normally increases, up to a certain temperature limit, which depends on the preparation conditions, and then the response decreases again when the temperature increases further. The thickness of the layer also influences the sensitivity, but a general correlation is difficult to make. The most porous films were obtained for pyrolysis temperatures between 400 and 450 °C and the layer thickness was between 20 and 170 nm.

The In_2O_3-based thin films manufactured by spray pyrolysis have a columnar structure, with almost parallel crystallites, generating a layer of about 200 nm thickness (25,92). The size of the primary particles depends to an important extent on the pyrolysis temperature. The sensing properties are displayed to reducing gases such as hydrogen and CO. The gas response increases slowly with the layer thickness. A detailed study on the influence of the thermal annealing process on the indium oxide stability (94) indicates an important increase of the grain size, especially in the case of particles obtained at higher pyrolysis temperatures. The response to reducing gases is not influenced by the crystallite size, whereas the exposure to oxidant gases (ozone) indicates a sharp decrease of the gas response with layer thickness.

15.5. PRACTICAL DEVICES AND COMMERCIAL APPLICATIONS

In this section, the most widely applied types of practical devices will be briefly exposed. To analyze complex mixtures of gases, advanced multiple sensor designs are suggested.

The metal-oxide resistive gas sensor design has turned from ceramic based on a thick catalytic layer (Figaro type) (95) to thin films. The modalities to improve the sensor fabrication process include development of new material systems, enhancing the fabrication process and manufacturing of "smart sensors," equipped with signal conditioning and filtration modules. The use of thin films as sensing materials brings important improvements of the sensitivity of the sensors.

Even though three working principles, resistors, diodes, and capacitors, of metal-oxide sensors are used to characterize these devices (96), resistors are almost exclusively used in commercial applications. The differences among the three devices concern the measured characteristics during interaction with the gas (see Figure 15.9) (96). In the case of resistors, the dependence between the voltage and the current intensity is linear and the slope depends on the charge-carrier density, which is directly related to the measured gas pressure. In the diode case, the I–U characteristic curve is displaced to a lower voltage value in the presence of a measured (reducing) gas. As the metal-oxide material exhibits semiconducting properties, the gas (oxidizing) interaction with the sensing surface introduces a depletion layer, which varies in size

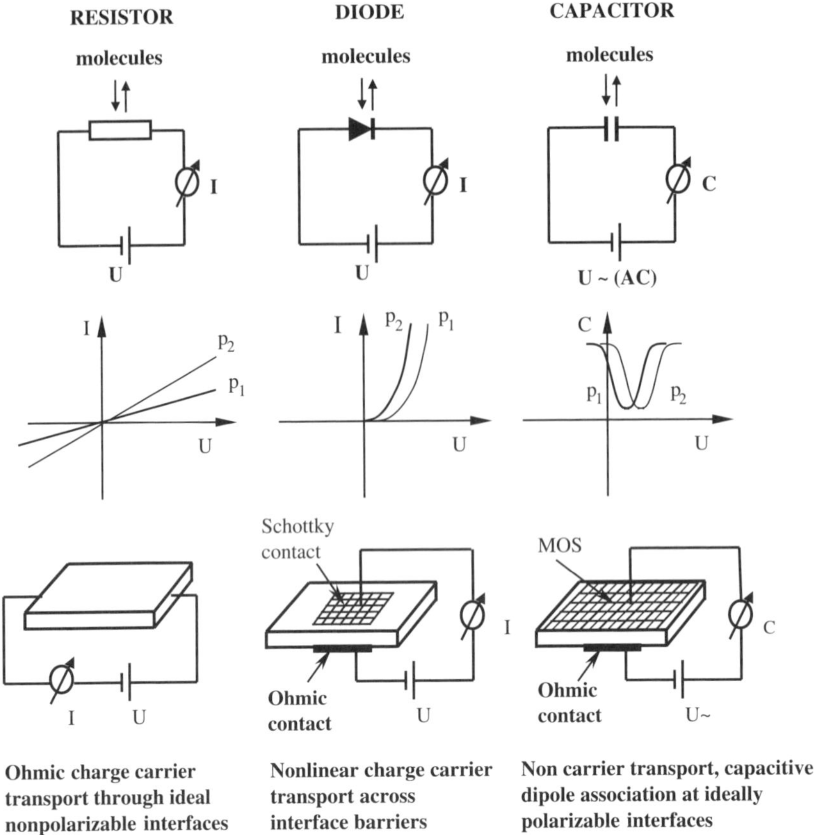

Figure 15.9. The resistor, diode, and capacitor device layout are used to investigate the electrical characteristics of metal oxide sensors. Reprinted from Ref. 96 with kind permission of Elsevier.

depending on nature and concentration of the gas, which is analyzed. Due to this also, the capacitive properties of a metal-oxide device will give information.

The yttria-stabilized zirconia (YSZ)-based solid electrolyte device is an oxygen ion conducting material used to measure excess or lack of oxygen in its simplest version (11). Several advanced modifications of this sensor have created, for example, a linear oxygen sensor and sensors for nitrogen oxides. Some of these will be described here.

Szabo and Dutta (97) used a special design of the sensor and of the entire mounting for detection of NO, NO_2, and their mixture. As sensing elements, they used chromium (III) oxide deposited on YSZ with a special shape (a single closed-end tube with a platinum reference electrode painted inside, near to the closed bottom). The gas to be analyzed was passed through a catalytic bed based on platinum deposited on zeolite Y previous to the analysis, in order to convert the NO to NO_2. The catalyst allowed attaining a conversion degree very close to the thermodynamic predictions. The value of the electromotive force (EMF) developed by the sensing process depended to a large extent on the temperature and on the nature of the nitrogen oxide. As expected from the electrochemical reactions involved in the sensing mechanism, the values of the EMF decreased in the case of NO with respect to the baseline registered in a $N_2 + 3\%O_2$ mixture, whereas in the case of NO_2, the values increased. In fact, logarithmic-type dependencies were found for the EMF 500 °C and for nitrogen oxide concentrations between 0 and 1000 ppm:

$$\mathrm{EMF}_{\mathrm{NO}_2} = -72.5 + 16.3 \ln[\mathrm{NO}_2] \tag{15.36}$$

$$\mathrm{EMF}_{\mathrm{NO}} = -32.3 - 2.6 \ln[\mathrm{NO}] \tag{15.37}$$

The mixtures of NO and NO_2 in different ratios led to an EMF intermediate between those of pure gases. This behavior is typical for mixed potential type sensors (27). The catalytic conversion of the gases prior to analysis led to essentially the same sensor response for all ratios between the two gases, indicating that the equilibrium was reached between the analyzed species during the contact with the catalyst. The sensor measurements performed on NO- and NO_2-containing gas flows at temperatures between 400 °C and 700 °C showed practically identical EMF values for both gases, for all ratios between the two species.

This association between catalytic conversion and gas sensing can be useful in simultaneous sensing of NO_x and CO. The Cr_2O_3 proved to be very sensitive to CO at concentrations of 0–1000 ppm. However, the catalytic converter totally transforms CO into CO_2, as confirmed by measurements. If a gas flow containing a nitrogen oxide and CO mixture is catalytically converted, the response of NO_x + CO is the same as for only NO_x. This allows measurement of 200–1000-ppm NO_x with this system independent of CO concentration.

The most widely practical applications of the metal-oxide chemical sensors concern:

1. A device, which starts an alarm when a toxic or dangerous gas species is present in the atmosphere surrounding the sensor.
2. Monitoring the composition of the exhaust gases from energy plants or of automotive exhaust gases and optimizing the combustion in order to minimize the level of pollutants emitted to the environment.
3. Evaluation the degree of pollution in the atmosphere by detecting gases such as ammonia and volatile organic compounds at very low concentrations in the air.
4. Sensing complex mixtures of scented substances in gas or liquid phase by instruments known as "electronic noses and tongues" (98). These devices find wide applications in the food and beverage industries.

The ***combustion control*** is nowadays a very important task, because the negative effects of gases such as CO, NO_x, and unburned hydrocarbons on the environment are well known (4). The control involves the development of sensitive and specific sensors for these species and correlating the information with a selective catalytic converter in a closed-loop control. The information from the sensor can be used to regulate the air-to-fuel ratio in order to minimize the amounts of unwanted and toxic gas compounds. As the nitrogen oxides come from the reaction between atmospheric nitrogen and oxygen at elevated temperatures in excess oxygen introduced to assure complete oxidation of hydrocarbons, establishing a stoichiometric ratio between the fuel and the air should be ideal. At the same time, a good regulation of the combustion should decrease the fuel consumption, thus minimizing the CO_2 emissions. At this composition ratio, the conversion efficiency of a catalyst employed to eliminate the toxic gases reaches a maximum. The sensor for this kind of systems should work at a gas composition close to the stoichiometric ratio between oxygen and fuel.

A typical example is the exhaust gas from automotives, which usually contains oxygen, hydrocarbons, carbon monoxide, hydrogen, and nitrogen monoxide. On the sensor surface, due to its catalytic potential, a series of reactions can occur:

$$2CO + O_2 \longrightarrow 2CO_2$$

$$CO + H_2O \longrightarrow CO_2 + H_2$$

$$2NO + 2CO \longrightarrow N_2 + 2CO_2$$

$$2NO + 2H_2 \longrightarrow N_2 + 2H_2O$$

$$2NO_2 + 4CO \longrightarrow N_2 + 2H_2O$$

$$2NO + 2H_2 \longrightarrow N_2 + 2H_2O$$

$$C_xH_y + (x + y/4)O_2 \longrightarrow xCO_2 + y/2H_2O$$

$$(2x + y/2)NO + C_xH_y \longrightarrow xCO_2 + y/2H_2O + (x + y/4)N_2$$

A complex indicator named the "lambda value" is defined by the ratio between the air-to-fuel mass ratio in the cylinder (real system) and at stochiometry (99):

$$\lambda = \frac{(m_{\text{air}}/m_{\text{fuel}})_{\text{engine}}}{(m_{\text{air}}/m_{\text{fuel}})_{\text{stoichiometric}}}$$

The information from the sensors is processed in a control unit that controls the air-to-fuel ratio. In practice, the oxygen-to-fuel ratio is allowed to oscillate around the stochiometric value, which increases further both the lifetime and the efficiency of the catalytic converter. The commercial lambda sensors currently used in automotive applications are based on oxygen-ion conducting YSZ.

Several lambda sensor configurations are developed today (100):

- As planar potentiometric sensors, based on the solid electrolyte properties of the zirconia, which remain electronically insulating. The potential difference over a ZrO_2 layer is directly related to the difference in oxygen concentration at the two sides of the zirconia layer.

- As oxygen pumping sensors, at temperatures above 600 °C, the transport of oxygen ions through the electrolyte can occur. The transport develops a pumping voltage proportional to the oxygen concentration gradient through the zirconia layer.
- As a dual cell with wide-range oxygen concentration sensing, made by a pumping cell and a sensing cell, separated by a porous diffusion barrier.
- Resistive-type sensors based on semiconducting metal oxide, reflecting in their electrical conductivity the equilibrium between the oxygen partial pressure in the measured gas and their bulk stoichiometry at temperatures above 700 °C.

Francioso et al. (101) manufactured a lambda sensor based on titania, relying on the change of the semiconductor resistance with oxygen partial pressure. The material was prepared by the sol-gel method and deposited on alumina substrates. The dependence of electrical resistance on the logarithm of the oxygen partial pressure was found to be linear for temperatures values of 400 °C, 500 °C, and 600 °C. A sudden increase of the resistance of the layer occurred between λ values of 1 and 1.1. For complex gas mixtures containing oxygen, carbon dioxide, nitrogen, monoxide, nitrogen, and methane, in ratios close to those in real exhaust mixtures, the resistance increased three times in a very narrow λ region close to 1.

Grilli et al. propose an YSZ sensor doped with tungsten oxide (102) as a pollutant detector at high temperatures. An optimal temperature value of 600 °C was found for the sensing of CO and NO_2 in the case of WO_3 prepared by thermal decomposition of ammonium tungstenate, whereas an optimum at 550 °C was noticed for the dopant powder prepared by precipitation. The tests performed with these sensors in exhaust gases of an engine bench test (103) showed good correlation with the results obtained using lambda sensors.

The monitoring of the environment quality can be performed by oxide-based chemical sensors. Indoor and outdoor air quality (104,105) control was performed by detection of volatile organic compounds (106,107) and emissions from wastewater treatment plants (108).

The typical indoor air contaminants are aldehydes, ozone, nitrogen dioxide, carbon monoxide, radon, solid particles, sulphur dioxide, lead, and water vapor (104). As, for example, water interferes in both NO_2 and CO detection and the discrimination between the different target species is achieved by using data processing and pattern recognition algorithms, which rely on the distinct response patterns provided by a sensor array. For CO detection, a $SnO_2 + SiO_2 + Au$ sensor material was used, whereas for NO_2 detection, the $SnO_2 + Au$ was preferred. The detection limits were 5 ppm for CO and 20 ppm for NO_2 and saturation occurred for NO_2 at about 100 ppm. These values are lower than the indoor threshold values, which after 8 hours of exposure were 9 ppm for CO and 53 ppm for NO_2, respectively. This means that these sensors exhibit excellent responses for indoor air monitoring. Furthermore, tests over a longer time period showed that the sensitivity was maintained for as long as 45 days. A certain drift with time is noticed, however, for both gases.

The outdoor air quality can be monitored by the same kind of sensors for detection and measurement of CO, NO_2, NO, and O_3 (105). The target detection limits of 3, 50, 100, and 20 ppm, respectively, were fulfilled by the recent research on SnO_2 sensors obtained by nanoparticle processing, allowing measurement ranges of up to 3, 15, 100, and 15 ppm, respectively.

The presence of volatile organic compounds in air can be monitored by the use of an SnO_2-based sensor array, one with undoped oxide and others with Pd, Pt, and Au doping (106). The data acquisition system is designed to measure the responses with eight sensors at a time. The organic compounds used for tests were 2-propanol, methanol, acetone, ethyl-methyl-ketone, hexane, benzene, and xylene. The different sensitivity to these species exhibited by the four types of sensors allows data processing to evaluate the contribution of each species in the system and thus monitoring the global composition of the analyzed gas mixture.

The ***applications of sensors in the food industry*** are a subject of great interest, because aroma is a specific quality indicator in this field. Flavor is generated by combinations of a big number of chemical compounds in certain ratios. For example, coffee aroma is made up of several hundreds of different molecules. The function of an electronic nose is based on the sensitivity of an array of semi-selective sensors, giving similar response patterns for similar aroma (so-called "aroma fingerprints") and different patterns for different aroma. The interaction between the array of sensors and the chemical species generates a series of signals that are processed by a computer connected online. The software contributes a specific pattern recognition program.

The results for expressing the composition of these complex mixtures are presented as graphical dependences in the form of "principal component analysis" (PCA) performed on normalized matrix of responses. In an array of "m" sensors, the sensitivity of a sensor "j" to a component "i" in a mixture can be expressed by the relation (109):

$$S_{ij} = \frac{|G_{\mathrm{vap}} - G_{\mathrm{air}}|}{\min(G_{\mathrm{vap}}, G_{\mathrm{air}})}$$

where G_{vap} and G_{air} are the conductance values after vapour exposure and in air respectively.

The normalized sensitivity in an individual sensor array is

$$S_{ij}^{N} = \frac{S_{ij}}{\sum_{i=1}^{m} S_{ij}}$$

The results are presented in planar plots having in abscissa the PC1 component and in the ordinate the PC2 or PC3. The "maps" thus obtained consist in areas with more or less grouped points corresponding to one chemical compound, each one being obtained with one of the sensors used in the measurement device. A typical graphic representation for such an analysis is shown in the Figure 15.10 (110).

Metal-oxide sensor manufacturing was used for very different kinds of product analysis: milk and cheese (110–113), wine (114), vegetable oils (115,116), coffee

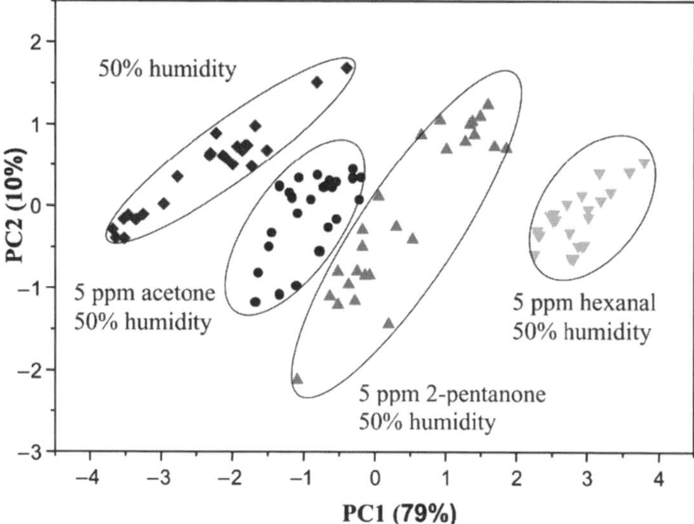

Figure 15.10. PCA plot of the detection of some solvents by an array of metal-oxide sensors. Reprinted from Ref. 110 with kind permission of Elsevier.

(117), or pharmaceutical oral solutions (118). Schaller et al. (119) and Ampuero and Bosset (120) review some food applications of the electronic nose in which metal oxides have been used as sensing layer.

15.6. CONCLUSIONS

The use of oxide nanomaterials in sensor manufacturing is a subject with a high scientific and practical impact, due to the wide range of possibilities displayed by the layer to interact with the molecules to be detected and measured from a gas mixture. The adsorption and the catalytic type interaction between the measured gas and the sensitive oxide layer can be fine-tuned by the proper choice of the chemical nature and the microscopic structure of the layer, the appropriate dopants, the stabilizing treatments, and so on. The modalities to disseminate the results are also multiple: electrical resistance measurements, ion conductivity through the solids, and field effects.

The main commercial applications are nowadays the environment monitoring and food quality analysis.

ACKNOWLEDGMENTS

Grants are acknowledged from CF, The Swedish Association of Graduate Engineers, and the Swedish Research Council.

REFERENCES

(1) Available at: http://www2.nose-network.org/members/2_Chemical_sensors.doc (accessed 2005_10_26).

(2) Available at: http://www.ipc.uni-tuebingen.de/weimar/research/maintopics/gassensors/overview.htm (accessed 2005_10_26).

(3) Meixner, H.; Lampe, U. Metal oxide sensors. *Sens. Actuat. B* **1996**, *33*, 198–202.

(4) Available at: http://www.ifm.liu.se/~spetz/Tutorial/ (accessed 2006_10_30).

(5) Doquier, N.; Candel, S. Combustion control and sensors: A review. *Proc. Energy Combust. Sci.* **2002**, *28*, 107–150.

(6) Anderson, G.L.; Hadden, D.M. *The Gas Monitoring Book*; Ickus Guides Avocet Press: New York, **1999**.

(7) Mizsei, J. How can sensitive and selective semiconductor gas sensors be made? *Sens. Actuat. B* **1995**, *23*, 173–176.

(8) Kohl, D. Function and applications of gas sensors. *J. Phys. D: Appl. Phys.* **2001**, *34*, R125–R149.

(9) Pal, B.N; Chakravorty, D. Electrical properties of composites with tin-tin oxide core-shell nanostructure and their sensing behaviour. *J. Phys. D: Appl. Phys.* **2005**, *38*, 3537–3542.

(10) Simon, I.; Barsan, N.; Bauer, M.; Weimar, U. Micromachined metal oxide gas sensors: Opportunities to improve sensor performance. *Sens. Actuat. B* **2001**, *73*, 1–26.

(11) Visser, J.H.; Soltis, R.E. Automotive exhaust gas sensing systems. *IEEE Trans. Instrument. Measur.* **2001**, *50(6)*, 1543–1550.

(12) Maskel, W.C. Progress in the development of zirconia gas sensors. *Solid State Ion.* **2000**, *134*, 43–50.

(13) Sanati, M.; Andersson, A. Kinetics and mechanism in the ammonoxidation of toluene over a TiO2(B)-supported vabadium oxide monolayer catalyst. 1. Selective reactions. *Ind. Eng. Chem. Res.* **1991**, *30(2)*, 312–320.

(14) Sanati, M.; Andersson, A. Kinetics and mechanism in the ammonoxidation of toluene over a TiO2(B)-supported vabadium oxide monolayer catalyst. 1. combustion reactions. *Ind. Eng. Chem. Res.* **1991**, *30(2)*, 320–326.

(15) Cabot, A.; Arbiol, J.; Morante, J.R.; Weimar, U.; Barsan, N.; Göpel, W. Analysis of the noble metal catalytic additives introduced by impregnation of as obtained SnO2 sol-gel nanocrystals for gas sensors. *Sens. Actuat. B* **2000**, *70*, 87–100.

(16) Osawa, T.; Ishiguro, Y.; Toyoda, K.; Nishimura, M.; Sasahara, T.; Doi, T. Detection of decomposed compounds from an early stage fire by an adsorption/combustion-type sensor. *Sens. Actuat. B* **2005**, *108*, 473–477.

(17) Lundström, I. Why bother about gas-sensitive field-effect devices? *Sens. Actuat. A* **1996**, *56*, 75–82.

(18) Forleo, A.; Francioso, L.; Epifani, M.; Capone, S.; Taurino, A.M.; Siciliano, P. NO_2-gas-sensing properties of mixed In_2O_3-SnO_2 thin films. *Thin Solid Films* **2005**, *490*, 68–73.

(19) Schmid, W.; Barsan, N.; Weimar, U. Sensing of hydrocarbons with tin oxide sensors: Possible reaction path as revealed by consumption measurements. *Sens. Actuat. B* **2003**, *89*, 232–236.

(20) Koziej, D.; Barsan, N.; Hoffmann, V.; Szuber, J.; Weimar, U. Complementary phenomenological and spectroscopic studies of propane sensing with tin dioxide based sensors. *Sens. Actuat. B* **2005**, *108*, 75–83.

(21) Korotcenkov, G. Gas response control through structural and chemical modification of metal oxide films: State of the art and approaches. *Sens. Actuat. B* **2005**, *107*, 209–232.

(22) Tiburcio-Silver, A.; Sanchez-Juarez, A. SnO_2:Ga thin films as oxygen gas sensor. *Mater. Sci. Eng. B* **2004**, *110*, 268–271.

(23) Baik, N.S.; Sakai, G.; Shimanoe, K.; Miura, N.; Yamazoe, N. Hydrothermal treatment of tin oxide solution for preparation of thin-film sensor with enhanced thermal stability and gas sensitivity. *Sens. Actuat. B* **2000**, *65*, 97–100.

(24) Williams, G.; Coles, G.S.V. Gas sensing properties of nanocrystalline metal oxide powders produced by a laser evaporation technique. *J. Mater. Chem.* **1998**, *8*(7), 1657–1664.

(25) Korotcenkov, G.; Macsanov, V.; Tolstoy, V.; Brinzari, V.; Schwank, J.; Faglia, G. Structural and gas response characterization of nano-size SnO_2 films deposited by SILD method. *Sens. Actuat. B* **2003**, *96*, 602–609.

(26) Golovanov, V.; Korotcenkov, G.; Brinzari, V.; Cornet, A.; Morante, J.; Arbiol, J.; Russinyol, E. CO-water interaction with SnO_2 gas sensors: Role of orientation effects. *Eurosensors XVI*; The 16th European Conference on Solid-State Transducers, Prague, Czech Republic, September **2002**.

(27) Korotcenkov, G.; Brinzari, V.; Cerneavschi, A.; Cornet, A.; Morante, J.; Cabot, A.; Arbiol, J. Crystallographic characterization of In_2O_3 films deposited by spray pyrolysis. *Sens. Actuat. B* **2002**, *84*, 37–42.

(28) Sasahara, T.; Kido, A.; Ishihara, H.; Sunayama, T.; Egashira, M. Highly sensitive detection of volatile organic compounds by an adsorption/combustion-type sensor based on mesoporous silica. *Sens. Actuat. B* **2005**, *108*, 478–483.

(29) Xu, C.N.; Miura, N.; Ishida, Y.; Matsuda, K.; Yamazoe, N. Selective detection of NH_3 over NO in combustion exhaust by using Au and MoO_3 doubly promoted WO_3 element. *Sens. Actuat. B* **2000**, *65*, 163–165.

(30) Sberveglieri, G. Recent developments in semiconducting thin-film gas sensors. *Sens. Actuat. B* **1995**, *23*, 103–109.

(31) Stankova, M.; Vilanova, X.; Llobet, E.; Calderer, J.; Bittencourt, C.; Pireaux, J.J.; Correog, X. Influence of the annealing and operating temperatures on the gas sensing properties of the rf sputtered WO_3 thin-film sensors. *Sens. Actuat. B* **2005**, *105*, 271–277.

(32) Gurlo, A.; Barsan, N.; Ivanovskaya, M.; Weimar, U.; Göpel, W. In_2O_3 and MoO_3-In_2O_3 thin film semiconductor sensors: Interaction with NO_2 and O_3. *Sens. Actuat. B* **1998**, *47*, 92–99.

(33) Einar Kruis, F.; Fissan, H.; Peled, A. Synthesis of nanoparticles in the gas phase for electronic, optical and magnetic applications—a review. *J. Aerosol Sci.* **1998**, *29(5/6)*, 511–535.

(34) Zakrzewska, K. Mixed oxides as gas sensors. *Thin Solid Films* **2001**, *391*, 229–238.

(35) Gleiter, H. Namostructured materials: Basic concepts and microstructure. *Acta Materialia* **2002**, *48*, 1–29.

(36) Szczuko, D.; Werner, J.; Ostwald, S.; Behr, G.; Wetzig, K. XPS investigations of surface segregation of doping elements in SnO_2. *Appl. Surf. Sci.* **2001**, *179*, 301–306.

(37) Korotcenkov, G.; Macsanov, V.; Boris, Y.; Brinzari, V.; Tolstooy, V.; Schwank, J.; Morante, J. Using of SILD technology for surface modifications of SnO_2 films for gas sensor applications. *Mater. Res. Soc. Symp. Proc.* **2003**, *750*, Y5.25.1-6.

(38) Timmer, B.; Olthuis, W.; Van den Berg, A. Ammonia sensors and their applications—a review. *Sens. Actuat. B* **2005**, *107*, 666–677.

(39) Wang, X.; Miura, N.; Yamazoe, N. Study of WO_3-based materials for NH_3 and NO detection. *Sens. Actuat. B* **2000**, *66*, 74–76.

(40) Yamaura, H.; Jinkawa, T.; Tamaki, J.; Moriya, K.; Miura, N.; Yamazoe, N. Indium oxide-based gas sensor for selective detection of CO. *Sens. Actuat. B* **1996**, *35–36*, 325–332.

(41) Galan-Vidal, C.A.; Munoz, J.; Dominguez, C.; Alegret, S. Chemical sensors, biosensors and thick-film technology. *Trends Analyt. Chem.* **1995**, *14(5)*, 225–231.

(42) Stankova, M.; Ivanov, P.; Llobet, E.; Brezmes, J.; Vilanova, X.; Garcia, I.; Cane, C.; Hubalek, J.; Malysz, K.; Correig, X. Sputtered and screen-printed metal oxide-based integrated micro-sensor arrays for the quantitative analysis of gas mixtures. *Sens. Actuat. B* **2004**, *103*, 23–30.

(43) Vaishnav, V.S.; Patel, P.D.; Patel, N.G. Preparation and characterization of indium tin oxide thin films for their application as gas sensors. *Thin Solid Films* **2005**, *487*, 277–282.

(44) Vaishnav, V.S.; Patel, P.D.; Patel, N.G. Indium tin oxide thin films gas sensors for detection of ethanol vapours. *Thin Solid Films* **2005**, *490*, 94–100.

(45) Dutta, P.K.; De Lucia, M.F. Correlation of catalytic activity and sensor response in TiO_2 high temperature gas sensors. *Sens. Actuat. B* **2006**, *115*, 1–3.

(46) Setkus, A.; Barato, C.; Comini, E.; Faglia, G.; Galdikas, A.; Kancleris, Z.; Sberveglieri, G.; Senuliene, D. Influence of metallic impurities on response kinetics in metal oxide thin film gas sensors. *Sens. Actuat. B* **2004**, *103*, 448–456.

(47) Lee, S.-M.; Lee, Y.-S.; Shim, C.-H.; Choi, N.-J.; Joo, B.-S.; Song, K.-D.; Huh, J.-S.; Lee, D.-D. Three electrodes gas sensor based on ITO thin film. *Sens. Actuat. B* **2003**, *93*, 31–35.

(48) Setkus, A. Heterogeneous reactionrate based description of the response kinetics in metal oxide gas sensors. *Sens. Actuat. B* **2002**, *87*, 346–357.

(49) Williams, D.E. Semiconductive oxides as gas-sensitive resistors. *Sens. Actuat. B* **1999**, *57*, 1–16.

(50) Miura, N.; Lu, G.; Yamazoe, N. Progress in mixed-potential type devices based on solid electrolyte for sensing redox gases. *Solid State Ionics* **2000**, *136/137*, 533–542.

(51) Stetter, J.R.; Penrose, W.R.; Yao, S. Sensors, chemical sensors, electrochemical sensors and ECS. *J. Electrochem. Soc.* **2003**, *150(2)*, S11–S16.

(52) Lundstrom, I.; Shivaraman, M.S.; Svensson, C.; Lundkvist, L. A hydrogen-sensitive MOS field-efect transisitor. *Appl. Phys. Lett.* **1975**, *26*, 55–57.

(53) Lloyd Spetz, A.; Nakagomi, S.; Savage, S. High temperature SiC-FET chemical gas sensors. In *Advances in Silicon Carbide Processing and Applications*; Saddow, S.E.; Agarwal, A. (Editors); Artech House: Boston, MA, **2004**, 29–67.

(54) Wallin, M.; Grönbeck, H.; Lloyd Spetz, A.; Skoglundh, M. Vibrational study of ammonia adsorption on Pt/SiO_2. *Appl. Surf. Sci.* **2004**, *235(4)*, 487–500.

(55) Wallin, M.; Grönbeck, H.; Lloyd Spetz, A.; Eriksson, M.; Skoglundh, M. Vibrational analysis of H_2 and D_2 adsorption on Pt/SiO_2. *J. Phys. Chem. B* **2005**, *109*, 9581–9588.

(56) Ekedahl, L.-G.; Eriksson, M.; Lundström, I. Hydrogen sensing mechanism of metal-insulator interfaces. *Acc. Chem. Res.* **1998**, *31(5)*, 249.

(57) Löfdahl, M.; Utaiwasin, C.; Carlsson, A.; Lundström, I.; Eriksson, M. Gas response dependence on gate metal morphology of field-effect devices. *Sens. Actuat. B* **2001**, *80*, 183–192.

(58) Wingbrant, H.; Svenningstorp, H.; Salomonsson, P.; Kubinski, D.; Visser, J.H.; Löfdahl, M.; Lloyd Spetz, A. Using a MISiC-FET sensor for detecting NH_3 in SCR systems. *IEEE Sens. J.* **2005**, *5(5)*, 1099–1105.

(59) Andersson, M.; Ljung, P.; Mattsson, M.; Löfdahl, M.; Lloyd-Spetz, A. Investigations on the possibilities of a MISiCFET sensor system for OBD and combustion control utilizing different catalytic gate materials. *Top. Catal.* **2004**, *30–31*, 365–368.

(60) Eriksson, M.; Salomonsson, A.; Lundström, I.; Briand, D.; Åbom, A.E. The influence of the insulator surface properties on the hydrogen response of field-effect gas sensors. *J. Appl. Physics* **2005**, *98*, 034903-1–034903-6.

(61) Lloyd Spetz, A.; Wingbrant, H.; Andersson, M.; Salomonsson, A.; Roy, S.; Wingqvist, G.; Katardjiev, I.; Eickhoff, M.; Uvdal, K.; Yakimova, R. New materials for chemical and biosensors. *Mat. Manufact. Proc.*, **2006**, 21, 253–256.

(62) Zhangooie, S.; Arwin, H.; Lundström, I.; Lloyd Spetz, A. Ozone treatment of SiC for improved performance of gas sensitive Schottky diodes. *Mat. Sci. Forum* **2000**, *338–342*, 1085–1088.

(63) Weidemann, O.; Hermann, M.; Steinhoff, G.; Wingbrant, H.; Lloyd Spetz, A.; Stutzmann, M.; Eickhoff, M. Influence of surface oxides on hydrogen-sensitive Pd:GaN Schottky diodes. *Appl. Phys. Lett.* **2003**, *83*(4), 773–775.

(64) Svenningstorp, H.; Tobias, P.; Lundstöm, I.; Salomonsson, P.; Mårtenson, P.; Ekedal, L.-G.; Lloyd Spetz, A. Influence of catalytic reactivity on the response of metal-oxide-silicon carbide sensor to exhaust gases. *Sens. Actuat. B* **1999**, *57*, 159–165.

(65) Wingbrant, H.; Lloyd Spetz, A. The influence of catalytic activity on the phase transition governer binary switch point of MISiC-FET lambda sensors. *Appl. Surf. Sci.* **2006**, *252*, 7473–7486.

(66) Salomonsson, A.; Roy, S.; Aulin, C.; Cerda, J.; Käll, P.-O.; Ojamäe, L.; Strand, M.; Sanati, M.; Lloyd Spetz, A. Nanoparticles for long-term stable, more selective MISiCFET gas sensors. *Sens. Actuat. B* **2005**, *107(2)*, 831–838.

(67) Salomonsson, A.; Roy, S.; Aulin, C.; Ojamäe, L.; Käll, P.-O.; Strand, M.; Sanati, M.; Lloyd Spetz, A. RuO2 and Ru nanoparticles for MISiC-FET gas sensors, *NSTI-Nanotech.* Anaheim, USA, May 8–12, **2005**, *2*, 269–272.

(68) Salomonsson, A.; Petoral Jr. R.M.; Uvdal, K.; Aulin, C.; Käll, P.-O.; Ojamäe, L.; Strand, M.; Sanati, M.; Lloyd Spetz, A. Nanocrystalline ruthenium oxide and ruthenium in sensing applications—an experimental and theoretical study. *J. Nanopart. Res.* DOI 10.1007/S11051-005-9058.

(69) Åbom, A.E.; Comini, E.; Sberveglieri, G.; Hultman, L.; Eriksson, M. Thin oxide films as surface modifiers of MIS field effect gas sensors. *Sens. Actuat. B* **2002**, *85*, 109–119.

(70) Wingbrant, H.; Svenningstorp, H.; Kubinski, D.; Visser, J.H.; Andersson, M.; Unéus, L.; Löfdahl, M.; Lloyd Spetz, A. MISiC-FET ammonia sensors for SCR control, in exhaust

and flue gases. In *Encyclopedia of Sensors*; Grimes, C.A.; Dickey, E.C. (Editors); in press.

(71) Wingbrant, H.; Lundén, M.; Lloyd Spetz, A. Modifications of the gate material to increase the long-term stability of field effect transistor lambda sensors. *Sens. Lett.* **2005**, *3*, 225–230.

(72) Comini, E.; Faglia, G.; Sberveglieri, G.; Calestani, D.; Zanotti, L.; Zha, M. Tin oxide nanobelts electrical and sensing properties. *Sens. Actuat. B* **2005**, *111–112*, 2–6.

(73) Ertl, G.; Knözinger, H.; Weitkamp, J. (Editors); *Handbook of Heterogeneous Catalysis, Vol. 1*; VCH Verlagsgesellschaft mbH: Weinheim, Germany, **1997**.

(74) Available at: http://www.psrc.usm.edu/mauritz/solgel.html (accessed 2005_10_26).

(75) Available at: http://www.chemat.com/html/solgel.html (accessed 2005_10_26).

(76) Tjong, S.C.; Chen, H. Nanocrystalline materials and coatings. *Mater. Sci. Eng* **2004**, *R 45*, 1–88.

(77) Hinds, W.C. *Aerosol Technology, Properties, Behaviour and Measurement of Airborne Particles*; John Wiley and Sons: New York, **1999**.

(78) Kennedy, M.K.; Kruis, F.E.; Fissan, H. Tailored nanoparticle films from monosized tin oxide nanocrystals: Particle synthesis, film formation and size-dependent gas-sensing properties. *J. Appl. Phys.* **2003**, *93(1)*, 551–560.

(79) Korotcenkov, G.; Brinzari, V.; Schwank, J.; Cerneavschi, A. Possibilities of aerosol technology for deposition of SnO_2-based films with improved gas sensing characteristics. *Mater. Sci. Eng* **2002**, *C19*, 73–77.

(80) Pratsinis, S.E. Flame synthesis of nanosize particles: Precise control of particle size. *J. Aerosol Sci.* **1996**, *27(suppl. 1)*, S153–S154.

(81) Patil, P.S. Versatility of chemical spray pyrolysis technique. *Mater. Chem. Phys.* **1999**, *59*, 185–198.

(82) Jidenko, N.; Borra, J.P.; Kasper, G.; Weber, A.P. Corona charging of agglomerated particles: Fragmentation or field-reinforcement? Abstracts of European Aerosol Conference **2005**; 138.

(83) Alonso, M.; Alguacil, F.J.; Perez, C.; Ramiro, E.; Sanchez, M. European Aerosol Conference **2005**; 137.

(84) Kruis, F.E.; Goossens, A.; Fissan, H. Synthesis of semiconducting nanoparticles. *J. Aerosol Sci.* **1996**, *27(suppl. 1)*, S165–S166.

(85) Chen, D.R.; Pui, D.Y.H.; Hummes, D.; Fissan, H.; Quant, F.R.; Sem, G.J.; Design and evaluation of a nanometer aerosol differential mobility analyzer (Nano DMA). *J. Aerosol Sci.* **1998**, *29(5/6)*, 497–509.

(86) Deppert, K.; Schmidt, F.; Krinke, T.; Dixkens, J.; Fissan, H. Electrostatic precipitator for homogeneous deposition of ultrafine particles to create quantum-dot structures. *J. Aerosol Sci.* **1996**, *27(suppl. 1)*, S151–S152.

(87) Gourari, H.; Lumbreras, M.; Van Landschoot, R.; Schoonman, J. Elaboration and characterization of SnO_2-Mn_2O_3 thin layers perpared by electrostatic spray deposition. *Sens. Actuat. B* **1998**, *47*, 189–193.

(88) Comini, E.; Ferroni, M.; Guidi, V.; Martinelle, G.; Sberveglieri, G. CO sensing properties of W-Mo and tin oxide RGTO multiple layers structures. *Sens. Actuat. B* **2003**, *95*, 157–161.

(89) Radecka, M.; Przewoznic, J.; Zarrewska, K. Microstructure and gas-sensing properties of (Sn, Ti)O_2 thin films deposited by RGTO technique. *Thin Solid Films* **2001**, *391*, 247–254.

(90) Szuber, J.; Uljanow, J.; Karczewska-Buczek, T.; Jakubik, W.; Waczynski, K.; Kwoka, M.; Konczak, S. On the correlation between morphology and gas sensing properties of RGTO SnO_2 thin films. *Thin Solid Films* **2005**, *490*, 54–58.

(91) Korotcenkov, G.; Brinzari, V.; Schwank, J.; di Batista, M.; Vasiliev, A. Peculiarities of SnO_2 thin film deposition by spray pyrolysis for gas sensor application. *Sens. Actuat. B* **2001**, *77*, 244–252.

(92) Korotcenkov, G.; Brinzari, V.; Cerneavschi, A.; Ivanov, M.; Cornet, A.; Morante, J.; Cabot, A.; Arbiol, J. In_2O_3 films deposited by spray pyrolysis: Gas response to reducing (CO, H_2) gases. *Sens. Actuat. B* **2004**, *98*, 122–129.

(93) Golovanonv, V.; Mäski-Jaskari, M.A.; Rantala, T.T.; Korotcenkov, G.; Brinzari, V.; Cornet, A.; Morante, J. Experimental and theoretical studies of indium oxide gas sensors fabricated by spray pyrolysis. *Sens. Actuat. B* **2005**, *106*, 563–571.

(94) Korotcenkov, G.; Brinzari, V.; Ivanov, M.; Cerneavschi, A.; Rodriguez, J.; Cirera, A.; Cornet, A.; Morante, J. Structural stability if indium oxide films deposited by spray pyrolysis during thermal annealing. *Thin Solid Films* **2005**, *479*, 38–51.

(95) Available at: http://www.figarosensor.com/products/common%281104%29.pdf (accessed 2006_10_30).

(96) Göpel, W. Ultimate limits in the miniaturization of chemical sensors. *Sens. Actuat. A* **1996**, *56*, 83–102.

(97) Szabo, N.F.; Dutta, P.K. Strategies for total NO_x measurements with minimal CO interference utilizing a microporous zeolitic catalytic filter. *Sens. Actuat. B* **2003**, *88*, 168–177.

(98) Gouma, P.; Sberveglieri, G. (Guest Editors); Novel materials and applications of electronic noses and tongues. *MRS Bull.* **2004**, *29(4)*, 697–731 (6 articles).

(99) Wingbrant, H.; Svenningstorp, H.; Salomonsson, P.; Tengström, P.; Moldin, D.; Ekedahl, L.-G.; Lundström, I.; Lloyd Spetz, A. Using a MISiCFET device as a cold start sensor. *Sens. Actuat. B* **2003**, *93(1–3)*, 295–303.

(100) Ivers-Tiffee, E.; Härdtl, K.H.; Menseklou, W.; Riegel, J. Principles of solid state oxygen sensors for lean combustion gas control. *Electrochimica Acta* **2001**, *47*, 807–814.

(101) Francioso, L.; Pressicce, D.S.; Taurino, A.M.; Rella, R.; Siciliano, P.; Ficarella, A. Automotive application of sol-gel TiO_2 thin film-based sensor for lambda measurement. *Sens. Actuat. B* **2003**, *95*, 66–72.

(102) Grilli, M.L.; Chevallier, L.; Di Vona, L.; Licoccia, S.; Di Bartolomeo, E. Planar electrochemical sensors based on YSZ with WO_3 electrode prepared by different chemical routes. *Sens. Actuat. B* **2005**, *111–112*, 91–95.

(103) Di Bartolomeo, E.; Grilli, M.L. YSZ-based electrochemical sensors: From materials preparation to testing in the exhaust of an engine bench test. *J. Eur. Ceram. Soc.* **2005**, *25*, 2959–2964.

(104) Zampolli, S.; Elmi, I.; Ahmed, F.; Passini, M.; Cardinali, G.C.; Nicoletti, S.; Dori, L. An electronic nose based on solid state sensor arrays for low-cost indoor air quality monitoring applications. *Sens. Actuat. B* **2004**, *101*, 39–46.

(105) Baraton, M.-I., Merhari, L. Nanoparticles-based chemical sensors for outdoor air quality monitoring microstations. *Mater. Sci. Eng. B* **2004**, *112*, 206–213.

(106) Srivastava, A.K. Detection of volatile organic compounds (VOCs) using SnO_2 gas-sensor array and artificial neural network. *Sens. Actuat. B* **2003**, *96*, 24–37.

(107) Morvan, M.; Talou, T.; Beziau, J.-F. MOS-MOSFET gas sensors array measurements versus sensory and chemical characterization of VOC's emissions from car seat foams. *Sens. Actuat. B* **2003**, *95*, 212–223.

(108) Nake, A.; Dubreuil, B.; Raynaud, C.; Talou, T. Outdoor in situ monitoring of volatile emissions from wastewater treatment plants with two portable technologies of electronic noses. *Sens. Actuat. B* **2005**, *106*, 36–39.

(109) Capone, S.; Siciliano, P.; Quaranta, F.; Rella, R.; Epifani, M.; Vasanelli, L. Analysis of vapours and foods by mean of an electronic nose based on a sol-gel metal oxide sensor array. *Sens. Actuat. B* **2000**, *69*, 230–235.

(110) Capone, S.; Epifani, M.; Quaranta, F.; Siciliano, P.; Taurino, A.; Vasanelli, L. Monitoring of rancidity of milk by means of an electronic nose and a dymanic PCA analysis. *Sens. Actuat. B* **2001**, *78*, 174–179.

(111) Eriksson, Å.; Persson Waller, K.; Svennersten-Sjaunja, K.; Augen, J.-E.; Lundby, F.; Lind, O. Detection of mastitic milk using a gas-sensor array system (electronic nose). *Int. Diary J.* **2005**, *15*, 1193–1201.

(112) O'Riordan, P.J.; Delahunty, C.M. Characterization of commercial Cheddar cheese flavour. 1. Traditional and electronic nose approach to quality assessment and market classification. *Int. Diary J.* **2003**, *13*, 355–370.

(113) O'Riordan, P.J.; Delahunty, C.M. Characterization of commercial Cheddar cheese flavour. 2. Study of Cheddar cheese discrimination by composition, volatile compounds and descriptive flavour assessment. *Int. Diary J.* **2003**, *13*, 371–389.

(114) Garcia, M.; Aleixandre, M.; Gutierrez, J.; Horrillo, M.C. Electronic nose for wine discrimination. *Sens. Actuat. B* **2006**, *113*(2), 911–916.

(115) Gonzalez Martin, Y.; Cerrato Oliveros, M.C.; Perez Pavon, J.L.; Garcia Pinto, C.; Moreno Cordero, B. Electronic nose based on metal oxide semiconductor sensors and pattern recognition techniques: Characterization of vegetable oils. *Anal. Chim. Acta* **2001**, *449*, 69–80.

(116) Cerrato Oliveros, M.C.; Perez Pavon, J.L.; Garcia Pinto, C.; Fernandez Laespada, M.E.; Moreno Cordero, B.; Forina, M. Electronic nose based on metal oxide semiconductor sensors as a fast alternative for the detection of adulteration of virgin olive oils. *Anal. Chimica Acta* **2002**, *459*, 219–228.

(117) Falasconi, M.; Pardo, M.; Sberveglieri, G.; Ricco, I.; Bresciani, A. The novel EOS^{835} electronic nose and data analysis for evaluating coffee ripening. *Sens. Actuat. B* **2005**, *110*, 73–80.

(118) Zhu, L.; Seburg, R.A.; Tsai, E.; Puech, S.; Misfud, J.-C. Flavor analysis in a pharmaceutical oral solution formulation using an electronic nose. *J. Pharm. Biomed. Anal.* **2004**, *34*, 453–461.

(119) Schaller, E.; Bosset, J.O.; Escher, F. "Electronic Noses" and their applications to food. *Lebensmittel Wissenscahften Technol.* **1998**, *31*, 305–316.

(120) Ampuero, S.; Bosset, J.O. The electronic nose applied to diary products: A review. *Sens. Actuat. B* **2003**, *94*, 1–12.

CHAPTER 16

Photovoltaic, Photoelectronic, and Electrochemical Devices Based on Metal-Oxide Nanoparticles and Nanostructures

JUAN BISQUERT

Departament de Ciències Experimentals, Universitat Jaume I, 12071 Castelló, Spain

16.1. INTRODUCTION

Nanostructured metal-oxide films formed by an assembly of nanoparticles sintered over a conducting substrate provide a large internal area that can realize different functionalities. Titanium dioxide and several other metal-oxides nanostructures (e.g., ZnO, SnO_2, and Nb_2O_5) have the advantage of easy processing and low production cost and have the potential to replace existing devices based on more expensive technologies and to give rise to new applications. In recent years, such nanostructures have been amply investigated for several applications, such as dye-sensitized solar cells (DSC) (1) photoelectrochromic windows (2) and electrical paint displays (3), and protein immobilization (4).

The most intensively investigated metal-oxide nanostructured device is the DSC, which was discovered in 1991 by Michael Grätzel (1). The DSC is formed with a network of TiO_2 nanoparticles sensitized to the solar spectrum with a monolayer of dye molecules and a redox electrolyte (5–8). Overall, 11% efficiency of sunlight energy conversion to electricity has been achieved in DSCs (6) but their large-scale production has not yet been reached mainly due to technical problems such as evaporation of the liquid ionic conductor. The DSC has also paved the way for several applications of metal-oxide nanomaterials in other devices based on similar principles of operation. Examples are solid-state nanostructured solar cells (9), electrochromic devices for displays (10), ultraviolet light-emitting diodes (LEDs) (11), and nanostructured WO_3 and Fe_2O_3 electrodes for hydrogen production by photocatalytic splitting of water

Synthesis, Properties, and Applications of Oxide Nanomaterials, Edited by José A. Rodríguez and Marcos Fernández-Garcia

(12,13). Energy storage devices such as batteries and supercapacitors can also benefit in many respects from nanostructuring (14).

In this chapter, we will review the main characteristics of operation of these nanostructured devices and the results that have been achieved so far. We will consider several topics related to device operation: the characteristics of carrier transport in mesoscopic metal-oxide films, the means of coating and functionalizing the surface, and the requirements of electrolytes and/or organic hole conductors that are usually employed for forming heterojunctions with nanostructured semiconductors. We will also highlight the physical properties of spatially regular nanostructures that are expected to give rise to new classes of devices. Quantum dots are small nanocrystalline particles where the strong confinement of the electron wave function provides discrete valence and conduction energy levels. The electrical and optical properties of quantum dots can be modified by controlling their dimension. Quantum dots can be assembled in highly ordered structures known as quantum dot solids, which may find, in the long term, applications in opto-electronic switches, LEDs, lasers, and solar cells (15,16). Carbon nanotubes, discovered in 1991, have become a milestone in nanomaterials research. Several types of oxidic nanowires are accessible now (17) and have attracted wide scientific and technological interest because of their novel structures and their potential applications in luminiscent devices and solar cells as well as in future nanoelectronic components, such as field-effect transistors (FETs), crossed junctions, and actuators.

16.2. GENERAL FEATURES OF NANOSTRUCTURED PHOTO-ELECTRONIC AND ELECTROCHEMICAL DEVICES

The fundamental structure of a nanostructured device is shown in Figure 16.1. Semiconductor nanoparticles can be prepared by wet chemical methods that lead to colloidal nanocrystals present in dispersion. The colloidal nanocrystals are usually deposited, e.g., by screen printing, onto a glass or flexible plastic support covered with a conducting layer. The film is thermally treated to form an electronically connected array of nanoparticles. The nanostructured film is filled with an ionic conductor or hole-conducting medium that is contacted with a counterelectrode. The global structure of the device consists of interpenetrating bicontinuous phases in which the simultaneous transport of several carriers in the different phases and interfacial processes must be optimized.

Electrons can be injected or extracted from the nanoparticulate film by external contact through the conductive support. Therefore, the whole internal area of the nanoparticulate network is electronically addressable from the substrate. This very large active area for device function that is obtained by a relatively simple preparation route is a main advantage of using metal-oxide nanomaterials for electronic and electrochemical devices. In photonic applications, the conducting substrate must be optically transparent, so that typically a thin-layer transparent conducting oxide (TCO) such as indium-doped tin oxide or fluor-doped tin oxide over glass is used.

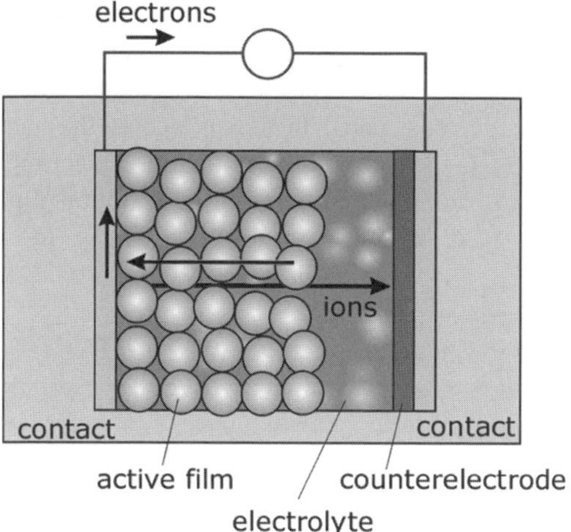

Figure 16.1. Scheme of an electroactive device formed by a nanoparticulate metal-oxide film and counterelectrode joined by an ionic conductor. Dark arrows indicate electrons pathways in the particulate network and in the external circuit, whereas the light arrow indicates ionic transport in the voids of the network up to the counterelectrode.

Although electrons move across the metal-oxide nanoparticulate network, the displacement of other charge carriers is needed in the medium filling the voids in the nanostructure of Figure 16.1, i.e., the ionic conductor that is addressed from the counterelectrode. Liquid electrolytes have the advantages that they immediately form a perfect electric junction in the internal surface, and that they provide a large ionic conductivity and wide potential window of stability. However, liquid electrolytes require efficient sealing, which is a major drawback for the long-term operation of devices. In general, an all solid-state construction is preferable with solid electrolytes that allow for more compact and mechanically flexible device designs. Organic conductors are convenient media for hole conduction and heterojunction formation in nanostructured devices, because they can be deposited in solution form and fill the pores homogeneously. However, the carrier mobilities are much lower than in inorganic semiconductors or liquid electrolytes. Recently, room-temperature ionic liquids have emerged as an excellent medium for ionic carrier in nanostructured devices (18,19), because they are nonvolatile and nonflammable, have high thermal stability, and show excellent electrochemical properties, namely, high ionic conductivity and wide potential windows of operation.

Metal-oxide nanoparticles are usually not intentionally doped and are therefore insulating. The high electronic conductivity required for addressing the internal surface from the substrate is obtained by the injection of electrons in the nanoparticles. These electrons may be electrically injected from the substrate or photogenerated in the nanoparticles surface by sensitizers such as organic dyes or quantum dots (20). (Doping of wide band gap metal-oxide nanoparticles for the sensitization to the visible spectrum has yielded so far very limited success; see below.) As a result the electron density varies by many orders of magnitude during device operation (see Figure 16.8). Consequently, a large density of negative charge is created in the nanoparticulate network, and in order to preserve overall charge neutrallity, the same amount of positive charge must occur in the surface of the nanoparticles, as it is suggested in Figure 16.2*a*. In many cases, to facilitate electron accumulation and transport by preserving the charge neutrality in the system is a primary function of

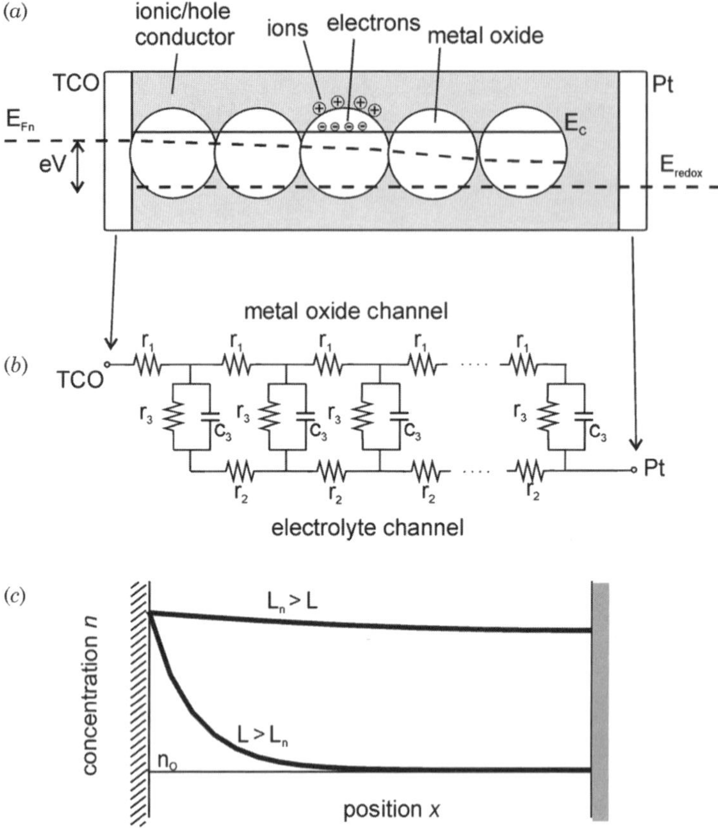

Figure 16.2. (*a*) Electron energy diagram of a nanostructured metal-oxide electrode in contact with a redox electrolyte (or hole conducting medium), illustrating the electrochemical potential of electrons E_{Fn} (Fermi level) when a potential V is applied to the substrate, and assuming that conduction band energy (E_c) remains stationary with respect to the redox level, E_{redox}. (*b*) The equivalent circuit (transmission line model) for a small periodic ac perturbation, which contains the transport resistance along the metal-oxide nanoparticles, r_1; the resistance in the hole/ion conducting medium, r_2; the charge transfer resistance at the metal-oxide/electrolyte interface, r_3; and the capacitance for charge accumulation in the metal-oxide particles and/or surface, c_3. (*c*) Distribution of electrons injected in the nanostructured film from the substrate, for the case in which the diffusion length is much larger than the nanoparticulate film thickness, $L_n > L$, and for the opposite case, $L_n < L$, as indicated. n_0 is the equilibrium concentration.

the medium that fills the pores. Positive carriers in the ionic medium should have a sufficiently high mobility not to limit the displacement of the electrons in the nanostructure. In addition, the average size of inorganic semiconductor units should be smaller (in the 10-nm range) than the Debye length, so that no internal electrical fields are built into one compact unit (21). In contrast to bulk semiconductors that show a spatial variation of energy levels in the space-charge region in the surface, in nanostructured networks filled with a conductive medium, there is no internal spatial distribution of the conduction band level in individual nanoparticles, and the whole nanostructure can be treated with a homogenous distribution of energy levels.

Let us discuss specific functions of carriers in some nanostructured devices with the configuration of Figure 16.1:

(a) *Charge Separation.* In DSC, positive and negative carriers are created by sunlight in the molecules adsorbed at the nanoparticles surface. Electrons are injected in the metal-oxide nanoparticles and travel all the way to the substrate. Meanwhile, a conjugated carrier travels in the medium filling the pores toward the counterelectrode. This second carrier may be a redox ion that reacts at the counterelectrode. It may be also an electronic carrier in an inorganic or organic hole conductor. The net effect of the overall process is the collection of a negative carrier at the substrate of the nanoparticulate film and a positive carrier at the counterelectrode.

(b) *Charge Recombination.* In the LEDs, electrons are injected from the substrate of the nanoparticulate film (the cathode) and holes are injected from the electrode contacting the other medium (the anode). Both carriers should travel in their respective media and meet at the surface of the nanostructure to recombine radiatively emitting light.

(c) *Charge Accumulation.* By modifying the carrier density in the nanostructure, the chemical potential of such a carrier is changed significantly. With accumulated charge, the device can supply a voltage and an electric current when the external circuit is connected to some load. This is the basic principle of operation of intercalation batteries and electrochemical supercapacitors (22).

A useful scheme for describing the transport and charge-transfer properties in nanostructured devices is the transmission line model shown in Figure 16.2. This model has been developed (23–25) for the interpretation of impedance spectroscopy results and has found wide application for understanding and characterizing the properties of electrochemical nanostructured devices (26–30) and DSC (31–33).

In the representation of Figure 16.2*a*, it is assumed that the medium filling the voids in the nanostructure possesses a large conductivity and a large carrier density (e.g., a liquid electrolyte). For this reason, the Fermi level/redox potential in the hole-conducting medium will not be significantly changed when the potential between the contacts is modified. The variation of potential mostly causes a displacement of the Fermi level (E_{Fn}) inside the metal-oxide nanostructure, from the equilibrium value, which equals the redox potential of the electrolyte, $E_{F0} = E_{\text{redox}}$, toward the conduction band. The gradient of the Fermi level indicates the direction of electron diffusion. The electron density decreases with the distance from the substrate due to charge transfer events from the semiconductor to acceptor species in solution.

In the transmission line model of Figure 16.2*b* there are different processes represented. In general the resistances describe carrier flux with respect to a difference of electrochemical potential, whereas capacitors represent carrier storage. The longitudinal resistance r_1 relates the local carrier flux to the variation of E_{Fn} with distance in the nanoparticulate structure, and it corresponds to the transport resistance; i.e., it is the inverse of the electronic conductivity (25). Similarly, r_2 describes the transport or carriers in the ion- or hole-conducting medium. In the situation indicated above for a liquid electrolyte, r_2 will be very low and provides no limitation for the device operation; however, in the case of organic conductors, r_2 cannot be neglected. The resistance r_3 describes the carrier flux by charge-transfer between the two phases and corresponds to recombination flux in DSC. Properties of the capacitance of the nanostructured network will be commented on in the next section.

In the characteristic structure of the transmission line of Figure 16.2*b*, the transport resistances, distributed in the spatial direction of each transport channel, are continuously interrupted by interphase elements r_3c_3. This combination represents the physical current flow; i.e., a carrier in the nanoparticle has a probability to move forward or to drop in the electrolyte. For the efficient operation of a device, the different elements governing transport and recombination must be correctly tailored and balanced. For example, in the presence of a large recombination flux (low r_3 and low conductivity (large r_1), carriers injected from the TCO will not move far from the substrate, and in this case, a large part of the device will not be operative.

Two main time constants describe more precisely such characteristics of a device: the lifetime of the carriers τ_n and the transit time over the film thickness L. If the carrier diffusion coefficient is D_n, the transit time is (up to a numerical constant of order 1)

$$\tau_{tr} = \frac{L^2}{D_n} \tag{16.1}$$

The relationship between the time constants can be written as

$$\frac{\tau_{tr}}{\tau_n} = \left(\frac{L}{L_n}\right)^2 \tag{16.2}$$

in terms of the diffusion length, which is defined as

$$L_n = \sqrt{D_n \tau_n} \tag{16.3}$$

When the quotient in Eq. 16.2 is $\ll 1$, the carriers have time to travel the entire thickness of the layer before recombining, which is an essential requirement for efficient solar cells. The diffusion length also gives a measure of the extent of the penetration of the carriers that are voltage-injected at the boundary of the layer.

Figure 16.2*c* indicates two extreme situations. The right boundary is considered completely blocking. In the case when the diffusion length is much larger than the film thickness, the carriers are nearly homogeneously distributed in the available volume. On the other hand in the high-reactivity case $L_n \ll L$, the carrier concentration decays rapidly close to the injecting boundary of the porous structure.

16.3. MEASUREMENT TECHNIQUES FOR CHARACTERIZATION OF NANOSTRUCTURED PHOTO-ELECTRONIC AND ELECTROCHEMICAL DEVICES

The most basic characteristic of a solar cell is the steady-state performance given by the current density-potential curve at fixed incident illumination. This curve gives the main parameters determining the solar cell performance: the short-circuit photocurrent (I_{sc}), the open-circuit potential (V_{oc}), and the fill factor (FF), which describes the quality of the diode behavior of the solar cell and determines the power output of the solar cell at the point of maximum power operation. Steady-state characteristics are obviuosly important for the performance of other devices such as LEDs. However, it is usually difficult to extract detailed device properties only from steady-state measurements. There are two main issues for investigation: (*1*) The determination of quantities such as the diffusion coefficient and the lifetime of the different carriers, which govern the fundamental electronic processes occurring in the device function. These time constants are also crucial characteristics for devices that require fast switching or fast power supply, such as elecrical paint displays and supercapacitors, respectively. (2) The separation of different effects in the global device response, for their separate optimization. These two issues can be addressed with a range of measurement methods in combination with rational variation of the device components.

The time constants mentioned above are obtained by specific small perturbation kinetic techniques that do not modify the steady state over which they are measured. Examples of the techniques are intensity modulated photocurrent spectroscopy (IMPS) (34–38) intensity modulated photovoltage spectroscopy (IMVS) (39), and small amplitude time transients (40–44).

The electrochemical impedance spectroscopy (24,31–33,45–47) measurement of nanostructured devices has the advantage of providing a determination of the time constants as well as spectrally separating different internal contributions of the overall device operation. However, when different processes are acting in combination in a nanostructured device, the interpretation of impedance spectroscopy results requires appropriate models for extracting the relevant information. As an important example, we discuss the impedance model of diffusion and reaction of the carriers in nanostructured electrodes, which has the following expression (24):

$$Z = \left(\frac{R_1 R_3}{1 + i\omega/\omega_3}\right)^{1/2} \coth\left[(\omega_3/\omega_1)^{1/2}(1 + i\omega/\omega_3)^{1/2}\right] \qquad (16.4)$$

where ω is the angular frequency of the perturbation. In Eq. 16.4, we use the total resistances for transport $R_1 = Lr_1$ and recombination $R_3 = r_3/L$, and the characteristic frequencies $\omega_1 = \tau_{tr}^{-1} = 1/R_1C_3$ and $\omega_3 = \tau_n^{-1} = 1/R_3C_3$, where $C_3 = Lc_3$ is the total capacitance. Equation 16.2 can be restated as

$$\frac{R_1}{R_3} = \frac{\omega_3}{\omega_1} = \left(\frac{L}{L_n}\right)^2 \tag{16.5}$$

The impedance pattern of Figure 16.3*a* corresponds to the case $R_1 \ll R_3$; i.e., the transport (diffusion) resistance is much lower than the recombination resistance. In this case, which indicates a large diffusion length, the impedance spectra show a small feature at high frequency that is the Warburg (diffusion) impedance, with an inclination of 45°; see Figure 16.3*a*. At lower frequencies, a large arc is obtained, which is due to

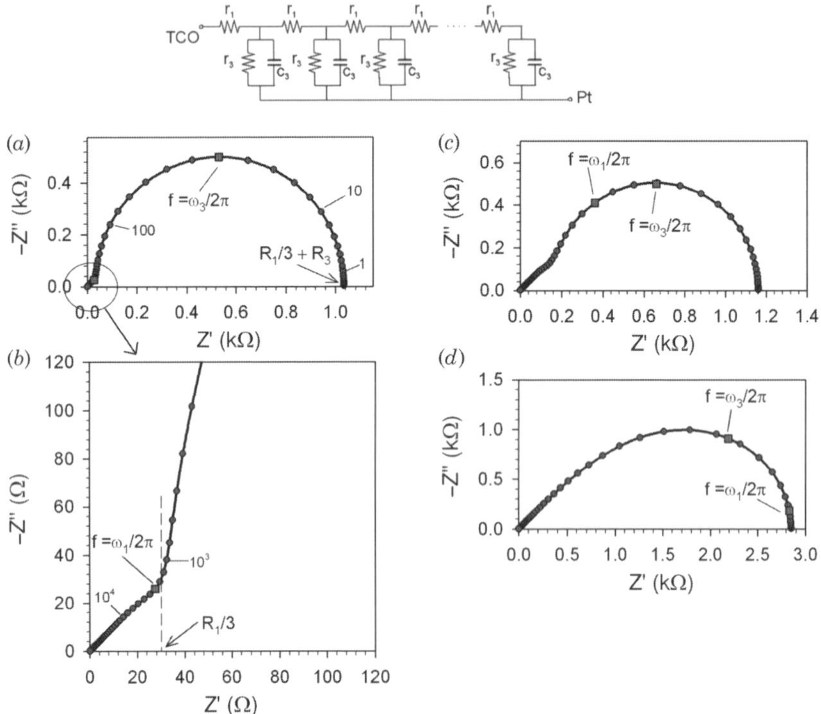

Figure 16.3. (*a*) Simulation of the diffusion-recombination impedance with reflecting boundary condition. Parameters are $R_1 = 10^2\ \Omega/\text{cm}^2$, $R_3 = 10^3\ \Omega/\text{cm}^2$ and $C_\mu = 5 \times 10^{-6}\ \text{F/cm}^2$. Shown are the frequencies in Hertz at selected points, the characteristic frequency of the low frequency arc (31.8 Hz, square point), related to the angular frequency ω_3, the low-frequency resistance, and the characteristic frequency of the turnover from Warburg behavior to low-frequency recombination arc (318 Hz, square point), related to the angular frequency ω_1. (*b*) Shows an enlargement of (*a*). (*c*) The case $R_1 = R_3$. (*d*) The case $R_3 \ll R_1$.

the combination of charge-transfer resistance and capacitance. In contrast, in the high-reactivity case, $R_3 \ll R_1$, or in other words, $L_n \ll L$, the impedance pattern changes to that of Figure 16.3*d*, which is known as the Gerisher impedance. Figure 16.3*b* shows the spectral shape obtained in the transition between the two previous extreme situations; i.e., $L_n \approx L$.

A representative measurement of impedance spectroscopy of nanostructured TiO_2 electrodes is shown in Figure 16.4 (48). The impedance pattern of Figure 16.3*a* is well realized, indicating that electron transport across the film length is faster than the rate of recombination. From such spectra, the different time constants τ_n and τ_{tr} can be obtained and the electron conductivity and diffusion coefficient. These quantities have been systematically determined as a function of bias potential (26,31).

It is interesting to mention the impedance behavior of nanostructured films in which there is no charge-transfer at all. This occurs in an inert electrolyte that realizes charge shielding but lacks any electron acceptors, which is a fundamental requirement of several capacitive devices such as Li-batteries and supercapacitors. The model obtained with $R_3 \to \infty$ in Eq. 16.4 is shown in Figure 16.5. The impedance of

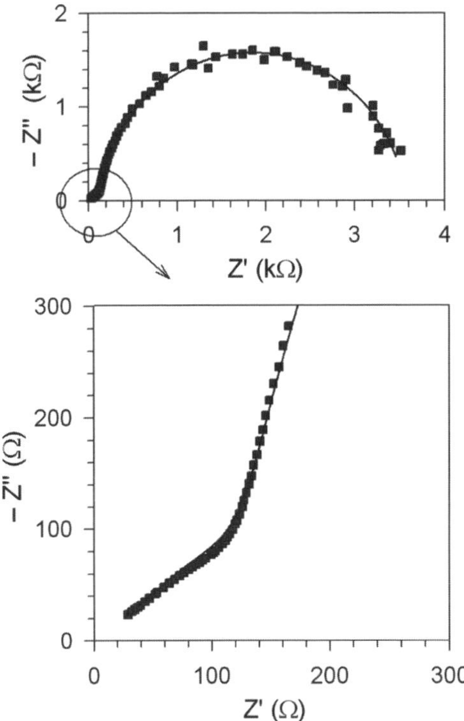

Figure 16.4. Impedance spectroscopy of a 8-μm-thick film of nanostructured TiO_2 (10-nm nanoparticles anatase) in aqueous solution at pH 2, with −0.250 V bias potential vs. Ag/AgCl under UV illumination. The line is a fit to the model of Eq. 16.4 (48).

unreactive porous electrodes shown in Figure 16.5 is characterized by a capacitive domain at low frequencies and a "Warburg" part at high frequencies that relates to the transport of the species flowing in the nanostructure (or in the electrolyte) in order to charge the whole film. For a perfect capacitor in the transverse branch, the impedance reduces to the classic transmission line of de Levie (49) for porous electrodes (50). But the capacitive response in many real systems does not show a perfect capacitor and is better described by a constant phase element (CPE) (51,52), with admittance

$$y_3 = q_3(\mathrm{i}\omega)^{\gamma_3}; \quad q_3 = Q_3/L \tag{16.6}$$

Equation 16.6 reduces to the perfect capacitor in the particular case $\gamma_3 = 1$. In general, the CPE in the transverse element of the transmission line changes the slopes of the Warburg and capacitive lines of the impedance spectra, depending on the extent that the exponent γ departs from the value of 1 of the ideal capacitor (23). The deviations from the ideal model in the low- and high-frequency lines have a common origin in the dispersive element Q_3, as seen in the example of Figure 16.5 for $\gamma_3 = 0.8$. Hence, the slopes are related by the high-frequency line having an exponent $\gamma_3/2$, whereas the low-frequency line is the full CPE; i.e., the latter exponent doubles the former one, as discussed in Ref. 23. This means that even in the presence of frequency dispersion, the behavior of a porous electrode can be spectrally recognized

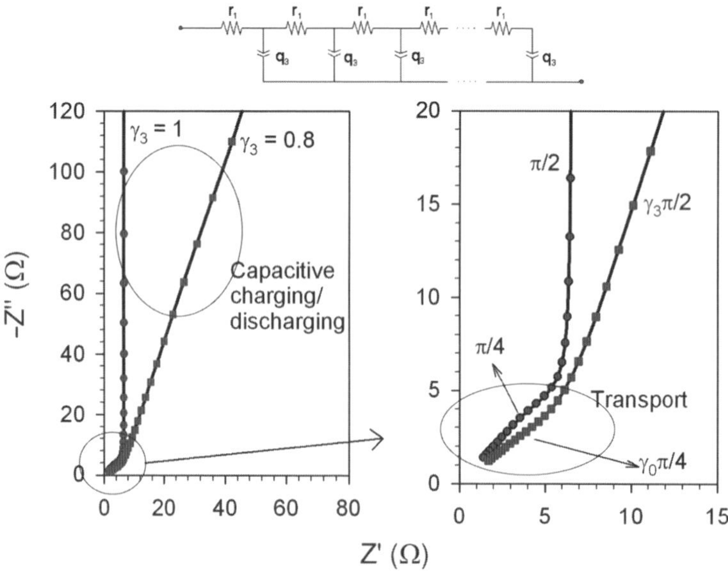

Figure 16.5. Simulation of the impedance spectra for the transmission line model indicated in the scheme with $R_1 = 20\,\Omega$, $Q_3 = 5\,\mathrm{mF\,s}^{1-\gamma_3}$, and two values of the CPE exponent γ_3 as indicated in the left diagram. Also indicated in the right diagram is the inclination of the low- and high-frequency lines, in radians.

by impedance measurements. An experimental instance of frequency dispersion can be observed in the less than 45° inclination of the spectrum of Figure 16.4 at high frequency, and another case is discussed below in Figure 16.16.

The spectra of the type of Figure 16.5 are very well realized in carbonaceous supercapacitors, for example (53). Independently of the presence of frequency dispersion, the spectra of Figure 16.5 indicate important limitations for nanostructured devices with large charge storage and fast transient response demands. In these devices, the amount of energy storage is governed by the total capacitance that is $C_3 = Lc_3$. If the specific capacitance of nanoparticles c_3 has been optimized, increasing the device capacitance requires increasing film thickness. However, one must take into account the charging region in the frequency domain that is observed in Figure 16.5 with the inclination $\leq\pi/4$. This region is onset at times shorter than $\tau_{tr} = L^2/D$. Therefore, increasing the film capacitance with larger thickness has the adverse effect of increasing (quadratically) the operation time of the device. Even if a highly conducting porous medium is used (such as carbon nanoparticles), the transient response will be governed by electrolyte diffusion coefficient. As an example, $D = 10^{-5}\,\text{cm}^2\,\text{s}^{-1}$ and $L = 10\,\mu\text{m}$ gives a limitation of $\tau = 0.1$ s for device operation.

In the situations discussed so far, we have focused on diffusive transport of carriers, granted that a large conductivity that anihilates any local space charge and shields electrical fields is available in at least one of the interpenetrating bicontinuous phases. Let us mention the other limiting case, consisting of basically *insulating* phases where carriers can be injected by suitable contacts. In the absence of intrinsic conductivity, application of a voltage requires the injected carriers to redistribute themselves to maintain a local field that drives the transport (usually by predominant drift mechanism). This is the space-charge-limited current transport (54) that is typically found, for example, in organic LEDs (55) and in some organic solar cells (56).

It is important to recognize that nanostructured devices are characterized by the abundance of interfaces of different kinds, which must be controlled for facilitating the proper device function. In particular, nanostructured solar cells are usually largely determined by the competition of kinetics of charge transfer at several interfaces (57), which governs the selectivity of contacts that is required in photovoltaic devices (17,58). The main (and larger) interface is that between the nanostructured metal oxide and the ionic/hole conductor, as it has already been described in terms of the resistance R_3 above. This interface is crucial for the diode characteristic of the solar cells (47) and usually receives a great deal of attention when material and device properties are selected. However, additional contacts should be correctly engineered to avoid unwanted resistances or internal short-circuits. Figure 16.6 shows as an example of additional impedances that may occur in the model of Figure 16.2*b* (59). Z_{c1} is an impedance at the electron injecting/extracting contact, which is determined by the interfacial barrier formed between the metal oxide and the conductive substrate. It is usually desired that this interface forms an ohmic contact in which $Z_{c1} \approx 0$ (short-circuit in Figure 16.2*b*) so that dissipative losses are avoided in device operation. Figure 16.6 also shows the impedance between the substrate and the ionic/hole-conducting medium. Ideally one usually needs this impedance to be totally blocking for electron transfer, i.e., $Z_{c2} \rightarrow \infty$ (open-circuit in Figure 16.2*b*),

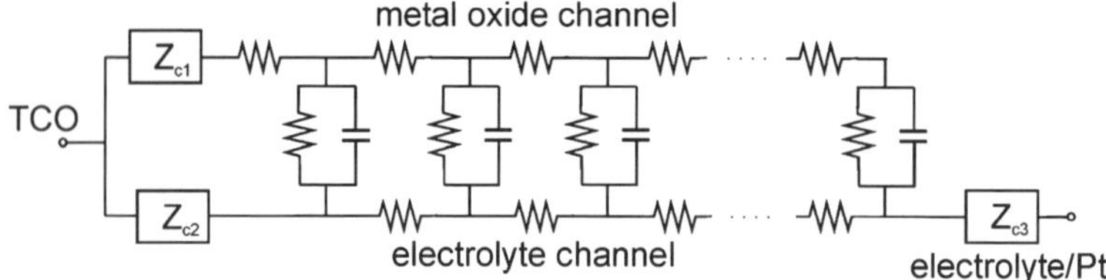

Figure 16.6. Transmission line equivalent circuit model, based in Figure 16.2*b*, and showing in addition the impedance at the metal oxide/subtrate interface, Z_{c1}, the impedance at the electrolyte/substrate interface, Z_{c2}, and the impedance at the electrolyte/counterelectrode interface, Z_{c3}.

to avoid the metal-oxide nanostructure from becoming short-circuited by the direct electronic pathway between the two external contacts across the ionic conductor. In DSC with an organic hole conductor the reactivity of the substrate is large and it must be pretreated with a thin metal-oxide layer that blocks the charge transfer to the hole-conducting medium while maintaining good electron–injection properties toward the nanoparticles. Finally, a resistance Z_{c3} relates to the facility of charge transfer at the counterelectrode. In summary, it should be appreciated that the proper device function requires the materials determining the three contact impedances in Figure 16.6 to be treated for reducing the equivalent circuit of Figure 16.6 to that of Figure 16.2*b*, which drives the current through the nanostructure area without side effects. We should also mention that in large area devices with a transparent conducting substrate, the transport of electric current along the TCO substrate often introduces major resistive losses.

Let us describe other techniques that provide relevant characteristics of the metal-oxide nanostructured films. The electrochemical transistor configuration, (Figure 16.7*a*) uses a conducting substrate that is divided into two separated regions, bridged by the nanostructured film (60), and allows us to determine the electronic conductivity as a function of electrode potential. The two regions of the conducting substrate can be controlled as an independent working electrode, so that the Fermi level in the film can be governed with a bias $U_1 \cong U_2$ (Figure 16.7*b*) while maintaining a small potential difference between the two sides, $\Delta U = U_2 - U_1$, which causes a current flow ΔI between WE_1 and WE_2 that enables us to measure the electronic conductivity σ. This method has been applied in several metal-oxide nanostructured films (60–63).

Figure 16.8 shows the nine-order-of-magnitude change of the electronic conductivity of nanocrystalline TiO_2 films immersed in solution when the Fermi level is displaced toward the conduction band. This figure illustrates the *global* transformation from insulator to conductor of the films by external potentiostatic control that is a genuine effect of the nanocrystalline structure, in comparison with the bulk material counterpart, which conductivity can only change in the surface.

Next we consider the special features of the capacitance of nanocrystalline films and its relationship to the density of states (DOS) (64). There are two main basic

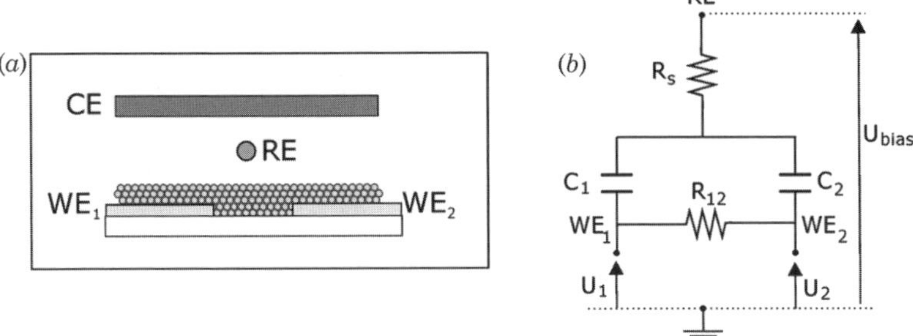

Figure 16.7. (*a*) Electrochemical transistor measurement configuration. The conducting substrate over which the film is deposited is divided in two regions, separated by an insulating gap. The separated regions serve as two working electrodes WE_1 and WE_2. (*b*) Equivalent circuit of the electrochemical transistor configuration. When WE_1 and WE_2 are shorted the film can be operated as in normal electrochemical cell with potential U_{bias}. The two working electrodes can also be operated independently with potentials U_1 and U_2 with respect to the reference electrode (RE). CE is the counter electrode.

mechanisms of accumulating charge with respect to voltage in electrochemical systems. The first one is that occurring in a standard dielectric capacitor where energy is stored in the electrical field, related to spatial charge separation. This capacitance is ubiquitous at interfaces with space charge such as Schottky barriers and Hemlholtz layers. In the other mechanism, which is predominant in many nanostructured and intercalation systems, the energy storage is associated with a change of the chemical potential of the accumulated species. When a voltage variation dV is applied to the conductive substrate of a nanostructured metal-oxide film, and the Fermi level is

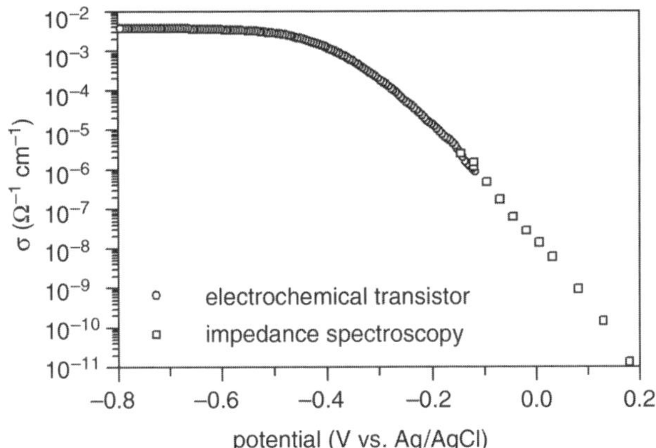

Figure 16.8. Electronic conductivity in nanostructured TiO_2 electrodes in solution at pH 2 as a function of the electrode potential, obtained by two different measurement methods as indicated (60).

displaced homogeneously in the film as $dE_{Fn} = -qdV$ (where q is the elementary charge), the electron density changes by a quantity dn. The electrochemical capacitance (per unit volume) relates the change of concentration of electrons n to the change of electrochemical potential:

$$C_\mu = q\frac{dn}{dE_{Fn}} \tag{16.7}$$

Assuming that the conduction band potential is stationary, the Fermi level inside the TiO_2 nanostructure is displaced toward the conduction band; i.e., the change of electrochemical potential implies a change of the chemical potential of electrons: $dE_{Fn} = d\mu_n$. Equation 16.7 gives a purely chemical capacitance (64). If $g(E)$ is the DOS in the band gap, the chemical capacitance is obtained integrating all contributions of occupied states:

$$C_\mu = q^2 \int_{-\infty}^{+\infty} g(E)\frac{df}{dE_{Fn}} dE \tag{16.8}$$

where f is the Fermi–Dirac distribution. To a good approximation near room temperature (65), Eq. 16.8 takes the form

$$C_\mu(E_{Fn}) = q^2\, g(E_{Fn}) \tag{16.9}$$

It is therefore found that the chemical capacitance is proportional to the DOS. The interpretation of Eq. 16.9 is that the extent of charging related to the perturbation dV corresponds to filling a slice of traps at the Fermi level. A common finding in nanostructured TiO_2 is an exponential distribution of localized states in the band gap as described by the expression:

$$g(E) = \frac{N_L}{k_B T_0} \exp\left[\frac{(E - E_c)}{k_B T_0}\right] \tag{16.10}$$

Here N_L is the total density and T_0 is a parameter with temperature units that determines the depth of the distribution.

One convenient method of measuring the chemical capacitance is the standard cyclic voltammetry technique (CV) (66), which monitors the current injected in the film as the potential varies at a constant speed, $s = dV/dt$. The electronic current flowing into a unit volume is

$$j = -q\frac{dn}{dt} = q^2\frac{dn}{dE_{Fn}}\frac{dV}{dt} = sC_\mu \tag{16.11}$$

Hence, $C_\mu(E_{Fn})$ is measured directly by cyclic voltammetry. Figure 16.9*c* illustrates that the shape of a CV reveals directly the shape of the exponential DOS of Eq. 16.10.

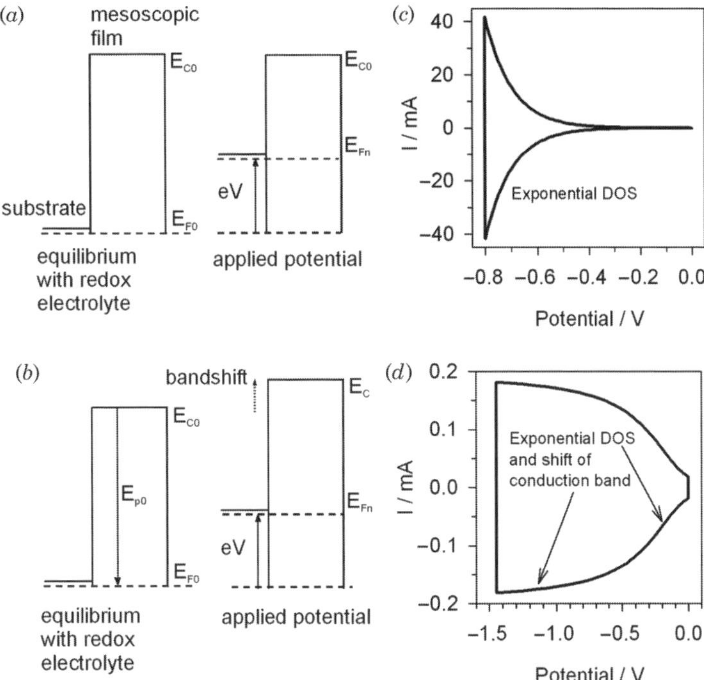

Figure 16.9. The left column shows the schematic energy diagram of a nanostructured semiconductor under application of a potential that rises homogeneously the Fermi level in the film for the case of (*a*) stationary energy levels and (*b*) shift of the energy levels by modification of the Helmhotz potential at the nanoparticles surface during electron injection. The right column shows the corresponding CVs for an exponential distribution of band gap states in the nanoparticles.

Limitations by transport that cause the film to depart from equilibrium during the measurement distort the shape of CV along the voltage axis, whereas the presence of cathodic charge transfer makes the anodic (positive current) peak much smaller than the cathodic (negative current) one (66). However, the quasi-equilibrium condition of the film can be recognized by an invariant shape of the CVs at different scan rate and a current that is proportional to s.

The CV method has been applied in several works for obtaining detailed information of nanostructured TiO_2 energy levels under different surface treatments (45,67) or electrolytes (68). This method is also important for studying the redox properties of organic molecules attached in the TiO_2 surface, as in the case of viologen molecules (27) and proteins (4,69). Provided that there is fast electron exchange between the semiconductor and the attached species, the functionalized electrode will show an excess of capacitance in the CV corresponding to the redox capacitance of the molecular species, in comparison with the bare TiO_2 electrode.

It has already been discussed that for maintaining charge neutrality into a nanostructured film, it is required that the increasing electron charge in the nanoparticles be accompanied by a positive ion charge at the semiconductor/electrolyte interface. As shown in Figure 16.9*b*, surface charging at increasing electron density in the nanoparticulate network changes the potential difference in the Helmholtz layer, producing an upward shift of the semiconductor energy levels (70). The combined effect of electron accumulation and partial band unpinning can be accounted for by a constant Helmholtz capacitance C_H connected in series to the chemical capacitance C_μ so that the total electrochemical capacitance becomes

$$C = \left(C_\mu^{-1} + C_H^{-1}\right)^{-1} \tag{16.12}$$

Hence, when the exponentially increasing C_μ becomes larger than C_H, the CV flattens to a constant value (Figure 16.9*d*), in consonance with the fact that the band shifts simultaneously with the displacement of the Fermi level, with a smaller rate of gain of electron density in the nanostructure.

16.4. DYE-SENSITIZED SOLAR CELLS

Since the first reports (1) of 1991, a major scientific and technical research effort has been devoted to the DSC (see Refs. 5–8), which now constitutes a credible chemical alternative to the already established silicon and thin-film photovoltaic technologies. In addition, the research on DSC has revealed many important general aspects of the principles of operation of nanostructured devices. In the following, we revise some properties of the DSC that have emerged from the research. Related devices will be described in the next section.

The main electronic processes occurring in a DSC are shown schematically in an energy diagram in Figure 16.10. The initial step of operation is the excitation of the dye molecules by absorption of sunlight. Ruthenium polypyridyl complexes have proved to be the most efficient TiO_2 sensitizers, with the *cis*-$RuL_2(SCN)_2$ (L = 2,2′-bipyridyl-4,4′-dicarboxylic acid) dye (N3 dye) demonstrating incident photon-to-electron conversion efficiencies (IPCEs) of up to 85% in the spectral region 400–800 nm corresponding to almost unity quantum yield (electrons per absorbed photon). The next steps produce the separation of the excitation of the dye into carriers in the different transport materials: the injection of an electron from a photoexcited dye to the conduction band of the TiO_2, and the transfer of an electron from the ionic conductor to the dye (71). The former process is usually completed within 200 ps, and the latter, the regeneration of the oxidized dye, is completed within the nanosecond time scale. When the charge carriers have been transferred to the separate electron and ionic/hole-conducting media, they propagate in their respective media toward the outer contacts. Electron transfer at the TiO_2/TCO interface is not thought to be a significant limiting factor in the DSC. On the other hand, the oxidized form of the redox carrier must be reduced at the counterelectrode, usually Pt-catalized TCO, as shown by the element Z_{c3} in the equivalent circuit of Figure 16.6.

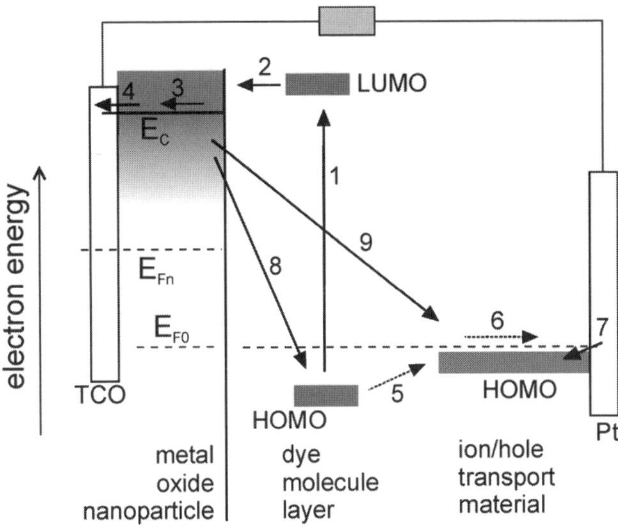

Figure 16.10. Schematic energy drawing of the electron (arrows) and hole (dashed arrows) transfer processes at the metal-oxide/dye layer/hole transport material in a DSC. The gray boxes indicate available electronic states, E_{F0} is the dark Fermi level, E_{Fn} is the Fermi level of electrons accumulated in TiO_2 nanoparticles, and E_c is the lower edge of the metal-oxide conduction band. (1) Photoexcitation. (2) Electron injection from lowest unoccupied molecular orbital (LUMO) of the dye to metal-oxide conduction band levels. (3) Electron diffusion in metal-oxide nanostructure. (4) Electron extraction at the TCO. (5) Hole transfer from highest occupied molecular orbital (HOMO) of the dye to hole transport material (HTM) HOMO. (6) Hole diffusion in HTM. (7) Electron transfer from the Pt counterelectrode to the oxidized ionic species, or to HTM HOMO. (8) Electron transfer from metal oxide to dye HOMO and to (a) HTM.

Nanostructured TiO_2 used in DSC is typically composed of anatase particles prepared by the hydrothermal method (63). They exhibit predominantly a bipyramidal shape, the exposed facets having (101) orientation, which is the lowest energy surface of anatase. The density of electronic states (DOS) in these nanostructures has been investigated using a variety of methods: cyclic voltammetry, as described above (66), impedance spectroscopy (31), voltage-decay charge-extraction method (72,73), and potential step-current integration (chronoamperometry) (63,74). All of these methods are based on the determination of the chemical capacitance. The results (31,66,72–74) agree in the existence of an exponential distribution of band gap states in nanostructured TiO_2, which is indicated in Figure 16.11*b*. However, the origin of such band gap states, and the influence of the surrounding media in the energetics of the electrons, remains unclear.

Let us look in more detail at the electronic processes occurring in the TiO_2 nanoparticle in the operation of the solar cell, following the schematic diagram of Figure 16.11*a*. Electrons photoinjected from the surface sensitizer accumulate in the TiO_2 nanostructure. Therefore, the electrochemical potential of the electrons in the

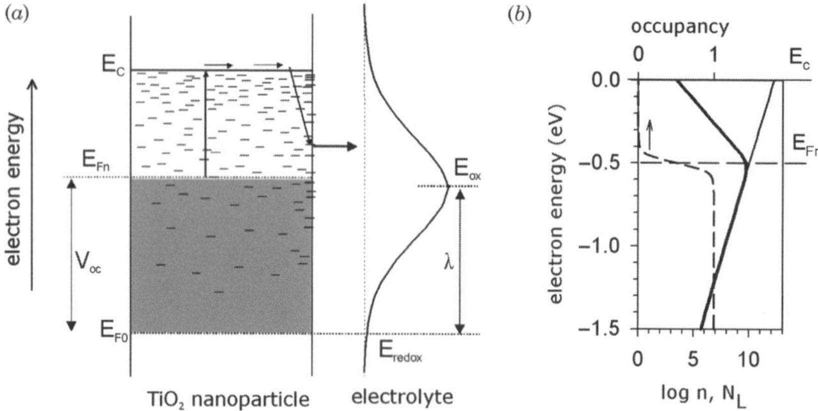

Figure 16.11. (*a*) Schematic energy diagram of electronic processes in a TiO_2 nanoparticle in a dye-sensitized solar cell. E_{F0} shows the position of the Fermi level in the dark, which is equilibrated with the redox potential (E_{redox}) of the iodide/triodide couple. E_{Fn} is the quasi-Fermi level of electrons accumulated in the nanoparticles, and E_c is the conduction band energy. The shaded region indicates the band gap states that are occupied with electrons (zero-temperature approximation of Fermi–Dirac distribution). The arrows show a sequence of processes in the following order: thermal promotion of an electron from a band gap localized state in the bulk of the nanoparticle, to the conduction band; electron diffusion in the conduction band states; trapping of an electron in a surface state; electron transfer from the surface state to the acceptor triiodide ions in solution, which form a distribution with an effective gaussian density of states, indicated in the right, with halfwidth λ, the reorganization energy (75). (*b*) Quantitative diagram of the thermal occupation at temperature $T = 300\,K$ (thick line) of an exponential DOS in the bandgap (thin line) with $T/T_0 = 0.25$, as determined by the Fermi-Dirac distribution function (dashed line).

TiO_2 nanoparticles (the quasi-Fermi level, E_{Fn}) increases. At open-circuit conditions, provided that there is sufficient electronic conductivity (produced by the injected electrons themselves, Figure 16.8), the Fermi level tends to be homogeneous in the nanostructure, and it produces an increase of the potential at the substrate, which is the photovoltage V_{oc}, typically 0.8 V. The number of electrons, determining the photovoltage, is established by dynamic equilibrium between the rate of gain of electrons (photoinjection) and the rate of their loss, which is caused by charge transfer at the surface of the nanoparticles.

The role of the redox mediator in the photovoltaic process is to reduce the oxidized dye, but the oxidized component of the redox couple (I_3^-) is present in the pores and is able to accept electrons from the TiO_2. In addition some injected electrons return to the oxidized dye molecules before they can be regenerated by the ionic conductor. These two charge transfer processes constitute the electron recombination in the DSC. It is important to recognize that the efficiency of the DSC relies on a strong differential kinetics of the redox couple at two interfaces (57). The transfer of electrons in the TiO_2 to oxidized ionic species must be slow, to facilitate electron storage, whereas electron transfer to the same species should be very fast at the counterelectrode, avoiding the requirement of a significant overpotential for drawing current (typically 15 mA cm^{-2}) from the cell under solar illumination. In Figure 16.11*a*, the recombination process is suggested for an electron that is trapped at a surface state in the band gap and subsequently is transferred to the fluctuating energy levels of the oxidized species in the electrolyte, following the Marcus–Gerisher model for electron transfer, as applied to DSC (75–77).

As the operation of the DSC is largely governed by the relative rates of several charge transfer steps, the energetics of the metal-oxide nanoparticle, relative to the molecular species in the surface and in solution, must be carefully controlled. There are several ways to modify the position of the semiconductor energy levels. Because there is no internal bandbending in individual nanoparticles (21), modifying the potential difference at the oxide/electrolyte interface *globally* displaces the energy levels suggested in Figure 16.11*a* with respect to the redox potential E_{redox}. An *upward shift* (to more negative potentials) of the energy levels in the semiconductor in the equilibrium condition of the solar cell, produces and increase of V_{oc}, because at the same electron density the E_{Fn} lies higher up, with respect to E_{redox}. These shifts can be achieved with agents such as 4-*tert*-butylpyridine (TBP) or ammonia that depronotate the TiO_2 surface (39). The TiO_2 energy levels can also be controlled by adsorption of a series of organic molecules with different dipole moments. The change of the conduction band position is illustrated by the displacement of the conductivity plot along the potential axis shown in Figure 16.12 when different molecular modificants are attached to the TiO_2 surface (78). Incidentally this plot shows the sensitivity of the conductivity measurement, described in Figure 16.7, to the changes of the TiO_2 surface, an effect that can be used for sensing as commented later on. In the DSC, a negative shift of the conduction band, however, hampers electron injection from the excited sensitizer (79), and it has been found that increasing the V_{oc} by these means usually produces a

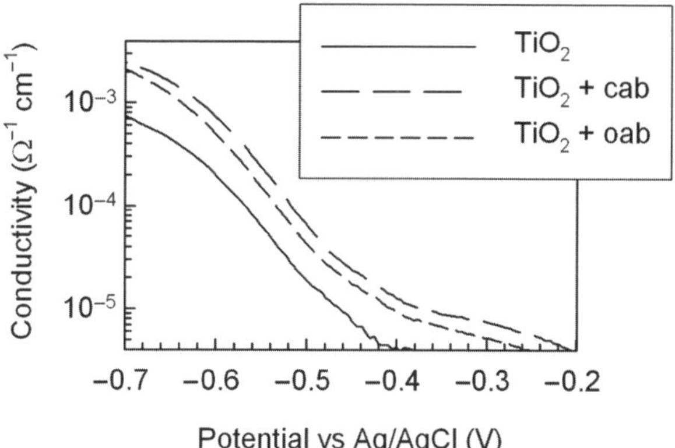

Figure 16.12. Conductivity plot of a bare, mesoporous TiO_2 film and molecular modified films with electrochemical deposited 4-methoxybenzenediazonium tetrafluoroborate (oab) and 4-cyanobenzenediazonium tetrafluoroborate (cab). Adapted from Ref. 78 with permission.

decrease of the photocurrent (39). In addition to these modifications, a light-induced shift of the conduction band at high electron density is possible as discussed before in Figure 16.9*b*.

Another route for improving the photovoltage in a DSC is to block the transference of the photoinjected electrons to the acceptor species in solution, provided that the blocking agents do not significantly decrease electron injection from the dye molecules. This involves the introduction of an insulating layer in the TiO_2 surface that is exposed to the solvent (80), which faces the complexity of mutual effects between the insulating layer and the dye, or with absorbants (e.g., amphiphilic molecules containing carboxylic or phosphonic end groups) that form a more compact monolayer, comprising the dye and coadsorbent, than the dye monolayer alone (81). Recently, it was found (45) that the coadsorption of 4-guanidinobutyric acid with the K-19 sensitizer (Ru (4,4′-dicarboxylic acid-2,2′-bipyridine) (4,4′-bis(p-hexyloxystyryl)-2,2′-bipyridine)$(NCS)_2$) remarkably increases the photovoltage, by a simultaneous shift of the TiO_2 conduction band and blocking of the recombination, without suffering significant decrease of the photocurrent. It has also been observed that one factor controlling the charge recombination dynamics in DSC is the spatial separation of the dye cation HOMO from the TiO_2 surface (82).

Composite material nanoporous electrodes formed by two materials that differ by their conduction band potential have been tested to improve the DSC performance (74,83–89). Arranging these materials in the correct geometry is expected to drive the electrons in the direction of the material having the lower conduction band. A coated matrix design, denoted as core shell, is achieved by the use of ultra-thin ($\leq$1 nm) conformal coating layers deposited onto the surface on the syntered nanocrystalline metal-oxide films. This design ensures free electron diffusion to the current collector by avoiding the introduction of energy barriers (the more negative material) in the transport pathways to the current collector. It has been found that the coating may have several different effects. For example, Nb_2O_5 coat, whose conduction band potential is 100 mV negative than that of the TiO_2 core particles, primarily forms of a “kinetic barrier” for recombination (84). On the other hand, $SrTiO_3$, that should form a 200-mV barrier on TiO_2, produces a shift of the conduction band of the core TiO_2 as a consequence of surface dipole generated at the $TiO_2/SrTiO_3$ interface, and consequently, the open-circuit photovoltage increases, whereas the short-circuit photocurrent decreases (86). It has been suggested that barrier coatings with high points of zero charge (basic coatings) are most effective for the retardation of the recombination dynamics (74). Impedance spectroscopy of Al_2O_3-coated TiO_2 films showed that the thin alumina layer mainly passivates the surface states in the TiO_2 and therefore reduces the recombination rate (90). Transient photovoltage (PV) spectroscopy results also (91) show that ultra-thin alumina layers prevent the formation of defects at the TiO_2 surface (91). It was found that the PV signal increases already at photon energies above 1 eV for the uncoated TiO_2, indicating strong absorption at defect states, whereas the TiO_2/Al_2O_3 films show PV signal only above 3.1 eV, the band gap of TiO_2. The authors suggested (91) that the alumina stabilizes the chemical potential of oxygen at the surface. Recently, Zaban et al. (92) analyzed a new type of high-surface-area TiO_2 electrode for DSCs, consisting of a transparent conductive nanoporous matrix that is coated with a thin layer of TiO_2. This design ensures a several nanometer distance between the TiO_2–electrolyte interface and the current collector throughout the nanoporous electrode, in contrast to several microns associated with the standard electrode. However, it was found that the photovoltaic performance of a TiO_2 coating thinner than 6 nm is rather inefficient (92).

The short-circuit photocurrent in a DSC is a central device parameter that depends mainly on two factors: light harvesting/electron injection and electron transport, provided that electrolyte processes are not limiting the DSC operation.

Strategies have been developed for improving the DSC conversion efficiencies by increasing the light harvesting, especially in the low-energy (red and near-IR) region of the solar spectrum. Dye molecules with high red absorbance have poor injection yields because of the low-lying excited state (LUMO). Increasing the thickness of the film beyond 10 μm in order to increase the absorbance in the red results in an increase in the electron transport length and the recombination rate, thereby decreasing the photocurrent. Furthermore, conventional nanocrystalline TiO_2 films formed by 10–30-nm nanoparticles are poor light-scatterers. The addition of 4–5-μm-thick light scattering layers of 400–nm TiO_2 particles to the conventional nanocrystalline layer enhances light harvesting in the red and near-IR spectral region by enhancing the scattering of light (93–95). TiO_2 photonic crystals and disordered scattering titania inverse opal structures have also been exploited to increase the light conversion efficiency of DSC, improving the conversion efficiency in the spectral range of 600–800 nm (96,97). A theoretical analysis shows improved photocurrent efficiency due to the mirror behavior of the colloidal superlattice (98).

Electron transport in DSC is driven mainly by diffusion, due to the effective electrolyte shielding of space charge. Detailed information on the physical parameters related to transport in DSCs has been obtained using small perturbation techniques at a fixed steady state such as IMPS (38,39) and impedance spectroscopy (26,31,90). It was found that both the effective electron diffusivity D_n and the effective electron lifetime τ_n (75,99) that are measured become a function of the steady state (34,38,43,67,75,100–104). This has usually been considered an effect of the distribution of band gap electronic states that influence the time constants. Variations of both diffusion coefficient and lifetime were attributed to the statistics of electrons in the material, which deviates from dilution, as described by thermodynamic factors (105).

The simplest approach to take trapping into account is the classic multiple trapping (MT) framework (106–109), which has been applied to DSC by several authors (34,38,110–112). In this approach, transport through extended states is slowed down by trapping–detrapping events, whereas direct hopping between localized states is neglected. Electronic states are composed of a transport level (usually identified with the lower edge of the conduction band) at the energy level E_c, with a diffusion coefficient D_0, and a density of localized states $g(E)$ distributed in the band gap. The total electron density is $n = n_c + n_L$, i.e., the sum of electron densities in conduction band (with an effective DOS N_c) and localized states. The mobility decreases rapidly below the level E_c defining the transport states, so that the motion of a bound electron is limited by the rate of thermal excitations to $E \geq E_c$. The varying D_n in this model was recognized as a chemical diffusion coefficient (112,113) that is given by

$$D_n = \left(\frac{\partial n_c}{\partial n_L} \right) D_0 \tag{16.13}$$

The prefactor in Eq. 16.13 is the relationship of free to a trapped number of electrons for a small variation of the Fermi level. This prefactor describes the delay of response of the chemical diffusion coefficient, with respect to the free electrons diffusion coefficient, by the trapping and detrapping process (105,114). The calculation of the chemical diffusion coefficient for an exponential distribution gives an exponential dependence on the Fermi-level position as follows (112,113):

$$D_n = \frac{N_c T_0}{N_L T} \exp \left[(E_{Fn} - E_c) \left(\frac{1}{k_B T} - \frac{1}{k_B T_0} \right) \right] D_0 \tag{16.14}$$

The functional dependence in Eq. 16.14 is a consequence of the exponential variation of the time constant for detrapping when the Fermi level scans the band gap in Figure 16.11.

To illustrate the connection between the DOS and the measured electron diffusion coefficient, we show in Figure 16.13 the chemical capacitance C_μ and the chemical diffusion coefficient D_n, obtained by impedance spectroscopy for three DSCs containing different species in the electrolyte (31). The two elements C_μ and D_n show exponential dependencies in the potential, in agreement with Eqs. 16.9, 16.10, and 16.14. Furthermore, for the different cells, these parameters display a global shift in the voltage axis for the different surface treatments, which demonstrates the global displacement of the TiO_2 energy levels in the energy axis, caused by the absorption of the indicated species, in the same way as in Figure 16.12. Fitting the capacitance of the reference cell, containing LiI electrolyte, in Figure 16.13*a*, in the intermediate domain where the chemical capacitance is observed separately from other capacitive contributions, $k_B T_0 = 62$ mV corresponding to $T/T_0 = 0.42$ at $T = 300$ K is obtained. This predicts an exponent $k_B T/(1 - T/T_0) = 44$ mV for the voltage-dependence of the chemical diffusion coefficient, and this in fact is the value obtained in Figure 16.13*b*. Therefore, the MT gives a consistent description of the observed features of electron transport and accumulation in anatase–TiO_2 DSC. However, the free electron displacement in the conduction band/transport level, predicted by the model when the Fermi level raises to fill most of the traps, has not been separately observed, possibly due to the difficulty of measuring at such negative potential by the bandshift induced by electron charging.

The measurement of the electron lifetime determines the time for the system to recover equilibrium under a small perturbation of the steady state, by removal of the excess carriers by recombination (99). In DSC, the lifetime can be determined by monitoring directly the variation of the position of the Fermi level with time (open-circuit photovoltage decay technique) (75,99). Assuming a process of trapping–detrapping in the bulk of a nanoparticle, and injection to the electrolyte from a single

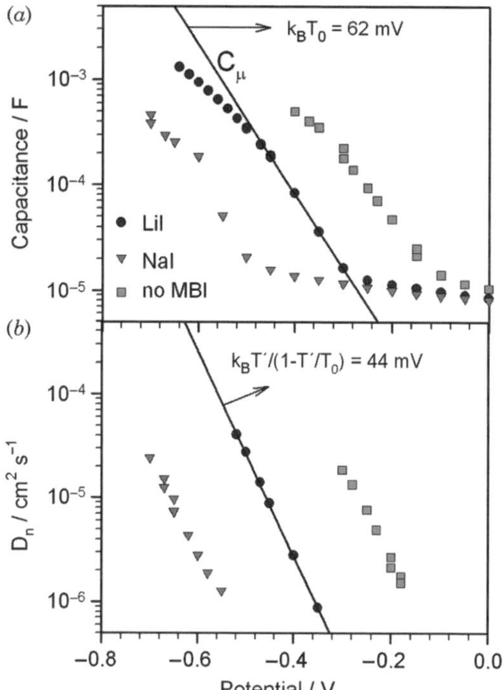

Figure 16.13. Parameters resulting from fit of experimental impedance spectroscopy spectra of three DSCs at different bias potentials in the dark (31). The electrolyte composition in the reference cell is 0.5M LiI, 0.05M I_2, 0.5 1-methylbenzimidazole (MBI) in 3-methoxypropionitrile (3-MPN). The second cell has 0.5M NaI instead of LiI, and the third cell has no MBI. (*a*) Capacitance of the cell without the contribution of the platinum counterelectrode capacitance. (*b*) Chemical diffusion coefficient of electrons. The lines are fits to the multiple trapping/exponential distribution (characteristic temperature T_0) model, and the parameters resulting from fits are indicated. T is the actual temperature of the cell, assumed 300 K; T' is the effective temperature obtained from the slope of the plots as indicated. Adapted from Ref. 31 with permission.

(conduction band) energy level, the time constant for the decay takes the form (99)

$$\tau_n = \left(\frac{\partial n_L}{\partial n_c}\right)\tau_{n0} \tag{16.15}$$

where τ_{n0} is the lifetime of the electron in the (conduction band) injection level. Equation 16.15 predicts an exponential dependence of the lifetime with the Fermi level that is indeed found in many reports (38,39,115). It is observed in Eqs. 16.13 and 16.15 that the factors $(\partial n_L/\partial n_c)$ in D_n and τ_n compensate when forming the diffusion length from measured quantities. The result is a constant

$$L_n = \sqrt{D_n\tau_n} = \sqrt{D_0\tau_{n0}} \tag{16.16}$$

The meaning of the compensation is clear when we note that the origin of the factor $(\partial n_L/\partial n_c)$ lies in carrier equilibration in the energy space, both for chemical diffusion coefficient in MT (D_n) and for measured lifetime (τ_n). Peter et al. (38,115) and Nakade et al. (43) have reported for DSCs the compensating behavior indicated in Eq. 16.16.

Measurements of the electron lifetime over a more extended voltage domain, indicated in Figure 16.14, show a more complex picture. In the important domain of the solar cell operation between 0.6 and 0.8 V, the lifetime depends exponentially on the potential, in agreement with the previously mentioned results. However, at lower potentials, there is a strong variation, and at higher potentials, the lifetime becomes constant. This behavior indicates a combination of several processes that govern electron recombination in DSC (75–77).

Besides the energy disorder that has been already emphasized for describing the diffusion coefficient, the extent of geometry disorder has been recognized as an important property of the nanocrystalline TiO_2 electrodes in DSC, due to the influence on long-range paths for electron transport (116,117). Geometrical disorder may become significant for open structures of the nanoparticulate network, owing to the existence of highly branched nanoparticle structures that influence electron transport dynamics. For example, for compact TiO_2 films (40% porosity) used in DSC, the average coordination number is about 6.6, whereas for open-structured films (80% porosity), the average number of particle interconnections is as low as 2.8 (116). Therefore, when increasing the porosity, the fraction of terminating particles (dead ends) in the TiO_2 film increases markedly, and this has the effect of increasing the average number of particles visited by electrons by 10-fold (117). Figure 16.15 shows

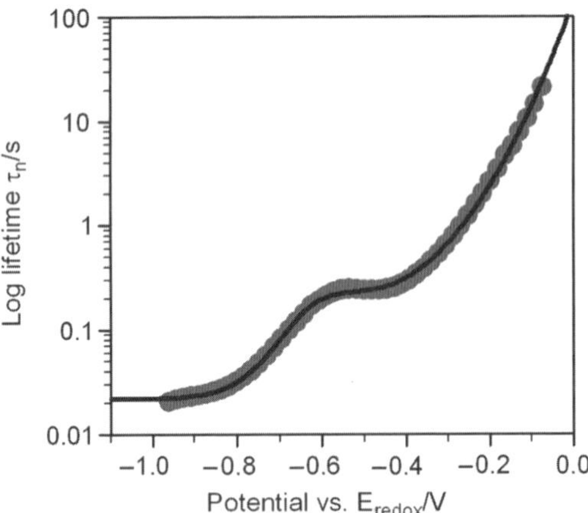

Figure 16.14. Electron lifetime determined from the decay of the cell potential in a DSC after application of a negative bias (>-1.2 V) in the dark. The line corresponds to the fit to a model that uses the quasi-static approximation for trapping in the bulk and a combination of charge transfer processes through conduction band and surface states. Adapted from Ref. 75 with permission.

the reported (44) evolution of the diffusion coefficient of electrons in nanostructured TiO_2, as a function of the excess porosity above the critical value for transport. Samples of different porosity are achieved by consecutive pressing the TiO_2 film, from 7 to 4 μm thickness. Using this method of changing the film thickness, the energy distribution and trapping factors are not significantly dependent on sample porosity. Therefore, the results show a change of the diffusion coefficient that is due to the geometrical effect of the coordination between nanoparticles. The power law behavior with porosity obtained in Figure 16.15 is related to electron diffusion along the percolation cluster in the nanoparticulate network as first suggested by Jao van de Lagemaat et al. (116,117).

Microscopic simulation using Monte-Carlo methods based on the continuous time random walk (CTRW) model has been applied to provide a detailed description of electron trapping dynamics in DSC (116–120). Simulations have described the results of fast optical pump-probe experiments that monitor the decay of the photoinduced dye cation excited by a laser pulse (121–123). Recently CTRW methods have provided valuable insights into the short-range diffusion and charge separation in dye-sensitized nanoscale TiO_2 layers measured by surface photovoltage transients (125,126).

Hole or ion conduction in the medium filling the pores can be realized in different configurations. The 11% conversion efficiency has been reached with the N3 dye, and a liquid electrolyte with iodine/iodide redox couples with a film thickness of over 15 μm. However, the long-term containment at elevated temperatures of the volatile solvent mixture employed remains a major challenge. Therefore, there has been an extensive quest for a solid or quasi-solid medium for replacing the volatile electrolyte while maintaining good performance of the DSC. Several options for the ionic/hole

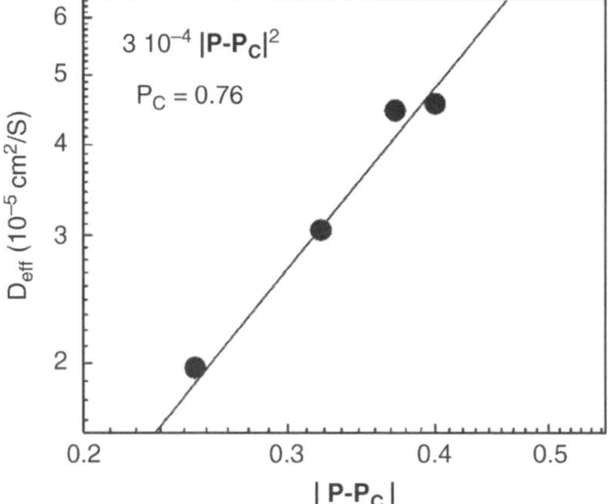

Figure 16.15. Plot of the effective (chemical) diffusion coefficient of electrons, obtained from transient photocurrents, as a function of the porosity of TiO_2 layers. The critical porosity was set to 0.76. Reused from Ref. 44 with permission.

transport material have been investigated (127):

- Inorganic *p*-type semiconductors (128).
- Organic hole-transport materials (HTM) such as spiro-OMeTAD (2,2′7,7′-tetrakis(N,N-di-p-methoxyphenyl-amine)-9,9′-spiro-bifluorene) (71,104,129,130).
- Polymer electrolytes.
- Ionic liquid conductors (19,131–133).

It seems that the I_3^-/I^- couple is unique in providing the strong kinetic asymmetry of recombination/regeneration reactions that gives so far the highest efficieny in DSC (57). Organic hole conductors generally exhibit lower differential kinetics than I_3^-/I^- couple and therefore provide a poorer performance of the solar cell. Room-temperature ionic liquids (RTILs) are solvent free redox systems with high stability and negligible vapor pressure, which are considered a key element for practical applications of the DSC (134,135). The very high density of ions present in these RTILs seems to facilitate the effective screening of the electric charges that are produced under illumination in the mesoporous films (127). DSC using RTIL have shown practical efficiencies in the range 6–7% and an improved stability at high temperatures. The limitation of the slow diffusion of iodide redox species in the viscous RTIL is currently being addressed by the development of novel sensitizers with an increased optical cross section allowing thinner TiO_2 films to be employed (95).

Complete models of the DSC have been developed based on porous battery theory (136,137), considering mass transport in the electrolyte (131,138–142), higher order recombination mechanisms (143) and the porosity effect (144). These methods allow the simulation of the photovoltaic performance of the DSC but involve a multitude of effects and complex calculation. Although V_{oc} and I_{sc} can be separately related to basic processes of the DSC, as discussed, the understanding of the diode characteristics, i.e., the FF, which is critical for device performance, has so far not been described in a simplified framework.

All-solid solar cells based on nanostructured or highly structured metal-oxide materials have been developed in different configurations (9,145–148), either with a thin light-absorber layer or with an absorber/hole transport material embedded in the nanostructured metal oxide. A nanocomposite solar cell, based on interpenetrating networks of TiO_2 and $CuInS_2$, has provided the highest energy conversion efficiency, of around 5%, of this type of solar cells (149).

16.5. NANOSTRUCTURED PHOTO-ELECTROCHEMICAL AND ELECTROCHEMICAL DEVICES BASED ON METAL-OXIDE NANOPARTICLES

The photocatalytic splitting of water into hydrogen and oxygen using solar light is a potentially clean and renewable source for hydrogen fuel. There has been extensive investigation into metal-oxide semiconductors such as TiO_2, WO_3, and Fe_2O_3, which can be used as photoanodes (12,13,150). The extension of the absorption of TiO_2 to the visible spectrum of the solar spectrum, by doping with carbon, nitrogen, or sulfur (151,152), has stimulated great interest, but the use of such materials has provided low conversion efficiencies so far. Nanoparticulate WO_3 electrodes exhibit larger photocurrents under solar illumination (12,150), due to the band gap energy of 2.5 eV that extends the photoresponse of WO_3 electrode into the blue part of the visible spectrum up to 500 nm. The photoelectrochemical behavior of nanostructured TiO_2 and WO_3 electrodes, concerning the photooxidation of organic compounds in water, has been studied in a number of works (153–159). The company Hydrogen Solar Ltd, UK, is trying to implement a tandem cell for the cleavage of water to hydrogen and oxygen by visible light based on work by M. Grätzel and J. Augustynski. The photoactive material in the top cell is a semiconducting oxide that absorbs the blue and green part of the solar emission spectrum and generates oxygen and protons from water with the energy collected. The not absorbed yellow and red light transmits the top cell and enters a DSC (157).

Electrochromism is a reversible process inducing optical transitions in a material via an electrochemical reaction (160). Electrochromism is of importance for applications in information displays with optical memory, smart windows, and anti-dazzling rear view mirrors for cars with variable reflectance. The most studied electrochromic material is WO_3, which turns from transparent to dark blue upon insertion of H, Li, and other guest atoms. Smart windows with switchable glazing allow dynamic control of solar energy gain, adapting the window optical properties to changing environmental conditions. Especially in temperate zone climates, solar radiation in summer conditions imposes an unwanted heat load on buildings, whereas passive heating by solar radiation is welcome in winter (161). One limitation of WO_3 intercalation films for implementation of a glass window with variable optical properties in a large scale is the requirement of external power sources to change the transmittance. To remove this limitation, self-powered smart windows combining a DSC with an WO_3 electrochromic layer counterelectrode have been suggested (2). A switchable photoelectrochromic device has been realized consisting on a dye-covered nanoporous TiO_2 layer, which is situated on a nanoporous WO_3 electrochromic layer (162–164). The layers can be kept thin, so that they are transparent. The transmittance of the device decreases when illuminated and can be increased under short-circuit conditions. Therefore, no external voltage source is needed to color/bleach the photoelectrochromic device.

Another approach to electrochromic devices that has proved successful for electrical displays applications was originally developed by M. Grätzel and D. Fitzmaurice (10). It consists of nanocrystalline TiO_2 films derivatized with a redox active molecule, such as a viologen endowed with an anchoring group to attach it firmly to the TiO_2 surface. Viologens (e.g., 1,1′-disubstituted 4,4′-bipyridinium dications) work as redox chromophores. The first reduction of the viologen dication is highly reversible and leads to the formation of the intensely deep blue-colored radical cation. The devices present rapid switching between the bleached and the colored states, high contrast ratios, and good cycling stability (27,165–169), due to the amplification of the optical phenomena by the high surface area of the nanocrystalline support, and fast interfacial electron transfer between the nanocrystalline oxide and the adsorbed modifier. Photoelectrochemical studies have shown (27) that the electronic charging of the semiconductor nanostructure is the key factor mediating between the electrode potential and coloration of the molecular monolayers anchored on the semiconductor surface. Displays based in this technology are close to commercialization by the company NTera, which announced in 2006 the successful production of working prototypes featuring the world's highest resolution naturally reflective electronic displays and plans to produce a high-contrast, monochromatic display that can replace the simple liquid crystal displays used in clocks, thermostats, and many other devices.

The optical and visible-sensitizing functions of molecular layers absorbed onto nanocrystalline metal oxides have been widely exploited in DSC and electrochromics, as commented above. Recently a new approach has been taken toward electric and electro-optic devices formed by molecular functionalized mesoscopic oxides, such as sensors and switchable molecular electronics, in which the molecular layer adsorbed in the surface serves instead as an electron or hole transport relay (170–174). It was shown that the molecular layer can be charged from the conductive substrate either with electrons or holes, depending on the applied bias potential (171,172). Figure 16.16 shows impedance spectroscopy results of mesoscopic Al_2O_3 and TiO_2 networks, covered with a monolayer of Ru-complex *cis*-RuLL$'$(NCS)$_2$ (L = 2,2$'$-bipyridyl-4,4$'$-dicarboxylic acid, $L' = 4,4'$-dinonyl-2,2$'$-bipyridyl) (Z907) (30). The results display the behavior of the model of Figure 16.5, at potentials in which injection of carriers in the respective semiconductor networks is forbidden. These results therefore show the injection, transport, and accumulation of electrons and holes in the molecular layers covering the surface of the mesoscopic oxides (30).

TiO_2 is biocompatible and combines selectively with some groups of biomolecules. Nanocrystalline TiO_2 has been employed for the immobilization of biomolecules (4,69,175), both for studying protein electrochemistry and for developing electrochemical biosensors, i.e., analytical devices consisting of a biological recognition element attached to the nanostructured TiO_2 that serves as a transduction element, relating the concentration of an analyte to a measurable response (176).

The conductivity of nanostructured TiO_2 in the gas phase (177–180) and as a function of ambient humidity has been described (181). The group of J. Augustynski has developed nanoparticles of semiconducting metallic oxides employed in gas sensors able to detect nitrogen oxides, carbon monoxide, and hydrocarbons. These nanocrystalline oxides (not disclosed, but apparently comprising WO_3 nanoparticles) are used by the Company MicroChemical Systems for manufacturing gas sensors for the automotive industry. The two types of microsensors measure changes in the electrical resistance of nanocrystalline films induced by road gas pollutants, resulting in improved sensitivity and response time of the microsensors fabricated by MicroChemical Systems, which has already delivered such microsensors in large quantities to car makers.

The use of nanosized positive or negative electrode materials (vs. Li) for Li batteries (14,182) offers attractive electrochemical characteristics over classic bulk materials, including very short diffusion length, low charge transfer resistance due

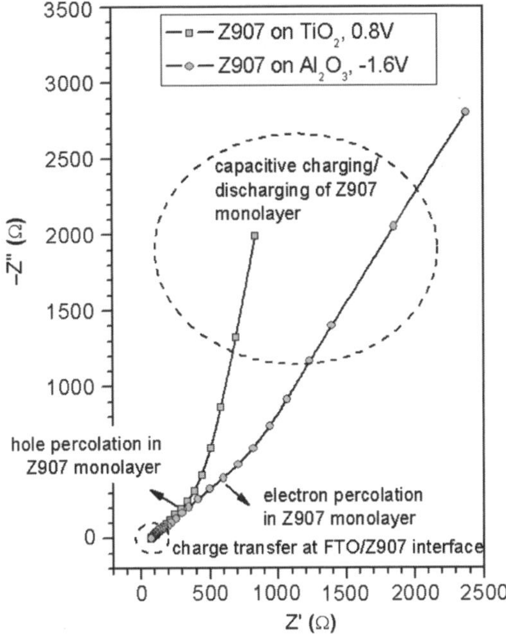

Figure 16.16. Experimental impedance plots of Z907-derivatized mesoscopic TiO_2 and Al_2O_3. Reprinted from Ref. 30 with permission.

to the high surface area of the active mass (182) and a better accommodation of the materials strains associated with the lithium insertion/extraction. The active material/electrolyte interface can be improved by acting on the current collector surface. Nano-architectured current collectors offer higher contact area than conventional 2D-substrate. As a result, for an equivalent thickness, the amount of active material is much higher on a nanostructured current collector. However, an adverse effect in primary nanoparticles is the possibility of increasing side reactions with the electrolyte, while the Li-reaction may be enhanced (183). TiO_2 nanotubes or nanowires have shown excellent Li-intercalation charge capacity and cyclability (184,185). The Li^+ intercalation renders the nanowires simultaneously electronically and ionically conducting (185). As in other devices commented on, the electrolyte is a crucial element in Li-batteries. It has been found that the addition of nanoparticle fillers, such as Al_2O_3 or TiO_2, increases several fold the conductivity of polymer electrolytes at 60–80°C, and prevents crystallization at room temperature (186).

Supercapacitors are high-power density devices that can be coupled with batteries to provide peak power and can replace batteries for memory backup. High-surface-area carbon-based materials are widely used for electrochemical double-layer supercapacitors. A nanostructured $Li_4Ti_5O_{12}$ has been developed as a negative electrode material suitable for use with activated carbon-positive electrodes (187,188). The combined effect of a high output voltage and an anode of greater specific capacity results in higher energy densities than carbon/carbon supercapacitors, while maintaining a high-power density and robustness.

16.6. ORDERED NANOPARTICULATE STRUCTURES

Quantum dots are nanocrystals of size roughly between 1 and 10 nm. In these crystals, the electron wave functions are delocalized in the limited space of the nanocrystal, and the electronic orbital has the same symmetry as in a conventional atom. The strong quantum confinement implies that insulating nanocrystals have a set of discrete atom-like valence and conduction energy levels. Their separation as well as the optical gap between the lowest conduction level and the highest valence level increases as the nanocrystal size decreases. Accordingly, electronic and electrooptic properties of quantum dots can be tailored by the dimensions.

Colloidal nanocrystals can be considered as building blocks for larger arquitectures. Nanocrystals with a narrow size dispersion can be assembled into two-dimensional or three-dimensional ordered arrays, which are termed nanocrystal superlattices, nanocrystal solids, or quantum dot solids. The particles are characterized by an inorganic core and stabilized by an organic surfactant, and both the core and the surfactant have a role in determining the superlattice crystallographic symmetry. As both the preparation of nanocrystals and the ordered assemblies is versatile, there is currently great interest in the investigation of the properties of the quantum dot solids (15,16,189).

The collective properties of electrons injected in a superlattice depend on individual electronic levels for a quantum dot, on the degree of electronic coupling between the dots, and on Coulomb interactions, both for multiple electrons in a quantum dot and in neighbor dots (16). Assemblies of quantum dots deposited on a conducting substrate, and filled with an ionic medium, allow the injection of a single kind of carrier in the superlattice. The Coulomb energy of electrons in a dot is considerably reduced by screening with electrolyte ions, as commented in connection to Figures 16.1 and 16.2. The energy for addition of electrons to a quantum dot contains several contributions (190,191): the self-energy of the electron interacting with its image charge at the surface of the particle (192); the electrostatic interactions, which are screened by the external medium; and the exchange energy. Characteristic capacitance spectra showing consecutive charging of different discrete energy levels have been reported for assemblies of CdSe (193,194) and ZnO colloids (195).

An ordered array of quantum dots could give rise to long-range coherent transport in delocalized bands, provided that sufficient electronic coupling exists between the same orbitals in neighbor dots. However, several factors lead to localization of carriers in the quantum dots (189). Colloidal nanoparticles prepared by wet chemical methods fluctuate in size. As the electronic response of an individual particle is determined by its size, there is some inherent disorder in an array of quantum dots, which produces a dispersion of the energy levels (189). Dispersion of energy levels can be also caused by a fluctuation of chemical composition (e.g., electronic surface states). The difference between energy levels in different quantum dots makes tunneling of electrons between quantum dots more difficult. Moreover, such dispersion leads to the random scattering of electronic waves. Thus, dispersion of quantum dots sizes and fluctuation of the chemical composition produce Anderson localization. Disorder in the spatial distribution of quantum dots also contributes to localization (196). The Coulomb repulsion between two electrons (holes) sitting on the same quantum dot introduces an energy gap between occupied and empty states (Hubbard insulator). Several characteristic phenomena for transport in disordered semiconductors have been reported in arrays of quantum dots, such as metal-insulator transitions (15,197) and Mott-like conduction gaps (62,196), the variable range hopping (VRH) transport, (198) temperature scaling in current-potential curves (199), and percolation thresholds depending on disorder (200). Experimental studies of arrays of monodisperse ZnO quantum dots, charge-compensated in electrolyte solution, have been reported (195,201,202). The influence of size dispersion over collective transport properties in these systems has been examined theoretically (203).

Ordered mesoporous materials with periodicity on an optical length scale can lead to materials that exhibit a full photonic band gap, i.e., a wavelength band in which photon propagation is forbidden (204). These photonic materials can be integrated into substrates of technological interest (205), and can be used in applications in which the manipulation of the flow of visible and near-IR light is required, as commented before for improving light-harvesting in DSC. TiO_2 is an attractive material for the fabrication of photonic band gap materials, especially the inverse opal, air spheres in TiO_2 matrix, due to the high refractive index, and good transparency of TiO_2 for the visible light. Inverse titania opal photonic crystals have shown how to effectively control the spontaneous emission from semiconductor quantum dots embedded in the photonic crystal (206). In these experiments, the crystal lattice parameter could be used to enhance the delay of the decay rates of light emission from the quantum dots. Dye-sensitized and solid-state solar cells have been fabricated on microporous titania inverse opal structures (207,208). The electrochemical behavior of these structures under Li intercalation was investigated (209).

16.7. NANOWIRES AND NANOTUBES

Nanowires and nanotubes are drawing tremendous attention due to their potential applications in various nanoscale devices. Among the oxide materials, those having a bulk crystalline structure in the form of layers or ribbons can be more easily induced to grow in the form of rods or tubes with a high aspect ratio and nanometric dimensions (17).

ZnO nanostructures such as nanowires, nanotubes, and nanorings have been grown successfully via a variety of methods, including chemical vapor deposition, thermal evaporation, and electrodeposition (210). Zinc oxide is a nontoxic n-type semiconductor that has favorable band energies for forming heterojunctions with hole-conducting polymers and can be grown as nanorod arrays with the appropriate dimensions for efficient nanorod-hole conductor devices. Vertically aligned ZnO nanorods, formed by electrodeposition on a transparent conducting oxide, have attracted much attention for applications in UV-optoelectronic devices such as light emitting diodes (11). Due to its near-cylindrical geometry and large refractive index, ZnO nanowire/nanorod is a natural candidate for optical waveguide and for optical resonant cavities that facilitate lasing in well-aligned ZnO nanowires (211).

With respect to solar cells based on nanoparticulate metal oxides, nanowire structures have the advantage of providing long and uninterrupted paths for electron transport while maintaining a high area density (212–214). ZnO nanowires have been used to form high heterojunction area solar cells (9,147,212,215); in particular, CdSe-sensitized CuSCN/nanowire ZnO heterojunctions have shown an encouraging photovoltaic performance (9). There has been recent progress in obtaining highly ordered transparent TiO_2 nanotube arrays for DSC with high electron lifetimes and excellent pathways for electron percolation (216). Carbon-doped TiO_2 ($TiO_{2-x}C_x$) nanotube arrays have also shown high photocurrent densities and more efficient water splitting under visible-light illumination than TiO_2 nanoparticles (217).

Recent literature results (218–220) have demonstrated that V oxides, such as V_2O_5, can be produced in the form of nanowires by means of different techniques, i.e., by hydrothermal, sol-gel, sol electrophoretic, or sol electrochemical deposition. V_2O_5 nanowires have many intrinsic merits, such as a uniform geometric cross section of $1.5\,nm \times 10\,nm$ and several micrometer length, which are important in nanofunctional materials. The heterojunction of V_2O_5 nanowires with carbon nanotube, field-effect transistors, and actuators as artificial muscles have been demonstrated as feasible electronic components (221). The fibers show a double layer structure, with each layer consisting of two V_2O_5 sheets. A V_2O_5 fiber is composed of only four atomic layers of vanadium, which therefore represents a wire with a thickness of molecular dimension. The conductivity of one V_2O_5 fiber was estimated to be $\sim$0.5 S/cm at room temperature (220). The electrochemical Li-intercalation behavior of nanoscale V_2O_5 was found to depend markedly on the morphology and defect structure (222).

Nanowires and nanotubes are also encouraging structures for chemical and biological sensors in which detection can be monitored electrically and/or optically. For example, when helium gas is absorbed physically into the interlayer or on the surface of soft V_2O_5 nanowires, it is possible to detect the helium gas from the variation of conductance through the nanowires (223). With molecular receptors or a selective membrane for the analyte of interest, the binding of a charged species to the surface of a nanowire or nanotube can lead to depletion or accumulation of carriers in the "bulk" of the nanometer diameter structure and increase sensitivity to the point that single-molecule detection is possible (224).

ACKNOWLEDGMENTS

The author is grateful to Jan Augustynski, Franco Decker, Michael Grätzel, Ian S. Plumb, Daniel Vanmaekelbergh, and Arie Zaban for their useful comments on nanostructured devices. The work was supported by Ministerio de Educación y Ciencia of Spain under project MAT2004-05168.

REFERENCES

(1) O' Regan, B.; Grätzel, M. *Nature* **1991**, *353*, 737.

(2) Bechinger, C.; Ferrer, S.; Zaban, A.; Sprague, J.; Gregg, B.A. *Nature* **1996**, *383*, 608.

(3) Cinnsealach, R.; Boschloo, G.; Rao, S.N.; Fitzmaurice, D. *Sol. Energy Mater. Sol. Cells* **1999**, *57*, 107.

(4) Topoglidis, E.; Palomares, E.; Astuti, Y.; Green, A.; Campbell, C.J.; Durrant, J.R. *Eletroanalysis* **2005**, *17*, 1035.

(5) Grätzel, M. *Nature* **2001**, *414*, 338.

(6) Grätzel, M. *J. Photochem. Photobiol. A: Chem.* **2004**, *164*, 3.

(7) Bisquert, J.; Cahen, D.; Rühle, S.; Hodes, G.; Zaban, A. *J. Phys. Chem. B* **2004**, *108*, 8106.

(8) Hagfeldt, A.; Grätzel, M. *Chem. Rev.* **1995**, *95*, 49.

(9) Lévy-Clément, C.; Tena-Zaera, R.; Ryan, M.-A.; Katty, A.; Hodes, G. *Adv. Mat.* **2005**, *17*, 1512.

(10) Grätzel, M. *Nature* **2001**, *409*, 575.

(11) Könenkamp, R.; Word, R.C.; Godinez, M. *Nano Lett.* **2005**, *5*, 2005.

(12) Wang, H.; Lindgren, T.; He, J.; Hagfeldt, A.; Lindquist, S.-E. *J. Phys. Chem. B* **2000**, *104*, 5686.

(13) Jorand-Sartoretti, C.; Alexander, B.D.; Solarska, R.; Rutkowska, I.A.; Augustynski, J.; Cerny, R. *J. Phys. Chem. B* **2005**, *109*, 13685.

(14) Aricò, A.S.; Bruce, P.; Scrosati, B.; Tarascon, J.M.; van Schalkwijk, W. *Nat. Mat.* **2005**, *4*, 366.

(15) Collier, C.P.; Vossmeyer, T.; Heath, J.R. *Ann. Rev. Phys. Chem.* **1998**, *49*, 371.

(16) Vanmaekelbergh, D.; Liljerorth, P. *Chem. Soc. Rev.* **2005**, *34*, 299.

(17) Patzke, G.; Krumeich, F.; Nesper, R. *Angew. Chem. Int. Ed.* **2002**, *41*, 2446.

(18) Huang, S.Y.; Kavan, L.; Exnar, I.; Grätzel, M. *J. Electrochem. Soc.* **1995**, *142*, L142.

(19) Wang, P.; Zakeeruddin, S.M.; Exnar, I.; Grätzel, M. *Chem. Commun.* **2002**, 2972.

(20) Zaban, A.; Micic, O.I.; Gregg, B.A.; Nozik, A.J. *Langmuir* **1998**, *14*, 3153.

(21) Bisquert, J.; Garcia-Belmonte, G.; Fabregat Santiago, F. *J. Solid State Electrochem.* **1999**, *3*, 337.

(22) Winter, M.; Brodd, R.J. *Chem. Rev.* **2004**, *104*, 4245.

(23) Bisquert, J.; Garcia-Belmonte, G.; Fabregat-Santiago, F.; Ferriols, N.S.; Bogdanoff, P.; Pereira, E.C. *J. Phys. Chem. B* **2000**, *104*, 2287.

(24) Bisquert, J. *J. Phys. Chem. B* **2002**, *106*, 325.

(25) Pitarch, A.; Garcia-Belmonte, G.; Mora-Seró, I.; Bisquert, J. *Phys. Chem. Chem. Phys.* **2004**, *6*, 2983.

(26) Fabregat-Santiago, F.; Garcia-Belmonte, G.; Bisquert, J.; Zaban, A.; Salvador, P. *J. Phys. Chem. B* **2002**, *106*, 334.

(27) García-Cañadas, J.; Fabregat-Santiago, F.; Kapla, J.; Bisquert, J.; Garcia-Belmonte, G.; Mora-Seró, I.; Edwards, M.O.M. *Electrochim. Acta* **2004**, *49*, 745.

(28) Gregg, B.A. *Coord. Chem. Rev.* **2004**, *248*, 1215.

(29) Grätzel, M. *Curr. Appl. Phy.* in press.

(30) Bisquert, J.; Grätzel, M.; Wang, Q.; Fabregat-Santiago, F. *J. Phys. Chem. B* in press.

(31) Fabregat-Santiago, F.; Bisquert, J.; Garcia-Belmonte, G.; Boschloo, G.; Hagfeldt, A. *Sol. Energy Mater. Sol. Cells* **2005**, *87*, 117.

(32) Wang, Q.; Moser, J.-E.; Grätzel, M. *J. Phys. Chem. B* **2005**, *109*, 14945.

(33) Hoshikawa, T.; Kikuchi, R.; Eguchi, K. *J. Electroanal. Chem.* **2006**, *588*, 59.

(34) de Jongh, P.E.; Vanmaekelbergh, D. *Phys. Rev. Lett.* **1996**, *77*, 3427.

(35) Vanmaekelbergh, D.; Iranzo Marín, F.; van de Lagemaat, J. *Ber. Bunsenges. Phys. Chem.* **1996**, *100*, 616.

(36) Vanmaekelbergh, D.; de Jongh, P.E. *Phys. Rev. B* **2000**, *61*, 4699.

(37) Dloczik, L.; Ileperuma, O.; Lauerman, I.; Peter, L.M.; Ponomarev, E.A.; Redmond, G.; Shaw, N.J.; Uhlendorf, I. *J. Phys. Chem. B* **1997**, *101*, 10281.

(38) Fisher, A.C.; Peter, L.M.; Ponomarev, E.A.; Walker, A.B.; Wijayanthan, K.G.U. *J. Phys. Chem. B* **2000**, *104*, 949.

(39) Schlichthörl, G.; Huang, S.Y.; Sprague, J.; Frank, A.J. *J. Phys. Chem. B* **1997**, *101*, 8141.

(40) Solbrand, A.; Lindström, H.; Rensmo, H.; Hagfeldt, A.; Lindquist, S.E.; Södergren, S. *J. Phys. Chem. B* **1997**, *101*, 2514.

(41) Solbrand, A.; Henningsson, A.; Sodergren, S.; Lindstrom, H.; Hagfeldt, A.; Lindquist, S.-E. *J. Phys. Chem. B* **1999**, *103*, 1078.

(42) Noack, V.; Weller, H.; Eychmüller, A. *J. Phys. Chem. B* **2002**,

(43) Nakade, S.; Saito, Y.; Kubo, W.; Kitamura, T.; Wada, Y.; Yanagida, S. *Electrochem. Commun.* **2003**, *5*, 804.

(44) Dittrich, T.; Ofir, A.; Tirosh, S.; Grinis, L.; Zaban, A. *Appl. Phys. Lett.* **2006**, *88*, 182110.

(45) Zhang, Z.; Zakeeruddin, S.M.; O'Regan, B.C.; Humphry-Baker, R.; Grätzel, M. *J. Phys. Chem. B* **2005**, *109*, 21818.

(46) Kern, R.; Sastrawan, R.; Ferber, J.; Stangl, R.; Luther, J. *Electrochim. Acta* **2002**, *47*, 4213.

(47) Mora-Seró, I.; Bisquert, J.; Fabregat-Santiago, F.; Garcia-Belmonte, G.; Zoppi, G.; Durose, K.; Proskuryakov, Y.; Oja, I.; Belaidi, A.; Dittrich, T.; Tena-Zaera, R.; Katty, A.; Lévy-Clement, C.; Barrioz, V.; Irvine, S.J.C. *Nano Lett.* in press.

(48) Fabregat-Santiago, F. *PhD Thesis*; Universitat Jaume I: Castelló, 2001.

(49) de Levie, R. In *Adv. Electrochem. Electrochem. Eng., Vol. 6*; Delahay, P. (Editor); Interscience: New York, 1967, 329.

(50) Raistrick, I.D. *Electrochim. Acta* **1990**, *35*, 1579.

(51) Pajkossy, T. *J. Electroanal. Chem.* **1994**, *364*, 111.

(52) Pajkossy, T.; Wandlowski, T.; Kolb, D.M.J. *J. Electroanal. Chem.* **1996**, *414*, 209.

(53) Kötz, R.; Carlen, M. *Electrochim. Acta* **2000**, *45*, 2483.

(54) Lampert, M.A.; Mark, P. *Current Injection in Solids*; Academic Press: New York, 1970.

(55) Blom, P.W.M.; Vissenberg, M.C.J.M. *Mat. Sci. Eng. R* **2000**, *27*, 53.

(56) Mihailetchi, V.D.; Wildeman, J.; Blom, P.W.M. *Phys. Rev. Lett.* **2005**, *94*, 126602.

(57) Peter, L.M. *J. Electroanal. Chem.* in press.

(58) Mora-Seró, I.; Bisquert, J. *Sol. Energy Mater. Sol. Cells* **2005**, *85*, 51.

(59) Bisquert, J. *Phys. Chem. Chem. Phys.* **2000**, *2*, 4185.

(60) Abayev, I.; Zaban, A.; Fabregat-Santiago, F.; Bisquert, J. *Phys. Stat. Sol. A* **2003**, *196*, R4.

(61) Meulenkamp, E.A. *J. Phys. Chem. B* **1999**, *103*, 7831.

(62) Roest, A.L.; Kelly, J.J.; Vanmaekelbergh, D. *Appl. Phys. Lett.* **2003**, *83*, 5530.

(63) Agrell, H.G.; Boschloo, G.; Hagfeldt, A. *J. Phys. Chem. B* **2004**, *108*, 12388.

(64) Bisquert, J. *Phys. Chem. Chem. Phys.* **2003**, *5*, 5360.

(65) Kytin, V.G.; Bisquert, J.; Abayev, I.; Zaban, A. *Phys. Rev. B* **2004**, *70*, 193304.

(66) Fabregat-Santiago, F.; Mora-Seró, I.; Garcia-Belmonte, G.; Bisquert, J. *J. Phys. Chem. B* **2003**, *107*, 758.

(67) Niinobe, D.; Makari, Y.; Kitamura, T.; Wada, Y.; Yanagida, S. *J. Phys. Chem. B* **2005**, *109*, 17892.

(68) Randriamahazaka, H.; Fabregat-Santiago, F.; Zaban, A.; García-Cañadas, J.; Garcia-Belmonte, G.; Bisquert, J. *Phys. Chem. Chem. Phys.* **2006**, *8*, 1827–1833.

(69) Li, Q.; Luo, G.; Feng, J. *Eletroanalysis* **2001**, *13*, 359.

(70) Bisquert, J.; Zaban, A. *Appl. Phys. A* **2003**, *77*, 507.

(71) Bach, U.; Lupo, D.; Comte, P.; Moser, J.E.; Weissörtel, F.; Salbeck, J.; Spreitzer, H.; Grätzel, M. *Nature* **1998**, *398*, 583.

(72) Peter, L.M.; Duffy, N.W.; Wang, R.L.; Wijayantha, K.G.U. *J. Electroanal. Chem.* **2002**, *524–525*, 127.

(73) Boschloo, G.; Hagfeldt, A. *J. Phys. Chem. B* **2005**, *109*, 12093.

(74) Palomares, E.; Clifford, J.N.; Haque, S.A.; Lutz, T.; Durrant, J.R. *Am. Chem. Soc.* **2003**, *125*, 475.

(75) Bisquert, J.; Zaban, A.; Greenshtein, M.; Mora-Seró, I. *J. Am. Chem. Soc.* **2004**, *126*, 13550.

(76) Bisquert, J.; Zaban, A.; Salvador, P. *J. Phys. Chem. B* **2002**, *106*, 8774.

(77) Salvador, P.; González-Hidalgo, M.; Zaban, A.; Bisquert, J. *Phys. Chem. B* **2005**, *109*, 15915.

(78) Ruhle, S.; Greenshtein, M.; Chen, S.-G.; Merson, A.; Pizem, H.; Sukenik, C.S.; Cahen, D.; Zaban, A. *J. Phys. Chem. B* **2005**, *109*, 18907.

(79) Lenzmann, F.; Krueger, J.; Burnside, S.; Brooks, K.; Grätzel, M.; Gal, D.; Rühle, S.; Cahen, D. *J. Phys. Chem. B* **2001**, *105*, 6347.

(80) Gregg, B.A.; Pichot, F.; Ferrere, S.; Fields, C.L. *J. Phys. Chem. B* **2001**, *105*, 1422.

(81) Wang, P.; Zakeeruddin, S.M.; Comte, P.; Charvet, R.; Humphry-Baker, R.; Grätzel, M. *J. Phys. Chem. B* **2003**, *107*, 14336.

(82) Clifford, J.N.; Palomares, E.; Nazeeruddin, M.K.; Grätzel, M.; Nelson, J.; Li, X.; Long, N.J.; Durrant, J.R. *J. Am. Chem. Soc.* **2004**, *126*, 5225.

(83) Kumara, G.R.R.A.; Tennakone, K.; Perera, V.P.S.; Konno, A.; Kaneko, S.; Okuya, M. *J. Phys. D: Appl. Phys.* 2001, *34*, 868.

(84) Zaban, A.; Chen, S.G.; Chappel, S.; Gregg, B.A. *Chem. Commun.* 2000, *22*, 2231.

(85) Chappel, S.; Chen, S.G.; Zaban, A. *Langmuir* **2002**, *18*, 3336.

(86) Diamant, Y.; Chen, S.G.; Melamed, O.; Zaban, A. *J. Phys. Chem. B* **2003**, *107*, 1977.

(87) Palomares, E.; Clifford, J.N.; Haque, S.A.; Lutz, T.; Durrant, J.R. *Chem. Commun.* **2002**, 1464.

(88) Lenzmann, F.; Nanu, M.; Kijatkina, O.; Belaidi, A. *Thin Solid Films* **2004**, *451–452*, 639.

(89) Zhang, X.T.; Liu H.-W., Taguchi, T.; Meng, R.-B.; Sato, O.; Fujishima, A.; *Sol. Energy Mater. Sol. Cells* **2004**, *81*, 197.

(90) Fabregat-Santiago, F.; García-Cañadas, J.; Palomares, E.; Clifford, J.N.; Haque, S.A.; Durrant, J.R.; Garcia-Belmonte, G.; Bisquert, J. *J. Appl. Phys.* **2004**, *96*, 6903.

(91) Dittrich, T.; Muffler, H.-J.; Vogel, M.; Guminskaya, T.; Ogacho, A.; Belaidi, A.; Strub, E.; Bohne, W.; Röhrich, J.; Hilt, O.; Lux-Steiner, M.C. *Appl. Surf. Sci.* **2005**, *240*, 236.

(92) Chappel, S.; Grinis, L.; Ofir, A.; Zaban, A. *J. Phys. Chem. B* **2005**, *109*, 1643.

(93) Usami, A.; *Chem. Phys. Lett.* **1997**, *277*, 105.

(94) Rothenberger, G.; Comte, P.; Grätzel, M. *Sol. Energy Mater. Sol. Cells* **1999**, *58*, 321.

(95) Kuang, D.; Ito, S.; Wenger, B.; Klein, C.; Moser, J.E.; Humphry-Baker, R.; Zakeeruddin, S.M.; Grätzel, M. *J. Am. Chem. Soc.* **2006**, *128*, 4146.

(96) Nishimura, S.; Abrams, N.; Lewis, B.A.; Halaoui, L.I.; Mallouk, T.E.; Benkstein, K.D.; van de Lagemaat, J.; Frank, A.J. *J. Am. Chem. Soc.* **2003**, *125*, 6306.

(97) Halaoui, L.I.; Abrams, N.; Mallouk, T.E. *J. Phys. Chem. B* **2005**, *109*, 6334.

(98) Mihi, A.; López-Alcaraz, F.J.; Míguez, H. *Appl. Phys. Lett.* **2006**, *88*, 193110.

(99) Zaban, A.; Greenshtein, M.; Bisquert, J. *ChemPhysChem.* **2003**, *4*, 859.

(100) Cao, F.; Oskam, G.; Meyer, G.J.; Searson, P.C. *J. Phys. Chem.* **1996**, *100*, 17021.

(101) Kambe, S.; Nakade, S.; Kitamura, T.; Wada, Y.; Yanagida, S. *J. Phys. Chem. B* **2002**, *106*, 2967.

(102) Nakade, S.; Kambe, S.; Matsuda, M.; Saito, Y.; Kitamura, T.; Wada, Y.; Yanagida, S. *Physica E* **2002**, *14*, 210.

(103) Nakade, S.; Saito, Y.; Kubo, W.; Kitamura, T.; Wada, Y.; Yanagida, S. *J. Phys. Chem. B* **2003**, *107*, 8607.

(104) Krüger, J.; Plass, R.; Grätzel, M.; Cameron, P.J.; Peter, L.M. *J. Phys. Chem. B* **2003**, *107*, 7536.

(105) Bisquert, J.; Vikhrenko, V.S. *J. Phys. Chem. B* **2004**, *108*, 2313.

(106) Tiedje, T.; Rose, A. *Solid State Commun.* **1981**, *37*, 49.

(107) Tiedje, T.; Cebulka, J.M.; Morel, D.L.; Abeles, B. *Phys. Rev. Lett.* **1981**, *46*, 1425.

(108) Orenstein, J.; Kastner, M. *Phys. Rev. Lett.* **1981**, *46*, 1421.

(109) Michiel, H.; Adriaenssens, G.J.; Davis, E.A. *Phys. Rev. B* **1986**, *34*, 2486.

(110) Kambili, A.; Walker, A.B.; Qiu, F.L.; Fisher, A.C.; Savin, A.D.; Peter, L.M. *Physica E* **2002**, *14*, 203.

(111) van de Lagemaat, J.; Frank, A.J. *J. Phys. Chem. B* **2000**, *104*, 4292.

(112) van de Lagemaat, J.; Kopidakis, N.; Neale, N.R.; Frank, A.J. *Phys. Rev. B* **2005**, *71*, 035304.

(113) Bisquert, J. *J. Phys. Chem. B* **2004**, *108*, 2323.

(114) Rose, A. *Concepts in Photoconductivity and Allied Problems*; Interscience: New York, **1963**.

(115) Peter, L.M.; Wijayantha, K.G.U. *Electrochem. Commun.* **1999**, *1*, 576.

(116) van de Lagemaat, J.; Benkstein, K.D.; Frank, A.J. *J. Phys. Chem. B* **2001**, *105*, 12433.

(117) Benkstein, K.D.; Kopidakis, N.; van de Lagemaat, J.; Frank, A.J. *J. Phys. Chem. B* **2003**, *107*, 7759.

(118) Anta, J.A.; Nelson, J.; Quirke, N. *Phys. Rev. B* **2002**, *65*, 125324.

(119) Nelson, J. *Phys. Rev. B* **1999**, *59*, 15374.

(120) van de Lagemaat, J.; Frank, A.J. *J. Phys. Chem. B* **2001**, *105*, 11194.

(121) Nelson, J.; Haque, S.A.; Klug, D.R.; Durrant, J.R. *Phys. Rev. B* **2001**, *63*, 205321.

(122) Tachibana, Y.; Haque, S.A.; Mercer, I.P.; Durrant, J.R.; Klug, D.R. *J. Phys. Chem. B* **2000**, *104*, 1198.

(123) Nogueira, A.F.; De Paoli, M.-A.; Montaneri, I.; Monkhouse, R.; Nelson, J.; Durrant, J.R. *J. Phys. Chem. B* **2001**, *105*, 7517.

(124) Mora-Seró, I.; Anta, J.A.; Garcia-Belmonte, G.; Dittrich T.; Bisquert, J. *J. Photochem. Photobiol. A: Chem.* in press.

(125) Mora-Seró, I.; Dittrich, T.; Belaidi, A.; Garcia-Belmonte, G.; Bisquert, J. *J. Phys. Chem. B* **2005**, *109*, 8035.

(126) Dittrich, T.; Mora-Seró, I.; Garcia-Belmonte, G.; Bisquert, J. *Phys. Rev. B* **2006**, *73*, 045407.

(127) Li, B.; Wang, L.; Kang, B.; Wang, P.; Qiu, Y. *Sol. Energy Mater. Sol. Cells* **2006**, *90*, 549.

(128) Tennakone, K.; Kumara, G.R.R.A.; Kottegoda, I.R.M.; Wijayantha, K.G.U.; Perera, V.P.S. *J. Phys. D: Appl. Phys.* **1998**, *31*, 1492.

(129) Krüger, J.; Plass, R.; Cevey, L.; Piccirelli, M.; Grätzel, M.; Bach, U. *Appl. Phys. Lett.* **2001**, *79*, 2085.

(130) Krüger, J.; Plass, R.; Grätzel, M.; Matthieu, H.-J. *Appl. Phys. Lett.* **2002**, *81*, 367.

(131) Papageorgiou, N.; Athanassov, Y.; Armand, M.; Bonhôte, P.; Petterson, H.; Azam, A.; Grätzel, M. *J. Electrochem. Soc.* **1996**, *143*, 3099.

(132) Murai, S.; Mikoshiba, S.; Sumino, H.; Katob, T.; Hayaseb, S. *Chem. Commun.* **2003**, 1534–1535.

(133) Kubo, W.; Kitamura, T.; Hanabusa, K.; Wada, Y.; Yanagida, S. *Chem. Commun.* **2002**, 374.

(134) Wang, P.; Zakeeruddin, S.M.; Moser, J.E.; Grätzel, M. *J. Phys. Chem. B* **2003**, *107*, 13280.

(135) Wang, P.; Zakeeruddin, S.M.; Humphry-Baker, R.; Grätzel, M. *Chem. Mater.* **2004**, *16*, 2694.

(136) Usami, A. *Chem. Phys. Lett.* **1998**, *292*, 223.

(137) Usami, A.; Ozaki, H. *J. Phys. Chem. B* **2001**.

(138) Papageorgiou, N.; Grätzel, M.; Infelta, P.P. *Sol. Energy Mater. Sol. Cells* **1996**, *44*, 405.

(139) Papageorgiou, N.; Barbé, C.; Grätzel, M. *J. Phys. Chem. B* **1998**, *102*, 4156.

(140) Ferber, J.; Stangl, R.; Luther, J. *Sol. Energy Mater. Sol. Cells* **1998**, *53*, 29.

(141) Stangl, R.; Ferber, J.; Luther, J. *Sol. Energy Mater. Sol. Cells* **1998**, *54*, 255.

(142) Kalaignan, G.P.; Kang, Y.S. *J. Photochem. Photobiol. A: Chem.* in press.

(143) Huang, S.Y.; Schilchthörl, G.; Nozik, A.J.; Grätzel, M.; Frank, A.J. *J. Phys. Chem. B* **1997**, *101*, 2576.

(144) Ni, M.; Leung, M.K.H.; Leung, D.Y.C.; Sumathy, K. *Sol. Energy Mater. Sol. Cells* **2006**, *90*, 1331.

(145) Ernst, K.; Engelhardt, R.; Ellmer, K.; Kelch, C.; Muffler, H.J.; Lux-Steiner, M.C.; Könenkamp, R. *Thin Solid Films* **2001**, *387*, 26.

(146) Könenkamp, R.; Dloczik, L.; Ernst, K.; Olesch, C. *Physica E* **2002**, *14*, 219.

(147) Lévy-Clément, C.; Katty, A.; Bastide, S.; Zenia, F.; Mora, I.; Munoz-Sanjose, V. *Physica E* **2002**, *14*, 229.

(148) Gavrilov, S.; Oja, J.; Lim, B.; Belaidi, A.; Bohne, W.; Strub, E.; Röhrich, J.; Lux-Steiner, M.C.; Dittrich, T. *Phys. Stat. Sol. A* **2006**, *203*, 1024.

(149) Nanu, M.; Schoonman, J.; Goossens, A. *Nano Lett.*, **2005**, *5*, 1716.

(150) Santato, C.; Ulmann, M.; Augustynski, J. *J. Phys. Chem.* **2001**, *105*, 936.

(151) Khan, S.U.M.; Al-Shahry, M.; Ingler, W.B. *Science* **2002**, *297*, 2243.

(152) Torres, G.R.; Lindgren, T.; Lu, J.; Granqvist, C.-G.; Lindquist, S.-E. *J. Phys. Chem. B* **2004**, *108*, 5995.

(153) Santato, C.; Ulmann, M.; Augustynski, J. *Adv. Mat.* **2001**, *13*, 511.

(154) Morand, R.; Lopez, C.; Koudelka-Hep, M.; Kedzierzawski, P.; Augustynski, J. *J. Phys. Chem.* **2002**, *106*, 7218.

(155) Solarska, R.; Santato, C.; Jorand-Sartoretti, C.; Ulmann, M.; Augustynski, J. *J. Appl. Electrochem.* **2005**, *35*, 715.

(156) Lana-Villarreal, T.; Bisquert, J.; Mora-Seró, I.; Salvador, P. *J. Phys. Chem. B* **2005**, *109*, 10355.

(157) Villarreal, T.L.; Gómez, R.; González, M.; Salvador, P. *J. Phys. Chem. B* **2004**, *108*, 20278.

(158) Villarreal, T.L.; Gómez, R.; Neumann-Spallart, M.; Alonso-Vante, N.; Salvador, P. *J. Phys. Chem. B* **2004**, *108*, 15172.

(159) Mora-Seró, I.; Lana-Villarreal, T.; Bisquert, J.; Pitarch, A.; Gómez, R.; Salvador, P. *J. Phys. Chem. B* **2005**, *109*, 3371.

(160) Granqvist, C.G. *Handbook of Inorganic Electrochromic Materials*; Elsevier: Amsterdam, 1995.

(161) Etzion, Y.; Erell, E. *Build. Environm.* **2000**, *35*, 433.

(162) Hauch, A.; Georg, A.; Baumgärtner, S.; Opara-Krasovec, U.; Orel, B. *Electrochim. Acta* **2001**, *46*, 2131.

(163) Krasovec, U.O.; Georg, A.; Georg, A.; Wittwer, V.; Joachim, L.; Topic, M. *Sol. En. Mater. Sol. Cells* **2004**, *84*, 369.

(164) Georg, A.; Georg, A.; Krasovec, U.O. *Thin Solid Films* **2006**, *502*, 246.

(165) Cummins, D.; Boschloo, G.; Ryan, M.; Corr, D.; Rao, S.N.; Fitzmaurice, D. *J. Phys. Chem. B* **2000**, *104*, 11449.

(166) Bonhote, P.; Gogniat, E.; Campus, F.; Walder, L.; Grätzel, M. *Displays* **1999**, *20*, 137.

(167) Edwards, M.O.M.; Gruszecki, T.; Pettersson, H.; Thuraisingham, G.; Hagfeldt, A. *Electrochem. Commun.* **2002**, *4*, 963.

(168) Petterson, H.; Gruszecki, T.; Johansson, L.H.; Norberg, A.; Edwards, M.O.M.; Hagfeldt, A. *SID Digest Tech. Papers* **2002**, 123.

(169) Garcia-Cañadas, J.; Peter, L.M.; Wijayantha, K.G.U. *Electrochem. Commun.* **2003**, *5*, 199.

(170) Bonhôte, P.; Gogniac, E.; Sophie, T.; Barbé, C.; Vlachopoulos, N.; Lenzmann, F.; Comte, P.; Grätzel, M. *J. Phys. Chem. B* **1998**, *102*, 1498.

(171) Wang, Q.; Zakeeruddin, S.M.; Cremer, J.; Bäuerle, P.; Humphry-Baker, R.; Grätzel, M. *J. Am. Chem. Soc.* **2005**, *127*, 5706.

(172) Wang, Q.; Zakeeruddin, S.M.; Zakeeruddin, M.K.; Humphry-Baker, R.; Grätzel, M. *J. Am. Chem. Soc.* in press.

(173) Trammell, S.A.; Meyer, T.J. *J. Phys. Chem. B* **1999**, *103*, 104.

(174) Westermark, K.; Tingry, S.; Persson, P.; Rensmo, H.; Lunell, S.; Hagfeldt, A.; Siegbahn, H. *J. Phys. Chem. B* **2001**, *105*, 7182.

(175) Paddon, C.A.; Marken, F. *Electrochem. Commun.* **2004**, *6*, 1249.

(176) Zhang, W.; Li, G. *Anal. Sci.* **2004**, *20*, 603.

(177) Könenkamp, R. *Phys. Rev. B* **2000**, *61*, 11057.

(178) Kytin, V.; Dittrich, T. *Phys. Stat. Sol. A* **2001**, *185*, 461.

(179) Kytin, V.; Dittrich, T.; Koch, F.; Lebedev, E. *Appl. Phys. Lett.* **2001**, *79*, 108.

(180) Kytin, V.; Dittrich, T.; Bisquert, J.; Lebedev, E.A.; Koch, F. *Phys. Rev. B* **2003**, *68*, 195308.

(181) Garcia-Belmonte, G.; Kytin, V.; Dittrich, T.; Bisquert, J. *J. Appl. Phys.* **2003**, *94*, 5261.

(182) Aurbach, D. *J. Power Sources* **2005**, *146*, 71.

(183) Poizot, P.; Laurelle, S.; Grugeon, S.; Dupont, L.; Tarascon, J.M. *J. Power Sources* **2001**, *97–98*, 235.

(184) Kavan, L.; Kalbaci, M.; Zukalova, M.; Exnar, I.; Lorenzen, V.; Nesper, R.; Grätzel, M. *Chem. Mater.* **2004**, *16*, 477.

(185) Armstrong, A.R.; Armstrong, G.; Canales, J.; García, R.; Bruce, P.G. *Adv. Mat.* **2005**, *17*, 862.

(186) Croce, F.; Appetecchi, G.B.; Persi, L.; Scrosati, B. *Nature* **1998**, *394*, 456.

(187) Amatucci, G.; Badway, Z.; Du Pasquier, A.; Zheng, T. *J. Electrochem. Soc.* **2002**, *148*, A930.

(188) Du Pasquier, A.; Plitz, A.; Gural, J.; Badway, Z.; Amatucci, G. *J. Power Sources* **2004**, *136*, 160–170.

(189) Remacle, F. *J. Phys. Chem. A* **2000**, *104*, 4739.

(190) Franceschetti, A.; Williamson, A.; Zunger, A. *J. Phys. Chem. B* **2000**, *104*, 3398.

(191) Franceschetti, A.; Zunger, A. *Appl. Phys. Lett.* **2000**, *76*, 1731.

(192) Brus, L.E. *J. Chem. Phys.* **1984**, *80*, 4403.

(193) Yu, D.; Wang, C.; Guyot-Sionnest, P. *Science* **2003**, *300*, 1277.

(194) Houtepen, A.J.; Vanmaekelbergh, D. *J. Phys. Chem. B* **2005**, *109*, 19634.

(195) Roest, A.L.; Kelly, J.J.; Vanmaekelbergh, D. *Appl. Phys. Lett.* **2003**, *83*, 5530.

(196) Remacle, F.; Levine, R.D. *Chem. Phys. Chem.* **2001**, *2*, 20.

(197) Romero, H.E.; Drndic, M. *Phys. Rev. Lett.* **2005**, *95*, 156801.

(198) Yu, D.; Wang, C.; Wehrenberg, B.L.; Guyot-Sionnest, P. *Phys. Rev. Lett.* **2003**, *92*, 216802.

(199) Parthasarathy, R.; Xiao-Min Lin, X.-M.; Jaeger, H.M. *Phys. Rev. Lett.* **2001**, *87*, 186807.

(200) Middleton, A.A.; Wingreen, N.S. *Phys. Rev. Lett.* **1993**, *71*, 3198.

(201) Roest, A.L.; Kelly, J.J.; Vanmaekelbergh, D. *Phys. Rev. Lett.* **2002**, *89*, 036801.

(202) Roest, A.L.; Houtepen, A.J.; Kelly, J.J.; Vanmaekelbergh, D. *Faraday Discuss.* **2004**, *125*, 55.

(203) van de Lagemaat, J. *Phys. Rev. B* **2005**, *72*, 235319.

(204) Yablonovitch, E. *Phys. Rev. Lett.* **1987**, *58*, 2059.

(205) Meseguer, F.; Míguez, H. *IEICE Trans. Electron.* **2004**, *E87*, 274.

(206) Lodahl, P.; van Driel, A.F.; Nikolaev, I.S.; Irman, A.; Overgaag, K.; Vanmaekelbergh, D.; Vos, W.L. *Nature* **2004**, *430*, 654.

(207) Huisman, C.L.; Schoonman, J.; Goossens, A. *Sol. Energy Mater. Sol. Cells* **2005**, *85*, 115.

(208) Somani, P.R.; Dionigi, C.; Murgia, M.; Palles, D.; Nozar, P.; Ruani, G. *Sol. Energy Mat. Sol. Cells* **2005**, *87*, 513.

(209) Kavan, L.; Zukalova, M.; Kalbaci, M.; Grätzel, M. *J. Electrochem. Soc.* **2004**, *151*, A1301.

(210) Fan, Z.; Lu, J.G. *J. Nanosci. Nanotech.* **2005**, *5*, 1561.

(211) Huang, M.H.; Mao, S.; Feick, H.; Yan, H.; Wu, Y.; Kind, H.; Weber, E.; Russo, R.; Yang, P. *Science* **2001**, *292*, 1897.

(212) Baxter, J.B.; Aydil, E.S. *Appl. Phys. Lett.* **2005**, *86*, 053114.

(213) Chae, W.-S.; Lee, S.-W.; Kim, Y.-R. *Chem. Mater.* **2005**, *17*, 3072.

(214) Greene, L.E.; Law, M.; Tan, D.H.; Montano, M.; Goldberger, J.; Somorjai, G.; Yang, P. *Nano Lett.* **2005**, *5*, 1231.

(215) Baxter, J.B.; Aydil, E.S. *Sol. Energy Mater. Sol. Cells* **2006**, *90*, 607.

(216) Mor, G.K.; Shankar, K.; Paulose, M.; Varghese, O.K.; Grimes, C.A. *Nano Lett.* **2006**, *6*, 215.

(217) Park, J.H.; Kim, S.; Bard, A.J. *Nano Lett.* **2006**, *6*, 24.

(218) Bullot, J.; Gallais, O.; Gauthier, M.; Livage, J. *Appl. Phys. Lett.* **1980**, *36*, 986.

(219) Kim, G.-T.; Waizmann, U.; Roth, S. *Appl. Phys. Lett.* **2001**, *79*, 3497.

(220) Muster, J.; Kim, G.T.; Krsticacute, V.; Park, J.G.; Park, Y.W.; Roth, S.; Burghard, M. *Adv. Mater.* **2000**, *12*, 420.

(221) Kim, G.T.; Muster, J.; Krstic, V.; Park, J.G.; Park, Y.W.; Roth, S.; Burghard, M. *Appl. Phys. Lett.* **2000**, *76*, 1875.

(222) Sun, D.; Kwon, C.W.; Baure, G.; Richmian, E.; MacLean, J.; Dunn, B.; Tolbert, S.H. *Adv. Mat.* **2004**, *14*, 1197.

(223) Yu, H.Y.; Kang, B.H.; Pi, U.H.; Park, C.W.; Choi, S.-Y.; Kim, G.T. *Appl. Phys. Lett.* **2005**, *86*, 253102.

(224) Yi Cui, Y.; Wei, Q.; Park, H.; Lieber, C.M. *Science* **2001**, *293*, 1289.

CHAPTER 17

Nanostructured Oxides in Photo-Catalysis

GERARDO COLÓN-IBÁÑEZ

Instituto de Ciencia de Materiales de Sevilla CSIC, C/Américo Vespucio 49, 41092-Sevilla, Spain

CAROLINA BELVER-COLDEIRA and MARCOS FERNÁNDEZ-GARCÍA

Instituto de Catálisis y Petroleoquímica CSIC, Campus Cantoblanco, 28049 Madrid, Spain

17.1. INTRODUCTION: SCOPE OF THE REVIEW

Photo-induced processes are studied in several industrial-oriented applications, which have been developed since their first descriptions in the scientific literature. Despite the difference in character and utilization, all photo-induced processes have the same origin. A semiconductor can be excited by light energy higher than the band gap inducing the formation of energy-rich electron-hole pairs. Chapter 16 describes the use of this energy in photovoltaic, photoelectrical, and photoelectrochemical devices, whereas here we will undertake its study in the photo-catalysis field. By photo-catalysis, it is commonly understood to be any chemical process catalyzed by a solid where the external energy source is an electromagnetic field with wavenumbers in the ultraviolet (UV)-visible range (1–6).

Customarily, photo-catalysts are solid oxide semiconductors, particularly TiO_2-anatase, but recent progress has expanded the chemical nature of photo-active systems by including (oxy)sulfides and (oxy)nitrides, doped-zeolites, and molecular entities embodied in zeotypes. As most of these catalysts have been specifically devised for application under sunlight, here we will differentiate between UV- and visible-active systems. This is, of course, a phenomenological way to classify photo-catalysts, but it appears most convenient from the point of view of enhancing new contributions to this work-field, not covered by previous reviews (1–6). Section 17.2 also includes a description of the reaction mechanism of photo-catalytic processes. Photo-catalysis is

Synthesis, Properties, and Applications of Oxide Nanomaterials, Edited by José A. Rodríguez and Marcos Fernández-Garcia

mainly involved in three areas concerning organic synthesis, degradation of pollutants and special reactions, like H_2Oreduction or N_2 fixation. All photo-induced chemical processes have in common the so-called "initial" steps, which include the absorption of light, diffusion, and trapping of charge and fate (recombination or phase transfer to the gas/liquid media) of charge carriers (7,8). These steps will be briefly introduced in Section 17.2 although the main point here, e.g., the influence of the nanostructure on such elemental steps, will be covered in the final part of this chapter. In any case, the main aim in describing reaction mechanisms in Section 17.2 is to give a glimpse into the potential chemistry evolving from solid surfaces under light activation. Section 17.3 will briefly discuss the field of potential assisted photo-catalysis.

The main subject of this review is, as mentioned, the study of the effect of the nanostructure on the photo-catalytic system; the world around this topic will be covered in Section 17.4 of this chapter. As an introduction, we will briefly discuss the structural and electronic properties of the solid catalysts influenced by size. As such properties have been thoroughly examined in Chapters 1 to 14, here we will summarize the specific points of interest for photo-catalysts. The review will continue by analyzing the influence of the nanostructure in the above-mentioned "initial steps" of the photo-catalytic reactions and finish by giving an overview of the most significant catalytic measurements as a function of primary particle size and morphological properties of nanostructured photo-catalysts.

17.2. GENERAL VIEW OF PHOTO-CATALYTIC SYSTEMS

When light with a characteristic wavelength in the energy range of interest for photo-catalysis interacts with a solid, both absorption and scattering phenomena occur concomitantly (9). Combined analysis of optical data obtained in transmittance and reflectance can be worked out to obtain a reasonable approximation to the absorption and scattering coefficients (10). Because the UV-vis wavelengths (550–300 nm) characteristic of excitation sources used in photo-catalytic processes are significantly larger than the primary size of nanosolids, ca. 10 nm, the radiation field response to the particle aggregates (with a characteristic dimension called secondary particle size) and scattering events is dominated by this morphological characteristic of the solid photo-catalyst (11). Absorption of light is obviously the initial step occurring on all photo-active systems used in catalysis. However, optical measurements in photo-active oxides prove that scattering is the dominant phenomenon and that absorption can occur after multiple scattering events, conditioning by reaction geometry (10), indicating that there is not an easy way to measure quantum yields (reaction rate per rate of absorbed photon). Additionally, quantum yields are spectral-dependent and are therefore specific for the excitation light source (with most of the times having several excitation lines) (12).

Solid semiconductors are characterized by the existence of a forbidden gap between the valence and the conduction band. As described, the process of photo-catalysis is relatively simple; is initiated by the absorption of a photon with energy equal to or greater than the band gap of the semiconductor (e.g., 3.2 eV for TiO_2), producing a charge carrier pair, that is an electron-hole (e^-/h^+) pair. Photo-catalytic semiconductors are typically oxides, TiO_2, ZnO, WO_3, CeO_2, or $A_xB_yO_z$, but also sulfides, MoS_x, WS_x, or BiS_x, or nitrides, TaN_x, SiN_x, or GeN_x. These pure phases are frequently doped with a hetero-atom, whether this is a cation or an anion, forming a solid solution, put in intimate contact with another semiconductor phase, or with ions or metals deposited on its surface. Thus, due to the band energy configuration of most used semiconductors, photo-excitation always requires photons with wavelength below ca. 550 nm and is typically in the UV- and/or visible spectral ranges. The corresponding systems working under UV-excitation will be detailed in Section 17.2.1, whereas those used under visible-light excitation will be covered in Section 17.2.2.

Upon irradiation, a photo-inducted process would thus take place. This will consist in the promotion of one valence band electron to the conduction band, leaving a hole in the valence band. Therefore, an electron-hole pair (exciton) is created with lifetimes in the femptosecond regime, which might undergo charge transfer to adsorbed species on the semiconductor surface by forming radicals that oxidize organic chemicals, or reduce metal species. Obviously, charge carriers may be involved in chemical reactions but also can recombine producing energy. Recombination is in fact the most likely process and can have a radiative or nonradiative nature (1,2,7,8). Basically, before an energy exchange with adsorbed species or a certain recombination steps can occur, the charge carriers would suffer trapping processes (1,2,7,8). Defects can capture such electrons and holes, by acting as electron (or hole) traps, through the following process:

$$D^{n+} + e^-_{cb} \longrightarrow D^{(n-1)+} \qquad \text{electron trapping} \qquad \textbf{(17.1)}$$

$$D^{n+} + h^+_{vb} \longrightarrow D^{(n+1)+} \qquad \text{hole trapping} \qquad \textbf{(17.2)}$$

If the pair $D^{n+}/D^{(n-1)+}$ is located below the conduction band edge (E_{cb}) and the energy level for $D^{n+}/D^{(n+1)+}$ above the valence edge (E_{vb}), the trapping of electron and holes would take place, affecting the lifetimes of charge carriers. The first/second pair originates the so-called donor/acceptor electronic levels. Defects can have a punctual nature, being intrinsic (like oxygen vacancies in nanostructured reducible oxides) or extrinsic (like dopants or impurities), or being line or plane defects. Their situation at the surface or the bulk of the material is critical as only the former can be involved in chemical reactions, being the latter mainly involved in recombination processes. The exact degree of localization of charge trapping in a defect is not known, but the terminology adopted discriminates between strongly localized, called deep traps, or shallow ones. Their energy characteristics are not well defined but should allow phenomenological differentiation between two main groups: shallow traps that exchange charge carriers with conduction/valence band species under thermal excitation at room temperature and deep traps that may not have this possibility or be strongly limited (1,2,7,8). The nature of trapping centers and main physicochemical characteristics have been analyzed in Chapters 5 and 6 of this book and will be briefly detailed for the most interesting photo-semiconductors in Section 17.4.1.

Within this situation, upon irradiation, the following electronic processes would take place: charge generation, charge trapping-charge release, recombination, and interfacial transfer (see detailed discussion in Section 17.4). The mechanisms for these electronic processes in a semiconductor are summarized in Scheme 17.1.

Photo-catalysis is mainly involved in three research areas, which are summarized in the following list:

- Organic synthesis:
  - Partial oxidation of linear and cyclic hydrocarbons
- Degradation of pollutants
  - Mineralization of organic compounds
  - Disinfection/destruction of biological materials
  - Detoxification of inorganic compounds and removal of ions
- Special reactions
  - Photo-fixation of nitrogen
  - Photo-reduction of CO_2 to organic compounds
  - Photo-splitting of water to produce H_2

EQUATION	ELECTRONIC STEPS
$TiO_2 + h\nu \rightarrow e^-_{CB} + h^+_{VB}$ $M^{n+} + h\nu \rightarrow M^{(n+1)+} + e^-_{CB}$ $M^{n+} + h\nu \rightarrow M^{(n-1)+} + h^+_{VB}$	**charge generation**
$Ti^{4+} + e^-_{CB} \rightarrow Ti^{3+}$ $M^{n+} + e^-_{CB} \rightarrow M^{(n-1)+}$ $M^{n+} + h^+_{VB} \rightarrow M^{(n+1)+}$ $OH^- + h^+_{VB} \rightarrow OH^\bullet$	**charge trapping**
$M^{(n-1)+} + Ti^{4+} \rightarrow M^{n+} + Ti^{3+}$ $M^{(n+1)+} + OH^- \rightarrow M^{n+} + OH^\bullet$	**charge release**
$e^-_{CB} + h^+_{VB} \rightarrow h\nu$ $Ti^{3+} + OH^\bullet \rightarrow Ti^{4+} + OH^-$ $M^{(n-1)+} + h^+_{VB} \rightarrow M^{n+}$ $M^{(n+1)+} + Ti^{3+} \rightarrow M^{n+} + Ti^{4+}$ $M^{(n-1)+} + OH^\bullet \rightarrow M^{n+} + OH^-$ $M^{(n+1)+} + e^-_{CB} \rightarrow M^{n+}$	**recombination**
$e^-_{CB}(or\ Ti^{3+}, M^{(n-1)+}) + O \rightarrow O^-$ $h^+_{VB}(or\ OH^\bullet, M^{(n+1)+}) + R \rightarrow R^+$	**interfacial charge transfer**

Scheme 17.1.

17.2.1. UV-Active Systems

As described, the process of photo-catalysis is relatively simple. Light energy from ultraviolet radiation (light) produces a charge carrier pair. Thus, due to the band energy configuration of most used semiconductors, UV-photo-excitation always requires photons of below 390 nm. Basically, the result of this energy change with chemical meaning is the formation of holes in the surface, and free electrons that are now available to form hydroxide (~OH), or other radicals, which can oxidize organic chemicals, or reduce metal species.

Semiconductors with photo-catalytic activity might show specific electronic, structural, surface, and morphologic features. A good photocatalyst should be

(*1*) photoactive,
(*2*) able to absorb visible and/or UV light,
(*3*) chemically and biologically inert and photostable,
(*4*) inexpensive, and
(*5*) nontoxic.

TiO_2, ZnO, $SrTiO_3$, CeO_2, WO_3, Fe_2O_3, CdS, and ZnS can act as photo-active material for redox processes due to their electronic structure, which is characterized by a filled valence band and an empty conduction band. Among these possible semiconductors, TiO_2 is the most used photo-catalytic material that fulfils all of these requirements and exhibits adequate conversion values (13). However, despite the high conversion values obtained for TiO_2, the calculated quantum yield for the studied reactions is appreciably low (less than 10% for most degradation processes) (14).

From the available semiconductors, ZnO is generally unstable in illuminated aqueous solutions, especially at low pH values. In liquid-phase photo-reactions, the photo-corrosion of ZnO is complete at pH lower than 4. The formation of Zn^{2+} is attributed to the oxidation of ZnO by h_{vb}^{+} (15). This phenomenon is considered as one of the main reasons resulting in the decrease of ZnO photo-catalytic activity in aqueous solutions (16,17). However, for applications in gas phase, this disadvantage should not exist (18). WO_3, like TiO_2, was found to be stable toward photo-corrosion. Even though it is reported to be a good candidate for oxygen evolution, the possibility of using this semiconductor for hydrogen evolution was attempted and succeeded (19). WO_3 absorbs only a portion of the visible radiation. Nevertheless, its absorption in the visible region could be increased by doping with transition metal ions. WO_3, although useful in the visible range, is generally less photo-catalytically active than TiO_2 (20). CeO_2 is an inexpensive and relatively harmless material that presents several characteristics that could be potentially advantageous for photo-catalytical applications, such as its stability and strong absorption under UV-illumination. Among others, CdS, ZnS, and iron oxides have been also tested (21–24).

The structural, electronic, and surface features play an important role in photo-catalytic applications. Although for heterogeneous catalysis, surface area is one of the conditioning characteristic of the catalysts, in the case of heterogeneous photo-catalysis, the activity is not necessary dependent on this feature, but on the availability of the active sites (25,26). Nevertheless, a large surface area can be determinant in certain photo-catalytic reactions in which the increase of the adsorption leads to enhance the reaction rate (27–29). However, it is also well known that materials presenting high values of surface area and porosity also exhibit a highly defective structure. A capital point is the presence of OH^- groups as they could have a significant contribution to the phase stability in the nanometer range (30,31) as well as define the acid/base character of the solid. The existence of crystalline defects like vacancies and/or low coordination centers in the structure turns detrimental for the photo-efficiency of the semiconductor, because they favor the electron-hole recombination (32,33). Crystallite size and concomitant phenomena (defects) are also determinant parameters to be considered (34–36). Nanosized semiconductors exhibit pronounced size effects on its electronic structure and related optical properties. These aspects will be detailed at length in Section 17.4.

Structural and electronic characteristics of the material also appear to be determinant. This is the case of ZnO and especially TiO_2. Monodispersed ZnO particles with well-defined morphological characteristics, such as spherical, ellipsoidal, needle, prismatic, and rod-like shapes have been obtained by means of different synthetic methods (37–40). It has been reported that the photo-catalytic activity of ZnO particles should strongly depend on the particle morphologies on which the specified crystal faces expose. This decisive difference in morphology probably influences the photo-catalytic activity because the exposed crystal faces or the ratios among exposed crystal faces are noticeably different for the ZnO particles composed with different crystallite forms (41,42).

TiO_2 occurs in nature in three crystallographic phases: rutile, anatase, and brookite, with the anatase the most commonly employed in photo-catalytic applications (6,13,43). Anatase is the polymorph of TiO_2 less thermodynamically stable, although from energy calculations this phase appears as the more likely phase when the grain size is around 10 nm (44). The crystalline structure can be described of terms TiO_6 octahedral chains differing by the distortion of each octahedron and the assembly pattern of octahedra chains. The Ti–Ti distances in the anatase structure are greater than in rutile, whereas Ti–O distances are shorter (45). These structural differences cause different mass densities and lead to a different electronic structure of the bands. Anatase phase is 9% less dense than rutile, presenting more pronounced localization of the Ti $3d$ states, and therefore, a narrower $3d$ band can be found in the anatase phase. Also the O $2p$–Ti $3d$ hybridation is different in the two structures (more covalent mixing in the rutile), with anatase exhibiting a valence and conduction band with more pronounced O $2p$–Ti $3d$ characters, respectively (46). These different structural features are presumably responsible for the difference in the mobility of charge carriers.

Although the anatase–TiO_2 presents the higher photoactivities reported (47), commercial Degussa P25, which has a mixture of anatase and rutile in a ratio 80:20, is one of the best TiO_2 photo-catalysts and frequently used as a reference. The relationship between the crystal structure of this commercial TiO_2 and its notably higher photo-efficiencies is still controversial. Bickely et al. reported a model based on anatase–rutile mixture and a fraction of amorphous phase (48). The anatase–rutile coexistence would lead to a synergetic effect between the two phases that improves the photoactivity of TiO_2, as has been reported by several authors for anatase–rutile mixtures prepared by different methods (49–51). This junction effect has been also described for the mixture anatase–brookite, which presents a high activity in gas-phase methanol photo-oxidation (52).

It is still necessary to develop a photo-catalytic system with improved structural or morphological properties that improve the charge separation and diffusion to the surface. Additionally, for liquid-phase photo-degradation processes, another requirement of the new photo-catalysts should be the ease to be separated from the reaction solution, a requisite of great importance for practical applications. Many efforts are actually performing in order to improve these two operational factors by the optimization of the catalyst features; crystal and electronic structure, surface acidity, smaller particle size, use of dopants (to extend the absorption range) and support materials, and/or designing immobilized photo-catalyst in different materials (glass, quartz, steel, membranes, fibers) (53–58) able to be used in liquid- and gas-phase configuration reactors.

17.2.1.1. Transition Metal Doping Another way to improve the efficiency of the photo-catalytic process is by doping with a metal ion the semiconductor photo-catalyst structure. Doping with suitable transition metal ions may allow extending the light absorption of large band gap semiconductors to the visible region (59). Additionally to the absorption spectrum expansion, doping with certain transition ions could act in certain cases as electron/hole trappers, leading to an increase of the charge carrier lifetimes and therefore avoiding the recombination processes (60–63). This occurs through the following process:

$$M^{n+} + e^-_{cb} \longrightarrow M^{(n-1)+} \qquad \text{electron trapping} \qquad \textbf{(17.3)}$$

$$M^{n+} + h^+_{vb} \longrightarrow M^{(n+1)+} \qquad \text{hole trapping} \qquad \textbf{(17.4)}$$

If the pair $M^{n+}/M^{(n-1)+}$ is located below the conduction band edge (E_{cb}) and the energy level for $M^{n+}/M^{(n+1)+}$ above the valence edge (E_{vb}), the trapping of electron and holes would take place, affecting the lifetimes of charge carriers.

It is clear that the photo-reactivities of doped semiconductors widely varies depending on the specific dopant considered. The origin of these photo-reactivities is clearly related to the efficiencies of the doping centers in trapping charge carriers and interceding in the interfacial transfer. Trapping either an electron or hole alone is ineffective for photo-degradation because immobilized charge species quickly recombines with a mobile counterpart (62).

The effect of doping a photo-catalytically active semiconductor is therefore a multifaceted problem, and the modifications induced in the photoactivity might be explained by the sum of changes occurring in the semiconductor matrix. Consequently, the relative efficiency of a metal ion dopant will depend on whether it serves as a mediator of interfacial charge transfer or as a recombination center. The ability of a doping ion to function as an effective trap is related to several factors, as the dopant concentration, its energy level within the semiconductor lattice, their *d* or *f* electronic configuration, the distribution within the particles, or the electron donor concentration. From literature review, an optimum doping content clearly exists for each selected dopant. In this sense Pleskov (64) reported that the value of the space charge region potential for the efficient separation of electron-hole pairs might be not lower than 0.2 eV. However, Pleskov theory is not applicable to nanostructured materials.

It is reported that doping quantum-sized TiO_2 with transition metal ions as Fe^{3+}, Mo^{5+}, Ru^{3+}, Os^{3+}, Re^{3+}, V^{4+}, and Rh^{3+} enhances the photo-reactivity and reduction processes (62). On the contrary, some authors state that the presence of a foreign metal species as Cr^{3+} is generally detrimental for the degradation of organic species in aqueous systems, although it accelerates the photo-reduction (65). Many examples indicate that the doping effect on TiO_2 photo-activity can be beneficial or detrimental for the same doping ion. Thus, although Grätzel et al. (66) found an inhibition of the electron-hole recombination on Mo^{5+} and V^{5+} doped TiO_2 based on EPR measurements, Luo et al. reported a significant diminution of the photo-activity for similar doped systems (67). Mu et al. (68) reported that doping with trivalent or pentavalent metal ions was detrimental for the photo-catalytic reactivity, whereas Karakitsou et al. (69) showed that doping with cations with valence higher than that of Ti^{4+} enhanced its photo-activity. Recently, lanthanide rare earth cations, such as La^{3+}, Ce^{4+}, Nd^{3+}, and Gd^{3+}, have been used for TiO_2 doping, leading to an increasing in the photo-activity values (70–73). By comparing the ionic radii of such cations with Ti^{4+}, it is clear that the incorporation of dopant cation into the TiO_2 matrix is not feasible. On the contrary, it could be possible that the Ti^{4+} ion might substitute for lanthanide ions in the lattice of the lanthanide oxide. As far as it concerns to lanthanide doping, it appears that dopant exhibiting half-filled electronic configuration induces higher photoactivity values (Gd^{3+}). The particular stability of such electronic configuration is well known. Thus, when dopant cation traps photo-excited electrons, the half-filled electronic configuration is shattered, decreasing its stability. In this situation, trapped electrons could easily be transferred to the oxygen molecule adsorbed in the surface as cation returns to the initial stable electronic situation. Within this scheme, it is clear that an enhancement in the charge transfer as well as an efficient separation of electrons and holes lead to a significant improvement in the photo-catalytic efficiency.

Special attention might be paid to iron doped TiO_2 due to the following attractive aspects:

(*1*) Iron(III) doped TiO_2 had been reported to enhance the lifetime of electron-hole pairs from nanosecond for naked TiO_2 colloids to minutes and even hours (74).

(*2*) It clearly enhances the absorption in the visible range.

(*3*) It has been proved for special photo-reactions of great interest as water splitting, N_2 photoreduction to ammonia, or oxidation reactions (75).

From the results reported about this system, it is clear that the deep knowledge of the structural features of this doped system is critical to understand its photo-catalytic behavior. In this sense, the role of iron dopant in TiO_2 is controversial. Some authors pointed that Fe^{3+} centers behave as electron/hole recombination centers (76–78), whereas some others postulated that its role is to perform the electron-hole separation, enhancing the photo-catalytic activity. As mentioned, the structural situation of a doped system drives its reactivity, and consequently, the preparation of iron–TiO_2 turns crucial. Aspects such as phase composition, solubility, location, and dispersion of iron in the different phases greatly vary in the specimens, depending on the preparation technique. Iron cations can be located substitutionally on the TiO_2 matrix because its ionic radius is similar to Ti^{4+}, however, depending on the preparation method, the iron doped system could form mixed oxides with other than anatase structure. Recombination processes can be the dominant route when the doping cation forms segregated phases (78). On the other hand, it is also known that the presence of iron in TiO_2 would catalyze the anatase to rutile transformation (79), leading to a diminution of its photo-catalytic activity. The enhancement of photo-activity is likely linked to an optimum iron loading (generally low concentrations $<1\%$). Regarding the origin of the photo-activity in these doped systems, it may be assumed that the iron doped system would give rise to a charge transfer transition from Fe^{3+} to Ti^{4+} levels (*d–d* transition $^2T_{2g} \rightarrow {}^2A_{2g}, {}^2T_{1g}$), creating new electronic levels within the TiO_2 band gap (bathocromic shift) (Figure 17.1). An additional charge transfer transition that would take place could be via the conduction band from one metal dopant to another identical center physically separated from the former ($Fe_a^{3+} + Fe_b^{3+} \rightarrow Fe_a^{4+} + Fe_b^{2+}$). As a result, formally, Fe^{3+} centers formally can act as electron and hole trapping centers (80). Bahnemann et al. (81) also describe a mechanism in which shallow trapping of the photo-electron by Fe^{3+} followed by rapid iron-catalyzed transfer of the electron to molecular oxygen is suggested as a tentative explanation for the enhancement of photo-catalytic activity by the Fe^{3+}-dopant. However, there must be a compromise between the carrier trap and recombination processes balanced by the amount of dopant (75).

Interesting results are reported by Araña et al. for carboxylic acid photodegradation using Fe^{3+} doped TiO_2 (82,83). In this case, surface iron in the TiO_2 participates in the photo-catalytic mechanism in a unexpected way. It seems that during photo-oxidation of the organic acid, Fe progresses toward solution by the formation of a Fe^{3+} complex. This iron complex is readily oxidized to CO_2 and Fe^{2+} species in solution. At this situation, would Fe^{2+} returns to its starting location at the

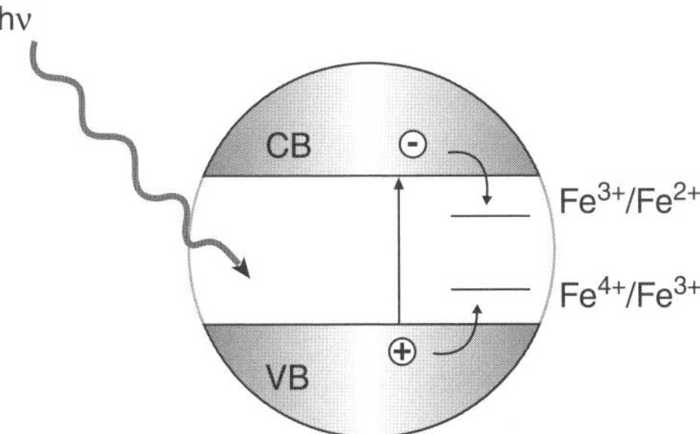

Figure 17.1. Schematic view of Fe role on Ti–Fe photocatalysts.

surface acting as a hole trapping center ($Fe^{2+} + h^{+} \rightarrow Fe^{3+}_{ads}$). Therefore, the heterogeneity of the photo-catalytic process would be doubtful, and an homogeneous photo-catalytic process would be considered.

As has been stated, the photo-physical mechanism of doped semiconductors is still controversial. In fact it is also reported that in same cases, the doping element acts as impurity, favoring the recombination by creation of structural defects (34). These contradictory results emerge because different doping methods leads to different morphological and crystalline situation in the doped photocatalyst. TiO_2 can be substitutionally or interstitially doped with different cations, forming a solid solution or a mixture of oxides. All metal dopants are conveniently substituted into TiO_2 lattice if their ionic radii are identical or nearly identical to that of the Ti^{4+} cation. Thus, the electronic diagram of the doped semiconductors will depend on the structural situation achieved.

17.2.1.2. Metal Coating Deposition of noble metals (NM) as Pt, Au, Ag, Pd, Rh, Ru, or Ir on semiconductor surface has also been considered to improve the photo-catalytic efficiency (84–89). As the photo-generated holes can react with adsorbed species, the electrons might be accumulated on the semiconductor particles, leading to an increase of the recombination process. To avoid this electron excess, oxygen is frequently used as electron scavenger. A similar role could be fulfilled by metal deposits acting as electron trapping. This way it is possible to enhance the charge carrier separation after photo-excitation. In the case of TiO_2 (properly an n-type semiconductor), a Schottky barrier is formed at the interface between TiO_2 and NM (Pt forms the highest barrier). A migration of the photo-generated electrons takes place toward NM deposited particles, leading to an electron-hole separation state (85,90). Moreover, the deposition of noble metal increases the electron transfer rate to oxygen and thereby the quantum yield. The basic principle is depicted in Figure 17.2.

To quantify the improvement achieved by the incorporation of the metal particle, Lee et al. (4) defined the *enhancement factor* (*E-factor*) for the photoprocess as

$$\text{E-factor} = \frac{\text{rate of process with NM}}{\text{rate of process without NM}} \tag{17.1}$$

Using this comparative parameter, they summarize some reactions improved by the deposition of Pt and Pd in several TiO_2 photo-catalysts. In most systems described, the NM loading on the semiconductor is invariable ca.1% w/w. Therefore, a higher amount of metal deposits leads to a decrease in the photo-efficiency of the reaction due to the electron-hole recombination enhancement or a UV shielding effect by the particles deposited, limiting the amount of light reaching the photo-catalyst. From the summarized results reported by Lee et al. arises that the enhancement of platinization is not always present. The nature of the pollutant to be photo-degraded is determinant in the effectiveness. Other parameters regarding the photo-catalyst are also decisive as the source and history of the semiconductor or the deposition method used. In this sense, *E-factors* reported are normally in the range 2–4. The large variation in the

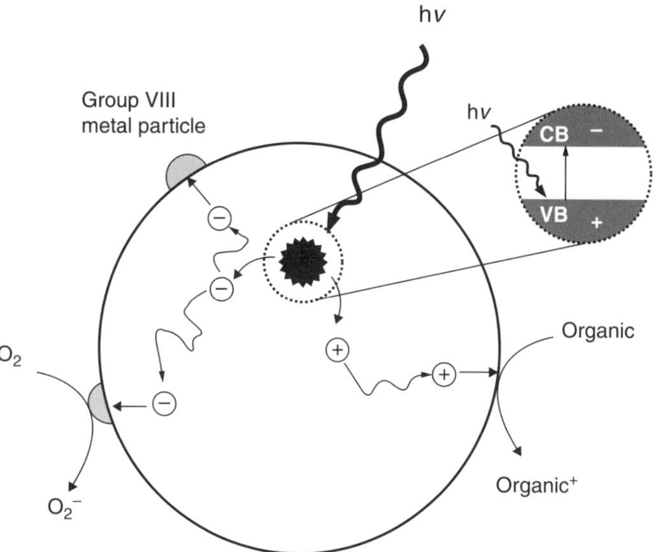

Figure 17.2. Schematic view of the charge separation process on NM/TiO_2 systems.

enhancement factor might be attributed to a combination of several parameters of the overall process.

17.2.1.3. Composite Photo-Catalytic Systems

17.2.1.3.1. Mixed Oxides and Heterojunctions To enhance the efficiency of photo-catalytic process, a number of composite systems formed by two or more different metal oxides has been reported. Mixed oxides and heterojunction composites systems have been presented as interesting alternatives to a single semiconductor photocatalyst.

Mixed oxides composed by a photoactive semiconductor dispersed on inert supports such as SiO_2 (91–93), Al_2O_3 (94), or ZrO_2 (95–97) as well as mixed oxides have also attracted interest in photocatalytic applications. A special case that will be discussed latter would be dispersion of the photo-active semiconductor on zeolite-like supports. The main scope of supporting the photo-catalyst on these supports is the stabilization of the structure and surface area. The increasing in the surface area and porosity values has a direct influence in the adsorptive properties of the photo-catalytic system. Thus, the dispersion of TiO_2 on silica supports reveals a certain stabilization of the anatase phase (91).

The mixture of two semiconductors, forming an electronic heterojunction, provides an approach to achieve a more efficient electronic process upon photo-induction (charge separation, an increased lifetime of the charge carriers, and an enhanced interfacial transfer to adsorbed substrates) (98–102). As these two semiconductors exhibit different energy levels for their corresponding conduction and valence bands, and therefore different band gap values, a synergetic effect in the photo-catalytic process would take place. Serpone et al. (98) first reported the special interaction between two photoactive semiconductors and the resulting interparticle electron transfer (IET) with the subsequent enhancement of the reductive properties of titania. This synergetic effect in the photoinduction could be considered in two ways, depending on the band gap values of the coupled semiconductors.

(*1*) Only one semiconductor undergoes in the photo-excitation, and the second is not photoactive (103). Depending of the relative values of band gap and the excitation energy, only one semiconductor would exhibit photoinduction. This is the case of SnO_2–TiO_2 ($E_g = 3.9$ vs. $E_g = 3.2$ eV, respectively) composite photo-catalysts when illuminated with UV light. It may be assumed that in this case, the photo-active semiconductor in a heterojunction system may act as similarly as a photosensitizer, but with the advantage that the driving force for the electron injection may be optimized through confinement effects (104).

(*2*) The two semiconductors experiment photo-excitation upon illumination (105). When band gap values are similar, or the photo-excitation energy is enough to promote electrons from the valence band in both systems, e.g., ZnO–TiO_2 or ZnO–SnO_2 (106).

17.2.1.3.2. Polyoxometalates The early transition metals (V, Nb, Ta, Mo, W) in their highest oxidation states can form metal-oxygen cluster anions, commonly referred to as polyoxoanions or polyoxometalates (POMs) (107). POMs could be classified in the following groups (108):

(*1*) isopoly compounds of the general formula, M_xO_{q-y}, produced by acid condensation of pure MoO_{2-4} or WO_{2-4}. The typical examples were Mo_7O_{6-24} (or W_7O_{6-24}) and $W_{10}O_{4-32}$;

(*2*) heteropoly compounds of the general formula $A_aM_xO_{q-y}$ produced by acid condensation of the mixed solution of MoO_{2-4} or WO_{2-4} and a heteroatom, A (A = P, Si, As, Fe, Co, etc.), for instance, Keggin ion, $PW_{12}O_{3-40}$, and Wells–Dawson ion, $P_2W_{18}O_{6-62}$;

(*3*) mixed heteropoly compounds that contain various ratios of Mo and W arising from a solution that contains both MoO_{2-4} or WO_{2-4} and a heteroatom, e.g., $P_2W_{15}Mo_3O_{6-62}$;

(*4*) Transition metal substituted POM (TMSP) produced by removing of M–O moieties from the saturated POM and replaced by transition metals with a varied of ligands, e.g., $[PW_{11}O_{39}Mn(H_2O)]^{6-}$.

POMs are efficient oxidants, exhibiting fast reversible multielectron redox transformations under mild conditions. They have several features in common with semiconductor metal-oxide clusters and can be considered as the analogous of the latter. Both classes of materials are constituted by d^0 transition-metals and oxide ions and exhibit similar electronic attributes, including well-defined HOMO–LUMO gaps (similar to semiconductor band gaps). The gaps inhibit the recombination of electrons and holes that are generated by the irradiation of the surface of the photo-catalysts with the light energy higher than the band gaps. Recently the ability of POM as electron acceptor has been applied to photo-catalytic systems, by adsorbing the POM species on to the semiconductor as polyoxometalate clusters (109–111), or even hybrid compounds consisting in an organic sensitizer molecule and a polyoxometalate unit (112). By accepting the photo-generated electrons from the semiconductor, these systems are able to retard the recombination process.

17.2.1.4. Adsorbent Supported Photo-Catalysts We have already described composite photo-catalytic systems mixed oxides. The improvement of these systems is referred in terms of the enhancement of the pollutant adsorption by an increment of the surface area and porosity. Recently, many attempts have been reported to develop photo-catalysts with special adsorptive properties. This is achieved by combining a photo-active material with specific adsorbents with mesoporous, microporous, or pillared structures such as activated carbon (113–117), clays (118,119), molecular sieves (120), or zeolites (121,122). The principle of the photocatalytic enhancement is based on the physicochemical adsorption of the pollutant or reaction intermediates in the inert adsorbent followed by the photo-catalytic process on the active site incorporated in the adsorbent matrix.

Among the mentioned supports, active carbon is the stronger adsorbent, with the TiO_2/C system the most studied one in either liquid- and gas-phase reactions. Carbon–TiO_2 can form different composites: titanium dioxide loaded on the activated carbon, and carbon coating the TiO_2 surface. In all cases, the enhancement of the photo-catalytic activity of TiO_2/C systems is explained by a synergetic effect between carbon and the photo-active species. Optimum adsorption strength is needed to improve the photo-activity. Structurally, a significant stabilization of the anatase phase is also observed. Surface modification as increasing of surface area and acidity are also reported (123,124). Additionally, in liquid-phase reactions, filtration seems to be aided because the decantability of photo-catalyst suspension is improved (114,115).

Zeolites and molecular sieves are inorganic materials characterized by a characteristic porosity and crystallinity, capable of complexing or adsorbing small- and medium-sized organic molecules. Just in the case of zeolites and molecular sieves, in addition to the adsorption enhancement, a shape selectivity of the substrate and products must be taken into consideration. In this sense, for liquid-phase reactions, it has been found that the hydrophobic–hydrophilic nature of catalysts significantly influences the catalytic reactivity and selectivity. Furthermore, the encapsulation of the photo-active semiconductor into heterogeneous zeolites or molecular sieves of high surface area would be effective in increasing the number of surface active sites due to the dispersion of the active species within the matrix structure and would produce quantum size effects.

Within this group, TiO_2–zeolite systems have been broadly studied. In this sense, it has been found that the control of both the dispersed state of TiO_2 and the porosity of the zeolite is important to improve the adsorption and catalytic activity (122). Titanium dioxide species anchored within the cavities or framework of the zeolite have revealed a unique local structure. Thus, from XANES spectra, it has been reported that Ti-species present a tetrahedral coordination instead of the typical octahedral coordination of TiO_2 (125,126). In this regard, the preparation method decisively influences the coordination state of Ti within the zeolite structure. Thus, in Ti-exchanged Y-zeolite, highly dispersed tetrahedral Ti-oxide species are found; meanwhile Ti-impregnated Y-zeolite involves the octahedrally coordinated Ti-oxide species. Highly dispersed TiO_4 species within the zeolite structure has been proposed to exhibit much higher and exceptional photocatalytic activity as compared with powdered TiO_2, especially on the decomposition of NO toward N_2 and O_2, or even for the CO_2 reduction to form CH_3OH and CH_4 (127,128). For this later reaction, it has been found that Ti-β zeolite catalyst having hydrophilic properties exhibited higher reactivity than hydrophobic Ti-β, attributed to formation of the charge-transfer excited complexes, $(Ti^{3+}–O^-)^*$. A charge transfer from the electron-trapped center (Ti^{3+}) into the π-antibonding orbital of NO and electron transfer from the p-bonding orbital of another NO into the trapped hole center (O^-) simultaneously occurs, leading to the decomposition of two NO molecules. On the other hand, a high selectivity for the formation of CH_3OH was observed on the Ti-β catalyst having hydrophobic properties. Therefore, the hydrophilic–hydrophobic properties of these zeolites seem to be the controlling factor in the reactivity and selectivity for the photo-catalytic reduction of CO_2 with H_2O (129).

Molecular sieves are mesoporous materials possessing large surface areas that can be used for harvesting light. The introduction of Ti-oxide species into the frameworks of mesoporous silicas has been found to provide highly efficient photo-catalysts. The channeling structure as well as the hydrophobic–hydrophilic character are the key features controlling their photo-catalytic activity. One example of the porous control of the photo-activity was provided by Yamashita et al. (130). Among the studied mesoporous silicas employed as TiO_2 support, the higher reactivity and higher selectivity for the formation of CH_3OH observed with the Ti-MCM-48 zeolite than with any other catalysts TS-1 has a smaller pore size (ca. 5.7 Å) and a three-dimensional channel structure; Ti-MCM-41 has a large pore size (>20 Å) but one-dimensional channel structure; as Ti-MCM-48 has both a large pore size (>20 Å) and three-dimensional channels. A combined contribution of the high dispersion state of the Ti-oxide species and large pore size having a three-dimensional channel structure was proposed.

Similarly, TiO_2 supported on pillared clays (mica, montmorillonite, and saponite) has also attracted much attention as a new type of photo-catalyst, because its superior adsorption property accelerates photo-catalytic reactions (131,132). TiO_2-pillared clay has a mesoporous structure due to its small TiO_2 particles, which are located as pillars between silicate layers, and it shows high adsorption ability due to its large specific surface area. As known, the interlayer surfaces of the pillared clay generally exhibit a hydrophobic character (133), which can be controlled by selection of raw clay. This constitutes an advantage to adsorb and enrich diluted hydrophobic organic compounds in water as toluene or trichloroethylene (134,135). In addition, it also shows photo-catalytic activity owing to nanosized pillars of TiO_2.

17.2.1.5. Immobilized Photo-Catalytic Systems Currently, most research in photo-catalysis is still devoted to powdered semiconductor catalyst. However, for any practical device, immobilized catalysts are preferred in liquid-phase reactions to avoid a costly separation step necessary after the reaction (136). Additionally, as the photo-catalyst dosage is increased to increase the conversion rates in liquid phases, the high turbidity created can decrease the depth of light penetration (137). Although immobilized systems have the problem that they are difficult to scale and most often have low interfacial surface areas, they exhibit interesting results in both liquid-phase and gas-phase applications. The following considerations might be taken into account when designing a photo-catalytic application using immobilized photo-catalysts:

(*1*) the exposed surface area can be considerably smaller than in the case of suspensions;

(*2*) support might be an inert with respect to the reaction media;

(*3*) a good adherence might be achieved to avoid abrasion and the loss of active mass;

(*4*) structural and surface interaction between the support and the immobilized semiconductor should be positive, or at least might produce a minimal decrease of the photo-catalyst performance.

Supports can be coated with photo-active films by different processes widely described in the literature, such as, chemical vapor deposition, physical vapor deposition, sputtering, electrophoretic deposition, and dip- and spin-coating with the desired suspensions. Generally speaking, vapor methods (CVD and PVD) need instrumental setup capable to control pressure and temperature, and therefore, their initial and running costs are normally high and the size of the substrate is limited. In this sense, the sol-gel methods of preparing films on the substrates have many advantages over other methods such as chemical vapor deposition, plasma spraying, anodization, and thermal oxidation.

On the other hand, the selection of the support material is critical because it directly affects the activity, homogeneity, and adhesion to the surface (138). Materials like ceramics tiles, paper, glass, fibreglass, and stainless steel have been employed, and other materials are currently being developed for possible future photo-catalytic applications. In the support choice, the possibility of introduction of foreign cationic impurities (Si^{4+}, Na^{+}, Cr^{3+}, Fe^{3+}) might be taken into account as a consequence of the thermal treatment applied for the enhancement of the adhesion, because these impurities would increase the electron-hole recombination processes (139).

Depending on its physical configuration, immobilized photo-catalytic systems can be divided in two groups: those photo-catalytic systems dispersed or embedded into a monolith acting as support, prepared by extrusion of a paste containing the photo-catalyst precursor together with natural silicates (140,141), and those disposed in films (137,142). However, most studied systems for photo-catalytic applications are those disposed in thin-film structures.

Within this configuration of thin film, it has been revealed that it should pay special attention to specific parameters that affect to the final photo-catalytic activity of the system. It is well known that the photo-catalytic activity of TiO_2 thin films strongly depends on the preparing method and post-deposition treatments, because they give decisive influence on the chemical and physical properties of TiO_2 included in the films. Therefore, it is necessary to choose adequate processing conditions to fabricate highly active photo-catalytic thin films (143). Thus, it has been pointed out that film thickness, porosity, and surface structure are crucial features to be considered in the immobilized photo-catalytic systems (144–146). Films with thickness smaller than the space charge layer have showed a larger photo-current than films with thickness larger than the space charge layer. The increase in photo-current is due to the effective electron-hole separation throughout the whole film thickness and the reduction of bulk recombination. Aggregation of TiO_2 particles in the interior region of thick films causes a decrease in the number of active sites and an increase in opacity and light scattering. Porosity and surface area also directly control the photo-activity of films. An appropriate roughness, and consequently high surface area and porosity, is proved to be beneficial for the enhancement of photo-activity of films (147–149). Finally, the selection of the film support is also of great significance. As mentioned, there is a great variety of support materials studied in photo-catalysis. The support choice is conditioned by its costs and its transparency to radiation. It is also worthy to note that in addition to the incorporation of impurities, the possible structural and electronic interactions between the coating and the support will affect the final photo-catalytic activity. Thus, in the case of TiO_2, the anatase–rutile transition temperature is found to be increased when TiO_2 is coating a glass support (borosilicate, ITO) (150) or stainless steel (151). Lassaletta et al. (152) describe the chemical interaction between TiO_2 and SiO_2 from the support by the formation of Si–O–Ti bondings. These bondings would induce a tetrahedral environment of Ti at the interface leading to a diminution of the electron mobility, and an increment of the band gap due to quantum effects.

17.2.2. Visible-Active Systems

Visible-light-driven photo-catalytic technology has attracted much attention and been widely studied, with the final aim of efficiently converting solar light energy into useful chemical energy. The effective utilization of the solar energy will lead to promising solutions not only for energy issues due to the exhaustion of natural energy sources but also for the many problems caused by environmental pollution. Semiconductor photo-catalysis is one of the most efficient technologies to eliminate undesired chemical substances for environmental conservation and to split water for green-energy hydrogen production (153–155). The development of new photo-catalytic materials seems to be a main research target in this field. In this sense, many efforts have been paid in recent years to develop new types of photo-catalyst materials active under sunlight mainly based on oxide semiconductors (154–157), which will be reviewed in this section.

The most viable nanoparticles for photo-catalysis are titanium dioxide, in its anatase phase, because of its high oxidative power, photo-stability, low cost, and nontoxicity (see Section 17.2.2) (2). Unfortunately, because of its large band gap of 3.2 eV, it can be activated only under UV light irradiation of wavelength less than 387 nm (158). UV light constitutes only a small fraction (5%) of the solar spectrum, so any improvement of the photocatalytic efficiency of TiO_2 by shifting its optical response to the visible range will have a profound positive effect. An important option to enhance the photo-response of titania consists of favoring the crystallization of other TiO_2 polymorphs, which have been described as susceptible materials to absorb visible radiation. The *ab initio* investigation of the electronic and optical properties of fluorite- and pyrite-type TiO_2 indicates that such cubic phases have important optical absorptive transitions in the region of the visible light. This fact is explained by the existence of a large amount of localized Ti *d* states at the bottom of the conduction band (159). On the other side, the rutile phase has a narrower band gap (3.0 eV) than anatase (3.2 eV) and hence should be easily shifted to the visible light region by suitable modification. The major obstacle for using rutile TiO_2 resides on the high calcination temperature necessary to synthesize it, which usually is obtained by calcining anatase samples. Controlling the synthesis method, it is possible to obtain an optimal rutile sample whose modification by sulfate promotion furthers the visible absorption (160). SO_4^{2-}-promoted titania incorporates the sulfate species into the TiO_2 network and occupies oxygen sites to form Ti–S bonds. The sulfur 3*p* states combine with the original titania valence band and contribute to the increased width of VB; this effect produces the decreasing of the band gap and consequently the visible absorption. Nevertheless, the main number of attempts dedicated to TiO_2 with the goal of improving their optical absorption and photo-catalytic performances is mainly focused on anatase modification (126,161). In this sense, the literature describes as the most feasible modifications processes such as doping (cationic or anionic), metal deposition, and surface sensitization (6,126), which will be described subsequently.

17.2.2.1. Anion and Cation Doping The cation doping has been described in Section 17.2.1.1 to explain its relevance in the UV-active systems. Nevertheless, doping the semiconductor with various transition metal ions may also lead to an enhanced photo-efficiency of the systems using solar irradiation, increasing the semiconductor efficiencies in the range 20–30% (6,126,162). Metal doping influences the photo-reactivity of TiO_2 by acting as electron or hole traps and by altering the e^-/h^+ pair recombination rate (see Figure 17.1). Metal incorporation introduces new energy levels in the titania band gap (163). These energy levels induce the red shift in the band gap transition and the visible light absorption through a charge transfer between the dopant and the semiconductor bands or through a *d–d* transition in the crystal field. The photo-physical mechanism of doped semiconductors is not always understood. Among others, unsolved problems related to the surface structure and to the contribution of the charge carrier dynamics are the main reasons for the complex study. All of these factors are the main cause for the controversy found in the literature concerning the doping effect. The transition metal ions more studied for titania doping are W (161,164), V (165), Fe (166), Cr (167,168), Au (169), and Mo (63,170) also, some studies exist in which the dopant agent are no transition cations such as earth alkaline metals (171) or even Bi (172). Their contribution on photo-catalytic activity under visible light is mainly related to the recombination rate (63), which in some cases is a positive effect, but in other cases, although the metal provokes an extension of absorption in the visible region, it does not enhance the photo-catalytic activity mainly focused on degradation reactions (62).

One new approach to induce visible light activated TiO_2 photo-catalysts is by doping with anions, such as N^{3-}, C^{4-}, and S^{4-}, or halides (F^-, Cl^-, Br^-, I^-) (163). It is suggested that these species substitute the oxygen lattice on TiO_2 and lead to a band gap narrowing, resulting in high visible absorption. Many works described the enhanced photo-catalytic properties of titania by nitrogen doping (173–176), although special care must be taken on the synthesis method (177–179). The nitrogen doping of TiO_2 was expected as a method for narrowing the band gap by changing the valence band structure without a change in the position of conduction band. Nevertheless, an open question in this approach is to know how the absorption shift occurs. The theory suggests that the narrowing of the band gap is due to mixing the N $2p$ and O $2p$ states (155,180), which identified the dominant transitions at the absorption edge from those from N $2p_\Pi$ to Ti d_{xy}, instead of from O $2p_\Pi$ as occurred in TiO_2 (181). Nevertheless, Irie et al. (182) suggest that the N $2p$ levels are separated (not mixed) from the valence band (formed by O $2p$ states), forming an isolated narrow band responsible for the visible light sensitivity of the oxynitride powders. On the other hand, some authors suggest that the N doping of TiO_2 is similar in properties to impurity sensitization (183). In this sense, the doped nitrogen exists as NO, and no Ti–N bonding is formed, contrary to the other cases where N is located at lattice O sites. Furthermore, doping with nitrogen not only modifies the electronic properties of titania but also can induce the formation of oxygen-defect sites (184). These species influence the photo-catalytic activity of N-doped TiO_2 because it can act as charge trapping centers; more details about this fact will be described in Section 17.4.

Coke-containing titanium dioxide is also active under visible radiation, showing a band gap narrowing comparable with the value observed for nitrogen-doped titania (185). However, the activity of $TiC_xO_{2-1/2x}$ is quietly lower than nitrogen doped systems probably due to the location of the valence band, the upper level of the BV shifts toward negative potentials, and thus, the oxidation power can decrease considerably (186). The characterization performed by density functional theory (DFT) methods suggests that carbon atoms substitute the oxygen sites, without structural changes, yielding a visible light response due to the appearance of an unoccupied impurity state occurring in the band gap. The consequently photo-catalytic activity depends on the oxygen vacancies formed during carbon doping because these defects fill the in-gap impurity states by emitting electrons inhibiting the photo-catalytic reaction (187). The synthesis method can also influence the TiO_2 band gap narrowing, which makes it possible to obtain systems with an optical gap of 2.32 eV (188) and 2.0 eV (189) by thermal treatment in air.

Other anionic dopants have been studied to enhance the photocatalytic activity of TiO_2 with visible light, as sulphur (190,191) and some halides (192–194). Although S has a larger ionic radius compared with N or C atoms, it has been possible to synthesize titanium oxysulfide where S bonds to Ti by substitution at O sites. Different band calculations indicate that the S 3*p* states mix with the valence band of TiO_2 increasing the width of the VB itself. This results in a decrease in the band gap energy and consequently leads to the photon-to-carrier conversion in the visible light region (195,196). A promoting effect on the photo-catalytic activity was also found in F-doped titania (192,193). Fluoride has been described as a substitutional component in TiO_2, especially at low concentrations (197). It has been described that F presence enhances the photo-catalytic performance by reduction of the recombination rate of the photo-generated charge carriers (198). This behavior is favored by the fact that the ionic radius of F^- ion is close to that of O^{2-} ion. Thus, an appropriate amount of F^- doping can fill the oxygen vacancies decreasing the recombination process (192,199). At this point, relevant to mention is a recent report that describes the iodine-doped titania as a novel visible-light photo-catalyst. Iodine incorporation causes an absorption in the visible light range with a red shift in the band gap transition (194).

These studies demonstrate that the doping process can provide an effective modification of the electronic structure of TiO_2 for the visible light absorption. The literature also shows the use of two dopants as another strategy to extend the visible response of titania samples (163). Recent works analyze the TiO_2 codoping process, showing optimum visible response with two different nonmetalic dopants, such as N/F (200) or N/C (201), with mixed metallic and nonmetallic dopants, such as La/N (202), or with different metal atoms, such as Sb/Cr (203).

17.2.2.2. Metal Deposition Metal coating has been described as an enhanced method to improve the photo-catalytic efficiency of a semiconductor under UV light (see Section 17.2.1.2.). The principal mechanism, described in Figure 17.2, consists of the photo-generated electron migration to the metal where it becomes trapped, avoiding the recombination process and favoring the hole diffusion to the surface (2). The utilization of these novel semiconductors under visible radiation has received some attraction, although their optimization must be investigated (4). One of the most popular metal coating systems has been Pt/TiO_2, which shows high conversion rates for different photo-catalytic reactions under UV light (see references from Section 17.2.1.2.). When this system is employed under solar or visible radiation, its efficiency is poor, as showed for the solar hydrogen production reported by Arakawa et al. (204). Better results have been recently obtained by Ag-deposited TiO_2 catalysts for the photo-catalytic degradation of rhodamine B under visible light (205). Compared with the pure TiO_2, the Ag-coated TiO_2 exhibits a significant increase in the photo-degradation rate, which results in an optimum Ag content (2%) to achieve the highest photo-efficiency. The authors suggest that Ag deposition enhances the photo-reaction according to the cooperative effects of three mechanisms. The first one supposes that Ag nanoparticles act as electron traps, enhancing the electron-hole separation and the subsequent transfer of the trapped electron to the surface. The second one considers the dye adsorption because the number of adsorbed molecules increases when titania is coated with silver. Thus, the presence of these molecules at the surface improves the photo-excited electron transfer from the visible-light sensitized dye to the conduction band and subsequently increasing the electron transfer to the surface.

The last mechanism involved is the plasmon-induced charge separation, which also has been described for Au-coated TiO_2 systems (206). This mechanism supposes that the visible light generates the photo-excited state of the metal nanoparticles based on the surface plasmon resonance. Then the photo-excited electrons are injected into the TiO_2 bulk improving the electron-hole separation (205). Simultaneously, the oxidized metal nanoparticles take electrons from a donor in the solution.

Other possible way to obtain a visible light sensitized photo-catalyst consists of TiO_2 impregnation with platinum group metal chlorides (4). Kisch et al. (207,208) studied TiO_2 impregnated with chlorides of Pt(IV), Pd(II), and Rh(III) for visible photo-degradation of 4-clorophenol, describing the TiO_2-chloroplatinate as the sample most active. The Pt complex is electronically excited by the absorption of visible light and undergoes homolytic cleavage of a Pt–Cl bond to generate intermediates of Pt(III) and an adsorbed chlorine atom. The former species injects an electron into the conduction band of the semiconductor, which subsequently reduces oxygen to superoxide and, eventually, leads to the generation of a hydroxyl radical, which both with the adsorbed chlorine atoms are assumed to oxidize the reactant. The complex content dependence was also investigated. A maximum rate was described around 3%, followed by a decrease at a higher platinum amount assigned tentatively to increasing Pt–Pt interactions (209). Gold and rhodium chlorides have also been described as sensitizers that operate via a mechanism similar to that indicated for platinum complex (209). These metal complexes suffer the problem of a slow but, ultimately limiting, photoinstability. Therefore, although metal coating has attracted a great deal of attention, its future as a route to sensitizing photo-catalytic reactions with visible light must be investigated.

17.2.2.3. Visible Light Sensitizers The solar energy improvement has motivated other fields of research, which consists of TiO_2 photosensitization by combining them with organic (210) and inorganic dyes (211–214) as well as with narrow band gap semiconductors (215). Dye-sensitized nanocrystalline TiO_2 promises to approach in the development of visible light responsive catalysts. Various photosensitizing dyes have been investigated (204,216). Grätzel et al. reported different dyes based on ruthenium complexes, such as (cis-di(thiocyanato)-*N*,*N*-bis(2,2′-bipyridyl dicarboxylate) Ru(II) and the "black dye" based on the trithiocyanato–ruthenium complex (211,214,217,218). Other charge transfer complexes capable of absorbing strongly in the visible spectrum have been formed on the TiO_2 surface after adsorption of transition metal cyanides, such as $[(Fe/Ru/Os)^{II}(CN)_6]^{4-}$, $Re^{III}(CN)_7^{4-}$, or $[(W/Mo)^{IV}(CN)_8]^{4-}$ (219). Reviewing the literature, the sensitizers of preferred choice are transition metal complexes derived from polypyridines (Ru(II)) (221–223), porphiryns, or phtalocyanines [Zn(II), Mg(II), Al(III)] as ligands.

In these systems, the dyes absorb visible light to form electronically excited states. From these states, electrons are injected into the conduction band of the semiconductor producing the visible photo-response of the catalyst. To work effectively, the excited state of the sensitizer needs to be located above the bottom of the semiconductor conduction band (224). A strong interaction between sensitizer and semiconductor is also required to have a fast and efficient electron injection. Consequently, the resulting effects of sensitization are

(*1*) the increasing of the efficiency of the excitation process, and

(*2*) the expansion of the absorption spectrum of the semiconductor through the excitation of the sensitizer. However, synthesis of these dyes involves several steps that increase their cost, and their thermal and photo-chemical stability is poor (162).

For these reasons, the attempts in photo-catalysis have been focused to replace the dye as sensitizer materials by semiconductor nanoparticles. More details about these sensitizers can be found in Chapter 16.

The association of TiO_2 with other semiconductors, possessing different energy levels from their corresponding conduction and valence bands, can enhance its photo-catalytic activity by increasing the charge separation. Two different cases can be distinguished, as described in Section 17.2.1.3 for UV-active systems: First, only one semiconductor is illuminated and the second is nonactivated; or second, both are illuminated. A proper placement of the individual semiconductors (e.g., convenient energy levels of the coupled photo-catalyst) and optimal thickness of the covering semiconductor are crucial for efficient charge separation. Undoubtedly, the geometry of particles, surface texture, and particle size also plays a significant role in the interparticle electron transfer (6). Different semiconductors have been studied to improve the solar absorption of titania, basically focused on metal chalcogenides. Bi_2S_3 is a promising candidate material because it has a fundamental absorption edge close to 800 nm. Different works have been made to control the synthesis of the sulphide nanoparticles and their interaction with titanium oxide, focusing on the particle distribution and tuning the conduction band as required (225). It is possible to obtain optimum Bi_2S_3/TiO_2 systems by a direct mixture of both semiconductors showing visible photo-degradation of different contaminants. Nevertheless, when the heterojunction is made by a precipitation method, no degradation is detected, which is attributed to titania covering by Bi_2S_3 films (100). CdS is also an important semiconductor owing to its unique electronic and optical properties and its potential applications in solar energy conversion and heterogeneous photocatalysis (226). The interaction CdS/TiO_2 yields optimum photo-catalytic activities under visible radiation. CdS can absorb visible light and generate electrons and holes. These electrons are injected into the conduction band of the TiO_2 particles where they can react with the surface reactants (228). These systems have been employed in different photo-catalytic reactions, such as contaminant photo-degradation (100) or water splitting (229). For many years, Mo and W dichalcogenides have been studied as potential electrodes for electrochemical solar conversion (230). However, the heterojunction of these semiconductors with titania has not been explored in detail because their conduction band energy levels are more negative than that of TiO_2. This means that the electrons cannot be transferred from them to TiO_2 energy levels. Even so, it has been reported that these chalcogenides exhibit quantum confinement effects, which increase their band gaps (231–233). This alternation allows employing MoS_2 and

WS_2 as photo-sensitizers for visible light. In this sense, Ho et al. (234) described MoS_2 and WS_2 nanocluster sensitized TiO_2 as an active photo-catalyst for the photo-degradation of contaminants under visible light. As described, the photo-generated electrons formed in MS_2 are transferred from the CB of chalcogenide into TiO_2 where they are accumulated at the lower lying CB of TiO_2, while holes accumulate at the VB of WS_2.

Although less studied, H_2O_2 has been described as a visible surface sensitizer that provides a novel approach for utilization of solar light to photo-degrade different pollutants (163). H_2O_2 can form complexes with valance-unfilled Ti(IV) on titania surface, exhibiting visible absorption. This light absorption produces photo-generated electrons that are transferred from the surface complexes to the CB of TiO_2, forming $^{\cdot}OOH$ or $^{\cdot}OH$ radicals that can produce the photo-oxidation of organic contaminants.

17.2.2.4. Titania Supported The synthesis and optimization of nanosized TiO_2 particles dispersed in porous supports have received attention to find new materials that can operate under visible light irradiation (162). The use of zeolites to improve the photo-efficiency of titania under UV light has been detailed in Section 17.2.1.4. These UV-active Ti/zeolites systems are not effective under visible light, but the metal implantation method enhances the visible absorption. In this sense, V-ion-implanted Ti/zeolites appears as optimum visible-photo-catalysts that led to high yield for decomposition reactions (235). In these systems, titanium oxide species are localized in tetrahedral coordination within the zeolite frameworks forming the Ti–O–V linkage with implanted V ions. This junction leads to the modification of the electronic properties of Ti-oxide species and enables the absorption of visible light to initiate reactions (236).

17.2.2.5. Metalates The strategies described in the literature to enhance the visible light absorption are focused on the modification of the conduction bands of stable oxide semiconductors. In general, these oxides are based on metal cations with d^0 or d^{10} configurations. The CB is formed by the empty orbitals of the metal cations (LUMOs), whereas the VB is based on O_{2p} orbitals. A way to modify the band gap consists of creating an electron donor level or a new valence band, which are effects described in the section above for the titania doping process. A new valence band could be also performed, employing elements with orbitals other than O_{2p}; in this sense, different metalates, described as $A_xB_yO_z$, have been studied as optimal materials for visible light absorption. The development of these materials is influenced by their photo-catalytic applications, with photo-degradation reactions and water splitting being the most common (43,154). Both types of reactions need semiconductors with an adequate band gap, but the last one is also important for the potentials of the conduction and valence bands; the bottom level of the CB must be more negative than the reduction potential of H^+/H_2, whereas the top level of the VB must be more positive than the oxidation potential of the O_2/H_2O (153,237). In this sense, different $A_xB_yO_z$ systems have been studied to obtain useful photo-catalysts for an optimum solar enhancement.

The presence of cations such as Bi^{3+}, In^{3+}, Sn^{2+} (s^2 configuration), or Ag^+ (d^{10} configuration) in an oxide system manages to elevate the valence band by means of the combination of their respective orbitals with the O_{2p} orbital from the oxygen, narrowing the band gap of the semiconductor. In this way, oxides such as $BiVO_4$, Bi_2WO_6, $AgNbO_3$, and $InMO_4$ (V, Nb, Ta) have been described as potential photo-catalysts for water splitting under visible radiation (153,154,238–243). The conducting properties of these materials depend on their electronic properties; as an example, it has been reported that the introduction of *s* orbital components increases the charge mobility and, at the same time, tunes the relative positions of CB and VB for $InMO_4$ systems (244). On the other side, the relevance of the structural properties is mainly due to the complex structures that these materials present, such as perovskites, scheelites, or spinels (245). The existence of chains or layers constituted by MO_6 octahedron favors the formation of a narrow conduction band and the delocalization of the charge carriers (246–248). As occurred in the case of TiO_2, the photo-response of these materials could be enhanced by the doping process choosing as dopants metal cations (246,249–251) or, even, some anions as nitrogen (252,253). Likewise, it has been confirmed that AMO_4 solids (M = Cr (254,255), Mo (256) , W (240,257,258); A = Ag, Bi, Pb, Sr, Ba) present optimum band gaps with related potentials to the reduction-oxidation of the water. Some $A_xB_yO_z$ have been reported as an effective photo-catalyst for contaminant degradation; important to mention are $CaBi_2O_4$ (259), $CaIn_2O_4$ (260–262), or Bi_2WO_6 (263), based on the modification of the valence band by ns^2 orbitals. Nevertheless, these materials present an important difficulty that consists of their low surface area and high particle size, mainly due to the synthesis method employed for their preparation that requires high temperatures. For this reason, important efforts are being performed to optimize the properties of these metalates to improve their photo-efficiency under visible light.

17.2.2.6. Oxynitrides, Nitrides, and Chalcogenides Most semiconductors employed in photo-catalytic reactions under solar irradiations are based on oxide compounds, just as has been described up to now. Nevertheless, other materials have been described as efficient semiconductors for photo-catalytic applications. Recently, (oxy)nitrides containing Ti^{4+} or Ta^{5+}, such as $LaTiO_2N$ (264,265), $MTaO_2N$ (266), TaON (267–270), and Ta_3N_5 (271,272), have been reported as potential visible-light-driven photo-catalysts. These systems have small band gap energies (e.g., TaON: 2.5 eV, $MTaO_2N$: 2.5–2.0 eV, Ta_3N_5: 2.05 eV) and absorb visible light at wavelengths 500–630 nm via the N_{2p} orbitals of the top of the valence bands (273). The chalcogenides have also been studied as new visible-light-driven photo-catalysts, mainly based on MoS_2, ZnS, and CdS semiconductors (232–276). These sulphides have narrow band gaps that are suitable for photo-catalytic reactions. However, they usually have poor thermal and chemical stability. For this reason, some efforts have been focused on developing an optimum chalcogenides photo-catalyst, obtaining satisfactory results with metal doped systems, such as Ni–ZnS (154,277).

17.2.3. Mechanism of Photo-Reactions: Liquid and Gas Phase

As can be inferred, heterogeneous photo-catalysis is a process in which a combination of photo-chemistry and catalysis is operable and implies that light and catalyst are necessary to bring out a chemical reaction. In this section, we attempt to illustrate possible photo-chemical phenomena that might be considered and that make possible the photo-catalytic reaction at the semiconductor interface. Thus, the primary steps involve the formation of the conduction band electrons and valence band holes and migration to the oxide surface. When carriers reach the surface, they can be ultimately tapped by intrinsic subsurface energy traps for holes and surface traps for electrons. Interfacial transfer lead to the final photo-reaction, oxidation in the case of hole transfer, and reduction in the case of electron transfer (278,279).

$$h^+ + (\mathrm{Red}_2)_{\mathrm{ads}} \longrightarrow (\mathrm{Ox}_2)_{\mathrm{ads}} \tag{17.5}$$

$$e^- + (\mathrm{Ox}_1)_{\mathrm{ads}} \longrightarrow (\mathrm{Red}_1)_{\mathrm{ads}} \tag{17.6}$$

Moreover, the secondary steps would also involve the reaction with chemisorbed O_2 and/or OH^-/H_2O molecules to generate a reactive oxygen species, such as $O_2^{\bullet -}$, $HOO^\bullet$ and $OH^\bullet$ radicals that will promote oxidative photo-reactions. In most experiments and applications with semiconductor photo-catalysis, oxygen is present and acts as an electron acceptor; thus, the formation of these radicals is aided. These oxygen species exhibit significantly high oxidant power, which can attack organic pollutants and lead to their partial or complete degradation. In addition to oxygen radicals, H_2O_2 can also be obtained from two different pathways in aerated aqueous solutions by means of photo-generated electrons and holes. This simple description of the chemical features in fact turns to a complicated concurrence of simultaneous reactions that drive the final photo-activity.

The simplified pathway for the photo-catalytic degradation of organic pollutant can be summarized as follows:

$$\text{Organic pollutant} + \mathrm{TiO_2} + h\nu \longrightarrow \mathrm{CO_2} + \mathrm{H_2O} + \text{mineral acid} \tag{17.7}$$

However, the detailed mechanism of the photocatalytic process is still not completely clear, particularly that concerning in the initial steps involved in the reaction of reactive oxygen species and the organic molecules. A simple summary of the secondary reaction steps could be as shown in Table 17.1.

TABLE 17.1. Surface Reactions Occurring in Photo-Catalytic Processes.

Surface Reactions Involving Holes	Surface Reactions Involving Electrons
	$O_2 + e^- \rightarrow -O_2^{-\bullet}$
$-OH^- + h^+ \rightarrow -OH^\bullet$	$-O_2^{-\bullet} + H^+ \rightarrow HOO^\bullet$
$-H_2O + h^+ \rightarrow -OH^\bullet + H^+$	$-O_2^{-\bullet} + 2H^+ \rightarrow H_2O_2$
$2H_2O + 2h^+ \rightarrow H_2O_2 + 2H^+$	$-O_2^{-\bullet} + e^- \rightarrow -O_2^{2-}$
	$H_2O_2 + e^- \rightarrow 2OH^\bullet$
	$H_2O_2 + -OH^\bullet \rightarrow HOO^\bullet + H_2O$

Regarding the oxidizing species (radical species versus hole), it has been stated that there are chemical evidences that both contribute to similar oxidation pathways. In this sense, for the photo-oxidation of phenol, the same intermediates have been detected by considering direct hole oxidation or radical oxidation (280).

A complete description of the photo-oxidation pathway for organic compounds is significantly complex. Detailed description of several pollutant photo-degradation in the presence of TiO_2 as catalysts has been reported by some authors (281–285). Commonly, the first step consists of the initial radical attack to the organic compound leading to double bond breaking, by means of electrophilic attack. Thus, in the case of aromatic molecules, the hydroxylation of the ring would lead to quinone and hydroquinone molecules. Following this scheme, further hydroxylation of formed intermediate molecules would lead to aliphatic intermediates.

On the basis of the photo-chemical processes described, various factors can affect the photocatalytic reaction rates in liquid-phase reactions. For example, the pH of the solution determines the surface charge on the semiconductor and the speciation of the substrate to be transformed. As it can be inferred from the above reaction schemes, the surface density of OH has an important role in the efficiency of the photo-catalytic process and strongly affects the rate of the oxygen evolution.

On the other hand, adsorption processes seem to have a particular importance because photo-catalytic reactions originate in the interface. It is logical to expect that the rate of pollutant degradation will be related to the adsorbed substrate concentration. Thus, the nature of substrate and its ability to be adsorbed clearly determine the photocatalytic efficiency (286). In this sense, Tanaka et al. found that the adsorption of dyes on TiO_2 was an important factor determining the final degradation rate (287). The kinetic regime depends on the substrate concentration, showing, in most cases, a Langmuirian behavior. Nevertheless, certain gas–solid adsorption–desorption equilibrium could be described by the Henry type relation, for which the initial photo-degradation rate was proportional to the square root of the adsorbed substrate concentration (288).

The photonic flux is also important because an excess of light promotes a faster electron-hole recombination. Ollis et al. (289) reported the effect of light intensity and stated that, at low intensities, the rate increased linearly with photon flux. At intermediate intensities, the rate depended on the square root of light intensity. And at high light intensities, the rate is independent of light intensity. Furthermore, Brusa et al. (290) recently reported that for the photocatalytic TiO_2 oxidation of cyclohexane, product selectivity and total efficiency of mono-oxygenated products critically depend on the irradiation conditions.

Particularly in gas-phase photo-reactions, the main factors affecting the photo-catalytic activity are reaction temperature, the presence of water, and the strong adsorption of intermediates, byproducts that could lead to the practical catalyst deactivation.

Temperature is one important factor in gas–solid heterogeneous photo-catalytic reactions. In the case of photo-catalytic oxidation applications in water, the narrow temperature range, which can be selected, can be a serious drawback, but in gas-phase applications, the operational range is wider. Catalytic oxidation with high yields is possible today at temperatures as low as 150 °C. High temperatures generally lead to higher rates because they provoke a more frequent collision between the substrate and the semiconductor. Generally speaking, an increase in the reaction temperature leads to an enhancement of the reaction rates, except for the photo-generation of electron-hole pairs (291). Thus, it has been reported that photo-catalytic degradation of trichloroethylene and methanol was more effective at a moderate temperature than at higher temperatures (292). Complete mineralization of trichloroethylene is significantly improved by increasing temperature up to a temperature limit of 125 °C using a TiO_2–monolith type photo-catalyst (293). Adsorption/desorption limitation during the photo-catalytic oxidation at temperatures above 125 °C is assumed to be the reason of the degradation drop above this temperature. At low temperature, the product desorption can be the rate-limiting (294) step; in contrast, adsorption of reactant would be the rate determining step at high temperature (292). Therefore, a compromise situation might be achieved in each specific case.

For gas-phase photo-degradation reaction, the substrate adsorption process is critical, and it has been reported to strongly depend on the relative humidity (295,296). The influence of water vapor on the photo-oxidation is complex because water plays an important role in the formation of the active species. In certain cases, the presence of water could be beneficial; meanwhile, in other cases, it has been reported to be detrimental. In general, it is assumed that water vapor has two different roles. At low humidity levels, water seems to aide to keep constant oxidation rates by the reposition of the surface hydroxyl groups exhausted during reaction. On the contrary, high concentrations of water vapor occur, a competition between water and an organic pollutant substrate for active sites would take place, and a significant decrease in the adsorbed substrate clearly influences the final photocatalytic degradation rate (297). Furthermore, the recombination process could also be favored by the presence of adsorbed water (298). However, as stated, a certain degree of humidity is required to keep the catalyst surface hydroxylation (295,299).

Deactivation processes seem to be more important in gas-phase reactions than in liquid phase. Some intermediates such as carboxylic acid formed during the photo-degradation of the organic pollutant can be chemisorbed at the active sites, leading to a progressive photo-catalyst deactivation. Up to date, there have been a few reports concerning the origin of the often observed decline in the photo-catalytic reactivity of TiO_2 photo-catalysts in both aqueous and gas-phase reaction systems (298,300–302). Cunningham et al. (303) reported this deactivation process in the gas-phase photo-catalytic reactions, pointing out three possible causes:

(*1*) the active site blocking by CO_2 as well as other carbonaceous deposit adsorption,

(*2*) the large consumption of the surface oxygen species, and

(*3*) the irreversible dehydroxylation of the surface.

17.2.4. Mechanism of Photo-Reactions: Disinfection

There are many circumstances in which it is necessary or desirable to remove or kill micro-organisms found in water, air, on surfaces, or even in a biological host.

Actually, the processes for pathogenic organism removing represent a top priority. Disinfection is required for water human consumption or in medical facilities in which biological contamination might be prevented. In this sense, many authors have shown the ability of the photo-catalytic process as a germicide method. By contrast to photo-catalytic processes, the water disinfection by UV light has been intensively studied for many years (304), and mechanisms of UV action and subsequent dark repair of cells has been proposed (305). Different mechanisms involved in the bactericidal action of TiO_2 photo-catalysis have been described (306,307), some of which were reviewed by Blake et al. (309). Results from the above studies suggest that the cell membrane is the primary site of reactive photo-generated oxygen species attack. Oxidative attack of the cell membrane leads to lipid peroxidation. The combination of cell membrane damage, and further oxidative attack of internal cellular components, ultimately results in cell death.

Photo-catalytic methods are unique in having several modes of action that can be brought to bear on disinfection. The target of disinfection processes are pathogenic organisms, including viruses, bacteria, fungi, protozoa, and algae. Ireland et al. (310) were the first to report inactivation of micro-organisms in waters using TiO_2 as a photo-catalyst. Further studies reported the photo-catalytic inactivation of bacteria such as *Escherichia coli*, *Bacillus pumilus*, *Salmonella typhimurium*, *Serratia marcescens*, *Pseudomonas stutzeri*, *Staphyloccus aureus*, *Streptococcus mutants*, *Lactobacillus acidophilus*, *Streptococcus rattus*, *Actinomyces viscosus*, *Streptococcus sobrinus*, *Bacillus subtilis*, spores of *Clostridium perfringens*, as well as lactobacillus phage PL-1, poliovirus, phage Q, and phage MS-2, virus (309). When water samples were irradiated with a UV-lamp (300–400 nm) in the presence of a semiconductor, a rapid cell death was observed. This effect was attributed to the generation of the strong oxidant $OH^{\bullet}$ radicals (311). Therefore, the conjunction of UV-light and a photo-active semiconductor were essential for the effective bacterial deactivation (312,313). The mechanism leading to the cell death is not yet fully understood. For a cell or virus in contact with TiO_2 surface, there may be a direct electron or hole transfer to the micro-organism or one of its components. Additionally, if TiO_2 particles have adequate size, they may penetrate into the cell and these processes could occur in the interior. The first proposed mechanism implies the oxidation of the intracellular coenzyme A, inhibiting the cellular living. On the contrary, it has also been proposed that cell death takes place by cell wall harm. This wall damage, followed by cytoplasmic membrane damage, leads to a direct intracellular attack as the sequence of events when micro-organisms undergo TiO_2 photo-catalytic attack. After eliminating the protection of the cell wall, the oxidative damage takes place on the underlying cytoplasmic membrane (314,315).

Many parameters are affecting the photo-catalytic disinfection process. In addition to the well-known operational parameters that influence the photo-catalytic process (pollutant concentration, pH, catalyst loading, ionic strength, photon flux, etc.) (283), photo-catalytic disinfection is affected by some others factors. Concerning photo-catalytic degradation of micro-organisms, Rincon et al. described the influence of several parameters on the photo-catalytic disinfection of water using TiO_2 as a photo-catalyst (316–318). They evaluated parameters such as light intensity, periodicity of irradiation, temperature, turbidity, and the amount of suspended and fixed catalyst.

Furthermore, the sensitivity of micro-organism to photo-catalytic treatment is clearly influenced by the chemical and microbiological characteristics of waters.

Bacterial reactivation after a photo-catalytic process is a serious problem that must be taken into account (319). In this sense, the rate of bacterial reactivation after photo-treatment is probably influenced by the generation of bioavailable organic and inorganic intermediates generated during photo-catalytic treatment and depends on the chemical characteristics of water. Illumination time is one of the most affecting parameters related to complete micro-organism inactivation. In this sense, it has been reported that in same cases, long irradiation time is required (316,320), whereas in some other cases, intermittent illumination is required. As a consequence, the effective disinfection time required for total inactivation of a micro-organism should be determined for the evaluation of the effectiveness of photo-catalytic disinfection to assure no regrowth process.

As mentioned, turbidity, caused by the presence of particulate matter, is an important parameter that affects negatively on the bacterial inactivation rate. This diminution in the disinfection process can be explained in different terms such as bacterial growth stimulation, oxygen diminution, competition of organic particles toward OH radical, or diminution of light penetration (316).

In the same way, temperature greatly influences photo-catalytic disinfection processes. Thus, micro-organism elimination by photo-catalysis is notably susceptible to this parameter, and narrow temperature ranges must be set to perform the best photo-catalytic disinfection rates. Additionally a critical temperature range exists for each kind of micro-organism that depends at the same time on other factors such as cell concentration or their physiological conditions.

Thus, the real application of photo-catalytic disinfection is limited by the above described factors, including long irradiation times to achieve safe inactivation rates and bacterial reappearance. From these considerations, the photo-catalytic inactivation of micro-organism is restricted to small volumes of water with a low microbial contamination.

17.3. POTENTIAL ASSISTED PHOTO-CATALYSIS

As stated in this chapter, photo-catalytic systems based on solid suspensions of semiconductor catalyst need a separation step at the end of the degradation process. To overcome this problem, immobilized systems have been proposed. However, in such systems, an additional diffusional limitation of reactives and products at the surface clearly leads to low quantum efficiency for the generation of surface hydroxyl groups. Low photocatalytic efficiency is therefore caused by a highest recombination rate with respect to charge transfer processes.

The first approach to enhance the efficiency of the photo-catalytic process is therefore the scavenging of photo-generated charge carriers and the enhancement of charge transfer of electrons and holes to their respective acceptors. Minimizing the recombination implies the increase of carrier availability for both oxidation and reduction processes that take place spatially separated at the surface. Thus, under this point of view, electrically assisted photo-catalysis combines the advantages of photo-catalysis and electro-catalysis. Electrochemically assisted photo-catalysis is a kind of photo-electrochemical combined technology that could improve the efficiency of photo-catalysis through an extra electric field. This concept of achieving charge separation in a semiconductor system with an electrochemical bias was first introduced by Honda and Fujishima (321). Using a single-crystal TiO_2, they were able to carry out the photo-electrolysis of water under the influence of an anodic bias.

Schematically, in the electrochemically assisted photo-catalytic degradation process, photo-generated electrons are driven out through an external circuit by the application of an electric field. Thus, oxidation and reduction reactions would take place at different electrodes of this photo-electrochemical device (Figure 17.3). Consequently, due to the electron scavenging process, the photo-reduction reaction will occur with lower activation energy and as a result the photo-oxidation rate will increase.

When the potential is modified externally, the electron concentration at the surface is also changed with respect to the equilibrium situation. For an n-type semiconductor, the electron density at the surface increases as applied potential is more negative than the flat band potential (U_{fb}). At the same time, the spatial charge region also varies with the applied potential.

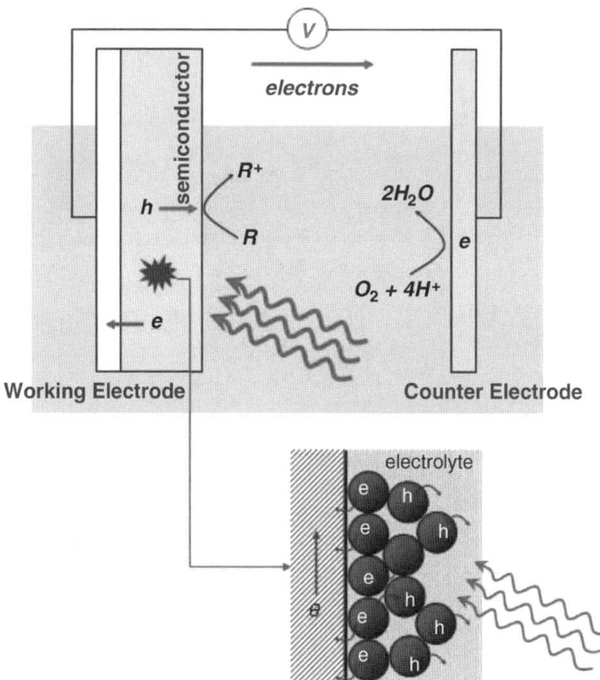

Figure 17.3. Reaction scheme of a typical potential-assisted photo-catalytic process. See color insert.

TABLE 17.2. Initial Steps of an Organic Molecule Photo-Catalytic Degradation Mechanism.

$TiO_2 + h\nu \rightarrow h^+_{vb} + e^-_{cb}$	$v_1 = k_1$
$-OH^- + h^+_{vb} \rightarrow -OH^\bullet$	$v_2 = k_2(-OH)(h^+_{vb})$
$-OH^\bullet + e^-_{cb} \rightarrow -OH^-$	$v_3 = k_3(-OH^\bullet)(e^-_{cb})$
$-OH^\bullet + R \rightarrow -OH^- + R^{\bullet +}$	$v_4 = k_4(R)(-OH^\bullet)$
$R + h^+_{vb} \rightarrow R^{\bullet +}$	$v_5 = k_5(R)(h^+_{vb})$
$R^{\bullet +} \rightarrow e^-_{cb} + R^{2+}$	$v_6 = k_6(R^{\bullet +})$
$R^{\bullet +} + e^-_{cb} \rightarrow R$	$v_7 = k_7(R^\bullet)(e^-_{cb})$

In a photo-electrochemical cell, in the absence of direct reactions between the oxidant and the organic substrate to be degraded (typical photo-catalysis process), the photo-current provides a direct measurement of the reaction rate (322). For each hole transferred in the semiconductor interface, an electron driving by the electric circuit must exist. For a typical photo-catalytic mechanism, the steps in Table 17.2 would be considered:

Thus, for a kinetic study of the reaction, the different step rates can be deduced:

$$d(h^+_{vb})/dt = v_1 - v_2 - v_5 \tag{17.2}$$

$$d(e^-_{cb})/dt = -I_{ph}/(FA) + v_1 - v_3 + v_6 - v_7 \tag{17.3}$$

$$d(R^\bullet +)/dt = v_4 + v_5 - v_6 - v_7 \tag{17.4}$$

$$d(OH^\bullet)/dt = v_2 - v_3 - v_4 \tag{17.5}$$

where I_{ph} is the photocurrent intensity, F is the Faraday constant, and A is the electrode area. At this point, we may suppose that the electroneutrality condition is achieved immediately $d(h^+_{vb})/dt = d(e^-_{cb})/dt$, and therefore, photo-current in a photo-electrocatalytical process can be defined in terms of partial reaction rates:

$$\frac{I_{ph}}{F \cdot A} = v_2 + v_5 - v_3 + v_6 - v_7 \tag{17.6}$$

As can be inferred from this expression, measured photo-current gives direct access to the photo-assisted process that is taking place. In practice, in a photo-electrochemical experiment, it is possible to measure separately the oxidation and reduction rates by means of the anodic and cathodic polarization curves. It can be stated that for the anodic polarization curve, a great dependence with illumination exists. On the contrary, cathodic curves do not experiment with this dependence. By combining the polarization curves, it is possible to determine the operational potential for which electron and hole flows are identical.

The efficiency of the photo-electrocatalytic devices for the degradation of organic pollutants has been demonstrated in recent contributions (323–327). Photo-electrochemically active semiconductor film devices have been prepared from colloidal ZnO, TiO_2, SnO_2, and WO_3 (324). Alternatively, other systems include coupled semiconductors SnO_2–TiO_2 (328) and recently TiO_2-based nanotubes (329) or single wall carbon nanotube films (330).

A special case to be considered is the application of photo-electrocatalytic devices for solar hydrogen production. Efficient hydrogen production by photo-electrochemical water splitting from sunlight is considered an important milestone in energy production since the work of Fujishima and Honda. Despite decades of research, today's best photo-electrocatalytic systems suffer from limitations that prevent their incorporation into the commercial energy market: Efficient material systems are typically expensive or unstable, whereas inexpensive and robust alternatives do not have adequate efficiency (331). However, some references make this objective a possible target within this hopeful field (332). In this sense, from literature data, efficiency values near 2% have been reported leading to average hydrogen evolutions of 400–700 μmol/h. A wide range of semiconductor materials has been proposed for such photo-electrochemical cells, including metal oxides, perovskites, and pyrochlore structures. Although these low hydrogen production efficiencies are reported, the growing number of studied systems opens up interesting possibilities in this field.

17.4. EFFECTS RELATED TO SIZE

The influence of the particle size over the photo-reactivity of semiconductors has been studied by several researches, with contradictory results (1). Some authors described that the quantum yield increases when the particle size decreases. This effect could be related to the increased surface area of the smaller diameter particles or to the enhanced redox potential of the conduction band electron, and may be due to the concentration of unsaturated surface sites (27,333). Nevertheless, other researchers described the inverse effect. They speculate that the quantum yield increases as particle size increases because the electron-hole recombination rates at the surface are slower due to the enhance separation of the photo-generated chargers (334). Also, this effect could be due to surface speciation and surface defect density, which are highly related to the preparation method (335,336). More recent studies suggest the existence of an optimum particle size for photo-catalytic reactions, due to competing effects of effective particle size on light absorption and scattering efficiency, charge-carrier dynamics, and surface area, which will be described at length in Section 17.4.3 (33,337). Before such section, Section 17.4.1 will give an overview of the structural and electronic properties of solid oxides of interest to explain photo-catalytic activity and Section 17.4.2 will describe results concerning the influence of size on the so-called "initial processes," e.g., the common steps to all photo-chemical reactions.

17.4.1. Effects on Structural and Electronic Properties

Structural and electronic properties of solid semiconductors have been thoroughly examined in Chapters 1 to 14, so in this section, we will just give a brief overview, including the specific points of interest for photo-catalysts.

17.4.1.1. Structural Effects Particle size holds a high influence over the structural properties of any material, e.g., its lattice symmetry and cell parameters. As described in Chapters 1 and 5, when the particle size decreases, the number of surface and interface atoms increases, generating strain/stress and structural perturbations, highly linked to the method used for synthesizing the solid (sol-gel process, hydrothermal synthesis, gas-phase thermal decomposition, etc.) (338). These modifications can lead to solids with different reactivity.

The structural effects with high influence over photo-catalysts are the crystalline phase, the nature and number of defect centers, and the surface chemical state. The importance of the defect centers is due to their behavior as trapping (electron or hole) or recombination centers, influencing the separation charge step. The surface chemical state is also important, because it influences charge transfer to molecules and the efficiency of photo-generated electrons and holes. We dedicate this section to describe the defects intimately linked to the particle-size effect, their nature, properties, and their influence on photo-reactions.

17.4.1.1.1. Nature of Defect Centers The size effect on the structure and stability of a particle is related to the energy balance between the surface and the bulk. The bulk structure of oxides is shaped by a close packing of the ion. Assuming thermodynamic control when the particle grows, the bulk structure extends by the lowest energy direction, creating surfaces that tend to minimize the surface energy by ion rearrangement or surface relaxation. This behavior causes structural differences between the bulk and the surface (339). When a nanosized solid is shaped, most ions are located at surface with unsatisfied coordination (340). Controlling the preparation method (reducing hydration and protonation effects), the energy minimization is reached by surface reconstruction, which increases the density of edges/corners and provides an increasing number of surface sites with respect to ideal-bulk terminology (341). The stability of a nanoparticle is also controlled by its stoichiometry, which can differ from the bulk to the surface. Metal-oxide nanoparticles usually present the metal ions with its maxima coordination, whereas the oxygen ions are located at the surface (339), typically in the form of hydroxyl groups, although the existence of bulk nonstoichiometries has sometimes been described (342). These effects produce the presence of defect sites in the nanoparticle surface that differs from those sited at the bulk.

The presence of defect sites and its difference between surface and bulk possess a relevant influence on the photo-induced processes that take place during a photo-catalytic reaction on a nanosized photo-catalyst. As described before, the photo-catalytic reaction began when a semiconductor is irradiated with photons whose energy is equal to or greater than its band gap energy ($E_{h\nu} > E_{\text{band gap}}$), leading to creation within the bulk of electron-hole pairs, which dissociate into free photo-electrons in the conduction band and photo-holes in the valence band. In the absence of scavengers, the energy is dissipated by recombination processes. If a suitable scavenger or surface defect state is available to trap the photo-generated charges, the recombination process is prevented and subsequent redox reactions may occur (1,343). It is in the charge trapping processes where the structural effects caused by the size effect

TABLE 17.3. Classification of Crystal Defects.

	Type of Defect
Point defect	Interstitial atoms
	Frenkel and Schottky defects
Line defect	Edge dislocation
	Screw dislocation
Plane defect	Lineage boundary
	Grain boundary
	Stacking fault

present their main influence. The main imperfections or defects detected in crystals are extensively described in crystallographic literature (344,345), and their presence in the main photo-catalytic materials will be described in Table 17.3.

As mentioned, a titanium dioxide semiconductor has been described as an ideal photo-catalyst due to their chemical stability, nontoxicity, and high photo-catalytic reactivity in the elimination of pollutants in air and water (126). The structural studies of the titanium–oxygen system show a phase diagram constituted by many stable phases (TiO_2, Ti_2O_3, Ti_3O_5, and TiO_{2-x} Magnéli phases) with a high diversity of structures. These phases show a high reducibility and oxygen mobility. The bulk reduction causes the formation of bulk defects such as doubly charged oxygen vacancies, Ti^{3+} or Ti^{4+} interstitials, and planar defects (342,346). The formation of these defects depends on the preparation method, from variables as temperature or pressure, complicating the analysis of the dominant defect.

The study of the surface defects in the anatase is not a simple one. The principal difficulty resides in the growth of adequate crystallites for analyzing the face thermodynamically stable. This fact is mainly governed by the synthesis method and the particle size effect. Different thermodynamic studies described the anatase phase as unstable compared with the rutile phase as bulk materials. Nevertheless, when the surface contribution to the energy is significant, the stability of the phases varies. To this respect, the thermodynamic work reported by Zhang et al. (340) shows that for particle sizes close to 14 nm, the free energy of rutile is higher than that of anatase, from which the anatase phase becomes more stable than the rutile phase. The authors indicate the higher rutile energy as a consequence of the higher unsatisfied charge densities at surface, which causes a higher surface-free energy. The study of the surface energy from the relaxed structures reveals its influence on the nanoparticle structure, although it also shows that the difference in stability between anatase and rutile phases is very small (347). Concerning the stability of nanophases, the work published by Li et al. can also be mentioned in which they demonstrate that the nature and stability of TiO_2 nanoparticles depend on additional factors such as the surface hydration controlled by the preparation method, with it being possible to synthesize stable rutile nanoparticles (31,348). Likewise, a recent work from Zhang et al. (349) provides instructions for preparing nanosized titania of a desired phase and of even mixed phases.

Several authors have analyzed the dominant faces present in anatase nanoparticles, establishing that the (101) surface is the most stable face, with smaller dominance of (001) surfaces (350). The (101) surface is very corrugated and shows O–Ti–O double chains along the [010] direction. This chains are connected by bridging oxygens, which also form rows along the same direction, creating a saw-tooth profile perpendicular to the [010] direction. This surface presents both fivefold- and sixfold-coordinated titanium atoms as well as twofold- and threefold-coordinated oxygens. The studies performed over the (001) surface only reveal the presence of fivefold-coordinated titanium atoms, next to twofold- and threefold-coordinated oxygen atoms (347,351,352). The presence of uncoordinated species at both surfaces can give rise to different crystal imperfections during the TiO_2 preparation process, with the most common as follows: step edges, line defects, impurities, and point defects, as the oxygen vacancies (342,346). Although much advanced research has been performed to learn about the bulk and surface of TiO_2 nanoparticles, contradictory results exist in the literature, among them, not alone as for the defects present in both but also concerning their role on the photo-catalytic activity (353). In this sense, special attention will be lent in the following section on the photo-catalytic behavior of titania defects.

In a more or less general view of photo-catalytic systems, we have described the efforts performed in the last seven years to enhance the photo-catalytic activity of semiconductors, which have mainly focused on the development of new capable materials to take advantage of the sunlight. One way to improve the efficiency of the photo-catalytic process consists of doping the photo-catalyst with suitable metal ions as described in detail in Sections 17.2.1.1 and 17.2.2.1. Some authors have reported a negative effect (80,354,355), but other researchers have described some metal-doped TiO_2 (Fe^{3+}, Mo^{5+}, Ru^{3+}, V^{4+}, or Rh^{3+}) as efficient photo-catalysts, with higher quantum yields than the pure titania (62,356). The metal-doped TiO_2 solids whose structural characteristics have been studied in more detail have been iron- and tungsten-doped TiO_2, lending special care on the particle size effect.

Iron-doped TiO_2 can form stable titanium–iron solid solutions, Fe_xTiO_y, with iron percents lesser than 5%. When these phases are formed, the iron ions are inserted into the titania structure and are located at interstitial sites or occupy some Ti^{4+} positions. The doping effect produces the formation of cationic vacancies, demonstrated by Rietveld refinement, whose concentration depends on the iron percent and the temperature of the synthesis treatment. The stability of these titanium lattice defects is explained by compensation for the extra hydroxyl ions present in the structure, indicated another time the influence of the synthesis process over the structural properties of nano-photo-catalysts (357).

The structural effect of the tungsten-doping into the anatase structure of nanoparticles has also been analyzed to understand its contribution over the TiO_2 photo-catalytic activity. W-doped TiO_2 forms mixed metal oxide, $Ti_{1-x}W_xO_{2+x}$, with the 20% being the solubility limit of W in the anatase network (161). The tungsten atoms occupy substitutional positions in anatase structure with the presence of cationic vacancies at their first cation–cation distance. The presence of vacancies was confirmed by EXAFS and Raman analysis to satisfy electroneutrality requirements. The W/Ti substitution and the cationic vacancies modify the anatase structural properties; the structure is distorted at a local level around W/Ti centers, and the lattice cell parameters are changed. The tungsten incorporation also affects the particle size, which decreases when the W content increases, probably ascribable to a limited stability of the mixed oxide nanoparticle with respect to the single oxide nanoparticle (358).

17.4.1.1.2. Defect Centers for Electron and Hole Trapping The photo-catalytic process comprises a complex mechanism that begins with the absorption of photons by the photo-catalyst and follows with the generation of electron-hole pairs. These photo-generated charges can recombine rapidly in the bulk although it is also possible to transfer to the semiconductor surface, where they can react with chemisorbed molecules or obviously recombine between them. The structural defects can act as trapping centers, with an important factor being their location, because the defects present at the surface could be involved in chemical reactions while the bulk defects are involved in recombination process. The small size of nanoparticles can also contribute to the recombination of the trapped charges themselves, due to the spatial proximity of the surface trapping centers. They are attracted by electrostatic forces, avoiding the stabilization of charge separation. To comprehend the behavior of trapping centers and to find the manner to block the recombination process, some works have analyzed the nature and properties of electron and hole trapping sites, with TiO_2 being the more studied semiconductor (8).

Electron and hole trapping sites have been investigated by several spectroscopic techniques. EPR has been extensively used because it provides information about the molecular environment and the electronic structure of paramagnetic intermediates in electron transfer reactions. The electron trapping centers detected in TiO_2 semiconductor are Ti^{4+} ions and oxide vacancies (8,359). The Ti^{4+} ions can be divided in two types of trapping sites: one of them located inside the nanoparticle and the other present at the surface (7). The bulk Ti^{4+} ions (defect sites) can interact with the photo-generated electron giving rise to Ti^{3+} center formation, which has been assigned by different authors as an interstitial ion. On the other side, the surface Ti^{4+} ions can trap the electrons when hole scavengers are present in the reaction medium, generating octahedral Ti^{3+} surface sites (8,360). Recently Hurum et al. (361) have identified a new trapping site specific to Degussa P25 titania, assigned to distorted four-coordinated interfacial sites. These data indicate that in this photo-catalyst, there are two different kinds of electron surface trapping sites: the anatase surface sites and the interfacial sites.

In many cases, titanium dioxide is prepared from titanium oxo-hydroxide by different techniques (sol-gel, hydrothermal, etc.). The presence of hydroxyl groups at TiO_2 surface has been studied to elucidate their influence over the electronic processes that take place during the photo-catalytic reaction. Some researchers have suggested that these surface hydroxyls, formed by the synthesis process, can be part of electron traps. Szczepankiewcz et al. (362) propose that photo-excited electrons are trapped at the surface as a Ti^{3+}–OH "complex." The formation of these species is facilitated by the ability of hydroxyl groups to dissipate the resulting energy through O–H related vibrations. The authors analyzed by infrared spectroscopy the electrons in the free state and both the electron trapping and the annihilation processes. The formation of these new hydroxylated species is described by the following scheme:

$$\mathrm{Ti^{IV}{-}OH} + e^-_{cb} \longrightarrow \mathrm{Ti^{III}{-}OH} \quad \textbf{(17.8)}$$

$$\mathrm{Ti^{III}{-}OH} + h^+_{vb} \longrightarrow \mathrm{Ti^{IV}{-}OH} \quad \textbf{(17.9)}$$

The photo-generated electron is trapped at the hydroxylated surface titania Ti^{4+}–OH (Reaction **17.8**), remaining in the particle in absence of a reducible species, although eventually is annihilated with an available hole (Reaction **17.9**) (362–364). These Ti^{3+}–OH species have also been detected when the titania activity was studied in the presence of water. In this case, their formation is attributed both to the electron trapping in Ti^{4+}–OH groups and to water dissociation at oxygen vacancies (365). These hydroxylated species suffer the Stark effect. The infrared studies show that the intensities of surface hydroxyl stretching bands are sensitive to electric fields caused by trapped charge carriers (366). This indicates, on the other hand, a strong localization of the trapped charge carrier.

As indicated, the oxygen vacancies ($Ti^{3+}V_oTi^{4+}$) have also been described as electron trapping centers, both located at the bulk or at the surface of nanoparticles, whose behavior is described as follows: $V_o^{2+} + e^-_{cb} \rightarrow V_o^+$ (8,367–369). The ionized oxygen vacancy formed (V_o^+) can subsequently interact with a valence hole band, yielding typically radiative recombination processes (80).

The study concerning hole trapping centers is more complex. The literature shows certain controversy about the nature of hole traps, mainly due to its dependence on the semiconductor surface properties. Because in an *n*-type semiconductor, as TiO_2, the holes are dominantly localized at the nanoparticle surface, their interaction with trapping centers depends on factors such as the synthesis method, the surface treatment, or its modification (8,359). Some reports assume that the photo-generated holes are trapped by surface hydroxyl groups, yielding hydroxyl radicals that were detected by the EPR technique (370–372).

$$\left\lfloor \mathrm{Ti^{IV}{-}O^{2-}{-}Ti^{IV}{-}OH} \right\rfloor + h^+_{vb} \longrightarrow \left\lfloor \mathrm{Ti^{IV}{-}O^{2-}{-}Ti^{IV}{-}{}^{\bullet}OH} \right\rfloor^+ \quad \textbf{(17.10)}$$

Nevertheless, Howe and Grätzel (373) detected by the same characterization technique a different signal assigned to the hole localized on the lattice oxygen atoms located not on the surface, but on the immediate subsurface layer of the hydrated anatase, yielding the following structure:

$$\left\lfloor \mathrm{Ti^{IV}{-}O^{2-}{-}Ti^{IV}{-}OH} \right\rfloor + h^+_{vb} \longrightarrow \left\lfloor \mathrm{Ti^{IV}{-}O^{\bullet -}{-}Ti^{IV}{-}OH} \right\rfloor^+ \quad \textbf{(17.11)}$$

Other researchers proposed that the holes produced by irradiation of titanium dioxide move to the surface, where they are trapped on oxygen radicals that are covalently bounded to Ti^{4+} ions (374):

$$\left\lfloor Ti^{IV}-O^{2-}-Ti^{IV}-OH \right\rfloor + h^{+}_{vb} \longrightarrow \left\lfloor Ti^{IV}-O^{2-}-Ti^{IV}-OH^{\bullet} \right\rfloor + H^{+} \quad \textbf{(17.12)}$$

More recently, Nakamura et al. have explained the dependence of the hole reactivity with the crystal-face (375). When a photo-generated hole reaches the surface, it can be trapped there, resulting in a surface trapped hole, named STH. The authors determine that the behavior of these STH depends on the crystal faces of rutile TiO_2. The (100) face poses its STH at the bottom of atomic grooves, covered by Ti–OH groups. Thus, these STH at this face are stable and can be accumulated in TiO_2 nanoparticles. The accumulation of these species can produce a high concentration of photo-generated electrons that can react with the STH by a radiative recombination, as it is indicated subsequently, being detected by using the photo-luminescence technique:

$$[\text{Ti–O–Ti}]_s + h^{+}_{vb} \longrightarrow [\text{Ti–O} \cdots \text{Ti}]^{+}_{s}(\text{STH}) \quad \textbf{(17.13)}$$

$$[\text{Ti–O} \cdots \text{Ti}]^{+}_{s} + e^{-}_{cb} \longrightarrow h\nu \quad \textbf{(17.14)}$$

On the other hand, the (110) face is exposed directly to the environment, and the STH formed can interact with H_2O or OH adsorbed at surface, giving rise to a new surface species. The difference between this situation and that mentioned can be observed by using photo-luminescence, because in the previous case, a radiative recombination was produced:

$$[\text{Ti–O–Ti}]_s + h^{+}_{vb} \longrightarrow [\text{Ti–O} \cdots \text{Ti}]^{+}_{s}(\text{STH}) \quad \textbf{(17.15)}$$

$$[\text{Ti–O} \cdots \text{Ti}]^{+}_{s} + \text{H}_2\text{O/OH} \longrightarrow [\text{Ti–O}^{\bullet}\ \text{HO–Ti}]_s + \text{H}^{+} \quad \textbf{(17.16)}$$

Similar experiments performed with ZnO as photo-catalyst show the same behavior just described for a TiO_2 semiconductor. Hydrated groups covering the surface of ZnO are trapping sites for the photo-generated holes, yielding radicals as follows: $\text{Zn–O} + h^{+}_{vb} \rightarrow \text{Zn}^{II}\text{O}^{\bullet}$ (374).

The presence of the charge trapping sites in TiO_2 nanoparticles is therefore related to the surface properties. It has been described that the surface modification of nanoparticles can result either in the formation of a new trapping sites or in the elimination of some trapping sites present in the surface. As shown, these surface species can trap holes yielding $HO^{\bullet}_{ads}$ and consequently free hydroxyl radicals, which can dimerize to form peroxide (373). There is abundant literature documenting the formation of hydroxyl radicals in photolysis of TiO_2 (372,373,376,377); many of those works study the hydroxyl formation by EPR using, as an example, nitroxide spin traps and nitroxyl radicals in their detection (372).

$$\text{HO}_{ads} + h^{+}_{vb} \longrightarrow \text{HO}^{\bullet}_{ads} \longrightarrow \text{HO}^{\bullet}_{free} \quad \textbf{(17.17)}$$

$$2\text{HO}^{\bullet}_{free} \longrightarrow \text{H}_2\text{O}_2 \quad \textbf{(17.18)}$$

The importance of adsorbed species has been described by Chen et al. to explain the photo-catalytic activity of nanocrystalline anatase (378). They described the surface oxygen vacancies ($Ti^{3+}V_oTi^{4+}$) as the sites where the oxygen is adsorbed, yielding to superoxide species O_2^- (346), as well as ones where the photo-generated electrons are trapped. Then, if a nanocrystalline TiO_2 contains a high amount of defects at the surface, it would present a good photo-catalytic activity. During the photo-catalytic reaction, along the transfer of photo-generated electrons at the intersurface of TiO_2, the holes are accumulated gradually, increasing the probability of electron-hole recombination, and thus, the catalytic activity falls. However, the presence of surface OH species can avoid the recombination process because these species act as hole trapping sites. The authors detailed the existence of a Ti^{3+}/OH optimum ratio for an effective photo-catalyst, with the value 1 being the best for its photo-catalytic study.

Other adsorbed species have also been described as charge trapping sites. Alcohols (ROH) and alkoxy (ROOH) groups have been described as hole trapping centers when they are anchored to the nanoparticle surface (376,379,380). In this way, an important survey has been performed by Shkrob et al. (380) analyzing the influence of different chemisorbed polyhydroxylated compounds (C_2–C_6 polyols) produces. They suggest that the polyol chemisorption involves an octahedrally coordinated Ti atom that is chelated by neighboring hydroxyl groups of the ligand. These binding sites serve as trapping sites for the photo-generated holes. The resulting trapped hole transfers a C–H proton to the environment, yielding a relatively stable titanium-bound-olyl radical Ti^{IV}–O–C$^{\bullet}$–RR'. These polyols complexes can compete for the hole both with surface defects and with electrons just because their reaction time (<100 fs) is not so far from the indicated hole scavengers. On the other side, important to mention is the behavior of the oxygen adsorbed as an electron trapping center. The influence of the adsorbed oxygen on the photo-catalytic activity has been extensively study and can be summarized in the scheme shows below. First of all, the photo-generated electron is transferred and stabilized on the adsorbed oxygen, which is reduced to a superoxide radical ion (Reaction **17.19**) (359,373,381). This radical can also capture protons, when available, generating O_2H radicals (Reaction **17.20**) or can be oxidized by other photo-generated electron yielding peroxide radicals (Reaction **17.21**) (373). These resulting trapped electrons can be involved in the photo-catalytic reaction, usually increasing the effectiveness of the photo-catalyst:

$$O_{2_{\mathrm{ads}}} + e^-_{cb} \longrightarrow O^-_{2_{\mathrm{ads}}} \qquad \textbf{(17.19)}$$

$$O^-_{2_{\mathrm{ads}}} + H^+ \longrightarrow HO^{\bullet}_{2_{\mathrm{ads}}} \qquad \textbf{(17.20)}$$

$$O^-_{2_{\mathrm{ads}}} + e^-_{cb} \longrightarrow O^{2-}_{2_{\mathrm{ads}}} \qquad \textbf{(17.21)}$$

The study of metal-doped TiO_2 photo-catalysts has demonstrated that dopant ions can be involucrated in electron and hole trapping processes. The influence of these species over the photo-catalytic activity depends on the nature and concentration of the dopant agent, as have been described in previous sections. The presence of metal ions in titania does not modify significantly the valence band edge of anatase; instead, it introduces new energy levels from the transition metal ions into the band gap of TiO_2 and/or modify the bottom of the conduction band (62). The energy levels created by the metal ion doping act as electron or hole trapping sites. When the energy level lies below but near the conduction band edge, it traps the excited electron, but when the energy level is above the valence band edge, the electrons in metal *d* orbitals can quench the photo-generated holes (80). From the energy level diagram of different metal-doped TiO_2 described in the literature, it is possible to study the behavior of the metal ions dopants, with iron, manganese, and vanadium being the most studied (382–384).

Iron ions in Fe-doped TiO_2 can act as either electron or hole trapping sites, being in both cases reversible processes. Fe^{3+} can trap holes yielding Fe^{4+} or can interact with the photo-generated electrons generating Fe^{2+} ions. Similar behavior has been described for vanadium-doped titania. Substitutional V^{4+} sites can trap holes and electrons yielding, respectively, V^{5+} and V^{3+}. Both reactions were difficult to distinguish, but the researchers described that both processes were occurring simultaneously, being reversible processes, as occurred in iron-doped titania. Nevertheless, molybdenum-doped titania showed differences from the iron and vanadium dopants. Interstitial Mo^{6+} has been described as an effective and irreversible electron traps, whereas the substitutional Mo^{5+} behaves as a reversible hole trapping site (66):

$$M^{n+} + h_{vb}^{+} \longrightarrow M^{(n+1)+} \tag{17.22}$$

$$M^{n+} + e_{cb}^{-} \longrightarrow M^{(n-1)+} \tag{17.23}$$

In metal-doped titania, it must be kept in mind that the trapping sites are not only associated with the metal ions. For example, in aluminium- and gallium-doped titania, Zwingel described the lattice oxygen ions as main hole trapping sites, following a similar behavior to the pure titania that can be described as follows (385):

$$\left\lfloor Ti^{4+}{-}O^{2-}{-}M^{3+} \right\rfloor + h_{vb}^{+} \longrightarrow \left\lfloor Ti^{4+}{-}O^{\bullet -}{-}M^{3+} \right\rfloor \tag{17.24}$$

17.4.1.1.3. Defect Distribution The nature, behavior, and local structural properties of defect sites in a nanosized photo-catalyst have been thoroughly examined in previous sections of this chapter. Now, it is important to describe the experimental techniques and theoretical methods that can give a detailed and full information about the structural characteristics of nanoparticles and the defect distribution properties, e.g., number, thermal, and energetic properties.

The **X-ray diffraction technique** has been extensively employed for obtaining structural information from crystalline solids. From the X-ray diffraction methods, it is possible to obtain information about different structural characteristics as solid phase, lattice parameters, crystallite size, microstrain, or defect concentration (386–388). First of all, the diffraction powder pattern can be qualitatively analyzed comparing the pattern obtained with the ICDD database of known compounds, with it being possible to identify the phase, analyze multiphase mixtures, and identify polymorphic mixtures (389). A quantitative analysis can also be made, trying to extract the structural data from the powder diffraction record (390). This kind of analysis presents some limitations, and care must be taken with multiphase mixtures, preferred orientation, and when phases have significantly different densities or crystallite sizes. The structure determination cannot be seen directly from a diffraction pattern, with it being necessary to follow different steps, which include experimental methods and *ab initio* structure solution, with the Rietveld refinement being a traditional way to obtain an accurate crystal structure (386,387,391). Other structural characteristics can be obtained analyzing the individual diffraction peaks, with it being possible to extract information about size, strain, and defects. Related to crystallite size, Scherrer (392) first observed that small crystallite size could give rise to line broadening and derived an equation where the crystallite size is inversely proportional to the broadening, which is called the *Scherrer Formula*.

The diffraction pattern can also be employed to obtain information about the lattice strain. Strain is a term used more often in engineering than in chemistry. Strain is defined as the deformation of an object divided by its ideal length, $\Delta\mathbf{d}/\mathbf{d}$ (388). In crystals, we can observe two types of strain: uniform strain and nonuniform strain. Uniform strain causes the unit cell to expand/contract in an isotropic way. This simply leads to a change in the unit cell parameters and shift of the peaks. No broadening is associated with this type of strain. However, nonuniform strain leads to systematic shifts of atoms from their ideal positions and to peak broadening. This type of strain arises from the following sources:

- Point defects (vacancies, site-disorder)
- Plastic deformation (cold worked metals, thin films)

The strain broadening was first analyzed by Stokes and Wilson (393). They observed that strained or imperfect crystals contained line broadening of a different sort than the broadening that arises from small crystallite size. Some years later, Williamson and Hall (394) proposed a method for deconvoluting size and strain broadening by looking at the peak width as a function of 2θ. As an example, we derived here the Williamson–Hall relationship for the Lorentzian peak shape, although it can be derived in a similar manner for the Gaussian peak shape:

$$\{\beta_{\text{obs}} - \beta_{\text{inst}}\} = \lambda/\{D_Y \cos\theta\} + 4\varepsilon_{\text{str}}\{\tan\theta\} \tag{17.7}$$

$$\{\beta_{\text{obs}} - \beta_{\text{inst}}\} \cos\theta = \lambda/D_V + 4\varepsilon_{\text{str}}\{\sin\theta\} \tag{17.8}$$

where D_V = the volume-weighted crystallite size, λ = the wavelength of the radiation, β = the integral breath of a reflection located at 2θ, and ε = the weighted average strain.

The Williamson–Hall analysis can be performed plotting $\{\beta_{\text{obs}} - \beta_{\text{inst}}\} \cos\theta$ on the y-axis versus $4 \sin\theta$ on the x-axis for the indexed peaks of the diffraction pattern. The strain value can be the result of stress in the lattice of the nanoparticles and reflects variations in cell dimension within the sample. These variations are present at the surface of the material with respect to the bulk, just because local symmetry and distances are different, but also attention must be paid on the variations produced by the presence of oxygen vacancies and other type of defects (described in 17.4.1.1.1) (338).

This type of study has been performed for anatase and rutile TiO_2 nanopowders to evaluate the presence of defects in these solids; important to mention is the influence of the synthesis process over these structural characteristics. Depero et al. (395), for example, calculated the particle size and microstrain for different titania samples by Fourier analysis and they studied the correlation between both properties, plotting $\langle \varepsilon^2 \rangle^{1/2} \langle \mathrm{M} \rangle$ versus $1/\langle \mathrm{M} \rangle$. Microstrain values were obtained for rutile nanoparticles although no significant changes were detected. This fact could suggest that the surface stress effect that usually induces an increase of the lattice spacing may balance the microstrain effect. The correlation between the parameters previously indicated fits to a straight line passing through the origin, which suggests that rutile microstrains depend only on the surface, because a crystallite free of microstrain is expected in the case of an infinite size. The same study performed over the anatase phase shows higher microstrain values for this phase comparing with the rutile phase, although a straight line is also described between both parameters. These results suggest that a finite and constant value of strain can be predicted for anatase crystallites at infinite size. This observation could be justified by the presence of intrinsic defects such as oxygen vacancies. The metal-doped titania has also been studied by the same authors. Their results suggested that microstrains depend on the specific cations introduced in the structure, as examples Ti–V compounds present larger strain values than powder containing Nb, Ta, Al, or Ga (395). Other authors have studied the defect concentration and cell parameters of iron-doped titania by Rietveld refinement (357). They observed that the iron content strongly affects the titanium defect concentration, increasing when the iron amount increased. They suggest that the formation of these lattice defects can be explained by compensation for the extra hydroxyl ions present in the crystal structures.

Impedance spectroscopy can also provide information about the defect distribution in titania samples, although not many works are related to this technique. The electrical conduction properties are studied by this method, showing an enhancement of the electronic conductivity when the particle size decreases (338). The theoretical explanation suggests that this behavior is due to depletion of oxygen vacancies and accumulation of conduction electrons. Knauth et al. evaluate the diffusion coefficient of titanium interstitials in nanocrystalline titania with the Nernst–Eistein relation indicated below, which relates the ionic conductivity with the concentration and diffusion coefficient of the majority of ionic defects (396):

$$\sigma = [Ti_i]D(Ti_i)F^2/RT \tag{17.9}$$

where $\sigma =$ the ionic conductivity; $F =$ is Faraday constant, $Ti_i =$ the titanium intersititials, and $D(Ti_i) =$ densityTi interstitials.

Raman spectroscopy is another powerful technique for nanoparticle characterization providing information about structural and morphological properties of oxide.

The effects of size on the phonon spectra of a variety of materials have been well established by using Raman scattering experiments on nanocrystals, in combination with theoretical phonon confinement models (338–399). Essentially, the theoretical background for the study of nanocrystalline materials is provided by the phonon confinement model. This factor is the main one responsible for the changes observed in the Raman spectrum, which are caused by the size effect. Nevertheless, other factors have been described that can contribute to Raman spectrum modification as the nonstoichiometry or the internal stress/surface tension.

The phonon confinement model links the q-vector selection rule for the excitation of Raman active optical phonons with long-range order and crystallite size (338,398,399). In an amorphous material, owing to the lack of long-range order, the q-vector selection rule breaks down and the Raman spectrum resembles the phonon density of states. Nanocrystals represent an intermediate behavior. For a nanocrystal of average diameter L, the strict "infinite" crystal selection rule is replaced by a relaxed version, with the result that a range of q vectors is accessible due to the uncertainty principle (338,401). The q-vector relaxation model can be used for the purpose of comparing experimental data with theoretically predicted phonon confinement. According to this model, for finite-sized crystals, the Raman intensity can be expressed using (338,398,401):

$$I(\omega) \cong \int_{\mathrm{BZ}} \exp(-q^2L^2/8)\frac{d^3q}{[\omega - \omega(\vec{q})] + [\Gamma_0/2]^2} \tag{17.10}$$

The $\rho(L)$ represents the particle size distribution, q is expressed in units of π/a_L (with a_L being the unit cell parameter), $\omega(\vec{q})$ is the phonon dispersion, and Γ_0 is the intrinsic linewidth of the bulk crystal. A spherically symmetric phonon dispersion curve is assumed and approximated by a simple linear chain model (401). For a given phonon mode, the slope of dispersion away from the BZ center determines the nature of the modification in the Raman line shape as a function of crystallite size: A negative slope, toward lower frequency, would produce a downshifted (red-shifted) Raman peak, whereas a positive slope would result in an upshifted (blue-shifted) Raman peak, in addition to an asymmetric peak broadening, as the crystallite size reduces. Usually for this kind of analysis, the most intense Raman mode is chosen for the solid studied.

The TiO_2 has been extensively studied by Raman spectroscopy to characterize the different phases of titanium oxide, the transitions between them, and the particle size. The results described in the literature are highly related to the synthesis method employed for preparing the samples. Different researchers have observed, as anticipated, an increased broadening and systematic shifts of the Raman peaks when going from the bulk to finer grained samples. In some works, the authors found a good agreement between the experimental Raman data and the phonon confinement model predictions, suggesting that spatial confinement of phonons in finite-sized nanocrystals is the major factor determining the Raman spectral characteristics of nanocrystalline anatase (402–404). Nevertheless, other authors cannot obtain a good explanation of the changes observed in the Raman spectrum considering only the phonon confinement. It has been reported that the surface strain affects the surface structure of nanoparticles and results in shifts and broadening of the Raman peaks (405,406). The surface strain combined with phonon confinement have been proposed by Xu et al. (407) to explain the blue shift and broadening of the E_g mode in TiO_2 nanoparticles prepared by a hydrothermal method. They employed a surface-active agent for improving the small size of the particles. This agent should produce a compressive stress on the first several layers of atoms of TiO_2 nanoparticles and make the surface atoms pack closely, resulting in higher vibrational wavenumbers. Nonstoichiometry can also be a significant factor contributing to the changes in the Raman spectrum of anatase nanoparticles (408,409). The influence of this factor has been observed by Zhang et al. (410). In their case, the phonon confinement appears as the dominant mechanism responsible for the blue shift of the Raman peak, and the nonstoichiometry has only a little influence on the blue shift. However, the theoretical calculation using the phonon confinement model gives smaller linewidths than the experimental results. The authors suggested that this deviation is caused by the nonstoichiometry effect that largely broadens the Raman line. Then, a combined mechanism involving phonon confinement and nonstoichiometry effects is proposed to explain the Raman lines in anatase samples. The influence of nonstoichiometry over the Raman spectrum has been calibrated by Parker et al. (409) for anatase and rutile TiO_2 phases. They obtained a relation between the O/Ti ratio and the peak position and full-width-at-half-maximum (FWHM) of the Raman modes. From this calibration, the presence of oxygen vacancies can be quantitatively determined by Raman scattering.

The presence and distribution of defect sites in titania phases can also be determined by **other methodologies**. Adsorption experiments appear as an important way to study the surface defect sites, as is demonstrated by the work of Liu et al. (411). They studied the interaction of methanethiol over different titanium dioxide surfaces, $TiO_2(110)$ and $TiO_{2-x}(110)$, using synchrotron based-photo-emission, thermal desorption mass spectroscopy (TDS), and first-principle density-functional (DF) slab calculations. The thermal desorption of adsorbed methanethiol over a perfect $TiO_2(110)$ substrate shows no evidence of CH_3SH dissociation. The molecule bonds to Ti sites via its S lone pairs and desorbs at temperatures in the range 206–160 K. Nevertheless, when defect sites are present, methanethiol is desorbed at a relatively high temperature, $\sim$288 K. This difference is explained by the authors by the presence of O vacancies in the oxide surface. These defect sites produce electronic states that facilitate the cleavage of the S–H bond and the deposition of CH_3S. The bond between CH_3S and O-vacancy sites is mainly covalent, but the bonding interactions are very strong and can induce the migration of O vacancies from the bulk to the surface of the oxide. In systems with a limited number of O vacancies, adsorbed CH_3S and H can recombine yielding a CH_3SH desorption detected into the gas phase. However, for surfaces with a large concentration of O vacancies and defects, there is extensive decomposition of CH_3SH at 100 K, producing a mixture of SO_x, CH_x, S, and CH_3S on the oxide. When the temperature is increased (250–750-K range), the C–S bond in adsorbed CH_3S breaks, desorbing CH_3 or CH_4 into the gas phase and leaving S and CH_x fragments on the surface. These results reflect that the different behavior of stoichiometry and nonstoichiometry titania surfaces with respect to methanethiol adsorption can be a fruitful way to analyze the defect density of titanium oxides.

Important to mention is the methodology developed by Ikeda et al. (412). To determine the molar amount of defective sites of TiO_2 powders. The authors examine the electron accumulation in a titania aqueous suspension produces by a photo-induced reaction in the presence of methylviologen (MV^{2+}) and sacrificial hole scavengers. First of all, they assume that the TiO_2 irradiation, with light of adequate energy ($E_{h\upsilon} > E_{\text{band gap}}$), produces reduced titanium species (Ti^{3+}) by trapping of electrons at defective sites (Reaction **17.25**). Thus, the accumulated electrons may reflect the number of defect sites, with the presence of a hole scavenger being necessary (e.g., methanol) to avoid the recombination pathway. From these considerations, if a deaerated aqueous MV^{2+} solution is injected into an aqueous suspension of TiO_2 containing these accumulated electrons (Ti^{3+}), the MV^{2+} species can be reduced by the trapped electrons by means of (Reaction **17.26**):

$$Ti^{4+}_{\text{defect site}} + e^-_{cb} \longrightarrow Ti^{3+} \qquad \textbf{(17.25)}$$

$$Ti^{3+} + MV^{2+} \longrightarrow Ti^{4+} + MV^{+\cdot} \qquad \textbf{(17.26)}$$

The molar amount of $MV^{+\cdot}$ can be determined by spectrophotometry using its absorption coefficient (413). This amount is thought to reflect the molar amount of defect sites (M_d) of TiO_2 samples, with this method being a way to analyze the number of defect sites in titania powders. As described in the next section, the same authors also define an equation that provides information about the (energy) distribution of the defective sites (412).

To finalize this section it is important to mention the **thermal behavior** of trapped charges. Berger et al. (414) report an important study about the nature and behavior of anatase trapping sites dependent on the temperature, employing for this purpose the combined measurements obtained by EPR and IR techniques. In accordance with the literature described in Section 17.4.1.1.2, the researchers detected at 90 K the following trapped charges: O^- as trapped holes at oxygen ions and Ti^{3+} as trapped electrons at titanium ions. Their formation takes place on the time scale of seconds to minutes. The study of the formation of these trapped charges demonstrates that the Ti^{3+} concentration never exceed 10% of the concentration of trapped holes. This suggests that most photo-excited electrons remain in the conduction band of the titanium dioxide at this temperature.

The same analysis performed at 140 K shows by EPR lower O^- formation, and no evidence of Ti^{3+} trapped electron, which implies that the trapped electrons recombine with holes at this temperature. The complementary analysis performed by IR spectroscopy demonstrates that, at this temperature, the electrons present in the conduction band remain stable and separated from their complementary holes. Nevertheless, when the temperature increases considerably, at 298 K, the infrared absorption decreased to a level close to the original baseline and the EPR experiments show no paramagnetic signals. These evidences imply that the trapped electrons and holes are lost; the charge carrier separation was reversed via thermally induced charge carrier recombination.

17.4.1.2. Electronic Effects The electronic structure of a solid is affected by size and altered from the continuous electronic levels forming a band to discrete-like or quantized levels. This is drastically observed when the particle size goes down to the nanometer range and is the origin of the so-called "quantum confinement" terminology referring to this phenomenon. From a solid state point of view, electronic states of confined materials can be considered as being a superposition of bulk-like states with a concomitant increase of the oscillator strength (415). The valence/conduction bandwidth observable of a solid material used in photo-catalysis, e.g., semiconductors, is a function of the crystal potential, and this, in turn, is perturbed by the effect of the size in two ways: a short-range effect induced by the presence of ions with a different coordination number and bond distance, and a large-range one, induced by changes in the Madelung potential of the oxide. Chapter 1 describes the most recent theoretical frameworks to deal with these physical phenomena, whereas Chapter 6 describes their influence in physico-chemical observables obtained by spectroscopical techniques. Here we will only give a brief summary of the most important effects with influence in photo-catalysis.

The natural consequences of these size-induced complex phenomena are the band narrowing and change of band energy. The thermodynamic limit (at 0 K) for a chemical reaction that can be carried out with the photo-generated charge carriers is given by the position of the band edges (the so-called flat-band potential). Both size-induced band-related changes modulate the onset energy where absorption occurs, e.g., the band gap energy, changing in this way the Fermi level of the system and the chemical potential of charge carrier species. A general, theoretical description has been implemented under the effective mass approximation, as described below. As will be explained in this chapter, this fact is typically invoked to interpret chemical activity changes as a function of size. An additional consequence of nanostructure is the presence of new electronic states located in the bulk band gap. As is well known, perfect crystals do not display such electronic features and are thus directly or indirectly related to a finite size in the nanometer range. Such mid-gap states are sometimes called "surface states" as the main structural sources of their origin are the presence of a different local symmetry at the surface and/or the presence of additional defects, among which, most typical in nano-oxides are punctual oxygen vacancies located at/near the surface (339,342). Energy distribution of vacancy-related mid-gap energy levels have been experimentally obtained (412,416,417), allowing the comparison between different samples. Roughly speaking, donor levels are energetically near the conduction band, whereas acceptor ones are located near the valence band and create Gaussian-like distributions with a width proportional to the two-thirds power of the number of defects (418). The trapping centers display distributions ranging from 0 to 0.6 eV from the corresponding flat-band potential with maxima (or depth) located sometimes at 60–100 meV (412,419) and others around 400–600 meV (416). The first class (called shallow traps) is believed to follow an exponential behavior after the work of Urbach, but this seems a rough picture in view of modern analyses. The corresponding electronic level distribution is strongly influenced by the difference in dielectric constants between the semiconductor nanoparticle and the neighboring medium (419). The latter (deep traps) may be related to the presence of defects like the uncoordinated Ti(IV) species. The presence of surface states can originate in important consequences under light excitation; it has been suggested that electrons trapped in such electronic levels do not generally equilibrate to the free electro's Fermi level, having thus a different chemical potential (420). This possibility has been, however, challenged on the basis of ultra-fast (femptosecond) transient analysis of charge carriers (379).

Apart from the above-mentioned electronic effects of size, a last general point to mention here is the absence in nanoparticulate materials of band bending under contact with the external media (in a heterogeneous photo-chemical reaction) (421). This infers an important property to all nanostructured photo-catalytic materials, as it facilitates the presence of both charge carrier species, electron and holes, at the surface of the particle, ready to be involved in subsequent chemical steps. This contrasts with the situation of bulk-like semiconductors where one charge carrier is typically depleted from the surface area.

17.4.2. Radiation–Matter Interaction: Excitation and De-excitation

The study of so-called "initial processes" of a photo-catalytic reaction and the influence of the characteristic size of the catalyst solid particles will be reviewed in this section. As outlined at the beginning of the preceding section, a natural consequence of the complex phenomena involved in the radiation–matter interaction (particularly those occurring in the initial steps) is that true photo-catalytic efficiencies are difficult to measure in practice. Despite this result, a significant amount of useful catalytic data has been published.

17.4.2.1. Absorption of Light: Charge Separation From basic quantum mechanism, it is known that absorption of light with energy higher than the band gap promotes electrons to the conduction band (CB), leaving holes into the valence band (VB), as sketched in Reaction **17.27**:

$$MO_x + h\nu \longrightarrow MO_x(e^-_{cb} + h^+_{vb}) \qquad \textbf{(17.27)}$$

where MO_x is an oxide semiconductor. This is the so-called charge separation phenomenon, which follows the well-known Fermi golden rule under the dipole approximation (418). The electron-hole pair formed after the absorption of light is called exciton. This phenomena occurs in the time scale of the femtosecond, below 200 fs (422,423). For TiO_2 and other oxides, absorption drives to the formation of singlet exciton states. Strong spin-orbit coupling with cation orbitals facilitates inter-system crossing to the triplet exciton states; this latter is longer lived and has the potential to undergo further physical/chemical transformations (8).

Due to quantum confinement, absorption of light becomes discrete-like and size-dependent. For nanocrystalline semiconductors, both linear and nonlinear optical (absorption included) properties arise as a result of transitions between electron and hole discrete or quantized electronic levels. Depending on the relationship between the radius of the nanoparticle (R) and the Bohr radius of the bulk exciton ($R_B = \varepsilon\hbar^2/\mu e^2$; μ being the exciton reduced mass and ε being the dielectric constant of the semiconductor), the quantum confinement effect can be divided into three regimes; weak, intermediate, and strong confinement regimes, which correspond to $R \gg R_B, R \approx R_B$, and $R \ll R_B$, respectively (424,425). In the first case, the energy of the exciton is larger than the quantization energy of both charge carrier species; the absorption spectrum is determined by the quantum confinement (transition energy and probabilities) of the exciton center of mass. The intermediate regime is precisely defined as the region where $R_e > R_B > R_h$ (where e and h stand for electron and hole, respectively), and here the hole is quasi-localized and the absorption spectrum comes from the oscillation movement of the hole around the center of the nanocrystal, suffering an average potential corresponding to the much faster electron movement. Finally, the optical absorption in the strong confinement regime is determined by the transitions between electron and hole quantized electronic levels. As explained in Chapter 6, the effective mass theory (EMA) (425) is the most elegant and general theory to explain the size dependence of the optical properties of nanometer semiconductors, although other theories as the free-exciton collision model (FECM) (426) or those described in Chapter 1 and based on the bond length–strength correlation (427) have been developed to account for several deficiencies of the EMA theory.

For the onset of light absorption, e.g., the optical band gap, as well as for all electronic transitions present in the optical absorption spectrum, the EMA theory predicts an r^{-2} dependence, with a main r^{-1} correction term in the strong confinement regime, whereas FECM gives a $\exp(1/r)$ behavior. Figure 17.4 depicts the optical band gap measured for TiO_2 (428–430), ZnO (430–432), and MoS_x (274) nanometer materials used for photo-catalysis. Bohr radius of TiO_2, ZnO, and MoS_2 are 1–2, 5–6,

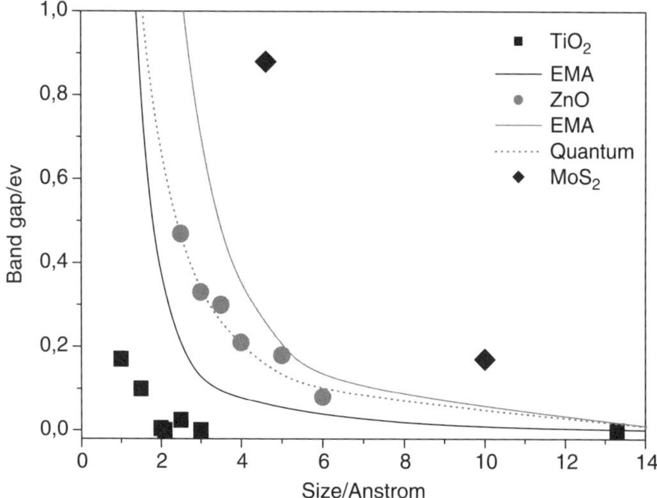

Figure 17.4. Band gap behavior versus size for TiO_2, ZnO, and MoS_2 materials.

and 1–2 nm, respectively. These figures indicate the size region where strong quantum confinement effects are expected in the band gap energy of such materials and, thus, in the chemical potential of charge carriers. This figure also includes predictions for the optical band gap by using the EMA theory or quantum chemical calculations. The comparison of the experimental and theoretical data reveals that the ZnO band gap behavior with size may be essentially described by the quantum confinement phenomenon, although several assumptions of the EMA theory make it to overestimate the size-dependent behavior. The overestimation of the EMA theory is in fact a general result for semiconductors and is due to several causes, among which the assumption of a size-invariant dielectric constant seems of major relevance for the latest version(s) of the theory (425).

Opposite to the ZnO case, TiO_2 displays strong differences with the behavior expected from quantum confinement, and significantly lower values than those corresponding to EMA or other calculations can be visualized in Figure 17.4. It should be however noted that there are samples showing apparent band gap energies approaching the EMA predictions (430); however, Serpone et al. (428) were able to rationalize this result by considering that such samples of apparent absorption onset do in fact correspond to direct electronic transitions in an otherwise indirect semiconductor. As explained in Chapter 6, metal oxides and other materials can be either direct or indirect semiconductors depending on whether the electronic transition is dipole allowed or forbidden, being in the latter case phonon-assisted. TiO_2 is an indirect semiconductor, but nanostructured samples may likely be direct ones. This is a general result as the confinement of charge carriers in a limited space causes their wave functions to spread out in momentum space, in turn increasing the likehood of direct (Frank–Condon type) transitions for bulk indirect semiconductors (433). The dominant contribution of one of the two types, e.g., direct and indirect, transitions as a function of size, preparation method, and other unknown variables may give a partial explanation to justify the large spread of result presented in the literature for nanometer TiO_2. The other phenomena that are also claimed to account for large differences with the EMA and related results are comprised in the catch-up term called "surface effects." This suggests that the mid-gap electronic states originated by the nanostructure and mentioned in Section 17.4.1.2 may strongly influence the optical behavior of nanometer TiO_2, but a detailed proof of such hypothesis has never been put forward (335). Despite this problem, it appears that the most detailed optical analysis of TiO_2 samples would give the steady-state behavior with size depicted in Figure 17.4. To date, no theoretical explanation has been given to interpret this anomalous behavior, which as mentioned is loosely explained as originated by "surface effects."

MoS_2 is a somewhat similar case to TiO_2, being a (bulk) indirect band gap semiconductor with a value of ca. 1.23 eV (990 nm) (294). The strong size effect displayed in Figure 17.4 is not adequately reproduced by using the EMA theory, but for these small clusters, the layered structure of this compound allows us to infer that the wave vector is not well defined and the application of the effective mass approximation is questionable. For nanometer layered compounds, only quantum chemical algorithms would predict the size-dependence, but such calculations have not been reported, to our knowledge, in the case of MoS_2.

In the last ten years, several oxides, sulfides, and nitrides have been used in a significant number of photo-catalytic processes, broadening the solid nature and chemical structures used in photo-catalysis, which were mainly limited to TiO_2 and, less frequently, ZnO in old times. Among the oxides, probably CeO_2 and WO_3 follow in importance to these two oxides. The size-dependence of the optical properties of cerium oxide is, however, a matter of debate; this latter one mainly comes from the fact that nanometer samples of the oxide typically present important quantities of the Ce(III) redox state, but it is not clear whether its presence in the solid can be justified by effect of the nanostructure and/or is an effect of the preparation procedure, as practically all specimens synthesized in the literature come from Ce(III) precursors (338). Obviously, the presence of the reduced state of the cation generates structural (strain) variation on lattice parameters and volume and several electronic effects, which strongly influence optical properties. Despite the large variation showed by the band gap energy values reported in the literature, it appears that a more or less band gap energy steady-state behavior with size is generally observed (338). The case of WO_3 also shares some of the difficulties pointed out above in the case of Ceria. Kubo et al. were able to show that the band gap of this oxide decreases with size from ca. 3.0 to 2.8 eV as a function of R^{-1} (434), but the presence of a variable number of oxygen defects, reduced W redox states, and mid-gap electronic states with size makes this an open question (435).

17.4.2.2. "Diffusion" and Trapping of Charge Once the electron-hole pair could be stabilized, in a triplet state in the TiO_2 case, both charge carriers should diffuse to the surface of the particle where they are normally trapped or localized to participate in chemical reactions. These phenomena occur in a broad time scale, ranging from picoseconds to mili- or even a fraction of second, as detailed below, and always compete with the charge recombination process. This latter one will be analyzed in Section 17.4.2.3, while here we will give an overview of the remaining processes.

Strictly speaking, diffusion of carriers should only occur where the particle radius (R) is significantly larger that the $\langle r \rangle$ expectation value for the e_{cb} and h_{vb} wave functions, which approach a radial dependence $r^{-3/2}\sin(\pi r/R)$ (425). Simple calculations indicate that the diffusion concept losses most of its meaning for nanometer particles. Despite this, a rough approximation to the diffusion coefficient (D) can be obtained by using the Fröhlich theory:

$$D = \mu kT/e \quad \mu = B(\exp(\theta/T - 1) \tag{17.11}$$

where μ is the charge carrier mobility, θ is the Debye temperature of the solid, k is the Boltzman's constant, and $B = B(m_e^*; \varepsilon; \theta)$ with m_e^* the effective electron mass and ε the dielectric constant of the solid. These calculations only considered the flat-band energy variations with size through the m_e^* parameter (436). From this, simple calculations of the time required ($t = R^2/\pi^2 D$) to reach the surface of spherical ZnO or TiO_2 nanoparticles with characteristic size below 10 nm indicate that the electron diffusion process will last, at most, for a few picoseconds, whereas the corresponding transition time for holes would be below 300 fs (430,437). So, size-dependence of this phenomenon hardly affects the photo-physic response of the solid in the nanometer range. On the other hand, we can mention that electron trapping time estimations are not only consistent with electron diffusion coefficient experimental values obtained for 2.5-nm TiO_2 nanoparticles (438) but also with TiO_2 samples with size in the 10–35-nm interval (437,439).

Apart from inherent limitations of the Frohlich theory, which will be briefly mentioned in Section 17.4.2.3, a main point is that electron and hole transport are known to be trap-limited in TiO_2, a fact not encountered from previous calculations with Frohlich theory (440,441). This means that one charge carrier may sample multiple trapping sites, depending of the thermal energy of the system, until it recombines. This seems a general result for any semiconductor used in catalysis. So, the real diffusion coefficient (D_r) is described by

$$D_r = Dk_d/k_t \tag{17.12}$$

where D is the free-like charge diffusion coefficient and k_t and k_d are pseudo-first-order rate constants for trapping and detrapping. Dependence of trapping nature and number of centers with particle size should then generate a dependence of the diffusion process of the carriers to the surface. Both characteristics of the solid trapping center distribution have been explored in Section 17.4.1. Here, we just point out that the assumption that trapping centers are mostly located at the surface or grain boundaries of the materials does not fully explain the experimental measurements of diffusion coefficients (439). A side point to mention is that photo-catalytic reactions are carried out in the presence of oxygen as an efficient electron scavenger and water (vapor for gas phase reaction or present in the medium for liquid reactions) to fully hydroxylate the surface; so, under reaction conditions, dominant electron and hole trapping processes occur with formation of oxygen radicals and OH-derived species, respectively (442–444), requiring an interfacial electron transfer step in the electron case. Additional discussion is also underway on whether free or trapped holes are the initial oxidant species, depending on the nature of the adsorbed molecule (1,5,7,8).

Despite the unclear chemical nature of trapping centers and their unknown dependence with experimental variables, particularly particle size and reaction conditions, nowadays progress has allowed us to set experimental methods that ensured the study of semiconductors systems having, on average, a single electron/hole pair per particle during a laser pulse, allowing the study of a trapping characteristic time constant in conditions resembling the real photo-catalytic measurements. Most of the studied have been performed in TiO_2 anatase samples as this is the "universal" system in photo-catalysis (7,379,422,423,445–448). Transition absorption experiments are used to follow the time-dependent response of the solid trapping centers, allowing the discriminate between charged species; although no general consensus exists, some strong indications suggest that trapped holes display an absorption maximum around 470–520 nm, whereas trapped electrons have this around 650–750 nm and free conduction band and/or shallow trapped electrons have a raising absorption intensity going from the far UV to the near IR regions. The most challenged question arises from the fact that the region around 700 nm can have significant contributions from both holes and electrons with (not well established) dominance of one of the species depending on several experimental variables (7,379,423). Trapped charges (shallow ones excluded) display cross sections about two-thirds orders of magnitude lower than that corresponding to free-like or shallow states (379). Measurements of trapping characteristic times by using absorption spectroscopy yield values below 1 ps for electrons and below 400 ps for holes (7,422,423,449). Furthermore, the capture of electrons by platinization of TiO_2 seems to have a time constant close to 2.3 ps (423,448). These time constants can be compared with those mentioned above for mobility/diffusion coefficient measurements and with those corresponding to the fixation of electrons through an interfacial charge transfer by adsorbed oxygen species (millisecond range) (450). The complete set of results would indicate, as mentioned above, that holes mainly interact (recombine) with multi-trapped electrons, as suggested by several authors (7,447), dominating the diffusion process of carrier species in TiO_2 under light excitation.

The assumption studies indicate that free-like/shallow-trapped electrons likely accumulate in the interior part of the particle, whereas both trapped charge species are mostly located at the surface of nanometer TiO_2 particles (361,447). This difference between electrons and holes obviously implicates a size-dependence of the trapping phenomena, favoring, presumably, the ease by which holes are trapped with respect to its charged counterpart with the decrease of particle size. On the other hand, considering this result and previous comments on the diffusion coefficients and trapping characteristic times dependence on morphological variables, it would seem that not only the main (surface) charge carrier trapping centers but also others, with specific features but with a minor contribution to the defect distribution, control the electron/hole transport properties. This question is further complicated with the already discussed point about whether the quasi-localized surface electronic states created by the solid defect distribution are in a chemical potential equilibrium with the unoccupied/occupied electronic band density of states.

A final paragraph can be here dedicated to the effect of metal dopands on the charge carrier trapping phenomena. As mentioned above, metal ions can act as electron acceptor or donor (e.g., hole acceptor) sites, depending on their electronic configuration. Recent studies with femtosecond time resolution of TiO_2 doped with Fe (cations that possess both electron acceptor and donor properties) and Cr (acceptor) indicate that a relatively larger amount of trapped holes is formed in both doped samples with respect to bare titania, with the effect being more important in the presence of Fe or the combined presence of Fe and Cr (448,451). Such trapping centers develop from 5 ps to 1 ns, without sensible time-dependent differences. The trapping phenomenon, on the other hand, does not seem to be strongly influenced by the metal percentage, at low levels of doping (below ca. 0.5 at. %) (451). The whole set of data presented in these contributions conclusively shows that doping introduces some imbalance in the electron-hole trapping, decreasing the hole trapping characteristic time and increasing the amount of trapped holes, with respect to the bare TiO_2, having consequences in the recombination process that will be analyzed in Section 17.4.2.3. The main point of this chapter concerning metal doping is that optimum metal content of the nanosize solid photo-catalyst is likely size dependent (33); when size becomes larger, the average path length of a charge carrier to the surface, where it should react with absorbed molecules, is longer. For a constant dopant concentration, the probability for electron-hole contact (e.g., charge recombination) would increase as does multiple trapping. Therefore, assuming no correlation with other physical phenomena governing defect distribution and number, to reduce the possibility of (excessive) multiple trapping, the dopant concentration should be decreased with increasing particle size. This point will be, however, fully addressed in the following sections.

17.4.2.3. Fate of Trapped Charge Species Electrons and holes subjected to a certain degree of stabilization could be subsequently involved in chemical processes after an interfacial charge transfer step. However, all charge processes involving trapping and interfacial transfer must compete with de-excitation or relaxation processes involving the recombination of charge carriers. In this section, we will study these two terminal processes that the charge carriers would undergo after the stabilization steps previously reviewed.

17.4.2.3.1. Electron-Hole Recombination Charge carriers formed in nanostructures oxides (MO_x) upon absorption of light (Reaction **17.27**) can relax or recombine in radiative or nonradiative ways, according to Reactions **17.28** to **17.31**:

$$e^-_{cb} + h^+_{vb} \longrightarrow MO_x + \text{energy} \qquad \textbf{(17.28)}$$

$$e^-_{\mathrm{Tr}} + h^+_{vb} \longrightarrow MO_x + \text{energy} \qquad \textbf{(17.29)}$$

$$e^-_{cb} + h^+_{\mathrm{Tr}} \longrightarrow MO_x + \text{energy} \qquad \textbf{(17.30)}$$

$$e^-_{\mathrm{Tr}} + h^+_{\mathrm{Tr}} \longrightarrow MO_x + \text{energy} \qquad \textbf{(17.31)}$$

where the subindexes CB, VB, and Tr stand for conduction and valence bands and trapped charge carriers, respectively. This was originally demonstrated by the rapid depletion of transient absorption spectra recorded during laser flash photolysis studies, in which decay behaviors mostly reflect the Reaction **17.28** (7). It should be noted that charge carrier energy relaxation or de-excitation in bulk semiconductors is dominated by nonradiative interactions with longitudinal optical phonons, leading to a fast (as described in the previous section, typically near or sub-picosecond) carrier cooling dynamics (452). This is the so-called Frohlich interaction (436), which was used to calculate diffusion coefficients in previous sections. However, in nanometer systems, no matter whether in a weak or strong confinement regime, the charge carrier relaxation mediated by interaction with phonons is drastically hindered because of restrictions imposed by energy and momentum conservation (leading to a phenomenon called "phonon Bottleneck"), and it appears that relaxation is in fact dominated by a "non-phonon" energy-loss mechanism mediated by Auger-type electron-hole energy transfer phenomena (452). Auger recombination in a bulk semiconductor is a nonradiative multiparticle process, leading to the electron-hole pair recombination via energy transfer to a third particle (electron or hole), which is re-excited to a higher energy state. In nanometer systems, confinement-induced enhancement in Coulomb interactions and relaxation in translation momentum conservation should lead to increased Auger rates with respect to bulk systems, whereas the atomic-like structure of energy levels should hinder the Auger process by a lower availability of final states (453). As a result, Auger recombination can only occur in nanometer systems with the participation of a phonon (as a four-particle process) or with involvement of a final state from the continuum of states outside the nanocrystal (454). Such complex behavior complicates a theoretical-analysis of quantum-confined Auger recombination and highlights the need for experimental mapping of the size-dependent Auger rates. Despite this result, the analysis gives an idea that a major fraction of electron and holes would still recombine by using radiativeless channels that might involve interparticle de-excitation and thus can be influenced not only by particle size but also by other morphological parameters as porosity and secondary particle size. It could be also mentioned here that the presence of dopants usually increases the likehood of nonradiative channels with respect to bare titania (80).

From the theoretical modeling and experimental analysis of recombination, we now know that several radiative processes are involved in the electron-hole recombination. Such processes would have to take into account steps taking place in femptosecond to subsecond time domains (7,423,445,455). Direct recombination models describe the reaction of the geminate ion pair in the lattice before any transfer to trapping sites (456). This type of recombination reduces efficiency as neither charge has a chance to migrate to the surface and takes place probably below the picosecond region (7,423,445). Serpone et al. found that this process could be the only (radiative) absorption decay in ca. 2-nm TiO_2 particles, having a 10-ns characteristic time, whereas it dominates (ca. 90%) in 10–25-nm particles, having a decay time decreasing with increasing particle size (445). The significant differences with the 2-nm sample can be grounded on the low crystallinity degree of the titania phase in such small particles.

Distinctive decay kinetics are observed after 1 ps, extending, depending on sample and experimental conditions, to the subsecond domain. Such recombination processes are called nongeminate recombination and include contributions with, at least, one previously trapped charge species. They may follow an average rate law switched from second order to first order when decreasing electron-hole pair number per solid particle (7,423,445,457). In any case, picosecond decay contributions are well explained by second-order kinetics, indicating similar initial number (density) of holes and electrons coming from trapping sites (7,423). First-order kinetics seem characteristic of micro- to subsecond processes (423), a fact that may be driven by the larger characteristic trapping times of holes, allowing the existence of long-lived electrons suffering multitrapping processes. The critical influence of several variables, among which the hydroxilation state of the surface appears as capital (458), has been shown to modulate the nature and importance of fast and slow nongeminate recombination processes and their decay kinetics, but this point still awaits rationalization in terms of the electronic and structural characteristics of samples, and particularly, no information about size dependence is available at the moment. A rather naive interpretation could be based on the fact that nongeminate recombination could occur by tunneling and thus displays a $\exp(-2r/\langle r_o \rangle)$ behavior, where r is the distance between trapping centers (assuming Reaction **17.31**, is the dominant process contributing to the nongeminate recombination) and $\langle r_o \rangle$ is the expectation value for the hydrogenoic wave function of the trapping carriers (see Section 17.4.2.2.). If the density of defects increases as size decreases, a shorter nongeminate recombination time would be expected as size decreases. However, if such defects are mainly located at the surface, although the density increases with size, the surface also increases with size as R^2, so it could be expected that surface density of defects would display a minimum for a certain particle size, which could give an optimum size having the larger recombination characteristic time.

In brief, considering that nonradiative processes are dominant and that among radiative ones, geminate recombination can account, at least, for 90% of the losses, it is clear that quantum yields of chemical reactions using light excitation sources can hardly reach 1%. The size-dependence of the overall recombination process is still not known but should be dominated by nonradiative processes, which frequently are not fully measured by researchers of nanometer systems.

A final comment concerns the influence of the metal dopants in recombination processes. For a constant level of doping, size-dependence should be mainly related to three points. First of all, the different nature of the nonradiative decay process with size and the influence of the metal in this dependence; has not been explored. Second, assuming the tunneling dependence of nongeminate recombination, a lower probability (e.g., an increase in characteristic time) is expected as size increases. A side point to be here mentioned is that some cations (like Fe, which is able to act as electron donor and acceptor, trapping both charge species in similar quantities) would increase the average nongeminate recombination time, but most of them would inevitably decrease it (62). Finally, a geometrical constraint derived from the distance of the dopant center to the surface will influence the relative probability of the trapped charge to be involved in interfacial charge transfer versus recombination. This latter one depends on the nature of the dopant (for example, some of them, like V, W, and Cr, typically would decrease the surface energy of titania, having thus a certain preference to be at the surface layer of the material) as well as on preparative conditions. Yet, there is not a general framework to interpret the size-dependence of the doping process for a photo-catalytic semiconductor.

17.4.2.3.2. Interfacial Charge Transfer In all chemical reactions and applications of semiconductor photo-catalysts, the charge carrier species must be in contact with the gas- or liquid-phase molecules to be degradated or chemically transformed. The chemical significance and characteristics of this process have been carefully reviewed in Section 17.2.3, and here we just want to give a short overview concerning the influence of the semiconductors size.

Oxygen is always present in the reaction media to act as the primary electron acceptor. Usually, free-like or previous trapped as Ti(III) (414) electrons are transferred to molecular oxygen adsorbed at the surface of the material yielding superoxide or hydrogen peroxide radicals as described in Section 17.4.1.1.2 (414,442–444,446,450,458).

Although the superoxide radical may also suffer disproportionation to yield neutral hydrogen peroxide and oxygen molecules by a reaction with two protons. In the liquid phase, such reactions depend on pH by the direct involvement of protons in the corresponding reactions (Reaction **17.20**) as well as the pH-dependence of the nanomater oxide flat-band potential, but also of additional variables, among which the presence of surface defects acting as adsorption sites are key ones. This latter one lends a strong size-dependence to oxygen activation steps, which however, is difficult to further elaborate from this simple explanation. A rate constant ca. 7.6×10^7 L mol^{-1} s^{-1} has been determined for Reaction **17.19** (446). Interface electron transfer is relatively slow with oxygen and neutral molecules, having characteristic time constants in the order of milliseconds, but can be notably faster in presence of ionic scavengers as NO_2^- (450). Also, ionic SCN^- is able to transfer an electron to a hole on titania in a way that effectively competes with fast electron-hole recombination processes (446). These comments just point out that the time domain of this elemental step is strongly influenced by the nature of the solid–molecule interaction. Some isotopic studies have been shown that in specific cases, oxygen radicals are additionally involved in several steps of photo-catalytic oxidation processes (2).

Photo-catalytic processes concerning oxidation of organic or inorganic molecules or ions may occur either via surface-bound hydroxyl radicals or valence-band hole before it is trapped within the particle or at its surface (414,444,446). The presence of Cl^- anions coming from residues of the Ti-precursor salt used to prepare titania has also been shown to form $Cl_2^{\cdot-}$ radicals upon a hole capture and subsequently oxidize phenol (459). As a simple and not yet well-established rationalization, it appears that ionic species, which can have fast adsorption rates on the surface of the semiconductor, can be oxidized by free-like holes, whereas adsorbed neutral species may be attacked by the two mentioned species (7). The competition between these two hole species makes the interfacial hole transfer size-dependent. Presumably, this would have significant impact on chemical reactions involving molecules adsorbed in a neutral way, e.g., gas/liquid-phase neutral molecules that do not suffer charge transfer from or to the solid surface.

17.4.3. Size Effects on Catalytic Properties

Catalytic phenomena occur at the surface of an important number of semiconductors described in Sections 17.2.1 and 17.2.2 after the interfacial charge transfer to oxygen and target molecules. A rationalization of the different mechanism involved in such steps is described in Section 17.2.3. A variety of kinetic mechanisms is involved in these chemical processes, but a fundamental question is raised on whether the "initial" photo-physical steps could be involved or control the reaction rate. Several groups have tried to establish correlations between reaction and radiative recombination rates, assuming this latter one is the main photo-physical event (63,356,412,451). The typical result is a somewhat scattered exponential decrease as the recombination rate increases. This is typically interpreted as if at relatively small recombination rates, the charge carrier recombination process may play a significant role on the overall process kinetics, which is nevertheless lost as the recombination rates grows. It can be noted that this type of behavior is not always observed (356). In any case, as described in Section 17.4.2.3.1, electron-hole de-excitation or relaxation is certainly a size-dependent phenomenon, but the significant number of radiative and nonradiative channels and, probably, the major importance of the latter type, makes it a difficult task to give a prediction concerning the effect of size. Despite this result, when the recombination step plays a kinetic role, size-effects can be expected.

Apart from that, it is obvious that the second general source of size-dependence comes from the corresponding band gap and flat-band energy variations. The behavior of these two observables with size has been described in Sections 17.4.1.2 and 17.4.2.1. As a general result, the blue shift in the band gap energy as size decreases will produce a small decrease in the photon absorption rate and carrier kinetic energy, but the flat-band position would change the redox potential of the electron-hole pair (1,2,5,7,8). As mentioned, the decrease of size in the nanometer range allows the appearance of mid-gap states, which may act as trapping centers, whose behavior depend on their energy relative position with respect to the valence and conduction band edge positions (412,416,417). The overall response of these two "electronic" effects is not really known, but as a general result, it appears that they would drive to an optimum influence corresponding to a certain size between ca. 3 and 12–15 nm. The inferior limit is established by the loss of the crystalline structure of the semiconductor and the dominant contribution of geminate recombination, whereas the upper one is given by the loss of significant influence of size in electronic properties.

Following the above simple interpretation of the main effects of size in parameters influencing chemical photo-activity of semiconductors, here we will review the literature reports displaying size-dependent behavior. These reports cover the use of sulfides, MS_x (M = Mo, Bi), ZnO, CeO_2, and TiO_2 as photo-catalysts, although the majority are devoted to the use of the latter.

Metallic sulfides have been used to eliminate phenol under visible light excitation, which is active for particles with a primary particle size ca. 4.5 nm (274). The large variation of band gap and edge energies allows photo-generated holes to oxidize water and create hydroxyl radicals for organic attack with a size below ca. 8 nm. MoS_s appears to maintain reasonable stability under reaction conditions, opening a way to use these chemical compounds in photo-catalytic elimination reactions. The binary MoS_x–TiO_2 (Degussa P25) system is also useful upon UV light excitation, displaying an activity rate enhancement of about 2 for sulfides particles with a 2-nm average particle size (274). The corresponding semiconductor junction is presumed to improve charge carrier separation, decreasing recombination rate and enhancing reaction rate. Such situation has also been shown to occur for BiS_x–TiO_2 systems upon adsorption of ionic molecules on the sulfide surface (225).

ZnO has been used as a photo-catalyst because it has similar band edge positions to that of TiO_2, but its self-induced oxidation in a number of applications in liquid phase limited its usefulness (460). Despite this result, the influence of size in the polymerization reaction of methyl methacrylate using the semiconductor as an photo-initiator indicates that optimum activity is obtained with size close to ca. 15 nm, with bulk-size ZnO being ineffective for such reaction (335). In this case, the quantum enhancement of the reaction is ascribed either to the influence of surface defects of the nanosolid or to a modification of the recombination rate. More modern papers use nanometer ZnO as part of binary ZnO/SnO_2 or ternary $ZnO/TiO_2/SnO_2$ systems in the photo-degradation of colorants (461) or H_2 production (462). In such cases, the electron-hole charge separation yielded by the physical contact between the two semiconductors allows the production of more or less stable systems, but no clear information about size-dependence of catalytic properties was reported. Very scarce reports have been devoted to CeO_2 photo-catalysts as a function of size. In fact, we are aware of a single publication that indicates that particles about 8 nm display optimum properties in the gas-phase elimination of toluene (19).

The anatase–TiO_2 system has been subjected to a large number of studies. Concerning size-dependence, the contributions can be classified into three different groups. The first group includes works that found an optimum primary size well below 10 nm. These are the gas-phase photo-oxidation of 1-butene (297), trichloroethylene (463), and toluene (464), and the liquid-phase hydrogenation of CH_3CCH (333), the photo-oxidation of rhodamine B (465), or the polymerization of a trisacrylate ester (466). Although not fully grounded, the studies infer that a strong quantum size effect dominates catalytic properties through two main factors. The first one is the change in redox potential suffered by both charge carriers as the band gap of the samples experienced a blue shift with decreasing size. The second one concerns the existence of mid-gap states associated with surface charge trapping centers positively affecting charge evolution. Both factors depend on primary size but also on other morphological parameters as the first is coupled with the different scattering versus absorption ratio presumably displayed by the samples and the second with grain boundaries between single particles, which are believed to have an important role as trapping centers. The influence not only of primary size but also of other morphological parameters clearly points out the difficulty in getting a general framework able to explain photo-catalytic behavior in cases where, apparently, size confinement plays a dominant role in the overall chemical reaction kinetics. The second group of works obtains a size-dependence with maximum ca. 10 nm; the gas-phase photo-oxidation of $CHCl_3$ (353), or the liquid-phase of acetone (467) are examples. Apart from the above-mentioned factors affecting photo-activity, one of the studies gives evidence of the importance of surface hydroxylation by analyzing the behavior of two samples with rather similar morphological properties (467). Finally, the third group comprises systems with photo-activity maximum well above a limiting size where quantum confinement effect can play a role. The gas-phase photo-oxidation of trichloroetane (468), and hexane (469), or the liquid-phase of 2-propanol (470), phenol (337), methyleneblue (35), and methylic orange (471), are illustrative examples. In these cases, the authors claim that the dominant effect of the recombination rate on the chemical reaction kinetics, which presumably should increase as size decreases, is not balanced by any of the positive effects coming from size confinement in the nanometer range.

ACKNOWLEDGMENTS

Financial support by projects CTQ2004-03409 and CTQ2004-05734-C02-02 are fully acknowledged. C.B. thanks MEC for JdC contract. Prof. J. Bisquert is also thanked for the critical reading of this chapter.

REFERENCES

(1) Hoffmann, M.R.; Martin, B.T.; Choi, W.; Bahnemann, D.W. *Chem. Rev.* **1995**, *95*, 69.
(2) Linsebigler, A.L.; Lu, G.; Yates, J.T. *Chem. Rev.* **1995**, *95*, 735.
(3) Serpone, N. *Sol. En. Mater. Sol. Cells* **1997**, *35*, 369.
(4) Lee, S.K.; Mills, A.M. *Platinum Metal Rev.* **2003**, *47*, 61.
(5) Bahnemann, D.W. *Sol. En.* **2004**, *77*, 445.
(6) Carp, O.; Huisman, C.L.; Rellr, A. *Prog. Solid State Chem.* **2004**, *32*, 33.
(7) Bahnemann, D.W. *Chemical Physics of Nanostructured Semiconductors*; NOVA Science: New York, 2003.
(8) Rajh, T.; Poluektov, O.G.; Thurnauer, M.C. *Chemical Physics of Nanostructured Semiconductors*; NOVA Science: New York, 2003.
(9) Pannkov, J.I. *Optical Processes in Semiconductors*; Dover: Mineola, New York, 1971.
(10) Cabrera, M.I.; Alfano, O.M.; Cassano, A.E. *J. Phys. Chem.* **1996**, *100*, 20043.
(11) Egerton, T.A.; Tooley, I.R. *J. Phys. Chem. B* **2004**, *108*, 5066.
(12) Emeline, A.V.; Frolov, A.V.; Ryabahuk, U.K.; Serpone, N. *J. Phys. Chem. B* **2003**, *107*, 7109.
(13) Fox, M.A.; Dulay, M.T. *Chem. Rev.* **1993**, *93*, 341.
(14) Blanco, J.; Malato, S. *Tecnología de Fotocatálisis Solar*; Cuadernos Monográficos: CIEMAT, 1996.
(15) Daneshvar, N.; Salari, D.; Khataee, A.R. *J. Photochem. Photobiol. A* **2004**, *162*, 317.
(16) Dijken, A.; Janssen, A.; Smitsmans, M.; Vanmaekelbergh, D.; Meijerink, A. *Chem. Mater.* **1998**, *10*, 3513.
(17) Dindar, B.; Icli, S. *J. Photochem. Photobiol. A* **2001**, *140*, 263.
(18) Yamaguchi, Y.; Yamazaki, M.; Yoshihara, S.; Shirakashi, T. *J. Electroanal. Chem.* **1998**, *442*, 1.
(19) Hernández-Alonso, M.D.; Hungría, A.B.; Martínez-Arias, A.; Fernández-García, M.; Coronado, J.M.; Conesa, J.C.; Soria, J. *Appl. Catal. B* **2004**, *50*, 167.
(20) Khalil, L.B.; Mourad, W.E.; Rophael, M.W. *Appl. Catal. B* **1998**, *17*, 267.
(21) Lin, Y.; Lin, R.; Yin, F.; Xiao, X.; Wu, M.; Gu, W.; Li, W. *J. Photochem. Photobiol. A* **1999**, *125*, 135.
(22) Han, M.; Huang, W.; Quek, C.H.; Gan, L.M.; Chew, C.H.; Xu, G.O.; Ng, S.C. *J. Mater. Res.* **1999**, *14*, 2092.
(23) Torres-Martínez, C.L.; Kho, R.; Mian, O.I.; Mehra, R.K. *J. Coll. Interf. Sci.* **2001**, *240*, 525.
(24) Yin, H.B.; Wada, Y.; Kitamura, T.; Yanagida, S. *Environ. Sci. Technol.* **2001**, *35*, 227.

(25) Campostrini, R.; Carturan, G.; Palmisano, L.; Schiavello, M.; Sclafani, A. *Mater. Chem. Phys.* **1994**, *38*, 227.

(26) Sclafani, A.; Herrmann, J.M. *J. Phys. Chem.* **1996**, *100*, 13655.

(27) Anpo, M.; Shima, T.; Kodama, S.; Kubokawa, Y. *J. Phys. Chem.* **1987**, *91*, 4305.

(28) Bahnemann, D.W.; Kholuiskaya, S.N.; Dillert, R.; Kulak, A.I.; Kokorin, A.I. *Appl. Catal. B* **2002**, *36*, 161.

(29) Colón, G.; Hidalgo, M.C.; Macías, M.; Navío, J.A.; Doña, J.M. *Appl. Catal. B* **2003**, *43*, 163.

(30) Chang, S.; Doong, R. *Chem. Mater.* **2005**, *17*, 4837.

(31) Li, G.S.; Li, L.; Boerio-Goates, J.; Woodfield, B.F. *J. Am. Chem. Soc.* **2005**, *127*, 8659.

(32) Tanaka, K.; Capule, M.F.V.; Hisanaga, T. *Chem. Phys. Lett.* **1991**, *187*, 73.

(33) Zang, Z.B.; Wang, C.C.; Zakaria, R.; Ying, Y.J. *J. Phys. Chem. B* **1998**, *102*, 10871.

(34) Colbeau-Justin, C.; Kunst, M.; Huguenin, D. *J. Mater. Sci.* **2003**, *38*, 2429.

(35) Toyoda, M.; Nanbu, Y.; Nakazawa, Y.; Hirano, M.; Inagaki, M. *Appl. Catal. B* **2004**, *49*, 227.

(36) Jung, K.Y.; Park, S.B.; Anpo, M. *J. Photochem. Photobiol. A* **2005**, *170*, 247.

(37) Chittofrati, A.; Matijevic, E. *Coll. Surf.* **1990**, *48*, 65.

(38) Trindade, T.; de Jesus, J.D.; O'Brien, P. *J. Mater. Chem.* **1994**, *4*, 1611.

(39) Auffrédic, J.; Boultif, A.; Langford, J.I.; Louër, D. *J. Am. Ceram. Soc.* **1995**, *78*, 323.

(40) Lu, C.H.; Yeh, C.H. *Ceram. Inter.* **2000**, *26*, 351.

(41) Bowker, M.; Houghton, H.; Waugh, K.C.; Giddings, T.; Green, M. *J. Catal.* **1983**, *84*, 252.

(42) Bolis, V.; Fubini, B.; Giamello, E. *J. Chem. Soc. Faraday Trans.* **1989**, *85*, 855.

(43) Fujishima, A.; Rao, T.N.; Tryk, D.A. *J. Photochem. Photobiol. C* **2000**, *1*, 1.

(44) Dietbold, U.; Ruzycki, N.; Herman, G.S.; Selloni, A. *Catal. Today* **2003**, *85*, 93.

(45) Burdett, J.K.; Hughbands, T.; Gordon, J.M.; Richarson, J.W.; Smith, J. *J. Am. Chem. Soc.* **1987**, *109*, 3639.

(46) Asahi, R.; Taga, Y.; Mannstadt, W.; Freeman, A.J. *Phys. Rev. B* **2002**, *65*, 224112.

(47) Bahnemann, D.W.; Bockelemann, D.; Goslich, R.; Hilgendorff, M.; Weichgrebe, D. *Photocatalytic Purification and Treatment of Water and Air*; Elsevier Science: Amsterdam, 1993.

(48) Bickley, R.I.; González-Carreño, T.; Lee, J.S.; Palmisano, L.; Tilley, R.J.D. *J. Solid State Chem.* **1991**, *92*, 178.

(49) Ohno, T.; Jarukawa, K.; Tokieda, M.; Matsumura, M. *J. Catal.* **2001**, *203*, 82.

(50) Sun, B.; Smirtionis, G. *Catal. Today* **2003**, *88*, 49.

(51) Wu, C.; Yue, Y.; Deng, X.; Hua, W.; Gao, Z. *Catal. Today* **2004**, *93–95*, 863.

(52) Ozawa, T.; Iwasaki, M.; Tada, H.; Akita, T.; Tanaka, K.; Ito, S. *J. Coll. Interf. Sci.* **2005**, *281*, 510.

(53) Hidalgo, M.C.; Sakthivel, S.; Bahnemann, D. *Appl. Catal. A* **2004**, *277*, 183.

(54) Byrne, J.A.; Eggins, B.R.; Brown, N.M.D.; McKinney, B.; Rouse, M. *Appl. Catal. B* **1998**, *17*, 25.

(55) Molinari, R.; Borgese, M.; Drioli, E.; Palmisano, L.; Schiavello, M. *Catal. Today* **2002**, *75*, 77.

(56) Molinari, R.; Grande, C.; Drioli, E.; Palmisano, L.; Schiavello, M. *Catal. Today* **2001**, *67*, 273.

(57) Liu, S.M.; Li, K. *J. Membrane Sci.* **2003**, *218*, 269.

(58) Sopyan, I.; Wanatabe, M.; Murasawa, S.; Hashimoto, K.; Fujishima, A. *J. Photochem. Photobiol. A* **1996**, *98*, 79.

(59) Anpo, M. *Catal. Surv. Jpn.* **1997**, *1*, 169.

(60) Carlson, T.; Griffin, G.L. *J. Phys. Chem.* **1986**, *90*, 5896.

(61) Sclafani, A.; Mozzanega, M.N.; Pichat, P.J. *J. Photochem. Photobiol. A* **1991**, *59*, 181.

(62) Choi, W.; Termin, A.; Hoffmann, M.R. *J. Phys. Chem. B* **1994**, *98*, 13669.

(63) Ikeda, S.; Sugiyama, N.; Pal, B.; Marci, G.; Palmisano, L.; Noguchi, H.; Uosaki, K.; Ohtani, B. *Phys. Chem. Chem. Phys.* **2001**, *3*, 267.

(64) Pleskov, Y.V. *Sov. Electrochem.* **1981**, *17*, 1.

(65) Martín, C.; Martín, I.; Rives, V.; Palmisano, L.; Schiavello, M. *J. Catal.* **1992**, *134*, 434.

(66) Grätzel, M.; Howe, R.F. *J. Phys. Chem.* **1990**, *94*, 2566.

(67) Luo, Z.; Gao, Q.H. *J. Photochem. Photobiol. A* **1992**, *63*, 367.

(68) Mu, W.; Herrmann, J.M.; Pichat, P. *Catal. Lett.* **1989**, *3*, 73.

(69) Karakitsou, K.E.; Verykios, X.E. *J. Phys. Chem.* **1993**, *97*, 1184.

(70) Ranjit, K.T.; Cohen, H.; Willner, I.; Bossmann, S.; Braun, A.M. *J. Mater. Sci.* **1999**, *34*, 5273.

(71) Wang, Y.; Cheng, H.; Zhang, L.; Hao, Y.; Ma, J.; Xu, B.; Li, W. *J. Mol. Catal. A* **2000**, *151*, 205.

(72) Xu, A.W.; Gao, Y.; Liu, H.Q. *J. Catal.* **2002**, *207*, 151.

(73) Xie, Y.; Yuan, C. *Mater. Res. Bull.* **2004**, *39*, 533.

(74) Moser, J.; Grätzel, M.; Gallay, J. *Helv. Chim. Acta* **1987**, *70*, 1596.

(75) Litter, M.I.; Navío, J.A. *J. Photochem. Photobiol. A* **1996**, *98*, 171.

(76) Bickely, R.I.; González-Carreño, T.; González-Elipe, A.R.; Munuera, G.; Palmisano, L. *J. Chem. Soc. Faraday Trans.* **1994**, *90*, 2257.

(77) Navío, J.A.; Testa, J.; Djedjeian, P.; Padrón, J.; Rodríguez, D.; Litter, M.I. *Appl. Catal. A* **1999**, *178*, 191.

(78) Beydoun, D.; Amal, R.; Low, G.K.C.; McEvoy, S. *J. Phys. Chem. B* **2000**, *104*, 4387.

(79) Iida, Y.; Ozaki, S. *J. Am. Ceram. Soc.* **1961**, *44*, 120.

(80) Nagaveni, K.; Hegde, M.S.; Madras, G. *J. Phys. Chem. B* **2004**, *108*, 20204.

(81) Wang, C.Y.; Böttcher, C.; Bahnemann, D.W.; Dohrmann, J.K. *J. Mater. Chem.* **2003**, *13*, 2322.

(82) Araña, J.; Díaz, O.G.; Miranda, M.; Rodríguez, J.M.D.; Melián, J.H.; Peña, J.P. *Appl. Catal. B* **2001**, *32*, 49.

(83) Araña, J.; Díaz, O.G.; Rodríguez, J.M.D.; Melián, J.H.; Cabo, C.G.I.; Peña, J.P.; Hidalgo, M.C.; Navío, J.A. *J. Mol. Catal. A* **2003**, *197*, 157.

(84) Litter, M.I. *Appl. Catal. B* **1999**, *23*, 89.

(85) Siemon, U.; Bahnemann, D.; Testa, J.J.; Rodríguez, D.; Litter, M.I.; Bruno, N. *J. Photochem. Photobiol. A* **2002**, *148*, 247.

(86) Hufschmidt, D.; Bahnemann, D.; Testa, J.; Emilio, C.; Litter, M. *J. Photochem. Photobiol. A* **2002**, *148*, 223.

(87) Liqqiang, J.; Baiqi, W.; Baifu, X.; Shudan, L.; Keying, S.; Weimin, C.; Honggang, F. *J. Solid State Chem.* **2004**, *177*, 4221.

(88) Tada, H.; Ishida, T.; Takao, A.; Ito, S. *Langmuir* **2004**, *20*, 7898.

(89) Emilio, C.A.; Testa, J.; Hufschmidt, D.; Colón, G.; Navio, J.A.; Bahnemann, D.W.; Litter, M.I. *J. Indust. Eng. Chem.* **2004**, *10*, 129.

(90) Furube, A.; Asahi, T.; Masuhara, H.; Yamashita, H.; Anpo, M. *Chem. Phys. Lett.* **2001**, *336*, 424.

(91) Xu, Y.; Zheng, W.; Liu, W. *J. Photochem. Photobiol. A* **1999**, *122*, 57.

(92) Murata, C.; Yoshida, H.; Hattori, T. *Stud. Surf. Sci. Catal.* **2002**, *143*, 845.

(93) Hu, C.; Tang, Y.; Yu, J.C.; Wong, P.K. *Appl. Catal. B* **2003**, *40*, 131.

(94) Ding, Z.; Hu, X.; Yue, P.L.; Lu, G.Q.; Greenfield, P.F. *Catal. Today* **2001**, *68*, 173.

(95) Navío, J.A.; Colón, G.; Herrmann, J.M. *J. Photochem. Photobiol. A* **1997**, *108*, 179.

(96) Schattka, J.H.; Schukin, D.G.; Jia, J.; Antonietti, M.; Caruso, R.A. *Chem. Mater.* **2002**, *14*, 5103.

(97) Colón, G.; Hidalgo, M.C.; Navío, J.A. *Appl. Catal. A* **2002**, *231*, 185.

(98) Serpone, N.; Borgarello, E.; Grätzel, M. *J. Chem. Soc. Commun.* **1983**, 342.

(99) Lo, S.C.; Lin, C.F.; Wu, C.H.; Hsieh, P.H. *J. Hazard. Mater.* **2004**, *114*, 183.

(100) Bessekhouad, Y.; Robert, D.; Weber, J.V. *J. Photochem. Photobiol. A* **2004**, *163*, 569.

(101) Sakhivel, S.; Geissen, S.U.; Bahnemann, D.W.; Murugesan, V.; Vogelpohl, A. *J. Photochem. Photobiol. A* **2002**, *148*, 283.

(102) Yin, H.B.; Wada, Y.; Kitamura, T.; Sakata, T.; Mori, H.; Yanagida, S. *Chem. Lett.* **2001**, *4*, 334.

(103) Tada, H.; Hattori, A.; Tokihisa, Y.; Imai, K.; Tohge, N.; Ito, S. *J. Phys. Chem.* **2000**, *104*, 4585.

(104) Vogel, R.; Hoyer, P.; Weller, H. *J. Phys. Chem.* **1994**, *98*, 3183.

(105) Li, D.; Haneda, H.; Ohashi, N.; Hishita, S.; Yoshikawa, Y. *Catal. Today* **2004**, *93–95*, 895.

(106) Wang, C.; Zhao, J.C.; Wang, X.M.; Mai, B.X.; Sheng, G.Y.; Peng, P.; Fu, J.M. *Appl. Catal. B*, **2002**, *39*, 269.

(107) Hill, C.L. *Chem. Rev.* **1998**, *98*, 1.

(108) Guo, Y.H.; Hu, C.W. *J. Clust. Sci.* **2003**, *14*, 505.

(109) Tanielian, C. *Coord. Chem. Rev.* **1998**, *178–180*, 1165.

(110) Yoon, M.; Chang, J.A.; Kim, Y.; Choi, J.R. *J. Phys. Chem. B* **2001**, *105*, 2539.

(111) Park, H.; Choi, W. *J. Phys. Chem. B* **2003**, *107*, 3885.

(112) Bonchio, M.; Carraro, M.; Scorrano, G.; Bagno, A. *Adv. Synth. Catal.* **2004**, *346*, 648.

(113) Herrmann, J.; Matos, J.; Disdier, J.; Guillard, C.; Laine, J.; Malato, S.; Blanco, J. *Catal. Today* **1999**, *54*, 255.

(114) Araña, J.; Rodríguez, J.M.D.; Tello, E.; Cabo, C.G.I.; Díaz, O.G.; Melián, J.H.; Peña, J.P.; Colón, G.; Navío, J.A. *Appl. Catal. B* **2003**, *44*, 161.

(115) Araña, J.; Rodríguez, J.M.D.; Tello, E.; Cabo, C.G.I.; Díaz, O.G.; Melián, J.H.; Peña, J.P.; Colón, G.; Navío, J.A. *Appl. Catal. B* **2003**, *44*, 153.

(116) Ao, C.H.; Lee, S.C. *J. Photochem. Photobiol. A* **2004**, *161*, 131.

(117) Janus, M.; Tryba, B.; Inagaki, M.; Morawski, A.W. *Appl. Catal. B* **2004**, *52*, 61.

(118) Shichi, T.; Tagaki, K. *J. Photochem. Photobiol. C* **2000**, *1*, 113.

(119) Shimizu, K.I.; Kaneko, T.; Fujishima, T.; Kodama, T.; Yoshida, H.; Kitayama, Y. *Appl. Catal. A* **2002**, *225*, 185.

(120) Moller, K.; Bein, T. *Chem. Mater.* **1998**, *10*, 2950.

(121) Ramamurthy, V. *J. Photochem. Photobiol. C* **2000**, *1*, 145.

(122) Anandan, S.; Yoon, M. *J. Photochem. Photobiol. C* **2003**, *4*, 5.

(123) Colón, G.; Hidalgo, M.C.; Navío, J.A. *Catal. Today* **2002**, *76*, 91.

(124) Colón, G.; Hidalgo, M.C.; Macías, M.; Navío, J.A.; Doña, J.M. *Appl. Catal. B* **2003**, *43*, 163.

(125) Yamashita, H.; Ichihashi, Y.; Anpo, M.; Louis, C.; Che, M. *J. Phys. Chem.* **1996**, *100*, 16041.

(126) Anpo, M.; Tackeuchi, M. *J. Catal.* **2003**, *216*, 505.

(127) Furube, A.; Asahi, T.; Masuhara, H.; Yamashita, H.; Anpo, M. *J. Phys. Chem. B* **1999**, *103*, 3120.

(128) Ikeue, K.; Mukai, H.; Yamashita, H.; Inagaki, S.; Matsuoka, M.; Anpo, M. *J. Synch. Rad.* **2001**, *8*, 640.

(129) Keita, I.; Hiromi, Y.; Anpo, M.; Takewaki, T. *J. Phys. Chem. B* **2001**, *105*, 8350.

(130) Yamashita, H.; Fujii, Y.; Ichihashi, Y.; Zhang, S.G.; Ikeue, K.; Park, D.R.; Koyano, K.; Tatsumi, T.; Anpo, M. *Catal. Today* **1998**, *45*, 221.

(131) Yoneyama, H.; Haga, S.; Yamanaka, S. *J. Phys. Chem.* **1989**, *93*, 4833.

(132) Ooka, C.; Akita, S.; Ohashi, Y.; Horiuchi, T.; Suzuki, K.; Komai, S.; Yoshida, H.; Hattori, T. *J. Mater. Chem.* **1999**, *9*, 2943.

(133) Yamanaka, S.; Malla, P.B.; Komarneni, S. *J. Coll. Interf. Sci.* **1990**, *134*, 51.

(134) Ooka, C.; Yoshida, H.; Suzuki, K.; Hattori, T. *Appl. Catal. A* **2004**, *260*, 47.

(135) Ilisz, I.; Dombi, A.; Mogyorosi, K.; Dekany, I. *Coll. Surf. A* **2003**, *230*, 89.

(136) Dijkstra, M.; Panneman, H.J.; Winkelman, J.; Kelly, J.J.; Beenackers, A. *Chem. Eng. Sci.* **2002**, *57*, 4895.

(137) Chang, H.T.; Wu, N.M.; Zhu, F. *Water Res.* **2000**, *34*, 407.

(138) Byrne, J.A.; Eggins, B.R.; Brown, N.M.; McKinney, B.; Rouse, M. *Appl. Catal. B* **1998**, *17*, 25.

(139) Fernández, A.; Lassaletta, G.; Jiménez, V.M.; Justo, A.; González-Elipe, A.R. *Appl. Catal. B* **1995**, *68*, 173.

(140) Avila, P.; Sánchez, B.; Cardona, A.I. *Catal. Today* **2002**, *76*, 271.

(141) Cardona, A.I.; Candal, R.; Sánchez, B.; Avila, P.; Rebollar, M. *Energy* **2004**, *29*, 845.

(142) Zhao, J.; Yang, X. *Build. Environ.* **2003**, *38*, 645.

(143) Kominami, H.; Kumamoto, H.; Kera, Y.; Ohtani, B. *Appl. Catal. B* **2001**, *30*, 329.

(144) Yu, J.; Zhao, X.; Zhao, Q. *J. Mater. Sci. Lett.* **2000**, *19*, 1015.

(145) Yu, J.; Zhao, X.; Zhao, Q. *Thin Solid Films* **2000**, *379*, 7.

(146) Yu, J.; Zhao, X. *Mater. Res. Bull.* **2000**, *35*, 1293.

(147) Fretwell, R.; Douglas, P. *J. Photochem. Photobiol. A* **2001**, *143*, 229.

(148) Kwon, C.H.; Shin, H.; Kim, J.H.; Choi, W.S.; Yoon, K.H. *Mater. Chem. Phys.* **2004**, *86*, 78.

(149) Guillard, C.; Debayle, D.; Gagnaire, A.; Jaffrezic, H.; Herrmann, J.M. *Mater. Res. Bull.* **2004**, *39*, 1445.

(150) Ha, H.Y.; Anderson, M.A. *J. Environ. Eng.* **1996**, *122*, 217.

(151) Kim, D.H.; Anderson, M.A.; Zeltner, W.A. *J. Environ. Eng.* **1995**, *121*, 590.

(152) Lassaletta, G.; Fernández, A.; Espinós, J.P.; González-Elipe, A.R. *J. Phys. Chem.* **1995**, *99*, 1484.

(153) Kudo, A.; Kato, H.; Tsuji, I. *Chem. Lett.* **2004**, *33*, 1534.

(154) Kudo, A. *Catal. Surv. Asia* **2003**, *7*, 31.

(155) Asahi, R.; Morikawa, T.; Ohwaki, T.; Aoki, K.; Ega, Y. *Science* **2001**, *293*, 269.

(156) Lettmann, C.; Hinrichs, H.; Maier, W.F. *Angew. Chem. Int. Ed.* **2001**, *40*, 3160.

(157) Malato, S.; Blanc, J.; Vidal, A.; Richter, C. *Appl. Catal. B* **2002**, *37*, 1.

(158) Fujishima, A.; Honda, K. *Nature* **1972**, *238*, 37.

(159) Mattesini, M.; Almeida, J.S.; Dubrovinsky, L.; Dubrovinskaia, N.; Johansson, B.; Ahuja, R. *Phys. Rev. B* **2004**, *70*, 115101.

(160) Yang, Q.; Xie, C.; Xu, Z.; Gao, Z.; Du, Y. *J. Phys. Chem. B* **2005**, *109*, 5554.

(161) Fuerte, A.; Hernández-Alonso, M.D.; Maira, A.J.; Martínez-Arias, A.; Fernández-García, M.; Conesa, J.C.; Soria, J.; Munuera, G. *J. Catal.* **2002**, *212*, 1.

(162) Anpo, M. *Bull. Chem. Soc. Jpn.* **2004**, *77*, 1427.

(163) Zhao, J.; Chen, C.; Ma, W. *Topics Catal.* **2005**, *35*, 267.

(164) Fuerte, A.; Hernández-Alonso, M.D.; Iglesias-Juez, A.; Martínez-Arias, A.; Conesa, J.C.; Soria, J.; Fernández-García, M. *Phys. Chem. Chem. Phys.* **2003**, *5*, 2913.

(165) Klosek, S.; Raftery, D. *J. Phys. Chem. B* **2001**, *105*, 2815.

(166) Chen, C.; Li, X.; Ma, W.; Zhao, J.; Hidaka, H.; Serpone, N. *J. Phys. Chem. B* **2002**, *106*, 318.

(167) Karvinen, S.; Lamminmäki, R.J. *Solid State Sci.* **2003**, *5*, 1159.

(168) Takeuchi, M.; Yamashita, H.; Matsuoka, M.; Anpo, M.; Hirao, T.; Itoh, N.; Iwamoto, N. *Chem. Lett.* **2000**, *67*, 135.

(169) Li, X.Z.; Li, F.B. *Environ. Sci. Tech.* **2001**, *35*, 2381.

(170) Fuerte, A.; Hernández-Alonso, M.D.; Maira, A.J.; Martínez-Arias, A.; Fernández-García, M.; Conesa, J.C.; Soria, J. *Chem. Commun.* **2001**, 2718.

(171) Zielinska, B.; Morawski, A.W. *Appl. Catal. B* **2005**, *55*, 221.

(172) Xu, X.H.; Wang, M.; Hou, Y.; Yao, W.F.; Wang, D.; Wang, H. *J. Mater. Sci. Lett.* **2002**, *21*, 1655.

(173) Sato, S. *Chem. Phys. Lett.* **1986**, *123*, 126.

(174) Gole, J.L.; Stout, J.D.; Burda, C.; Lou, Y.; Chen, X. *J. Phys. Chem. B* **2004**, *108*, 1230.

(175) Sakatani, Y.; Nunoshige, J.; Ando, H.; Okusako, K.; Koike, H.; Takata, T.; Kondo, J.N.; Hara, M.; Domen, K. *Chem. Lett.* **2003**, *32*, 1156.

(176) Li, D.; Haneda, H.; Hishita, S.; Ohashi, N. *Mater. Sci. Eng. B* **2005**, *117*, 67.

(177) Burda, C.; Lou, Y.; Chen, X.; Samia, A.C.S.; Stout, J.; Gole, J.L. *Nano Lett.* **2003**, *3*, 1049.

(178) Sano, T.; Negishi, N.; Koike, K.; Takeuchi, K.; Matsuzawa, S. *J. Mater. Chem.* **2004**, *14*, 380.

(179) Kuroda, Y.; Mori, T.; Yagi, K.; Makihata, N.; Kawahara, Y.; Nagao, M.; Kittaka, S. *Langmuir* **2005**, *21*, 8026.

(180) Morikawa, T.; Asahi, R.; Ohwaki, T.; Aoki, A.; Taga, Y. *Jpn. J. Appl. Phys.* **2001**, *40*, L561.

(181) Asahi, R.; Taga, Y.; Mannstadt, W.; Freeman, A.J. *Phys. Rev.* **2000**, *61*, 7459.

(182) Irie, H.; Watanabe, Y.; Hashimoto, K. *J. Phys. Chem. B* **2003**, *107*, 5483.

(183) Sato, S.; Nakamura, R.; Abe, S. *Appl. Catal. A* **2005**, *284*, 131.

(184) Ihara, T.; Miyoshi, M.; Iriyama, Y.; Matsumoto, O.; Sugihara, S. *Appl. Catal. B* **2003**, *42*, 403.

(185) Sakthivel, S.; Kisch, H. *Angew. Chem. Int. Ed.* **2003**, *42*, 4908.

(186) Irie, H.; Watanabe, Y.; Hashimoto, K. *Chem. Lett.* **2003**, *32*, 772.

(187) Kamisaka, H.; Adachi, T.; Yamashita, K. *J. Chem. Phys.* **2005**, *123*, 084704.

(188) Khan, S.; Al-Shahry, M.; Ingeler, W.B. *Science* **2002**, *297*, 2243.

(189) Barborini, E.; Conti, A.M.; Kholmanov, I.; Piseri, P.; Podestà, A.; Milani, P.; Cepek, C.; Sakho, O.; Macovez, R.; Sancrotti, M. *Adv. Mater.* **2005**, *17*, 1842.

(190) Umebayashi, T.; Yamaki, T.; Tanaka, S.; Asai, K. *Chem. Lett.* **2003**, *32*, 330.

(191) Ohno, T.; Mitsui, T.; Matsumura, M. *Chem. Lett.* **2003**, *32*, 364.

(192) Hattori, A.; Yamamoto, M.; Tada, H.; Ito, S. *Chem. Lett.* **1998**, *27*, 707.

(193) Yu, J.C.; Yu, J.G.; Ho, W.K.; Jiang, Z.T.; Zhang, L.Z. *Chem. Mater.* **2002**, *14*, 3808.

(194) Hong, X.; Wang, Z.; Cai, W.; Lu, F.; Zhang, J.; Yang, Y.; Ma, N.; Liu, Y. *Chem. Mater.* **2005**, *17*, 1548.

(195) Umebayashi, T.; Yamaki, T.; Itoh, H.; Asai, K. *Appl. Phys. Lett.* **2002**, *81*, 454.

(196) Umebayashi, T.; Yamaki, T.; Yamamoto, S.; Miyashita, A.; Tanaka, S.; Sumita, T.; Asai, K. *J. Appl. Phys.* **2003**, *93*, 5156.

(197) Yamaki, T.; Sumita, T.; Yamamoto, S. *J. Mater. Sci. Lett.* **2002**, *21*, 33.

(198) Subbarao, S.N.; Yun, Y.H.; Kershaw, R.; Dwight, K.; Wold, A. *Inorg. Chem.* **1979**, *18*, 488.

(199) Hattori, A.; Shimoda, K.; Tada, H.; Ito, S. *Langmuir* **1999**, *15*, 5422.

(200) Li, D.; Haneda, H.; Hishita, S.; Ohashi, N. *Chem. Mater.* **2005**, *17*, 2588.

(201) Noguchi, D.; Kawamata, Y.; Nagatomo, T. *J. Electrochem. Soc.* **2005**, *152*, D124.

(202) Miyauchi, M.; Takashio, M.; Tobimatsu, H. *Langmuir* **2004**, *20*, 232.

(203) Kato, H.; Kudo, A. *J. Phys. Chem. B* **2002**, *106*, 5029.

(204) Arakawa, H.; Sayama, K. *Res. Chem. Intermed.* **2000**, *26*, 145.

(205) Sung-Suh, H.M.; Choi, J.R.; Hah, H.J.; Koo, S.M.; Bae, Y.C. *J. Photochem. Photobiol. A* **2004**, *163*, 37.

(206) Tian, Y.; Tatsuma, T. *J. Am. Chem. Soc.* **2005**, *127*, 7632.

(207) Burgeth, G.; Kisch, H. *Coord. Chem. Rev.* **2002**, *230*, 40; and references therein.

(208) Macyk, W.; Kisch, H. *Chem. Eur. J.* **2001**, *7*, 1862.

(209) Zang, L.; Macyk, W.; Lange, C.; Maier, W.; Antonius, C.; Meissner, D.; Kisch, H. *Chem. Eur.* **2000**, *6*, 379.

(210) Kamat, P.V.; Fox, M.A. *Chem. Phys. Lett.* **1983**, *102*, 379.

(211) O'Regan, B.; Grätzel, M. *Nature* **1991**, *353*, 737.

(212) Kalyanansundaram, K.; Grätzel, M. *Coord. Chem. Rev.* **1998**, *177*, 347.

(213) Benko, G.; Kallioinen, J.; Myllyperkio, P.; Trif, F.; Korppi, J.; Yartsev, A.P.; Sundstrom, V. *J. Phys. Chem. B* **2004**, *108*, 2862.

(214) Bauer, C.; Boschloo, G.; Mukhtar, E.; Hagfeldt, A. *J. Phys. Chem. B* **2001**, *105*, 5585.

(215) Vogel, R.; Pohl, K.; Weller, H. *Chem. Phys. Lett.* **1990**, *174*, 241.

(216) Hara, K.; Sugihara, H.; Tachibana, Y.; Islam, A.; Yanagida, M.; Sayama, K.; Arakawa, H. *Langmuir* **2001**, *17*, 5992.

(217) Nazeeruddin, M.K.; Péchy, P.; Grätzel, M. *Chem. Commun.* **1997**, 1705.

(218) Nazeeruddin, M.K.; Kay, A.; Rodicio, I.; Humpbry-Baker, R.; Miller, E.; Liska, P.; Vlachopoulos, N.; Grätzel, M. *J. Am. Chem. Soc.* **1993**, *115*, 6382.

(219) Vrachnou, E.; Grätzel, M.; McEvoy, A.J. *J. Electroanal. Chem.* **1989**, *258*, 193.

(220) Vrachnou, E.; Vlachopoulos, N.; Grätzel, M. *J. Chem. Soc. Chem. Commun.* **1987**, 868.

(221) Choi, Y.M.; Choi, W.Y.; Lee, C.H.; Hyeon, T.; Lee, H.I. *Environm. Sci. Technol.* **2001**, *35*, 966.

(222) Cheung, S.; Fung, A.; Lam, M. *Chemosphere* **1998**, *36*, 2461.

(223) Fung, A.; Chiu, B.; Lam, M. *Water Res.* **2003**, *37*, 1939.

(224) Zhang, J.Z. *J. Phys. Chem. B* **2000**, *104*, 7239.

(225) Peter, L.M.; Wijayantha, K.G.U.; Riley, D.J.; Waggett, J.P. *J. Phys. Chem. B* **2003**, *107*, 8378.

(226) Hu, K.; Brust, M.; Bard, A. *Chem. Mater.* **1998**, *10*, 1160.

(227) Weller, H. *Angew. Chem. Int. Ed. Engl.* **1993**, *32*, 41.

(228) Sant, P.A.; Kamat, P.V. *Phys. Chem. Chem. Phys.* **2002**, *4*, 198.

(229) Hirai, T.; Suzuki, K.; Komasawa, I. *J. Coll. Interf. Sci.* **2001**, *244*, 262.

(230) Calabrese, G.S.; Wrighton, M.S. *J. Am. Chem. Soc.* **1981**, *103*, 6273.

(231) Kohtani, S.; Hiro, J.; Yamamoto, N.; Kudo, A.; Tokumura, K.; Nakagaki, R. *Catal. Commun.* **2005**, *6*, 185.

(232) Wilcoxon, J.P. *J. Phys. Chem. B* **2000**, *104*, 7334.

(233) Huang, J.M.; Laitinen, R.A.; Kelley, D.F. *Phys. Rev. B* **2000**, *15*, 10995.

(234) Ho, W.K.; Yu, J.C.; Lin, J.; Yu, J.; Li, P. *Langmuir* **2004**, *20*, 5865.

(235) Anpo, M.; Takeuchi, M.; Ikeue, K.; Dohshi, S. *Curr. Opin. Solid State Mater. Sci.* **2002**, *6*, 381.

(236) Yamashita, H.; Anpo, M. *Catal. Surv. Asia* **2004**, *8*, 35.

(237) Zou, Z.G.; Ye, J.H.; Arakawa, H. *Int. J. Hydrogen Energ.* **2003**, *28*, 663.

(238) Kudo, A.; Ueda, K.; Kato, H.; Mikami, I. *Catal. Lett.* **1998**, *53*, 229.

(239) Tokunaga, S.; Kato, H.; Kudo, A. *Chem. Mater.* **2001**, *13*, 4624.

(240) Kudo, A.; Hiji, S. *Chem. Lett.* **1999**, *28*, 1103.

(241) Ye, J.; Zou, Z.; Arakawa, H.; Oshikiri, M.; Shimoda, M.; Matsushita, A.; Shishido, T. *J. Photochem. Photobiol. A* **2002**, *148*, 79.

(242) Kato, H.; Kobayashi, H.; Kudo, A. *J. Phys. Chem. B* **2002**, *106*, 12441.

(243) Ye, J.; Zou, Z.; Oshikiri, M.; Matsushita, A.; Shimoda, M.; Imai, M.; Shishido, T. *Chem. Phys. Lett.* **2002**, *356*, 221.

(244) Oshikiri, M.; Boero, M.; Ye, J.; Zou, Z.; Kido, G. *J. Chem. Phys.* **2002**, *117*, 7313.

(245) Takata, T.; Tanaka, A.; Hara, M.; Kondo, J.N.; Domen, K. *Catal. Today* **1998**, *44*, 17.

(246) Zou, Z.; Arakawa, H. *J. Photochem. Photobiol. A* **2003**, *158*, 145.

(247) Zou, Z.; Ye, J.; Arakawa, H. *Chem. Phys. Lett.* **2000**, *332*, 271.

(248) Zou, Z.; Ye, J.; Arakawa, H. *Mater. Res. Bull.* **2001**, *36*, 1185.

(249) Domen, K.; Hara, M.; Kondo, J.N.; Takata, T.; Kudo, A.; Kobayashi, H.; Inoue, Y.; Korean, J. *Chem. Eng.* **2001**, *18*, 862.

(250) Zou, Z.; Ye, J.; Sayama, K.; Arakawa, H. *J. Photochem. Photobiol. A* **2002**, *148*, 65.

(251) Zou, Z.; Ye, J.H.; Sayama, K.; Arakawa, H. *Nature* **2001**, *414*, 625.

(252) Matsushima, S.; Obata, K.; Nakamura, H.; Arai, M.; Kobayashi, K. *J. Phys. Chem. Solid* **2003**, *64*, 2417.

(253) Ji, S.M.; Borse, P.H.; Kim, H.G.; Hwang, D.W.; Jang, J.S.; Bae, S.; Lee, J. *Phys. Chem. Chem. Phys.* **2005**, *7*, 1100.

(254) Yin, J.; Zou, Z.; Ye, J. *Chem. Phys. Lett.* **2003**, *378*, 24.

(255) Wang, D.; Zou, Z.; Ye, J. *Chem. Phys. Lett.* **2003**, *373*, 191.

(256) Kato, H.; Matsudo, N.; Kudo, A. *Chem. Lett.* **2004**, *33*, 1216.

(257) Saito, N.; Kadowaki, H.; Kobayashi, H.; Ikarashi, K.; Nishiyama, H.; Inoue, Y. *Chem. Lett.* **2004**, *33*, 1452.

(258) Tang, J.; Zou, Z.; Ye, J. *J. Phys. Chem. B* **2003**, *107*, 14265.

(259) Tang, J.; Zou, Z.; Ye, J. *Angew. Chem. Int. Ed.* **2004**, *43*, 4463.

(260) Tang, J.; Zou, Z.; Yin, J.; Ye, J. *Chem. Phys. Lett.* **2003**, *382*, 175.

(261) Tang, J.; Zou, Z.; Yin, J.; Ye, J. *Chem. Mater.* **2004**, *16*, 1644.

(262) Tang, J.; Zou, Z.; Ye, J. *Catal. Today* **2004**, *93–95*, 891.

(263) Tang, J.; Zou, Z.; Yin, J.; Ye, J. *Catal. Lett.* **2004**, *92*, 53.

(264) Kasahara, A.; Nukumizu, K.; Hitoki, G.; Takata, T.; Kondo, J.N.; Hara, M.; Kobayashi, H.; Domen, K. *J. Phys. Chem. A* **2002**, *106*, 6750.

(265) Kasahara, A.; Nukumizu, K.; Takata, T.; Kondo, J.N.; Hara, M.; Kobayashi, H.; Domen, K. *J. Phys. Chem. B* **2003**, *107*, 791.

(266) Yamasita, D.; Takata, T.; Hara, M.; Kondo, J.N.; Domen, K. *Solid State Ion.* **2004**, *172*, 591.

(267) Hara, M.; Hitoki, G.; Takata, T.; Kondo, J.N.; Kobayashi, H.; Domen, K. *Catal. Today* **2003**, *78*, 555.

(268) Hitoki, G.; Takata, T.; Kondo, J.N.; Hara, M.; Kobayashi, H.; Domen, K. *Chem. Commun.* **2002**, 1698.

(269) Nakamura, R.; Tanaka, T.; Nakato, Y. *J. Phys. Chem. B* **2005**, *109*, 8920.

(270) Hara, M.; Takata, T.; Kondo, J.N.; Domen, K. *Catal. Today* **2004**, *90*, 313.

(271) Hitoki, G.; Ishikawa, A.; Takata, T.; Kondo, J.N.; Hara, M.; Domen, K. *Chem. Lett.* **2002**, 736.

(272) Zhang, Q.; Gao, L. *Langmuir* **2004**, *20*, 9821.

(273) Chun, W.J.; Ishikawa, A.; Fujisawa, H.; Takata, T.; Kondo, J.N.; Hara, M.; Kawai, M.; Matsumoto, Y.; Domen, K. *J. Phys. Chem. B* **2003**, *107*, 1798.

(274) Thurston, T.R.; Wilcoxon, J.P. *J. Phys. Chem. B* **1999**, *103*, 11.

(275) Kudo, A.; Sekizawa, M. *Catal. Lett.* **1999**, *58*, 241.

(276) Lee, J.K.; Lee, W.; Yoon, T.J.; Park, G.S.; Choy, J.H. *J. Mater. Chem.* **2002**, *12*, 614.

(277) Kudo, A.; Sekizawa, M. *Chem. Commun.* **2000**, 1371.

(278) Martin, S.T.; Herrmann, H.; Choi, W.; Hoffmann, M.R. *Trans. Faraday Soc.* **1994**, *90*, 3315.

(279) Martin, S.T.; Herrmann, H.; Hoffmann, M.R. *Trans. Faraday Soc.* **1994**, *90*, 3323.

(280) Pelizzetti, E.; Minero, C. *Electrochim. Acta* **1993**, *38*, 47.

(281) Teurich, J.; Linder, M.; Bahnemann, D.W. *Langmuir* **1996**, *12*, 6368.

(282) Dieckmann, M.S.; Gray, K.A. *Water Res.* **1996**, *30*, 1169.

(283) Herrmann, J.M. *Catal. Today* **1999**, *53*, 115.

(284) Vautier, M.; Guillard, C.; Herrmann, J.M. *J. Catal.* **2001**, *201*, 46.

(285) Houas, A.; Lachheb, H.; Ksibi, M.; Elaloui, E.; Guillard, C.; Herrmann, J.M. *Appl. Catal. B* **2001**, *31*, 145.

(286) Subramanian, V.; Pangarkar, V.G.; Beenackers, A.A. *Clean Prod. Pro.* **2000**, *2*, 149.

(287) Tanaka, K.; Padarmpole, K.; Hisanaga, T. *Water Res.* **2000**, *34*, 327.

(288) Demeestere, K.; Visscher, A.D.; Dewulf, J.; Leeuwen, M.; Langenhove, H.V. *Appl. Catal. B* **2004**, *30*, 261.

(289) Ollis, D.F.; Pelizzetti, E.; Serpone, N. *Environ. Sci. Technol.* **1991**, *25*, 1523.

(290) Brusa, M.A.; Grela, M.A. *J. Phys. Chem. B* **2005**, *109*, 1914.

(291) Zeltner, W.A.; Hill, C.G.; Anderson, M.A. *Chemtech.* **1993**, *23*, 21.

(292) Kim, S.B.; Hwang, H.T.; Hong, S.C. *Chemosphere* **2002**, *48*, 437.

(293) Sánchez, B.; Cardona, A.I.; Romero, M.; Avila, P.; Bahamonde, A. *Catal. Today* **1999**, *54*, 369.

(294) Pichat, P.; Herrmann, J. *Photocatalysis: Fundamentals and Applications*; Wiley & Sons: New York, 1989.

(295) Coronado, J.M.; Zorn, M.E.; Tejedor, I.; Anderson, M.A. *Appl. Catal. B* **2003**, *43*, 329.

(296) Zorn, M.E.; Tompkins, D.T.; Zeltner, W.A.; Anderson, M.A. *Environ. Sci. Technol.* **2000**, *4*, 5206.

(297) Cao, L.; Huang, A.; Spiess, F.J.; Suib, S.L. *J. Catal.* **1999**, *188*, 48.

(298) Park, D.R.; Zhang, J.; Ikeue, K.; Yamashita, H.; Anpo, M. *J. Catal.* **1999**, *185*, 114.

(299) Jo, W.K.; Park, H.K. *Chemosphere* **2004**, *57*, 555.

(300) Phillips, L.A.; Raupp, G.B. *J. Mol. Catal.* **1992**, *77*, 297.

(301) Obee, T.N.; Brown, R.T. *Environ. Sci. Technol.* **1995**, *29*, 1223.

(302) Peral, J.; Ollis, D.F. *J. Catal.* **1992**, *136*, 554.

(303) Cunninham, J.; Hodnett, B.K.; Ilyas, M.; Leahy, E.L. *Faraday Discuss. Chem. Soc.* **1981**, *72*, 283.

(304) Harm, W. *Biological Effects of Ultraviolet Radiation*; Cambridge University Press: Cambridge, 1980.

(305) Young, R.A.; Bjorn, L.O.; Moan, J.; Nultsch, W. *Environmental UV Photobiology*; Cambridge University Press: New York, 1993.

(306) Matsunaga, T.; Tomoda, R.; Nakajima, T.; Wake, H. *FEMS Microbiol. Lett.* **1985**, *29*, 211.

(307) Saito, T.; Iwase, T.; Horie, J.M.T. *J. Photochem. Photobiol. B* **1992**, *14*, 369.

(308) Sunada, K.; Watanabe, T.; Hashimoto, K. *J. Photochem. Photobiol. A* **2003**, *156*, 227.

(309) Blake, D.M.; Maness, P.C.; Huang, Z.; Wolfrim, E.; Huang, J.; Jacoby, W. *Separ. Purif. Method.* **1999**, *28*, 1.

(310) Ireland, J.C.; Klostermann, P.; Rice, E.; Clark, R. *Appl. Environ. Microbiol.* **1993**, *59*, 1668.

(311) Ireland, J.C.; Valinierks, J. *Chemosphere* **1992**, *25*, 383.

(312) Zhang, P.; Scrudato, R.; Germano, G. *Chemosphere* **1994**, *28*, 607.

(313) Wei, C.; Lin, W.; Zainal, Z.; Williams, N.E.; Zhu, K.; Kruzic, A.P.; Smith, R.L.; Rajeshwar, K. *Env. Sci. Technol.* **1994**, *28*, 934.

(314) Maness, P.C.; Smolinski, S.L.; Blake, D.M.; Huang, Z.; Wolfrum, E.J.; Jacoby, W.A. *Appl. Environ. Microbiol.* **1999**, *65*, 4094.

(315) Zheng, H.; Maness, P.C.; Blake, D.M.; Wolfrum, E.J.; Smolinski, S.L.; Jacoby, W.A. *J. Photochem. Photobiol. A* **2000**, *130*, 163.

(316) Rincón, A.G.; Pulgarín, C. *Appl. Catal. B* **2003**, *44*, 263.

(317) Rincón, A.G.; Pulgarín, C. *Appl. Catal. B* **2003**, *49*, 99.

(318) Rincón, A.G.; Pulgarín, C. *Appl. Catal. B* **2004**, *51*, 283.

(319) Herrera Melián, J.A.; Doña Rodríguez, J.M.; Viera Suárez, A.; Tello Rendón, E.; Campo, C.A.; Araña, J.; Pérez Peña, J. *Chemosphere* **2000**, *41*, 323.

(320) Gourmelon, M.; Cillard, J.; Pommepuy, M. *J. Appl. Bacterol.* **1994**, *77*, 105.

(321) Fujishima, A.; Honda, K. *Nature* **1972**, *238*, 37.

(322) Mandelbaum, P.A.; Regazzoni, A.E.; Blesa, M.A.; Bilmes, S.A. *J. Phys. Chem. B* **1999**, *103*, 5505.

(323) Vinodgopal, K.; Hotchandani, S.; Kamat, P.V. *J. Phys. Chem. B* **1993**, *97*, 9040.

(324) Vinodgopal, K.; Kamat, P.K. *Sol. Energy Mat. Sol. Cells* **1995**, *38*, 401.

(325) Vinodgopal, K.; Stafford, U.; Gray, K.A.; Kamat, P.V. *J. Phys. Chem.* **1994**, *98*, 6797.

(326) Calvo, M.E.; Candal, R.J.; Bilmes, S.A. *Environ. Sci. Technol.* **2001**, *35*, 4132.

(327) Egerton, T.A.; Kosa, S.A.M.; Christensen, P.A. *Phys. Chem. Chem. Phys.* **2006**, *8*, 398.

(328) Vinodgopal, K.; Kamat, P.K. *Environ. Sci. Technol.* **1995**, *29*, 841.

(329) Yang, S.G.; Liu, Y.Z.; Sun, C. *Appl. Catal. A: Gen.* **2006**, *301*, 284.

(330) Barazzouk, S.; Hotchandani, S.; Vinodgopal, K.; Kamat, P.V. *J. Phys. Chem.* **2004**, *108*, 17015.

(331) Bak, T.; Nowotny, J.; Rekas, M.; Sorrell, C.C. *Int. J. Hydrogen En.* **2002**, *27*, 991.

(332) Aroutiounian, V.M.; Arakelyan, V.M.; Shahnazaryan, G.E. *Sol. En.* **2005**, *78*, 581.

(333) Anpo, M.; Shima, T.; Kodama, S.; Kubokawa, Y. *J. Phys. Chem.* **1987**, *91*, 4305.

(334) Grela, M.A.; Colussi, A.J. *J. Phys. Chem.* **1996**, *100*, 18214.

(335) Hoffmann, A.J.; Mills, G.; Yee, H.; Hoffmann, M.R. *J. Phys. Chem.* **1992**, *96*, 5540.

(336) Hoffmann, A.J.; Mills, G.; Yee, H.; Hoffmann, M.R. *J. Phys. Chem.* **1992**, *96*, 5546.

(337) Almquist, C.B.; Biswas, P. *J. Catal.* **2002**, *212*, 145.

(338) Fernández-García, M.; Martínez-Arias, A.; Hanson, J.C.; Rodríguez, J.A. *Chem. Rev.* **2004**, *104*, 4063.

(339) Al-Abadleh, H.A.; Grassian, V.H. *Surf. Sci. Rep.* **2003**, *52*, 63.

(340) Zhang, H.; Banfield, J.F. *J. Mater. Chem,* **1998**, *8*, 2073.

(341) Lucas, E.; Decker, S.; Khaleel, A.; Seitz, A.; Fultz, S.; Ponce, A.; Li, W.; Carnes, C.; Klabunde, K.J. *Chem. Eur. J.* **2001**, *7*, 2505.

(342) Dietbold, U. *Surf. Sci. Rep.* **2003**, *48*, 53.

(343) Herrmann, J.M. *Top. Catal.* **2005**, *34*, 49.

(344) Kelly, A.; Groves, G.W.; Kidd, P. *Crystallography and Defects*; John Wiley & Sons: New York, 2000.

(345) Bloss, F.D. *Crystallography and Crystal Chemistry*; Holt, Reinehart & Winston: New York, 1971.

(346) Thompson, T.I., Yates, J.Y. *Top. Catal.* **2005**, *35*, 197.

(347) Lazzeri, M.; Vittadini, A.; Selloni, A. *Phys. Rev. B* **2001**, *63*, 155409.

(348) Li, G.S.; Li, L.P.; Boerio-Goates, J.; Woodfield, B.F. *J. Mater. Res.* **2003**, *18*, 2664.

(349) Zhang, H.; Banfield, J.F. *Chem. Mater.* **2005**, *17*, 3421.

(350) Penn, R.L.; Banfield, J.F. *Geochim. Cosmochim. Acta* **1999**, *63*, 1549.

(351) Vittadini, A.; Selloni, A.; Rotzingerand, F.P.; Grätzel, M. *Phys. Rev. Lett.* **1998**, *81*, 2954.

(352) Lazzeri, M.; Vittadini, A.; Selloni, A. *Phys. Rev. B* **2001**, *65*, 119901.

(353) Wang, C.C.; Zhang, Z.; Ying, J.Y. *Nano struct. Mater.* **1997**, *9*, 583.

(354) Serpone, N.; Lawless, D.; Disdier, J.; Herrmann, J.M. *Langmuir* **1994**, *10*, 643.

(355) Sivalinga, G.; Nagaveni, K.; Hegde, M.S.; Madras, G. *Appl. Catal. B: Environ.* **2003**, *45*, 23.

(356) Di Paola, A.; García-López, E.; Ikeda, S.; Marcí, G.; Ohtani, B.; Palmisano, L. *Catal. Today* **2002**, *75*, 87.

(357) Wang, J.A.; Limas-Ballesteros, R.; López, T.; Moreno, A.; Gómez, R.; Novaro, O.; Bokh, X. *J. Phys. Chem. B* **2001**, *105*, 9692.

(358) Fernández-García, M.; Martínez-Arias, A.; Fuerte, A.; Conesa, J.C. *J. Phys. Chem. B* **2005**, *109*, 6075.

(359) Nakaoka, Y.; Nosaka, Y. *J. Photochem. Photobiol. A* **1997**, *110*, 299.

(360) Howe, R.F.; Grätzel, M. *J. Phys. Chem.* **1985**, *89*, 4495.

(361) Hurum, D.C.; Kimberly, A.G.; Rajh, T.; Thurnauer, M.C. *J. Phys. Chem. B* **2005**, *109*, 977.

(362) Szczepankiewicz, S.H.; Moss, J.A.; Hoffmann, M.R. *J. Phys. Chem. B* **2002**, *106*, 2922.

(363) Szczepankiewicz, S.H.; Colussi, A.J.; Hoffmann, M.R. *J. Phys. Chem. B* **2000**, *104*, 9842.

(364) Corrent, S.; Cosa, G.; Scaiano, J.C.; Galletero, M.S., Alvaro, M.; Garcia, H. *Chem. Mater.* **2001**, *13*, 715.

(365) Henderson, M.A.; Epling, W.S.; Peden, C.H.F.; Perkins, C.L. *J. Phys. Chem. B* **2003**, *107*, 534.

(366) Szczepankiewicz, S.H.; Moss, J.A.; Hoffmann, M.R. *J. Phys. Chem. B* **2002**, *106*, 7654.

(367) Kolle, U.; Moser, J.; Grätzel, M. *Inorg. Chem.* **1985**, *24*, 2253.

(368) Ghosh, A.M.; Wakim, F.G.; Addis, R.R. *Phys. Rev.* **1969**, *184*, 979.

(369) Avudaithai, M.; Kutty, T.R.N. *Mater. Res. Bull.* **1988**, *23*, 1675.

(370) Anpo, M.; Shima, T.; Kubokawa, Y. *Chem. Lett.* **1985**, 1799.

(371) Kasinski, J.J.; Gomez-Jahn, L.A.; Faran, K.J.; Gracewski, S.M.; Dwayne Miller, R.J. *J. Chem. Phys.* **1989**, *90*, 12536.

(372) Jaeger, C.D.; Bard, A.J. *J. Phys. Chem.* **1979**, *83*, 3146.

(373) Howe, R.F.; Grätzel, M. *J. Phys. Chem.* **1987**, *91*, 3906.

(374) Micic, O.I.; Zhang, Y.; Cromack, K.R.; Trifunac, A.D.; Thurnauer, M.C. *J. Phys. Chem.* **1993**, *97*, 7277.

(375) Nakamura, R.; Ohashi, N.; Imanishi, A.; Osawa, T.; Matsumoto, Y.; Koinuma, H.; Nakato, Y. *J. Phys. Chem. B* **2005**, *109*, 1648.

(376) Rabani, J.; Yamashita, K.; Ushida, K.; Stark, J.; Kira, A. *J. Phys. Chem. B* **1998**, *102*, 1689.

(377) Coronado, J.M.; Maira, A.J.; Conesa, J.C.; Yeung, K.L.; Augugliaro, V.; Soria, J. *Langmuir* **2001**, *17*, 5368.

(378) Chen, C.Q.; Liu, H.B.; Gu, G.B. *Mater. Chem. Phys.* **2005**, *91*, 317.

(379) Shkrob, I.A.; Sauer, M.C. *J. Phys. Chem. B* **2004**, *108*, 12497.

(380) Shkrob, I.A.; Sauer, M.C.; Gosztola, D. *J. Phys. Chem. B* **2004**, *108*, 12512.

(381) Anpo, M.; Che, M.; Fubini, B.; Garrone, E.; Giamello, E.; Paganini, M.C. *Top. Catal.* **1999**, *8*, 189.

(382) Mizushima, K.; Tanaka, M.; Iid, S. *J. Phys. Soc. Jpn.* **1972**, *32*, 1519.

(383) Mizushima, K. *Phys. Chem. Solids* **1979**, *40*, 1129.

(384) Bin-Daar, G.; Dare-Edwards, M.P.; Goodenough, J.B.; Hamnett, A. *J. Chem. Soc. Faraday Trans.* **1983**, *79*, 1199.

(385) Zwingel, D. *Solid State Comm.* **1978**, *26*, 775; Zwingel, D. *Solid State Comm.* **1976**, *20*, 397.

(386) *Defect and Microstructure Analysis by Diffraction*; Snyder, R.L.; Fiala, J.; Bunge, H.J. (Editors); Oxford University Press: New York, 1999.

(387) Smart, L.; Moore, E. *Solid State Chemistry*; Chapman & Hall: New York, 1995.

(388) Cullity, B.D. *Elements of X-ray Diffraction*; Addison-Wesley: London, 1978.

(389) International Centre for Diffraction Data: Newtown Square, PA.

(390) *Structure Determination from Powder Diffraction Data*; David, W.; Shankland, K.; McCusker, L.; Baerlocher, C. (Editors); Oxford University Press: New York, 2002.

(391) *The Rietveld Method*; Young, R.A. (Editor); Oxford University Press: New York, 1995.

(392) Scherrer, P. *Nachr. Gott.* **1918**, *2*, 98.

(393) Stokes, A.R.; Wilson, A. *Proc. Phys. Soc.* **1944**, *56*, 174.

(394) Williamson, G.K.; Hall, W.H. *Acta Metallurgica* **1953**, *1*, 22.

(395) Depero, L.E.; Sangaletti, L.; Allieri, B.; Bontempi, E.; Marino, A.; Zocchi, M. *J. Crystal Growth* **1999**, *198/199*, 516.

(396) Knauth, P.; Tuller, H.L. *J. Appl. Phys.* **1999**, *85*, 897.

(397) Nemanich, R.J.; Solin, S.A.; Martin, R.M. *Phys. Rev. B* **1981**, *23*, 6348.

(398) Richter, H.; Wang, Z.P.; Ley, L. *Solid State Commun.* **1981**, *39*, 625.

(399) Doss, C.J.; Zallen, R. *Phys. Rev. B* **1993**, *21*, 15626.

(400) Tiong, K.K.; Amirtharaj, P.M.; Pollak, F.H.; Aspnes, D.E. *Appl. Phys. Lett.* **1984**, *44*, 122.

(401) Nelly, S.; Pollak, F.H.; Tomkiewicz, M. *J. Phys. Chem. B* **1997**, *101*, 2730.

(402) Swamy, V.; Kuznetsov, A.; Dubrovinsky, L.S.; Caruso, R.A.; Shchukin, D.G.; Muddle, B.C. *Phys. Rev. B* **2005**, *71*, 184302.

(403) Barborini, E.; Kholmanov, I.; Piseri, P.; Ducati, C.; Bottani, C.; Milani, P. *Appl. Phys. Lett.* **2002**, *81*, 3052.

(404) Bersani, D.; Lotticia, P.P.; Ding, X.Z. *Appl. Phys. Lett.* **1998**, *72*, 73.

(405) Lee, M.H.; Choi, B.C.K. *J. Am. Ceram. Soc.* **1991**, *74*, 2309.

(406) Ma, W.; Lu, Z.; Zhang, M. *Appl. Phys. A* **1998**, *66*, 621.

(407) Xu, C.Y.; Zhang, P.X.; Yan, L. *J. Raman Spectrosc.* **2001**, *32*, 862.

(408) Parker, J.C.; Siegel, R.W. *J. Mater. Res.* **1990**, *5*, 1246.

(409) Parker, J.C.; Siegel, R.W. *Appl. Phys. Lett.* **1990**, *57*, 943.

(410) Zhang, W.F.; He, Y.L.; Zhang, M.S.; Yin, Z.; Chen, Q. *J. Phys. D* **2000**, *33*, 912.

(411) Liu, G.; Rodríguez, J.A.; Chang, Z.; Hrbek, J.; González, L. *J. Phys. Chem. B* **2002**, *106*, 9883.

(412) Ikeda, S.; Sugiyama, N.; Murakami, S.; Kominami, H.; Kera, Y.; Noguchi, H.; Uosaki, K.; Torimoto, T.; Ohtani, B. *Phys. Chem. Chem. Phys.* **2003**, *5*, 778.

(413) Watanabe, T.; Honda, K. *J. Phys. Chem.* **1982**, *86*, 2617.

(414) Berger, T.; Sterrer, M.; Diwald, O.; Knözinger, E.; Panayotov, D.; Thompson, T.L.; Yates, J.T. *J. Phys. Chem. B* **2005**, *109*, 6061.

(415) Moriarty, P. *Rep. Prog. Phys.* **2001**, *64*, 297.

(416) Boschloo, G.; Fitsmaurice, D. *J. Phys. Chem. B* **1999**, *103*, 228.

(417) Durrant, J.R. *J. Photchem. Photobiol. A* **2002**, *148*, 5.

(418) Cohen, M.M. *Introduction to the Quantum Theory of Semiconductors*; Gordon: Amsterdam, 1972.

(419) Movilla, J.L.; García-Belmonte, G.; Bisquert, J.; Planelles, J. *Phys. Rev. B* **2005**, *72*, 153313.

(420) Mora-Seró, I.; Busquet, J. *Nanoletters* **2003**, *3*, 945.

(421) Grätzel, M. *Heterogeneous Photochemical Electron Tranfer*; CRC: Boca Raton, FL, 1989; 106–108.

(422) Colombo, D.P. Jr.; Bowman, R.M. *J. Phys. Chem.* **1996**, *100*, 18445; and references therein.

(423) Iwata, K.; Takaya, T.; Hamaguchi, H.; Yamakata, A.; Ishibashi, T.; Unishi, H.; Kuruda, H. *J. Phys. Chem. B* **2004**, *108*, 20233.

(424) Efros, A.I.L.; Efros, A.L. *Sov. Phys. Semicond.* **1982**, *16*, 772.

(425) Yoffre, I. *Adv. Phys.* **2001**, *50*, 1.

(426) Glinka, Y.D.; Lin, S.H.; Hwang, L.P.; Chen, Y.T.; Tolk, N.H. *Phys. Rev. B* **2001**, *64*, 085421.

(427) Pan, L.K.; Sun, C.Q. *J. Appl. Phys.* **2004**, *95*, 3819.

(428) Serpone, N.; Lawless, D.; Khairutdinov, R. *J. Phys. Chem.* **1995**, *99*, 16646.

(429) Monticone, S.; Triferi, R.; Kanaev, A.V.; Scolan, E.; Sánchez, C. *Appl. Surf. Sci.* **2000**, *162–163*, 565.

(430) Enright, B.; Fritzmaurice, D. *J. Phys. Chem.* **1996**, *100*, 1027.

(431) Liu, Y.L.; Liu, Y.C.; Liu, Y.X.; Shen, D.Z.; Lu, Y.M.; Zhang, J.Y.; Fa, X.W. *Physica B* **2002**, *322*, 31.

(432) Visnawatha, R.; Sapra, S.; Satpati, B.; Satyam, P.V.; Der, B.N.; Serma, D.D. *J. Mater. Chem.* **2004**, *14*, 681.

(433) Iyer, S.S.; Xie, Y.-H. *Science* **1993**, *260*, 40.

(434) Kubo, T.; Nishikitani, Y. *J. Electrochem. Soc.* **1998**, *145*, v1729.

(435) Karazhanov, S.Zh.; Zhang, Y.; Wang, L.-W.; Masacarenas, A.; Deb, S. *Phys. Rev. B* **2003**, *68*, 233204.

(436) Frohlich, H.; Mott, N. *Proc. R. Soc. London* **1939**, *A171*, 496.

(437) Kroeze, J.A.; Savenije, T.J.; Warman, J.M. *J. Am. Chem. Soc.* **2004**, *126*, 7608; and references therein.

(438) Grätzel, M.; Frank, A.J. *J. Phys. Chem.* **1982**, *86*, 2964.

(439) Nakade, S.; Saito, Y.; Kubo, W.; Kitamura, T.; Wada, Y.; Yanigida, S. *J. Phys. Chem. B* **2003**, *107*, 8607.

(440) Advicensses, G.J. *Phys. Rev. B* **1995**, *51*, 9661.

(441) Van meekelberh, D.; de Jongh, P.E. *Phys. Rev. B* **2000**, *61*, 4649.

(442) Hirakawa, T.; Nosaka, Y. *Langmuir* **2002**, *18*, 3247; and references therein.

(443) Goto, H.; Hamada, Y.; Ohno, T.; Matsumura, M. *J. Catal.* **2005**, *225*, 223.

(444) Cornu, C.J.G., Colussi, A.J.; Hoffmann, M.R. *J. Phys. Chem. B* **2003**, *107*, 3156.

(445) Serpone, N.; Lawless, D.; Khairutdinov, R. *J. Phys. Chem.* **1995**, *99*, 16555.

(446) Bahnemann, D.W.; Hildendorff, M.; Memming, R. *J. Phys. Chem. B* **1997**, *101*, 4265.

(447) Yoshihara, T.; Katoh, R.; Furube, A.; Tamaki, Y.; Mrai, M.; Hara, K.; Murata, S.; Arakawa, H.; Tachiya, M. *J. Phys. Chem. B* **2004**, *108*, 2817.

(448) Yang, X.; Tamai, N. *Phys. Chem. Chem. Phys.* **2001**, *3*, 3393.

(449) Turne, G.M.; Beard, M.C.; Schumuttenmauer, C.A. *J. Phys. Chem. B* **2002**, *106*, 11716.

(450) Gao, R.; Safrany, A.; Rabani, J. *Radiat. Phys. Chem.* **2003**, *67*, 25.

(451) Salmi, M.; Tkachenko, N.; Lamminmaki, R.-J.; Karvinen, S.; Vehmanen, V.; Lemmetynien, H. *J. Photochem. Photoboil. A* **2005**, *175*, 8.

(452) Klimov, V.I. *J. Phys. Chem. B* **2000**, *104*, 6112.

(453) Chepic, D.; Efros, A.I.L.; Ekimov, A.; Ivanov, M.; Kharchenko, V.A.; Kudriavtsev, I. *J. Lumin.* **1990**, *47*, 113.

(454) Efros, A.I.L.; Rosen, M. *Phys. Rev. Lett.* **1997**, *78*, 1110.

(455) Yamakata, A.; Ishibashi, T.-A.; Onishi, H. *Chem. Phys. Lett.* **2001**, *333*, 271.

(456) Bisquert, J.; Zaban, A.; Salvador, P. *J. Phys. Chem. B* **2002**, *106*, 8774.

(457) Rothenberger, G.; Moser, J.; Gratzel, M.; Serpone, N.; Sharma, D.K. *J. Am. Chem. Soc.* **1985**, *107*, 8054.

(458) Herderson, M.A.; White, J.M.; Uetsuka, H.; Onishi, H. *J. Am. Chem. Soc.* **2003**, *125*, 14974.

(459) Grabner, G.; Li, G.; Quint, R.M.; Getoff, N. *J. Chem. Soc. Faraday Trans.* **1991**, *87*, 1097.

(460) Kukuta, N.; White, J.M.; Bard, A.J.; Campion, A.; Fox, M.A.; Webber, S.E.; Finlayson, M. *J. Phys. Chem.* **1985**, *89*, 48.

(461) Wang, C.; Wang, X.; Xu, B.-Q.; Zhao, J.; Mai, B.; Peng, P.; Sheng, G.; Fu, J. *J. Photochem. Photobiol. A* **2004**, *168*, 47.

(462) Tennakone, K.; Bandara, J. *Appl. Catal. A* **2001**, *208*, 335.

(463) Maira, J.; Yeung, K.L.; Lee, C.Y.; Yue, P.L.; Chan, C.K. *J. Catal.* **2000**, *192*, 185.

(464) Maira, J.; Yeung, K.L.; Soria, J.; Coronado, J.M.; Belver, C.; Lee, C.Y.; Augugliaro, V. *Appl. Catal. B* **2001**, *29*, 327.

(465) Hao, W.C.; Zheng, S.K.; Wang, C.; Wang, T.M. *J. Mater. Sci. Lett.* **2002**, *21*, 1627.

(466) Damm, C.; Voltzke, D.; Abicht, H.-P.; Israel, G. *J. Photochem. Photoboil. A* **2005**, *174*, 171.

(467) Wu, L.; Yu, C.J.; Zhang, L.; Wang, X.; Ho, W. *J. Solid State Chem.* **2004**, *177*, 2584.

(468) Rivera, P.; Tanak, K.; Hisanaga, T. *Appl. Catal. B* **1993**, *3*, 37.

(469) Deng, X.; Yue, Y.; Gao, Z. *Appl. Catal. B* **2002**, *39*, 135.

(470) Ohtani, B.; Ogawa, Y.; Nishimoto, S.-I. *J. Phys. Chem. B* **1997**, *101*, 3746.

(471) Chen, G.; Luo, G.; Yang, X.; Sun, Y.; Wang, J. *Mater. Sci. Eng. A* **2004**, *380*, 320.

CHAPTER 18

Oxide Nanomaterials for the Catalytic Combustion of Hydrocarbons

ILENIA ROSSETTI and LUCIO FORNI

Dipartimento di Chimica Fisica ed Elettrochimica, Università degli Studi di Milano, v. C. Golgi 19, I-20133 Milano, Italy

18.1. INTRODUCTION

The impact of human activities on the environment has recently gained growing attention over the last few decades. After a basic assessment of the major environmental concerns, such as the Greenhouse Effect and energy efficiency, attention was paid to the need of a worldwide answer to these problems, to be combined with the expected improvement of the quality of life and economic development (1,2). The key points of this philosophy are the control and limitation of the emissions from anthropogenic sources and the increase of energy efficiency, especially when using fossil fuels as an energy source. Particular attention is devoted to the fields of transportation and power generation, which are considered two of the major responsibles for pollutants emission. Although different innovative technologies based on renewable energy sources are under study, an immediate short-term answer can be found by improving the existing technologies for automotive and stationary power generation. The topic is highly demanding and of primary importance, particularly in view of the booming economic growth of some major developing countries. As an example, the increase of energy consumption during the last years is reported in Figure 18.1 (3). It is clear that despite a worldwide increase of energy demand, North America (in particular, the United States) and mostly Asia (in particular, China and India) play a leading role, the latter having exceeded the energy need of the most industrialized countries in 2003 (Figure 18.1). As energy production is mainly based on fossil fuels combustion (in gasoline or diesel motor engines and in thermoelectrical power generation units), a similar trend is shown by CO_2 worldwide emissions (Figure 18.2).

Synthesis, Properties, and Applications of Oxide Nanomaterials, Edited by José A. Rodríguez and Marcos Fernández-Garcia

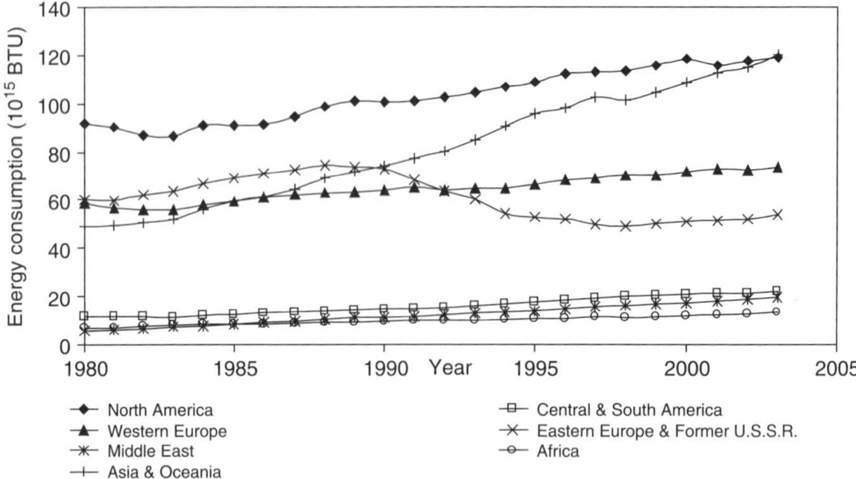

Figure 18.1. World total primary energy consumption by geographic areas (1980–2003). North America's data include Mexico. Data expressed as quadrillion (10^{15}) Btu.

The situation is even more critical from this point of view, due to the use of obsolete, less energy-efficient technologies in the developing countries.

Emissions from mobile sources are mainly constituted of partially unburnt hydrocarbons (UHC), CO, NO_x, SO_x, and of course, CO_2 and H_2O. In addition, some nitrous oxide (N_2O), a powerful greenhouse gas, can form through the reduction of NO and NO_2 by catalytic converters in the case of "cold start". Indeed, mobile

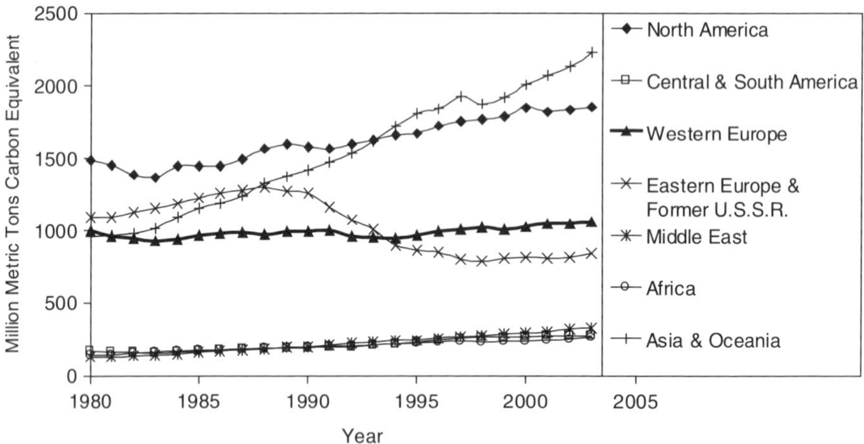

Figure 18.2. World carbon dioxide emissions from the combustion and flaring of fossil fuels (1980–2003). North America's data include Mexico. Data expressed as million (10^6) metric tons carbon equivalent.

combustion is responsible for ca. 20% of total U.S. anthropogenic N_2O emissions (4). As for stationary combustion processes, e.g., power generation units, the same pollutants can be found in effluent gases, with more or less relevant amounts of SO_2, depending on the fuel nature and its pretreatment.

According to Ref. 5, three different mechanisms account for NO_x formation during combustion:

(1) "Thermal NO_x" form through oxygen and nitrogen coupling, following the Zeldovich mechanism (6). The rate of thermal NO_x production increases nearly linearly with residence time and strongly depends on flame temperature.
(2) "Prompt NO_x" form in fast coupling of nitrogen with radicals formed at the flame front, a mechanism favored by moderate temperatures and fuel-rich conditions.
(3) "Fuel NO_x" form by oxidation of nitrogen containing compounds in the fuel. Their formation depends moderately on flame temperature and is favored by fuel-lean conditions.

Table 18.1 reports emissions from energy consumption for electricity production in the United States in the last decade. Despite the increase of CO_2 release due to the overall increase of electricity production, SO_2 emissions slightly decreased through the years, due to more efficient fuel desulphurization. An even stronger decrease of NO_x release can be noticed due to more sophisticated control of thermal plants (7) and mostly to more efficient postcombustion treatments, such as the selective catalytic reduction (SCR) process (8–11). In addition, N_2O can form in stationary combustion units as a result of chemical interactions between nitrogen oxides, mostly NO_2, and other combustion products. It is interesting to observe that N_2O emissions from anthropogenic sources have been reduced in the agricultural and industrial fields, whereas they are still increasing when considering energy production and waste management (Table 18.2), primarily due to combustion

TABLE 18.1. Gaseous Emission Connected with Electricity Production in the United States (1993–2003) (4). Data Reported as Thousand Metric Tons.

	CO_2	SO_2	NO_x
1995	2,079,761	11,898	7,885
1996	2,155,453	12,908	6,281
1997	2,223,347	13,524	6,324
1998	2,313,013	12,509	6,235
1999	2,326,558	12,445	5,732
2000	2,429,394	11,297	5,380
2001	2,379,603	10,966	5,045
2002	2,397,937	10,515	4,802
2003	2,408,961	10,594	4,396

TABLE 18.2. N_2O Emissions from Anthropogenic Sources in the United States (1990–2003) (4).

	Thousand Metric Tons N_2O		Percent Change	
Source	1990	2003	1990–2003	2002–2003
Energy	183	228	24.4%	1%
Agriculture	814	788	−3.2%	−0.8%
Industrial	96	45	−53.0%	−11.8%
Waste management	16	20	23.7%	0.9%

processes. Moreover, in the case of stationary power generation units, methane, a harmful greenhouse gas (1 ton CH_4 equals 23.5 tons CO_2), is often released due to incomplete combustion.

Energetic efficiency is a must from the environmental point of view, but in the last decades, it has also become of primary importance from the economical one. Indeed, the world's energy needs nowadays to depend essentially on fossil fuel availability, raising many strategic problems. Hence, one of the main tasks of the current century is the increase of energetic efficiency of the existing power generation technologies also in view of the decreasing availability of fuels and of their increasing cost.

18.2. COMBUSTION TECHNOLOGIES

Thermal methods for power generation are based on combustion of oil, natural gas, or coal to produce electricity. Some of the most representative currently available technologies are as follows.

18.2.1. Steam Power (ST) Plants

The conventional steam power plant is a mature technology widely used throughout the world. It consists of a steam generator (boiler), a steam turbine, and a particulate removal equipment, which is typically an electrostatic precipitator (ESP). Although steam units up to 1300 MW and supercritical steam conditions of 24 MPa have been tried, in recent years, most units range in size between 300 and 900 MW, with 16 MPa 550 °C/565 °C steam outlet conditions, and overall efficiency ranging between 34% and 38%. Supercritical steam units have been in use in some countries (Denmark and Japan) and have achieved efficiencies up to 42%. All types of coals, oil, natural gas, and biomass, separately or in combination, can be used as fuels in steam plants. However, the power plant needs to be designed for the available fuel (12).

18.2.2. Integrated Gasification Combined Cycle (IGCC)

The first step of this technology is coal gasification. IGCC combines both steam and gas turbines ("combined cycle"), and 40–42% efficiency may be achieved depending on the level of integration of the various processes. The fuel gas leaving the gasifier must be cleaned up from sulphur compounds and particulate, reducing the overall process efficiency and increasing costs. IGCC plants can achieve 99% SO_2 removal and NO_x emissions below 50 ppm. The IGCC technology being in a demonstrative phase, cost projections range from 1200 to 1400 US$/kW. The primary constraints to the application of gasification and IGCC in developing countries are that the technology needs further demonstration, the costs are higher than those of competing technologies, and the fact that environmental regulations in developing countries do not require the so high SO_2 removal and low-NO_x emission achieved by IGCC.

18.2.3. Gas Turbines (GT)—Simple Cycle

The gas turbine apparatus is based on a compressor, a combustor, a power turbine, and a generator. Ambient air is compressed to high pressure in a multistage compressor and fed to the combustion chamber together with the fuel, usually natural gas. The combustion gas powers an axial turbine that drives the compressor and the generator before exhausting to atmosphere. The advantage of this technology is that there is no need of heat transfer to a water/steam cycle to power a steam turbine. The latest gas turbines are designed for inlet temperature of 1500 °C and compression ratio as high as 30:1 (for aero-derivatives models), giving thermal efficiency of 35% or more for a simple-cycle gas turbine (13). The temperature of turbine exhaust gas is about 540 °C, providing the opportunity of heat recovery in a steam boiler (cogeneration facility). Many different technological implementations were introduced to further improve the overall efficiency of the system. Control of NO_x emission is particularly difficult, and energy output and efficiency, as well as NO_x formation, increases with increasing firing temperature. However, the use of lean gas mixtures of combustor design able to better average out temperature and of SCR equipments allow us to reduce NO_x emission to 25 ppm with respect to 300 ppm for natural gas-fuelled gas turbines with no corrective measures. However, a drawback of lean combustion is that CO and UHC increases significantly at reduced power (14).

18.2.4. Combined Cycles

These units combine the Rankine (steam turbine) and Brayton (gas turbine) thermodynamic cycles by using heat recovery boilers to capture the energy in the GT exhaust gases for steam production to feed another turbine. In GT design, the firing temperature, compression ratio, mass flow, and centrifugal stress are the factors limiting both unit size and efficiency. For example, each 55 °C increase in firing temperature gives a 10–13% output increase and a 2–4% efficiency increase. The most critical areas determining the engine efficiency and life lie within the hot gas path, i.e., the combustion chambers and the turbine first stage stationary nozzles and rotating buckets. A typical gas turbine compression ratio is 16:1, but it can raise up to 30:1 in aero-derivative models, with roughly 50% of the total turbine power being required just to drive the compressor. The GT has the inherent disadvantage that reduced air density due to high ambient temperature or high elevation over sea level causes a significant reduction in power output and efficiency, because the mass flow through the gas turbine is reduced. Today the largest aero-derivative power gas turbine is probably the General Electric's 40-MW LM6000 engine, with a 40% simple-cycle efficiency and a weight of 6 tons. Gas turbines of about 150 MW are already in operation, manufactured by at least four separate groups (General Electric and its licensees, Asea Brown Boveri, Siemens, and Westinghouse/Mitsubishi). These groups are also developing, testing, and/or marketing gas turbine sizes of about 200 MW. Combined-cycle units are made up of one or more such GTs, each with a waste heat steam generator arranged to supply steam to a single steam turbine. Typical costs are ca. 600 US$/kW (15). Combined-cycle efficiency is already over 50% and research aimed at 1370 °C turbine inlet temperature, which may make 60% efficiency possible. GTs burn mainly natural gas and light oil. Crude oil, residual oil, and some distillates contain corrosive components and require a fuel treatment equipment. In addition, ash deposits from these fuels result in gas turbine deratings of up to 15% and in an increase of the frequency of maintenance stops. A lot of application examples can be found in Ref. 15.

18.3. CATALYTIC COMBUSTION

The flameless combustion of flammable fuel-air mixtures on platinum wires was discovered in the nineteenth century and led to the invention of the miner's safety lamp by Davy in 1817 (16) and of the pneumatic gas lighter by Döbereiner in 1823 (17). Catalytic combustion is a means for burning a gaseous fuel without flame, thus avoiding massive generation of NO_x. Moreover, it imposes less restrictive constraints concerning flammability limits and reactor design. This is especially important in power generation by gas turbines (18,19). Replacement of a conventional burner with a catalytic stage allows us to carry out the oxidation of methane at lower temperature (<1200 °C) with respect to homogeneous flames (1800–2000 °C), in which large amounts of NO_x form (>200 ppm). Efficient burner design and control allow us to reduce NO_x emissions to 25 ppm (at 15% O_2), but catalytic combustion attains levels as low as 2–3 ppm. Moreover, this approach is cheaper than the various effluent clean-up methods. Additional advantages are better flame stability and fewer pulsations with respect to the conventional lean premix burners (20). As for the fuel, methane combustion releases the lowest amount of CO_2 per unit of produced energy. In view of ever more stringent regulations on CO_2 emissions, this makes natural gas, whose main component is CH_4 (80–90%), even more attractive as a fuel for energy production, provided that methane is not released.

In principle, total methane combustion under fuel lean conditions can be achieved at an even lower temperature, 550–800 °C depending on catalyst activity. However, the amount of fuel used in practical application would lead to unacceptable adiabatic temperature raise, igniting the catalyst and destroying it. Therefore, a hybrid design concept is usually adopted, providing one or two catalytic beds, which convert only a part of the fuel, and a subsequent conventional burner zone, where the remaining fuel (and, when needed, additional air) are added and homogeneous flame combustion takes place. This allows us to keep temperature lower and much better controlled, reducing substantially the formation of NO_x in the flame zone. A schematic representation of the concept and of the typical temperature of each step is represented in Figure 18.3.

Air is compressed and mixed with the fuel before the catalytic step. Pressure ratios were originally low (up to 12:1), with low adiabatic temperature (<1300 °C). The improvement of the materials for the most critical hot zones and the design of new cooling systems allowed us to noticeably increase both pressure ratio (up to 30:1) and adiabatic flame temperature (ca. 1500 °C). Pressure is a critical parameter, affecting the overall yield of the turbine, the energy needed for compression and the catalyst behavior. The catalytic reaction is not directly affected by pressure, but the risk of homogeneous ignition is strongly pressure dependent and must be avoided (20). By contrast, elevated pressures allow us to start up the homogeneous combustion at relatively low temperature (800–900 °C) and to complete fuel oxidation with limited NO_x formation (14). If the temperature reached after compression is not enough for the catalytic reaction to light on, an auxiliary preburner provides the necessary heat. This depends mainly on the activity of the catalyst and on its possible thermal deactivation during use, leading to progressively higher light-off temperature. However, preburners should be avoided, whenever possible, because they form some NO_x and can cause uneven temperature and gas velocity distribution at a catalyst bed inlet.

Catalyst design has to be tailored on the specific fuel to be used. Indeed, GTs are usually fed with natural gas, which is mainly constituted of CH_4, the less reactive

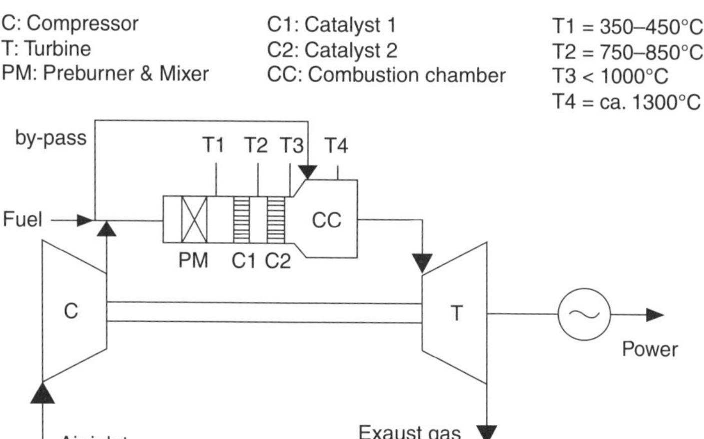

Figure 18.3. Scheme of the hybrid catalytic combustion-gas turbine design.

component of the gas mixture. If significant quantities of higher hydrocarbons are present, homogenous ignition can take place within the catalyst bed, deactivating it. The original design (21) of catalytic GT systems was based on the catalytic combustion of the entire fuel/air mixture, provoking catalyst destruction due to the adiabatic temperature raise occurring in the process. This pushed to the "hybrid" concept, in which part of the fuel bypasses the catalytic section to be fed directly in the homogenous burner (22–25). Dimensioning of the bypass section depends on the enthalpy of combustion, so to keep below 1000 °C the temperature in the catalytic step.

An alternative way to limit hot spots and ignition in the catalytic section is a proper catalyst design during deposition on honeycombs (26,27). Active phase deposition can be e.g., tailored so to have alternate active–inactive channels. This allows a fairly good control of temperature raise in the catalyst monolith, but it prevents the use of different catalyst forms, such as foams or tortuous channel supports. A further possibility is to employ a fuel-rich feeding mixture, in order to convert it in the catalytic step to a mixture of partial and complete oxidation products, to be completely oxidized in the postcatalytic section (28). In this way, the limited extent of complete oxidation reaction prevents the catalyst from overheating (28,29). Moreover, the presence of CO and H_2, highly reactive species, improves flame stability, allowing us to decrease NO_x and UHC emissions (30,31).

To predict, control, and optimize catalyst performance in GT applications, sophisticated modeling techniques are employed to simulate both temperature profiles along the catalyst bed and in the homogeneous sections. Therefore, computational tools were recently developed that provide opportunities to analyze the elementary chemical processes occurring at the gas–solid interface and couple them to the surrounding gas phase. In contrast to previous simulations (32,33), the new numerical codes are also able to take into account transient phenomena. Details about reaction and burner modeling can be found in Refs. 34 through 45.

A final reference concerns the application of deep oxidation to the abatement of methane emissions from natural gas or methane combustion devices, catalytic or not. A typical application would be the clean-up of the exhaust gas from stationary combustion units or from lean-burn natural gas vehicles (NGVs). The latter are increasingly diffusing in urban areas, especially in the form of heavy-duty vehicles, such as buses. Indeed, the use of natural gas offers a lot of environmental advantages also in the transportation field, showing much higher fuel efficiency and lower NO_x emission. The EuroIV standards (October 2005) impose the limit of 3.5 g/kWh for NO_x emission for heavy-duty vehicles. This limit will be cut to 2.0 g/kWh at the end of 2008 (EuroV). Currently, NO_x emissions from diesel buses are ca. 5 g/kWh, meeting the requirement of the EuroIII regulation, whereas only 2 g/kWh are found with lean-burn NG bus, so fully meeting the more stringent future regulations. Moreover, very low amounts of SO_x and of particulate form during combustion of natural gas. The main concern about NGVs emissions is unburnt methane, which is a powerful greenhouse gas, as mentioned. Without any clean-up treatment, methane emission from a NG bus is ca. 3–4 g/kWh, much higher than the 1.1 g/kWh required by the EuroIV and EuroV standards. Hence, highly active catalysts are needed for postcombustion abatement of methane (46,47).

18.4. CATALYSTS FOR THE FLAMELESS COMBUSTION OF METHANE

Catalyst design for the catalytic flameless combustion (CFC) of methane is a highly demanding task. Indeed, high catalytic activity is needed, to ensure low light-off temperature, together with high thermal stability, to withstand the thermal shocks due to the extremely high reaction temperatures. These two requirements are hardly attainable simultaneously. Following the possible hybrid design concept (Figure 18.4), the problem is faced by using two catalytic steps. In the first one, the key point is the light-off temperature (maximum 350–450 °C), and hence, catalytic activity must be sufficiently high. In the second step, the key point is thermal stability, activity being less important, due to the higher reaction temperature (ca. 700–900 °C). Different solutions are possible, but the current trend is to use noble metal-based catalysts in the first step and more thermal-resistant catalysts such as perovskites or hexaaluminates in the second step (49,50). Nanotechnology and nanoscience can be powerful tools for both catalyst design and improvement of catalytically active species.

18.4.1. Supported Noble Metals

Many different examples of noble metal-based catalysts are reported in the literature and recently reviewed in Refs. 5, 46, and 51.

Platinum and palladium show low light-off temperatures in oxidation of hydrocarbons and other organic chemicals (52–56), palladium being generally more active than platinum in the combustion of methane (57–61). The resistance of palladium to thermal deactivation is usually better than that of platinum (62), but its resistance to poisoning by either sulphur- or lead-containing species is worse (54,63,64). However, besides these technical problems, one of the main drawbacks related to the use of noble metal-based catalysts is their high and continuously fluctuating cost (Figure 18.4).

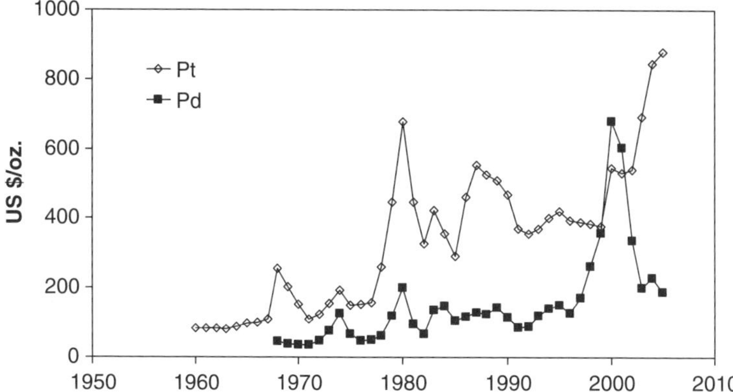

Figure 18.4. Noble metals' price fluctuation over the years. Data based on London PM fix quotations (48).

Furthermore, noble metals availability raises some concern in the case of wide-scale application at high loading, such as the current one. Hence, the research has moved toward mixed metal oxides, which can show satisfactory activity and suitable thermal stability.

18.4.2. Mixed Metal Hexaaluminates

Mn-substituted Ba-hexaaluminates are reported to be very active catalysts for CFC of methane (65–71). The main problem with such catalysts is that activity is usually strongly dependent on surface area, and the latter is commonly low in mixed oxides prepared through conventional techniques. To overcome the problem, microemulsion preparation techniques have been developed to control the oxide particle size, ranging, e.g., from 10 to 1000 nm and hence to control surface area (72). However, although surface areas in excess of 50 m^2/g were obtained, depending on the preparation conditions, catalytic activity was not optimal, showing light-off temperature always higher than 400 °C and never attaining methane full conversion below 650 °C. In addition, neither a deep study of catalyst stability nor of structured catalysts have been done yet.

18.4.3. Perovskite-Like Mixed Oxides

Among the mixed oxides, those with perovskite-like structure are perhaps the most widely studied. They have general formula ABO_3, where A is usually a large-radius metallic cation (e.g., a lanthanide or alkali-earth metal ion) and B is a lower size transition metal ion. The same crystal structure can be obtained with carbides, halides, nitrides, and hydrides, although oxides are by far the most common. One of the main advantages of this family of oxides is the high versatility of their composition, which can be further expanded through the partial substitution of the cations at both A and B position to form $A_{1-x}A'_xB_{1-y}B'_yO_{3\pm\delta}$ compounds, with widely different properties and catalytic behavior (73). Another interesting feature is the exceptional stability of the perovskitic structure, the crystal field force of which can force some ions to assume unusual oxidation states. A typical example is Cu in Ba-substituted lanthanum cuprate, in which Cu is present in both the 2+ and the 3+ oxidation states. This allowed us to also develop an interesting set of superconducting materials working at a relatively high temperature (74,75).

The ideal perovskitic structure is cubic, the smaller B ion being octahedrally coordinated to 6 oxygen ions and the larger A ion attaining a coordination number of 12 (Figure 18.5). The following geometry-based relationship between ionic radii holds for the ideal structure (73):

$$r_A + r_O = 2^{1/2}(r_B + r_O)$$

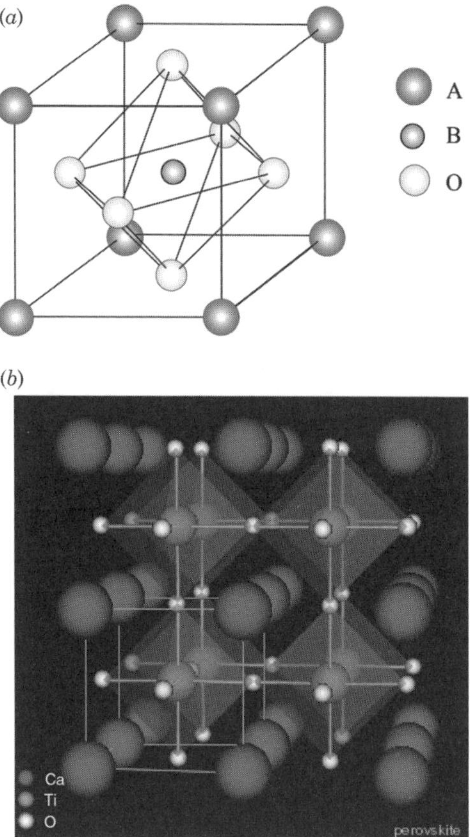

Figure 18.5. Example of perovskite-like structure: (*a*) unit cell (76) and (*b*) example of $CaTiO_3$ tri-dimensional array (77). See color insert.

However, some distortion is possible without destroying the perovskitic lattice. A tolerance factor t is then introduced:

$$t = (r_A + r_O)/(2^{1/2}(r_B + r_O))$$

More or less distorted, the perovskitic structure can be obtained for t values ranging from 0.75 to 1 (78).

One of the most interesting properties of perovskites is their nonstoichimetry (73,79–81). Indeed, the whole structure has to be neutral, so cation charges have to be distributed to give an overall 6+ sum: e.g., $A^{1+}B^{5+}O_3$, $A^{2+}B^{4+}O_3$, or $A^{3+}B^{3+}O_3$. However, the presence of defects, such as cationic or anionic vacancies, is very common in real structures. Anionic vacancies are the most frequent and can extend up to a whole oxygen layer, leading to a different family of compounds, the brownmillerites (81,82) (e.g., $Ca_2Fe_2O_5$ and $La_2Ni_2O_5$), in which up to one sixth of the oxygen atoms are vacant. Sometimes an excess of oxygen can be also found. This is the case of $LaMnO_{3+\delta}$ (83–87), which shows a very good catalytic activity for total oxidation reactions.

Some perovskites show interesting magnetic properties, which are connected with the B ion properties. Paramagnetic species can also form, due to coupling with structural or adsorbed oxygen, which in turn may be involved in oxidation reactions. Hence, EPR can be a powerful technique to investigate the nature of the possible active species (88–92).

As for electric conductivity, no generalization is possible, due to the dependence of the electrical properties on the nature of the B metal. Indeed, conductors, semiconductors, and insulators can be found, depending on the possibility to form a delocalized orbitals array in which electrons can freely move. Semiconducting materials, e.g., titanates, can be very interesting for photo-electrocatalytic application, but no significant dependence of activity for full oxidation reaction on electric conductivity was ever found (93).

More interesting from this point of view is ionic conductivity. This is due to ion transport through the lattice, which is made easier by the relatively open perovskitic structure. Ionic mobility is mainly due to oxygen ions, which can more or less easily move through oxygen vacancies. Indeed, the activation energy for O^{2-} jumping between oxygen vacancies is usually low. This phenomenon offers interesting application possibilities, e.g., in the development of ceramic materials for semipermeable membranes. Moreover, some perovskites characterized by high ionic mobility, such as $La_{(1-x)}Sr_xCo_{(1-y)}Fe_yO_{3-\delta}$, have been used as cathode materials for fuel cells (94–96). Generally speaking, oxygen mobility depends on both the facility of oxygen transport through the bulk phase and the exchange properties at the gas–solid interphase. The first parameter plays a leading role and depends mainly on the size of A and B ions, i.e., on the tolerance factor t. A minimum in the activation energy for oxygen transport was found for $t \approx 0.81$ (97). The reason of this phenomenon is lattice relaxation around the vacancy, a sort of expansion of the vacancy permitting the decrease of the electrostatic repulsion of the surrounding cations, leading to an easier oxygen transit (73). Hence, in principle, it would be possible to increase oxygen mobility by increasing oxygen vacancies concentration, of course, respecting the mentioned structural requirements. This can be obtained, e.g., by catalyst doping with an A′ ion of lower valence with respect to A (e.g., $La_{(1-x)}Sr_xMO_{3-\delta}$) and/or by thermal treatment at high temperature in inert gas, which can lead to oxygen release. Although oxygen mobility is a key point, catalytic activity strongly depending on it (*vide infra*), cationic mobility can also be an important parameter. It represents the ability of A and/or B ions to move through the lattice. Although not directly connected with catalytic activity, it represents an important parameter during the synthesis of the mixed oxides, e.g., through a solid state reaction, allowing the rearrangement of the ions in the correct position and, mainly, improving the stability of the structure. Indeed, segregation of the single oxides (A_xO_y and B_xO_y) is possible, especially at a high temperature, leading to partial or complete phase destruction. An easier segregation can be observed for some dopants, mainly at A position, possibly vanishing their effect (98). Hence, caution is needed to prevent structural modifications during a high-temperature application, such as the CFC of methane. Similarly to other properties, a generalization is difficult. It can only be said that cation diffusion depends on ion size and charge (99), but of course structural factors and the overall material stability play an important role.

Oxygen adsorption and release can be a good index of oxygen mobility. Furthermore, it can offer interesting information, especially when the catalyst is employed for oxidation reactions. A powerful technique to collect such data is temperature-programmed desorption (TPD) of preadsorbed O_2. It consists of sample presaturation in air or pure O_2, followed by heating in inert gas, while monitoring the release of both adsorbed and structural oxygen. Indeed, in a typical TPD pattern, two oxygen desorption peaks can be detected, called α and β. The α peak occurs at relatively low temperature ($<500\,°C$), and it is usually attributed to surface O_2 adsorbed on anionic vacancies. By contrast, the β peak appears at higher temperature ($>600\,°C$), and it is usually ascribed to the release of structural oxygen (100,101). Moreover, the β signal is tightly connected to the thermal decomposition temperature of the B ion oxide (relative to the oxidation state in which B is present in the perovskite) (98,102). Hence, the presence and the intensity of both the α and β peaks depend on the nature of the B ion (thermal decomposition temperature and easiness to form anionic vacancies) and on the presence of dopants, which can modify the vacancies concentration and/or alter the B–O bond energy, affecting the release of structural oxygen (98,100,101,103–107). As an example, in Figure 18.6, the results are reported of the TPD analysis carried out on some lanthanum-based perovskites differing for the nature of B ion or for the presence of an A′ dopant (98,103).

These features are connected with oxygen mobility. In addition, the energy of interaction between oxygen and B ion is intuitively tied to the ease of oxygen transport

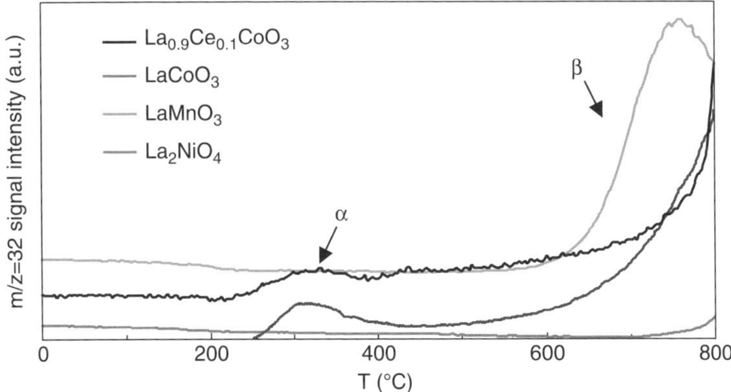

Figure 18.6. Typical TPD pattern of some perovskites. The signal refers to O_2 release. Labels indicate the so-called α and β peaks.

through the oxide crystal lattice. These parameters are in turn correlated with catalytic activity for deep oxidation reactions, such as the CFC of methane.

Catalyst doping with A′ ions of different valence with respect to A can either lead to an increase of vacancies concentration and/or to oxidation or reduction of the B ion, depending on the system under study. In the previously mentioned example (Figure 18.6), the effect of Ce^{4+} and Eu^{2+} on $LaCoO_3$ catalysts prepared by flame-hydrolysis (FH) (*vide infra*), was studied by TPD (98); the β TPD peak appearing at ca. 750 °C was attributed to the release of bulk oxygen and it was related to the temperature of thermal decomposition of the cobalt(III) oxide (895 °C), due to the partial Co^{3+} reduction to Co^{2+}. By contrast, the α peak was connected with the nature and amount of the A′ dopant. Indeed, with Ce^{4+} substituting for La^{3+}, the α peak appeared at ca. 350–400 °C and its intensity was directly dependent on the substitution degree. With Eu^{2+} as dopant (not reported in Figure 18.6), a similar increase of the α peak intensity with increasing concentration of this ion in the lattice was observed, together with a shift toward higher onset temperature (ca. 500 °C). Eu^{2+} substitution for La^{3+} can provoke two main effects: an increase of vacancies concentration and/or the oxidation of the B ion to a higher oxidation state. The latter is very unlikely, with Co^{3+} being yet in its highest oxidation state, so oxygen vacancies formation should be expected. It should be underlined that the effect would be different with other B metal ions, such as Mn^{3+}, which would likely be oxidized to Mn^{4+} (108,109). By contrast, Ce^{4+} doping would provoke a partial Co^{3+} reduction to Co^{2+}, in order to save the overall electro-neutrality of the system. It must be taken into account that a limit for CeO_2 solubility into the perovskitic lattice exists. Indeed, with high substitution degree (e.g., 20%), segregation of some CeO_2 usually occurs, determining the formation of cationic vacancies (98).

Upon high-temperature treatment in inert gas (He), followed by saturation with air during cooling, oxygen vacancies form and fill in and possible oxidation of Co^{2+} to Co^{3+} occurs. Ce^{4+} doping leads to the formation of less tightly bound oxygen species, i.e., released at lower temperature, which can be efficiently exploited during the oxidation reaction. A clear relationship between the temperature of the α peak maximum and the concentration of the dopant was indeed observed (98) and correlated with catalytic activity. However, the onset temperature and intensity of both the α and the β peaks also depend on the B ion. Indeed, the set of FH-prepared $LaBO_3$ catalysts (B = Co, Fe, Mn) and $La_2NiO_{4\pm\delta}$, studied by TPD (103) showed a different behavior (Figure 18.6). $La_2NiO_{4\pm\delta}$ showed a relatively intense α peak around 300 °C, indicating a high concentration of oxygen vacancies and of adsorbed oxygen, relative to this defective structure, whereas the onset temperature of the β peak was ca. 650 °C. $LaMnO_{3\pm\delta}$ did not show any α desorption peak, whereas the onset temperature of the β peak was perceptibly lower than in the previous case, leading to a lower and broader maximum of the β desorption peak at 700–750 °C. $LaCoO_{3\pm\delta}$ showed a very small, almost undetectable α peak around 450 °C and a β peak, starting over 700 °C (*vide supra*). Finally, $LaFeO_{3\pm\delta}$ showed neither α nor β peaks within the temperature range investigated. The onset of the β peak can be correlated to that of thermal decomposition of the pure B ion oxide also in the case of the Mn- and Ni-based samples (102,110). The absence of the α peak for the undoped $LaBO_3$ samples is in line with the above-mentioned discussion on the role of doping at A position to form such defective sites. The presence of the β peak and its onset temperature can be adopted as a qualitative description of catalyst reducibility and, together with the α peak, of oxygen mobility within the oxide bulk. Both of these parameters can be correlated with catalytic activity (98,103,111).

Methane full oxidation on perovskite-like catalysts can be interpreted through a Mars–van Krevelen mechanism, based on a redox cycle centered on the B ion. Hence, the substrate is oxidized by the lattice oxygen, reducing the B ion to a lower oxidation state. The latter is oxidized back to the original oxidation state by gaseous oxygen co-fed with the reactant. Clearly, this reaction mechanism involves oxygen mobility (i.e., the easiness of oxygen transport through the bulk phase) and the B ion redox properties. The activation energy for oxygen transport is higher for Mn- than for Co-substituted perovskites (112). However, this reaction mechanism is usually active at a relatively high temperature, at which oxygen mobility can be effective. According to Voorhoeve et al. (113), this is called the *intrafacial* mechanism and it plays a leading role at a temperature higher than 500–600 °C. It can be accompanied by a low-temperature reaction mechanism, called *suprafacial*, which directly involves surface adsorbed oxygen, and it is connected with the presence of oxygen vacancies. Hence, the adsorption properties of the catalyst are also directly involved in the reaction mechanism. To summarize, catalytic activity at both low and high temperatures are tightly connected to the presence of anionic vacancies, the latter increasing both oxygen adsorption and the subsequent desorption at low temperature and fast oxygen mobility at high temperature. In addition, structural parameters (affecting oxygen mobility) and B ion redox properties mainly affect the high-temperature catalytic performance.

The above reported conclusions are supported by the comparison of TPD data with catalytic activity for methane oxidation (98,103). Indeed, the catalysts that did not show any α peak, i.e., possessing a low concentration of oxygen vacancies, were very active at low temperature through the suprafacial mechanism. This confirmed the low oxygen mobility of $LaFeO_{3\pm\delta}$ and $LaMnO_{3\pm\delta}$ samples. By contrast, the catalysts characterized by the presence of a more or less intense α peak, such as $La_2NiO_{4\pm\delta}$, and, to a minor extent, $LaCoO_{3\pm\delta}$, were exclusively or predominantly active through the intrafacial mechanism (103). An example of catalytic activity dependence on doping at A position (i.e., possible formation of oxygen vacancies) and on the nature of the B ion is given in Figure 18.7 (98,103). The low-temperature suprafacial mechanism is evident for the Mn and Fe containing samples, which lead to ca. 15% and 30% CH_4 conversion, respectively, at a temperature as low as 300–350 °C. The easier reducibility–oxidizability of Co, Mn, and Fe with respect to Ni can be invoked to explain their higher catalytic activity, together with the uneasy formation of a pure perovskitic lattice for La_2NiO_4, in the absence of high-temperature additional treatments (103,114). The beneficial effect of Ce doping is also evident, lowering noticeably the light-off temperature (ca. 230 °C, with respect to ca. 280 °C of the undoped $LaCoO_3$ sample) and similarly decreasing the temperature of full methane conversion (559 °C vs. 590 °C). By contrast, lower activity has been observed for Sr(II) and Eu(II) doped cobaltites (98,106), in which the effect of the dopant would

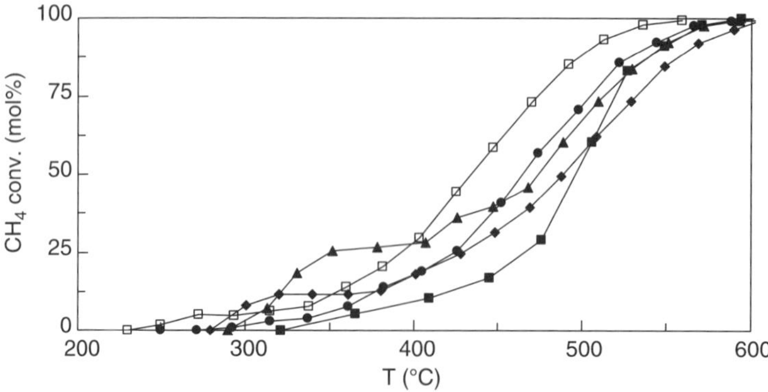

Figure 18.7. Catalytic activity of some selected samples. (♦) $LaMnO_3$; (□) $La_{0.9}Ce_{0.1}CoO_{3\pm\delta}$; (•) $LaCoO_3$; (■) La_2NiO_4; (▲) $LaFeO_3$.

be an increase of oxygen vacancies and, hence, in principle, an increase of catalytic activity. A possible explanation was found through EPR analysis, which evidenced the formation of spin-glass, observed (115) in $La_{0.9}Sr_{0.1}CoO_{3-\delta}$, which would account for the low oxygen exchange ability of the sample and therefore for its lower activity with respect to the Ce-doped homologue, where spin-glass systems were not observed.

Partial substitution at the B position can strongly influence both oxygen mobility and catalytic activity. A recent investigation on the activity of mixed Co- and Fe-based catalysts for the CFC of methane allowed us to further underline the deep connection between these two parameters (93). Indeed, composite materials such as $La_{1-x}Sr_xCo_{1-y}Fe_yO_{3-\delta}$ gained growing attention due to their high mixed conductivity. At ca. 800 °C, their electronic conductivity can attain $10^2\ \Omega^{-1}cm^{-1}$ and ionic conductivity ranges between 10^{-2} and $1\ \Omega^{-1}cm^{-1}$. These compositions find application in different fields, mainly as cathodes for solid oxides fuel cells (SOFCs) (94–96), but also as membranes for oxygen separation and for membrane reactors for syngas production and oxidation of hydrocarbons (116–119). Both doping at A position with higher or lower valence A′ ions and at B position deeply affect the TPD pattern and hence the catalytic performance. Unfortunately, many frequently counteracting phenomena accompany the substitution of both the A- and the B-ion, so that it is very difficult to predict *a priori* the catalytic behavior of the resulting material.

A further complication arises from the dependence of structural, conducting, and catalytic properties on the preparation method. In fact, there is a strong dependence of defects formation, phase purity, and particle size on the preparation procedure adopted; all of these features of course affecting the catalytic performance and thermal stability of the material (*vide infra*).

A final remark concerns another important property, surface area, which is usually very important in catalytic applications. As this parameter strictly depends on the temperature attained during catalyst preparation, the conventional preparation routes usually lead to low-surface-area materials (1–20 m^2/g), often characterized by poor activity. With the samples essentially being nonporous, the contribution to surface area is substantially given by the external surface of particles. Hence, decreasing particle size, i.e., moving from microsized toward nanostructured oxides, is a powerful method to increase surface area and thus catalytic performance.

18.5. METHODS FOR THE PREPARATION OF NANOSIZED PEROVSKITIC OXIDES

The application of perovskitic mixed oxides here analyzed in detail is the CFC of hydrocarbons, using methane as a sample reactant, due to its wide use and to its stability, which allows us to extend the results to the higher homologues through a conservative approach. The reaction imposes restrictive requirements for catalyst design. Indeed, the catalyst must show the highest possible activity, to offer sufficiently low light-off temperature (at least $<400\,°C$), avoiding the need of any pre-burner (see paragraph 3). Moreover, high activity has to be coupled with sufficiently high resistance to deactivation, due, e.g., to poisoning by sulphur compounds, but mainly to thermal sintering. Unfortunately, these two features are hardly obtainable simultaneously.

The first reported procedure for the preparation of perovskites is based on a solid-state reaction. The precursor oxides are mixed, ground, and calcined at high temperature, usually $>1000\,°C$, several cycles of calcination-milling (CM) being repeated until the desired perovskitic phase is obtained (50,120). This route leads to highly sintered materials, generally characterized by satisfactory thermal resistance, however accompanied by very low surface area ($<2\,m^2/g$) and thus by poor activity and sometimes by low phase purity. Hence, this procedure is commonly applied to the synthesis of ceramic materials, but it is unsuitable for the preparation of perovskites for the current catalytic application. A variation of the method is the solid-state reaction of the precipitated single hydroxides. This reaction is faster than with oxides, requiring lower calcination temperature (ca. 900 °C). However, despite the higher phase purity of the obtained product, the surface area remains too low.

To increase surface area, many different techniques were proposed. Among these, the co-precipitation–thermal decomposition of different precursors (carbonates, nitrates, oxalates, etc.) leads to somewhat better results (up to $10\,m^2/g$), mainly due to the lower calcination temperature during the decomposition of precursors (73,121–124). In addition, some mechanical methods based on high-energy ball-milling were proposed to overcome the need of a final high-temperature calcination (125–127). Unfortunately, the difficulty in obtaining a complete conversion of precursors strongly limits the possibility of their practical application (128). A better improvement are the sol-gel (SG) type techniques, based on the use of a complexing agent (citric, malic or tartaric acid, or polyethylene glycol, just to mention a few) to avoid excessive growth of the primary particles forming the gel. These methods allow us to obtain good perovskite-like phase purity by calcination at relatively low temperature (commonly 650–750 °C), permitting satisfactory values of surface area (up to $30\,m^2/g$) (88,89,93,106,109,128–140). However, the increase of surface area, and hence of activity, due to the lower calcination temperature, is obtained at the expenses of a poor thermal stability. Thus, these preparation routes are suitable for a relatively low-temperature application only, such as pollutants abatement like de-NO_x processes and CO oxidation. A detailed comparison between different preparation procedures has been recently reported (128).

A possible way to combine the high thermal stability conferred by high temperature calcination with satisfactory catalytic activity (i.e., high surface area and phase purity) comes from the observation that particle sintering depends on both calcination temperature and time. Hence, a sensible decreasing of calcination time would limit any extensive sintering, and thus, it can lead to high-surface-area nanosized materials.

A few years ago, we set up a method based on flame hydrolysis (FH) for the preparation of perovskitic mixed oxides (88,89,98,103,128,141,142). This preparation technique is well known for the synthesis of very fine particles of some oxides, such as silica or titania. However, to the best of our knowledge, the FH method was applied to single oxides only and required the availability of a volatile precursor, such as $SiCl_4$ in the case of SiO_2 (143,144). We substantially modified this procedure by using an aqueous solution of precursors, instead of operating in gas phase, so in principle extending its potentiality to the preparation of many different single or mixed oxides.

Our FH apparatus (141) was based on a properly shaped nebulizer, fed with an aqueous solution of the perovskite precursor salts. The latter is prepared by dissolving the soluble salts (nitrates, acetates, or citrates) into the desired ratios either in pure water or in a HNO_3 solution. The concentration of precursors in the resulting solution is to be kept at 5–6 wt%, in order to reach a compromise between particle size (and hence surface area), powder crystallinity, and productivity. The addition of a complexing agent, such as citric acid, helps in obtaining a clear solution and to increase flame temperature, the organic compound also acting as additional fuel. A similar effect was noticed when using organic anions of the precursor salts. However, it has to be underlined that an excessive enhancement of flame temperature may cause an unacceptable drop of surface area, so the amount of organic material in the solution has to be calibrated. The resulting solution is fed into a preformed $H_2 + O_2$ flame by means of a nebulizer, whose orifice diameter determines the size of the droplets and, eventually, the particle size. Moreover, the air flow needed for the nebulization may show a quenching effect on flame temperature, when excessive.

Powder collection was achieved by means of a properly designed electrostatic precipitator working at 10 kV (141), consisting of an AISI 316 stainless steel multipin effluviator, surrounded by a coaxial cylindrical collector, allowing us to recover over 80% of the product. An example of the powder particle size obtained by this method is given in Figure 18.8*a*.

A wide variety of perovskitic mixed oxides were so synthesized, some of the most representative being collected in Table 18.3. One key point was the versatility of the procedure, which ensured satisfactory phase purity with widely different compositions, even in the preparation of ternary oxides. The better results as for phase purity and particle size uniformity were obtained with La-based samples. The $SrTiO_3$ samples often showed some segregation of $SrCO_3$ as a consequence of the high stability of the latter, which can form during the combustion of organic precursors.

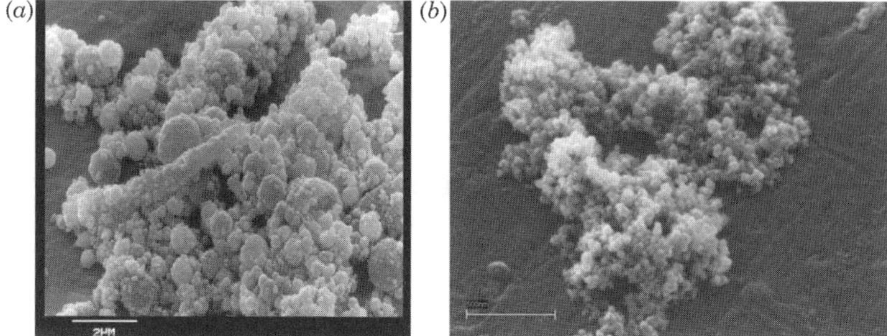

Figure 18.8. Example of powder particle size of perovskitic mixed-oxide catalysts prepared by (*a*) FH and (*b*) FP.

Furthermore, for the La-based samples, particle size distribution showed more uniform and particles smaller than 100 nm were usually obtained, independently of B ion nature and of the presence of A′ dopants. Consequently, surface area was rather high (considering the high preparation temperature), lying around 15–25 m^2/g. As a result of high phase purity and high surface area, the catalytic activity attained and, in some cases, exceeded that of the SG-prepared catalysts (Table 18.3). However, the main result achieved by the FH technique was the high thermal resistance of materials. This parameter was measured by keeping the working sample at the temperature T_f of maximum conversion for at least 100 h, looking for any worsening of catalytic performance. After this preliminary test, at least three cycles of rapid deactivation were performed, consisting in overheating the working sample at 800 °C for 1 h; after which, the temperature was brought back to T_f to check the residual methane conversion. An example of thermal resistance data for samples prepared through different methods is reported in Figure 18.9.

Another set of samples prepared by FH are titanates, interesting semiconducting materials, which can also be employed for the current application in the absence or presence of dopants, such as Ag ions. In this case, the particle size distribution of these catalysts was less uniform than for La-based perovskites and surface area showed a bit lower. However, a higher calcination temperature was usually needed to also obtain a sufficient phase purity through the SG preparation procedure, leading to a low surface area (Table 18.3). As for doping at A position, Ag^{n+} incorporation into the perovskite lattice was never complete, independent of the preparation procedure. The latter, however, determined more or less extensively the kind of active species (among them even Ag^{2+}-based active species were found) responsible for catalyst activity (88,89,145). In addition, the presence of some extra-framework, segregated metallic Ag helped in increasing thermal resistance, acting as a spacer and so preventing any extensive sintering of the mixed oxide even after treatment at very high temperature (Figure 18.9) (89).

The FH preparation procedure, although it is very versatile and powerful in combining high surface area and thermal resistance, showed as main drawback a very low productivity (e.g., a few mg/h with our apparatus), essentially due to the very low concentration of the solution and to the nebulization method.

TABLE 18.3. Chemical–Physical Properties and Activity for the CFC of Methane of Some Perovskite Samples Differing in Composition and/or Preparation Procedure.

Catalyst	Crystal Phases[a]	Particle Size (nm)[b]	SSA_{BET} (m^2/g)[c]	T_b (°C)[d]	$T_{1/2}$ (°C)[d]	T_f (°C)[d]	Ref.
$LaCoO_3$-FH	P	20–80	17.9	>290	466	602	98,141
$La_{0.9}Ce_{0.1}CoO_{3\pm\delta}$-FH	P	20–80	24.0	<250	438	560	98,141
$La_{0.9}Eu_{0.1}CoO_{3\pm\delta}$-FH	P	20–80	18.3	360	462	580	98,141
$LaMnO_3$-FH	P	20–60	19.3	270	489	605	103
$LaFeO_3$-FH	P	20–60	15.6	290	472	596	103
La_2NiO_4-FH	La_2NiO_4	20–60	22.8	320	497	594	103
$La_{0.9}Ce_{0.1}CoO_{3\pm\delta}$-CM	P	>1000	4.6	390	550	>600	98
$La_{0.9}Ce_{0.1}CoO_{3\pm\delta}$-SGC	P	400–600	22.0	350	520	>600	98
$LaMnO_3$-SGC	P	400–600	25.5	<290	395	453	128
$LaMnO_3$-EDTA	A	n.d.	n.d.	n.d.	n.d.	n.d.	128
$LaMnO_3$-BP1	A	n.d.	n.d.	350	569	>650	128
$LaMnO_3$-BP2	A	n.d.	3.0	350	561	>650	128
$LaCoO_3$-SGC	P	400–600	12.2	270	390	477	128
$LaCoO_3$-EDTA-500 °C	A	n.d.	n.d.	n.d.	n.d.	n.d.	128
$LaCoO_3$-EDTA-700 °C	P	n.d.	11.6	300	442	530	128
$SrTiO_3$-FH	P + $SrCO_3$	40–200	12.0	<250	540	650	88
$SrTiO_3$-SGC-950 °C	P	n.d.	1.0	400	575	>650	88
$SrTiO_3$-FP	P	30–60	107	360	520	>650	89
$Sr_{0.9}Ag_{0.1}TiO_{3\pm\delta}$-SGC-5[e]	P + Ag	30–100	24	250	400	500	89
$Sr_{0.9}Ag_{0.1}TiO_{3\pm\delta}$-FH	P + Ag	50–800	12	350	542	650	89

$Sr_{0.9}Ag_{0.1}TiO_{3\pm\delta}$-FP	P + Ag	30–60	67	250	402	503	89
$Sr_{0.9}K_{0.1}TiO_{3\pm\delta}$-FH	P	50–500	5	345	550	>650	145
$Sr_{0.9}K_{0.1}TiO_{3\pm\delta}$-SGC-850 °C	P + $SrCO_3$	n.d.	5	n.d.	n.d.	n.d.	145
$Sr_{0.9}Gd_{0.1}TiO_{3\pm\delta}$-FH	P	20–400	10	350	575	>650	145
$Sr_{0.9}Gd_{0.1}TiO_{3\pm\delta}$-SGC-850 °C	P + $SrCO_3$	100–500	2	n.d.	n.d.	n.d.	145

Notes: Catalytic activity determined under the following conditions: 0.2 g of catalyst (0.15–0.25 mm particle size), diluted with 1.3 g of quartz powder; sample activation in 20 cm^3/min of air, temperature ramp of 10 °C/min up to 600 °C, then kept for 1 h; activity tests carried out by feeding 10 cm^3/min of 1.04% CH_4 in He + 10 cm^3/min of air, heating by 2 °C/min from 250 up to 600 °C. The acronym refers to the preparation method: FH = flame hydrolysis, CM = calcination-milling, SGC = sol-gel citrates, EDTA = sol-gel with EDTA as complexing agent, BP = ball-milling in planetary mill, and FP = flame pyrolysis. Details on the preparation can be found in the corresponding reference (last column).

[a] Crystal phases revealed by XRD analysis. P = pure perovskite; A = amorphous. Reference data in Ref. 146.

[b] Determined by SEM analysis.

[c] Specific surface area (SSA) calculated through the BET method.

[d] T_b = temperature of light-off; $T_{1/2}$ = temperature of 50% CH_4 conversion; T_f = temperature of 100% CH_4 conversion.

[e] Heating rate during calcination (°C/min).

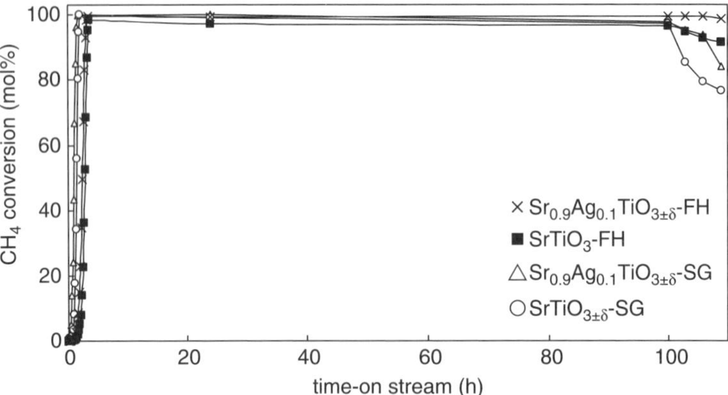

Figure 18.9. Thermal stability of some $Sr_{1-x}Ag_xTiO_{3\pm\delta}$ ($x = 0, 0.1$) samples prepared through the sol-gel (SG) method or by flame hydrolysis (FH). The last three points of each curve refer to the rapid deactivation cycles (see text).

Many research groups reported different flame-based synthesis techniques for the preparation of single or, in some cases, mixed oxides, based on aerosol or liquid-feed flame pyrolysis (FP) (147–166). Among the methods proposed, the spray FP seems the most interesting for the production of perovskitic oxides, because it does not require volatile precursors. It is based on a specially designed vertical nozzle (147–158,166) fed with oxygen and with an organic solution of the precursors, the solvent acting as the fuel for the flame. High productivity and phase purity may be obtained, together with nanometer-size particles (Figure 18.8*b*) and hence exceptionally high surface area, even over $100\,m^2/g$. The relatively high temperature of the flame in principle should also ensure thermal stability, once the main operating parameters have been optimized. Hence, we recently set up and optimized an FP apparatus for the preparation of perovskitic mixed oxides, at first using $LaCoO_3$ as reference material and then extending the procedure to different compositions (88,89,167,168).

A detailed description of the apparatus and of the effect of the main operating parameters on the properties of the obtained catalysts is given elsewhere (167). Briefly, the apparatus is composed of three sections:

(*1*) the flame reactor (burner), depicted in Figure 18.10;
(*2*) the feeding rate control devices for gaseous and liquid reagents; and
(*3*) the catalyst powder collection system, which is exactly the same adopted for the FH apparatus (*vide supra*).

The burner includes a capillary tube (inner diameter 0.6 mm) through which the precursors solution is fed by means of a syringe pump. This tube ends in the center of the nozzle, where oxygen acts both as an oxidant and as a dispersing agent, to form very small droplets of solution. The main flame is ignited and supported by a ring of 12 premixed $O_2 + CH_4$ flamelets. Oxygen linear velocity and pressure drop across

Figure 18.10. Particular of the FP burner.

the nozzle can be varied by selecting a proper feeding rate and/or by adjusting the nozzle discharge cross section and geometry.

The main steps of the FP synthesis are

(*1*) droplets formation;
(*2*) solvent evaporation and burning;
(*3*) precursors decomposition;
(*4*) particle nucleation; and
(*5*) particle growth due to coalescence, condensation, agglomeration, and sintering.

Droplets size depends mainly on liquid and oxygen flow rate and linear velocity and can be modeled through correlations between proper adimensional parameters (166). Better dispersion and short droplet lifetime were observed by using oxygen rather than air as dispersant/oxidant in the main nozzle. Indeed, large droplets can have a lifetime longer than the residence time in the hottest zone of the flame, leading to poor particle size homogeneity (150). Furthermore, if the solvent readily evaporates, product formation proceeds via monomer or clusters formation as in aerosol processes (see, e.g., ZrO_2 synthesis reported in Ref. 152). By contrast, if solvent evaporation is too slow, egg-shell hollow particles can initially form and then explode, due to the rapid evaporation of the residual solvent inside them (149,152,155,159).

Once the primary particles have formed through nucleation, their growth is due to coalescence, depending on collisions between the particles and sintering. The former phenomenon can be controlled by decreasing particles concentration in the flame, however at the expenses of a lower productivity. Moreover, extensive sintering can be avoided by lowering flame temperature and particle residence time in the hottest zone of the flame. This parameter can be modulated in two different ways:

(*1*) by shortening flame height through a proper selection of O_2 feeding flow rate and

(*2*) by increasing O_2 linear velocity.

In principle, to obtain high-surface-area dense particles, the operating parameters leading to small particle size should be selected. This is often uneasy, the phenomena depending on many factors, sometimes contrasting with each other. Some of the most representative are

(*1*) O_2 and CH_4 flow rate to the supporting flamelets;

(*2*) organic liquid solution flow rate and concentration;

(*3*) O_2 flow rate to the main nozzle;

(*4*) pressure drop across the nozzle;

(*5*) nozzle geometry; and

(*6*) nature of the organic solvent and of metal precursors.

As for the supporting flamelets, when air is used as an oxidizing agent, a poor stability of the main flame can be observed, the optimum being a stoichiometric O_2/CH_4 mixture. The overall flow rate of the reagent is not critical from the point of view of the properties of the obtained material (i.e., phase purity, surface area, and particle size), provided that a stable flame can be maintained.

Productivity can be increased by increasing the liquid solution flow rate. However, the higher the latter parameter, the higher the concentration of primary particles in the flame. This increases the frequency of particle collision, therefore, leading to often unacceptable particle growth. Hence, if the goal is to obtain high surface area (i.e., poorly agglomerated products), a compromise has to be reached between productivity and particle size.

Liquid flow rate can be varied by keeping constant the O_2/liquid feeding ratio. This keeps unchanged the flame height, without significant change of product surface area. Indeed, the increase of particles concentration in the flame can be counterbalanced by the decrease of their residence time within it. Alternatively, the liquid flow rate can be increased at constant, over-stoichiometric, O_2 flow rate. This leads to increased productivity and higher flame height, accompanied, however, by a progressive drop of surface area, due to the increase of particles concentration within the flame and to the increased residence time within it, favoring particle sintering.

Similar considerations can be addressed as for the effect of precursors concentration in the feeding solution. Indeed, to increase productivity, highly concentrated solutions are preferable, of course, within the precursors solubility limits. However, also in this case, the increased concentration of primary particles in the flame entrains an increase of particles collision frequency and hence of particle growth. In addition, particles necking due to enhanced sintering in gas phase can be observed, favored by the higher concentration of precursors in the flame. Hence, to prepare nanosize, high-surface-area particles, diluted solutions are preferable. In this case, productivity can be enhanced by increasing the liquid flow rate at constant O_2/liquid feeding ratio (*vide supra*) (167).

A detailed study of nozzle geometry and of its effect on the product properties is reported in Ref. 167. In our design, nozzle configuration from convergent into convergent–divergent can be varied by moving its inner body. For a convergent nozzle configuration, the highest O_2 discharge velocity is sonic, i.e., ca. 341 m/s, whereas for the convergent–divergent geometry, supersonic discharge conditions can be attained. If O_2 velocity is increased at constant O_2 flow rate, i.e., by narrowing the nozzle discharge section and increasing the pressure drop across it, a significant effect on surface area can be obtained. Indeed, an increase of surface area was noticed with increasing O_2 linear velocity under subsonic discharge conditions. However, when sonic velocity was reached, a further increase of O_2 pressure drop did not further increase surface area, which remained constant, with the discharge linear velocity being constant. When nozzle geometry changed into convergent–divergent, a further increase of O_2 discharge velocity was possible (i.e., supersonic discharge conditions), bringing about a further decrease of particle size and hence an increase of surface area. This behavior can be explained in terms of residence time of the primary particles within the flame: the lower the residence time, the lower the particles sintering. So, to decrease particle size, O_2 discharge linear velocity has to be increased. Needless to say that for the current application, as CFC catalysts, this effect has to be balanced with the need of phase purity (i.e., sufficient residence time to allow the complete conversion of the precursors into the perovskite-like phase and the complete burning of the organic residua) and, even more important, of thermal stability, which is tightly connected with calcination time, besides temperature (167). In addition, O_2 velocity has a secondary effect on particle residence time into the flame. In fact, a decrease of flame height was observed with increasing of the latter parameter. Moreover, a noticeable raising of discharge velocity could result in a transition of the flow regime from laminar to turbulent. This can in turn affect particle size, because the more efficient mixing characterizing the turbulent flow provokes a more uniform particle size distribution (around 20 nm) and limits any extensive sintering, with a consequent increase of surface area (157,167).

If O_2 flow rate is increased at a constant nozzle cross-section area, an increasing pressure drop across the nozzle is observed, which, under subcritical (i.e., below sonic velocity) discharge conditions, leads to increasing O_2 linear velocity. The surface area of the powder was found to rapidly increase with increasing O_2 flow rate (and velocity) until a sonic regime is attained. Then, a further increase of O_2 flow rate provokes a further increase of surface area, but it is much less rapid than under subsonic conditions. The reason for this behavior can be found in the above-reported dependence of particle size on O_2 linear velocity, but also in the diluting effect of excess oxygen. Indeed, increasing oxygen flow rate at a constant liquid feeding rate leads to a higher dispersion of the particles within the flame, decreasing the probability of collision and hence of particle growth. Furthermore, the quenching effect due to the higher oxygen flow rate, lowering the flame temperature, and hence the particle sintering rate, contributes to the decrease of particle size. The latter two effects can be separated from that of O_2 linear velocity by increasing oxygen flow rate under sonic conditions. The quenching effect can be particularly critical, because it can inhibit the formation of a crystalline perovskitic phase.

Together with calcination time (i.e., residence time of the particles in the flame), flame temperature is an important parameter in determining particle growth. Flame temperature can be roughly estimated on the basis of the fuel adiabatic combustion enthalpy, but it is hard to predict its real value through modeling. Hence, the best practice is to measure its distribution, e.g., though high-resolution thermal maps. As an example, the effect of oxygen pressure drop on flame temperature is reported in Table 18.4. It is evident that an increase of pressure drop causes a quenching of flame temperature, which becomes more dramatic with increasing the combustion enthalpy of the solvent mixture. The latter represents another crucial point, directly affecting flame temperature. It is intuitive that an increase of the combustion enthalpy of the solvent causes the increase of flame temperature (when keeping constant any other operating parameter), hence provoking a decrease of surface area and an increase of particle size due to sintering. An example of the effect of solvent composition on flame temperature is also given in Table 18.4. As expected, this reflects on catalytic activity, which progressively decreases with increasing flame temperature (i.e., decreasing surface area) and on thermal stability, which shows the opposite trend.

From the results above summarized, it can be concluded that the FP technique is very versatile. Indeed, its operating parameters can be properly tuned so to obtain either very small particle size, and hence high surface area (even higher than 100 m^2/g, as for sample $SrTiO_3$-FP in Table 18.3), or more sintered materials. Of course, this affects catalytic activity for the current application. Provided that a pure perovskitic phase is obtained, a decrease of particle size of the FP samples usually brings about an increase of catalytic activity. An example is given in Table 18.5, where the parameter considered is pressure drop through the nozzle and hence O_2 linear velocity at the

TABLE 18.4. Dependence of FP Flame Temperature on Pressure Drop (ΔP) Across the Nozzle and on Solvent Nature.

ΔP (bar)	Solvent	ΔH_c (kJ/cm^3)	T_{max} (°C)	H_{Tmax} (cm)
1.5	PA	−21	824	1.8
0.3	PA	−21	897	1.1
1.5	PA + Xi (8:2)	−27	895	1.1
0.3	PA + Xi (8:2)	−27	1527	1.1
1.5	PA + Xi (2:8)	−45	1089	2.5
0.3	PA + Xi (2:8)	−45	1910	2.3

Abbreviations: PA = propionic acid; Xi = xylene (figures indicate the volumetric ratio); ΔH_c = combustion enthalpy of the solvent; T_{max} = max flame temperature in the main flame zone; H_{Tmax} = distance from nozzle mouth.

TABLE 18.5. Surface Area, Catalytic Activity, and Thermal Stability of FP-Prepared $LaCoO_3$ Samples (167).

		Catalytic Activity			Thermal Stability		
ΔP (bar)	SSA[a] (m^2/g)	T_b (°C)[b]	$T_{1/2}$ (°C)[b]	T_f (°C)[b]	CH_4 Conv. (mol%) 1st Cycle	CH_4 Conv. (mol%) 2nd Cycle	CH_4 Conv. (mol%) 3rd Cycle
0.4	47.3	279	430	527	94.8	93.3	92.3
1.5	54.2	276	404	497	80.0	78.2	76.1
6.0	65.4	274	383	471	57.9	45.6	45.2

Notes: Operation parameter varied: pressure drop (ΔP) across the nozzle. O_2 discharge linear velocity: subsonic ($\Delta P = 0.4$ bar), sonic ($\Delta P = 1.5$ bar), and supersonic ($\Delta P = 6.0$ bar). Thermal stability reported as residual CH_4 conversion at T_f after the rapid deactivation cycles at 800 °C.

[a]Specific surface area (SSA) calculated through the BET method.

[b]T_b = temperature of light-off; $T_{1/2}$ = temperature of 50% CH_4 conversion; T_f = temperature of 100% CH_4 conversion.

nozzle discharge section. The transition from subsonic to sonic and finally to supersonic oxygen velocity is accompanied by a progressive increase of surface area and, in a parallel way, of catalytic activity. However, from the previous considerations, we can also conclude that the preparation conditions characterized by a shortening of residence time into the flame or a decrease of flame temperature lead to less thermally resistant materials. This is well evidenced in the same Table 18.5, in which a higher residual methane conversion is shown by the more sintered material, i.e., that obtained with the lower pressure drop across the nozzle. The activity data of the FP prepared samples (Table 18.5) should be compared with those of the samples prepared through other routes (Table 18.3). It may be concluded that even under preparation conditions, which are less favorable from the point of view of activity, very active catalysts can be obtained, competing with those obtained through traditional preparation techniques. Moreover, even if some thermal deactivation occurs, mainly due to the high surface energy connected with the particle nanometric size, such a deactivation tends to level off. Indeed, even for the worst resistant catalyst, after an initial loss of ca. 50% in methane conversion, the latter levels off without any further significant decrease. A different behavior is shown by the SG-prepared samples, which further significantly deactivate after the second and third overheating cycles (Figure 18.9).

In conclusion, both the FP and the FH preparation methods can be satisfactorily used for the preparation of nanometer-size perovskitic catalysts for the flameless combustion of methane. Moreover, these routes show intrinsic practical advantages with respect to the traditional preparation procedures, being one-step, continuous techniques. In addition, the FP procedure is even more versatile than FH and allows us to substantially increase productivity with respect to FH, although the more complex set of parameters affecting the properties (particle size, surface area, thermal stability, etc.) of the desired material requires a careful optimization of the procedure.

18.6. STRUCTURED CATALYSTS FOR THE CFC OF METHANE

For practical application in the CFC of hydrocarbons, the catalyst needs to be supported. Indeed, the pressure drop along a pelletized or extruded catalyst bed would be unacceptable, leading to a decrease of the inlet pressure in the GT, with a consistent reduction of its efficiency. To decrease the pressure drop, different support configurations can be used, such as ceramic of metallic foams. These supports are characterized by a tortuous pore network, allowing good gas mixing and, hence, better heat transfer, avoiding the formation of pronounced hot spots (or, in the worst case, ignition) with the consequent catalyst destruction (169–173). However, honeycomb monoliths are by far the most commonly used supports. They are usually made of cordierite, a Mg–Al–Si-based ceramic material, but they can also be metallic, depending on working temperature. Ceramic monoliths are well suited for the current application, due to their low thermal expansion coefficient, which is preserved from thermal shocks (174,175). They are usually obtained by extrusion, with either square, triangular, or hexagonal channel cross sections of a different opening size. Channel size is an important parameter, determining the ratio between homogeneous gas-phase reaction and surface reaction. Larger channels allow lower pressure drop across the monolith, but they expose a lower available surface for the reaction. Common commercial channel density is about 400 channels per square inch (cpsi).

In some recent studies, the active phase has also been directly mixed with the cordierite precursors paste before extrusion (176–179). However, the technique usually adopted for the deposition of the active phase is the dip-coating, or wash-coating, which consists of the immersion and withdrawal of the honeycomb from a solution-suspension of the active-phase precursors, followed by drying and calcination, to decompose the precursor and obtain the desired active-phase formation and dispersion (174,180–182). The calcination temperature to obtain the perovskite-like phase is much higher (700–800 °C) than for noble metal-based catalysts. However, this method is not suitable to obtain a good perovskite dispersion in the absence of a primer (*vide infra*). Indeed, the poor anchoring of the active phase on the bare monolith surface causes the easy surface migration of the oxide, with consequent formation of very large (20–100 μm) and poorly active particles, as evidenced by scanning electron microscopy micrographs (Figure 18.11*a*). Some better results were obtained by deposition of the preformed perovskite from a suspension. However, the thermal stability of the catalyst in this case is usually insufficient (174).

In general, uniform distribution of a thin catalyst layer can be achieved by controlling the hydrodynamics of the system, the rheology of the suspension, and the speed of dipping-withdrawal of the monolith. When substrate speed and liquid viscosity are moderate, film thickness (h) can be calculated through the Landau–Levich equation (183):

$$h = \frac{0.94 \cdot (\eta U)^{2/3}}{\gamma_{LV}^{1/6} \cdot (\rho g)^{1/2}}$$

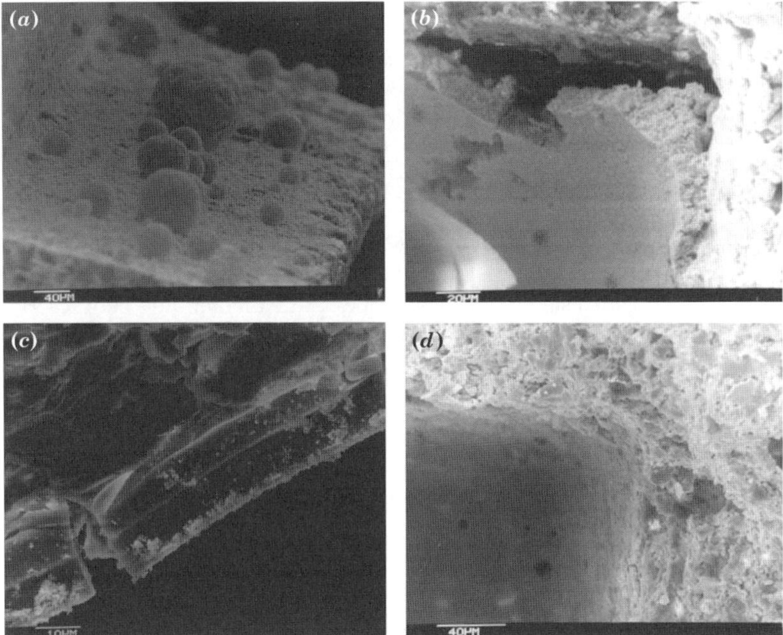

Figure 18.11. SEM micrographs of honeycomb-supported samples. (*a*) Perovskite directly synthesized on the honeycomb from nitrate precursors, by impregnation and calcination at 800 °C. (*b*) Excessive loading of $LaFeO_3$ on La_2O_3-coated cordierite, after calcination at 800 °C. (*c*) $La_{0.9}Ce_{0.1}CoO_{3\pm\delta}$ loaded on γ-Al_2O_3-coated cordierite and calcined at 800 °C. (*d*) Example of compact perovskite coating (obtained with La_2O_3 as primer from nitrate precursor calcined at 500 °C) after calcination at 800 °C.

where U is the withdrawal speed, η is the viscosity, γ_{LV} is the surface tension of the liquid, and ρ is the density. This model applies to Newtonian liquids, and viscosity is considered constant.

To favor a better adhesion of the mixed oxide-type active phases to the monolith, a primer is usually needed. The most commonly used primer is γ-Al_2O_3, which is loaded onto the support by dip-coating from a stable suspension of boehmite.

During the preparation of the alumina suspension, pH has to be carefully controlled. The addition of nitric acid is very important to improve the stability of the suspension and to govern the thickness of the deposited layer (174,180,184). Indeed, a low-pH solution favors the peptization of the boehmite through the following reaction (185):

$$\mathrm{AlOOH + 3H^+ \longrightarrow Al^{3+} + 2H_2O}$$

whereas at higher pH, an equilibrium is established between the solid surface and the solution:

$$\mathrm{{=}Al{-}OH + H^+ \longleftrightarrow {=}Al{-}OH_2^+}$$

Furthermore, Al^{3+} hydrated ions are involved in the following equilibria (186):

$$[Al(H_2O)_6]^{3+} \longleftrightarrow [Al(H_2O)_5(OH)]^{2+} + H^+$$

$$[Al(H_2O)_6]^{3+} \longleftrightarrow [Al(H_2O)_4(OH)_2]^{+} + 2H^+$$

which contrast the increase of pH accompanying the dissolution of a part of the boehmite and the protonation of the surface of solid particles. To minimize these phenomena and obtain a stable suspension, the optimal molar ratio between H^+ and Al^{3+} lies around 0.05 (186).

Similar considerations can also be referred to other possible primer materials, such as lanthanum oxide. Indeed, at pH = 6.5, the precipitation of lanthanum hydroxide prevents the formation of a stable suspension and leads to a poorly uniform coating. On the other hand, at too low a pH, the La precursor cannot be easily adsorbed on the cordierite acidic surface. Hence, the optimal pH of the solution/suspension of La^{3+} should be around a value of 6 (174).

To obtain a strong and uniform grafting of both primer and perovskite layers on the cordierite support, the speed of the honeycomb withdrawal during the dip coating has to be carefully controlled. A high speed usually leads to an excessive and uneven deposition, the excess fluid accumulating at the bottom of the monolith. On the other hand, a low speed causes the adhesion of a too thin liquid layer to the surface, imposing several dip-coating cycles to obtain the desired loading. Furthermore, primer and perovskite loadings depend mostly on the concentration of the solution/suspension, while being almost independent of immersion time (174,180).

The loading of both primer and active phases should be optimized with respect to different parameters, such as coating adhesion [measurable through ultrasonic treatment and determination of monolith weight loss (187)], activity, and thermal stability. In the case of perovskite supporting, ca. 4–5 wt% of primer and ca. 2 wt% of perovskite is allowed to obtain satisfactory activity and thermal stability (174,180,188). Indeed, a too high loading usually entrains poor mechanical and thermal resistance of the deposited layer, with consequent formation of cracks and exfoliation of active phase after use at high temperatures (Figure 18.11*b*).

The primer plays a very important role for both catalytic activity and thermal stability in the current application, and it must be properly selected according to the active phase nature. In fact, when noble metals are used, proper dispersion has to be ensured, so the primer choice will be addressed toward high-surface-area materials, such as alumina. Furthermore, proper additives can be required to improve its properties, such as oxygen buffers and stabilizers. By contrast, in perovskite supporting, the catalyst constitutes the bulk phase, the primer favoring its adhesion only. Hence, selection must be done on the basis of thermal properties and inertness toward the active phase. Indeed, due to the high temperature attained during the CFC of methane, possible solid-state reactions can occur between the primer and the active perovskitic layer. This is the case of $LaCoO_3$ samples, which are very active catalysts for the current application both as such or doped with A′ substituents. They can transform into other less active oxides through interaction with the alumina primer layer. Indeed, $CoAl_2O_4$ with spinel structure can form at relatively low temperatures (ca. 500 °C) (174,189–191), favored by the presence of water, supplied by the reaction, which enhances the migration of Co toward alumina. The same transformation was observed at higher temperatures (ca. 800 °C) even in dry conditions (192,193). This phenomenon was not observed with different B metal ions, such as Mn or Fe. Furthermore, La segregation from a $LaCoO_3$ perovskite may also occur, with formation of $LaAlO_3$ (194). Besides these chemical interactions, another key point is the difference in thermal expansion coefficients of cordierite, alumina, and the oxidic active phase. Indeed, if this difference is too high, as in the above reported example, problems can arise during heating and cooling cycles, bringing about the formation of deep cracks (Figure 18.11*c*), with consequent exfoliation of the coating layer (174).

From the previous considerations, it can be concluded that alumina does not represent the best primer for honeycomb supporting of La-based perovskites. An alternative primer can be La_2O_3 (114,174,188), wash-coated on the honeycomb starting from a solution/suspension of a La precursor, showing very satisfactory results from both the points of view of activity and thermal stability. A recent study (114,188) on the properties of the primer material and on its modifications during catalytic performance allowed us to conclude that the best results can be achieved by loading the primer from its nitrate precursor, followed by calcination at relatively low temperature (500 °C). During calcination, $La(NO_3)_3$ undergoes a series of complex transformations, leading to several hydroxy–nitrate intermediates, which melt in their crystallization water and progressively lose NO_x and water to give the final La_2O_3 primer layer. After calcination of the primer precursor at 500 °C, the conversion into the oxide is still incomplete and the morphology of the layer resembles that of a needle carpet. When the perovskite powder is loaded onto it, a rough surface is available, favoring a better grafting of the active phase. Furthermore, when the sample is finally calcined at higher temperature, the complete conversion of the primer precursor into La_2O_3 occurs, partially incorporating the active phase, due to the change of the needle-like morphology into a compact layer (Figure 18.11*d*).

A final remark should be addressed to the possible physical modifications occurring to some active phases during honeycomb supporting. Indeed, some physical or chemical transformations can take place during the various loading steps, such as possible acid leaching, phase transitions during calcination, or possible interactions with the La_2O_3 primer, with consequent modification of the perovskite-like phase stoichiometry, as in the case of La_2NiO_4 samples (114).

18.7. CONCLUSIONS

The growing demand of energy imposes progressively more restrictive criteria on its production, coupled with the selection of technologies characterized by higher efficiency and lower emission of harmful species. The catalytic flameless combustion of hydrocarbons, mainly of natural gas, is an efficient route to supply heat in gas turbines for stationary power generation and for the clean-up of the exhaust gas from both stationary and mobile applications. Its inherently lower temperature ($<1000\,^{\circ}C$), with respect to homogeneous combustion, allows us to almost completely suppress NO_x emissions (<3 ppm), meeting the ever more stringent environmental regulations.

Catalyst design for the flameless combustion of hydrocarbons is a highly demanding task. Indeed, high activity and high thermal stability are usually required, to obtain low light-off temperature coupled with resistance under the very severe operating conditions. Supported noble metal oxides, especially PdO, proved to be very active catalysts, but their often unsatisfactory thermal resistance and high cost raise some problems. Mixed metal oxides, especially of a perovskite-like structure, can be efficiently used for the current application.

The flame-based preparation techniques can be effective routes to combine activity and stability of perovskites, allowing us to obtain nanostructured materials, characterized by high phase purity and exceptionally high surface areas, leading to very active and stable catalysts.

Finally, the practical application of these nanosized materials requires proper supporting on preshaped inert monoliths. Furthermore, a primer is usually needed to ensure a proper grafting and good dispersion of the active phase. The selection of the primer has to meet the characteristics of the active phase. For instance, Al_2O_3 is usually adopted for PdO loading on monoliths, whereas La_2O_3 was very efficient for supporting La-based perovskites on cordieritic honeycombs.

REFERENCES

(1) Available at: http://unfccc.int/resource/docs/convkp/kpeng.html.

(2) Available at: http://www.un.org/documents/ga/conf151/aconf15126-1annex1.htm.

(3) Available at: http://www.eia.doe.gov/pub/international/iealf/tablee1.xls.

(4) Available at: http://www.eia.doe.gov/fueloverview.html.

(5) Ciuparu, D.; Lyubovsky, M.R.; Altman, E.; Pfefferle, L.D.; Datye, A. *Catal. Rev.* **2002**, *44(4)*, 593.

(6) Zeldovich, Y. *J. Phys. (U.S.S.R.)* **1946**, *10*, 321.

(7) Available at: http://www.worldenergy.org/wec-geis/publications/default/tech_papers/17th_congress/1_3_08.asp.

(8) Hong, H.; Yunbo, Y. *Catal. Today* **2005**, *100(1–2)*, 37–47.

(9) Gomez-Garcia, M.A.; Pitchon, V.; Kiennemann, A. *Environ. Int.* **2005**, *31(3)*, 445–467.

(10) Kung, M.C.; Kung, H.H. *Top. Catal.* **2004**, *28(1–4)*, 105–110.

(11) Amiridis, M.D.; Mihut, C.; Maciejewski, M.; Baiker, A. *Top. Catal.* **2004**, *28(1–4)*, 141–150.

(12) Stratos Tavoulareas, E.; Charpentier, J.-P. *Clean Coal Technologies for Developing Countries*; World Bank Technical Paper No. 286, Energy Series, July 1995.

(13) Williams, R.H.; Larson, E.D. *Aero-derivative Turbines for Stationary Power*; The Center for Energy and Environmental Studies: Princeton University, May 1988.

(14) Kuper, W.J.; Blaauw, M.; van der Berg, F.; Graaf, G.H. *Catal. Today* **1999**, *47*, 377.

(15) Available at: www.worldbank.org/html/fpd/em/power/EA/mitigatn/thermpow.stm.

(16) Knight, D. *Humphry Davy: Science and Power*; Cambridge University Press: Cambridge, 1998.

(17) Kauffman, G.B. *Platinum Metals Rev.* **1999**, *43*, 122.

(18) Available at: http://www.energy.ca.gov/pier/energy_update/2002-08_FACTSHEET_CATALYTIC.PDF.

(19) Cocchi, S.; Modi, R.; Nutini, G.; Spencer, M.J.; Sarento, G.N. In *Catalytic Combustion, Vol. 1*; Forzatti, P.; Groppi, G.; Ciambelli, P.; Sannino, D. (Editors); Polipress: Milano, 2005; 27.

(20) Carroni, R.; Schmidt, V.; Griffin, T. *Catal. Today* **2002**, *75*, 287.

(21) Pfefferle, W.C. *Belgian Patent* 814,752; 1974.

(22) Furuya, T.; Sasaki, T.; Hanakata, Y.; Mitsuyasu, K.; Yamada, M.; Tsuchiya, T.; Furuse, Y. *Proc. International Workshop on Catalytic Combustion*, Tokyo, 1994; 162.

(23) Ozawa, Y.; Fujii, T.; Kikumoto, S.; Sato, M.; Fukuzawa, H.; Saiga, M.; Watanabe, S. *Proc. International Workshop on Catalytic Combustion*, Tokyo, 1994; 166.

(24) Forzatti, P. *Catal. Today* **2003**, *83*, 3.

(25) Dalla Betta, R.A. *Catal. Today* **1997**, *35*, 129.

(26) Young, W.E.; Carl, D.E. Eur. Patent 0304707 A1; 1989.

(27) Carroni, R.; Griffin, T.; Mantzaras, I.; Reinke, M. *Catal. Today* **2003**, *83*, 157.

(28) Lyubovsky, M.; Smith, L.L.; Castaldi, M.; Karim, H.; Nentwick, B.; Etemad, S.; LaPierre, R.; Pfefferle, W.C. *Catal. Today* **2003**, *83*, 71.

(29) Cimino, S.; Lisi, L.; Russo, G.; Landi, G. In *Catalytic Combustion, Vol. 1*; Forzatti, P.; Groppi, G.; Ciambelli, P.; Sannino, D. (Editors); Polipress: Milano, 2005; 65.

(30) Tavazzi, I.; Maestri, M.; Beretta, A.; Groppi, G.; Tronconi, E.; Forzatti, P. In *Catalytic Combustion, Vol. 1*; Forzatti, P.; Groppi, G.; Ciambelli, P.; Sannino, D. (Editors); Polipress: Milano, 2005; 77.

(31) Karim, H.; Lyle, K.; Etemad, S.; Smith, L.; Pfefferle, W.; Dutta, P.; Smith, K. ASME Paper GT-2002-30083.

(32) Warnatz, J.; Allendorf, M.A.; Kee, R.J.; Coltrin, M.E. *Combust. Flame* **1994**, *96*, 393.

(33) Deutschmann, O.; Behrendt, F.; Warnatz, J. *Catal. Today* **1994**, *21*, 461.

(34) Mantzaras, J. In *Catalytic Combustion, Vol. 1*; Forzatti, P.; Groppi, G.; Ciambelli, P.; Sannino, D. (Editors); Polipress: Milano, 2005; 13.

(35) Appel, C.; Mantzaras, J.; Schaeren, R.; Bombach, R.; Inauen, A. *Combust. Flame* **2005**, *140*, 70.

(36) Appel, C.; Mantzaras, J.; Schaeren, R.; Bombach, R.; Kaeppeli, B.; Inauen, A. *Proc. Combust. Inst.* **2002**, *29*, 1031.

(37) Hayes, R.E.; Kolaczkowski, S.T. *Catal. Today* **1999**, *47*, 295.

(38) Groppi, G.; Tronconi, E.; Forzatti, P. *Catal. Rev.-Sci. Eng.* **1999**, *41*, 227.

(39) Young, L.C.; Finlayson, B.A. *AIChE J.* **1976**, *22(2)*, 343.

(40) Groppi, G.; Belloli, A.; Tronconi, E.; Forzatti, P. *Chem. Eng. Sci.* **1995**, *50*, 2705.

(41) Gupta, N.; Balakotaiah, V. *Chem. Eng. Sci.* **2001**, *56*, 4771.

(42) Di Benedetto, A.; Marra, F.S.; Russo, G. *Chem. Eng. Sci.* **2003**, *58*, 1079.

(43) Dogwiler, U.; Mantzaras, J.; Appel, C.; Benz, P.; Kaeppeli, B.; Bombach, R.; Arnold, A. *Proc. Combust. Inst.* **1998**, *27*, 2275.

(44) Raja, L.L.; Kee, R.J.; Deutschmann, O.; Warnatz, J.; Schmidt, L.D. *Catal. Today* **2000**, *59*, 47.

(45) Deutschmann, O.; Schwiedernoch, R.; Maier, L.I.; Chatterjee, D. *Stud. Surf. Sci. Catal.* Iglesia, E.; Spivey, J.J.; Fleisch, T.H. (Editors); **2001**, *136*, 251.

(46) Gélin, P.; Primet, M. *Appl. Catal. B: Environ.* **2002**, *39*, 1.

(47) Fino, D.; Russo, N.; Saracco, G.; Specchia, V. In *Catalytic Combustion, Vol. 1*; Forzatti, P.; Groppi, G.; Ciambelli, P.; Sannino, D. (Editors); Polipress: Milano, 2005; 175.

(48) Available at: www.kitco.com.

(49) Groppi, G.; Cristiani, C.; Forzatti, P. In *Specialist Periodical Reports, Vol. 13*; Spivey, J.J. (Editor); The Royal Society of Chemistry: London, 1998.

(50) Zwinkels, M.F.M.; Järås, S.G.; Menon, P.G.; Griffin, T. *Catal. Rev.-Sci. Eng.* **1993**, *35*, 319.

(51) Centi, G. *J. Molec. Catal. A: Chem.* **2001**, *173*, 287.

(52) Papaefthimiou, P.; Ioannides, T.; Verykios, X.E. *Appl. Catal. B: Environ.* **1997**, *13*, 175.

(53) Veser, G.; Ziauddin, M.; Schmidt, L.D. *Catal. Today* **1999**, *47*, 219.

(54) Burch, R.; Crittle, D.J.; Hayes, M.J. *Catal. Today* **1999**, *47*, 229.

(55) Aryafar, M.; Zaera, F. *Catal. Lett.* **1997**, *48*, 173.

(56) Paulis, P.; Gandia, L.M.; Gil, A.; Sambeth, J.; Odriozola, J.A.; Montes, M. *Appl. Catal. B: Environ.* **2000**, *26*, 37.

(57) Forzatti, P.; Groppi, G. *Catal. Today* **1999**, *54*, 165.

(58) Betta, R.A.D.; Ezawa, N.; Tsurumi, K.; Schlatter, J.C.; Nickolas, S.G. U.S. Patent 5,183,401; 1993; to Catalytica.

(59) Anderson, R.A.; Stein, K.C.; Feenan, J.J.; Hofer, L.J. *Ind. Eng. Chem.* **1961**, *53*, 809.

(60) Groppi, G.; Cristiani, C.; Lietti, L.; Forzatti, P. *Stud. Surf. Sci. Catal.* **2000**, *130*, 3801.

(61) McCarty, J.G.; Gusman, M.; Lowe, D.M.; Hildenbrand, D.L.; Lau, K.N. *Catal. Today* **1999**, *47*, 5.

(62) Summers, J.C.; Monzoe, D.R. *Ind. Eng. Chem. Prod. Res. Dev.* **1981**, *20*, 23.

(63) Meeyoo, V.; Trimm, D.L.; Cant, N.W. *Appl. Catal. B: Environ.* **1998**, *16*, L101.

(64) Schelef, M.; Otto, K.; Otto, N.C. *Adv. Catal.* **1978**, *27*, 311.

(65) Machida, M.; Eguchi, K.; Arai, H. *J. Catal.* **1990**, *123*, 477.

(66) Groppi, G.; Bellotto, M.; Cristiani, C.; Forzatti, P.; Villa, P.L. *Appl. Catal. A: Gen.* **1993**, *104*, 101.

(67) Wang, J.; Tian, Z.; Xu, J.; Xu, Y.; Xu, Z.; Lin, L. *Catal. Today* **2003**, *83*, 213.

(68) Xu, J.; Tian, Z.; Xu, Y.; Xu, Z.; Lin, L. *Stud. Surf. Sci. Catal.* **2004**, *147*, 481.

(69) Teng, F.; Xu, J.; Tian, Z.; Wang, J.; Xu, Y.; Xu, Z.; Xiong, G.; Lin, L. *Chem. Comm.* **2004**, *16*, 1859.

(70) Long, Z.-R.; Hu, R.-S.; Gao, G.-J.; Zhang, X.-H.; Bai, Y. In *Catalytic Combustion, Vol. 2*; Forzatti, P.; Groppi, G.; Ciambelli, P.; Sannino, D. (Editors); Polipress: Milano, 2005; 39.

(71) Rawlinson, D.A.; Sermon, P.A. In *Catalytic Combustion, Vol. 2*; Forzatti, P.; Groppi, G.; Ciambelli, P.; Sannino, D. (Editors); Polipress: Milano, 2005; 146.

(72) Tian, Z.; Xu, J.; Xu, Y.; Lin, L. In *Catalytic Combustion, Vol. 1*; Forzatti, P.; Groppi, G.; Ciambelli, P.; Sannino, D. (Editors); Polipress: Milano, 2005; 99.

(73) Peña, M.A.; Fierro, J.L.G. *Chem. Rev.* **2001**, *101*, 1981.

(74) Bednorz, J.G.; Muller, K.A. *Angew. Chem.-Engl. Ed.* **1988**, *100*, 757.
(75) Labbe, J.; Bok, J. *J. Europhys. Lett.* **1987**, *3*, 1225.
(76) Available at: http://journals.iucr.org/j/issues/2004/02/00/hx5000/hx5000fig5.html.
(77) Available at: http://jcrystal.com/steffenweber/gallery/StructureTypes/st5.html.
(78) Goldschmidt, V.M. *Skr. Nor. Viedenk. Akad., Kl. 1 Mater. Naturvidensk.* **1926**, *8*.
(79) Rao, C.N.R.; Gopalakrishnan, J.; Vidyasagar, K. *Indian J. Chem. A* **1984**, *23A*, 265.
(80) Smith, D.M. *Ann. Rev. Mater. Sci.* **1985**, *15*, 329.
(81) Smith, D.M. In *Properties and Applications of Perovskite Type Oxides*; Tejuca, L.G.; Fierro, J.L.G. (Editors); Dekker: New York, 1993; 47.
(82) Sayagues, M.J.; Vallet-Regi, M.; Caneiro, A.; Gonzales Calbet, J.M. *J. Solid State Chem.* **1994**, *110*, 295.
(83) Tofield, B.C.; Scott, W.R. *J. Solid State Chem.* **1974**, *10*, 183.
(84) Voorhoeve, R.J.H.; Remeika, J.P.; Trimble, L.E.; Kooper, A.S.; Disalvo, F.J.; Gallagher, P.K. *J. Solid State Chem.* **1975**, *14*, 395.
(85) Vogel, E.M.; Johnson Jr., D.W.; Gallagher, P.K. *J. Am. Ceram. Soc.* **1977**, *60*, 31.
(86) Taguchi, H.; Sugita, A.; Nagao, M.; Tabata, K. *J. Solid State Chem.* **1995**, *119*, 164.
(87) Kamegashira, N.; Miyazaki, Y.; Yamamoto, H. *Mater. Chem. Phys.* **1984**, *11*, 187.
(88) Oliva, C.; Bonoldi, L.; Cappelli, S.; Fabbrini, L.; Rossetti, I.; Forni, L. *J. Molec. Catal. A: Chem.* **2005**, *226*, 33.
(89) Fabbrini, L.; Kryukov, A.; Cappelli, S.; Chiarello, G.L.; Rossetti, I.; Oliva, C.; Forni, L. *J. Catal.* **2005**, *232*, 247.
(90) Forni, L.; Oliva, C.; Vatti, F.P.; Kandala, M.A.; Ezerets, A.M.; Vishniakov, A.V. *Appl. Catal. B: Environ.* **1996**, *7*, 269.
(91) Oliva, C.; Termignone, G.; Vatti, F.P.; Forni, L.; Vishniakov, A.V. *J. Mater. Sci.* **1996**, *31*, 6333.
(92) Oliva, C.; Forni, L.; Ezerets, A.M.; Mukovozov, I.E.; Vishniakov, A.V. *J. Chem. Soc. Faraday Trans.* **1998**, *94*, 587.
(93) Campagnoli, E.; Tavares, A.C.; Fabbrini, L.; Rossetti, I.; Dubitsky, Yu.A.; Zaopo, A.; Forni, L. *J. Mater. Sci.* **2006**, *41*(15), 4713.
(94) Park, S.; Vohs, J.M.; Gorte, R.J. *Nature* **2000**, *404*, 265.
(95) Hibino, T.; Hashimoto, A.; Inoue, T.; Tokuno, J.; Yoshida, S.; Sano, M. *Science* **2000**, *288*, 203.
(96) Baker, R.T.; Metcalfe, I.S.; Middleton, P.H.; Steele, B.C.H. *Sol. State Ion.* **1994**, *72*, 328.
(97) Minh, N.Q. *J. Am. Ceram. Soc.* **1993**, *76*, 563.
(98) Leanza, R.; Rossetti, I.; Fabbrini, L.; Oliva, C.; Forni, L. *Appl. Catal. B: Environ.* **2000**, *28*, 55.
(99) Islam, M.S. *J. Mater. Chem.* **2000**, *10*, 1027.
(100) Yamazoe, N.; Teraoka, Y.; Seiyama, T. *Chem. Lett.* **1981**, 1767.
(101) Seiyama, T.; Yamazoe, N.; Eguchi, K. *Ind. Eng. Chem. Prod. Res. Dev.* **1985**, *24*, 19.
(102) Teraoka, Y.; Yoshimatsu, M.; Yamazoe, N.; Seiyama, T. *Chem. Lett.* **1984**, 893.
(103) Rossetti, I.; Forni, L. *Appl. Catal. B: Environ.* **2001**, *33*, 345.
(104) Nakamura, T.; Misono, N.; Yoneda, Y. *Bull. Chem. Soc. Jpn.* **1982**, *55*, 394.
(105) Nitadori, T.; Misono, M. *J. Catal.* **1985**, *93*, 459.

(106) Ferri, D.; Forni, L. *Appl. Catal. B: Environ.* **1998**, *16*, 119.

(107) Ponche, S.; Peña, M.A.; Fierro, J.L.G. *Appl. Catal. B: Environ.* **2000**, *24*, 193.

(108) Oliva, C.; Forni, L.; Pasqualin, P.; D'Ambrosio, A.; Vishniakov, A.V. *Phys. Chem. Chem. Phys.* **1999**, *1(2)*, 355.

(109) Marchetti, L.; Forni, L. *Appl. Catal. B: Environ.* **1998**, *15(3–4)*, 179.

(110) Iwamoto, M.; Yoda, Y.; Yamazoe, N.; Seiyama, T. *J. Phys. Chem.* **1978**, *82*, 2564.

(111) Petunchi, J.O.; Nicastro, J.L.; Lombardo, E.A. *J. Chem. Soc. Chem. Commun.* **1980**, 467.

(112) Islam, M.S.; Cherry, M.; Catlow, C.R.A. *J. Solid State Chem.* **1996**, *124*, 230.

(113) Voorhoeve, R.J.H.; Remeika, J.P.; Johnson, D.W. *Science* **1973**, *180*, 62.

(114) Fabbrini, L.; Rossetti, I.; Forni, L. *Appl. Catal. B: Environ.* **2006**, *63*(1-2), 131.

(115) Oliva, C.; Forni, L.; Vishniakov, A.V. *Spectrochim. Acta A* **2000**, *56*, 301.

(116) Zhang, H.M.; Yamazoe, N.; Teraoka, Y. *J. Mater. Sci. Lett.* **1998**, *8*, 995.

(117) Benson, S.J.; Waller, D.; Kilner, J.A. *J. Electrochem. Soc.* **1999**, *146*, 1305.

(118) Tsai, C.Y.; Dixon, A.G.; Ma, Y.H.; Moser, W.R.; Pascucci, M.R. *J. Am. Ceram. Soc.* **1998**, *81*, 1437.

(119) Xu, Q.; Huang, D.; Chen, W.; Lee, J.; Wang, H.; Yuan, R. *Scripta Materialia* **2004**, *50*, 165.

(120) Berndt, U.; Maier, D.; Keller, C. *J. Solid State Chem.* **1975**, *13*, 131.

(121) Keshavaraja, A.; Ramaswamy, A.V. *Indian J. Eng. Mater. Sci.* **1994**, *1*, 229.

(122) Jain, A.N.; Tiwary, S.K.; Singh, R.N.; Chartier, P. *J. Chem. Soc. Faraday Trans.* **1995**, *91*, 1887.

(123) Teraoka, Y.; Nanri, S.; Moriguchi, I.; Kagawa, S.; Shimanoe, K.; Yamazoe, N. *Chem. Lett.* **2000**, 1202.

(124) Nakajima, Y.; Ogura, T.; Kinoshita, M.; Komirama, T. *Chem. Express* **1989**, *4*, 225.

(125) Szabo, V.; Bassir, M.; Van Neste, A.; Kaliaguine, S. *Appl. Catal. B: Environ.* **2002**, *37*, 175.

(126) U.S. Patent 6,017,504; 2000, to Université Laval (Canada).

(127) Szabo, V.; Bassir, M.; Van Neste, A.; Kaliaguine, S. *Appl. Catal. B: Environ.* **2003**, *43*, 81.

(128) Campagnoli, E.; Tavares, A.; Fabbrini, L.; Rossetti, I.; Dubitsky, Yu.A.; Zaopo, A.; Forni, L. *Appl. Catal. B: Environ.* **2005**, *55(2)*, 133.

(129) Ho, S.; Park, I.H. *J. Korean Chem. Soc.* **1994**, *38*, 276.

(130) Taguchi, H.; Yoshioka, H.; Naguo, M. *J. Mater. Sci. Lett.* **1994**, *13*, 891.

(131) Li, X.; Zhang, H.; Chi, F.; Li, S.; Xu, B.; Zhao, M. *Mater. Sci. Eng.*, **1993**, *B18*, 209.

(132) Park, H.B.; Kweon, H.J.; Kim, S.J.; Kim, K. *J. Korean Chem. Soc.* **1994**, *38*, 852.

(133) Shimizu, Y.; Murata, T. *J. Am. Ceram. Soc.* **1997**, *80*, 2702.

(134) Tiwaris, K.; Chartier, P.; Singh, N.R. *J. Electrochem. Soc.* **1995**, *142*, 148.

(135) Xiong, G.; Zhi, Z.L.; Yang, X.; Lu, L.; Wang, X. *J. Mater. Sci. Lett.* **1997**, *16*, 1064.

(136) Wang, X.; Li, D.; Lu, L.; Wang, X. *J. Alloy Compounds* **1996**, *237*, 45.

(137) Cifà, F.; Dinka, P.; Viparelli, P.; Lancione, S.; Benedetti, G.; Villa, P.L.; Viviani, M.; Nanni, P. *Appl. Catal. B: Environ.* **2003**, *46*, 463.

(138) Viparelli, P.; Villa, P.L.; Basile, F.; Trifirò, F.; Vaccari, A.; Nanni, P.; Viviani, M. *Appl. Catal. A: Gen.* **2005**, *280*, 225.

(139) Ferri, D.; Forni, L.; Dekkers, M.A.P.; Nieuwenhuys, B.E. *Appl. Catal. B: Environ.* **1998**, *16*, 339.

(140) Svensson, E.; Nassos, S.; Boutonnet, M.; Järås, S.G. In *Catalytic Combustion, Vol. 1*; Forzatti, P.; Groppi, G.; Ciambelli, P.; Sannino, D. (Editors); Polipress: Milano, 2005; 104.

(141) Giacomuzzi, R.A.M.; Portinari, M.; Rossetti, I.; Forni, L. *Stud. Surf. Sci. Catal.* **2000**, *130*, 197.

(142) Oliva, C.; Forni, L.; D'Ambrosio, A.; Navarrini, F.; Stepanov, A.D.; Kagramanov, Z.D.; Mikhailichenko, A.I. *Appl. Catal. A: Gen.* **2001**, *205*, 245.

(143) Long, J.; Teichner, S.J. *Rev. Hautes Temper. et Refract.* **1965**, *2*, 47.

(144) Moser, W.R. (Editor); *Advanced Catalysts and Nanostructured Materials*; Academic Press: New York, 1996.

(145) Oliva, C.; Cappelli, S.; Rossetti, I.; Kryukov, A.; Bonoldi, L.; Forni, L. *J. Molec. Catal. A: Chem.* **2006**, *245*, 55.

(146) *Selected Powder Diffraction Data*; J.C.P.D.S.: Swarthmore, PA, Minerals.

(147) Stark, W.J.; Mädler, L.; Pratsinis, S.E. European Patent 1,378,489 A1; 2004 to ETH, Zurich.

(148) Schultz, H.; Stark, W.J.; Maciejewski, M.; Pratsinis, S.E.; Baiker, A. *J. Mater. Chem.* **2003**, *13*, 2979.

(149) Mädler, L.; Pratsinis, S.E. *J. Am. Ceram. Soc.* **2002**, *85(7)*, 1713.

(150) Heine, M.C.; Pratsinis, S.E. *Ind. Eng. Chem. Res.* ASAP Article, web release date: March 24, 2005.

(151) Strobel, R.; Stark, W.J.; Mädler, L.; Pratsinis, S.E.; Baiker, A. *J. Catal.* **2003**, *213*, 296.

(152) Mueller, R.; Jossen, R.; Kammler, H.K.; Pratsinis, S.E.; Akhtar, M.K. *AIChE J.* **2004**, *50(12)*, 3085.

(153) Tani, T.; Takatori, K.; Pratsinis, S.E. *J. Am. Ceram. Soc.* **2004**, *87(3)*, 365.

(154) Strobel, R.; Pratsinis, S.E.; Baiker, A. *J. Mater. Chem.* **2005**, *15*, 605.

(155) Stark, W.J.; Mädler, L.; Maciejewski, M.; Pratsinis, S.E.; Baiker, A. *Chem. Commun.* **2003**, 588.

(156) Mueller, R.; Jossen, R.; Pratsinis, S.E.; Watson, M.; Akhtar, M.K. *J. Am. Ceram. Soc.* **2004**, *87(2)*, 197.

(157) Stark, W.J.; Wegner, K.; Pratsinis, S.E.; Baiker, A. *J. Catal.* **2001**, *197*, 182.

(158) Jossen, R.; Pratsinis, S.E.; Stark, W.J.; Mädler, L. *J. Am. Ceram. Soc.* **2005**, *88*, 1388.

(159) Jung, K.Y.; Kang, Y.C. *Mater. Lett.* **2004**, *58*, 2161.

(160) Kim, S.; Gislason, J.J.; Morton, R.W.; Pan, X.Q.; Sun, H.P.; Laine, R.M. *Chem. Mater.* **2004**, *16*, 2336.

(161) Marchal, J.; John, T.; Baranwal, R.; Inklin, T.; Laine, R.M. *Chem. Mater.* **2004**, *16*, 822.

(162) Johannessen, T.; Koutsopoulos, S. *J. Catal.* **2002**, *205*, 404.

(163) Mäkelä, J.M.; Keskinen, H.; Forsblom, T.; Keskiken, J. *J. Mater. Sci.* **2004**, *39*, 2783.

(164) Kilian, A.; Morse, T.F. *Aerosol Sci. Tech.* **2001**, *34*, 227.

(165) Seo, D.J.; Park, S.B.; Kang, Y.C.; Choy, K.L. *J. Nanopart. Res.* **2003**, *5*, 199.

(166) Mädler, L.; Kammler, H.K.; Mueller, R.; Pratsinis, S.E. *J. Aerosol Sci.* **2002**, *33*, 369.

(167) Chiarello, G.L.; Rossetti, I.; Forni, L. *J. Catal.* **2005**, *236*, 251.

(168) Chiarello, G.L.; Rossetti, I.; Forni, L. In *Catalytic Combustion, Vol. 1*; Forzatti, P.; Groppi, G.; Ciambelli, P.; Sannino, D. (Editors); Polipress: Milano, 2005; 165.

(169) Beebe, W.; Cairns, K.D.; Pareek, V.K.; Nickolas, S.G.; Schlatter, J.C.; Tsuchiya, T. *Catal. Today* **2000**, *59*, 95.

(170) Thevenin, P.O.; Ersson, A.G.; Kusar, H.M.J.; Menon, P.G.; Järås, S.G. *Appl. Catal. A: Gen.* **2001**, *212*, 189.

(171) Gibson, L.J.; Ashby, M.F. *Cellular Solids, Structures and Properties*; Pergamon Press: Oxford, 1988.

(172) Thevenin, P.O.; Menon, P.G.; Järås, S.G. *CATTECH*, **2003**, *7*, 10.

(173) Maione, A.; Demoulin, O.; Ruiz, P. In *Catalytic Combustion, Vol. 2*; Forzatti, P.; Groppi, G.; Ciambelli, P.; Sannino, D. (Editors); Polipress: Milano, 2005; 147.

(174) Fabbrini, L.; Rossetti, I.; Forni, L. *Appl. Catal. B: Environ.* **2003**, *44*, 107.

(175) Cimino, S.; Lisi, L.; Pirone, R.; Russo, G.; Turco, M. *Cat. Today* **2000**, *59*, 19.

(176) Isupova, L.A.; Alikina, G.M.; Snegurenko, O.I.; Sadykov, V.A.; Tsybulya, S.V. *Appl. Catal. B: Environ.* **1999**, *21*, 171.

(177) Ciambelli, P.; Palma, V.; Tikhov, S.F.; Sadykov, V.A.; Isupova, L.A.; Lisi, L. *Cat. Today* **1999**, *47*, 199.

(178) Tretyakov, V.F.; Sadykov, V.A.; Rozovskii, A.Ya.; Lunin, V.V. *Heterogeneous Catalysis, Proc. IX Intern. Symp.*; Petrov, L.; Bonev, Ch.; Kadinov, G. (Editors); Institute of Catalysis, Bulgarian Academy Science: Sofia, 2000; 623.

(179) Schneider, R.; Kiessling, D.; Wendt, G.; Burckhardt, W.; Winterstein, G. *Catal. Today*, **1999**, *47*, 429.

(180) Forni, L.; Rossetti, I. *Appl. Catal. B: Environ.* **2002**, *38*, 29.

(181) Isupova, L.A.; Kulikovskaya, N.A.; Amosov, Yu.I.; Saputina, N.F.; Plyasova, L.M.; Rudina, N.A.; Sadykov, V.A.; Ovsyannikova, I.A.; Tretyakov, V.F.; Burdeinaya, T.N. In *Catalytic Combustion, Vol. 2*; Forzatti, P.; Groppi, G.; Ciambelli, P.; Sannino, D. (Editors); Polipress: Milano, 2005; 34.

(182) Cimino, S.; Lisi, L.; Pirone, R.; Russo, G. In *Catalytic Combustion, Vol. 2*; Forzatti, P.; Groppi, G.; Ciambelli, P.; Sannino, D. (Editors); Polipress: Milano, 2005; 53.

(183) Levich, V.G. *Physicochemical Hydrodynamics*; Prentice-Hall: Englewood Cliffs, NJ, 1962.

(184) Valentini, M.; Groppi, G.; Cristiani, C.; Levi, M.; Tronconi, E.; Forzatti, P. *Catal. Today* **2001**, *69*, 307.

(185) Laiti, E.; Oehman, L.O. *J. Coll. Interf. Sci.* **1996**, *183*, 441.

(186) Morgado, E.J.; Lam, Y.L.; Menezes, S.M.C.; Nazar, L.F. *J. Coll. Interf. Sci.* **1995**, *176*, 432.

(187) Fino, D.; Russo, N.; Saracco, G.; Specchia, V. In *Catalytic Combustion, Vol. 1*; Forzatti, P.; Groppi, G.; Ciambelli, P.; Sannino, D. (Editors); Polipress: Milano, 2005; 175.

(188) Fabbrini, L.; Rossetti, I.; Forni, L. *Appl. Catal. B: Environ.* **2005**, *56(3)*, 221.

(189) Garbowski, E.; Guenin, M.; Marion, M.C.; Primet, M. *Appl. Catal.* **1990**, *64*, 209.

(190) Jongsomjit, B.; Panpranot, J.; Goodwin, J.C. *J. Catal.* **2001**, *204*, 98.

(191) Bolt, P.H.; Habraken, F.H.P.M.; Geus, J.W. *J. Solid State Chem.* **1998**, *135*, 59.

(192) Chen, C.; Bouwmeester, H.J.M.; Kruidhof, H.; ten Elshof, J.E.; Burggraaf, A.J. *J. Mater. Chem.* **1996**, *6*, 815.

(193) Xue, P.; Shen, Y.; Sun, Y.; Hu, R. *Fenzi Cuihua* **1998**, *12*, 424; *Chem. Abstr.* **1999**, *130*, 72050.

(194) Munakata, F.; Tanimura, M.; Takamoto, K.; Kaneko, H.; Yamaguchi, H.; Inoue, Y.; Akimune, Y. *J. Ceram. Soc. Japan* **1995**, *103*, 1041.

CHAPTER 19

Nanostructured Oxides in DeNO$_x$ Technologies

MARCOS FERNÁNDEZ-GARCÍA and ARTURO MARTÍNEZ-ARIAS

Instituto de Catálisis y Petroleoquímica CSIC, Campus Cantoblanco, 28049 Madrid, Spain

JAVIER PÉREZ-RAMÍREZ

Institut Catalá d'Investigació Química Av. Paisos Catalans 16 Campus Univesitary, Tarragona, Spain

19.1. INTRODUCTION

N-containing molecules are typical pollutants directly related to industrial activities in many cases. The combustion of fossil fuels to meet the society requirement of energy and the needs of the chemical industry is the main source for the presence of nitrogen oxides (NO_x and N_2O) in the atmosphere, whereas N-containing molecules present in aqueous media are mainly originated from agricultural activities. These pollutants are responsible for severe environmental problems such as acid rain, smog, global warming, ozone layer depletion, and drinking water deterioration. Many NO_x-controlling technologies have been developed, but there are still not enough to meet the more stringent regulations to come in the near future (1).

To face these problems, emission control and post-treatment technologies have been implemented and improved constantly. Oxides are key components in some of these technologies and, particularly, in the two main methods used currently for the removal of emission gases: three-way catalysts (TWCs) developed for mobile sources that use gasoline as fuel and the selective catalytic reduction of NO_x with ammonia (SCR), which is applied in the case of emissions from stationary sources such a power plants. Together with these, nowadays other technologies have reached or are reaching maturity to treat NO_x-related environmental problems in the gas and liquid phases. As an incomplete list, we can mention the use of adsorbents or scrubber solutions, selective non-catalytic reduction with aqueous ammonia, catalytic wet

Synthesis, Properties, and Applications of Oxide Nanomaterials, Edited by José A. Rodríguez and Marcos Fernández-Garcia

air oxidation, NO_x-storage and reduction catalysts (NSRs), catalytic reduction of liquid-phase nitrates, or photo-catalysis (2–9). Of course, oxide-based technologies are not the only ones, and for example, the use of biological systems for treating water is now extensively used and actively investigated. Here, we will describe an overview of oxide-based technologies with special emphasis in analyzing the existent information concerning size influence in the performance of these systems. To rationalize the work currently available in the open literature, we have differentiated between processes occurring in the gas and the liquid phase. Sections 19.2 and 19.3 will be devoted to the study of NO_x and N_2O decomposition and reduction technologies in the gas phase, whereas Section 19.4 describes the liquid-phase treatments. To conclude, Section 19.5 introduces other technologies of interest.

19.2. DIRECT DECOMPOSITION

19.2.1. NO_x

As is well known, NO accounts for 95% of all nitrogen oxide emissions mainly resulting during processes of combustion of fossil fuels (10). From a thermodynamic point of view, NO is unstable toward decomposition into molecular nitrogen and oxygen:

$$NO \longrightarrow \tfrac{1}{2}N_2 + \tfrac{1}{2}O_2; \quad \Delta G_f^0 = -86\,\text{kJ mol}^{-1} \qquad \textbf{(19.1)}$$

However, this decomposition reaction presents high activation energy (ca. $350\,\text{kJ mol}^{-1}$) (11,12), with the use of a catalyst required to facilitate the process. Many oxide systems have shown different degrees of ability to catalyse such a reaction (10,13,14). These include simple or mixed transition metal oxides, alkaline-earth oxides, rare-earth oxides, and zeolites exchanged with metals (Co, Ce, Sm, and most particularly, Cu) (10,13–16). An early study by Winter examined different types of metallic oxides for this reaction indicating that its rate depends strongly on the rate of the O_2 desorption step (15). This was observed to also correlate with rates of oxygen exchange, which in both cases is generally related to the strength of the metal–oxygen bond in the oxide (15,16). These arguments are generally employed to explain the higher activities, among simple oxides, of CuO, Co_3O_4, or MnO_2 as well as the significant enhancement observed upon doping of Co_3O_4 with Na^+ cations (18). Concerning the nature of active sites and the reaction mechanism, it was proposed that the reaction involved the reductive chemisorption of two NO molecules on adjacent anion vacancies, reaction of the neighboring chemisorbed molecules to yield N_2, and final desorption of oxygen for recovery of the vacancies active center (15). A similar type of clustered vacancies surface center (more recently directly observed by STM) (19,20) was proposed to be active, in contrast to single vacancy center, for low-temperature NO decomposition (to N_2 or N_2O and with gradual deactivation as a consequence of reoxidation of the anionic vacancies centers) over reduced nanosized cerium oxide (21). Hyponitrite species were proposed as intermediates in the NO decomposition process on the basis of infrared evidence (21). Similar proposals with respect to the nature of the centers active and intermediate involved in the decomposition process have been done in the case of La_2O_3 (22). In agreement with these proposals, modifications of the cerium oxide nanostructure as a consequence of supporting it on alumina (yielding low size ca. 6-nm ceria particles, probably affected by an epitaxial relationship with the underlying γ-Al_2O_3) (23) has been shown to be detrimental to NO reduction (24), in correlation with the strong hindering of formation of clustered vacancies centers in such supported ceria configuration (23).

The NO decomposition reaction over the oxides has been observed to be (as over metals (10)) first order with respect to NO pressure and to be significantly inhibited by the presence of oxygen (15,25). This latter result limits its practical application, although oxygen tolerance can be enhanced by the addition of appropriate promoters, as shown by Hamada in the case of silver-promoted cobalt oxide (26). Active sites composed by a complex of Ag and Co oxides and the low affinity of silver for oxygen are proposed as main points to explain the higher oxygen tolerance (26). Perovskite-type oxides have also been tested for the NO decomposition reaction with the aim of achieving lower inhibition by oxygen; this has been based on the fact that defective perovskite configurations could permit easy desorption of oxygen from the bulk (10,27–31). Indeed, maximum NO decomposition activity is usually achieved for intermediate levels of M^{2+}–M^{3+} substitution, although inhibition by oxygen is typically observed (29,30). Moreover, oxygen inhibition is also generally found for perovskite oxides (10,32). Some progress in this respect has been reported to occur upon weak doping with silver of $La_{0.6}Ce_{0.4}CoO_3$, in which oxygen is shown to promote significantly the decomposition activity (33).

A significant enhancement of the NO decomposition catalytic activity with respect to simple or mixed oxide systems was observed upon use of copper-exchanged zeolites (13,14). In particular, the overexchanged Cu-ZSM-5 form (with MFI structure) displays unique properties in this respect (14), being able to decompose NO at temperatures between 673 and 773 K (34). Different reasons have been pointed out to explain this exceptional activity, including optimized textural, geometrical, and acid–base properties, efficient oxygen handling (in line with previous arguments), and optimum configuration of the active copper sites (13,14,35). Concerning this latter reason, certain controversy still appears to exist concerning the amount of copper cations forming the active site (13,36), although dimeric cationic copper or dispersed copper oxide clusters configurations appear most likely to be involved (37,38).

19.2.2. N_2O

Nitrous oxide (N_2O), also known as laughing gas, has been long considered as a relatively harmless species and has suffered from a lack of interest from scientists, engineers, and politicians. However, during the last decade, a growing concern has been noticed because N_2O is a harmful gas in our atmosphere, contributing to the Greenhouse Effect and ozone layer depletion (39,40). Biological processes in soils and oceans are the primary natural source of N_2O. Human activities are believed to be responsible for half or more of the total annual generation of N_2O. Agriculture, through soil cultivation and the use of nitrogen-fertilizers, contributes to 57% of the total emission in the European Union (EU). Chemical production and the burning of organic material and fossil fuels are also important sources. A rapidly increasing source that also doubled between 1990 and 1998 is the transport sector, as a side-effect of the introduction of the catalytic converter to control NO_x. N_2O emissions that can be reduced in the short term are associated with chemical production and the energy industry (~35% in the EU). Emissions from the chemical industry mainly apply to adipic acid and nitric acid production plants and, in general, processes using nitric acid as oxidizing agent or involving ammonia oxidation.

In the mid-1990s, adipic acid producers fulfilled a voluntary reduction of N_2O emissions (from 600 kton per year in 1994 to <100 kton per year currently) through the development of thermal and catalytic measures for N_2O decomposition (41). Several catalysts have been developed, such as the DuPont and UOP system (ElimiNox) based on a binary oxide supported system ($CoO–NiO/ZrO_2$). At this moment, nitric acid plants have represented the largest single source of N_2O in the chemical industry (400 kton per year). The 75 operative nitric acid plants in the EU-15 emit 130 kton of N_2O per year (40 Mton CO_2-eq), representing 11% of the total greenhouse gas emissions from industry worldwide (42). Extrapolation of de-N_2O technology from adipic acid to nitric acid plants proved ineffectual, and as a consequence, extensive research programs for developing technology in nitric acid production were initiated in the late 1990s. Abatement options in different locations of the process have been investigated, as analyzed in a recent review (43). Catalytic N_2O decomposition below the noble metal gauzes in the ammonia burner (referred to as process-gas decomposition or high-temperature option) and in the tail-gas train at a lower temperature (referred to as tail-gas decomposition) were concluded as the most cost-effective abatement measures, ranging from 0.2 to 1 euro per ton CO_2-eq. Commercial systems for in process-gas N_2O decomposition include supported oxide systems, such as CuO/Al_2O_3 (BASF) and Co_2AlO_4/CeO_2 (Yara International), as well as bulk mixed oxides such as the lanthanum cerium cobalt perovskite ($La_{0.8}Ce_{0.2}CoO_3$, Johnson Matthey). For tail-gas abatement, Uhde is commercializing the EnviNOx process for the combined removal of N_2O and NO_x in a single fixed-bed reactor using iron-containing zeolites.

Many solid catalysts have been reported for direct N_2O decomposition (Reaction **19.2**) and include supported and unsupported metals, pure and mixed oxides, and zeolitic systems (44). However, most of the studies in the literature do not have direct applicability because testing conditions differ strongly as to what is found in industrial N_2O sources, particularly regarding realistic effects by certain gases in the feed. Here we summarize the work with oxide-based systems attending to the scope of the book.

$$N_2O \longrightarrow N_2 + \tfrac{1}{2}O_2;\ \ \Delta H_f^0 = -82\,\text{kJ mol}^{-1} \qquad \textbf{(19.2)}$$

Pure oxides for N_2O decomposition have been reviewed elsewhere (44–48). The highest activities are exhibited by the oxides of the transition metals of group VIII

(Rh, Ir, Co, Fe, Ni), by CuO and by some rare-earth oxides (La_2O_3) (45,47,49,50). High activities per unit surface area are also claimed for CaO, SrO, V_2O_3, and HfO_2 (46,47,51). Moderate activities are found for elements of groups III–VII (Mn, Ce, Th, Sn, Cr) and of group II (Mg, Zn, Cd). Figure 19.1 provides a comparison of specific activities (per unit surface area) of various oxides. The valence of an element is also important. For manganese, which can have various oxidation states, the specific activity order was MnO < MnO_2 < Mn_3O_4 < Mn_2O_3 (52), so 3+ appears as the optimal oxidation state. For vanadium, V_2O_3 is much more active than the nearly inactive V_2O_5 (53). It should be mentioned that depending on the experimental conditions, some oxides are not stable and are partially converted, like MnO_2, MnO (52), Cu_2O (54), and CoO (55). The apparent activation energies range between 80 and 170 kJ mol^{-1}. The rate is usually proportional to the partial N_2O pressure or has a slightly lower order due to the inhibition of produced oxygen. The order in partial O_2 pressure for strong inhibition amounts to −0.5. Some oxides seem not to be affected by the presence of oxygen, which holds for Ca, Sr, La, Ce, Zn, and Hf (48).

Much work has been done on *mixed oxide systems*, like doped oxides or solid solutions, spinels, and perovskites, not only for the N_2O decomposition reaction as such, but also predominantly for a better mechanistic understanding of catalytic phenomena over oxidic transition metal (TM) in general (56–63). The latter are of high fundamental value and provide guidelines for the recent development of industrial catalysts. Cimino, Stone, and others systematically studied the effect of various transition metal ion concentrations in relatively inert oxide matrices like MgO, Al_2O_3, and $MgAl_2O_4$ for N_2O decomposition. The catalytic activity already develops strongly in very dilute solutions (<1 TM ion per 100 cations), where the activity per TM ion

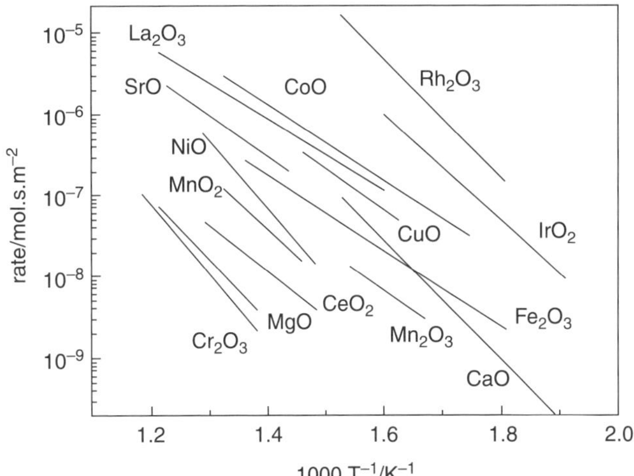

Figure 19.1. N_2O decomposition rates of selected pure oxides, illustrated as Arrhenius plots calculated for 10 kPa N_2O and 0.1 kPa O_2 pressure (46,47,52,53). Reprinted with permission from Elsevier.

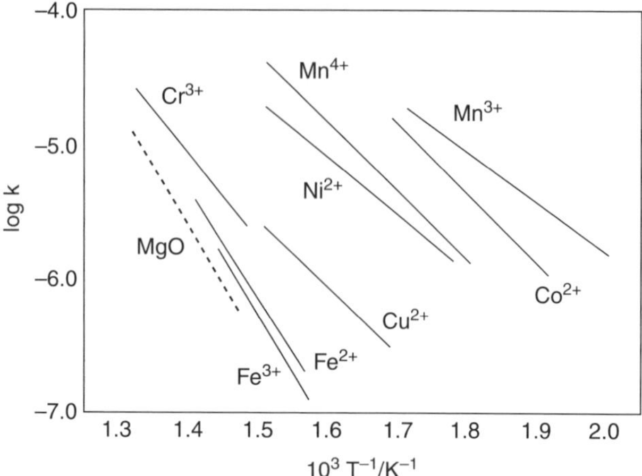

Figure 19.2. Relative activities of 3*d* TM ion solutions in MgO (1 at.-%), illustrated as Arrhenius plots (44,56). Reprinted with permission from Elsevier.

is the highest at the lowest dilution. For example, Figure 19.2 compares the activity of several transition metals at 1 atomic% concentration in MgO.

The TM ions act specifically; i.e., the activity of different oxidation states varies widely. An example of the former is the study of Ciminio and Indovina (64), who demonstrated that Mn^{3+} ions dispersed in a MgO matrix had the most active oxidation state compared with Mn^{2+} and Mn^{4+}, in agreement with the results for the pure oxides (52). The activation energies of the higher concentrations TM resemble those of the pure oxides of the same oxidation state; this unequivocally shows that a size effect influences chemical activity. For Ni, Cr, and Co ions in MgO, the activation energy decreases with increasing dilution, which is ascribed to a weaker bonding of the adsorbed oxygen (56,57,59,62,65). Although molecular oxygen can adsorb on these systems and exhibits an inhibiting effect, this adsorption is much less for the diluted system. Cr has been studied further in magnesium aluminate and α-alumina, with similar results as for magnesia. Co-aluminate has been reported as the most active system of the pure TM aluminates (66).

Perovskites and other mixed oxides (e.g., spinels, pyrochlores) are of a similar interest as the solid solution systems. Perovskites can be represented by the general formula ABO_3 (67,68). Also, double (and multiple) structures exist, as indicated by $AA'BB'O_6$. The A ion is generally the larger one, often a rare-earth element, mostly La, and catalytically relatively inactive, whereas B is mostly a 3*d*-transition metal element like Cu, Cr, Fe, Co, Ni, Mn, and Ti, mostly responsible for the catalytic activity, although the A ion influences this. Partial substitution of the A and/or B ion with an ion of another valence gives rise to abnormal valencies of the B-site cation (e.g., Cu^{3+} and Ni^{3+}) and/or to oxygen vacancies. These structures can be represented by $A'_xA_{1-x}B'_yB_{1-y}O_{3-\lambda}$, where λ is the oxygen vacancy to compensate for the cation charges. Sr and other alkaline-earth elements and Ce are often applied for A-site substitution, whereas B-site substitution is studied to a lesser extent.

Thus, a wide variety of structures can be formed, explaining the large amount of studies on these materials. Perovskites are known for their structural and thermal stability, due to their high-temperature preparation, resulting in low specific surface areas ($<10\,m^2/g$). The wide compositional variety control over the oxygen defects and valence of metal ions makes them suitable model compounds to study the relation between solid-state chemistry and catalytic activity, including the N_2O decomposition reaction (67–70).

Despite their low specific surface area, the reactivities of these catalysts is relatively high. Upon variation of the B-site in $LaMO_3$ (M = Co, Ni, Mn, Fe, Cr) (67,70), good correlations were found between the activation energy for the decomposition reaction (35–130 kJ/mol), and the oxygen binding bond strength plays an important role. The activity order is Co > Ni, Cu > Fe > Mn > Cr (Figure 19.3). The oxide matrix has a considerable effect on the activity of the TM ions as is apparent by comparing Figures 19.1–19.3. The effect of oxygen varies considerably. The reaction at 6.6-kPa N_2O over Fe is not, over Ni weakly and over Co-systems strongly inhibited by oxygen (70). The reaction is first order in partial N_2O pressure and −0.5 in partial O_2 pressure for strong inhibition.

Variation of the A-site ion in $MMnO_3$ (M = La, Nd, Sm, Gd) resulted in a decreasing activation energy from 105 to 30 kJ/mol, which is also explained by an increased electron density on the Mn site, resulting in a facilitated desorption of oxygen (67). For the related structure M_2CuO_4 (M = La, Pr, Nd, Sm and Gd), the activation energies ranged nonsystematically between 45 and 85 kJ/mol at 6.6 kPa, whereas at 26 kPa, the values dropped to between 45 and 85 kJ/mol. Oxygen inhibited strongly.

Partial substitution of the three-valent La by the two-valent Sr in La_2CuO_4 induces an increasing average Cu valence oxidation state (up to ca. +2.3) and the presence

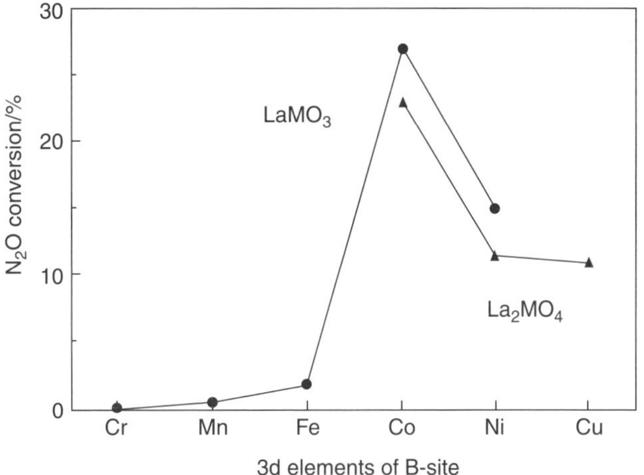

Figure 19.3. Comparison of activities for N_2O decomposition of mixed oxides of the type $LaMO_3$ and La_2MO_4. Conditions 723 K, 0.1 kPa N_2O, and $W/F_{N2O} = 7200$ g s/mmol (44,70). Reprinted with permission from Elsevier.

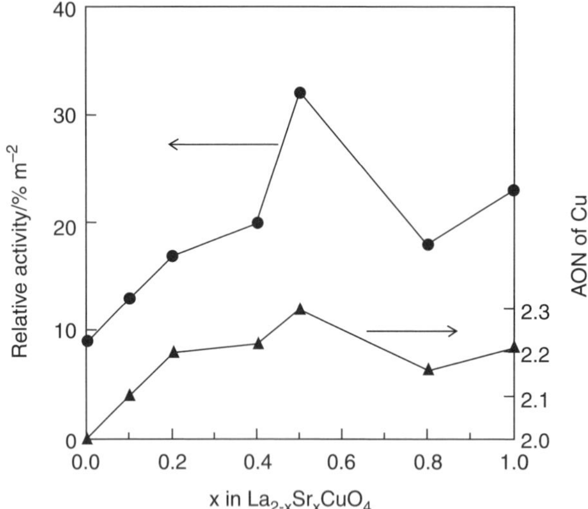

Figure 19.4. N_2O conversion, normalized per unit surface area, over La_2CuO_4 doped with Sr as a function of the fraction Sr for reaction at 723 K, 0.1 kPa N_2O, and $W/F_{N_2O} = 7200$ g s/mmol. Included is the average oxidation number (AON) of copper (44,70). Reprinted with permission from Elsevier.

of oxygen defects (70–72). The highest activity is obtained at an optimal fraction of Sr of 0.5, where the cooper has its highest average oxidation state (Figure 19.4). No correlation with the oxygen vacancy concentration is found, which is apparently not the activity-determining factor. The oxygen inhibition is the strongest at low pressures, but at higher partial O_2 pressure(>1 kPa), the rate does not decrease further, like that observed for Pt catalysts.

Very active mixed-oxide catalysts were reported, and prepared from thermal decomposition (ca. 700 K) of transition metal (Co, Cu, Ni, Rh, Ru, Pd, La) containing hydrotalcites, belonging to the class of clay minerals. Some of them are even more active than the zeolitic catalysts (73–75). A typical reaction order found is Co-Rh > Co-La > Co-Mg > Co-ZSM-5. Conversion of N_2O already occurs below 500 K. The apparent activation energies amount to 45–55 kJ/mol, and the reaction is first order in partial N_2O pressure. The Co-containing calcined samples exhibit a sustained life at temperatures above 900 K in a wet and oxygen containing atmosphere with 10 vol.% N_2O (74), keeping promises for practical application.

19.3. NO$_X$ REDUCTION REACTIONS

Reduction reactions of nitrogen oxides and the corresponding catalytic systems are somewhat unspecific in respect to the nitrogen-to-oxygen ratio of the target molecule, and current technologies deal simultaneously with NO and N_2O molecules in the so-called stoichiometric (equal amount of reductant and oxidant molecules), reducing or fuel-rich gas mixtures, whereas NO and NO_2 appear simultaneously in oxygen-rich gas mixtures (3–6). Although the exact conditions under which these molecules are eliminated differ from one case to another in terms of the reaction temperature and contact time, catalyst formulation and other important variables, a key characteristic of the reaction is the nature of the reductant molecule used to reduce nitrogen oxides and to obtain (*vide infra*) N_2. For this reason, this section is divided into subsections that deal with the most important results concerning the use of CO, hydrocarbons, hydrogen, and ammonia.

19.3.1. CO

The reduction of NO by CO ($NO + CO \rightarrow \frac{1}{2}N_2 + CO_2$) is one of the most important reactions taking place in automotive converters. Although such a reaction is mainly catalysed by highly active dispersed noble metals in modern automotive converters (5,14), base metal-oxide catalysts were among the first investigated for such process, whereas perovskite oxides also display an ability for the process (14,31). Early work by Shelef et al. revealed distinct activities and selectivities for various supported transition metal oxides, as summarized in Table 19.1 (76,77). These were mainly explained on the basis of the abilities of the different oxides to handle NO in the general context of redox-type mechanisms (NO reoxidation of active sites, proposed as a rate determining step, or facility to accommodate NO molecules in neighboring sites) (76). The results reveal a general increase of the selectivity to N_2 formation with reaction temperature, although selectivity differences between the catalysts are apparent. In this respect, N_2O was proposed as a gas-phase intermediate in the reaction on the basis of parallel tests for the N_2O–CO reaction and kinetic measurements, whereas various intermediates are proposed to explain the catalytic features observed (14,77). Nanosized Co_3O_4 (average size of 26 nm) has also been recently tested in the CO–NO reaction (78). Overall, a 50% NO conversion is observed in this case at

TABLE 19.1. Catalytic Parameters of Various Metal-Oxide Catalysts Supported on 95% Alumina-5% Silica for the CO–NO Reaction (1.2 mol% CO and 2.0 mol% NO).

Catalyst	50% CO Isoconversion Temperature (K)	Temperature of Maximum N_2O Yield (K)	N_2O Concentration at the Maximum (mol %)
Fe_2O_3	418	473	0.84
$CuCr_2O_4$	428	453	0.36
Cu_2O	448	493	0.28
Cr_2O_3	493	513	0.62
NiO	523	573	0.68
Co_3O_4	623	623	0.12
Al_2O_3–SiO_2 (support)	698	733	0.34
MnO	708	–	0.00
V_2O_5	833	–	0.00

383 K (and full conversion below 573 K); note, however that experimental conditions are different than those employed by Shelef et al. (Table 19.1), and therefore, direct comparison is not possible. The reaction has been shown to take place simultaneously with a reduction of Co_3O_4 into CoO. A redox mechanism between Co^{3+} and Co^{2+} ions based on NO decomposition was proposed.

An aspect of practical interest with respect to employment of CO as reductant concerns the influence of having oxygen in the reactant mixture (i.e., the respective ability of O_2 and NO for catalytically oxidizing CO). In this respect, the main focus of the work by Shelef et al. was put on analyzing differences between the oxides with respect to catalytic performances for individual reactions CO–NO versus CO–O_2 (76). It was interesting in this respect that most oxides displayed higher activity for the latter reaction, with Fe_2O_3 and Cr_2O_3 being the only exceptions among the examined systems. Nevertheless, analysis of Cr_2O_3 in a reactant mixture containing the three components revealed a significant inhibition of NO reduction in the presence of O_2 (76). Similar results have been observed in the case of highly dispersed copper oxide catalysts supported on nanostructured ceria (or a structurally related Ce–Zr mixed oxide) (79). Basic aspects pointed out as most relevant to explain such results have been related to the facility to achieve partially reduced states in the dispersed CuO phase, in agreement with proposals for other supported-CuO systems (14), and the particular redox properties of the CuO–nanoceria interface (79). In this respect, considering operation within a redox-type mechanism, oxygen is demonstrated to be a stronger oxidant of both copper and ceria interfacial components than NO, thus explaining the difficulties to reduce NO in the presence of oxygen in the reactant mixture (79).

On the other hand, the relevancy of generating the adequate configuration of surface defects for NO reduction is illustrated by differences in the catalytic activities of two different nanostructured configurations of ceria under CO–O_2–NO mixtures (24). The considerably higher activity for NO reduction observed over a bulk CeO_2 sample (with ca. 15-nm average size on the basis of analysis of the X-ray diffractogram (23)) in comparison with an Al_2O_3-supported CeO_2 sample (in which average ceria size is of ca. 6 nm, according to TEM and XRD (23); consider also that alumina does not apparently participate in the catalytic activity (24)) has been explained on the basis of the higher facility for generating associated vacancies centers (active for decomposing NO, as mentioned (21)) in the former with respect to the latter (24). It must, however, be noted that this must not be analyzed merely as a particle size effect because such active configuration of surface defects can also appear in CeO_2 particles of sizes lower than 6 nm (23). Morphological changes as a consequence of an epitaxial relationship with the underlying alumina appear thus to play an important role on specific activities of the ceria nanoparticles.

19.3.2. Hydrocarbons

The employment of hydrocarbons (HC's) as reductants of NO_x is of high relevancy from an environmental point of view because both types of pollutants usually appear associated in the waste of processes involving combustion of fossil fuels. In this respect, the different situations can be classified as a function of the chemical nature and/or source of the hydrocarbon involved. Methane is usually considered a useful reductant for emissions produced in stationary power plants using natural gas as fuel, whereas it is the most chemically stable hydrocarbon. In turn, higher hydrocarbons are more reactive and considered useful reductants of NO in mobile sources, in accordance with the nature of the exhaust gases or the type of fuel employed (80). In any of the cases, the catalytic requirements for the materials to be employed in each case are basically imposed by the composition of the gas mixture. Thus, for emissions produced as a consequence of stoichiometric combustion conditions (giving rise to redox balanced stoichiometric emission), three-way catalysts are employed to deal with NO–HC reactions, in which active sites are essentially related to noble metals (with nanostructured ceria-related oxides acting as the activity promoter) (5,80). However, oxide systems can play a more relevant role for treatment of lean-burn emissions in which NO_x can strongly compete with oxygen for the hydrocarbon reductant according to a process in which selective catalytic reduction (SCR) of the NO_x (competing with hydrocarbon combustion) is desirable:

reductants	oxidants		main products
	$\mathbf{NO_x(N + NO_2)}$		$\mathbf{N_2 + CO_2 + H_2O}$
HC's +	O_2	$\xrightarrow{\text{SCR catalyst}}$	
	etc.		(undesired) subproducts
			$N_2O, CO, C_xH_yO_z$, HCN, etc.

Taking these aspects into account, this section will be essentially dedicated to analysis of oxide materials for such SCR processes.

19.3.2.1. CH_4 Several oxidic systems have displayed interesting properties as catalysts for NO_x-SCR with methane. In general terms, these can be classified into two groups: metal-exchanged zeolites and simple (promoted or not) oxides. Among the former, unique performance has been revealed by Co-exchanged zeolites, most particularly using topologies such as those in Co-ZSM-5 and Co-ferrierite (81,82). These (along with other metal-exchanged zeolites, among which Ga- and In-exchanged are probably most active) apparently present important advantages with respect to Cu-exchanged zeolites, active for other deNO_x processes as shown throughout this chapter, in terms of oxygen tolerance (81). This has been related to the respective activities for methane combustion as well as to differences in the way Cu^{2+} and Co^{2+} comparatively activate NO (83). Highly dispersed Co^{2+} ions exchanged in the zeolitic matrix (undergoing changes within the Co^{2+}/Co^{3+} redox couple) are considered to be the active sites for the reaction (82,84); in turn, NO seems to be activated through formation of an adsorbed intermediate (of the nitrate form) in whose formation zeolitic protons can also have a role while NO oxidation (to NO_2) can also compete with the SCR process (84).

The main drawback of these zeolitic systems is that the activity of the catalysts is significantly inhibited by steam (81), although this effect can be limited, for instance,

TABLE 19.2. Summary of Catalytic Properties of Various Oxide Systems Under CH_4–NO–O_2 Mixture at 773 K (16).

	Rate of N_2 Formation (μ mol s^{-1} $\times 10^3$)		
Catalyst	Per Gram	Per m^2	TOF (s^{-1} $\times 10^3$)
La_2O_3	30	8.5	2.2
Sr/La_2O_3	22	13	2.3
CeO_2	0.75	0.30	0.17
Nd_2O_3	23	6.7	2.54
Sm_2O_3	25	3.5	0.77
Sr/Sm_2O_3	21	3.2	1.0
Tm_2O_3	2.7	2.9	0.49
Lu_2O_3	18	7.3	1.4
Li/MgO	0.84	0.24	0.32
Co-ZSM-5	1280	6.6	2.3

by mixing the zeolite catalysts with metal oxides such as manganese oxide (85). However, the zeolitic systems are hydrothermally unstable due to dealumination and loss of the active phase (36). This has motivated the search for new catalysts, among which simple (or promoted) oxide systems have shown interesting properties, although in general, their activity window appears at a higher temperature than for the zeolitic systems (81). As an example, Ga_2O_3/Al_2O_3 has been claimed to be more active and selective than Ga-ZSM-5 and Co-ZSM-5 at high temperature as a consequence of its lower hydrothermal degradation (86). On the other hand, the concept of favoring CH_4 activation has led Vannice et al. to test catalysts known as active in methane oxidative coupling (16,87,88). Comparison of a series of these oxide systems with Co-ZSM-5 is shown in Table 19.2, in which it can be appreciated that some oxide systems are comparable with Co-ZSM-5 on a turnover frequency (TOF) basis. They also have the advantage of exhibiting no activity bendover at high temperatures. However, their low surface area is a serious drawback, and development of stable nanostructured systems appears interesting in this respect. Indeed, dispersing La_2O_3 on alumina (and promoting with Sr) has been shown to improve significantly the activity not only on a weight but also on a TOF basis, suggesting new interesting properties for the nanostructured configurations of these systems (89). Other oxides like CaO (promoted with La) or perovskite systems have also shown interesting properties for this reaction (90,91).

19.3.2.2. Higher Hydrocarbons Following the pioneering work by Iwamoto et al. as well as that by Held et al., which showed the activity of Cu/ZSM-5 for the process (13,92), many systems have been tested for the NO_x-SCR with higher hydrocarbons, taking into account the possibilities of the process to treat diesel engine lean-burn emissions (5). A classification of these systems can be established on the basis of their catalytic behavior as platinum group metal and base metal-oxide systems, whereas systems of metal exchanged on microporous materials (mainly zeolites) apparently provide optimum configuration of the catalytic components although, as mentioned in the former section, their thermal stability is usually low, which limits their practical application (13,93). Platinum group metals catalysts (in which active sites are essentially believed to be constituted by zero-valent states of the metals) (5,92) are the most active for the process, although they have the drawback of forming generally substantial amounts of N_2O (a greenhouse gas—about 200 more powerful than CO_2−) and are ineffective at high temperature (5).

Base metal-oxide systems are the least active (they only show activity at relatively high temperature), although they are usually fully selective to N_2 and can present improved thermal stability (5). In general terms, the extent of the selective reduction of NO_x depends on a series of factors like the nature of the metal oxide or of the hydrocarbon (or another co-reductant or gaseous promoter like H_2, CO, etc.) employed as reductant, the nature of the support employed to disperse the metal oxide, as well as preparation parameters (metal precursor employed, type of pretreatment, etc.) (92). More or less dispersed oxides of Co, Ni, Cu, Fe, Sn, Ga, Au, In, and Ag have been tested for this process employing mainly Al_2O_3, TiO_2, and ZrO_2 to support them (13,92). Among them, probably the most promising results have been obtained for the silver system supported on Al_2O_3 (5,92). The main catalytic properties of this system are illustrated in Figure 19.5 in which the catalytic performance of systems differing in the silver loadings is shown (94). Optimum performance is usually achieved for low or intermediate silver loadings. This has been related to the unique properties of the highly dispersed Ag^+ cations that appear in a tetrahedral-like local coordination closely resembling that of β-$AgAlO_2$ (95). Subtle differences with respect to this aluminate structure were related to XAFS observation of a significant local disorder around Ag centers as well as a noticeable reduction of the average Ag–O first-shell coordination distance with respect to those present in the β-$AgAlO_2$ material. For higher loadings, segregation of Ag_2O entities (and/or further surface silver reduction) could occur upon contact with the reducing agent at reaction temperatures. Such highly loaded systems then become much more active for hydrocarbon oxidation (Figure 19.5) or even for N_2O formation (5,92,94,95).

Concerning the reaction mechanism, it appears that both the NO and the hydrocarbon require a first step of activation through formation of chemisorbed nitrates and oxygenated hydrocarbon species (acrylate or acetate, in the case of propene) in the case of silver or tin oxide systems (5,94,96), followed by generation of nitrogen-containing hydrocarbon intermediates that could finally decompose to the reaction products via intermediates like cyanide or isocyanate with influence of gas-phase reaction determining N-selectivity (92,94,97). Other hypotheses, like NO preactivation to NO_2, have been proposed for supported gold systems in which significant promotion of the SCR activity is observed upon addition of Mn_2O_3 (98). Acid–base properties of the support are also proposed to play a relevant role on these activating steps (92,94,95).

Concerning the composition of the reactant mixture in terms of the type of hydrocarbon employed, the effectiveness of the HC in the SCR of NO_x during the competition with the burning reaction with oxygen (see Scheme in Section 19.3.2) increases with the HC molecular weight; this is probably due to the parallel increase in the heat of adsorption and the decrease of the C–H bond strength (99). It is

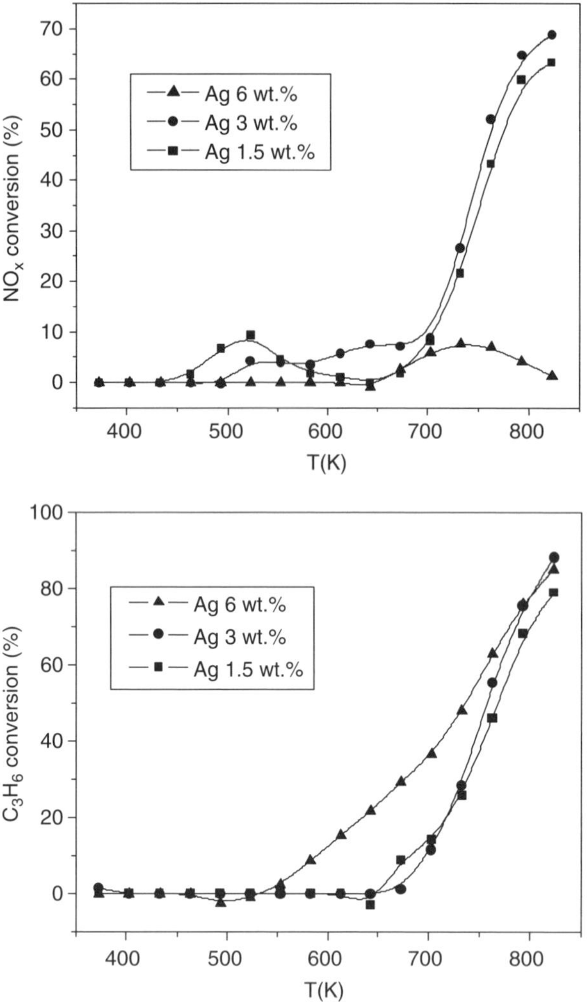

Figure 19.5. Conversion of NO$_x$ and propene obtained during tests under 0.1% NO, 0.1% C_3H_6, 5% O_2 and 3% H_2O (N_2 balance) at a fixed space velocity of 4×10^4 h^{-1} over different alumina-supported silver catalysts (95).

interesting that in the case of employment of higher hydrocarbons, the possibility of operation under a mechanism combining heterogeneous and surface-mediated homogeneous radical reactions has also been pointed out (100). When comparing between alkanes, alkenes, and oxygenated hydrocarbons with the same number of carbons, NO$_x$ reduction is typically most difficult with the former and easiest with the latter (92). Another interesting approach to enhance NO$_x$-SCR activity consists in adding a gaseous promoter like H_2 or CO to the reactant mixture (92). Thus, in the case of alumina-supported silver catalysts, an important increase of the NO$_x$ conversion is observed upon H_2 addition (101). The absence of such promotion upon CO addition and the similarity with H_2 concerning the effects induced on active silver species (clustering) suggests that hydrogen could have a direct chemical effect on the reaction mechanism (101).

19.3.3. H_2

Employment of hydrogen as a NO reductant ($NO + H_2 \rightarrow \frac{1}{2}N_2 + H_2O$) is interesting from the point of view of exhaust treatment, given the formation of hydrogen (usually associated with CO as a consequence of incomplete hydrocarbon combustion) during fossil fuel combustion. N_2O and NH_3 can also be formed as subproducts of such a reaction (14). Different oxide materials (base oxides, perovskites, metal exchanged zeolites) have been tested for this reaction (14). Shelef and Gandhi first investigated the reduction of NO with hydrogen on copper chromite catalysts (102). On these oxides, NH_3 is one of the main products of the reaction in a wide range of temperatures, although the product distribution strongly depends on the inlet concentration: generally speaking, the selectivity to NH_3 decreases monotonically with the NO concentration, whereas the NH_3 yield reaches a maximum at intermediate inlet concentrations. Similar results have been obtained with other oxides (14). The influence of water on the reactant mixture appears also highly relevant, apparently inducing the opposite effects related to inhibition of NO reduction while favoring NH_3 formation (14,102). Acke and Panas have examined the positive influence of Na and Fe impurities present in CaO on the catalytic activity, which are attributed to modifications in the Arrhenius pre-exponential factor rather than to changes in the apparent activation energy within a reaction mechanism basically involving two redox steps with surface oxygen abstraction and reoxidation; such difference is proposed to be due to the presence of defects (like O^-) induced by the impurities (103). The same work examines the activity for this reaction over a series of alkaline-earth simple oxides observing a linear increase in the apparent activation energy with the lattice parameter (103). A positive effect of doping an oxide for this reaction has also been reported in the case of Fe-doped ZrO_2; the spillover of hydrogen species (as hydroxyls) on the support is proposed to play an important role on the activity enhancement (104).

The activity of $La_{0.9}A_{0.1}BO_{3+\delta}$ (A = Ce, Eu; B = Mn, Co) and $La_{0.8}Sr_{0.2}BO_{3+\delta}$ (B = Mn, Fe, Co, Ni) perovskites for this reaction was analyzed by Ferri et al. (105). They found Co as the most active among cations in the B position, which is attributed to the presence of oxygen vacancies suitable for NO adsorption along with the lower oxidation state of the B-site metal cation within a redox type mechanism. Substitution at A-site with a bivalent (Eu) versus tetravalent (Ce) metal cation also influences the catalytic activity observing enhanced activity for cobaltates, when Ce is substituted for La, and for manganites when Eu is introduced, which is attributed to differences in the type of defects present in each case. In the $La_{0.8}Sr_{0.2}BO_3$ series, the stability of the B ionic species seems the key factor for the order of activity of the catalysts.

19.3.4. Ammonia

The selective catalytic reduction (SCR) process is based on the following reactions:

$$4NH_3 + 4NO + O_2 \longrightarrow 4N_2 + 6H_2O \quad (\mathbf{19.3})$$

$$2NH_3 + NO + NO_2 \longrightarrow 2N_2 + 3H_2O \quad (\mathbf{19.4})$$

$$8NH_3 + 6NO_2 \longrightarrow 7N_2 + 12H_2O \quad (\mathbf{19.5})$$

Reaction **19.3** proceeds very rapidly on a catalyst at 523 > T > 723 K in excess oxygen for the overall stoichiometry of the process ($NH_3/NO = 1/1$) because NO_x typically consist in more than 90% NO on stationary plan gas exhausts. The selective term refers to the ability of ammonia to react selectively with NO_x instead of being oxidized by oxygen to form N_2, N_2O, and NO. Mechanistic details of this reaction can be found in Ref. 106. Common applications of this technology are coal-, oil-, and gas-fired power plants and cogeneration plants, but also it has been applied to industrial and municipal waste incinerators, and in the glass, steel, and cement industries. Conventional SCR catalysts are employed in the form of honeycomb monoliths, plates, and coated metal monolights (106–108).

In addition to Reactions **19.3** to **19.5**, potential side reactions, depending on the catalyst and operational variables, impact the working conditions of SCR and downstream units. At low temperature, nitrite/nitrate salt formation by the $NO_2 + NH_3$ reaction may generate explosion risk problems. Nitrous oxide (N_2O) may be formed in specific conditions, particularly in the absence of moisture. When sulfur is present in the fuel of power plants, SO_2 and minor percentages of SO_3 can be formed. SO_2 can be oxidized to SO_3 on the catalyst with further reaction with ammonia and generation of sulfuric acid and sulfates (NH_4HSO_4; $(NH_4)_2SO_4$) in the presence of water. Sulfates can be deposited and accumulated on the catalysts and other parts, affecting performance. Additional poisons are Na, K, and As coming from the fuel and/or lubricating oils in coal-fired applications, whereas in municipal solid waste treatment, alkaline metals and halogenidric acids can be detected (107–109).

Three types of catalysts have been developed for SCR application and mainly differentiated by the temperature range of operation. At low temperature, 432–573 K, Pt-based systems were originally used as they combine high NO reduction and CO oxidation activity, although their replacement by metal (Ni, Cu, etc.) oxides has been implemented in some cases due to problems related to the noble metal high activity in ammonia oxidation (108). More recently, the use of activated carbon or carbon fibers promoted with transition (Mn, Ni) metal oxides/hydroxides has been extensively studied (110). Concerning the oxide role on the SCR reaction, lack of size-related information can be noticed, although NO usually contributes to the dispersion or limits aggregation of the active metal under the specific SCR reaction conditions, whereas other reactants, like CO, may volatilize the metal by forming carbonyl entities, eliminating them from the solid catalyst over an extended period of operation. This potential dynamic behavior of the system under reaction conditions makes it difficult to work out size-dependence phenomena (106). In the case of carbon-supported systems, the selectivity to N_2 and concomitant presence of N_2O as a product seem connected with the crystalline size of MO_x entities, with smaller sizes of the active phase particles producing less N_2O (110).

At medium temperatures, 523–723 K, the VO_x/TiO_2 system is widely used. Among hundreds of catalyst formulations investigated, those based in vanadia supported on titania in the anatase form, and promoted with tungsta and molybdena, were found to present genuine advantages in NO reduction and SO_2 oxidation reactions (106–108). The use of V as the active element and its superior activity and stability were well established by the 1970s, whereas the choice of TiO_2 anatase relies on the fact that the oxide is only weakly and reversibly sulfated in presence of $SO_2 + O_2$ while it also can accomplish the role of a promoter of catalytic activity. Vanadia is responsible for the activity in NO reduction but also for the undesired SO_2 oxidation. Accordingly, V content is generally kept low and below 1 wt% in the presence of high SO_2 concentration streams, with the maximum loading value close to 6 wt%, a value near the monolayer surface coverage. WO_x and MoO_x, approximately 5–10 wt%, are employed to increase acidity, activity, and thermal stability besides limiting SO_2 oxidation. Additionally, MoO_x prevents deactivation by As. Finally, silico-aluminates and fiberglass are used as ceramic additives to improve mechanical properties. Novel formulations, not deactivated by SO_2, are based on VO_x/carbon (111).

The structural and electronic properties of V species have been reviewed in several dedicated publications (112–114). As detailed below, the capital chemical importance of the V oxide nature and structural and electronic properties evolves from the fact that oxygen handling in the reaction involves V species. Bulk V_2O_5 posseses platelet morphology, whereas supported V species are strongly dependent on the support nature coverage and hydration state (115). V-supported species usually anchor the oxide substrate by preferentially interacting with the basic hydroxyls. Below monolayer surface coverage, isolated $[VO_4]$ units and/or polymeric, raft-like species are observed in all systems, using supports such as TiO_2, CeO_2, ZrO_2, and Al_2O_3, except in the SiO_2 case. Independently of the structure, V is in a V(V) dominant state under oxidant gas mixtures and V(IV) and less significantly V(III) only get a concentration about traces under the influence of net-reducing gas atmospheres. These isolated/polymeric species appear in real VO_x/TiO_2 catalysts, and obviously, their chemistry is well differentiated from a bulk phase. Both contain two oxygen species, one shared with the support phase and other corresponding to a $V = O$ double bond. For polymeric entities, additional V–O–V species are present. Differences between the chemical behavior of such species can be expected as activity may be strongly influenced by acidity properties at high temperature, but redox properties may be key variables at low temperature, so oxygen lability and vacancy replacement are likely key steps in the reaction scheme (106,108,115). On the other hand, above monolayer surface coverage, V_2O_5 crystallites are readily detected.

A somewhat similar structural behavior in terms of species occurring below monolayer surface coverage is observed with W and Mo oxide entities. The main differences concern the existence of variable quantities of tetrahedral-like $[MO_4]$ and distorted octahedral-like $[MO_x]$ ($x = 5,6$) as isolated species and a more complex linkage, from a structural point of view, of M–O–M units. Both cations have a single double M=O bond in all isolated and polymeric species (116,117).

Unfortunately, the influence of the support size is still to be investigated as this subject has only been attempted by using $VO_x/TiO_2/SiO_2$ systems where, under the appropriate preparation conditions, V selectively interacts with titania (118,119). The titania size is determined by varying the corresponding loading. Results do not, however, allow us to extract conclusive information as, as said before, the uncoupling of the size and the TiO_2–SiO_2 interaction effect is difficult to perform on experimental observables.

Finally, above 673 K and using a maximum temperature of 873 K, zeolite-based systems have been used (107,108). Zeolites in the acid form and exchanged with transition metal ions (e.g., Fe) can be used at temperatures where previous systems lack thermal stability. The thermal stability of the zeolite system is increased, and its tendency to sulfidation is diminished by increasing the Si/Al ratio through the removal of aluminium cations from the crystal matrix. The Fe cationic species present in a zeolite and particularly on ZSM-5 are strongly dependent on the preparation method, but below or nearly complete exchange, mono- and dimeric (oligomeric) species are usually obtained (120,121). The reactants interaction with these species is completely different from the single oxide; such differences are mainly detected in the unselective oxidation of ammonia, which is favored in the presence of FeO_x aggregates and, maybe, oligomeric species, with respect to isolated species (122).

19.4. LIQUID PHASE DeNO$_X$

Groundwater pollution by harmful organic compounds is a growing concern in advanced societies. Although current liquid-phase depollution technologies may not include broadly applied and important DeNO$_x$ technologies, we may give here a brief overview to liquid-phase treatments closely related to gas-phase DeNO$_x$ processes but carried out in the liquid phase. These are the elimination of nitrates in drinking water and of NO-containing organic molecules using catalytic wet air oxidation processes.

Nitrates are one of the most problematic and widespread of the vast number of groundwater contaminants. Although their elimination by using separation (ion exchange, reverse osmosis, electrodialysis) and destruction (electrochemical reduction, biological denitrification) technologies is well developed, there is still concern in terms of increasing the treating capacity, reducing the time of treatment, and favoring the disposal of remainings (123). Recently, the development of the selective heterogeneous catalytic hydrogenation to N_2 using hydrogen or formic acid as reducing agents has been shown as a potential alternative. After the work of Volpov et al. (124), the combination of noble metals (Pd, Pt) with transition ones (Cu, Ni, Sn, In) has displayed promising activity and selectivity to nitrogen (125–132). Warna et al. (133) proposed a kinetic model for catalytic nitrate reduction essentially accepted by the community and which has been completed by many others authors in subsequent contributions (127,132). The promoting transition metal seems responsible for nitrate reduction to nitrite by a redox reaction, whereas the main function of the noble metal could be to regenerate the metallic state of the active phase by means of activated (chemisorbed) hydrogen and to further reduce nitrite to nitrogen. Ammonia is detected as a byproduct evolving from over-reduction reactions. It has been shown that the kinetics and the selectivity to nitrogen of the process are severely limited by diffusion (134,135), and several strategies have been devoted to overcoming this problem by operating under poor surface hydrogen concentration with the help of membranes (135) or by taking advantage of a reversible electrostatic interaction with layered supports (128).

Concerning the contribution of oxide components to this reaction, most interesting papers are related to the use of reducible oxides like SnO_2 (134,136), or TiO_2 (137). It would appear that the metal–support interaction developed in the pre-reduction treatment of the catalyst may produce the existence of oxygen vacancies and help in the electron flow control, influencing in this way the selectivity of the reduction process and allowing the use of monometallic (Pd, Pt) systems with improved properties (selectivity to N_2) with respect to bimetallic systems supported on alumina, carbon, or other typical supports. In the case of Pd, this is strongly related to the behavior of the Pd beta-hydride phase formed in the presence of hydrogen and the control of the yield of over-reduced products. It can be envisaged that methods that control the existence of oxygen vacancies may influence the reaction by affecting hydrogen–hydride chemical activity through metal–support interactions (137). Size and doping are therefore variables that await a careful scanning by interested researchers. A last contribution to the reduction of nitrates is the use of Ag/TiO_2 photo-catalytic systems, which in the presence of adequate sacrificial hole scavengers display promising activity (138).

The *catalytic wet air oxidation* process is typically used to treat toxic and/or non-biodegradable organic pollutants. In the context of this chapter, it is used to eliminate NO-containing molecules (139) or N_2O (140), but the main problems are related to the stability of heterogenous metal-supported catalysts by the leaching of active components in the reaction media. The wet oxidation key steps seem to involve oxygen activation and handling and the direct electron transfer (redox reaction) to the NO-containing molecule; both steps are, as illustrated in other chapters, size-dependent but to our knowledge have not been explored to date.

19.5. OTHER STRATEGIES FOR NO_X REMOVAL

19.5.1. NO_x-Sorbing Materials

A general view of absorbent materials and the influence of particle size in their behavior has been described in Chapter 14. Sorptive NO_x removal uses adsorption, absorption or bulk-like type reactions, and/or solid–gas/liquid reactions and, typically, is used as a previous step in several technological processes or in combination with biological technologies (3–8). Oxide-type materials are investigated for the sorption of mainly NO and NO_2 in the gas phase. For reviewing, the existent technologies are based on materials that can be classified into three main categories. The first contains oxides corresponding to alkaline, transition, and rare-earth oxides. The second embraces oxygen-containing materials as exchanged zeolites, spinels, perovskites, and double-layered cuprates. Finally, a third one will be shortly reviewed that will englobe non-oxidic materials such as carbonaceous adsorbents.

19.5.1.1. Metal Oxides

Alkaline and Alkaline-Earth Oxides. Thermodynamic and reaction data demonstrate that the basicity of the bulk-like oxide drives the NO_x absorption or trapping performance by being directly related to the corresponding calculated equilibrium constants [e.g., $MO_x + NO_x = M(NO_3)_y$ for nitrate production, the main product although nitrites are often observed]. At 625 K, the performance order is $K > Ba \geq Sr > Na > Li \geq Mg$ (141,142). At higher temperatures, desorption of NO_x species may take place for some oxides, limiting the usefulness of these oxides for real applications (4). These oxides mainly absorb NO_2, and in the presence of O_2, only the help of a metallic component makes the reaction with NO and further generation of nitrates possible. The presence of other gases, e.g., H_2O, CO, CO_2, and O_2, typically present in energy production emission streams, drastically changes the NO_x-uptake of the oxide. At least in the case of Ba, the presence of CO_2 in a gas mixture dominantly generates Ba carbonate at $423 > T > 873$ K, with this compound essentially being stable in all oxidizing conditions. The BaO to $BaCO_3$ transformation occurs at room temperature for samples exposed to air in extended periods typical for storing industrial absorbents/catalysts. Other gases, like H_2O or SO_2, limit NO_x-uptake but typically do not fully suppress it (4,142). Other possible initial phases of applied systems, where alumina or other oxides are used as a carrier, like $BaAl_2O_4$ or $BaSnO_3$ display distinctive properties, having a selective interaction with NO_2 in the presence of NO and other gases (CO_2, H_2O) (143,144) but breaking Ba-M (M = Al, Sn) bonds and forming Ba sulfates in the presence of SO_2 (145).

No information is available for the size behavior of alkaline oxides, but their deposition on alumina allows us to obtain some data. As said, binary MO_x/Al_2O_3 materials are used for technological applications related to exhaust gas cleaning in automobiles (4). Of course, oxide–Al_2O_3 interactions would obscure the size-related information as there is no easy way to uncouple both effects from experimental observables. Additionally, the heterogeneity of the initial system can be noteworthy. After calcining in air, Ba may form a BaO-type phase for surface coverage below the monolayer while two less-interacting (with alumina) types of Ba carbonates can be additionally detected. Most apparent differences with bulk-types phase behavior appear for one of these Ba carbonates, which may absorb NO_x species and form nitrates by displacing carbonates groups, whereas XRD or Raman-detected, alumna-supported carbonates seem stable and mostly form NO_x-surface adspecies, maintaining their characteristic bulk $BaCO_3$ structure (146,147). The presence of a noble metal in storage and reduction catalysts (NSRs) in the vicinity of the MO_x moities has profound implications on the absorptive behavior of the system (4), and NO_x-storage properties seem to be optimized for Ba quantities corresponding to coverages near or slightly below the monolayer but, as said, is strongly dependent of the Pt–Ba degree of interaction under operation conditions (148). The behavior under contact with hydrocarbons has been analyzed in Section 19.3.2.2.

Transition Metal Oxides. NO-uptake on transition metal oxides has been typically measured on Al_2O_3 or Al_2O_3;SiO_2 supported samples (149–151). Table 19.3 gives an overview of main results as summarized by Gómez-Varela et al. (3). Some oxides, like Fe, Cr, or Mn, are in a partially reduced chemical state by effect of a pre-reduction treatment. These show a significantly higher sorption capacity than the fully oxidized samples. It is established that NO chemisorption on Cr(III), Co(II), and Mn(II) is particularly large due to the formation of metal–adsorbate covalent bonds. It is, however, clear that such chemical species may not be stable in an O_2-containing stream; in fact, Fe(III) and Mn(II) are known to be oxidized by NO alone. Cu is the only element with a higher sorption capacity in the fully oxidized state than for the reduced form (Cu_2O). The reasonable sorption capacity showed by Cu cannot be directly used in certain practical applications because CuO is also a good SO_2 scavenger via formation of copper sulfate (150). This fact, however, has allowed the development of CuO/Al_2O_3 systems, sometimes promoted by other oxides, for NO_x–SO_2 "simultaneous" removal using the help of NH_3 as reductant; this system appears as a serious competitor and/or can be used together with the technology based on $Na(OH)/Na_2O$ materials (152,153). Again, lack of the NO-uptake size-dependence can be noticed in the literature concerning the transition metal oxide's behavior.

Table 19.3 also includes unsupported Pt oxide particles with a primary particle size of ca. 30 nm, showing a low sorption capacity per weight of the material but relatively high per square meter when compared with the rest of the systems. Pd surfaces also present strong chemisorption (154), indicating that noble metal and particularly Pt may be good NO-adsorbers. They can be used with NO_2, NO, or $NO + O_2$ as they catalyze NO oxidation to NO_2, but they also easily oxidize SO_2 to SO_3 with fast sulfate deposition and poisoning. This problem has been attempted to be solved by addition of V, Nb, or other oxides (4). Size-dependence of NO-sorption on noble metal

TABLE 19.3. NO Sorption Capacity (mg NO g^{-1}; in parenthesis mg NO m^{-2}) for Supported Metal Oxides on Alumina-Silica 95:5 w/w at 1 mbar of NO and 298 K.

Metal Oxide	Supported Amount (wt%)	Sorption Capacity
Fe_3O_4	8.5	20.7 (0.23)
α-Fe_2O_3	8.5	5.4 (0.06)
CuO	8.9	12.5 (0.12)
NiO	6.9	11.0 (0.15)
Cr_2O_3	10.0	6.0 (0.12)
Co_3O_4	–	4.6 (0.23)
Pt (surface oxidized)	–	1.7 (0.20)

particles is related to the NO adsorption versus dissociation probabilities, disfavoring the adsorption as the particle grows in the nanometer range (155).

Rare-Earth Oxides. The NO and N_2O interactions with bulk-like rare-earth oxides have been analyzed by Rosynek (156). The study indicates that oxygen vacancies play a significant role in the interaction, giving raise to reduced (N_2/N_2O) products. Further studies were able to show that hyponitrites are intermediates in the reductive pathways (21). The presence of other molecules, like water, decreases the probability of obtaining N_2/N_2O by eliminating vacancies, but there is a structure sensitivity factor as (111) terminations are more resistant to being oxidized (157). The inherent presence of vacancies in the nanometer range (158) would favor the relative activity of small nanoparticles in the formation of reductive pathways under real operation conditions, but this has not been fully examined up to date. Other sorption products are formed by oxidation reactions, giving raise to nitrites and nitrates mainly present at surface layers (21,156) NO_2, on the other hand, mildly oxidizes oxygen vacancies present at the oxide surface, giving rise to the presence of nitrites and nitrates at the surface (156).

Recently, a lot of work has been devoted to the study of binary and ternary rare-earth-based oxides with fluorite-type structure, the structure typical of several rare-earth oxides (Ce, Pr, Tb) (159–161). Alio-valent (M(IV)) cations improve oxygen handling (e.g., mobility and exchange) properties but may not strongly influence NO-uptake properties until the basicity of the materials is strongly affected. However, under the appropriate reaction mixture or temperature ($T > 673$ K), the ease by which oxygen vacancies are formed strongly influences NO-uptake and behavior (159). According to XRD–Rietveld analyses, the presence of M cations with valence below +4 yields the stoichiometric quantity of oxygen vacancies to achieve electroneutrality (161). Due to the significantly large number of such vacancies comparing with those related to nanoscale size, even for a small weight percentage of the M cation, these solid solutions are likely to increase significantly NO-uptake and favor reductive pathways with respect to the corresponding single-oxide parent system. The presence of an M cation influences oxygen mobility and the surface versus bulk vacancy equilibrium, also modifying the thermal behavior of the NO–solid interaction. A particular case could be Mn–Ce samples, as Mn is able to significantly enhance the oxidation of NO to nitrites and nitrates and to store these products with the cooperative help of Ce(IV) cations (162). To the best of our knowledge, no detail comparison has been published for NO-uptake capacity of rare-earth oxide solid solutions, although in some cases, like Ce-Zr, the NO–solid interaction has been analyzed (163).

19.5.1.2. Other Oxide-Type Materials

Ion-Exchanged Zeolites. Zeolite-based systems can be viewed as the limit case in size-dependent studies as they can represent, if adequately synthesized, systems where M cations are isolated or forming part of rather size-limited (e.g., dimers and trimers) entities. Since the pioneering work of Iwamoto et al. (13,164) zeolite studies have been extensively reported in the literature as potentially useful materials for some $DeNO_x$ processes. The NO_x-sorption properties of ZSM-5 exchanged with various metals at 273 K were examined by Zhang et al. (165). ZSM-5 seems to offer significant advantages over other zeolitic supports for the specific case of the NO-uptake (166). Adsorption occurs both through a reversible (via NO^+) and an irreversible (via NO_2^+, NO_2^-, NO_3^-) way. As summarized in Table 19.4, zeolite exchanged with alkaline and rare-earth elements as Ce, La, or Na, take up NO only weakly. The highest sorption capacity was measured with Co > Cu > Ni > Mn and, in some cases, is comparable with the uptake displayed by the pure oxides, reaching values between 10 and 15 mg NO g^{-1} at the conditions mentioned in Table 19.1. However, contrary to pure oxides, the presence of oxygen has a strong influence, decreasing significantly the NO sorption capacity except in the case of Na where formation of a similar N_2O_3 intermediate is reported in the presence/absence of oxygen (167). The detrimental effect is also observed with H_2O and CO and seems to affect both the reversible and the irreversible ways of adsorption, whereas SO_2 mainly decreases the irreversible one. Fe-ZSM-5 was sometimes noted as a system able to work under complex gas mixtures due to a specific interaction with NO, but this result may lack consistency due to problems in reproducing the synthesis of solid catalysts (120). The reversible/irreversible ways are important in all cases studied and make the NO-uptake process sensitive to the level of exchange, presumably by the different proportion of mono-, bi-coordinated, and oligomeric species and the significant differences in their interaction with NO (168,169).

TABLE 19.4. NO Sorption Capacity (g NO g^{-1}) for Metal-Exchanged ZSM-5 Under 10% NO/He at 273 K.

Cation	Cation Content (wt%)	Degree of Exchange (%)	Sorption Capacity
	Transition Metals		
Co	3.1	90	0.795
Cu	5.9	157	0.720
Ni	2.4	68	0.288
Mn	4.2	127	0.235
Ag	10.9	90	0.148
Fe	2.1	62	0.134
Cr	0.9	41	0.060
Zr	3.8	96	0.056
	Other Metals		
Ca	1.3	54	0.126
Sr	5.4	105	0.109
Ba	6.4	80	0.109
Mg	0.7	46	0.034
Ce	0.4	8	0.028
La	0.4	7	0.002
Na	2.8	100	0.001

TABLE 19.5. NO_2 Sorption Capacity ($mg\ NO_2\ g^{-1}$) for Perovskite Materials During Temperature Ramps up to 873 K.

Structure	Sorption Capacity
$BaSnO_3$	17
$SrSnO_3$	13
$CaSnO_3$	9
$BaZrO_3$	12
$SrZrO_3$	8
$CaZrO_3$	3
$BaTiO_3$	3

Other *oxide-ion-containing systems* are spinels (AB_2O_4), perovskites (ABO_3), and double-layer cuprates $La_{2-x}SrCu_2O_6$. Although some perovskites like $ZnCo_2O_4$ with Co(III) ions on octahedral positions display important sorption capacity (149,170), some samples with (Co, Ni, Mg) on tetrahedral coordination (A site) could decompose N_2O on N_2/O_2 at 475–573 K in the absence/presence of O_2/H_2O, a fact with potential practical use (171). However, generally speaking, perovskites seem better alternatives, and Hodjati et al. have shown their potential as NO_2 sorption materials (Table 19.5), with the formation of bulk nitrates and a partial breaking of the solid structure in a process surprisingly reversible (172). NO sorption properties seem nevertheless significantly inferior (173). Finally, Marchida et al. reported the NO–solid reaction based on intercalation into substituted double-layered cuprates (174).

19.5.1.3. Other Materials Apart from oxides, there are other attractive alternatives for NO sorption materials. Among those studied, heteropolyacid and carbonaceous solids are most relevant and the information concerning these systems has been reviewed in Ref. 3. As these materials are not covered by the main subject of this chapter, we just mention that doped carbon fibers (with metal oxides and/or organic compounds) display high NO sorption values, but the oxidation state stability and the capture of SO_2 seem real drawbacks for the practical implementation of a carbon-based technology. The doping with metal oxides/hydroxides improves the system performance in the SCR of NO_x, particularly the selectivity to N_2, presumably by affecting NO-related reaction steps, but the size-dependence of this phenomenon has been barely studied (175). Heteropolyacids show bulk and surface NO sorption capacity even in the presence of CO_2 and SO_2, with this potential being useful in the future if the thermal stability of this solid type is improved.

19.5.2. Miscellaneous Strategies

To end this chapter, we would like to briefly mention other strategies that have been shown potential to solve DeNO$_x$ problems at specific situations. These are mainly based on photo-catalytic processes. Although there are publications in several DeNO$_x$

fields, the only promising photocatalytic DeNO$_x$ process seems to be the NO reduction with CO, which uses TiO_2 (176) or Co on SiO_2 or MCM (mobil composition of matter) supported (177,178) catalysts. In the first case, the presence of surface hydroxyl groups appear important for catalytic performance, whereas in the second, the presence of light excited [Mo^{5+}–O^-] entities, which further react with CO, may be key intermediates in a step-like process that follows a NO → N_2O → N_2 reductive pathway. The size-dependence study of the catalytic properties awaits implementation, but at least for TiO_2, the strong correlation of surface acidity–basicity with size would suggest a significant influence.

REFERENCES

(1) Environmental Assessment Report Environment in Europe. EEA report; Copenhagen, April 24, 2003.

(2) Ramachandran, B.; Hermann, R.G.; Choi, S.; Stenger, H.G.; Lyman, C.E.; Sale, J.W. *Catal. Today* **2000**, *55*, 281.

(3) Gómez-Varela, M.A.; Pitchon, V.; Kienemann, A. *Environ. Int.* **2005**, *31*, 445.

(4) Epling, W.S.; Campbell, L.E.; Yezerets, A.; Currier, N.W.; Parks II, J.E.; *Catal. Rev. Sci. Eng.* **2004**, *46*, 163.

(5) Martínez-Arias, A.; Fernández-García, M.; Conesa, J.C.; Anderson, J.A. In *Supported Metals in Catalysis*; Anderson, J.A.; Fernández-García, M. (Editors); Imperial College Press: London, 2005; Chapter 8.

(6) Heck, R.M. *Catal. Today* **1999**, *53*, 519.

(7) Hermann, J.M. *Top. Catal.* **2004**, *46*, 163.

(8) Pethaiyan, J.; Klabunde, K. *Adsorbents*; In Rodriguez, J.; Fernandez-Garcia, M. (Editors); Wiley: New York, in press, Chapter 14.

(9) Glevov, L.S.; Zakivova, A.G.; Tretyakov, O.F.; Burdeinaya, T.N.; Akopova, G.S. *Petrol. Chem.* **2002**, *42*, 143.

(10) Fritz, A.; Pitchon, V. *Appl. Catal. B* **1997**, *13*, 1.

(11) Glick, H.; Klien, J.J.; Squire, W. *J. Chem. Phys.* **1957**, *27*, 850.

(12) Howard, C.H.; Daniels, F. *J. Phys. Chem.* **1958**, *62*, 360.

(13) Iwamoto, M. *Stud. Surf. Sci. Catal.* **2000**, *130*, 23.

(14) Parvulescu, V.I.; Grange, P.; Delmon, B. *Catal. Today* **1998**, *46*, 233.

(15) Winter, E.R.S. *J. Catal.* **1971**, *22*, 158.

(16) Zhang, X.; Walters, A.B.; Vannice, M.A. *J. Catal.* **1995**, *155*, 290.

(17) Boreskov, G.K. *Disc. Faraday Soc.* **1966**, *41*, 263.

(18) Park, P.W.; Kil, J.K.; Kung, H.H.; Kung, M.C. *Catal. Today* **1998**, *42*, 51.

(19) Nörenberg, H.; Briggs, G.A.D. *Phys. Rev. Lett.* **1997**, *79*, 4222.

(20) Esch, F.; Fabris, S.; Zhou, L.; Montini, T.; Africh, C.; Fornasiero, P.; Comelli, G.; Rosei, R. *Science*, **2005**, *309*, 752.

(21) Martínez-Arias, A.; Soria, J.; Conesa, J.C.; Seoane, X.L.; Arcoya, A.; Cataluña, R. *J. Chem. Soc. Faraday Trans.* **1995**, *91*, 1679.

(22) Huang, S.-J.; Walters, A.B.; Vannice, M.A. *J. Catal.* **2000**, *192*, 29.

(23) Martínez-Arias, A.; Fernández-García, M.; Salamanca, L.N.; Valenzuela, R.X.; Conesa, J.C.; Soria, J. *J. Phys. Chem. B* **2000**, *104*, 4038.

(24) Cataluña, R.; Arcoya, A.; Seoane, X.L.; Martínez-Arias, A.; Coronado, J.M.; Conesa, J.C.; Soria, J.; Petrov, L.A. *Stud. Surf. Sci. Catal.* **1995**, *96*, 215.

(25) Amirnazmi, A.; Benson, J.E.; Boudart, M.; *J. Catal.* **1973**, *30*, 55.

(26) Hamada, H.; Kintaichi, Y.; Sasaki, M.; Ito, T. *Chem. Lett.* **1990**, 1069.

(27) Shin, S.; Ogawa, K.; Shimomura, K.; *Mater. Res. Bull.* **1979**, *14*, 133.

(28) Tabata, K. *J. Mater. Sci. Lett.* **1988**, *7*, 147.

(29) Teraoka, Y.; Fukuda, H.; Kagawa, S. *Chem. Lett.* **1990**, 1.

(30) Halasz, I.; Brenner, A.; Shelef, M.; Ng, K.Y.S. *Catal. Lett.* **1991**, *11*, 327.

(31) Peña, M.A.; Fierro, J.L.G. *Chem. Rev.* **2001**, *101*, 1981.

(32) Tofan, C.; Klvana, D.; Kirchnerova, J. *Appl. Catal. A* **2002**, *226*, 225.

(33) Liu, Z.; Hao, J.; Fu, L.; Zhu, T. *Appl. Catal. B* **2003**, *44*, 305.

(34) Iwamoto, M.; Furukawa, H.; Mine, Y.; Uemura, F.; Mikuriya, S.; Kagawa, S. *J. Chem. Soc. Chem. Commun.* **1986**, 1272.

(35) Li, Y.; Hall, W.K. *J. Catal.* **1991**, *129*, 202.

(36) Shelef, M. *Chem. Rev.* **1995**, *95*, 209.

(37) Lei, G.D.; Adelman, B.J.; Sarkany, J.; Sachtler, W.M.H. *Appl. Catal. B* **1995**, *5*, 245.

(38) Moretti, G.; Ferraris, G.; Fierro, G.; Lo Jacono, M.; Morpurgo, S.; Faticanti, M. *J. Catal.* **2005**, *232*, 476.

(39) Crutzen, P.J. *J. Geophys. Res.* **1971**, *76*, 7311.

(40) Donner, L.; Rantanen, V. *J. Atmos. Sci.* **1980**, *37*, 119.

(41) Reimer, R.A.; Slaten, C.S.; Seapan, M.; Lower, M.W.; Tomlinson, P.E. *Environ. Progr.* **1994**, *13*, 134.

(42) Jensen, T.K. *Proc. Int. Fert. Soc.* **2004**, *538*, 1.

(43) Pérez-Ramírez, J.; Kapteijn, F.; Schöffel, K.; Moulijn, J.A. *Appl. Catal. B* **2003**, *44*, 117.

(44) Kapteijn, F.; Rodríguez-Mirasol, J.; Moulijn, J.A. *Appl. Catal. B* **1996**, *9*, 25.

(45) Godolets, G.I. *Stud. Surf. Sci. Catal.* **1983**, *15*, 200.

(46) Winter, E.R.S. *J. Catal.* **1969**, *15*, 144.

(47) Winter, E.R.S. *J. Catal.* **1970**, *19*, 32.

(48) Winter, E.R.S. *J. Catal.* **1974**, *34*, 431.

(49) Li, Y.; Armor, J.N. *Appl. Catal. B* **1992**, *1*, L31.

(50) Zhang, X.; Walters, A.B.; Vannice, M.A. *Appl. Catal. B* **1994**, *4*, 237.

(51) Soete, G.G.D. *Rev. Inst. Franc. Petr.* **1993**, *48*, 413.

(52) Yamashita, T.; Vannice, M.A. *J. Catal.* **1996**, *164*, 254.

(53) Volpe, M.L.; Reddy, J.F. *J. Catal.* **1967**, *7*, 76.

(54) Dell, R.M.; Stone, F.S.; Tiley, P.F. *Trans. Faraday Soc.* **1953**, *49*, 201.

(55) Amphlett, C.B. *Trans. Faraday Soc.* **1953**, *50*, 273.

(56) Stone, F.S. *J. Solid State Chem.* **1975**, *12*, 27.

(57) Cimino, A.; Bosco, R.; Indovina, V.; Schiavello, M. *J. Catal.* **1966**, *5*, 271.

(58) Keenan, A.G.; Iyengar, R.D. *J. Catal.* **1966**, *5*, 301.

(59) Cimino, A.; Pepe, F. *J. Catal.* **1972**, *25*, 362.

(60) Egerton, T.A.; Stone, F.S.; Vickerman, J.C. *J. Catal.* **1974**, *33*, 299.

(61) Egerton, T.A.; Stone, F.S.; Vickerman, J.C. *J. Catal.* **1974**, *33*, 307.

(62) Cimino, A.; Indovina, V.; Pepe, F.; Schiavello, M. *J. Catal.* **1969**, *14*, 69.

(63) Indovina, V.; Cordischi, D.; Occhiuzzi, B.; Arieti, A. *J. Chem. Soc. Faraday Trans.* **1979**, *75*, 2177.

(64) Ciminio, A.; Indovina, V. *J. Catal.* **1970**, *17*, 54.

(65) Cimino, A.; Indovina, V.; Pepe, F.; Stone, F.S. *Gazz. Chim. Ital.*, **1973**, *103*, 935.

(66) Saito, Y.; Yoneda, Y.; Makishima, S. *Actes du 2iéme Congrès International sur la Catalyse*; Technip: Paris, 1961; 1937–1953.

(67) Swamy, C.S.; Christopher, J. *Cat. Rev. -Sci. Eng.* **1992**, *34*, 409.

(68) Yamazoe, N.; Teraoka, Y. *Catal. Today* **1990**, *8*, 175.

(69) Ramanujachari, K.V.; Kameswari, N.; Swamy, C.S. *J. Catal.* **1984**, *86*, 121.

(70) Wang, J.; Yasuda, H.; Inumaru, K.; Misono, M. *Bull. Chem. Soc. Jpn.* **1995**, *68*, 1226.

(71) Yasuda, H.; Nitadori, T.; Mizuno, N.; Misono, M. *Bull. Chem. Soc. Jpn.* **1993**, *66*, 3492.

(72) Christopher, J.; Swamy, C.S. *J. Mol. Catal.* **1990**, *62*, 69.

(73) Kannan, S.; Swamy, C.S. *Appl. Catal.* **1994**, *3*, 109.

(74) Armor, J.N.; Braymer, T.A.; Farris, T.S.; Li, Y.; Petrocelli, F.P.; Weist, E.L.; Kannan, S.; Swamy, C.S. *Appl. Catal. B* **1996**, *7*, 397.

(75) Swamy, C.S.; Kannan, S.; Li, Y.; Armor, J.N.; Braymer, T.A. U.S. Patent 5,407,652; 1995.

(76) Shelef, M.; Otto, K.; Gandhi, H. *J. Catal.* **1968**, *12*, 361.

(77) Shelef, M.; Otto, K. *J. Catal.* **1968**, *10*, 408.

(78) Zhang, Z.; Geng, H.; Zheng, L.; Du, B. *J. All. Comp.* **2005**, *392*, 317.

(79) Martínez-Arias, A.; Fernández-García, M.; Hungría, A.B.; Iglesias-Juez, A.; Gálvez, O.; Anderson, J.A.; Conesa, J.C.; Soria, J.; Munuera, G. *J. Catal.* **2003**, *214*, 261.

(80) Lox, E.S.J.; Engler, B.H. In *Environmental Catalysis*; Ertl, G.; Knözinger, H.; Weitkamp, J. (Editors); Wiley-VCH: Weinheim, 1999; 1.

(81) Armor, J.N. *Catal. Today* **1995**, *26*, 147.

(82) Kaucky, D.; Vondrova, A.; Dedecek, J.; Wichterlova, B. *J. Catal.*, **2000**, *194*, 318.

(83) Desai, A.J.; Kovalchuk, V.I.; Lombardo, E.A.; d'Itri, J.L. *J. Catal.* **1999**, *184*, 396.

(84) Resini, C.; Montanari, T.; Nappi, L.; Bagnasco, G.; Turco, M.; Busca, G.; Bregani, F.; Notaro, M.; Rocchini, G. *J. Catal.* **2003**, *214*, 179.

(85) Misono, M. *CATTECH* **1998**, *2*, 183.

(86) Shimizu, K.; Satsuma, A.; Hattori, T. *Appl. Catal. B* **1998**, *16*, 319.

(87) Zhang, X.; Walters, A.B.; Vannice, M.A. *J. Catal.* **1994**, *146*, 568.

(88) Vannice, M.A.; Walters, A.B.; Zhang, X. *J. Catal.* **1996**, *159*, 119.

(89) Shi, C.; Walters, A.B.; Vannice, M.A. *Appl. Catal. B*, **1997**, *14*, 175.

(90) Costa, C.N.; Anastasiadou, T.; Efstathiou, A.M. *J. Catal.* **2000**, *194*, 250.

(91) Belessi, V.C.; Costa, C.N.; Bakas, T.V.; Anastasiadou, T.; Pomonis, P.J.; Efstathiou, A.M. *Catal. Today* **2000**, *59*, 347.

(92) Burch, R.; Breen, J.P.; Meunier, F.C. *Appl. Catal. B* **2002**, *39*, 283; and references therein.

(93) Traa, Y.; Burger, B.; Weitkamp, J. *Microp. Mesop. Mater* **1999**, *30*, 3.

(94) Martínez-Arias, A.; Fernández-García, M.; Iglesias-Juez, A.; Anderson, J.A.; Conesa, J.C.; Soria, J. *Appl. Catal. B* **2000**, *28*, 29.

(95) Iglesias-Juez, A.; Hungría, A.B.; Martínez-Arias, A.; Fuerte, A.; Fernández-García, M.; Anderson, J.A.; Conesa, J.C.; Soria, J. *J. Catal.* **2003**, *217*, 310.

(96) Lee, J.; Yezerets, A.; Kung, M.C.; Kung, H.H. *Chem. Comm.* **2001**, 1404.

(97) Eränen, K.; Klingstedt, F.; Arve, K.; Lindfors, L.; Murzin, D.Y. *J. Catal.*, **2004**, *227*, 328.

(98) Ueda, A.; Haruta, M. *Appl. Catal. B* **1998**, *18*, 115.

(99) Shimizu, K.; Satsuma, A.; Hattori, T. *Appl. Catal. B* **2000**, *25*, 239.

(100) Eränen, K.; Lindfors, L.; Klingstedt, F.; Murzin, D.Y.; *J. Catal.* **2003**, *219*, 25.

(101) Wichterlová, B.; Sazama, P.; Breen, J.P.; Burch, R.; Hill, C.J.; Capek, L.; Sobalík, Z. *J. Catal.* **2005**, *235*, 195.

(102) Shelef, M.; Gandhi, H.S. *Ind. Eng. Chem. Prod. Res. Dev.* **1974**, *13*, 80.

(103) Acke, F.; Panas, I. *J. Phys. Chem. B*, **1998**, *102*, 5127.

(104) Guglielminotti, E.; Boccuzzi, F. *Appl. Catal. B* **1996**, *8*, 375.

(105) Ferri, D.; Forni, L.; Dekkers, M.A.P.; Nieuwenhuys, B.E. *Appl. Catal. B* **1998**, *16*, 339.

(106) Parvulescu, V.I.; Grange, P.; Delmon, B. *Catal. Today* **1998**, *46*, 233.

(107) Heck, R.M. *Catal. Today* **1999**, *53*, 519.

(108) Forzatti, P. *Appl. Catal. A* **2001**, *222*, 221.

(109) Lisi, L.; Lasorella, G.; Mallogi, S.; Riso, G. *Appl. Catal. B* **2004**, *50*, 251.

(110) Grzybek, T.; Klinik, J.; Dutka, B.; Papp, H.; Suprun, V. *Catal. Today* **2001**, *101*, 93.

(111) García-Bordejé, E.; Lázaro, M.J.; Moliner, R.; Galindo, J.F.; Sotres, J.; Baró, A.M. *J. Catal.* **2004**, *223*, 395.

(112) Bond, G.C.; Vedrine, J.C. *Catal. Today* **1994**, *20*, papers therein.

(113) Trifiro, F.; Grzybowska, B. *Appl. Catal. A* **1997**, *157*, papers therein.

(114) Vedrine, J.C. *Catal. Today* **2000**, *56*, papers therein.

(115) Wachs, I.E. *Catal. Today* **2005**, *100*, 79.

(116) Wachs, I.E. *Catal. Today* **1996**, *27*, 437.

(117) Weckhuisen, B.M.; Meng, J.-M.; Wachs, I.E. *J. Phys. Chem. B* **2000**, *104*, 7382.

(118) Gao, X.; Bare, S.R.; Fierro, J.L.G.; Wachs, I.E. *J. Phys. Chem. B* **1999**, *103*, 618.

(119) Keranen, J.; Guimon, C.; Auroux, A.; Iiskola, E.I.; Niinistro, L. *Phys. Chem. Chem. Phys.* **2003**, *5*, 5333.

(120) Hall, W.K.; Feng, X.B.; Dumesic, J. *Catal. Lett.* **1998**, *52*, 13.

(121) Capek, L.; Krlibrich, V.; Derecek, J.; Grygar, T.; Wicherlowa, B.; Martens, J.A.; Brosius, R.; Tokarova, V. *Microp. Mesop. Mat.* **2005**, *80*, 279.

(122) Schwilder, M.; Kumar, M.S.; Klementiev, K.; Pohl, M.M.; Bruckner, A.; Grunet, W. *J. Catal.* **2005**, *231*, 314.

(123) Canter, L.W. *Nitrates in Groundwater*; CRC Press: Boca Raton, FL, 1996.

(124) Schmidt, J.; Vorlop, D. *Proc. 4th European Congress on Biotechnology*, Vol. 1; Elsevier: Amsterdam, 1987; 155.

(125) Horold, S.; Vorlop, K.D. *Environ. Technol.* **1993**, *14*, 21.

(126) Chollier-Brym, M.J.; Gavagnin, R.; Struckul, G.; Marella, M.; Tomaselli, M.; Ruiz, P. *Catal. Today* **2002**, *75*, 49.

(127) Gauthard, F.; Epron, F.; Barbier, J. *J. Catal.* **2003**, *220*, 182.

(128) Palomares, A.E.; Prato, J.G.; Rey, F.; Corma, A. *J. Catal.* **2004**, *221*, 62.

(129) Pintar, A.; Batista, J.; Muzariz, I. *Appl. Catal. B*, **2004**, *52*, 49.

(130) Sa, J.; Vinek, V. *Appl. Catal. B*, **2005**, *57*, 247.

(131) Matatov-Meytal, V.; Sheintuch, M. *Catal. Today*, **2005**, *102-3*, 121.

(132) Barrabés, N.; Just, J.; Dafinov, A.; Medina, F.; Fierro, J.L.G.; Sueiras, J.E.; Salangre, P.; Cesteros, Y. *Appl. Catal. B* in press.

(133) Warna, J.; Turunen, I.; Selmi, T.; Maunula, T. *Chem. Eng. Sci.* **1994**, *49*, 5763.

(134) Gavagnin, R.; Biasetto, L.; Pinna, F.; Strukul, G. *Appl. Catal. B* **2002**, *38*, 91.

(135) D'Arino, M.; Pinna, F.; Strukul, G. *Appl. Catal. B*, **2004**, *53*, 191.

(136) Strukul, G.; Gavagnin, R.; Pinna, F.; Modiferri, E.; Perathoner, S.; Centi, G.; Marella, M.; Tomaselli, M. *Catal. Today* **2000**, *55*, 139.

(137) Sa, J.; Berger, T.; Fottinger, K.; Riss, A.; Anderson, J.A.; Vinek, H. *J. Catal.* **2005**, *234*, 282.

(138) Zhang, F.; Jin, R.; Chen, J.; Shao, C.; Gao, W.; Li, L.; Guan, N. *J. Catal.* **2005**, *232*, 424.

(139) Tsumura, R.; Higashimoto, S.; Matsuoka, M.; Yamashita, H.; Che, M.; Coluccia, S. *J. Catal.* **1999**, *184*, 390.

(140) Chekchook, M.; Deibe, G.; Foussard, J.N.; Debellefontained, H. *Environ. Tech.* **1995**, *16*, 645.

(141) Kobayashi, T.; Yamada, T.; Kayano, K. *SAE* Technical Papers, Series 970745.

(142) Breen, J.P.; Marella, M.; Pistarino, C.; Ross, J.R.H. *Catal. Lett.* **2002**, *80*, 123.

(143) Hodjati, S.; Vaezzadeh, K.; Petit, C.; Pitchon, V.; Kienemann, A. *Appl. Catal. B*, **2000**, *27*, 117.

(144) Despress, J.; Koebel, M.; Krocher, O.; Elsener, M.; Wokaun, A. *Appl. Catal. B*, **2005**, *43*, 389.

(145) Courson, K.; Khalfi, A.; Mahzoul, H.; Hodjati, S.; Moral, N.; Kienemann, A. *Catal. Comm.* **2002**, *3*, 471.

(146) Piacentini, M.; Maciejewski, M.; Baiker, A. *Appl. Catal. B* **2005**, *59*, 187.

(147) Anderson, J.A.; Liu, Z.; Fernández-García, M. *Catal. Today* in press.

(148) Hendershot, R.; Rogers, W.B.; Snively, C.M.; Ogunnakide, B.A.; Lauterbach, J. *Catal. Today* **2004**, *98*, 375; *Appl. Surf. Sci.* **2006**, *252*, 2588.

(149) Shelef, M. *Catal. Rev. Sci. Eng.* **1975**, *11*, 1.

(150) Kill, J.H.A.; Prins, W.; van Swaaij, W.P.N. *App. Catal. B* **1992**, *1*, 13.

(151) Eguchi, K.; Watabe, M.; Machida, M.; Arai, H. *Catal. Today* **1996**, *27*, 297.

(152) Wey, M.Y.; Lu, C.Y.; Tseng, H.H.; Fu, C.H. *J. Waste Manag. Ass.* **2002**, *52*, 449.

(153) Liu, Q.; Liu, Z.; Huang, Z.; Xie, G. *Catal. Today*, **2004**, *93–94*, 833.

(154) Loffred, D.; Simon, D.; Sautet, P. *J. Catal.* **2005**, *213*, 211.

(155) Martínez-Arias, A.; Hungría, A.B.; Fernández-García, M.; Iglesias-Juez, A.; Conesa, J.C. *J. Catal.* **2004**, *223*, 85.

(156) Rosynek, M.P. *Catal. Rev. Sci. Eng.* **1977**, *16*, 111.

(157) Henderson, M.A.; Perkins, C.L.; Engelhard, M.H.; Thevuthasan, S.; Peden, C.H. *Surf. Sci.* **2003**, *526*, 1.

(158) Hernández-Alonso, M.D.; Hungría, A.B.; Martínez-Arias, A.; Coronado, J.M.; Fernández-García, M. *Phys. Chem. Chem. Phys.* **2004**, *6*, 3524.

(159) Trovarelli, A. *Catalysis by Ceria and Related Materials*; Imperial College Press: London, 2002.

(160) Di Monte, R.; Kaspar, J. *J. Mater. Chem.*, **2005**, *15*, 633.

(161) Fernández-García, M.; Wang, X.; Belver, C.; Iglesias-Juez, A.; Hanson, J.C.; Rodriguez, J.A.; *Chem. Mater.* **2005**, *17*, 4181.

(162) Machida, M.; Uto, M.; Kuraga, D.; Kijima, T. *Chem. Mater.* **2000**, *12*, 3165.

(163) Daturi, M.; Bion, N.; Saussey, J.; Lavalley, J.C.; Hadouin, C.; Seguelong, T. *Phys. Chem. Chem. Phys.* **2001**, *3*, 252.

(164) Iwamoto, M. *Stud. Surf. Sci. Catal.* **1990**, *54*, 121.

(165) Zhang, Y.; Yahiro, M.; Mizuno, N.; Izumi, J.; Iwamoto, M. *Lagmuir* **1993**, *9*, 2337.

(166) Arai, H.; Marchida, M. *Catal. Today* **1994**, *22*, 97.

(167) Sultana, A.; Loenders, M.; Monticelli, O.; Kiraohhock, C.; Jackobs, P.A.; Martens, J.A. *Angew. Chem. Int. Ed.* **2000**, *39*, 2934.

(168) Li, Y.; Armor, J.N. *Appl. Catal. B* **1993**, *3*, 55.

(169) Truex, T.J.; Searles, R.A.; Sun, D.C. *Platinum Metal Rev.* **1999**, *36*, 2.

(170) Pnayotov, D.; Matyshak, V.; Skyaror, V.; Vlasenko, V.; Mehandjev, D. *App. Catal.* **1986**, *24*, 37.

(171) Yan, L.; Ren, T.; Wang, X.; Ji, D.; Suo, J. *Appl. Catal. B* **2003**, *45*, 85.

(172) Hodjati, S.; Petit, C.; Pitchon, V.; Kiennemann, A. *Appl. Catal. B* **2000**, *27*, 117.

(173) Tascón, J.M.D.; Tejuca, L.C.; Rochester, C.H. *J. Catal.* **1985**, *95*, 558.

(174) Machida, M.; Murakan, H.; Kijima, T. *Appl. Catal. B* **1998**, *17*, 195.

(175) Grzybek, T.; Klinik, J.; Czajka, P.; Papp, H.; Suprun, V. *Polish, J. Environ. Stud.* **2004**, *13*, 14.

(176) Bowering, N.; Walker, G.S.; Harrison, P.G. *Appl. Catal. B* **2005**, *62*, 208.

(177) Susbotina, I.R.; Shelimov, B.N.; Kazansky, V.B.; Lisachenko, A.; Che, M.; Coluccia, S. *J. Catal.* **1999**, *184*, 390.

(178) Tsumura, R.; Higashimoto, S.; Matsuoka, M.; Yamashita, H.; Che, M.; Anpo, M. *Catal. Lett.* **2000**, *68*, 101.

CHAPTER 20

Chemistry of SO_2 and $DeSO_x$ Processes on Oxide Nanoparticles

JOSÉ A. RODRIGUEZ

Department of Chemistry, Brookhaven National Laboratory, Upton, NY 11973, USA

20.1. INTRODUCTION

The chemistry of sulfur dioxide (SO_2) on oxide nanoparticles is receiving a lot of attention due to its importance in the industrial production of sulfuric acid (1) and environmental catalysis (2–4). Sulfur-containing molecules are common impurities present in coal and crude oil. SO_2 is one of the major air pollutants released to the atmosphere as a result of the combustion of fuels in power plants, factories, houses, and transportation (2). It contributes to the generation of smog and constitutes a serious health hazard for the respiratory system (2). After its oxidation and reaction with water in the atmosphere, it is responsible for the acid rain that kills vegetation and corrodes buildings and monuments in modern cities (2). In addition, the SO_2 produced by the combustion of sulfur-containing fuels in automotive engines poisons the catalysts that are used for the removal of CO and NO in exhaust catalytic converters ($2CO + O_2 \rightarrow 2CO_2$; $2CO + 2NO \rightarrow 2CO_2 + N_2$) (4,5). When SO_2 is present in the catalytic converter at high concentrations, it dissociates on the precious-metal component of the catalyst (Rh, Pd, or Pd), blocking active sites and reducing the overall activity of the system through medium- or long-range electronic effects (6,7). At the levels of 5 to 20 ppm currently present in the typical automotive exhaust, SO_2 interacts primarily with the ceria component of the catalytic converter, and the poisoning of this oxide is a major concern nowadays (8–10).

Governments are constantly tightening regulations to limit the production of SO_2 and emission of sulfur compounds into the air (3–5). Over the past 30 years, several processes have been proposed and developed for the removal of SO_2 from exhaust

---

Synthesis, Properties, and Applications of Oxide Nanomaterials, Edited by José A. Rodríguez and Marcos Fernández-Garcia

systems and $DeSO_x$ operations (3–5,11). There is still not a universally acceptable solution to this problem. Due to their low cost, oxides are frequently used as sorbents or scrubbers for trapping the SO_2 molecule in industrial processes (11). Also, there is a general interest (3,4,11–15) in using oxides as catalysts for the Claus reaction ($SO_2 + 2H_2S \rightarrow 2H_2O + 3S_n$) and the reduction of sulfur dioxide by CO ($SO_2 + 2CO \rightarrow 2CO_2 + S_n$). In these reactions, the rupture of the S–O bonds on the oxide catalyst is one of the most difficult steps.

This chapter presents an overview of recent studies that focus on the chemistry of SO_2 and $DeSO_x$ processes on oxide nanoparticles. In principle, the unique electronic and structural properties of oxide nanostructures can lead to a high activity for the cleavage of S–O bonds. A comparison with studies performed using single crystals like ZnO(0001), MgO(100), or $CeO_2(110)$ allows a detailed analysis of structure sensitivity and the impact of defects and O vacancies. This chapter is organized as follows. First, a brief general description of the behavior of SO_2 on bulk oxide systems is presented. This is followed by a more detailed description of the interaction of SO_2 with nanoparticles of CaO, MgO, SrO, Al_2O_3, Al_2O_3/MgO, Fe_2O_3, Fe_2O_3/CaO, CeO_2, $Ce_{1-x}Zr_xO_2$, and $Ce_{1-x}Ca_xO_{2-y}$. Approaches useful for facilitating or promoting the dissociation of SO_2 on oxide nanoparticles are discussed.

20.2. CHEMISTRY OF SO_2 ON BULK OXIDES

A study of the interaction of SO_2 with CeO_2 is interesting for two basic reasons. First, ceria doped with copper or other metals can catalyze the reduction of SO_2 by CO (14,15). And second, the SO_2 formed during the combustion of fuels in automotive engines can affect the performance of the CeO_2 or $Ce_{1-x}Zr_xO_2$ present in catalysts used for reducing CO and NO_x emissions (8–10). The species responsible for ceria deactivation is mainly attributed to cerium sulfate, which blocks the Ce^{3+} sites for the redox cycle in the process of oxygen storage/release (8–10). Reaction of SO_2 with CeO_2 powders and polycrystalline ceria films supported on Pt(111) at 25 °C shows sulfate (SO_4) as the main surface species as evidenced by a combination of X-ray absorption near-edge structure (XANES), temperature programmed desorption (TPD), and high-resolution photoemission (16). Photoemission studies for the adsorption of SO_2 on $CeO_2(111)$ and $Ce_{1-x}Zr_xO_2(111)$ point to the formation of a SO_x species on the surface that could be either SO_3 or SO_4 (17,18). The identification of this species on the basis of only photo-emission is not conclusive (17). To address this issue, XANES was used to study the interaction of SO_2 with $CeO_2(111)$ and $Ce_{1-x}Zr_xO_2(111)$ surfaces (19). Figure 20.1 shows S K-edge spectra for the adsorption of SO_2 on $CeO_2(111)$ and $Ce_{0.7}Zr_{0.3}O_2(111)$ surfaces at room temperature. A comparison with the corresponding peak positions for sulfates and sulfites (19) indicates that SO_4 is the main species formed on the oxide surfaces with a minor concentration of SO_3. There is no dissociation of the adsorbate.

The top layer of $CeO_2(111)$ and $Ce_{0.7}Zr_{0.3}O_2(111)$ contains only O atoms, see Figure 20.2. The adsorption of SO_2 on these O atoms would yield directly sulfite or

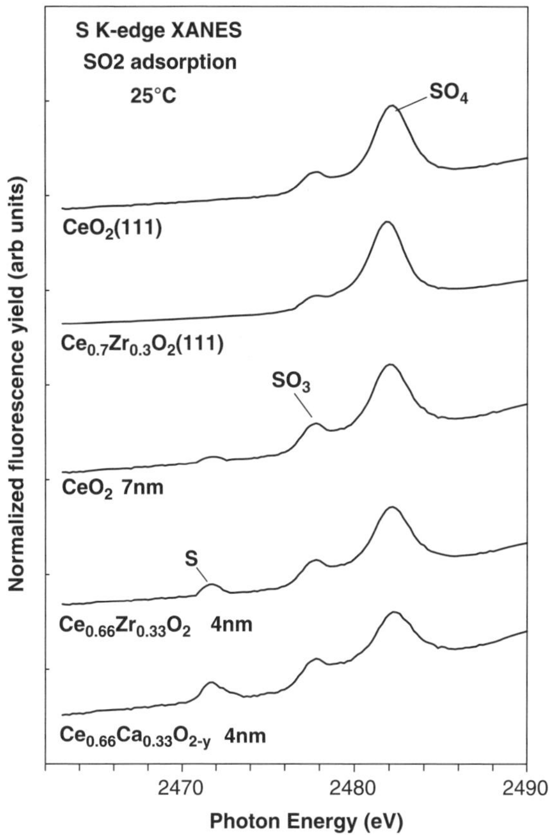

Figure 20.1. S K-edge spectra taken after dosing SO_2 to $CeO_2(111)$ and $Ce_{0.7}Zr_{0.3}O_2(111)$ surfaces, and nanoparticles of CeO_2, $Ce_{0.66}O_{0.33}O_2$ and $Ce_{0.66}Ca_{0.33}O_{2-y}$. The samples were exposed to 0.1 Torr of SO_2, for 5 min at 25 °C. Reprinted from Ref. 19.

sulfate species:

$$SO_{2,\mathrm{gas}} + O_{\mathrm{lattice}} \longrightarrow SO_{3,\mathrm{ads}} \qquad \mathbf{(20.1)}$$

$$SO_{3,\mathrm{ads}} + O_{\mathrm{lattice}} \longrightarrow SO_{4,\mathrm{ads}} \qquad \mathbf{(20.2)}$$

"Holes" in the top layer of $CeO_2(111)$ and $Ce_{0.7}Zr_{0.3}O_2(111)$ expose Ce and Zr cations in the second layer; see Figure 20.2. These cations have all their O neighbors (eight in total) and interact very weakly with an adsorbed SO_2 molecule (17).

Figure 20.3 displays photo-emission data for the adsorption of SO_2 at 300 K on a MgO(100) crystal, bottom panel, and a MgO(100) epitaxial film grown on a Mo(100) substrate, top panel (20). The (100) face of MgO consists of a 50%–50% mixture of Mg and O atoms. The photo-emission data indicate that only the O atoms interact strongly with SO_2 forming a mixture of SO_3 and SO_4 species on the oxide surface

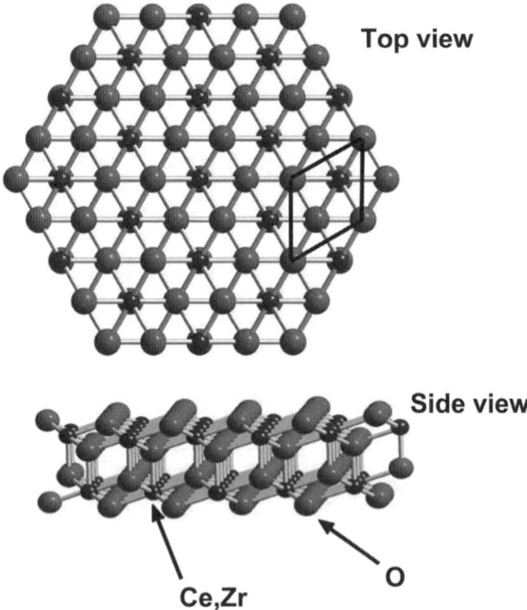

Figure 20.2. Top and side views of an oxygen-terminated $Ce_{1-x}Zr_xO_2(111)$ surface ($x < 0.4$). The large spheres represent O atoms, and the small spheres correspond to Ce or Zr atoms in a solid solution.

(20). An identical result was observed with XANES after exposing MgO powders to moderate pressures of SO_2 (21). The metal centers of MgO interact weakly with SO_2 and are not able to dissociate the molecule (20,21). In general, bulk powders of pure stoichiometric oxides (MgO, Al_2O_3, TiO_2, Cr_2O_3, Fe_2O_3, NiO, CuO, ZnO, ZrO_2, V_2O_5, MoO_3, $CoMoO_4$, and $NiMoO_4$) do not decompose SO_2 (12,13,16–26). Well-defined surfaces of oxides that expose metal cations with a high coordination number {MgO(100), ZnO(0001), $TiO_2(110)$, NiO(001), $V_2O_5(0001)$, $Fe_2O_3(0001)$} also do not dissociate SO_2 (17,18,27–31). Thus, stoichiometric oxides can be very good as sorbents (forming SO_3 or SO_4 species), but in general, they will not be active as catalysts for reactions that involve S–O bond cleavage.

At a theoretical level, the bonding between SO_2 and the cations of oxides has been investigated using extended two-dimensional slabs {MgO, CaO, SrO, TiO_2 (21,24,32)}. In all theoretical calculations, the frontier molecular orbitals of SO_2 mix poorly with electronic states located on the metal centers of the oxides. Figure 20.4 compares the band energies of a common oxide (MgO) and the molecular orbital energies of SO_2 (33,34). For the oxide, the empty and occupied bands are indicated by dotted and solid lines, respectively. The lowest unoccupied molecular orbital (LUMO) of SO_2 is S–O anti-bonding (32,34). In typical oxides, the occupied states of the metal centers are too stable for interacting or transferring electron density into the LUMO of SO_2 (i.e., no effective band-orbital mixing). This leads to small SO_2 adsorption

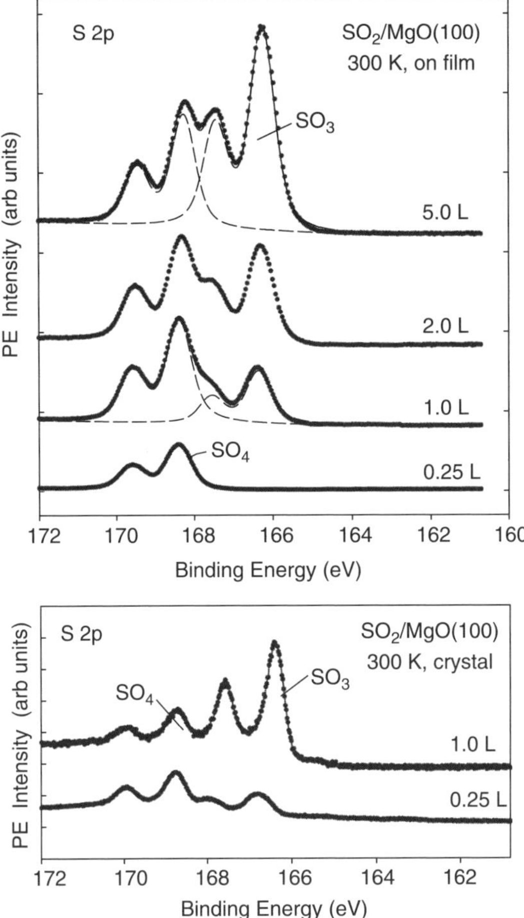

Figure 20.3. S 2*p* photo-emission spectra for the adsorption of SO_2 on a MgO(100) single crystal and a MgO(100) epitaxial film grown on a Mo(100) substrate. Reprinted from Ref. 20.

energies on the cations and prevents dissociation of the molecule (21,24,32). To dissociate SO_2 on an oxide surface, one needs to create occupied metal states above the valence band of the oxide. This can be accomplished by the introduction of O vacancies or other structural defects on the surface. This type of defect is frequent on oxide nanoparticles and tend to move up, the occupied levels or down the empty levels (19); see Figure 20.4.

The presence of O vacancies in $CeO_2(111)$ and $Ce_{0.7}Zr_{0.3}O_2(111)$ induces the interaction of SO_2 with the metal cations and dissociation of the molecule (17,18). The right-side panel in Figure 20.5 shows S 2*p* photo-emission results for the adsorption of SO_2 on several ceria systems (16). The left-panel displays the corresponding valence photo-emission spectra for the ceria systems *before* the adsorption of SO_2 (16). In the case of CeO_2, the valence spectrum shows no signal in the region between 4 and

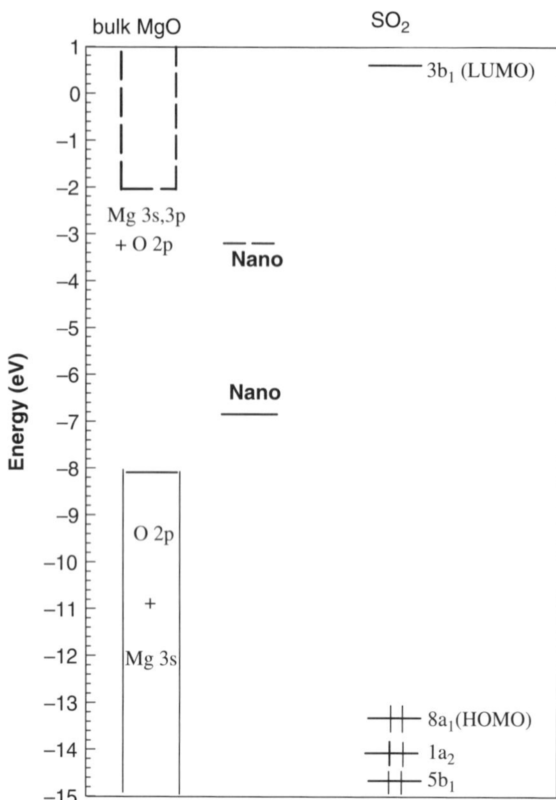

Figure 20.4. Energy position for the valence and conduction bands of bulk MgO. Empty states are shown as doted lines, whereas solid lines denote occupied states. For comparison, we also include the molecular orbital energies of SO_2 and the type of energy shift that can occur for the states of a nanoparticle. Such a shift facilitates interactions with the LUMO of SO_2. The zero of energy is the vacuum level. Reprinted from Refs. 33 and 34.

0 eV, where Ce^{3+} appears. The features between 8 and 4 eV contain O 2*p* character (main component) *and* metal character (16–18). On this system, the adsorption of SO_2 mainly produces SO_4 (16). O atoms can be preferentially removed from CeO_2 by Ar^+ sputtering (16–18). In Figure 20.5, the valence spectra for CeO_{2-x} and Ce_2O_{3+x} are characterized by a Ce^{3+} peak near 2 eV (16,17). Thus, the introduction of O vacancies creates occupied metal states above the valence band of CeO_2. This phenomenon should make the oxide active for the dissociation S–O bonds (see above). And, indeed, the S 2*p* data exhibit features between 164 and 162 eV that come from the full decomposition of the SO_2 molecule on CeO_{2-x} and Ce_2O_{3+x}. CeO_2 is useful as a catalysts for the reduction of SO_2 by CO only at elevated temperatures, when CO can create O vacancies and associated Ce^{3+} sites in the oxide (16,33,35).

MgO(001), TiO_2(110), NiO(001), V_2O_5(0001), and Fe_2O_3(0001) also become active for the decomposition of SO_2 after the creation of O vacancies by ion sputtering

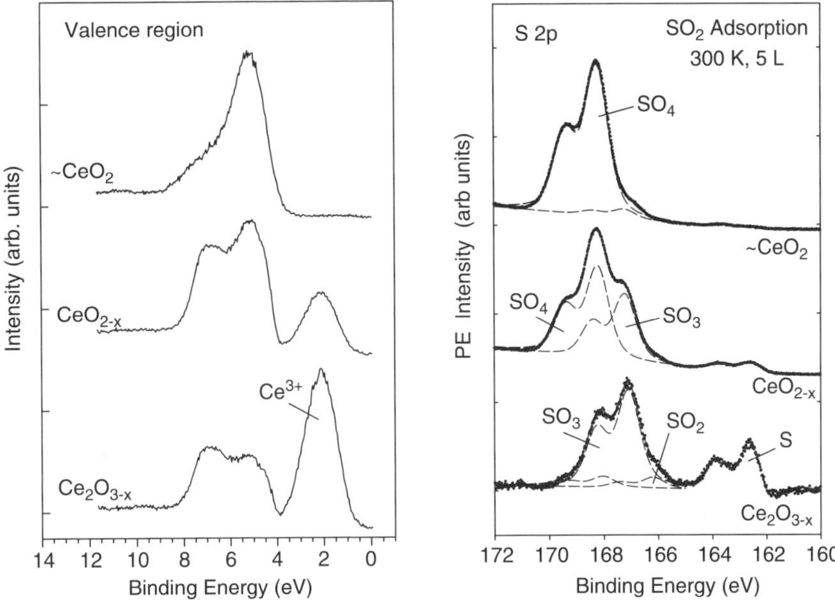

Figure 20.5. Right: Valence photo-emission spectra for a series of ceria systems. Left: S 2*p* spectra taken after dosing 5 langmuir of SO_2 at 300 K to the ceria surfaces. Reprinted from Ref. 16.

(21,27–32). Theoretical calculations show that, on these defects, the adsorption energy of SO_2 is much larger than on flat stoichiometric surfaces with a substantial weakening and elongation of the S–O bonds (21,32). In principle, oxides that can be reduced by CO to form O vacancies could be active catalysts for the $2CO_{gas} + SO_{2,gas} \rightarrow 2CO_{2,gas} + S_{ads}$ reaction. This is the case of CuO and CeO_2 (14,15,16,35). To favor the formation of O vacancies and active centers in these oxides, the reaction is usually carried out at high temperatures with a CO/SO_2 ratio in the feed >2. In the case of ceria, doping with a second metal helps the formation of O vacancies (36,37) and increases catalytic activity (14,15,35,38).

20.3. CHEMISTRY OF SO_2 AND $DeSO_X$ PROCESSES ON OXIDE NANOPARTICLES

"Size effects" have not been studied for many oxides in a systematic way (39). For a vast number of oxide compounds, this comes out from the fact that the preparation procedures are not able to give a size distribution approaching the delta function (39). Detailed studies have appeared in the literature examining the interaction of SO_2 with only a few types of oxide nanoparticles: CaO, MgO, SrO, BaO, Al_2O_3, Al_2O_3/MgO, Fe_2O_3, Fe_2O_3/CaO, CeO_2, and $Ce_{1-x}Zr_xO_2$. The focus has been mainly on oxides of the alkaline-earth elements, frequently used as $DeSO_x$ scrubbers (11), and ceria based materials, useful as $DeSO_x$ catalysts (14,15).

20.3.1. Interaction with CaO, MgO, and SrO

Nanoparticles of CaO, MgO, and SrO usually prefer to adopt a nearly perfect or somewhat distorted cubic shape, exposing the (100) face of a rocksalt crystal structure (40,41). Nanoparticles exhibiting (110) and (111) faces are much less common and frequently are not stable at high temperatures. For example, when Mg metal is burned in air or oxygen, the MgO smoke particles that are formed are almost perfect cubes having (100) faces (42). Special procedures to prepare MgO nanoparticles exhibiting (110) and (111) faces have been partially successful (43), but in general, they tend to facet to surfaces containing (100) planes (44).

An important aspect to consider when dealing with MgO nanoparticles is the possible presence of O vacancies (45). These can have a tremendous influence on the electronic and chemical properties of the nanoparticles. The anionic vacancies in MgO are known as F centers; depending on the charge, one can have F, F^+, and F^{2+} centers that correspond to the removal of a neutral O atom, an O^-, or an O^{2-} anion, respectively (45). The F centers can be described as an electron pair trapped in the cavity left by the missing oxygen. They can produce electronic states localized well above the valence band of MgO (45). The F^+ centers consist of a single electron associated with the vacancy and give rise to a typical signal in electron paramagnetic resonance (EPR). Finally, F^{2+} centers are strongly electron deficient and have a tendency to ionize bonded molecules.

The adsorption of SO_2 on nanoparticles of MgO at 300 K mainly produces SO_4 groups, $SO_2(gas) + 2O(oxide) \rightarrow SO_4(adsorbed)$, with a very small amount of SO_3 and without cleavage of S–O bonds (i.e., no deposition of atomic sulfur on the Mg cations) (44). In contrast, the adsorption of SO_2 on a MgO(100) crystal under similar conditions mainly yields SO_3 with SO_4 as a secondary product (see above) (44). Calculations based on the Hartree–Fock method and density functional theory (DFT) indicate that the formation of SO_4 on MgO(100) is not spontaneous and requires a major reconstruction of the surface with the participation of defect sites (21,24). In the calculations, the interaction of SO_2 with MgO(100) produces a SO_3-like species (21,24). For SO_2 on the MgO nanoparticles, the existence of corner and edge sites in the oxide substrate facilitates the structural changes necessary for the formation of SO_4 (46).

For several industrial applications, MgO is doped with small amounts of a transition metal. Such doping can induce structural transformations and be used to stabilize MgO nanoparticles that expose (110) or (111) faces (44). The doping also can lead to perturbations in the electronic properties of the nanoparticles by favoring the formation of O vacancies or by introducing new occupied states above the valence band of MgO as shown in Figure 20.6 (47). The position of the new occupied states depends on the nature of the dopant element. This phenomenon is particularly important when the doping is done with metals like Fe or Cr that induce states 2–3 eV above the MgO

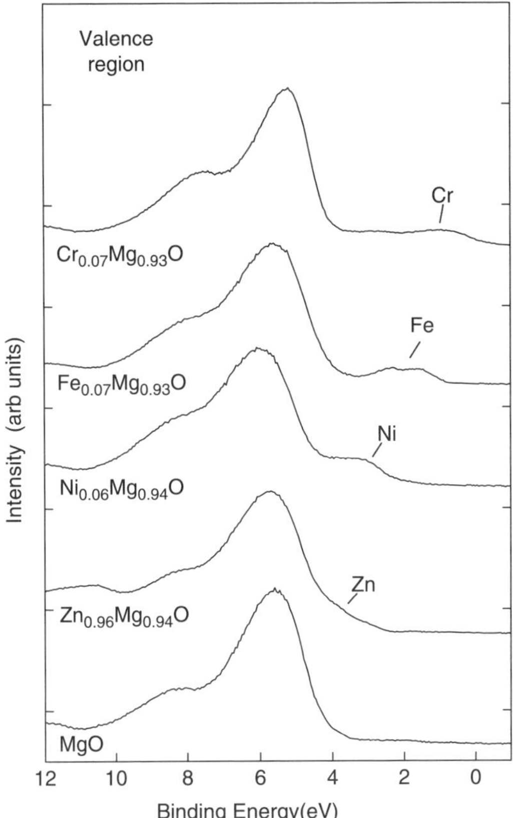

Figure 20.6. Valence photo-emission spectra for pure and doped magnesium oxide. Reprinted from Refs. 39 and 48.

valence band. In general, the $TM_xMg_{1-x}O$ systems (TM = Ni, Fe, Mn, Cr) exhibit electronic and chemical properties different from those of pure MgO (44,47).

The doping of the MgO nanoparticles with Cr produces a system that is extremely efficient for the destruction of SO_2 (44,48). The $Mg_{1-x}Cr_xO_2$ nanoparticles are able to cleave the S–O bonds at temperatures below 300 K, whereas bulk MgO and Cr_2O_3 only adsorb the molecule to form SO_3 or SO_4 species (44,48). The Cr atoms in the $Mg_{1-x}Cr_xO_2$ nanoparticles are trapped in a "+2" formal oxidation state and have occupied electronic states that appear well above the MgO valence band (see Figure 20.6). These properties facilitate interactions with the LUMO of SO_2 and make the nanoparticles more chemically active than bulk MgO or Cr_2O_3 (44,48).

Depending on the exact procedure followed for the preparation, nanoparticles of MgO and CaO with polyhedral or hexagonal shapes can be prepared, but they also contain OH groups (40,49). In these morphological shapes, the nanoparticles posses more defects than expected for the typical cubic shape of MgO and CaO (49). Such defects could be of the Frenkel or Schottky type (vacancies) or be manifested as unusual configurations of edges, corners, or crystal planes (40). Adsorption data for SO_2 and other molecules are conclusive that the polyhedral or hexagonal nanocrystals of MgO and CaO are more reactive than cubic microcrystals of these oxides (49). This has been mainly attributed to morphological differences, including the concentration of defects. However, intrinsic electronic effects due purely to "smallness" (confinement) could not be ruled out (49).

A comparison of the reactivity of CaO particles with sizes of 7.3 and 14.6 nm shows that the smaller particles are ~1.5 times more efficient for the trapping of SO_2 than the bigger particles (50). This can be attributed to a larger concentration of corner and edge sites in the 7.3-nm particles, sites that are important for the adsorption of SO_2 (51). Both sizes of CaO nanoparticles are more reactive than the bulk oxide. Furthermore, the pure nanoparticles of CaO exhibit a higher activity for S–O bond cleavage than nanoparticles of pure MgO (44,50,51). A mixture of calcium sulfite, calcium sulfate, and calcium sulfide is observed on the CaO upon the adsorption of SO_2. These species may be formed through two different mechanisms (50):

$$CaO + SO_2 \longrightarrow CaSO_3 \tag{20.3}$$

$$CaO + SO_2 \longrightarrow \tfrac{3}{4}CaSO_4 + \tfrac{1}{4}CaS \tag{20.4}$$

The SO_2/CaO system is complex, and cation-to-cation exchanges of the $SO_3^{2-} \leftrightarrow O^{2-}$, $SO_4^{2-} \leftrightarrow O^{2-}$, and $S^{2-} \leftrightarrow O^{2-}$ type may be occurring (50). One reason why the nanocrystals display such a large efficiency for trapping SO_2 is that the small size of the particles permits a shorter ion migration distance to the core of the particle than for that of conventional calcium oxide. With the shorter ion migration distances, it is easier to react with the entire CaO particle by effectively "eating out" the core (50).

Nanoparticles of SrO also can be used to mitigate atmospheric pollution and sequester SO_2 (52). A particularly large efficiency was seen for nanoparticles with sizes under 10 nm. The results of extended X-ray absorption five structure (EXAFS) indicate that this large efficiency is associated with a substantial degree of disorder in the lattice structure (52).

20.3.2. Interaction with Al_2O_3 and Al_2O_3/MgO

Nanocrystals of Al_2O_3 and Al_2O_3/MgO were produced through a modified aerogel synthesis (53). The resulting oxides were in the form of powders having crystallites of about 2 nm or less in dimension. They exhibited a reactivity toward SO_2 much larger than that of bulk aluminas or MgO (53). This is thought to be due to morphological differences, whereas larger crystallites have only a small percentage of reactive sites on the surface, smaller crystallites possess much higher surface concentration of such sites per unit surface area.

Figure 20.7 shows infrared spectra collected after exposing the nanoparticles (NCs) and commercial powders (CMs) to SO_2 (20 Torr, room temperature) followed

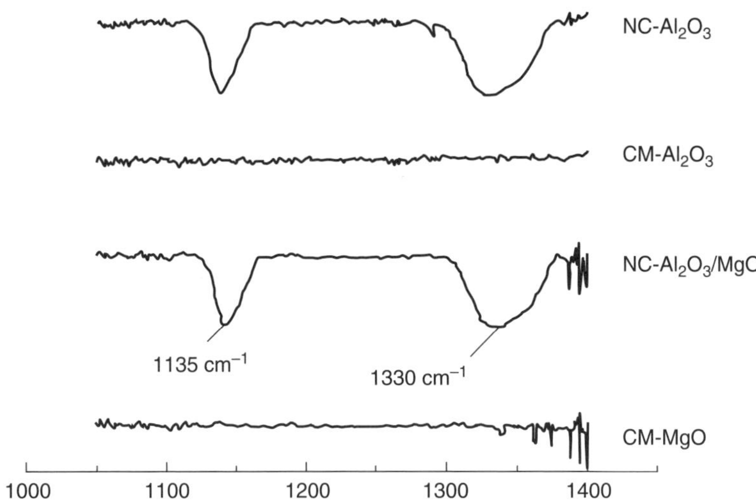

Figure 20.7. Infrared spectra collected after exposing nanoparticles of Al_2O_3 and Al_2O_3/MgO to 20 Torr of SO_2 at room temperature followed by a two 2-hour evacuation. "NC" and "CM" refer to nanocrystalline and commercial, respectively. Reprinted from Ref. 53.

by a 2-hour evacuation (53). The CM–MgO and CM–Al_2O_3 showed no adsorbed species. On the other hand, both NC–Al_2O_3 and NC–Al_2O_3/MgO show new peaks at 1135 and 1130 cm^{-1} that correspond to chemisorbed monodentate SO_2. Clearly the nanoparticles have a special chemical activity. The nanoparticles of Al_2O_3/MgO were found to be the best SO_2 sorbents among the examined samples (53). A significant feature is that, by a cogellation synthesis, Al_2O_3 and MgO have been intermingled, which engenders enhanced reactivity/capacity over the pure forms of nanoscale Al_2O_3 or MgO (53).

20.3.3. Interaction with Fe_2O_3, Fe_2O_3/CaO, and Fe_2O_3/SrO

Nanoparticles of Fe_2O_{3-x} with sizes in the range of 2–4 nm can be used for sequestering SO_2 and do have a moderate activity for the $SO_2 + 2CO \rightarrow 2CO_2 + S_n$ reaction (3,44). S K-edge spectra taken after the adsorption of SO_2 at room temperature point to the presence of S, SO_2, and SO_3 on the surface of the Fe_2O_{3-x} nanoparticles (44). They are much more reactive than bulk Fe_2O_3 (3,44), probably due to the presence of O vacancies and surface defects (44). Measurements of X-ray powder diffraction show a lot of stress, caused by imperfections, in the lattice of the Fe_2O_{3-x} nanoparticles (44).

Coating with Fe_2O_3 enhances the ability of CaO and SrO nanoparticles to adsorb SO_2 (50,52,54). Only small amounts of Fe_2O_3 are necessary to see this phenomenon (see Figure 20.8). The supported Fe_2O_3 acts as a facilitator, and there is a direct reaction of the calcium or strontium oxide with SO_2. The reaction does not stop at the surface, and the CaO and SrO behave as a stoichiometric reagents (50,54).

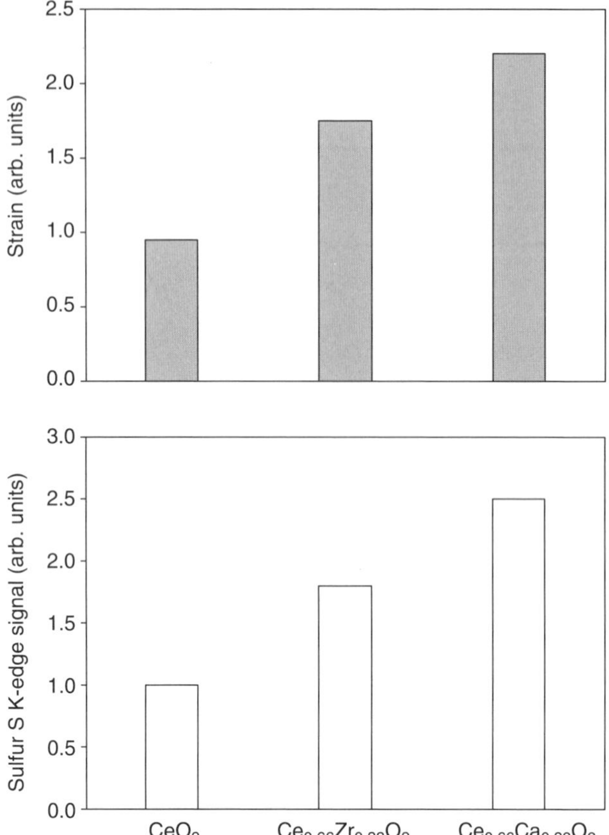

Figure 20.8. Top panel: Lattice strain for CeO_2, $Ce_{0.66}Zr_{0.33}O_2$, and $Ce_{0.66}Ca_{0.33}O_{2-y}$ nanoparticles (36,37). Bottom panel: Amount of atomic sulfur deposited on CeO_2, $Ce_{0.66}Zr_{0.33}O_2$, and $Ce_{0.66}Ca_{0.33}O_{2-y}$ nanoparticles as a consequence of the dissociation of SO_2. The samples were exposed to 0.1 Torr of SO_2, for 5 min at 25 °C. Then, the gas was evacuated and S K-edge spectra were collected. Reprinted from Ref. 19.

Table 20.1 compares the reaction efficiencies for the trapping of SO_2 by calcium oxide nanoparticles pure or coated with Fe_2O_3 (50). The CaO nanoparticles have sizes of 7.3 and 14.6 nm (SP and MP particles, respectively, in our notation). The extent of the reaction is indicated by the breakthrough number and the number of moles of SO_2 that are adsorbed per mole of CaO (50). The breakthrough number is defined as the number of 1-mL injections that are made to the reaction cell until the first trace of excess SO_2 is eluted from the bed of the adsorbent. In Table 20.1, it is obvious the tremendous effect of Fe_2O_3 on the efficiency of the system. MP–CaO particles coated with Fe_2O_3 are more active than SP–CaO particles, and when the SP–CaO particles are coated with Fe_2O_3, the efficiency of the system is very close to the maximum limit (i.e., 0.94 versus 1.0). Systematic studies indicate that the enhancement in reactivity is a kinetic phenomenon (50).

TABLE 20.1. Breakthrough Numbers and Reaction Efficiencies for the Trapping of SO_2 by CaO and Fe_2O_3/CaO Nanoparticles.[a]

Sample	Breakthrough Number	Reaction Efficiencies (mol of SO_2/mol of CaO)[b]
MP–CaO[c]	11	0.36
Fe_2O_3/MP–CaO	23	0.55
SP–CaO[d]	19	0.51
Fe_2O_3/SP–CaO	44	0.94

[a]From Ref. 50.
[b]Theoretical maximum would be 1.0.
[c]MP = 14.6 nm in size.
[d]SP = 7.3 nm in size.

The structure of Fe_2O_3/SrO nanoparticles before and after exposure to SO_2 was investigated using Sr and Fe K-edge EXAFS measurements (52). In the fresh Fe_2O_3/SrO nanoparticles, the first Sr–O peak is almost unchanged with respect to that in bulk SrO, but the second Sr–Sr peak is dramatically decreased. This points to a disordered oxide lattice in the Fe_2O_3/SrO nanoparticles and may help to enhance their reactivity toward SO_2 (52). The Fe K-edge EXAFS data for the fresh Fe_2O_3/SrO nanoparticles indicate that the Fe_2O_3 coating has a large degree of disorder, and some cations may be reduced to metallic iron (52). Upon interaction with SO_2, the characteristic Fe–S peak for iron sulfide was not seen. Unfortunately, it was not possible to discriminate between disordered Fe_2O_3 and $Fe_2(SO_4)_3$ because they have a similar first Fe–O peak (52).

20.3.4. Interaction with CeO_2 and Doped-Ceria

Figure 20.1 compares S K-edge XANES spectra collected after dosing SO_2 at 25 °C to $CeO_2(111)$ and $Ce_{0.7}Zr_{0.3}O_2(111)$ surfaces and nanoparticles of CeO_2, $Ce_{0.66}Zr_{0.33}O_2$, and $Ce_{0.66}Ca_{0.33}O_{2-y}$ (19). The spectra for the extended surfaces display a peak for SO_4 as a dominant feature with a minor peak for SO_3. For the oxide nanoparticles, again one finds that SO_4 is the main sulfur-containing species present on the surface, but in addition, features are seen at photon energies between 2470 and 2472 eV that denote the existence of metal-S bonds (19) as a consequence of the full dissociation of sulfur dioxide ($SO_2 \rightarrow S + 2O$) on the cerium cations. In principle, clusters and nanoparticles of $Ce_{1-x}Zr_xO_2$ probably have metal cations at corner and edge sites that can interact well with the SO_2 molecule. On some of these special sites that are very reactive, SO_2 decomposes. In addition, there may be O vacancies in the surface of the $Ce_{0.66}Zr_{0.33}O_2$ and $Ce_{0.66}Ca_{0.33}O_{2-y}$ nanoparticles that facilitate S–O bond cleavage (36). In Figure 20.1, the $Ce_{0.66}Ca_{0.33}O_{2-y}$ system has the largest concentration of O vacancies (36) and the highest reactivity for the dissociation of SO_2.

From the X-ray diffraction (XRD) data of the nanoparticles, one can get a strain parameter that is a measurement of the lattice stress existing in the materials because of surface defects (differences in local symmetry and distances with respect to the bulk) and/or the crystal imperfections (O vacancies, other point defects, line defects, and plane defects) (36,37,39). The top panel in Figure 20.8 shows the strain parameters for the CeO_2, $Ce_{0.66}Zr_{0.33}O_2$, and $Ce_{0.66}Ca_{0.33}O_{2-y}$ nanoparticles (36,37). Pure ceria nanoparticles exhibit a larger lattice strain than bulk ceria (36). Clearly, the introduction of an alien species like Zr or Ca leads to extra forces that increase the strain in the lattice of the nanoparticles. The effects of Ca are more significant because Ca likes to form oxides with a relatively low content of O and the oxidation states of Ca and Ce are different (36). In Figure 20.8, a qualitative correlation can be found between the amount of S deposited on the nanoparticles, as a consequence of the dissociation of SO_2 (19), and the strain parameter of the nanoparticles. The nanoparticles that have more lattice imperfections, $Ce_{0.66}Ca_{0.33}O_{2-y}$, are the more active for S–O bond cleavage.

Figure 20.9 shows the effect of the temperature on the sulfate (SO_4) signal for the CeO_2 and $Ce_{1-x}Zr_xO_2$ systems in Figure 20.1 (19). As the temperature is raised, SO_4 decomposes. In the case of the $CeO_2(111)$ and $Ce_{0.7}Zr_{0.3}O_2(111)$ surfaces, the adsorbed SO_4 transforms into SO_2 gas. On the other hand, in the case of the nanoparticles, most decomposed SO_4 yields SO_2 gas, but a fraction undergoes complete decomposition depositing S on the oxide substrate. The SO_4 adsorbed on the nanoparticles is somewhat more stable than that present on the (111) surfaces. For both types of systems, the presence of Zr seems to induce an increase in the thermal stability of the adsorbed sulfate. The Zr cations also enhance the thermal stability of SO_4 species formed on partially reduced $Ce_{1-x}Zr_xO_{2-y}(111)$ surfaces (17).

Nanoparticles of $Ce_{1-x}D_xO_{2-y}$(D = Sr, Sc, La, Gd, Ni, Cu) display catalytic activity for the reduction of SO_2 with carbon monoxide ($SO_2 + 2CO \rightarrow 2CO_2 + S$) or methane ($2SO_2 + CH_4 \rightarrow CO_2 + 2H_2O + 2S$) (55,56). High catalytic activity was observed when ceria was doped with copper or nickel. It is well established that copper facilitates the reduction of ceria nanoparticles by CO (57). A redox mechanism has been proposed to explain the reduction of SO_2 by CO on the $Ce_{1-x}D_xO_{2-y}$ nanoparticles (55):

$$\text{cat-O} + CO \longrightarrow \text{cat-[ ]} + CO_2 \qquad \textbf{(20.5)}$$

$$\text{cat-[ ]} + SO_2 \longrightarrow \text{cat-O} + SO \qquad \textbf{(20.6)}$$

$$\text{cat-[ ]} + SO \longrightarrow \text{cat-O} + \qquad \textbf{20.7)}$$

First, an oxygen vacancy is created as a surface capping oxygen is removed by CO. Then, SO_2 donates one of its oxygens to the vacancy to form SO. The SO is mobile on the surface until it finds another vacancy to donate its oxygen or a vacancy may migrate to a neighboring site to accept its oxygen. High oxygen mobility in the catalyst will facilitate the oxygen transfer from one site to another on the surface or from the bulk to the surface. A dopant may help in this respect. However, surface reduction by CO is still the key step to initiate the reaction (55). Under reaction conditions, CO

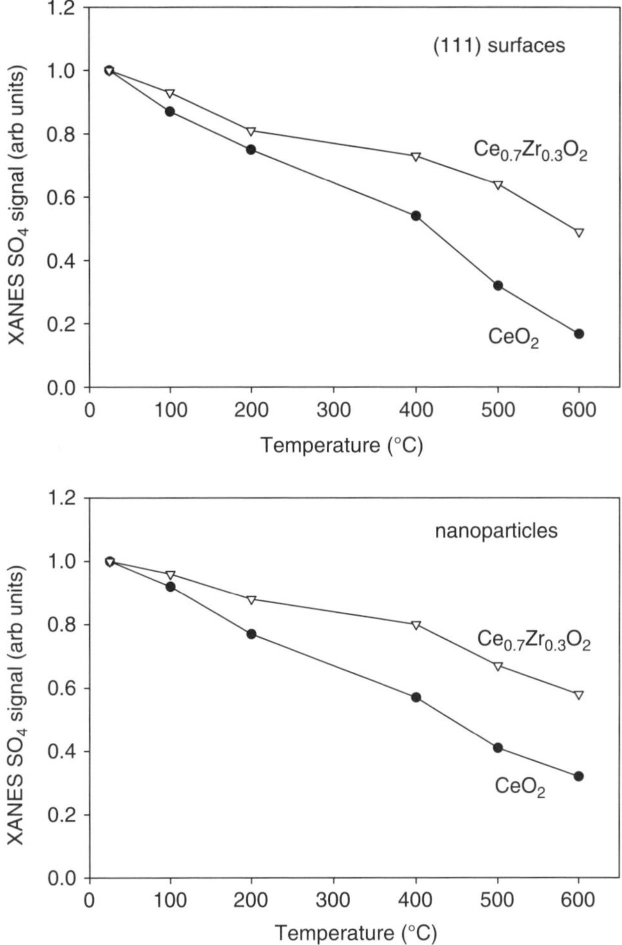

Figure 20.9. Effect of temperature on the XANES signal for the SO_4 formed on the CeO_2 and $Ce_{1-x}Zr_xO_2$ systems of Figure 20.1. The top panel shows the results for the (111) surfaces, whereas the bottom panel contains the corresponding results for the nanoparticles. Reprinted from Ref. 19.

and SO_2 compete for the surface oxygen. Reaction with CO produces an O vacancy (Reaction **20.5**) whereas reaction of SO_2 with the surface oxygens forms sulfite or sulfate species that are strongly bound and hinder the redox reaction.

The Cu–Ce(La)–O nanocatalysts displayed a better performance than Ce(La)–O_2 or CuO (55). Thus, a synergistic effect was observed for the Cu-dopped ceria. Copper and cerium oxide probably play different roles in the redox mechanism (55). Cerium oxide comprises the matrix of the catalyst and provides the oxygen and oxygen vacancy sources, whereas Cu cations promote the reducibility of cerium oxide and provide surface sites for CO adsorption (55,57).

20.4. CONCLUSIONS

On bulk stoichiometric oxides, SO_2 mainly reacts with the O centers to form SO_3 or SO_4 species that decompose at elevated temperatures. Adsorption on the metal cations occurs below 300 K and does not lead to cleavage of the S–O bonds. In bulk oxides, the occupied cation bands are too stable for effective bonding interactions with the LUMO of SO_2. The effects of quantum confinement on the electronic properties of oxide nanoparticles and the structural defects that usually accompany these systems in general favor the bonding and dissociation of SO_2. Thus, nanoparticles of MgO, CaO, SrO, Al_2O_3, Fe_2O_3, and CeO_2 are all more efficient for sequestering SO_2 than the corresponding bulk oxides. Structural imperfections in pure or metal-doped ceria nanoparticles accelerate the reduction of SO_2 by CO by facilitating the formation and migration of O vacancies in the oxide surface.

ACKNOWLEDGMENTS

The author is grateful for financial support from the U.S. Department of Energy (Divisions of Chemical and Materials Science) under Contract DE-AC02-98CH10086.

REFERENCES

(1) Adlkofer, J. *Handbook of Heterogeneous Catalysis*, Vol. 4; Ertl, G.; Knözinger, H.; Weitkamp, J. (Editors); Wiley-VCH: New York, 1997; 1776.

(2) Stern, A.C.; Boubel, R.W.; Turner, D.B.; Fox, D.L. *Fundamentals of Air Pollution, 2nd ed.*; Academic Press: Orlando, FL, 1984.

(3) Piéplu, A.; Saur, O.; Lavalley, J.-C.; Legendre, O.; Nédez, C. *Catal. Rev. Sci. Eng.* **1998**, *40*, 409.

(4) *Environmental Catalysis*; Armor, J.N. (Editor); ACS Symposium Series No 552 American Chemical Society: Washington, DC, 1994.

(5) Shelef, M.; McCabe, R.W. *Catal. Today* **2000**, *62*, 35.

(6) Bartholomew, C.H.; Agrawal, P.K.; Katzer, J.R. *Adv. Catal.* **1982**, *31*, 135.

(7) Rodriguez, J.A.; Hrbek, J. *Acc. Chem. Res.* **1999**, *32*, 719.

(8) Beck, D.D.; Sommers, J.W.; DiMaggio, C.L. *Appl. Catal. B* **1997**, *11*, 273.

(9) Beck, D.D. *Catal. Deact.* **1997**, *111*, 21.

(10) Gorte, R. Private communication.

(11) Slack, A.V.; Holliden, G.A. *Sulfur Dioxide Removal from Waste Gases, 2nd ed.*; Noyes Data Corporation: Park Ridge, NJ, 1975.

(12) Centi, G.; Passarini, N.; Perathoner, S.; Riva, A. *Ind. Eng. Chem. Res.* **1992**, *31*, 1947.

(13) Waqif, M.; Saur, O.; Lavalley, J.C.; Perathoner, S.; Centi, G. *J. Phys. Chem.* **1991**, *95*, 4051.

(14) Zhu, T.; Kundakovic, K.; Dreher, A.; Flytzani-Stephanopoulos, M. *Catal. Today* **1999**, *50*, 381.

(15) Liu, W.; Wadia, C.; Flytzani-Stephanopoulos, M. *Catal. Today* **1996**, *28*, 391.

(16) Rodriguez, J.A.; Jirsak, T.; Freitag, A.; Hanson, J.C.; Larese, J.Z.; Chaturvedi, S. *Catal. Lett.* **1999**, *62*, 113.

(17) Liu, G.; Rodriguez, J.A.; Chang, Z.; Hrbek, J.; Peden, C.H.F. *J. Phys. Chem. B* **2004**, *108*, 2931.

(18) Overbury, S.H.; Mullins, D.R.; Huntley, D.; Kundakovic, L.J. *J. Phys. Chem. B* **1999**, *103*, 11308.

(19) Rodriguez, J.A.; Wang, X.; Liu, G.; Hanson, J.C.; Hrbek, J.; Peden, C.H.F.; Iglesias-Juez, A.; Fernández-García, M. *J. Molec. Catal. A: Chem.* **2005**, *228*, 11.

(20) Rodriguez, J.A.; Pérez, M.; Jirsak, T.; González, L.; Maiti, M. *Surf. Sci.* **2001**, *477*, L279.

(21) Rodriguez, J.A.; Jirsak, T.; Freitag, A.; Larese, J.Z.; Maiti, A. *J. Phys. Chem. B* **2000**, *104*, 7439.

(22) Rodriguez, J.A.; Jirsak, T.; Chaturvedi, S.; Kuhn, M. *Surf. Sci.* **1999**, *442*, 400.

(23) Rodriguez, J.A.; Jirsak, T.; Chaturvedi, S.; Dvorak, J. *J. Molec. Catal. A* **2001**, *167*, 47.

(24) Schneider, W.F. *J. Phys. Chem. B* **2004**, *108*, 273.

(25) Waqif, M.; Saad, A.M.; Bensitel, M.; Bachelier, J.; Saur, O.; Lavalley, J.C. *J. Chem. Soc., Faraday Trans.* **1992**, *88*, 2931.

(26) Rodriguez, J.A.; Hanson, J.C.; Chaturvedi, S.; Brito, J.L. *Stud. Surf. Sci. Catal.* **2000**, *130*, 2795.

(27) Warburton, D.R.; Pundie, D.; Muryn, C.A.; Prahakaran, K.; Wincott, P.L.; Thorton, G. *Surf. Sci.* **1992**, *269/270*, 305.

(28) Muryn, C.; Purdie, D.; Hardman, P.; Johnson, A.L.; Prakash, N.S.; Raiker, G.N.; Thorton, G.; Law, D. *Faraday Discuss. Chem. Soc.* **1990**, *89*, 77.

(29) Kurtz, R.L.; Henrich, V.E. *Phys. Rev. B* **1987**, *36*, 3413.

(30) Zhang, Z.; Henrich, V.E. *Surf. Sci.* **1990**, *225*, 47.

(31) Li, X.; Henrich, V.E. *Phys. Rev. B* **1993**, *48*, 17486.

(32) Rodríguez, J.A.; Liu, G.; Jirsak, T.; Hrbek, J.; Chang, Z.; Dvorak, J.; Maiti, A. *J. Am. Chem. Soc.* **2002**, *124*, 5242.

(33) Rodriguez, J.A.; Jirsak, T.; Hrbek, J. *J. Phys. Chem. B* **1999**, *103*, 1966.

(34) Rodriguez, J.A.; Chaturvedi, S.; Kuhn, M.; Hrbek, J. *J. Phys. Chem. B* **1998**, *102*, 5511.

(35) Tschope, A.; Liu, W.; Flytzani-Stephanopoulos, M.; Ying, J.Y. *J. Catal.* **1995**, *157*, 42.

(36) Rodríguez, J.A.; Wang, X.; Hanson, J.C.; Liu, G.; Iglesias-Juez, A.; Fernández-García, M. *J. Chem. Phys.* **2003**, *119*, 5659.

(37) Wang, X.; Hanson, J.C.; Liu, G.; Rodriguez, J.A.; Iglesias-Juez, A.; Fernández-García, M. *J. Chem. Phys.* **2004**, *121*, 5434.

(38) de Carolis, S.; Pascual, J.L.; Petterson, L.G.M.; Baudin, M.; Wojcik, M.; Hermansson, K.; Palmqvist, A.E.C.; Muhammed, M. *J. Phys. Chem. B* **1999**, *103*, 7627.

(39) Fernández-García, M.; Martínez-Arias, A.; Hanson, J.C.; Rodríguez, J.A. *Chem. Rev.* **2004**, *104*, 4063.

(40) Klabunde, K.J.; Stark, J.; Koper, O.; Mobs, C.; Park, D.G.; Decker, S.; Jiang, Y.; Lagadic, I.; Zhang, D. *J. Phys. Chem.* **1996**, *100*, 12142.

(41) Tasker, P.W. *Adv. Ceram.* **1984**, *10*, 176.

(42) Moodie, A.F.; Warble, C.E. *J. Crystal Growth* **1971**, *10*, 26.
(43) Mackrodt, W.C.; Tasker, P.W. *Chem. Brit.* **1985**, *21*, 13.
(44) DeSantis, E.; Ferrari, J. Private communication.
(45) Pacchioni, G.; Pescarmona, P. *Surf. Sci.* **1998**, *412/413*, 657.
(46) Pacchioni, G.; Clotet, A.; Ricart, J.M. *Surf. Sci.* **1994**, *315*, 337.
(47) Rodriguez, J.A. *Catal. Today* **2003**, *85*, 177.
(48) Rodriguez, J.A.; Jirsak, T.; Pérez, M.; Chaturvedi, S.; Kuhn, M.; González, L.; Maiti, A. *J. Am. Chem. Soc.* **2000**, *122*, 12362.
(49) Lucas, E.; Decker, S.; Khaleel, A.; Seitz, A.; Futlz, S.; Ponce, A.; Li, W.; Carnes, C.; Klabunde, K.J. *Chem. Eur. J.* **2001**, *7* 2505.
(50) Decker, S.; Klabunde, K.J. *J. Am. Chem. Soc.* **1996**, *118*, 12465.
(51) Pacchioni, G.; Ricart, J.M.; Illas, F. *J. Am. Chem. Soc.* **1994**, *116*, 10152.
(52) Moscovici, J.; Michalowicz, A.; Decker, S.; Lagadic, I.; Latreche, K.; Klabunde, K. *J. Synchrotron Rad.* **1999**, *6*, 604.
(53) Carnes, C.L.; Kapoor, P.N.; Klabunde, K.J. *Chem. Mater.* **2002**, *14*, 2922.
(54) Decker, S.P.; Klabunde, J.S.; Khaleel, A.; Klabunde, K.J. *Environ. Sci. Technol.* **2002**, *36*, 762.
(55) Liu, W.; Wadia, C.; Flytzani-Stephanopoulos, M. *Catal. Today* **1996**, *28*, 391.
(56) Flytzani-Stephanopoulos, M.; Zhu, T.; Li, Y. *Catal. Today* **2000**, *62*, 145.
(57) Wang, X.; Rodriguez, J.A.; Hanson, J.C.; Gamarra, D.; Martínez-Arias, A.; Fernández-García, M. *J. Phys. Chem. B* **2006**, *110*, 428.

CHAPTER 21

H_2 Production and Fuel Cells

XIANQIN WANG

Institute for Interfacial Catalysis, Pacific Northwest National Laboratory, Richland, WA, 99352 (USA)

JOSÉ A. RODRIGUEZ

Chemistry Department, Brookhaven National Laboratory, Upton, NY, 11973 (USA)

21.1. INTRODUCTION

The finding of new sources of energy is perhaps the most important problem that faces humanity today (1). The crucial energy problem is caused by the decrease in fossil fuel reserves due to world population growth, technological developments, and increasing energy demand; the global climate change due to the increase of carbon dioxide concentration from the burning of fossil fuels in the atmosphere; and the conflicts and wars due to fluctuations in energy prices, economic recessions, decrease in living standards, and increase in unrest among countries (2). The energy policy of the developed countries is directed by the need for a secure energy supply and by the wish for sustainable growth. Therefore, the new challenges are to create alternative fuels, clean the environment, deal with the causes of global warming, and keep us safe from toxic substances and infectious agents (3).

There are many different kinds of alternative fuels: liquefied petroleum gas, natural gas, methanol, dimethylether, ethanol, bio-diesel, synfuel, and hydrogen (4). Among these alternative fuels, hydrogen is a potential solution for satisfying many of our energy needs while reducing (and eventually eliminating) carbon dioxide and other greenhouse gas emissions (5–8). In the early 1970s, the potential use of hydrogen was discussed at the beginning of the energy crisis. Nowadays, it is high on the political agenda and on the priorities of agencies funding research. Hydrogen is expected to play an important role in future energy scenarios, replacing to some degree fossil fuels and becoming the preferred portable energy carrier for vehicles and in stationary

Synthesis, Properties, and Applications of Oxide Nanomaterials, Edited by José A. Rodríguez and Marcos Fernández-Garcia

applications as well. Fuel cells will play a key role for both situations, as will be discussed below.

Hydrogen can be manufactured via a variety of routes as shown in Figure 21.1. Hydrogen, the most plentiful element available, can be extracted from water by electrolysis. One can imagine capturing energy from the sun and wind and/or from the depths of the earth to provide the necessary power for electrolysis. Alternative energy sources such as these are the promise for the future, but for now they are not feasible for the power needs across the globe. A transitional solution is required to convert certain hydrocarbon fuels to hydrogen (9). Renewable energy sources such as hydraulic, solar, wind, geothermal, wave, and solid waste energy can be used to generate hydrogen from hydrocarbons and/or water. The use of natural gas represents a partial solution because of its high content of hydrogen. Biomass may be converted into hydrogen via gasification or by pyrolysis, followed by liquid-phase "reforming" at subcritical conditions. Currently, nearly 95% of the hydrogen supply is produced from the reforming of crude oil, coal, natural gas, wood, organic wastes, and biomass (10,11). Hydrogen is manufactured today primarily by reacting steam with natural gas. This is the cheapest route, at least for the demand known today, and the process is highly efficient (4).

Fuel cells play a key role for the use of hydrogen in stationary and mobile applications. A fuel cell is a device that uses hydrogen (or a hydrogen-rich fuel) and oxygen to cleanly and efficiently produce electricity by an electrochemical process with water and heat as byproducts. Fuel cells are unique in terms of the variety of their potential applications; they can provide energy for systems as large as a utility power station and as small as a laptop computer. Fuel cells have several benefits over conventional

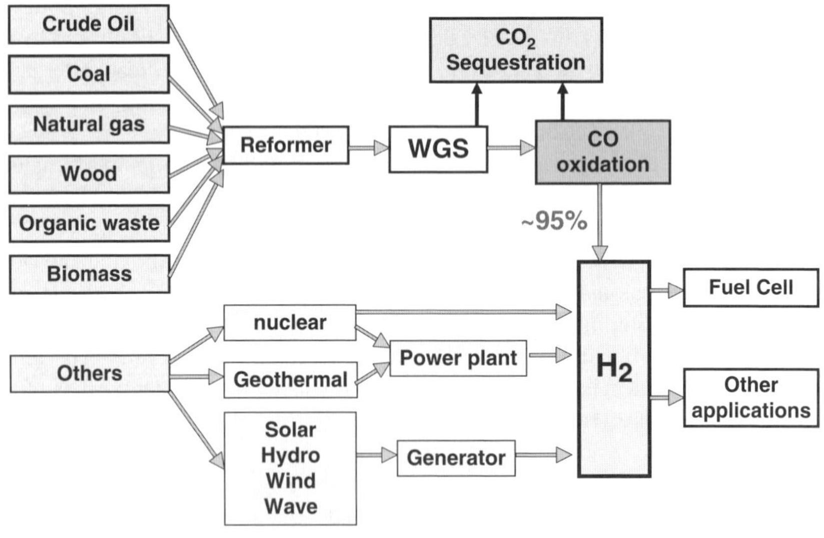

Figure 21.1. H_2 production pathways.

combustion-based technologies currently used in many power plants and passenger vehicles. They produce much smaller quantities of greenhouse gases and none of the air pollutants that create smog and cause health problems. If pure hydrogen is used as a fuel, fuel cells emit only heat and water as a byproduct.

A single fuel cell consists of an electrolyte and two catalyst-coated electrodes (a porous anode and cathode). Although there are different fuel cell types as shown in Figure 21.2, all work on the same principle (12,13):

- Hydrogen, or a hydrogen-rich fuel, is fed to the anode where a catalyst separates hydrogen's negatively charged electrons from positively charged ions (protons).
- At the cathode, oxygen combines with electrons and, in some cases, with species such as protons or water, resulting in water or hydroxide ions, respectively.
- For polymer electrolyte membrane and phosphoric acid fuel cells, protons move through the electrolyte to the cathode to combine with oxygen and electrons, producing water and heat.
- For alkaline, molten carbonate, and solid oxide fuel cells, negative ions travel through the electrolyte to the anode where they combine with hydrogen to generate water and electrons.
- The electrons from the anode side of the cell cannot pass through the electrolyte to the positively charged cathode; they must travel around it via an electrical circuit to reach the other side of the cell. This movement of electrons is an electrical current.

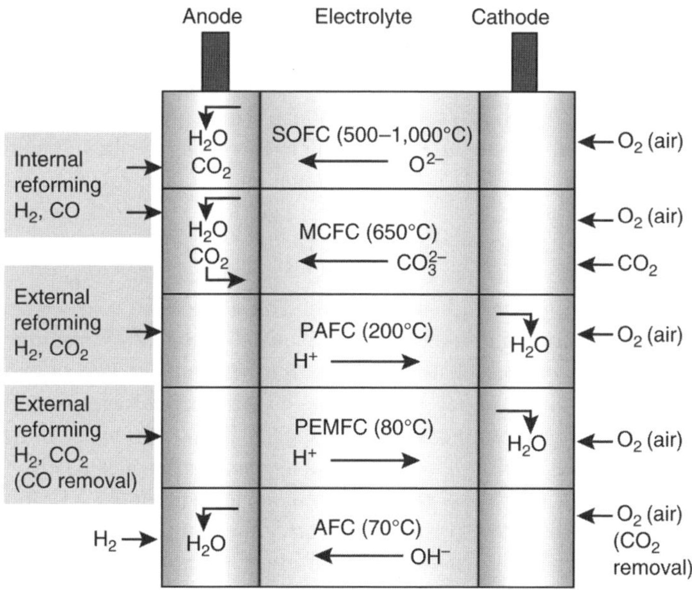

Figure 21.2. Different types of fuel cells (taken from Ref. 12). See color insert.

In most fuel cells, electrochemical oxidation of hydrogen takes place at the anode. In high-temperature fuel cells, it is possible to convert the fuel to hydrogen inside the cell by using the heat from the electrochemical reaction, but otherwise it is necessary to convert the primary fuel outside the stack into a hydrogen-rich gas that is fed to the anode. The coupling of fuel processing with the fuel cell operation is essential to achieve high plant efficiency. It is necessary to integrate the fuel processing system and the fuel cell stack to use the waster heat generated in the fuel cell stack itself. The aim is to reduce the amount of heat generated in the fuel cell power plant, because this can be transferred into electricity only via the Carnot cycle (14).

The primary objective of this chapter is to discuss the key role of free and supported oxide nanomaterials in clean hydrogen production and in the fabrication of fuel cells. Clean hydrogen production from the water–gas-shift (WGS) reaction will be discussed in detail in the second part of this chapter. As an example, we will show recent in situ X-ray diffraction (XRD) and X-ray absorption fine structure (XAFS) studies for the WGS on ceria-based nanocatalysts. Potential applications of oxide nanomaterials in fuel cells will be covered in the third part of this chapter, making emphasis on proton-exchange membrane fuel cells (PEMFCs) and solid oxide fuel cells (SOFCs).

21.2. PRODUCTION

21.2.1. Oxide-Based Nanocatalysts and the Water–Gas-Shift Reaction

As mentioned, hydrogen is manufactured primarily by the steam reforming process today: $C_nH_m + nH_2O \rightarrow nCO + (n - m/2)H_2$. The reformed fuel contains 1–10% CO, which degrades the performance of the Pt electrode in fuel cell systems (15,16). To get clean hydrogen for fuel cells and other industrial applications, the WGS reaction ($CO + H_2O \rightarrow CO_2 + H_2$) and preferential CO oxidation ($2CO + O_2 \rightarrow 2CO_2$) processes are critical (Figure 21.1) (10,11,17). To accelerate them, heterogeneous catalysts are frequently used (18). A fundamental understanding of the configuration and properties of the active sites for the WGS reaction is a prerequisite for designing catalysts with a high activity or efficiency (18).

In industrial applications, the classic catalyst formulations contain a mixture of Fe and Cr oxides or Cu and Zn oxides (18). There has been a lot of controversy about what is the active phase and the mechanism of the WGS reaction on these catalysts (19). Most of the debate resides in the mechanistic area on whether the mechanism is *associative*, taking place through intermediates such as formates, or *regenerative*, via redox reactions involving special forms of copper. In addition to formates, other species such as carbonates and hydroxycarbonates may be participating in the reaction (18,19). A generally accepted mechanism for the WGS reaction yet fails to exist, and this is not surprising considering the complex nature of the catalyst surface. The need for in situ characterization under reaction conditions is clear for these systems (19).

The design of new WGS catalysts is aimed at obtaining systems with a high activity and a lower temperature of operation (20). At the same time, one has to find a compromise between activity and cost. CeO_2-based nanocatalysts have been reported to be very promising for the WGS reaction. Among these materials, the two most studied systems currently are Cu- and Pt-based catalysts (19–22). It is anticipated that, with proper development, metal-promoted ceria catalysts should realize much higher CO conversions than even commercial CuZnO catalysts (23). However, the roles played by ceria and the metal in the WGS reaction are still a matter of debate. The redox properties and oxygen storage capacity of ceria are usually considered important, whereas the metal plays a direct role in the mechanism of the WGS reaction.

21.2.1.1. WGS Reaction on Cu-Ceria Nanocatalysts Two synchrotron-based techniques, time-resolved XRD and XAFS, have been used for examining the behavior of Cu-ceria nanocatalysts in situ under the WGS reaction. The properties of $Ce_{1-x}Cu_xO_2$ catalysts prepared by a novel microemulsion method (24–27) can be compared with those of conventional CuO_x/CeO_2 and AuO_x/CeO_2 catalysts prepared by incipient impregnation methods (28–31). The Cu atoms embedded in ceria have an oxidation state higher than those of the cations in Cu_2O or CuO (32). The lattice of the Ce_{1-x} Cu_xO_2 systems still adopts a fluorite-type structure, but it is highly distorted with multiple cation–oxygen distances in contrast to the single cation–oxygen bond distance seen in pure ceria. The doping of CeO_2 with copper introduces a large strain into the oxide lattice and favors the formation of O vacancies. Cu approaches the planar geometry characteristic of Cu(II) oxides but with a strongly perturbed local order. The chemical activities of the $Ce_{1-x}Cu_xO_2$ nanoparticles were tested using the reactions with H_2 and O_2 as probes (32). During the reduction in hydrogen, an induction time was observed and became shorter after raising the reaction temperature. The fraction of copper that can be reduced in the $Ce_{1-x}Cu_xO_2$ oxides also depends strongly on the reaction temperature. A comparison with data for the reduction of pure copper oxides indicates that the copper embedded in ceria is much more difficult to reduce. The reduction of the $Ce_{1-x}Cu_xO_2$ nanoparticles is reversible, and no significant amounts of CuO or Cu_2O phases are generated during reoxidation (32). This reversible process demonstrates the unusual structural and chemical properties of the Cu-doped ceria materials. The chemical properties of these unusual materials were tested for the hydrogen production via the WGS reaction.

A typical set of time-resolved XRD patterns is shown in Figure 21.3*a* for a $Ce_{0.95}Cu_{0.05}O_2$ catalyst during the WGS reaction at different temperatures. Only the diffraction lines for ceria are seen (33,34), indicating that most copper is embedded in the host oxide forming a solid solution. In Figure 21.3*b*, one can see a big increase in catalytic activity for the production of H_2 when going from 200 °C to 300 °C and then to 400 °C. CO_2 production followed the same curve as H_2 production and is not presented in Figure 21.3*b*. This turn-on of the catalytic activity did not produce diffraction lines for CuO_x or Cu species. As we will see in Figure 21.4, the Cu in $Ce_{0.95}Cu_{0.05}O_2$ is fully reduced under WGS conditions, but it is dispersed within the

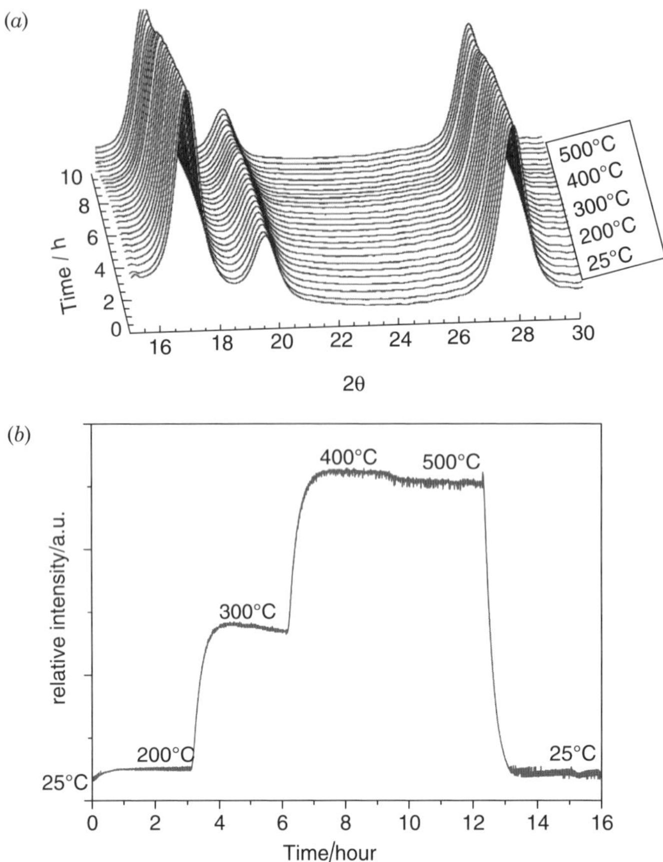

Figure 21.3. (*a*) Time-resolved X-ray diffraction patterns for a $Ce_{0.95}Cu_{0.05}O_2$ catalyst during the WGS reaction at different temperatures ($\lambda = 0.922$ Å). (*b*) The relative concentration of H_2 produced as a function of time at different temperatures during the WGS reaction as shown in (*a*).

oxide probably forming nanoparticles that have sizes under 2 nm and do not yield diffraction peaks.

Similar experiments were carried out with in situ time-resolved XAFS. Figure 21.4*a* displays Cu K-edge X-ray absorption near-edge structure (XANES) spectra collected over the $Ce_{0.95}Cu_{0.05}O_2$ catalyst cooled down to room temperature after the WGS reaction at 25, 100, 200, 300, and 400 °C. The corresponding Fourier transforms for the extended XAFS (EXAFS) region are shown in Figure 21.4*b*. The XANES data for temperatures above 200 °C, when substantial WGS activity was seen, show a line shape typical of pure metallic copper (35,36). In the Fourier transform for the EXAFS, the position of the main peak located between 1.1 and 2.0 Å at room temperature is similar to the first Cu–O coordination shell in CuO (37,38). The Cu–O peak disappeared at high temperatures with a simultaneous increase of the Cu–Cu peak for metallic copper. The product curves from the in situ XAFS experiments are

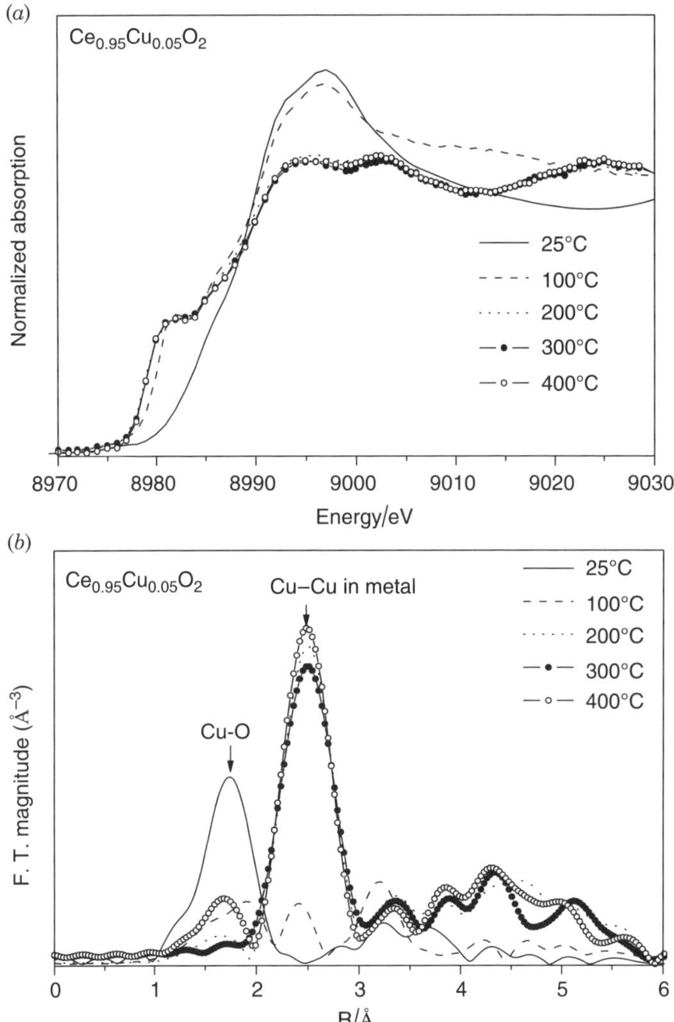

Figure 21.4. (*a*) Cu K-edge XANES and (*b*) EXAFS spectra collected over a $Ce_{0.95}Cu_{0.05}O_2$ catalyst cooled down to room temperature after the WGS reaction at different temperatures.

not shown here because they were similar to those for the in situ TR–XRD experiments. These results demonstrate the important role of metallic copper in the WGS reaction. Another set of experiments for a $Ce_{0.8}Cu_{0.2}O_2$ catalyst, Cu K-edge XANES spectra in the top panel of Figure 21.5, also indicates that Cu is fully reduced when the catalyst displays high WGS activity.

Figure 21.5*b* compares the Ce L_3-edge XANES spectra from the $Ce_{0.8}Cu_{0.2}O_2$ sample in the WGS reaction at different temperatures with the spectrum for a $Ce(NO_3)_3 \cdot 6H_2O$ reference, in which the cerium atoms are trivalent. The two main peaks in the spectrum of the sample at room temperature are separated by

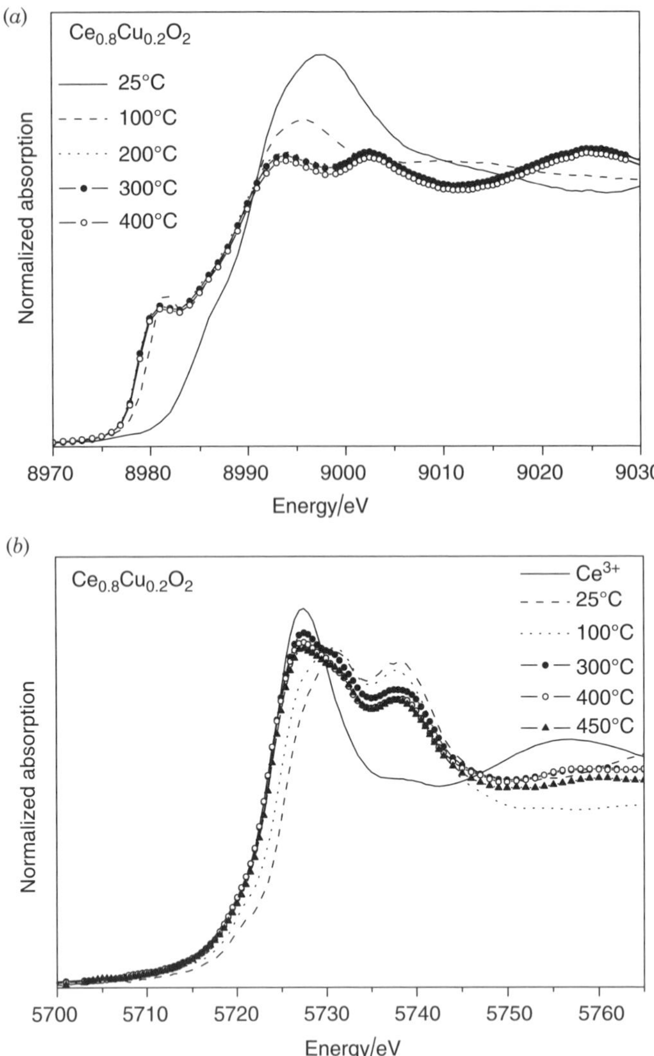

Figure 21.5. (*a*) Cu K-edge XANES spectra collected over a $Ce_{0.8}Cu_{0.2}O_2$ catalyst cooled down to room temperature after the WGS reaction at the indicated temperatures. (*b*) Ce L_3-edge XANES spectra collected over $Ce_{0.8}Cu_{0.2}O_2$ during the WGS reaction at different temperatures.

approximately 7 eV, in agreement with previous results (39). Based on the comparison of the intensities of the spectra near the Ce^{3+} position, it is clear that oxygen vacancies and Ce^{3+} were formed during the WGS reaction. The amount of oxygen vacancies and Ce^{3+} is seen to increase with the raise of reaction temperature up to 300 °C, but decreases at higher temperatures, especially above 400 °C. When the $Ce_{0.8}Cu_{0.2}O_2$ catalyst was exposed to 5% CO in He, the amount of Ce^{3+} formed was bigger than for the experiments in Figure 21.5*b* and increased continuously with the temperature of exposure. These results show that ceria was oxidized by H_2O under the WGS reaction, with oxygen vacancies and Ce^{3+} being eliminated, especially at high temperature (40,41).

As Ce^{3+} is bigger than Ce^{4+}, a simple reduction leads to a ceria lattice cell expansion. Changes in the ceria lattice parameter can be directly related to the concentration of oxygen vacancies and Ce^{3+} cations in this oxide (42–44). Figure 21.6*a* shows the

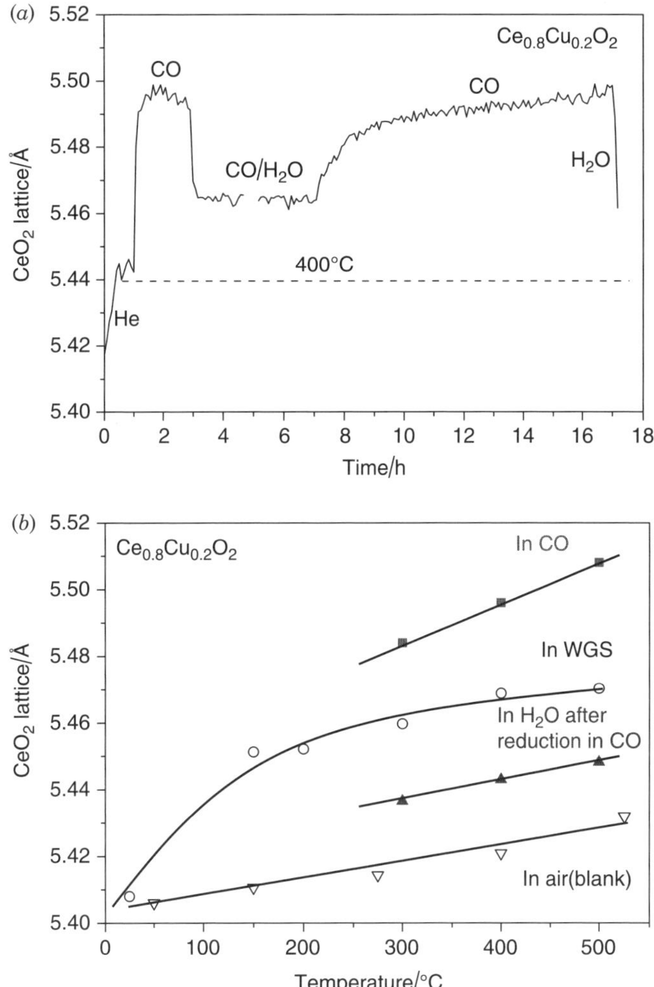

Figure 21.6. (*a*) Ceria lattice parameters during gas-switch experiments over a $Ce_{0.8}Cu_{0.2}O_2$ catalyst at 400 °C. (*b*) Ceria lattice parameters of $Ce_{0.8}Cu_{0.2}O_2$ in the gas-switch experiments as a function of temperature.

lattice parameters for ceria determined from (111) diffraction peaks of TR–XRD patterns for $Cu_{0.2}Ce_{0.8}O_2$ under different gases at 400 °C. The sample was first heated to 400 °C in He. The ceria lattice displayed a significant increase after exposure to CO and a decrease in H_2O, indicating that CO reduced ceria while H_2O oxidized it. In other words, CO created oxygen vacancies while H_2O eliminated them.

In the bottom panel of Figure 21.6 are summarized the results of gas-switch experiments for $Cu_{0.2}Ce_{0.8}O_2$ at different temperatures. The ceria lattice parameter under the WGS reaction reflects a combination of the effects of CO reduction and H_2O oxidation, implying that oxygen vacancies on the fluorite phase are involved in the chemistry of the process.

The activity and behavior of ceria impregnated with copper oxides, CuO_x/CeO_2, during the WGS reaction has also been investigated (32). Figure 21.7 compares the XRD patterns for the $Ce_{1-x}Cu_xO_2$ and CuO_x/CeO_2 systems. The mixed metal oxides exhibit only the diffraction lines of ceria because they are essentially solid solutions. On the other hand, diffraction peaks for CuO and Cu_2O are observed for the CuO_x/CeO_2 catalysts. In experiments of time-resolved XRD, it was found that the diffraction lines for CuO and Cu_2O disappeared under the WGS reaction, whereas diffraction lines for Cu appeared (32). This $CuO_x \rightarrow Cu$ transformation is also clear in the Cu K-edge XANES spectra in Figure 21.8. The 5%CuO_x/CeO_2 catalyst initially

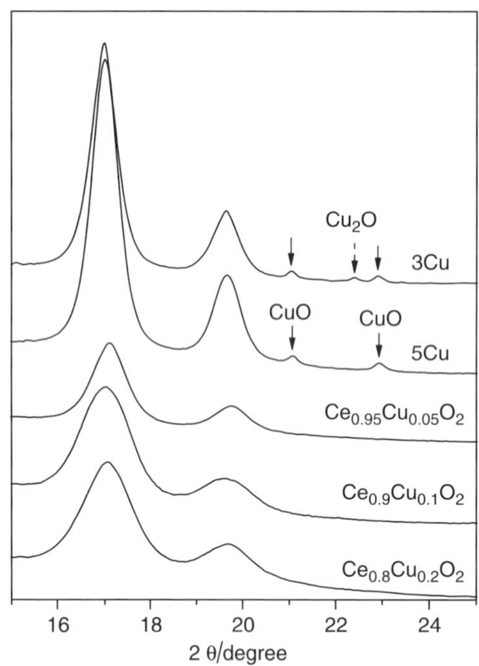

Figure 21.7. XRD patterns for ceria impregnated with copper oxides (top two traces, 3% and 5% Cu in weight) and $Ce_{1-x}Cu_xO_2$ samples (bottom three traces) ($\lambda = 0.922$ Å), (taken from Ref. 32).

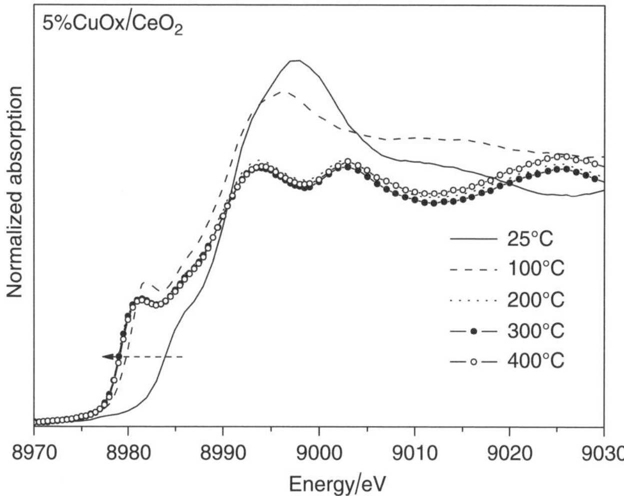

Figure 21.8. Cu K-edge XANES spectra collected over a 5%CuOx/CeO_2 catalyst cooled down to room temperature after the WGS reaction at the indicated temperatures.

exhibits the spectrum typical of CuO, but the line shape of metallic Cu was observed at temperatures above 200 °C when the production of H_2 and CO_2 was significant. This trend corroborates that Cu, and no CuO_x, is the species present in the active sites for the WGS reaction.

Ce L_3-edge XANES spectra (not shown) demonstrated the partial reduction of ceria in 5%CuO_x/CeO_2 during the WGS, but the formation of Ce^{3+} was not as pronounced as seen in Figure 21.5*b* for $Cu_{0.2}Ce_{0.8}O_2$, probably because the copper in the impregnated sample had a lower dispersion within the oxide support than in the doped sample. Figure 21.9 summarizes the behavior of the ceria lattice in 5%CuO_x/CeO_2 under different gases at 400 °C. Exposure to CO produces the formation of Ce^{3+} and expands the ceria lattice. The adsorption of H_2O on the O vacancies removes the Ce^{3+} and contracts the lattice. These are in agreement with those changes for the doped samples shown in Figure 21.4*a*.

21.2.1.2. WGS Reaction on Au-Ceria Nanocatalysts During the past decade, many studies have established that Au on reducible oxides has a remarkable catalytic activity for many important industrial reactions, including the low-temperature oxidation of CO, the WGS reaction, hydrocarbon oxidation, NO reduction, and the selective oxidation of propylene to propylene oxide (45–50). Au-CeO_2 nanocatalysts have recently been reported to be very promising catalysts for the WGS reaction (47,50). As prepared, these catalysts contain nanoparticles of pure gold and gold oxides (AuO_x) dispersed on a nano-ceria support. The ceria support may not be a simple spectator in these systems (51). Pure ceria is a very poor WGS catalyst, but the properties of this oxide were found to be crucial for the observed activity of the Au–CeO_2 catalysts (46,52).

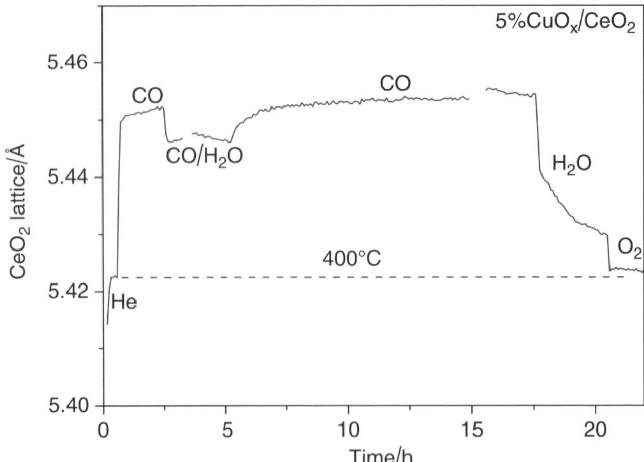

Figure 21.9. Ceria lattice parameters in gas-switch experiments over a 5%CuOx/CeO_2 sample at 400 °C.

Several models have been proposed for explaining the unusual catalytic properties of gold supported on oxides (44,45,48): From special chemical properties resulting from the limited size of the active gold particles (usually less than 5 nm), to the effects of metal:support interactions. In particular, it has been suggested by several authors that the active phase in Au–CeO_2 catalysts are AuO_x nanoparticles or more specifically cationic $Au^{\delta+}$ species (47,53). Similar $Au^{\delta+}$ centers have been proposed as the active centers for the oxidation of CO on Au/MgO catalysts (54). In contrast, Haruta et al. have indicated that the active species are small metallic gold particles (46,49). In studies dealing with a Au/TiO_2(110) model catalyst, Goodman et al. found that for catalytic activity, metallic gold is indispensable (48,55,56). However, the catalytic activity could be further enhanced by perturbing the electronic structure of gold ($Au^{\delta-}$) by a strong interaction between the metal and O vacancies in the oxide support (45,46,49,56–58). In principle, the active sites for the catalytic reactions could be located only on the supported Au particles or on the perimeter of the gold–oxide interface (45,49,57–60).

To address these issues, an in situ time-resolved characterization of the Au–CeO_2 catalysts under operational conditions is critical. Synchrotron-based in situ time-resolved X-ray absorption spectroscopy (XAS) (50,61–63) and time-resolved XRD (64,65) were employed to study the WGS reaction over nanostructured Au–CeO_2 catalysts. In addition, the oxidation and reduction of Au nanoparticles supported on a rough ceria film or a CeO_2(111) single crystal were investigated using photoemission (50,51).

We will begin by examining results of in situ TR–XRD experiments for the WGS reaction over a 2.4wt%Au–Ce(10at%Gd)O_x catalyst (50). A typical set of results is shown in Figure 21.10. The catalyst was held at temperatures of 300, 400, and 500 °C. The XRD pattern for the as-prepared catalyst, bottom of Figure 21.10, showed the typical peaks for nano-ceria (50) and no features that could be attributed to nanoparticles of Au or AuO_x. The gold species were present as aggregates or clusters that had sizes below 2 nm. Exposure of the $Au–CeO_2$ catalyst to a mixture of CO/H_2O at temperatures of 300–500 °C led to the formation of H_2 and CO_2 without inducing the appearance of diffraction peaks for AuO_x or Au species. The catalytically active gold particles had a very small size (<2 nm) and were in close contact with the ceria support because they did not agglomerate at temperatures as high as 500 °C.

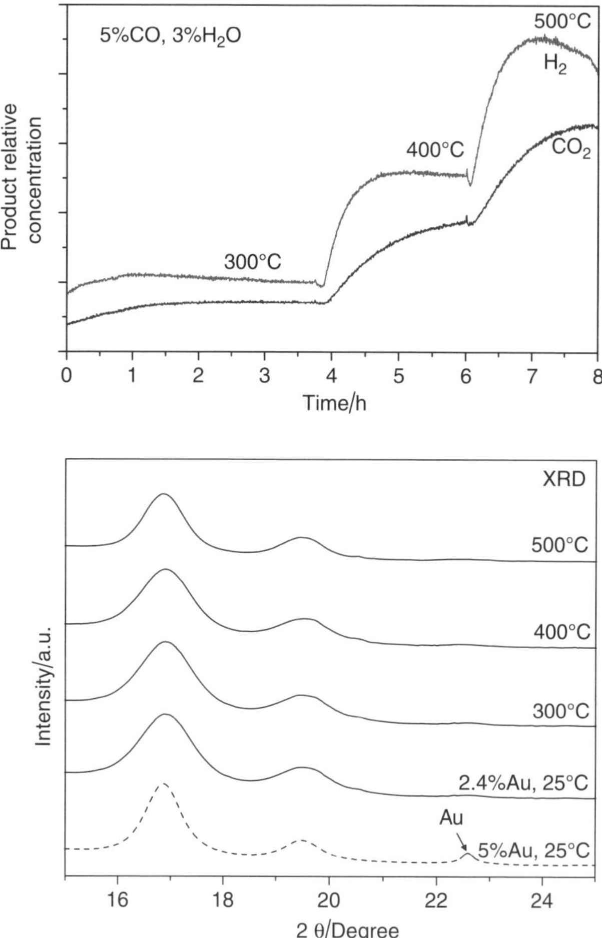

Figure 21.10. Top panel: Relative amounts of H_2 and CO_2 formed during the WGS reaction on a 2.4wt%Au–Ce(10at%Gd)O_x catalyst. A mixture of 5%CO and 3%H_2O in He (total flow ~10 mL/min) was passed over the catalyst at 300, 400, or 500 °C. Bottom panel: X-ray diffraction patterns collected in situ during the WGS on the 2.4wt%Au–Ce(10at%Gd)O_x catalyst. For comparison, we are also including the XRD pattern for a fresh catalyst with 5wt%Au, dashed trace.

The chemical state of gold during the WGS was determined by means of in situ XANES. Figure 21.11 shows Au L_3-edge XANES spectra collected at room temperature for fresh catalysts with a Au content of 0.5wt% (dashed trace) or 2.4wt% (solid traces) (50). The line shape of these two spectra is very similar and shows a clear feature at ~2.5 eV above the edge that is not seen for metallic gold and is characteristic of gold oxides (66,67). The intensity of this peak is higher than that observed for Au_2O and closer to that seen in Au_2O_3 (66,67). Once the 2.4wt%Au–Ce(10at%Gd)O_x catalyst was exposed to a mixture of CO/H_2O at elevated temperatures, the XANES features for gold oxide disappeared. At temperatures above 200 °C, when significant WGS activity was detected, the line shape of the Au L_3 edge resembled that of pure gold (50). The XANES spectra in Figure 21.11 were obtained under a reaction mixture of 5% CO and 3% H_2O in He (total flow ~10 mL/min). Similar results were found when using a 1% CO and 3% H_2O in He reaction mixture (50). The AuO_x in the Au–CeO_2 catalysts was also fully reduced upon exposure to CO or H_2 at 200–500 °C. The in situ TR–XANES data indicate that cationic $Au^{\delta+}$ species *cannot be* the key sites responsible for the WGS activity in Figure 21.10, because they do not exist under reaction conditions. Small Au aggregates (<2 nm in size) supported on ceria exhibit a WGS activity not seen for pure gold or ceria.

In a separate set of experiments, the interaction of O_2 and mixtures of CO/H_2O with gold nanoparticles supported on a well-defined $CeO_2(111)$ surface or rough ceria films was investigated. These experiments were done using a ultra-high vacuum

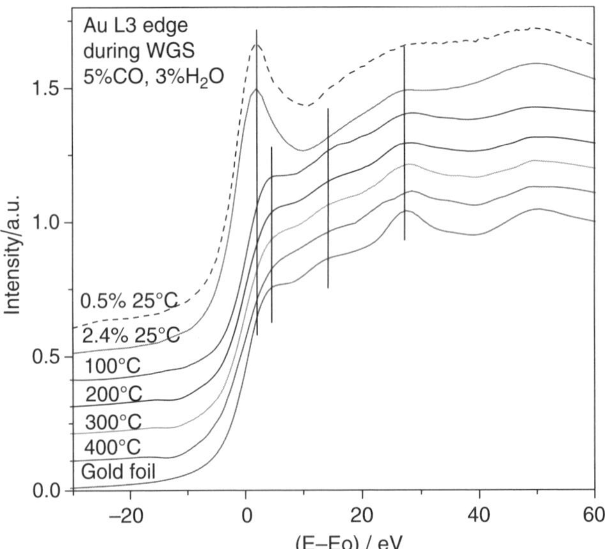

Figure 21.11. Au L_3-edge XANES spectra collected in situ during the WGS reaction over a 2.4wt%Au–CeO_2 catalyst. The sample was exposed to a mixture of 5%CO and 3%H_2O in He (total flow ~10 mL/min) at the indicated temperatures. For comparison, we also include the spectra for a fresh 0.5wt%Au–CeO_2 catalyst, dashed trace, and a gold foil.

(UHV) chamber that has attached a high-pressure cell or batch reactor (51,68–70). The sample could be transferred between the reactor and the vacuum chamber without exposure to air. Small coverages of gold (0–0.2 monolayer) were vapor-deposited on the $CeO_2(111)$ surface at 25 °C. At these coverages, the admetal grows forming three-dimensional (3D) islands on the $CeO_2(111)$ substrate that have sizes in the range of 0.5–3 nm (50). The $Au/CeO_2(111)$ surfaces were exposed in the batch reactor to a mixture of H_2O (10 Torr) and CO (pressures varying from 2 to 20 Torr) at temperatures of 300–400 °C. The production of H_2 and CO_2 through the WGS was observed. After pumping the gases, photo-emission spectra (Au 4*f* and Ce 3*d* core levels) indicated that Au remained in a zero-valent state (i.e., no formation of AuO_x), and there was a partial reduction of the $CeO_2(111)$ support.

Interfaces of AuO_x/CeO_2 were prepared by first vapor-depositing small amounts of Au (<0.2 monolayer) on a Ce film supported on Pt(111) (70). The Au–Ce/Pt(111) systems were then oxidized by reaction with 500 Torr of O_2 in a high-pressure cell. This produced AuO_x/CeO_2 interfaces that have the typical Ce 3*d* features of ceria (50,71) and Au $4f_{7/2}$ peaks located in between those expected for Au^{+1} and Au^{+3} (50). As in the case of $Au/CeO_2(111)$, the AuO_x/CeO_2 interfaces were exposed to mixtures of CO/H_2O at 300–400 °C in a reaction cell. Figure 21.12 displays typical results of photo-emission. Initially, Au^{3+} and some Au^0 were present in the AuO_x/CeO_2 system. Exposure to a mixture of 5 Torr of CO and 10 Torr of H_2O at 30 °C led to a full reduction of the gold oxide and partial reduction of ceria (Ce 3*d* spectrum, not shown). The gold oxide in the AuO_x/CeO_2 interfaces was fully reduced upon exposure to CO, H_2, or reaction mixtures that had a CO/H_2O ratio varying from 0.2 to 2 (50).

The top panel in Figure 21.13 compares the WGS activity of $Au/CeO_{2-x}(111)$ and Au/CeO_{2-x} catalysts that contained the same amount of gold (~0.15 monolayer). These catalysts were preconditioned by exposing the fresh $Au/CeO_2(111)$ and

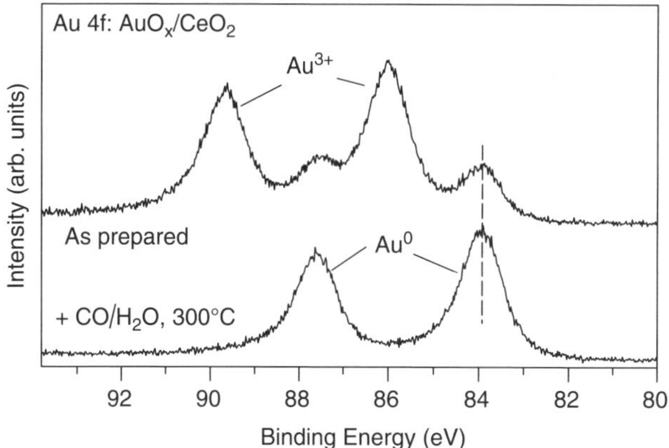

Figure 21.12. Au 4f core-level photo-emission spectra collected before and after exposing a AuO_x/CeO_2 interface ($\theta_{Au} \sim 0.15$ monolayer) to a mixture of 5 Torr of CO and 10 Torr of H_2O at 300 °C for 5 min.

AuO_x/CeO_2 systems to a mixture of 20 Torr of CO and 10 Torr of H_2O for 5 minutes at 350 °C. Photo-emission indicated that the catalysts contained only pure Au and a quantitative analysis of the line shape of the Ce 3*d* core levels (72,73) gave stoichiometries of $Au/CeO_{1.94}(111)$ and $Au/CeO_{1.88}$. These compositions remained essentially constant during the experiments in the top panel of Figure 21.13. In these experiments,

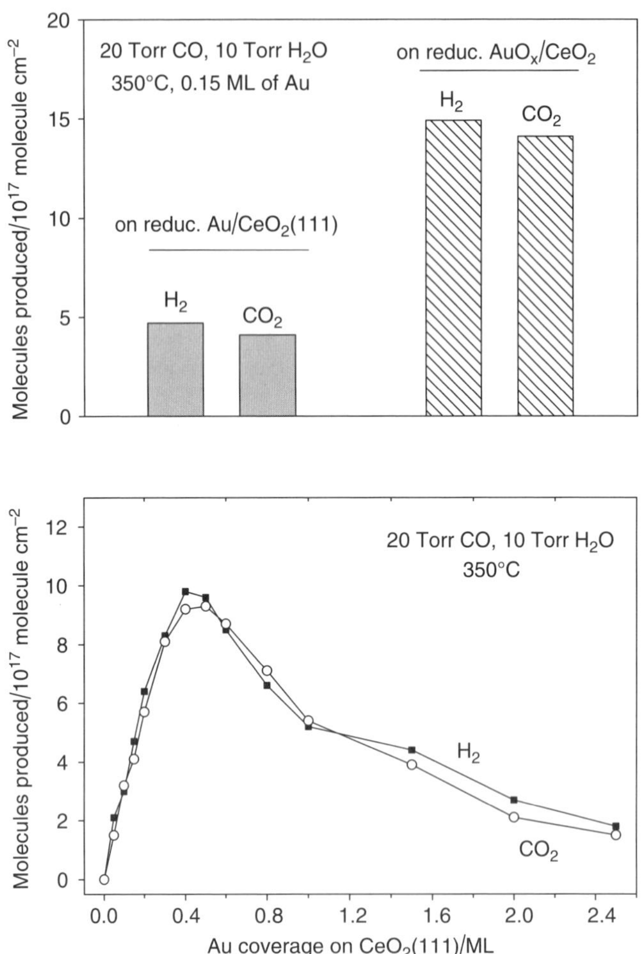

Figure 21.13. Top panel: Amounts of H_2 and CO_2 produced during the WGS reaction on reduced $Au/CeO_2(111)$ and AuO_x/CeO_2 ($\theta_{Au} \sim 0.15$ monolayer) catalysts. The catalysts were preconditioned as described in the text. Then, they were exposed to a mixture of 20 Torr of CO and 10 Torr of H_2O at 350 °C for 5 min in a batch reactor. The amounts of H_2 and CO_2 produced during this time period are plotted in the graph. Bottom panel: Amounts of H_2 and CO_2 produced during the WGS reaction on reduced $Au/CeO_2(111)$ catalysts with different Au coverages. After preconditioning (see text), each surface was exposed to a mixture of 20 Torr of CO and 10 Torr of H_2O at 350 °C for min.

the $Au/CeO_{1.94}(111)$ and $Au/CeO_{1.88}$ catalysts were under a mixture of 20 Torr of CO and 10 Torr of H_2O for 5 minutes at 350 °C. Then, the amount of H_2 and CO_2 produced through the WGS was measured. Both catalysts have a much higher activity than that observed for Cu(110) and Cu/ZnO catalysts under similar reaction conditions (74). The catalyst generated from AuO_x/CeO_2 was clearly the more active, probably because it contained a larger amount of surface defects and O vacancies than the catalyst generated from $Au/CeO_2(111)$ (50).

The bottom panel in Figure 21.13 shows the effect of Au coverage on the catalytic activity of $Au/CeO_{2-x}(111)$. These surfaces were preconditioned as described above, and then we measured the catalytically produced H_2 and CO_2 after exposure to a mixture of 20 Torr of CO and 10 Torr of H_2O for 5 minutes at 350 °C. A maximum in the catalytic activity is seen for Au coverages of 0.4–0.5 monolayer. The drop in the catalytic activity at high Au coverage is probably an effect of a growth of the Au nanoparticles (>3 nm in size) (50,75) and agrees with the fact that there is no increase in WGS activity when going from a 2.4%wtAu– to a 5.0%wtAu–Ce(10at%Gd)O_x catalyst (47).

The results of photo-emission and in situ TR-XAS indicate that the active phase of the gold/ceria catalysts for the WGS contained pure Au and CeO_{2-x}. This is consistent with studies that show that Au as an admetal facilitates the reduction of ceria (76). The generation of Ce^{3+} in ceria produces an expansion in the unit cell of the oxide that can be monitored with TR–XRD (50). Figure 21.14 shows the lattice parameters for ceria determined from (111) diffraction peaks of TR–XRD patterns for 2.4wt%Au–Ce(10at%Gd)O_x under different gases at 500 °C. The sample was first heated from 25 to 500 °C in He, producing an increase of ~0.02 Å in the ceria lattice due to thermal expansion (50). The oxide lattice varied significantly after exposure

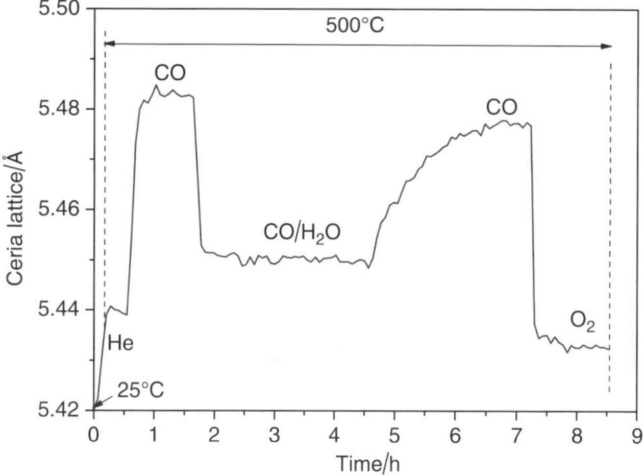

Figure 21.14. Ceria lattice parameters determined from TR–XRD patterns for a 2.4wt%Au–Ce(10at%Gd)O_x catalyst under different gases at 500 °C: pure He, 5% CO in He, 5% CO and 3% H_2O in He, again 5% CO in He, and finally 5% O_2 in He.

to CO, an ~0.04 Å increase as a consequence of the formation of O vacancies and Ce^{3+}, or H_2O, an ~0.03 Å decrease probably as a result of the decomposition of this adsorbate on the O vacancies. Under a mixture of CO/H_2O, O vacancies were formed that could be removed only upon exposure to O_2. The changes of ceria in Au–CeO_x catalysts were consistent with those changes observed for ceria in Cu–CeO_x catalysts during similar gas-switch experiments, indicating the same role played by ceria for both systems.

From the data in Figures 21.11–21.14, it appears that the rate determining steps for the WGS at temperatures above 250 °C occur at the gold–ceria interface, with the active phase involving small gold clusters (<2 nm) and O vacancies. A strong interaction between the admetal and the support prevents the generation of big particles (>2 nm) of gold and is probably a key issue for determining the catalytic properties of the gold/ceria system. It is not necessary to invoke a special catalytic activity for nonmetallic gold nanoparticles (AuO_x) (50).

Figure 21.15 compares the conversion of CO during the WGS reaction over the same amount (~5 mg) of 2.4%AuO_x/CeO_2, 5%CuO_x/CeO_2, $Ce_{0.8}Cu_{0.2}O_2$, $Ce_{0.95}Cu_{0.05}O_2$, and CuO at 400 °C. No activity was observed over pure nano-ceria under the current operating conditions. The 2.4%AuO_x/CeO_2, 5%CuO_x/CeO_2, and $Ce_{0.95}Cu_{0.05}O_2$ catalysts have a similar activity. Figure 21.15 clearly shows that the catalyst based on pure CuO has little activity. This catalyst certainly had the largest concentration of copper sites among the studied samples. These results indicate that interactions between copper and ceria enhance the activity of the metal in Cu–CeO_2 catalysts, suggesting that the WGS reaction may take place preferentially on the metal/ceria interface. The same may be also valid for the Au–CeO_2 catalyst because

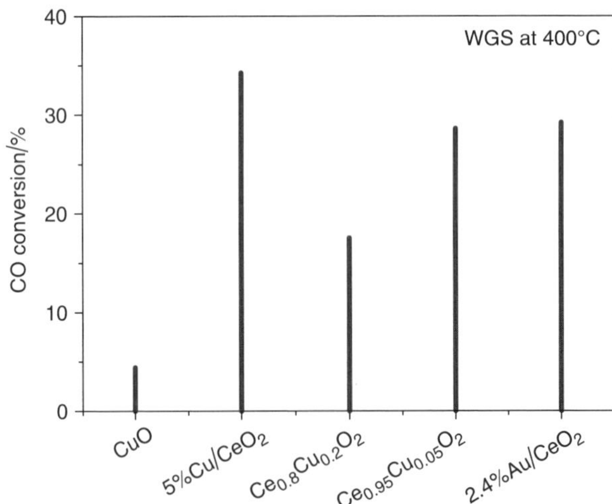

Figure 21.15. CO conversions for the WGS reaction at 400 °C over CuO, 5%CuOx/CeO_2, $Ce_{0.8}Cu_{0.2}O_2$, $Ce_{0.95}Cu_{0.05}O_2$, and 2.4wt%AuOx/CeO_2 catalysts with the same operating conditions (Space velocity ~1.5×10^5/h, ~5 mg of catalyst).

this system is more active than either pure gold or pure ceria (47). The moderate activity of the $Ce_{0.8}Cu_{0.2}O_2$ catalyst may be a consequence of the formation of a large amount of strongly bound carbonates on its surface (32). The CuO_x and AuO_x species present in the fresh Cu–CeO_2 and Au–CeO_2 nanocatalysts of Figure 21.15 do not survive above 200 °C under the WGS conditions (32,50). Data of time-resolved XRD and XANES indicate that metallic copper or gold and oxygen vacancies in ceria are involved in the generation of the catalytically active sites.

21.2.1.3. WGS on Other Oxide-Based Catalysts There has been an interest in the Pt/CeO_2 system since the early 1980s, when in connection with the development of the three-way catalysts for reducing exhaust engine pollution, it was discovered that such a system enhances the rate of the WGS (77). Academic research on Pt, Pd, Rh, Ni, Fe, and Co-ceria as catalysts for the WGS involved mechanistic studies and the effect of the ceria oxygen-storage properties on the WGS activity (78,79). The performance of the catalysts was strongly affected by the method used for the pretreatment of the ceria support (78). It was proposed that the WGS occurred primarily through a bifunctional mechanism in which CO adsorbed on the precious metal is oxidized by ceria, which in turn is oxidized by water (78,79). A reaction mechanism that involves a surface formate intermediate cannot be ruled out (80). Enhancement of the WGS activity of Pt/CeO_2 has been observed after doping of the oxide with Zr (81). The doping with zirconia decreases the temperature for the surface reduction step in the WGS process (81). Pt/MgO and Pt/ZrO_2 are active catalysts for the WGS reaction (80), but they are not as good as Pt/CeO_2.

Ruthenium deposited on α-Fe_2O_3 has been mentioned in the literature as giving promising WGS conversions (82). Non-precious-metal formulations for the WGS are sought because of their low cost. The WGS activity of alkaline-promoted basic oxides based on Mg, Al, La, Nd, Pr, Mn, and Mg has been tested and found satisfactory at elevated temperatures (>400 °C) (83). Mixtures of Co- or Ni-promoted Mo, V, or W oxides are also catalytically active under similar conditions (83).

21.2.2. Production of H_2 Through Other Methods

Hydrogen can be produced by the steam reforming, $C_nH_m + nH_2O \rightarrow nCO + (n - m/2)H_2$, or the partial oxidation, $C_nH_m + n/2\ O_2 \rightarrow nCO + (m/2)H_2$, of hydrocarbons. Natural gas is the major feedstock for the production of hydrogen via syngas (84). Because CH_4 is a very stable molecule, it has to be processed under very severe conditions (84). The steam reforming of methane is a highly endothermic reaction ($\Delta H^\circ = 206\,kJ/mol$), whereas the partial oxidation is slightly exothermic ($\Delta H^\circ = -36\,kJ/mol$). The traditional methane steam-reforming catalysts for the production of hydrogen and synthesis gas contain Ni/NiO or CoO_x supported on alumina (85) or combined with a rare-earth oxide like ceria (86). The catalysts are frequently promoted with alkali and alkali-earth oxides to improve their stability and facilitate coke gasification (85).

Partial oxidation of methane is a much faster reaction than steam reforming, offering therefore the advantage of smaller reactors and higher throughputs (87). Catalysts include mixtures of NiO–MgO, nickel modified hexa-aluminates, and platinum group metals deposited on alumina or ceria-containing supports (87–89). Examples of catalysts for the oxidative reforming of methane based on ceria are Ni/CeO_2 and Ni/CeO_2–ZrO_2 (90), Pt/CeO_2–ZrO_2 (89,91), and Ru/CeO_2–ZrO_2 (89). The activity of these catalysts seems to be enhanced by metal-support interactions and the oxygen storage capacity (OSC) of ceria under oxidizing-reducing conditions (42,89,92). More systematic studies need to be done investigating the application of oxide nanocatalysts in the production of hydrogen through steam reforming or the partial oxidation of hydrocarbons.

Hydrogen can be obtained directly from methanol following three different processes: (*1*) steam reforming, (*2*) decomposition, and (*3*) partial oxidation (84). Typical catalysts for methanol steam reforming are based on Cu–ZnO–Al_2O_3 (93). Sometimes zirconia is added to these catalysts to enhance its activity or to improve its selectivity toward carbon dioxide (94). NiO and CuO_x supported on high-surface-area carriers and promoted with other metal oxides have been used for methanol decomposition: $CH_3OH \rightarrow CO + 2H_2$ (95). Small nanoparticles (3–8 nm) of copper oxide undergo partial reduction before becoming catalytically active for methanol decomposition (96). Partial oxidation of methanol over Cu-containing catalysts produces mainly carbon dioxide and hydrogen: $CH_3OH + 0.5O_2 \rightarrow CO_2 + 2H_2$. Cu–ZnO–$Al_2O_3$ catalysts used for methanol synthesis display a very good performance in methanol partial oxidation (84).

Other possible routes for the production of H_2 are thermochemical water splitting and photo-assisted water splitting (2,5,84). Currently, these processes are not efficient enough for the industrial production of hydrogen. Photo-chemical water splitting to H_2 and O_2 has been successfully carried out using ultraviolet irradiation of aqueous suspensions (or thin-film photo-anodes) of various photo-catalysts, including TiO_2 (97), transition-metal doped TiO_2 (98,99), $SrTiO_3$ (100), and transition-metal loaded $SrTiO_3$ (101). Among these photo-catalysts, TiO_2 is the one most frequentely employed because of its low cost, nontoxicity, and structural stability (99). However, pure titania is only activated by ultraviolet light (97). Recently, N-doped titania ($TiO_{2-x}N_x$) has been synthesized and uses as a photo-catalyst that uses visible light (>420 nm) (102–104). This material absorbs solar light in a region where TiO_2 or $SrTiO_3$ do not due to their wide band gaps. Such a behavior is very important for practical applications because solar light contains less than 5% of ultraviolet radiation. In principle, a dopant element can induce energy levels within the band gap of TiO_2 that may increase the yield for electron-hole pair formation under illumination with visible light (97,98,102). Two different models have been proposed to explain the good performance of $TiO_{2-x}N_x$ as a photo-catalyst (102,105,106). In one model, hybridization of the N 2*p* states of the dopant with the O 2*p* valence band of TiO_2 leads to band narrowing (102). Optical absorption studies on N-doped TiO_2 point to a gap narrowing (102), but no band gap narrowing has been observed in photoemission studies (107). In the second model, the dopant N 2*p* levels form localized states within the band gap just above the O 2*p* valence band maximum for titania

(105). For a detailed discussion of the photo-chemical properties of pure and doped nanoparticles of TiO_2, the reader is referred to Chapter 16.

21.3. FUEL CELLS

In this section, we will describe two types of fuel cells that make use of the H_2 produced through the processes described in the previous section. These fuel cells have found a place in technological applications, and they may use either H_2 or hydrogen-rich fuels in their operation. However, if pure hydrogen is used as a fuel, the fuel cells emit only heat and water as a byproduct.

21.3.1. Polymer Electolyte Membrane Fuel Cell (PEMFC)

The PEMFC, more commonly known as the proton-exchange membrane fuel cell, was first used in the 1960s as an auxiliary power source in the Gemini space flights. Advances in this technology were stagnant until the late 1980s when the fundamental design underwent significant reconfiguration. Nowadays, the PEMFC is one of the most promising fuel cell types for widespread use because it offers several advantages for transport and several other applications. Its low-temperature operation, high power density, fast start-up, system robustness, and low emissions have ensured that the majority of motor manufacturers are actively pursuing PEMFC research and development. Already in Europe with demonstration buses and passenger vehicles in California, for example, a first market introduction of fuel cell vehicles will be seen in the near future (108).

As indicated in Figure 21.16 (109), a single proton-exchange membrane (PEM) cell comprises three types of components: a membrane-electrode assembly (MEA), two bipolar plates, and two seals. In its simplest form, the MEA consists of a membrane, two dispersed catalyst layers, and two gas diffusion layers (GDLs). The membrane separates the half reactions allowing protons to pass through to complete the overall reaction. The electron created on the anode side is forced to flow through an external circuit, thereby creating current. The GDL allows direct and uniform access of the fuel and oxidant to the catalyst layer, which stimulates each half reaction. In a fuel cell stack, the bipolar plates are vital components. Each bipolar plate supports two adjacent cells. The bipolar plates typically have four functions:

(*1*) to distribute the fuel and oxidant within the cell,
(*2*) to facilitate water management within the cell,
(*3*) to separate the individual cells in the stack, and
(*4*) to carry current away from the cell.

In the absence of dedicated cooling plates, the bipolar plates also facilitate heat management. Different combinations of materials, fabrication techniques, and layout

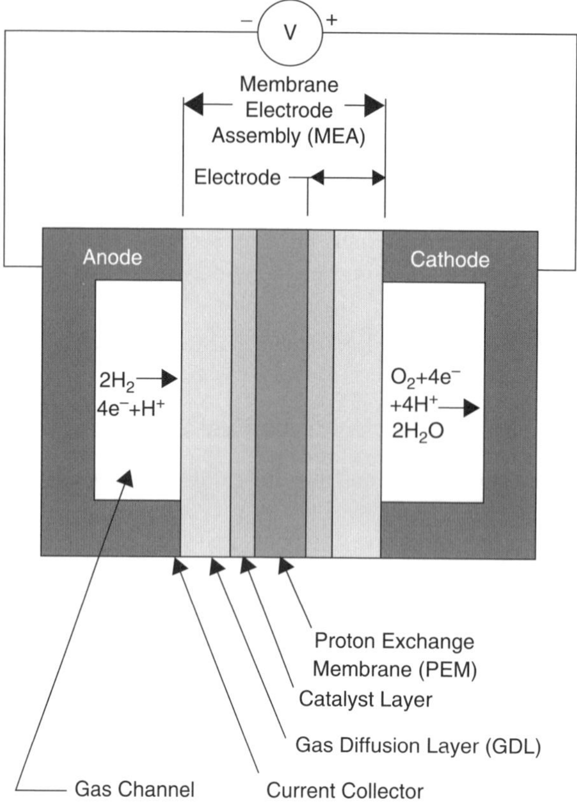

Figure 21.16. PEMFC schematic diagram (taken from Ref. 109).

configurations have been developed for these plates to achieve high performance and optimal cost (108).

Mehta and Cooper reviewed the material needs for designing and manufacturing PEM fuel cells for vehicle applications (110). In their article, 16 polymer electrolyte membranes, 2 types of GDLs, 8 types of anode catalysts, 4 types of cathode catalysts, and over 100 bipolar plate designs are recommended for further study. The prime requirements of fuel cell membranes are high proton conductivity, low methanol/water permeability, good mechanical and thermal stability, and moderate price (110). Membranes and their operating parameters have a profound influence on the performance of PEMFCs. Membranes built from perfluorinated ionomers, hydrocarbon and aromatic polymers, and acid–base complexes have been described by Smitha et al. (111). Membranes containing nano-arrays of NiO, TiO_2, and titanium silicates have been prepared and tested (112,113). Promising avenues for further research in this area have been identified (110–113).

Farrauto et al. reviewed methods for the production of hydrogen for the PEM fuel cells (114). Technologies including heterogeneous catalysts for the WGS reaction, the reforming of methanol (either autothermal or steam reforming), the preferential oxidation of CO, and anode tail gas combustion were discussed in detail in this article (114). The rapid development in recent years of the PEM fuel cell technology has stimulated research in all areas of fuel processor catalysts for hydrogen generation. The principal aim is to develop more active catalytic systems that allow for the reduction in size and increase the efficiency of fuel processors. The overall selectivity in generating a low CO content hydrogen stream as needed by the PEM fuel cell is dependent on the efficiency of the catalysts used in each segment of the fuel processor. The hydrogen produced via steam reforming and partial oxidation of hydrocarbons always contains CO (83,84). As discussed in the previous section, oxide-based nanocatalysts can be very efficient for the removal of CO and the production of nearly pure H_2 through the WGS reaction.

In a PEM fuel cell, the type of fuel used dictates the appropriate type of anode or cathode catalysts needed to facilitate the chemical transformations inside the cell. Within this topic, tolerance to carbon monoxide is an important issue. The amount of CO present in the feed can be greatly reduced through the WGS, but CO always will be present at the ppm level. It has been determined that the PEM fuel cell must be capable of tolerating a CO concentration of at least 100 ppm in order to reduce the size of the hydrogen production/cleaner unit (115). An approach is to alloy the anode catalyst to minimize its sensitivity toward CO; one (a binary catalyst) or sometimes two elements (a ternary catalyst) are added to the base catalyst, usually Pt/C. Mehta and Cooper summarized over 26 different kinds of catalysts for the MEA component of PEM fuel cells (110). The anode catalysts were divided into three groups based on the number of metal elements used: single metal catalysts (Pt/C), binary catalysts (Pt–Ru/C, Pt–Mo/C, Au–Pd/C, etc.), and tertiary catalysts (Pt–Ru–Mo/C, Pt–Ru–W, Pt–Ru–Al_4, and Pt–Re–MgH_2) (110). In addition to Pt/C, Pt–Ni/C, and Pt–Co/C were reported for cathode applications (116). Anode and cathode catalysts built from oxide nanostructures that have a reasonable high conductivity and resistance to CO are also under consideration and being tested (112,113).

The design of electrodes for PEMFCs is a delicate balancing of transport media (109). Conductance of gas, electrons, and protons must be optimized to provide efficient transport to and from the electrochemical reactions. This is accomplished through careful consideration of the volume of conducting media required by each phase and the distribution of the respective conducting network (109). In addition, the issue of electrode flooding cannot be neglected in the electrode design process. New fabrication methods, which have now become conventional, were adopted and optimized to a high degree. Possibly, the most significant barrier that PEM fuel cells have to overcome is the costly amount of platinum required as a catalyst. The large amount of platinum in the original PEM fuel cells was one of the reasons why fuel cells were excluded from commercialization (109,110). Thus, the reconfiguration of the PEM fuel cell was targeted directly on the electrodes employed and, more specifically, on reducing the amount of platinum in the electrodes. This continues to be a driving force for further research on PEM fuel cell electrodes (109), and low-cost conducting oxide nano-arrays could be an alternative (112,113).

21.3.2. Solid Oxide Fuel Cell (SOFC)

The solid oxide fuel cell is an all-solid device that converts the chemical energy of gaseous fuels, such as hydrogen and natural gas, to electricity through electrochemical processes (117,118). These fuel cells provide many advantages over traditional energy conversion systems, including high efficiency, reliability, modularity, fuel adaptability, and very low levels of NO_x and SO_x emissions. The efficiency of SOFCs is inherently high as it is not limited by the Carnot cycle of a heat engine (117). A solid oxide fuel cell essentially consists of two porous electrodes separated by a dense, oxygen ion conducting electrolyte (117,118). They can be arranged in a planar or tubular configuration (Figure 21.17) (118). The cathode or air electrode operates at elevated temperatures (usually 800–1100 °C) and participates in the oxygen reduction reaction:

$$0.5O_2(g) + 2e \longrightarrow O^{2-}(s)$$

i.e., oxygen in the gas phase is reduced to oxide ions, consuming two electrons in the process. The air electrode frequently consist of lanthanum manganite doped with alkaline and rare-earth elements (118,119). Lanthanum manganite ($LaMnO_3$) is a *p*-type perovskite oxide and shows reversible oxidation–reduction behavior. Recently, a composite cathode was fabricated using well-dispersed nanosize grains of $La(Sr)MnO_3$ and Y_2O_3 stabilized ZrO_2 (120). The composite cathode had a unique morphology that led to high electrochemical activity at 800 °C, suggesting that intermediate temperatures ($<$800 °C) of operation can be achieved by using composite arrays of oxide nanoparticles (120).

SOFCs are based on the concept of an oxygen-ion conducting electrolyte through which the oxide ions (O^{2-}) migrate from the air electrode (cathode) side to the fuel electrode (anode) side where they react with the fuel (H_2, CH_4, etc.) to generate an electrical voltage. Fluorite-structured oxide materials such as yttria-stabilized zirconia

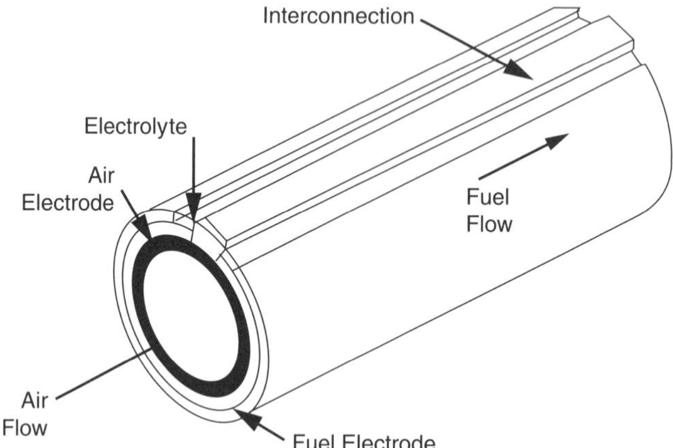

Figure 21.17. Tubular configuration of an SOFC (taken from Ref. 118).

(YSZ), rare-earth doped ceria, and rare-earth doped bismuth oxide are widely used for electrolytes in SOFCs (118,119). The oxygen-ion conducting capabilities of these oxides in many cases improve when going from the bulk to the nanoscale (see Chapter 13), also improving their performance in SOFCs (113,117,120,121).

In an SOFC, the anode is the electrode for the electrochemical oxidation of the fuel. Hydrogen oxidation is one of the most important electrode reactions in SOFCs. The process for the electrochemical oxidation of H_2 can be written as

$$H_2(g) + O^{2-}_{\text{oxide}} \longrightarrow H_2O(g) + 2e$$

where O^{2-}_{oxide} is an oxygen ion in the electrode. To minimize the polarization losses of the H_2 oxidation reaction, anode materials should have a high electronic conductivity and electrocatalytic activity (117,121). Nanostructured oxides can be particularly useful in this respect (113,120). Current activities in the area of materials development for SOFCs are increasingly focused on decreasing the operating temperatures of the cell from 1000–1100 °C to 500–800 °C (118–120). The most common anodes for SOFCs contain a metal-oxide cermet: Ni/ Y_2O_3–ZrO_2 (the so-called Ni/ YSZ cermet) (118,121). Conventional Ni/ YSZ cermet anodes are usually made of commercial NiO and YSZ powders, which are then homogenized by mechanical mixing and milling. The NiO/ YSZ ink is applied to the YSZ electrolyte and sintered to form a porous Ni/ YSZ cermet electrode (118). The performance of these cermet anodes depends critically on the microstructure and distribution of the Ni and YSZ phases (118). The use of nanosize grains for the oxide precursors leads to a better dispersion/ mixing and a subsequent improvement in the performance of the Ni/ YSZ cermets (120,121). The Ni/ YSZ cermets have very good electrocatalytic properties for cells that use H_2 as a fuel, but they suffer a series of drawbacks in systems where natural gas is the fuel, notably the sulfur poisoning and carbon deposition caused by the cracking of C2 and higher hydrocarbons (121). Theoretical studies of more conventional supported nickel catalysts have focused on ensemble size effects, nickel crystallite size effects, and control of highly active step and edge sites to reduce carbon formation (122–124). The acidity of the support has also been considered important, with acidic supports such as Al_2O_3 to be avoided, and basic supports such as MgO or $MgAl_2O_4$ being preferred (125). With these latter supports, it is believed that adsorption of H_2O is increased, facilitating the gasification of carbon that is deposited on the surface (126). Takeguchi et al. investigated the carbon deposition during the steam reforming of methane on Ni/ YSZ anodes modified with the addition of MgO, CaO, SrO, or CeO_2 (127,128). The presence of nanoparticles of CaO, SrO, and CeO_2 suppressed the carbon deposition, whereas the addition of MgO nanoparticles promoted it and decreased the steam-reforming activity of the anodes. A high content (1–2 wt%) of CeO_2 and SrO in the Ni/ YSZ cermets also led to a substantial decrease in the steam-reforming activity. However, other literature showed that MgO has unique properties in its interaction with NiO, due to its ability to form a solid solution. On reduction, Ni appears to be stabilized relative to sintering and carbon formation, as it maintains a strong interaction with the MgO or a MgO–NiO solid solution (125). Sample preparation methods and catalyst compositions played important roles in the different behaviors of the MgO containing system and need to be investigated more thoroughly in the future (125). Additional studies showed that the addition of Ru and Pt promotes the reforming activity and suppresses the carbon deposition (127,128).

Several studies have been carried out on the use of electronic or mixed conducting oxides to make alternative metal/oxide cermets such as ceria-stabilized zirconia (CeO_2–ZrO_2), calcium-doped ceria (Ca–CeO_2), yttria-doped ceria (YDC), samaria-doped ceria (SDC), and praseodymium oxide (PrO_x) (121). Mixed results were reported, and the performance of the electrode depends strongly on the preparation method (121). Eguchi et al. observed that the activity of Ni/oxide cermets for hydrogen oxidation increased following the sequence (129):

$$\mathrm{Ni/YSZ} < \mathrm{Ni/CeO}_x < \mathrm{Ni/SDC} < \mathrm{Ni/PrO}_x$$

The improved performance was attributed to the increase of the effective reaction area due to the ionic conduction of the oxides (126). A comparison of the hydrogen–oxidation activity of Ni/CeO_2 cermets prepared with bulk NiO and CeO_2 or with nano-NiO (3–6 nm) and CeO_2 (5–8 nm) shows that the second type of electrode is ~four times more active than the first (130).

Gorte et al. focused their efforts on the development of Cu/CeO_2/YSZ-based anodes (121,131–133). Copper is much less reactive than nickel. Copper is inert to hydrogen or hydrocarbon oxidation and has no catalytic activity for the formation of C–C bonds, thus suppressing the carbon deposition. The primary function of copper in the cermet anodes is to provide a path for the conduction of electrons (129,130). A pure Cu/YSZ anode has very low electrochemical activity for both H_2 and CH_4 oxidation reactions. Ceria catalyst was added to the Cu/YSZ cermet to improve the electrode activity (128,129).

Pure or doped titanates (TiO_2, $SrTiO_3$, $ZrTiO_4$) and lanthanum chromites ($LaCrO_3$) have been tested as anode materials (117,121). Chromite/titanate-based perovskite anodes generally show much poorer electrochemical activity for H_2 oxidation as compared with that of Ni/YSZ cermet anodes despite the high electrical conductivity of some perovskites (121). The performance of these oxides can be improved when going from the bulk to the nanoscale (113,127), but then there are issues with their long-term stability (119). In general, they do not show clear advantages when combining activity and stability and comparing with respect to the properties of the Ni/YSZ system.

21.4. CONCLUSIONS

Oxide nanosystems play a key role as components of catalysts used for the production of H_2 via the steam reforming or the partial oxidation of hydrocarbons and for the WGS reaction. The behavior seen for Cu-ceria and Au-ceria WGS catalysts indicates that the oxide is much more than a simple support. The special chemical properties of the oxide nanoparticles (defect rich, high mobility of oxygen) favor interactions with the reactants or other catalyst components. More in situ characterization and mechanistic studies are necessary for the optimization of these nanocatalysts.

The use of oxide nanomaterials for the fabrication of PEMFCs and SOFCs can lead to devices with a high practical impact. One objective is to build electrodes with low-cost conducting oxide nano-arrays. The electron and oxygen-ion conducting capabilities of many oxides improve when going from the bulk to the nanoscale. Furthermore, one can get a more homogeneous surface morphology and an increase of the effective reaction area. Much more fundamental and practical research needs to be done in this area.

ACKNOWLEDGMENT

The work done at Brookhaven National Laboratory was financed through Contract DE-AC02-98CH10086 with the U.S. Department of Energy (DOE), Division of Chemical Sciences. The work done at Pacific Northwest National Laboratory operated by Battelle Memorial Institute was financed by the U.S. Department of Energy under Contract DE-AC05-76RL01830.

REFERENCES

(1) Smalley, R.E. Talk at Brookhaven National Lab, Rice University, October 2004.

(2) Midilli, A.; Ay, M.; Dincer, I.; Rosen, M.A. *Renew. Sustain. Energy Rev.* **2005**, *9*, 255–271.

(3) Energy Policy Act (EPAct); 1992.

(4) Rostrup-Nielsen, J.R. *Catal. Rev.* **2004**, *46(3–4)*, 247.

(5) President's Hydrogen Initiative; 2003; Jacobson, M.Z.; Colella, W.G.; Golden, D.M. *Science*, **2005**, *308*, 1901.

(6) Energy Policy Act (EPAct); 1992.

(7) President's Hydrogen Initiative; 2003.

(8) Jacobson, M.Z.; Colella, W.G.; Golden, D.M. *Science*, **2005**, *308*, 1901.

(9) Farrauto, R.; Hwang, S.; Shore, L.; Ruettinger, W.; Lampert, J.; Giroux, T.; Liu, Y.; Ilinich, O. *Ann. Rev. Mater. Res.* **2003**, *33*, 1.

(10) Spivey, J.J. *Catal. Today* **2005**, *100*, 171.

(11) Tonkovich, A.Y.; Zilka, J.L.; LaMont, M.J.; Wang, Y.; Wegeng, R.S. *Chem. Eng. Sci.* **1999**, *54*, 2947.

(12) Steele, B.C.H.; Heinzel, A. *Nature* **2001**, *414*, 345.

(13) Available at: http://www.eere.energy.gov/.

(14) Rostrup-Nielsen, J.R.; Sehested, J.; Nørskov, J.K. *Adv. Catal.* **2002**, *47*, 65.

(15) Liu, Y.; Fu, Q.; Flytzani-Stephanopoulos, M. *Catal. Today* **2004**, *93–95*, 241.

(16) Suh, D.J.; Kwak, C.; Kim, J.H.; Kwon, S.M.; Park, T.J. *J. Power Sources* **2005**, *142*, 70.

(17) Song, C. *Catal. Today* **2002**, *77*, 17.

(18) Li, Y.; Fu, Q.; Flytzani-Stephanopoulos, M. *Appl. Catal. B* **2000**, *27*, 179.

(19) Rhodes, C.; Hutchings, G.J.; Ward, A.M. *Catal. Today* **1995**, *23*, 43.

(20) Koryabkina, N.A.; Phatak, A.A.; Ruettinger, W.F.; Farrauto, R.J. Ribeiro, F.H. *J. Catal.* **2003**, *217*, 233.

(21) Liu, Y.; Fu, Q.; Stephanopoulos, M.F. *Catal. Today* **2004**, *93–95*, 241.

(22) Fu, Q.; Saltsburg, H.; Flytzani-Stephanopoulos, M. *Science* **2003**, *301*, 935.

(23) Jacobs, G.; Chenu, E.; Patterson, P.M.; Williams, L.; Sparks, D.; Thomas, G.; Davis, B.H. *Appl. Catal. A: Gen.* **2004**, *258*, 203.

(24) Fernández-García, M.; Martínez-Arias, A.; Guerrero-Ruiz, A.; Conesa, J.C.; Soria, J. *J. Catal.* **2002**, *211*, 326.

(25) Iglesias-Juez, A.; Hungría, A.B.; Gálvez, O.; Fernández-García, M.; Martínez-Arias, A.; Guerrero-Ruiz, A.; Conesa, J.C.; Soria, J. *Stud. Surf. Sci. Catal.* **2001**, *138*, 347.

(26) Fernández-García, M.; Martínez-Arias, A.; Hungría, A.B.; Iglesias-Juez, A.; Conesa, J.C.; Soria, J. *Phys. Chem. Chem. Phys.* **2002**, *4*, 2473.

(27) Hungria, A.B.; Martínez-Arias, A.; Fernández-García, M.; Iglesias-Juez, A.; Guerrero-Ruiz, A.; Calvino, J.; Conesa, J.C.; Soria, J. *Chem. Mater.* **2003**, *15*, 4309.

(28) Enache, D.I.; Knight, D.W.; Hutchings, G.J. *Catal. Lett.* **2005**, *103*, 43.

(29) Zheng, X.; Zhang, X.; Wang, X.; Wang, S.; Wu, S. *Appl. Catal. A: Gen.* **2005**, *295*, 142.

(30) Zhang, S.M.; Huang, W.P.; Qiu, X.H.; Li, B.Q.; Zheng, X.C.; Wu, S.H. *Catal. Lett.* **2002**, *80*, 41.

(31) Martínez-Arias, A.; Fernández-García, M.; Soria, J.; Conesa, J.C. *J. Catal.* **1999**, *182*, 367.

(32) Wang, X.Q.; Rodriguez, J.A.; Hanson, J.C.; Gamarra, D.; Martínez-Arias, A.; Fernández-García, M. *J. Phys. Chem. B* **2006**, *110*, 428.

(33) Zhang, F.; Jin, Q.; Chan, S.-W. *J. Appl. Phys.* **2004**, *95*, 4319.

(34) PDF #34-0394, JCPDS Powder Diffraction File; Int. Center for Diffraction Data: Swarthmore, PA, 1993.

(35) Kau, L.-S.; Solomon, D.J.; Penner-Haln, J.E.; Hodgson, K.O.; Solomon, E.I. *J. Am. Chem. Soc.* **1987**, *109*, 6433.

(36) Skårman, B.; Grandjean, D.; Benfield, R.E.; Hinz, A.; Andersson, A.; Wallenberg, L.R. *J. Catal.* **2002**, *211*, 119; and references therein.

(37) Ma, X.D.; Yokoyama, T.; Hozumi, T.; Hashimoto, K.; Ohkoshi, S. *Phys. Rev. B* **2005**, *72*, 094107.

(38) Günter, M.M.; Ressler, T.; Jentoft, R.E.; Bems, B. *J. Catal.* **2001**, *203*, 133.

(39) Fernández-García, M. *Catal. Rev.-Sci. Eng.* **2002**, *44*, 59.

(40) Otsuka, K.; Hatano, M.; Morikawa, A. *J. Catal.* **1983**, *79*, 493.

(41) Sadi, F.; Duprez, D.; Gérard, F.; Miloudi, A. *J. Catal.* **2003**, *213*, 226.

(42) Trovarelli, A. *Catal. Rev. Sci. Eng.* **1996**, *38*, 439.

(43) Yao, H.C.; Yao, Y.F. *J. Catal.* **1984**, *86*, 254.

(44) Engler, B.; Koberstein, E.; Schubert, P. *Appl. Catal.* **1989**, *48*, 71.

(45) Rodriguez, J.A. In *Dekker Encyclopedia of Nanoscience and Nanotechnology*; Dekker: New York, 2004; 1297–1304.

(46) Haruta, M. *Catal. Today* **1997**, *36*, 153.
(47) Fu, Q.; Saltsburg, H.; Flytzani-Stephanopoulos, M. *Science* **2003**, *301*, 935.
(48) Chen, M.S.; Goodman, D.W. *Science* **2004**, *306*, 252.
(49) Campbell, C.T. *Science* **2004**, *306*, 234.
(50) Wang, X.Q.; Rodriguez, J.A.; Hanson, J.C.; Pérez, M.; Evans, J. *J. Chem. Phys.* **2005**, *123*, 221101.
(51) Rodriguez, J.A.; Pérez, M.; Evans, J.; Liu, G.; Hrbek, J. *J. Chem. Phys.* **2005**, *122*, 241101.
(52) Fu, Q.; Deng, W.; Saltsburg, H.; Flytzani-Stephanopoulos, M. *Appl. Catal. B: Environ.* **2005**, *56*, 57.
(53) Carrettin, S.; Guzman, J.; Corma, A. *Angew. Chem. Int. Ed.* **2005**, *44*, 2242.
(54) Guzman, J.; Gates, B.C. *J. Am. Chem. Soc.* **2004**, *126*, 2672.
(55) Valden, M.; Lai, X.; Goodman, D.W. *Science* **1998**, *281*, 1647.
(56) Goodman, D.W. *Catal. Lett.* **2005**, *99*, 1.
(57) Rodriguez, J.A.; Liu, G.; Jirsak, T.; Hrbek, J.; Chang, Z.; Dvorak, J.; Maiti, A. *J. Am. Chem. Soc.* **2002**, *124*, 5242.
(58) Vijay, A.; Mills, G.; Metiu, H. *J. Chem. Phys.* **2003**, *118*, 6536.
(59) Häkkinen, H.; Abbet, S.; Sanchez, A.; Heiz, U.; Landman, U. *Angew. Chem. Int. Ed.* **2003**, *42*, 1297.
(60) Remediakis, I.N.; Lopez, N.; Nørskov, J.K. *Angew. Chem. Int. Ed.* **2005**, *44*, 1824.
(61) Wang, X.Q.; Hanson, J.C.; Liu, G.; Rodriguez, J.A.; Iglesias-Juez, A.; Fernández-García, M. *J. Chem. Phys.* **2004**, *121*, 5434.
(62) Wang, X.Q.; Hanson, J.C.; Frenkel, A.I.; Kim, J.-Y.; Rodriguez, J.A. *J. Phys. Chem. B* **2004**, *108*, 13667.
(63) Wang, X.Q.; Hanson, J.C.; Liu, G.; Rodriguez, J.A.; Belver, C.; Fernández-García, M. *J. Chem. Phys.* **2005**, *122*, 154711.
(64) Norby, P.; Hanson, J.C. *Catal. Today* **1998**, *39*, 301.
(65) Chupas, P.J.; Ciraolo, M.F.; Hanson, J.C.; Grey, C.P. *J. Am. Chem. Soc.* **2001**, *123*, 1694.
(66) Schwartz, V.; Mullins, D.R.; Yan, W.; Chen, B.; Dai, S.; Overbury, S.H. *J. Phys. Chem. B* **2004**, *108*, 15782.
(67) Jacobs, G.; Ricote, S.; Patterson, P.M.; Graham, U.M.; Dozier, A.; Khalid, S.; Rhodus, E.; Davis, B.H. *Appl. Catal. A: Gen.* **2005**, *292*, 229.
(68) Rodriguez, J.A.; Wang, X.Q.; Liu, G.; Hanson, J.C.; Hrbek, J.; Peden, C.H.F.; Iglesias-Juez, A.; Fernández-García, M. *J. Molec. Catal. A: Chem.* **2005**, *228*, 11.
(69) Campbell, C.T.; Daube, K.A. *J. Catal.* **1987**, *104*, 109.
(70) Rodriguez, J.A.; Jirsak, T.; Freitag, A.; Hanson, J.C.; Larese, J.Z.; Chaturvedi, S. *Catal. Lett.* **1999**, *62*, 113.
(71) Ferriz, R.M.; Gorte, R.J.; Vohs, J.M. *Catal. Lett.* **2002**, *82*, 123.
(72) Liu, G.; Rodriguez, J.A.; Hrbek, J.; Dvorak, J.; Peden, C.H.F. *J. Phys. Chem. B* **2001**, *105*, 7762.
(73) Shyu, J.Z.; Weber, W.H.; Gandhi, H.S. *J. Phys. Chem.* **1988**, *92*, 4964.
(74) Nakamura, J.; Campbell, J.M.; Campbell, C.T. *J. Chem. Soc. Faraday* **1990**, *86*, 2725.

(75) Oyama, S.; Ito, T.; Rodriguez, J.A.; Nakamura, J. In publication.
(76) Jacobs, G.; Chenu, E.; Patterson, P.M.; Williams, L.; Sparks, D.; Thomas, G.; Davis, B.H. *Appl. Catal. A: Gen.* **2004**, *258*, 203.
(77) Mendelovici, L.; Steinberg, M. *J. Catal.* **1985**, *96*, 285.
(78) Bunluesin, T.; Gorte, R.J.; Graham, G.W. *Appl. Catal. B: Environ.* **1998**, *15*, 107.
(79) Hilaire, S.; Wang, X.; Luo, T.; Gorte, R.J.; Wagner, J. *Appl. Catal. A: Gen.* **2001**, *215*, 271.
(80) Chenu, E.; Jacobs, G.; Crawford, A.C.; Keogh, R.A.; Patterson, P.M.; Sparks, D.E.; Davis, B.H. *Appl. Catal. B: Environ.* **2005**, *59*, 45.
(81) Ricote, S.; Jacobs, G.; Milling, M.; Ji, Y.; Patterson, P.M.; Davis, B. *Appl. Catal. A: Gen.* **2006**, *303*, 35.
(82) Basinska, A.; Kepinski, L.; Domka, F. *Appl. Catal. A: General* **1999**, *183*, 143.
(83) Ghenciu, A.F. *Curr. Opin. Solid State Mater. Sci.* **2002**, *6*, 389.
(84) Peña, M.A.; Gómez, J.P.; Fierro, J.L.G. *Appl. Catal. A: Gen.* **1996**, *144*, 7.
(85) Rostrup-Nielsen, J.R. *Catal. Today* **1993**, *18*, 305.
(86) Craciun, R.; Daniel, W.; Knozinger, H. *Appl. Catal. A: Gen.* **2002**, *230*, 153.
(87) Bharadwaj, S.S.; Schmidt, L.D. *Fuel Proc. Technol.* **1995**, *42*, 109.
(88) Ashcroft, A.T.; Cheetham, A.K.; Green, M.L.H.; Grey, C.P. *Nature* **1990**, *344*, 319.
(89) Pantu, P.; Kim, K.; Gavalas, G.R. *Appl. Catal. A: Gen.* **2000**, *193*, 203.
(90) Dong, W.S.; Jun, K.W.; Ro, H.S.; Liu, Z.W.; Park, S.E. *Catal. Lett.* **2002**, *78*, 13.
(91) Sadykov, V.A.; Kuznetsova, T.G. *React. Kin. Catal. Lett.* **2002**, *76*, 83.
(92) Bedrane, S.; Descorme, C.; Duprez, D. *Catal. Today* **2002**, *75*, 401.
(93) Velu, S.; Suzuki, K.; Kapoor, M.P.; Ohashi, F.; Osaki, T. *Appl. Catal. A: Gen.* **2001**, *213*, 47.
(94) Breen, I.P.; Roos, J.R.H. *Catal. Today* **1999**, *51*, 521.
(95) Marczewski, M.; Peplonski, R. *React. Kinet. Catal. Lett.* **1983**, *22*, 241.
(96) Kreese, J.P.; Antok, A.; Smith, R.T. Private communication.
(97) Fujishima, A.; Honda, K. *Nature* **1972**, *238*, 37.
(98) Grätzel, M. *Nature* **2001**, *414*, 338.
(99) Arakawa, H.; Sayama, K. *Catal. Surv. Jpn.* **2000**, *4*, 75.
(100) Bolts, J.M.; Wrighton, M.S. *J. Phys. Chem.* **1976**, *80*, 2641.
(101) Domen, K.; Kudo, A.; Onishi, H. *J. Catal.* **1986**, *102*, 92.
(102) Asahi, R.; Morikawa, T.; Ohwaki, T.; Aoki, K.; Taga, Y. *Science* **2001**, *293*, 269.
(103) Chen, X.; Burda, C. *J. Phys. Chem. B* **2004**, *108*, 15446.
(104) Nakamura, R.; Tanaka, T.; Nakato, Y. *J. Phys. Chem. B* **2004**, *108*, 10617.
(105) Di Valentin, C.; Pacchioni, G.; Selloni, A. *Phys. Rev. B* **2004**, *70*, 085116.
(106) Lee, J.Y.; Park, J.; Cho, J.-H. *Appl. Phys. Lett.* **2005**, *87*, 011904.
(107) Batzill, M.; Morales, E.; Diebold, U. *Phys. Rev. Lett.* **2006**, *96*, 026103.
(108) Li, X.; Sabir, I. *Int. J. Hydr. En.* **2005**, *30*, 359.
(109) Litster, S.; McLean, G. *J. Power Sources* **2004**, *130*, 61.
(110) Metha, V.; Cooper, J.S. *J. Power Sources* **2003**, *114*, 32.
(111) Smitha, B.; Sridhar, S.; Khan, A.A. *J. Membrane Sci.* **2005**, *259*, 10.

(112) Fukui, S.; Naito, T.; Li, Q. Private communication.

(113) Schoonman, J. *Solid State Ion.* **2000**, *135*, 5.

(114) Farrauto, R.; Hwang, S.; Shore, L.; Ruettinger, W.; Lampert, L.; Giroux, T.; Liu, Y.; Ilinich, O. *Annu. Rev. Mater. Res.* **2003**, *33*, 1.

(115) Blomen, L.J.M.J.; Mugerwa, M.N. *Fuel Cell Systems*; Plenum Press: New York, 1993.

(116) Ross, P.N.; Markovic, N.M.; Schmidt, T.J.; Stammenkovic, V. *New Electrocatalysts for Fuel Cells*, Transportation Fuel Cell Power System, Annual Progress Report, 2000; 115.

(117) Singhal, S.C.; Kendall, K. *High Temperature Solid Oxide Fuel Cells: Fundamentals. Design and Applications*; Elsevier: Oxford, 2003.

(118) Singhal, S.C. *Solid State Ion.* **2000**, *135*, 305.

(119) Badwal, S.P.S. *Solid State Ion.* **2001**, *143*, 39.

(120) Fukui, T.; Ohara, S.; Naito, M.; Nogi, K. *J. Nanopar. Res.* **2001**, *3*, 171.

(121) Jiang, S.P.; Chan, S.H. *J. Mater. Sci.* **2004**, *39*, 4405.

(122) Besenbacher, F.; Chorkendorff, I.; Clausen, B.S.; Hammer, B.; Molenbroek, A.M.; Norskov, J.K.; Stensgaard, I. *Science* **1998**, *279*, 1913.

(123) Rostrup-Nielsen, J.R. Catalyst deactivation. In *Studies in Surface Science and Catalysis, Vol. 68*; Bartholomew, C.H.; Butt, J.B. (Editors); Elsevier: Amsterdam, 1991.

(124) Bengaard, H.S.; Norskov, J.K.; Sehested, J.; Clausen, B.S.; Nielsen, L.P.; Molenbroek, A.M.; Rostrup-Nielsen, J.R. *J Catal.* **2002**, *209*, 365.

(125) Hu, Y.H.; Ruckenstein, E. *Catal. Rev.* **2002**, *44(3)*, 423.

(126) Trimm, D.L. *Catal. Today* **1999**, *49*, 3.

(127) Takeguchi, T.; Kikuchi, R.; Yano, T.; Eguchi, K.; Murata, K. *Catal. Today* **2003**, *84*, 217.

(128) Takeguchi, T.; Yano, T.; Kani, Y.; Kikuchi, R.; Eguchi, K. In *SOFC-VIII, Vol. 7*; Singhal, S.C.; Dokiya, M. (Editors); Electrochemical Society: Pennington, NJ, 2003; 704.

(129) Eguchi, K.; Setoguchi, T.; Okamoto, K.; Arai, H. In *Proc. of the International Fuel Cell Conf*; Makihari: Japan, 1992; 373.

(130) Park, S.; Kim, J. To be published.

(131) Park, S.D.; Vohs, J.M.; Gorte, R.J. *Nature* **2000**, *404*, 265.

(132) Kim, H.; Lu, C.; Worrell, W.L.; Vohs, J.M.; Gorte, R.J. *J. Electrochem. Soc.* **2002**, *149*, A247.

(133) Gorte, R.J.; Kim, H.; Vohs, J.H. *J. Power Sources* **2002**, *106*, 10.

CHAPTER 22

Oxide Nanomaterials in Ceramics

VICENTE RIVES, RAQUEL TRUJILLANO, and MIGUEL A. VICENTE
Departamento de Química Inorgánica. Universidad de Salamanca, Spain

22.1. INTRODUCTION

The word *ceramics* is usually associated with art, dinnerware, pottery, tiles, brick, and toilets. These products, however, should be referred to as traditional or silicate-based ceramics. Despite these traditional products have been, and continue to be, important to society, a new class of ceramics is emerging. These advanced or technical ceramics are being used for applications simply unexpected (or even unknown) some years ago, such as space shuttle tiles, engine components, artificial bones and teeth, computers and other electronic components, or cutting tools.

Broadly speaking, "ceramics" can be defined as inorganic compounds, constituted by a couple of elements, one metallic and the other nonmetallic, with markedly different electronegativities (as alumina Al_2O_3 or zirconia ZrO_2), typically produced from clays and other natural minerals or chemically processed powders. However, some ceramic compounds are formed between two nonmetallic elements neighboring in the Periodic Table (as silicon nitride Si_3N_4, silicon carbide SiC, or boron nitride BN), thus having strongly covalent properties, and other compounds are formed by more than two elements (as aluminium titanate or sialons). Ceramics are typically crystalline in nature. Glass is often considered a subset of ceramics, although it is somewhat different than ceramics in that it is amorphous or lacks long-range crystalline order (1).

All "classic" ceramics contain clays; actually, the word ceramics is related to the Greek word for designating clays ($\kappa\varepsilon\rho\alpha\mu\iota\kappa\eta$). Although some common ceramics, mostly those found in homeware, are still related to clay materials, a new class of ceramics, usually denoted as *advanced ceramics* and not containing clays, have become very important in the last few decades.

Advanced ceramics are distinguished from traditional ceramics by their larger strength, higher operating temperatures, improved toughness, and tailorable properties. Advanced ceramics, also known as engineering or technical ceramics, refer

Synthesis, Properties, and Applications of Oxide Nanomaterials, Edited by José A. Rodríguez and Marcos Fernández-Garcia

to materials that exhibit superior mechanical properties, corrosion/oxidation resistance, and thermal, electrical, optical, or magnetic properties. Advanced ceramics can be classified as structural ceramics, electrical and electronic ceramics, ceramic coatings, and chemical processing and environmental ceramics. An important advantage of advanced ceramics is that they can replace metals in applications where materials with lower density and higher melting points, together with high resistance to corrosive environments such as acids, alkalis, and organic solvents, are needed, such as engine components, cutting tools, valves, bearings, and chemical process equipment (2).

Structural ceramics are useful for applications such as industrial wear parts, bioceramics, cutting tools, and engine components. Electronic applications for ceramics with low thermal expansion coefficient and high thermal conductivity include superconductors, substrates, magnets, capacitors, and transducers. Electronic ceramics, which have the largest share of the advanced ceramic market, includes capacitors, insulators, substrates, integrated circuits packages, piezoelectrics, magnets, and superconductors. Ceramic coatings find applications in engine components, cutting tools, and industrial wear parts. Chemical processing and environmental ceramics include filters, membranes, catalysts, and catalyst supports.

We review in this chapter the chemical and structural properties of the main ceramic compounds, classified from their composition. Thus, we may consider zirconia, boron nitride, silicon carbide, silicon nitride, sialon, boron carbide, and aluminum titanate. We describe their structures and their main preparation methods, in some cases based on synthesis reactions and in others on purification of natural ores, also paying special attention to the chemical aspects of the processes. Although we describe some peculiarities of these syntheses, we will not discuss in detail other important steps of the preparation of these materials, such as sintering and densification, discussed in other chapters of this book.

22.2. MAIN CERAMIC COMPOUNDS

Table 22.1 lists several ceramic materials currently included under the heading of advanced structural ceramics. The most important advanced structural materials being developed at this time are alumina, zirconia, and beryllia (all of them oxides) and the covalent materials silicon carbide and silicon nitride. In addition, cordierite is also important because of its use in heat exchangers. Table 22.2 lists some important mechanical properties of several of these materials.

Aluminas exhibit good mechanical performances at temperatures as high as 1900 °C provided they are not exposed to thermal shock, impact, or highly corrosive atmospheres. Above 2000 °C, the strength of alumina drops; consequently, many applications are under steady-state, high-temperature conditions, but not where abrupt temperature changes would cause failure arising from thermal shock. Aluminas have good creep resistance up to about 1800 °C, above which other ceramics show a better behavior. In addition, aluminas are susceptible to corrosion from strong acids, steam, and sodium.

TABLE 22.1. Chemical Formulas of Some Advanced Structural Ceramics (3).

Generic Name	Chemical Formula
Alumina	Al_2O_3
Zirconia	ZrO_2
Zircon	$ZrSiO_4$ or $ZrO_2 \cdot SiO_2$
Spinel	$MgAl_2O_4$ or $MgO \cdot Al_2O_3$
Mullite	Al_6SiO_{11} or $3Al_2O_3 \cdot SiO_2$
Cordierite	$Mg_2Al_4Si_5O_{18}$ or $2MgO \cdot 2Al_2O_3 \cdot 5SiO_2$
Silicon carbide	SiC
Silicon nitride	Si_3N_4
SiAlON	$Si_{6-z}Al_zO_zN_{8-z}$
Boron carbide	B_4C
Aluminum nitride	AlN
Glass ceramics	Al_2O_3, Li_2O, and SiO_2 with TiO_2 or ZrO_2 as nucleating agents

Beryllia (BeO, with the hexagonal wurtzite-like structure; Figure 22.1) ceramics are efficient heat dissipaters and excellent electrical insulators. They are used in electrical and electronics applications, such as microelectric substrates, transistor bases, and resistor cores. Beryllia has excellent thermal shock resistance (some grades can withstand 800 °C/s changes), a very low thermal expansion coefficient, and a high thermal conductivity. It is expensive, however, and is an allergen to which some persons are sensitive.

TABLE 22.2. Engineering Ceramic Materials and their Main Properties (3).

Property	Si_3N_4	SiC	ZrO_2	Al_2O_3 (+ZrO_2)	Al_2TiO_5
Density, g/cm^3	3.2	3.2	6	3–4	3
Bending strength, MPa	200–1000	500	500 to >1000	300–600	40
Bending strength at 1400 °C, MPa	200–600	200–400	100	100–400	50
Hardness, GN/m^2	14–17	25–30	12	18–23	
Fracture toughness K_{IC}, MPa m$^{0.5}$	3–10	3–5	6–15	3 (8)	1
Young's modulus	200–300	400	200	400 (300)	20
Thermal expansion coefficient 20–1200 °C, (10^{-6} K^{-1})	3	4.5	10	8	2
Thermal conductivity, W m^{-1} K^{-1}	10–40	100–140	2	30	2
Thermal shock resistance	high	high	medium	low (medium)	very high
Abrasion resistance	very good	very good	good	good	no

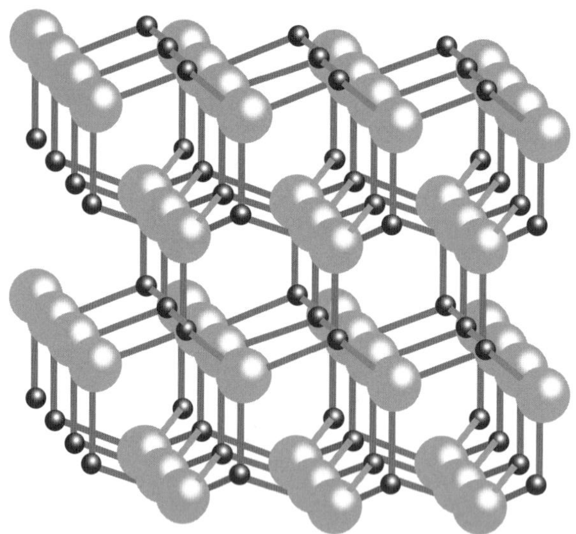

Figure 22.1. Idealized structure of beryllia, BeO. Black spheres correspond to $Be^{2+}$and large, gray spheres to O^{2-} (4).

Zirconia is used primarily for its extreme inertness to most metals. Zirconia ceramics are very strong even near their melting point, well over 2200 °C, the highest of all ceramics. Applications for fused or sintered zirconia include crucibles and furnace bricks.

Transformation-toughened zirconia ceramics are among the strongest and toughest ceramics even made. These materials are of three main types: Mg–PSZ (zirconia partially stabilized with magnesium oxide), Y–TZP (Yttria-stabilized tetragonal zirconia polycrystals), and ZTA (zirconia-toughened alumina) (5).

Applications of Mg–PSZ ceramics are mainly under low- and moderate-temperature abrasive and corrosive environments—pump and valve parts, seals, bushings, impellers, and knife blades. Y–TZP ceramics (stronger than Mg–PSZ but less flaw tolerant) are used for pump and valve components requiring wear and corrosion resistance in room-temperature service. Lower dense ZTA ceramics have a better thermal shock resistance, and they are cheaper than the other two; they are used in transport equipment where they need to withstand corrosion, erosion, abrasion, and thermal shock.

Many engineering ceramics have multioxide crystalline phases. An especially useful one is *cordierite* ($Al_3Mg_2AlSi_5O_{18}$ in the dehydrated form, constituted by alternate layers of SiO_4 tetrahedra, on one side, and AlO_4 tetrahedra and MgO_6 octahedra, on the other; Figure 22.2), which is used in cellular ceramic form as a support for a washcoat and catalyst in catalytic converters in automobile emission systems. Its low thermal expansion coefficient accounts for its resistance to thermal fracture.

Glass ceramics are prepared from molten glass and subsequently crystallized by heat treatment. They are constituted by several oxides forming complex, multiphase microstructures. Glass ceramics do not exhibit the strength-limiting porosity of conventional sintered ceramics, but other properties can be tailored by controlling

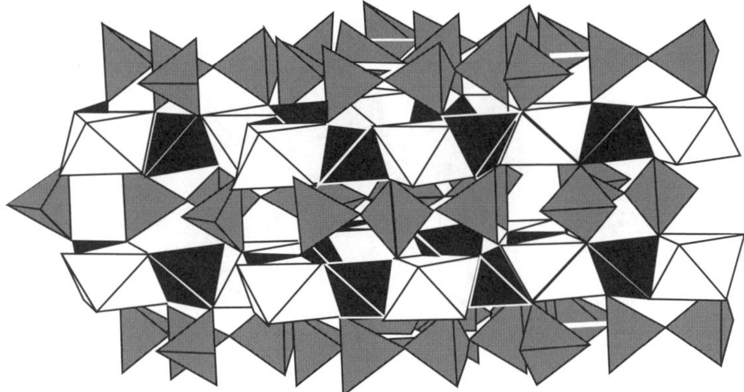

Figure 22.2. Idealized structure of cordierite (4). The gray tetrahedra correspond to SiO_4 units, the black tetahedra to AlO_4 units, and the white octahedra to MgO_6 units.

their crystalline structure in the host glass matrix. Major applications are as cooking vessels, tableware, smooth cooktops, and several technical products such as radomes.

The three common glass ceramics, lithium-aluminum-silicate (LAS, or β-spodumene, Figure 22.3), magnesium-aluminum-silicate (MAS, or cordierite), and

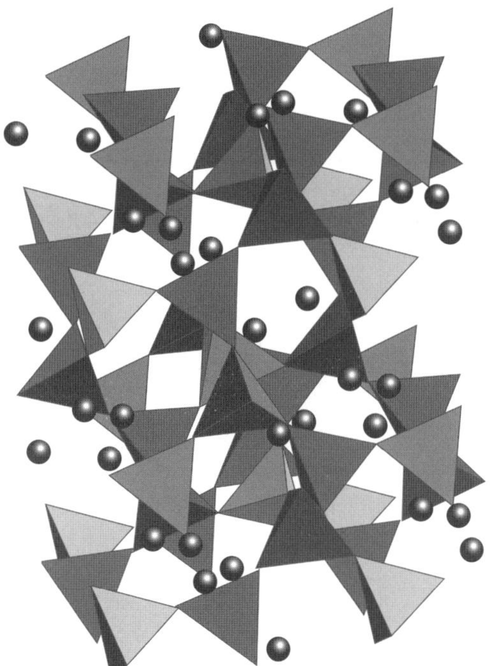

Figure 22.3. Idealized structure of β-spodumene (4). The tetrahedra correspond to SiO_4 and AlO_4 units, and the black sphere to Li^+ ions.

Figure 22.4. Idealized structure of keatite (4).

aluminum-silicate (AS, or aluminous keatite, a silica polymorh; Figure 22.4), are stable at high temperatures, have near-zero thermal expansion coefficients (LAS and AS have essentially nonmeasurable thermal expansion up to 400 °C), and are resistant to high-temperature corrosion, especially oxidation. The high silica content of LAS is responsible for its low thermal expansion, but the presence of silica decreases strength; regarding its chemical resistance, LAS is sensitive to sulphur and sodium attack.

Several *metal carbides and nitrides* also qualify as engineering ceramics. Most commonly used are boron carbide and nitride, silicon carbide and nitride, and aluminum nitride.

Boron carbide is known for its very large hardness and low density, making it useful for lightweight, bulletproof armor plates. The material has the best abrasion resistance of any ceramics, so it is also specified for pressure-blasting nozzles and similar high-wear applications. A use limitation of boron carbide is due to its low strength at high temperatures.

Despite their high prices, silicon carbide (SiC), aluminum nitride (AlN), and boron nitride (BN) are challenging alumina, particularly for the more critical applications. BN, for example, has a high dielectric strength and near-zero thermal expansion in some temperature ranges.

Silicon carbide and silicon nitride can be considered the high-temperature, high-strength superstars of the engineering ceramics. These are the strongest structural ceramics for high-temperature oxidation-resistant service. However, SiN and SiC primary particles do not easily self-bond. Consequently, many processing variations have been devised to fabricate parts from these materials, creating several tradeoffs in cost, fabricability, and properties. As ceramic powders can be consolidated by hot pressing, under simultaneous high temperature and pressure and, in some cases, the presence of additives acting as bond-forming catalysts, fully dense material can be formed.

On the other hand, Si_3N_4 and SiC particles can be bonded without pressure by several processes, namely reaction bonding, recrystallization (for silicon carbide), or reaction sintering. With these processes, "green" parts can be dry or isostatically pressed, extruded, slip-cast, or, in some cases, formed by conventional plastic molding techniques such as injection molding, and then sintered. Complex shapes, close to finished size, can be produced by these techniques, but the ceramic is only about 80% as dense as the hot-pressed counterpart and has lower strength and poorer thermal shock resistance.

Silicon carbide, either hot pressed or reaction bonded, is not as strong as silicon nitride up to about 1400 °C; silicon nitride grain boundaries soften, or creep, and strength drops. Above 1400 °C, silicon carbide is the strongest ceramic known. At 1300 °C, however, strength of hot-pressed ceramics nearly equals that of reaction-bonded ceramics (2).

22.3. TYPES OF ADVANCED MATERIALS

According to their applications, advanced ceramics can be classified into structural ceramics and electronic ceramics or electroceramics. In the last few years, a new important family has arised, the so-called bioceramics.

22.3.1. Structural Ceramics

Structural ceramics are those ones in which the mechanical properties (namely, strength, modulus, toughness, wear resistance, and hardness) are the most important clues controlling their performance and applications. These ceramics constitute about 10% of the market nowadays, but this market quota may highly increase if they become effective to be used in engines. Their hardness, low density, and resistance—both to high temperature and to corrosive environments—favor their use, whereas their low fracture toughness and low reliability are some drawbacks. In any case, as mentioned, it is expected that ceramics with better properties to be used in engines and automotive components will be hopefully developed.

From the chemical point of view, structural ceramics are mainly composed of inorganic materials and often possess nonmetallic properties, such as good thermal and electrical insulators. Some components include ionic–covalent materials (alumina, zirconia, titania, and magnesia) as well as covalent materials (diamond, cubic boron nitride, silicon nitride, silicon carbide, aluminum nitride, cordierite, and glass ceramic). Structural ceramics are used in industries such as Aerospace, Electronics, Heat-treating, High-temperature Furnaces, Instrumentation, Lighting, Medical, Semiconductor, and Welding and Plasma Cutting.

World demand for advanced ceramics will strongly increase in the next years. According to the predictions from Freedonia Group, world demand for advanced ceramics will increase over 7% annually, driven by the manufacture of a wide variety of electronic components. Alumina will remain the primary material used to produce advanced ceramics, whereas ferrites, beryllia, and zirconia ceramics continue to erode the alumina market share. This study analyzes the US$24 billion world advanced ceramics industry, with historical data (1989–1999), and forecasts to 2004 and 2009 by material (e.g., alumina, titanate, ferrite, zirconate, cordierite, beryllia, silicon nitride) and by end-use market, as well as for 6 geographic regions and 11 key countries. The study also examines the market environment, details industry and market share, and profiles over 40 key companies, including KEMET, Kyocera, Murata, NGK Insulators, Saint-Gobain, TDK, Toshiba, and Vishay Intertechnology (6).

Another well-known marketing company, Business Communications Company (BCC), estimates in its December 2005 report that the North American market of advanced structural ceramics will reach US$2.3 billion in 2005 and will expand at an average annual growth rate of 5.8% to reach US$3 billion by 2010. This study is based on an overview of the various advanced ceramic materials and their production technologies and applications, on the determination of the current market size and future growth for applications in tool inserts, wear-resistant components, energy and high temperature resistant components, aircraft and aerospace components, military, bioceramics and finally, auto and turbine engine components, and on a determination of the current market size and future growth of alumina-based ceramic, zirconia, silicon carbide, boron carbide, silicon nitride, titanium diboride, glass and glass-ceramics, and other categories (7).

22.3.2. Electroceramics

Electroceramics are widely present today in our daily lives. Essential, and often critical, components in devices ranging from the humble gas lighter to the sophisticated cellular phone contain these ceramics. They may have insulating, dielectric, piezoelectric, magnetic, optical, or superconducting properties, allowing them a widespread use in electrical and electronic devices. Chemically, electroceramics are mainly composed of mixed oxides, sometimes with the spinel or perovskite structures.

Piezoelectric lead *zirconate titanate* (PZT) is one of the best known electroceramics and is widely used in gas lighters; telephones and autofocus cameras; capacitors made of barium titanate ceramic in televisions, radios, and almost all electronic equipment; as well as the microwave dielectric ceramics used in highly selective filters for cellular phones and satellite communication systems. *Soft ferrites* are used in sensitive radio antennas and transformers; *hard ferrites* are used in many small electric motors—as many as 20 such motors are used in some models of automobiles for functions such as power seats and door locks. Fuel cells, which may replace the internal combustion engine in cars in the near future, contain ceramics such as zirconia working as solid electrolytes to generate power from fuels without combustion.

A class of electroceramics that show diverse phenomena and have wide applications are the ferroelectric ceramics. Similar to ferromagnets, these ceramics can be electrically polarized permanently and the direction of polarization can also be reversed. They have potential applications as memories in computers, mechanical sensors, and transducers and as room temperature infrared sensors.

Although ceramics have traditionally been used for their mechanical and thermal stability, their unique electrical, optical, and magnetic properties have become of increasing importance in many key (and even strategic) technologies, including communications, energy conversion and storage, electronics, and automation. Such materials are now classified under electroceramics as distinguished from other functional ceramics such as advanced structural ceramics.

Historically, developments in the various subclasses of electroceramics have paralleled the growth of new technologies, examples including ferroelectrics (high dielectric capacitors, nonvolatile memories), ferrites (data and information storage), solid electrolytes (energy storage and conversion), piezoelectrics (sonar), and semiconducting oxides (environmental monitoring).

22.3.3. Bioceramics

A market has also developed for ceramics called "bioceramics," which are used as hip and bone transplants, as supports for directed delivery of limited and controlled amounts of drugs to the affected areas, and as components in implant devices such as pace makers. Their inertness to body fluids and adequate mechanical strength makes some ceramics the ideal materials for these applications.

22.4. CHEMICAL NATURE OF ADVANCED CERAMICS

As noted, some of the most important advanced ceramics currently used are composed of alumina, zirconia, silicon nitride, silicon carbide, boron carbide, or boron nitride. In some cases, mixtures of these compounds are used, whereas in others, small amounts of other compounds are added to improve their properties. Without paying special attention to the effects of the addition of dopants to a given oxide, we review here some important properties of these compounds that make them available for ceramic uses.

22.4.1. Alumina (Al_2O_3)

Alumina is the most versatile engineered ceramic because it can be used up to very high temperature, and because of its chemical, electrical, and mechanical properties. Its relatively low cost and easy fabrication also enhance its use. Alumina products range from relatively low-calcined grades of polishing aluminas to the extremely hard, fused alumina and synthetically produced sapphire.

The main properties that make alumina valuable in ceramic applications are its high melting point (2050 °C), hardness (9 on the Mohs scale), strength, dimensional stability, chemical inertness, and electrical insulating ability. Together with its availability in large quantities at moderate prices, these properties have led to extensive and varied uses of alumina as a ceramic material (8).

Alumina is currently the most highly developed engineering ceramic. A wide range of high Al_2O_3 ceramics is produced under different trade names, the Al_2O_3 content usually ranges from 92% to 99.8%, with trace components, particularly silica, enhancing its properties. Properties of 99.8% alumina are listed in Table 22.3. Strength and temperature capability increase with Al_2O_3 content, but shaping capability decreases with increasing Al_2O_3 content. SiO_2, MgO, and other additives are added to aid shaping and to control sintering temperature. The microstructure of high-Al_2O_3 ceramic also contains some glassy phase arising from the sintering additives. One drawback for its use is a poor thermal shock resistance and the difficulty of machining sintered materials to obtain high-dimensional accuracy of complex-shaped components. Indeed, because of its high hardness, α-Al_2O_3 (corundum, Figure 22.5) has long been used in abrasives, either in the form of loose grains or bonded material in grinding wheels. Retention of hardness at high temperature is an important characteristic because the tool temperature can reach values as high as 1000 °C during the machining of ductile materials such as steel. Silicon carbide whiskers can be added to Al_2O_3, increasing

TABLE 22.3. Some Mechanical, Thermal, and Electrical Properties of 99.8% Al_2O_3 alumina (3).

Property (conditions, unities)	Value
Bulk density (20 °C, g/cm^3)	3.96
Tensile Strength (20 °C, MPa)	220
Flexural (Bending) Strength (20 °C, MPa)	410
Elastic Modulus (20 °C, GPa)	375
Hardness (20 °C, kg/mm^2)	14
Fracture Toughness (20 °C, MPa•$m^{1/2}$)	4–5
Porosity (20 °C, %)	0
Maximum working temperature (-, °C)	1700
Thermal Expansion Coefficient (10^{-6}/°C)	25–300 °C: 7.8 25–1000 °C: 8.1
Thermal Conductivity (20 °C, W/m K)	28
Dielectric Strength (2.5 mm tk, ac-kv/mm)	10
Dielectric Constant (1 MHz, -)	9.7
Volume Resistivity (Ohm cm)	20 °C: $>10^{14}$ 300 °C: 10^{10} 1000 °C: 10^6
Loss Factor (1 MHz, -)	0.009
Dissipation Factor (1 MHz, -)	0.0001

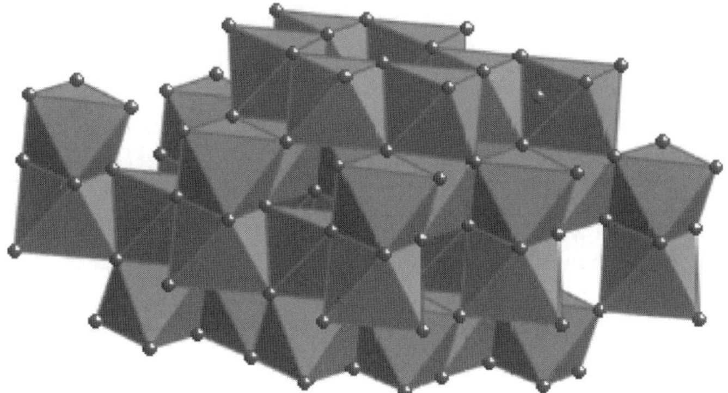

Figure 22.5. Idealized structure of corundum, α-Al_2O_3 (4). The centers of the octahedra are occupied by Al^{3+} ions (two out of three octahedra in each layer), and the corners are occupied by O^{2-} ions, here shown as small spheres.

its thermal conductivity and, consequently, the resistance to thermal shock. They also significantly increase toughness, which has led to improved tool performance (8,9).

Sintered Al_2O_3 is also used for missile radomes, external containers for radar, and other equipment. The radome must be transparent to radar signals, thus requiring some particular dielectric properties; in addition, as an external surface, it is also exposed to aerodynamic and thermal effects during flight. It is also widely used for crucibles, tubes, and rods for high-temperature use and for a large number of wear-resistant and corrosion-resistant specialized items. Perhaps the most important single products are spark plugs and the optically translucent polycrystalline alumina lamp hulls for high-temperature sodium vapor street lamps.

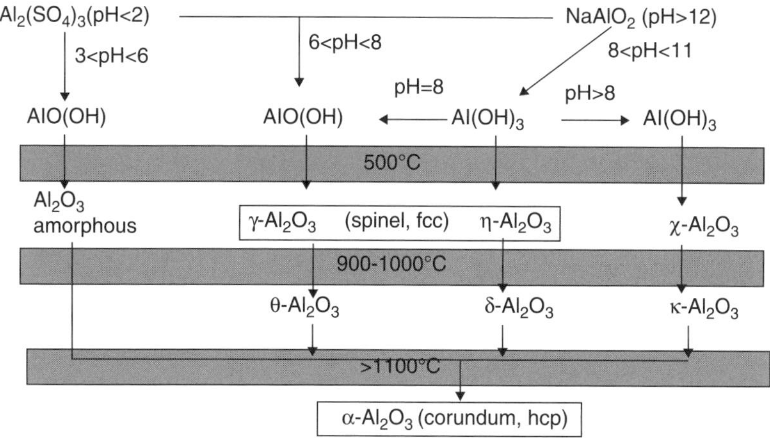

Figure 22.6. Flowchart showing transformations among Al–O compounds.

When aluminum hydroxides or oxo-hydroxides are heated in air at atmospheric pressure, they undergo a series of compositional and structural changes (Figure 22.6), before ultimately being converted to α-Al_2O_3. Several aluminium hydroxides and oxo-hydroxides are known. A general classification of aluminum hydroxides includes well-defined *crystalline forms* as the three trihydroxides $Al(OH)_3$: *gibbsite, bayerite*, and *nordstrandite*, as well as two modifications of aluminum oxide hydroxide AlO(OH): *boehmite* and *diaspore* (10). Besides these well-defined crystalline phases, several other forms have been described in the literature. However, there is still a controversy as to whether they are truly new phases or simply forms with distorted lattices containing adsorbed or interlamellar water and impurities (β-alumina actually corresponds to the chemical formula $NaAl_{11}O_{17}$; Figure 22.7).

Gelatinous hydroxides may consist of predominantly X-ray indifferent aluminum hydroxide or pseudoboehmite. The X-ray diffraction pattern of the latter shows broad bands in positions roughly coincident with those for strong reflections of the well-crystallized oxide hydroxide boehmite.

The aluminum hydroxides broadly distributed in nature are gibbsite, boehmite, and diaspore. *Gibbsite* and *bayerite* have similar structures. Their lattices are built of layers of anion octahedra (that for bohemite is shown in Figure 22.8), in which aluminum cations occupy two thirds of the octahedral holes. In the gibbsite structure, the layers are somewhat displaced relative to one another in the direction of the

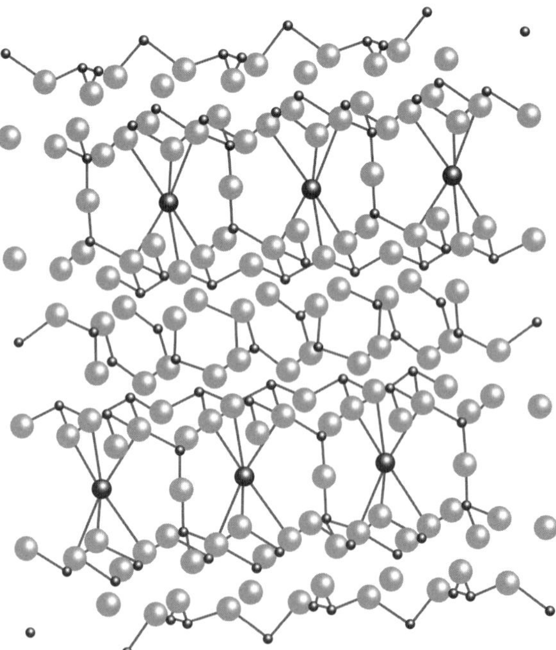

Figure 22.7. Idealized structure of β-Al_2O_3 (4). Small black spheres correspond to Al^{3+} ions, large black spheres to Na^+ ions, and gray spheres to O^{2-} ions.

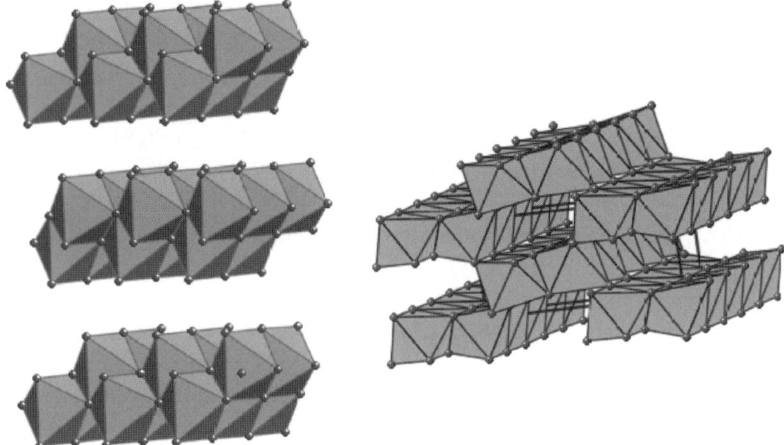

Figure 22.8. Idealized structure of (left) bohemite and (right) diaspore, showing the stacking of octahedra layers occupied by Al^{3+}; the corners are occupied by O^{2-} or OH^- ions, here shown as small spheres (4).

a-axis. As a consequence, the hexagonal symmetry of this lattice type (brucite type) is lowered to monoclinic.

In bayerite, the layers are arranged in approximately hexagonal close packing. Because of shorter distances between the layers, the density is higher than in the case of gibbsite. The crystal class and space group of bayerite have not yet been established clearly.

The individual layers of hydroxyl octahedra in both the gibbsite and the bayerite structures are linked to one another only through weak hydrogen bonds. Bayerite does not form large single crystals; the most commonly observed growth forms are spindle- or hourglass-shaped somatoids. The long axis of these somatoids stands normal to the basal plane; i.e., the somatoids consist of stacks of $Al(OH)_3$ layers. The effect of alkali ions on the structures of $Al(OH)_3$ types was investigated by several workers, and intercalation of Li^+ ions in empty octahedral holes transforms gibbsite into the hydrotalcite-like structure.

Nordstrandite, the third form of $Al(OH)_3$, was described by van Nordstrand and others. The structures of nordstrandite and bayerite were investigated and compared with those of monoclinic and triclinic gibbsite, which had been determined previously.

The lattice of nordstrandite is built of the same, electrically neutral $Al(OH)_3$ octahedral layers that form gibbsite and bayerite. The lattice period amounts to 1.911 nm in the direction normal to the layer. This corresponds to the sum of identical layer distances of bayerite plus gibbsite. The identical nordstrandite structure consists of alternating double layers, in which the OH octahedra are arranged once in the packing sequence of bayerite, and then in that of gibbsite. Materials containing continuous transitions from bayerite through nordstrandite to gibbsite have been prepared through a proper selection of precipitation conditions.

The thermal transformations from the hydroxides or the oxo-hydroxides into alumina are topotactic. Despite a loss of 34 or 15% of mass for the trihydroxides or oxide hydroxides, respectively, the habit of the primary crystals and crystal aggregates is only slightly changed. This leads to a considerable internal porosity, which may increase the specific surface area of the material to several hundreds of square meters per gram. Several structural forms develop that, albeit thermodynamically unstable, are reproducibly formed and are characteristic for a given temperature range and starting material. The hexagonally closest packed α-Al_2O_3 (corundum) modification is the only stable oxide in the Al_2O_3–H_2O system. Corundum is a common mineral in igneous and metamorphic rocks. Red and blue varieties of gem quality are known as ruby and sapphire, respectively. The lattice of corundum is composed of hexagonally closest packed oxygen ions forming layers parallel to the (0001) plane (Figure 22.5). Only two thirds of the octahedral holes are occupied by aluminum ions. The structure may be described roughly as consisting of alternating layers of Al and O ions.

The simplest transformation yielding to corundum is that from diaspore, (Figure 22.8). As the structures of these two compounds are very similar to each other, nucleation of α-Al_2O_3 requires only a minor rearrangement of the oxygen lattice after the hydrogen bonds are broken, and complete transformation can be attained at temperatures as low as 860–870 K. The newly formed corundum grows epitaxially on the decomposing diaspore, with the (0001) plane of Al_2O_3 parallel to the (010) plane of AlO(OH). Transformation to the corundum (α-Al_2O_3) proceeds through an intermediate α'-Al_2O_3 phase (8).

The thermal transformation, at ambient pressure, of boehmite and the trihydroxides to α-Al_2O_3 requires considerably more structural rearrangements than from disapore and is generally not completed until the temperature reaches at least 1100–1125 °C. The first step in the reaction sequence is the diffusion of protons to adjacent OH groups and the subsequent formation and removal of water. This process begins at a temperature close to 200 °C. If the water molecules cannot diffuse rapidly out of large trihydroxide particles, hydrothermal conditions may develop locally, resulting in the formation of γ-AlO(OH). As water loss proceeds, a large internal porosity develops. The lattice voids left by the escaping water are not readily healed because of the slow diffusion in this low-temperature range. The voids are oriented parallel and perpendicular to the basal plane of the trihydroxide crystals.

Transition oxides formed at lower temperatures are mostly two-dimensional, formed by short-range ordered domains within the texture of the decomposed hydroxides. Extensive three-dimensional ordering begins at ca. 775 °C. Until completely converted to corundum, the solid retains considerable amounts of hydroxyl ions; most likely protons are retained to maintain electroneutrality in cation-deficient areas. Therefore, the presence of protons may retard reordering of the cation sublattice. The high surface area ($>75\ m^2/g$) of γ-Al_2O_3 has been shown to provide thermodynamic stability. The addition of fluorine to the furnace atmosphere removes protons. As a result, rapid transition to α-Al_2O_3 occurs at temperatures as low as 875 °C. Markedly tabular corundum crystals form, possibly because the preceding transition alumina is mostly two-dimensionally ordered.

Alumina is prepared by the Bayer process (11). The method is based on the fact that, due to their amphoteric properties, boehmite, gibbsite, and diaspore can be dissolved in NaOH solutions under moderate hydrothermal conditions; although the solubility of Al_2O_3 in NaOH is temperature-dependent, most other components of natural bauxite are quite inert in the process (e.g., silica forms a nearly insoluble byproduct). These features lead to formation of a sodium aluminate solution, from which impurities are removed by filtration, and finally precipitation of pure $Al(OH)_3$ from the cooled solution (or after gentle decrease of pH by bubbling CO_2). The final operation in production of alumina is calcination, for which the temperature is raised above 1110 °C.

This reaction can take several pathways, and several transition forms may appear. The end of all pathways is α-alumina (Figure 22.6). In European practice, a major portion of the alumina is converted to the α-phase, sometimes by the addition of a fluoride salt to lower the transformation temperature. In American practice, calcination always has been less severe; so, normally, only 20% of the alumina is in the α-phase.

22.4.2. Zirconia (ZrO_2)

Among the families of engineering ceramic materials, zirconias provide the best combination of mechanical properties; for instance, they show the highest mechanical strength and toughness at room temperature. At higher temperatures where the mechanical strength of metals would plummet, the drop-off in the strength of engineering ceramics is more gradual. Zirconias have better wear-resistant properties than metals, are usually resistant to corrosion, and possess a thermal expansion coefficient close to those of many metals.

The use of zirconium oxides in technical ceramics is well established, notably for their electrical, wear, and heat-resistant properties. Zirconia and yttria-stabilized zirconia materials are vital ingredients in a huge range of industrial and domestic products, including capacitors, microwave telecommunications, piezoelectrics, wear part ceramics, molten metal filters, fibre optic ferrules, and oxygen sensors.

Zirconia is very resistant to acids and bases, but it slowly dissolves in concentrated hydrofluoric acid or hot concentrated sulphuric acid. It is also resistant to many fluxes, molten glasses, or melts, silicate, phosphate, or borate, but it is attacked by fluoride or alkaline melts. Zirconium oxide and alkaline oxides or caustic substances can be fired together to form solid-solution oxides, known as zirconate compounds, or a mixture of both. Because of its high melting temperature (2764 °C), zirconia can be used for structural applications at higher temperatures than alumina (12).

Zirconium oxide is extracted from two commercial ores: baddeleyite (natural form of zirconia, containing ~2% HfO_2 and other impurities) and zircon ($ZrO_2 \cdot SiO_2$; Figure 22.9). Various undesired elements should be removed from these ores, silica in the case of zircon and low-level impurities of other compounds, such as titanium and iron oxides, from baddeleyite.

The breakdown of zircon and baddeleyite through reaction with sodium hydroxide is the most common method for chemical production of the pure oxide. Sodium hydroxide and zircon react above 600 °C to form sodium zirconate, sodium zirconate

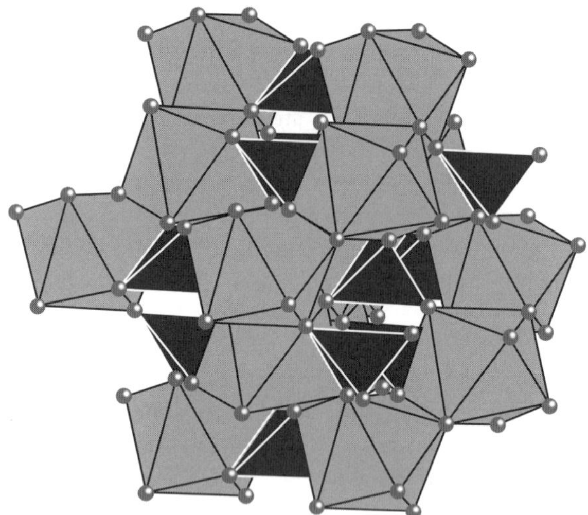

Figure 22.9. Idealized structure of zircon, $ZrSiO_4$ (4). The centers of the gray polyhedra are occupied by Zr^{4+} ions, those of black tetrahedra by Si^{4+} ions, and the O^{2-} ions (spheres) are in the corners.

silicate, and sodium silicate. Careful control of the zircon/sodium hydroxide ratio, temperature, and reaction conditions (particularly the surrounding atmosphere) can yield nearly complete conversion. When sodium carbonate is used, reactions require higher temperatures and depend on the sodium carbonate to zircon ratio.

Three polymorphs of ZrO_2 are stable at atmospheric pressure: the cubic one above 2370 °C, the tetragonal phase above 1170 °C, and the monoclinic one below 1000 °C. The cubic form has the fluorite structure, and the tetragonal and monoclinic structures are also known (Figure 22.10). Transformation of the monoclinic phase to the tetragonal phase begins at ca. 1050 °C and is completed at ca. 1170 °C. This transformation

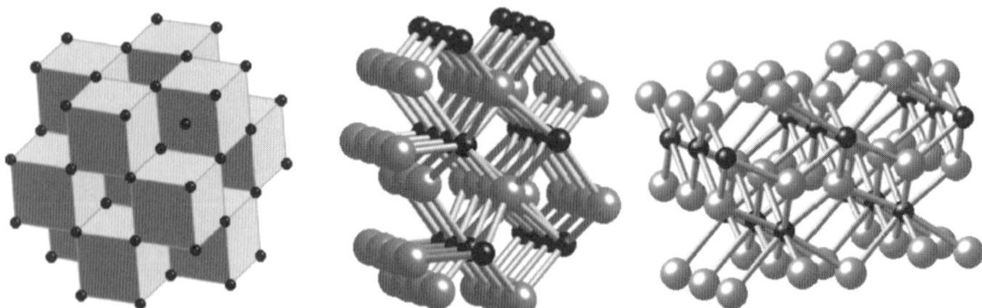

Figure 22.10. Idealized structures of ZrO_2 polymorphs (4): (left) cubic, (center) monoclinic, and (right) tetragonal. In the first case, the centers of the cubes are occupied by Zr^{4+} ions, shown as small black spheres in the other two structures. The other spheres correspond to O^{2-} ions.

is accompanied by a volume shrinkage of 3–5% and shows thermal hysteresis: On gradual cooling, the tetragonal phase is stable to ca. 1000 °C and the conversion to the monoclinic one is only completed at ca. 800 °C. Upon rapid quenching, the tetragonal phase is metastable to room temperature.

Only the tetragonal phase provides ceramic materials with satisfactory properties, but the mechanical behavior limits its use. The cubic phase shows moderate mechanical properties, whereas the monoclinic phase weakens the mechanical performance and simultaneously reduces the cohesion among the particles and thus the density. Consequently, the percentage of the monoclinic phase in a mixture should be as small as possible for structural ceramics. Some properties of these phases are summarized on Table 22.4.

The monoclinic–tetragonal transformation is accompanied by a volume change of ca. 7% together with considerable hysteresis, resulting in stress and cracking in pure ZrO_2 ceramics. Consequently, such ceramics cannot be manufactured. The addition of suitable metal oxides (mainly CaO, MgO, and Y_2O_3 in solid solution with ZrO_2 for refractory purposes) leads to stabilization of the cubic modification down to room temperature. A fully stabilized ZrO_2 can be achieved by the addition of a 5–10% stabilizer. Partially stabilized zirconia raw materials with a 3–6% stabilizer added are now used in most cases. This leads to smaller volume changes on heating and provides the additional advantage of creating microcracks, which increases the thermal shock resistance of the brick (1).

The volume expansion during the tetragonal-to-monoclinic phase transition can be used as an advantage to improve both the strength and the toughness of ceramics: Zirconia particles within a critical size range are introduced into the ceramic, and the constraining pressure exerted by the surrounding matrix retains zirconia in the tetragonal structure. If a crack grows in such a ceramic, the tensile stress around the advancing crack tip allows the tetragonal–monoclinic transformation to take place. The resulting expansion introduces a compressive stress around the crack tip, thereby reducing its tendency to propagate. The addition of 15 vol % ZrO_2 to Al_2O_3 improves fracture toughness to 12 Mpa $m^{1/2}$; this material is used commercially in modern Al_2O_3 cutting tools. Transformation toughening in ZrO_2 is achieved by isothermally aging a partially stabilized (3% MgO) ZrO_2 at 1400 °C to develop small tetragonal precipitates within the cubic matrix. This material is used for cutting and slitting industrial materials; the insulating potential in automotive engine parts, where the low

TABLE 22.4. Some Properties of Zirconia Polymorphs (12).

	Monoclinic	Tetragonal	Cubic
Space Group	P21/c	P42/nmc	Fm3m
a (Å)	5.156	5.094	–
b (Å)	5.191	5.177	
c (Å)	5.304		
β (°)	98.9		
Density (g/cm^3)	5.830	6.100 (calculated)	6.090 (calculated)

thermal conductivity is advantageous, is currently being evaluated. The low thermal conductivity of ZrO_2 has also resulted in its use in coating jet engine components (1).

As indicated, to improve its properties, zirconia is extensively modified by the addition of elements that stabilize the cubic structure (13). The amount of these oxides added may produce a partially stabilized zirconia (PSZ) or form a zirconia modification that maintains the cubic structure from its melting point to room temperature, namely, fully stabilized zirconia (FSZ). The oxides more widely used to stabilize zirconia are MgO, CaO, Y_2O_3, and rare-earth oxides, giving rise to novel and innovative ceramic materials that have brought about considerable technological change. The acronyms frequently used to denote zirconia ceramic alloys are summarized in Table 22.5.

Usually, the molar percentage and symbol of the added, stabilizing element used are also written in the acronym. Stabilized zirconium oxides contain from 3% yttria to 8% calcia. They have the same crystal structure from room temperature to melting, avoiding the catastrophic mechanical failure shown by pure zirconia ceramic parts on transforming from the tetragonal to monoclinic phase while cooling.

Partially stabilized zirconias can be thermally cycled to precipitate metastable tetragonal zirconia within the grains of cubic zirconia. These materials have higher toughness than fully stabilized zirconias. The increased toughness of PSZ is the result of stress-induced martensitic transformation of the metastable tetragonal grains to the monoclinic form in the stress field of a propagating crack. This leads to development of an entirely tetragonal zirconia, known as tetragonal zirconia polycrystal (TZP). The orthorhombic structure can be stabilized at atmospheric pressure by addition of >12 mol % of niobia or tantala or a mixture thereof (2).

Y–TZP is very hard, and it has been the ceramic of choice for most uses. However, this material is prone to environmental degradation under humid conditions. Also Mg–TZP and Ca–TZP undergo continuous leaching of the Mg^{2+} and Ca^{2+} cations in aqueous environments, in addition to requiring critical control during annealing. These problems have led us to pay more attention to Ce–TZP, which is more stable under environmental conditions and less prone to leaching and degradation of mechanical properties in aqueous environments. This is possible because the ceria content is larger than 10 mole %. Besides, Ce–TZP has an attractive yellowish-green lustre and is cheaper than Y–TZP (2).

To prepare zirconium oxides with satisfactory stability, the presence of yttrium oxide is necessary, the ideal content being 5.15 wt.%. The most common impurity when preparing zirconia is alumina, which should be less than 0.5 wt.%. The ISO

TABLE 22.5. Acronyms for Zirconia Ceramics Alloys.

TZP	tetragonal zirconia polycrystals
PSZ	partially stabilized zirconia
FSZ	fully stabilized zirconia
TTC	transformation toughened ceramics
ZTA	zirconia toughened alumina
TTZ	transformation toughened zirconia

TABLE 22.6. Chemical Composition and Some ISO Properties for Zirconia Materials (14).

Property	Value
$ZrO_2 + HfO_2$	>94.05%
HfO_2	≤5%
Y_2O_3	4.95 ± 0.45%
Al_2O_3	<0.5%
Others	<0.5%
Density	≥6.00 g/cm^3
Average grain size	≤0.6 μm
Bending strength	≥800 MPa
Biaxial strength	≥500 MPa
Sphericity tolerance	10 μm
Surface roughness	0.02 μm

requirements for ceramic zirconia, both for chemical composition and for some properties, are summarized in Table 22.6. When using zirconia for special optical and ophthalmic glasses, the requirements are even more strict.

Some applications of zirconia ceramics have been already indicated, and others are listed in Table 22.7. Although stabilized zirconias are insulators at room temperature, at elevated temperatures, the vacancies in the anion lattice allow O^{2-} ions to diffuse and the zirconia becomes a solid electrolyte with applications in oxygen sensors and high-temperature fuel cells. Some stabilized zirconias can also be used as resistors or susceptors. Inductively heated yttria-stabilized zirconia cylinders are used as heat sources to melt quartz boules for the drawing of quartz optical fibres.

Zirconia is a constituent of lead zirconate titanate (PZT) used in piezoelectric ceramics for applications as gas furnace and barbeque igniters, microphone and phonograph crystals, ultrasonic transducers for medical ultrasound imaging, for agitation in cleaning tanks, and for underwater sonar. With the further addition of lanthanum (PLZT), ferroelectric optically active transparent ceramics have become available (12).

22.4.3. Boron Nitride (BN)

Boron and nitrogen form the 1:1 compound BN, which has allotropes isostructural to those of carbon. This is not surprising considering that these elements are neighboring carbon in the Periodic Table, and the average number of electrons per atom in BN is the same as in a carbon atom. The allotropes of BN (Figure 22.11) are as follows:

α-BN, hexagonal modification with a layered structure similar to graphite, sometimes called white graphite.

β-BN, high-pressure diamond-like modification, cubic zinc blende structure.

γ-BN, dense hexagonal modification, wurtzite structure.

TABLE 22.7. Some Applications of ZrO_2 and the Benefit Thereof Induced (12).

Application	Benefit
Adhesives	Reduces water sensitivity and improves adhesion of aqueous adhesive formulations.
Antiperspirants	Improves efficacy of antiperspirant formulations.
Printing inks	Non-yellowing adhesion promoter to enhance adhesion, thermal, solvent, and scratch-resistance properties of inks on difficult substrates such as polymer films and metals.
Oil industry	Clay stabilizer for secondary oil recovery, rheological control of cementing applications, fracturing fluids.
Paint	Through drier for alkyd based paints, thixotropic agent for flat emulsion paints.
Pigment coating	Surface treatment of titanium dioxide and other pigments to improve durability by reducing UV degradation of coating binders.
Metal treatment	Improves adhesion of surface treatment to aluminium.
Waterproofing	Component of waterproofing formulation to improve adhesion to textile substrates.
Ceramic colours	For all zircon and zirconia pigments.
Electronic ceramics	Dielectrics, sensors, and piezoelectrics and capacitors.
Engineering ceramics	Tough wear-resistant technical ceramics.
Glass	For glass toughening and optical property control.
Solid oxide fuel cells	Custom-made products for the electrolyte and anode.
Ceramic colours	For all zircon and zirconia pigments.
Electronic ceramics	Dielectrics, sensors, and piezoelectrics and capacitors.
	Paper Industry
Adhesives	Promotes greater printing ink adhesion to plastics and metals.
Crosslinkers	In printing inks and polymers.
Paper coatings	Strength giving, water-resistant surface coatings for paper and paperboard.
Pigment coating	For better weathering of paint films.
Polymer adhesives	Various additives including crosslinkers.

The most common form of boron nitride is the graphite-like hexagonal one, used increasingly because of its unique combination of properties: low density (2.27 g/cm^3), high-temperature stability (melting point close to 3000 °C), chemical inertness (especially its resistance to acids and molten metals and its stability in air up to 1000 °C and in inert gases like N_2, CO, and Ar up to 270 °C), stability to thermal shock, workability of hot-pressed shapes, extremely low electrical conductivity—which makes it an excellent insulator—and high thermal conductivity.

The cubic form of boron nitride, β-BN, resembles diamond in its crystal structure and in some of its properties. Pure β-BN is colorless and a good electrical insulator. Doping with Li_3N or other elements (Be, Si, C, or P) turns β-BN yellow to black

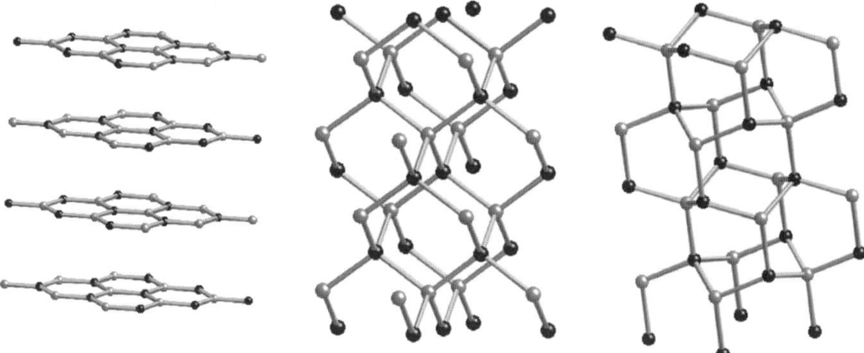

Figure 22.11. Idealized structures of BN polymorphs (4): (left) α, (center) β, and (right) γ.

and modifies other properties, such as conductivity or toughness. The calculated density of β-BN is 3.48 g/cm^3. Because β-BN is stable in air up to 1400 °C, higher than diamond, which burns away at only 800 °C, it is an effective high-temperature abrasive (15).

Wurtzite-like boron nitride, γ-BN, which is metastable at the pressures and temperatures where β-BN is prepared, has a density close to that of cubic β-BN. Above 5.5 GPa and 1300 °C, the γ-BN to β-BN transformation takes place.

Hexagonal α-BN is prepared mainly by two processes on an industrial scale (16):

1. Reaction of boric oxide or boric acid with ammonia in the presence of a carrier, often tricalcium orthophosphate:

$$B_2O_3 + 2NH_3 \longrightarrow 2BN + 3H_2O \ (900\,^\circ C)$$

   A second heat treatment for purification and stabilization is carried out at >1500 °C under N_2.

2. Reaction of boric acid or borax with organic nitrogen compounds, e.g., urea or melamine:

$$B_2O_3 + CO(NH_2)_2 \longrightarrow 2BN + CO_2 + 2H_2O \ (>1000\,^\circ C)$$

The first method usually leads to formation of crystalline α-BN as thin hexagonal platelets with a thickness of about 0.1–0.5 μm and a diameter of up to 5 μm. The second method can produce a "turbostratic" boron nitride, a hexagonal structure characterized by partial or complete absence of three-dimensional order among its lamellae. A third method, known as self-propagating high-temperature synthesis, has been developed recently (17), which is based on a solid–solid reaction, strongly exothermic, where the intense heating of the reacting mass is self-propagating by way of a combustion front or wave; depending on the process, the temperatures in this front can reach values as high as 4000 °C, creating pressures up to 25 Mpa. The main advantages of this technology are the simplicity of the process and the low energy consumption, the high purity of the products formed, and the high productivity (15).

Properties of boron nitride ceramics are listed in Table 22.8. Cubic boron nitride is the second hardest material known after diamond. In fact, it is considered not a form a boron nitride but a different compound. It was first introduced commercially in 1969; it is known by the initials CBN and commercialized with the trade names of Borazon and Amber Boron Nitride, among others.

This abrasive may be made in any of the high-pressure, high-temperature apparatuses used for diamond production. By using different raw materials and manufacturing conditions, a variety of products having a range of friabilities may be produced. Color ranges from almost colorless through shades of yellow, red, and black. As with diamonds, the producers supply CBN in sized grains and flours.

CBN uses are similar to those of diamond, i.e., machining and drilling. Grinding wheels made from resin- or metal-bonded β-BN grits are in many cases superior and less expensive than diamond wheels because of their better chemical resistance and greater toughness. Sintered dense compacts of β-BN are being used increasingly as cutting tools for hard steels and tough nickel-based superalloys, as wire drawing dies, and as cutting heads in rock drills (15).

TABLE 22.8. Properties of Boron Nitride Ceramics (15).

	Hot-Pressed BN	Hot Isostatically Pressed BN	Hot-Pressed Composite Ceramics BN/ZrO_2
Bulk density, g/cm^3	2.0	2.2	2.8–3.6[b]
Porosity, vol %	<7	<1	<7
Modulus of rupture (4-point),[a] MPa	100/80	50	130/70
Young's modulus,[a] GPa	70/35	30	80/35
Thermal conductivity, $Wm^{-1}K^{-1}$	20 °C 65/45	50	35–25/20–18[b]
	400 °C 50/30	40	31–21/17–15
	700 °C 30/20	30	28–18/15–13
	1000 °C 15/10	20	25–15/13–18
Thermal expansion coefficient, $10^{-6}\,K^{-1}$ (at 20–1000 °C)	1.2/8.0	4.0	4.5–9.0/8.5–11.0[b]
Specific heat, J/gK (at 20 °C)	0.8	0.8	0.7
Electrical resistivity, Ωcm (at 20 °C)	$>10^{12}$	$>10^{12}$	$>10^{12}$
Dielectric strength, kV/mm (at 20 °C)	>6	>6	>6

[a]Test orientation vertical/parallel to pressing direction.
[b]Dependent on BN/ZrO_2 ratio.

22.4.4. Silicon Carbide (SiC)

Silicon carbide was the first synthetic abrasive invented and commercialized. It was discovered by Acheson in 1891, when he tried to make diamonds by an electric-arc heating process. As Acheson wrongly thought that this compound was a combination of carbon and corundum, he called it "carborundum," a name maintained even after knowing its real chemical composition.

Its chemical formula is SiC, with a composition of 70.05 wt % Si and 29.95 wt % C. Its density is $3.21\ g/cm^3$. It does not exhibit a congruent melting point. In a closed system, i.e., under equilibrium conditions at a total pressure of 0.1 MPa, SiC decomposes at $2830 \pm 40\,°C$ into carbon and a silicon-rich melt; in an open system decomposition, it begins at ca. 2300 °C (18).

Silicon carbide has special chemical and physical properties, especially a high chemical resistance, a great hardness, and semiconducting properties. It is consequently used in the production of abrasives, refractory materials, heating elements, voltage-dependent electrical resistors, light-emitting diodes, and structural ceramic components. Silicon carbide also has the ability to dissolve in molten iron, and it is therefore used in ferrous metallurgy as an alloying and deoxidizing agent.

Silicon carbide has the diamond structure, with C and Si atoms located in alternated positions in the structure, each one having a tetrahedral coordination of atoms of the other type (Figure 22.12).

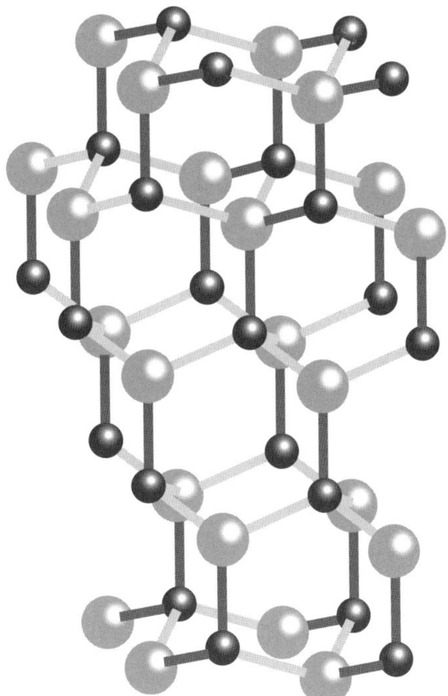

Figure 22.12. Idealized structure of SiC (4) showing the wurtzite structure, with Si and C atoms tetrahedrally bonded to the opposite ones.

Silicon carbide exists as both polymorphs and polytypes. As indicated, coordination of both C and Si atoms is tetrahedral. Depending on the plane packing, two forms of silicon carbide may occur: cubic β-SiC and α-SiC, which includes a large number of hexagonal and rhombohedral crystal types. In addition, different polytypes are formed, such as 3C, 4H, 6H, and 15R. β-SiC is metastable and is formed initially in SiC production. It changes into α-SiC above 1900 °C (19). Different packing patterns for these polytypes are shown in Figure 22.13.

Silicon carbide is produced industrially from silicon dioxide and carbon, which react according to the overall equation:

$$SiO_2 + 3C \longrightarrow SiC + 2CO$$

The reaction is strongly endothermic with $\Delta H_{298K} = +618.5\,kJ/mol$ (20). The reaction in fact takes place in several stages. The initiation step

$$SiO_2 + C \longrightarrow SiO + CO$$

is thermodynamically possible above ca. 1700 °C and in fact starts when SiO_2 melts.

The next steps are

$$SiO + C \longrightarrow Si + CO$$
$$Si + C \longrightarrow SiC$$

The SiC thus formed appears as an incrustation on the carbon grains, preventing further formation of SiC, and ultimately causing the reaction to halt. The SiC then

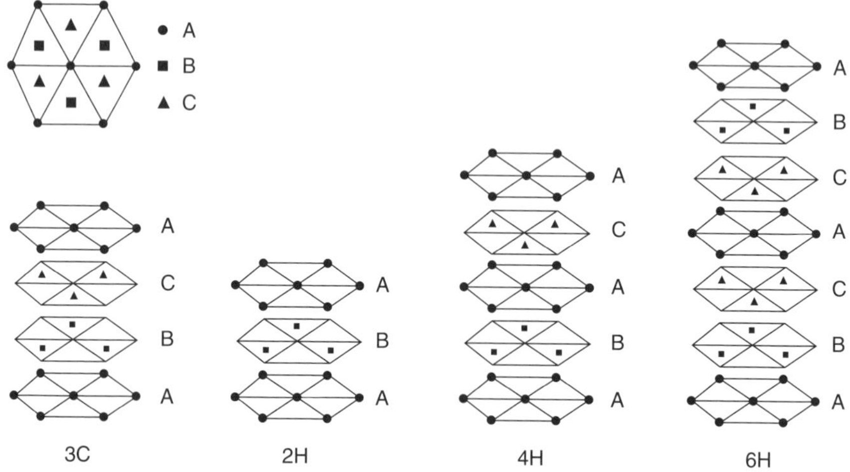

Figure 22.13. Different packing patterns for SiC polytypes.

reacts further

$$SiC + 2SiO_2 \longrightarrow 3SiO + CO$$

$$SiC + SiO \longleftrightarrow 2Si(g) + CO$$

The yield and quality of silicon carbide are seriously affected by any impurities existing in the raw materials. Therefore, high-purity raw materials should be used.

The source of silicon is generally very pure, natural silica sand or ground quartz. The SiO_2 content is 99.0–99.9%, preferably >99.7%, depending on the quality of SiC required. The usual impurities are Fe_2O_3 (0.01–0.05%) and Al_2O_3 (0.02–0.50%) (21).

The preferred carbon source is low-ash petroleum coke, with a typical composition 93.7% C_{fix}, 6.2% volatiles, and 0.1% ash. Charcoal and anthracite are also used to produce silicon carbide, but these materials are not sufficiently pure to produce the highest quality SiC grades such as green SiC.

An important new process is the direct manufacture of sinterable β-SiC powders from the gas phase by decomposing gaseous silicon halides in the presence of hydrocarbons in a plasma jet reactor. In another CVD process, the starting materials are organosilicon compounds. Another method of producing a sinterable β-SiC powder is the carbothermic reduction of finely divided SiO_2.

In addition to its hardness, an important property of silicon carbide is its chemical resistance: It is resistant to organic solvents, alkalis, acids, salt solutions, and even *aqua regia* and fuming nitric acid.

The behavior of silicon carbide toward oxygen and oxygen-containing gases is of industrial importance. Oxidation of pure SiC begins at ca. 600 °C, forming a SiO_2 coating on the surface of the SiC that prevents further oxidation. This is named *passive* oxidation, and it may be distinguished from the *active* oxidation, which takes place under oxygen deficient conditions above 1000 °C, leading to decomposition of SiC and formation of SiO, through the following reaction (22):

$$SiC + O_2 \longrightarrow SiO + CO$$

Silicon carbide is attacked and decomposed by oxidizing agents if the SiO_2 protective layer is removed, thereby enabling the reaction to proceed unhindered. SiC is decomposed in this way by fused alkalis such as $Na_2CO_3 + Na_2O_2$ or $Na_2CO_3 + KNO_3$. In the presence of oxygen, fused lead borate causes complete decomposition of SiC. Pb_3O_4 or a mixture of $K_2Cr_2O_7$ with $PbCrO_4$ reacts very vigorously on heating with silicon carbide.

A mixture of hydrofluoric acid, nitric acid, and sulphuric acid slowly attacks silicon carbide. However, complete dissolution only takes place under pressure and at a high temperature if SiC is very finely divided. Chlorine reacts with SiC above 800 °C, forming silicon tetrachloride and carbon. Silicon carbide behaves toward molten metals in various ways: It is not attacked by molten zinc or zinc vapor, whereas molten aluminum attacks SiC slowly, forming Al_4C_3 and Si (23).

SiC is sensitive to oxygen and oxygen-containing gases such as water vapor. In an oxidizing atmosphere, SiC begins to decompose to SiO_2 and CO or CO_2 at ca. 900 °C. Above ca. 1200 °C, the SiO_2 layer formed on the SiC grains becomes glassy and retards further oxidation. Oxidation can be largely prevented by a dense protective coating (e.g., an Al_2O_3 surface layer). SiC is also sensitive to alkali melts, basic slags, chlorine gas, and metal melts other than lead, zinc, and copper. Alkali resistance can be increased by using a nitride bond.

SiC is used in blast furnaces, kiln furniture in the ceramic industry (maximum service temperature, 1500 °C), incinerators, retorts for zinc distillation, aluminum industry (submerged burner tubes), and in tubes for ceramic regenerators (3,18).

22.4.5. Silicon Nitride (Si_3N_4)

Silicon and nitrogen form a covalent compound with chemical formula Si_3N_4, whose structure is shown in Figure 22.14. In the cubic and tetragonal forms, each silicon atom is tetrahedrally linked to four nitrogen atoms and each nitrogen atom is in the center of a triangle, whose corners are occupied by silicon atoms, whereas in the hexagonal form, the coordination is tetrahedral for both atoms.

Silicon nitride exists in two basic forms: porous and dense. The porous form, *reaction-bonded silicon nitride* (*RBSN*), is produced by controlled nitridation of a compact silicon powder in a nitrogen atmosphere at 1400 °C. *Dense silicon nitride* is made by several processes, including hot pressing and pressureless sintering.

RBSN has the lowest strength over most of the temperature range. The strength depends on bulk density, which can be controlled in the range 2.0–2.8 g/cm^3, compared with the calculated density of ca. 3.2 g/cm^3. Pore morphology is an additional strength-controlling factor. Flexure strength is relatively low, but it is maintained to >1200 °C because the absence of sintering aids means that no glassy phases are present in the microstructure. One manufacturing advantage is that little shrinkage occurs on nitridation. Thus, the pressed and machined silicon powder shape can often be made close to the final dimensions. Reaction-bonded silicon nitride is used in light

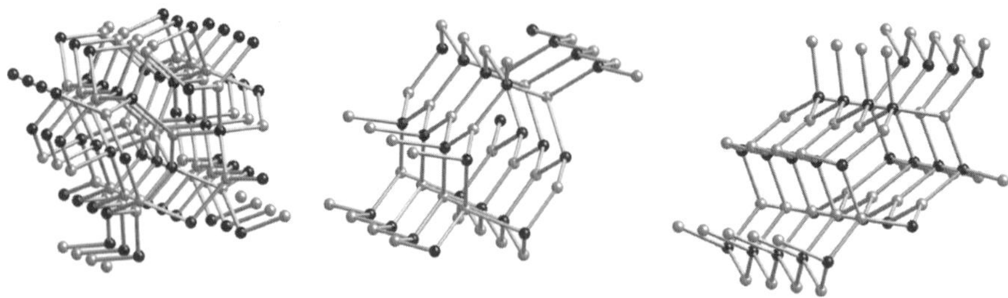

Figure 22.14. Idealized structures of three Si_3N_4 polymorphs (4): (left) cubic, (center) tetragonal, and (right) hexagonal. The black spheres correspond to Si atoms and the gray ones to N atoms.

alloy foundries for pouring spouts, continuous casting dies, and so on, and in heat treatment and brazing fixtures.

The production of dense silicon nitride requires the incorporation of sintering aids. Sintering may be carried out at atmospheric or slight overpressure to produce sintered material (sintered silicon nitride, SSN). Sintering takes place during hot pressing (hot pressed silicon nitride, HPSN) or hot isostatic pressing; the amount of sintering aid required generally decreases with applied pressure. Oxides such as MgO, Y_2O_3, and Al_2O_3 are incorporated as sintering aids, and densification occurs at 1600–1800 °C by dissolution and precipitation of β-Si_3N_4 from the liquid produced by the sintering aids. The resultant microstructure contains two or more phases depending on the aids used. The first widely used additive was MgO, which reacts with SiO_2 to develop a low-viscosity glassy phase that limits temperature capability. A more refractory glass is produced by addition of Y_2O_3; Y_2O_3-containing materials, therefore, possess higher temperature capability.

Dense silicon nitride has been used for various static and rotating turbine components and has been built into automobile turbocharger rotors.

The properties of silicon nitride are strongly dependent on the additives it contains. Table 22.9 compares the properties of silicon nitride hot pressed with MgO additive, silicon nitride sintered with Y_2O_3 additive, reaction bonded silicon nitride, and Sialon. Generally speaking, higher strengths are achieved both at low and high temperatures by hot pressing; however, this is not an easily commercialized process for mass production. Reaction-bonded silicon nitride, made by exposing a compact silicon powder to nitrogen at high temperatures, has poor low-temperature strength but good high-temperature strength (3).

22.4.6. Sialons ($Si_{6-z}Al_zO_zN_{8-z}$)

A compound related to silicon nitride that has found successful commercialization is *SiAlON*, an aluminum silicon oxide nitride. The most common form has the formula $Si_{6-z}Al_zO_zN_{8-z}$, where z ranges from 0 to 4. Compounds with z near zero have good

TABLE 22.9. The Effects of Additives on the Properties of Silicon Nitride (24).

Material	Bending Strength, Four Point, MPa			Elasticity Modulus GPa	Thermal Expansion Coefficient, $10^{-6}\,K^{-1}$	Thermal Conductivity, $W\,m^{-1}\,K^{-1}$
	RT	100 °C	1375 °C			
Hot pressed (MgO additive)	690	620	330	317	3.0	30–15
Sintered (Y_2O_3 additive)	655	585	275	276	3.2	28–12
Reaction bonded (2.45 g/cm^3)	210	345	380	165	2.8	6–3
SiAlON (sintered)	485	485	275	297	3.2	22

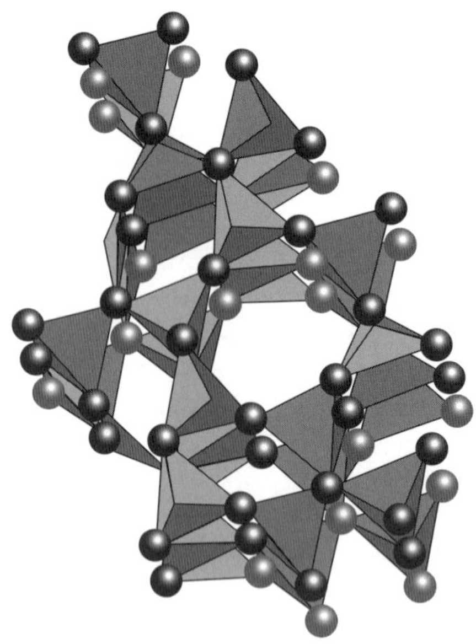

Figure 22.15. Idealized structure of $Si_3Al_3O_3N_5$ (4). The centers of the tetrahedra are occupied by Si or Al atoms, whereas the corners (spheres) correspond to O or N atoms.

high-temperature strength, creep resistance, and oxidation resistance, whereas those with z approaching 4 have good low-temperature properties, i.e., strength, toughness, and abrasion resistance; the structure for $z = 3$ is shown in Figure 22.15.

Sialons are based on a solid solution of Al_2O_3 in Si_3N_4, where nitrogen is partially replaced by oxygen and silicon is partially replaced by aluminum. They may be produced by firing under a nitrogen-containing atmosphere a green mixture composed of Si–SiC, which may also include Al_2O_3 or other alumina-bearing compounds. They also contain additives such as Y_2O_3, because they are made by hot pressing or sintering. The second phase in such materials can be crystallized to $Y_3Al_5O_{12}$. The result is a higher temperature capability because this is not a glassy phase. Sialon containing a glassy second phase has been used for machining nickel superalloy turbine disks at a surface speed of 5 m/s, compared with a maximum of 0.25 m/s for tungsten carbide inserts (3).

22.4.7. Boron Carbide (B_4C)

Recent studies (25) using density functional theory have shown that the structure of boron carbide, B_4C, can be described as a arrangement of 3-atom linear chains and 12-atom icosahedra (Figure 22.16); all chains have a CBC structure and although most of the icosahedra have a $B_{11}C$ structure—the C atom placed in a polar site—a few percentage have a B_{12} structure or a $B_{10}C_2$ structure with the two C atoms placed in two antipodal sites.

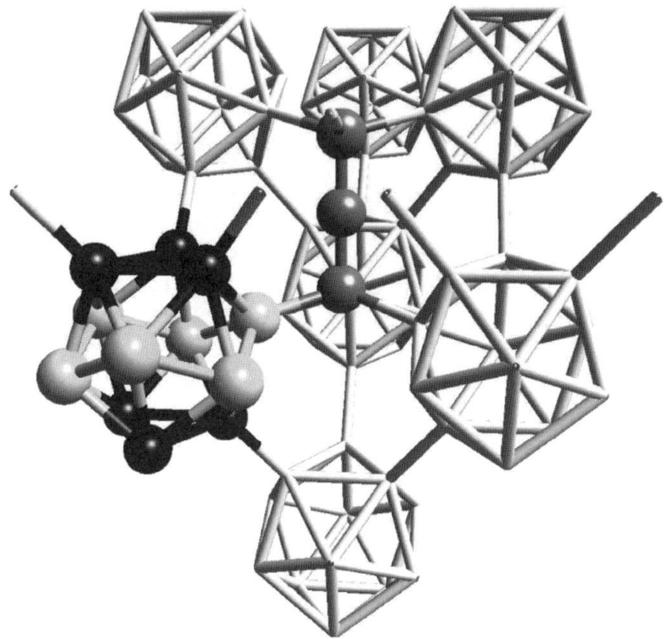

Figure 22.16. Structure of B_4C as concluded from DFT studies. Reprinted from Ref. 25 with permission.

Boron carbide is produced by carbothermic reduction from a charge of high-purity boron oxide glass and high-purity coke according to the following reaction, usually carried out in arc or Acheson furnaces (26):

$$2B_2O_3 + 7C \longrightarrow B_4C + 6CO$$

It can be also prepared by reduction of boric oxide with magnesium in the presence of carbon, at high temperature, with the reaction being:

$$2B_2O_3 + 6Mg + C \longrightarrow B_4C + 6MgO$$

This abrasive is unsuited for use in bonded or coated products. However, for loose-grain applications, such as the lapping of cemented carbides and other hard materials, it is used alone or mixed with silicon carbide.

22.4.8. Aluminium Titanate (Al_2TiO_5)

The chemical formula of this compound can be written as Al_2TiO_5 or as $Al_2O_3{\cdot}TiO_2$. It has the psuedobrookite (Fe_2TiO_5) crystal structure.

TABLE 22.10. Typical Physical Properties for Aluminium Titanate (27).

Property	Value
Density	$3–3.4\,g.cm^{-3}$
Modulus of Rupture (RT)	30 MPa
Modulus of Rupture (1000 °C)	60 MPa
Young's modulus	20 GPa
Thermal expansion (20–600 °C)	$0–1 \times 10^{-6}\,K^{-1}$
Thermal expansion (600–1000 °C)	$1–2 \times 10^{-6}\,K^{-1}$
Thermal conductivity (RT–1000 °C)	$<2 W.m.K^{-1}$
Maximum service temperature	1000 °C continuous 1100 °C intermittent
Thermal shock resistance	Excellent
Resistance to molten metals	Good

Aluminium titanate is synthesized by solid-state reaction (ceramic method) between alumina and titania, with a temperature higher than 1350 °C being needed. The reaction product is sintered in air at 1400–1600 °C.

Ceramics manufactured from aluminium titanate exhibit extremely good resistance to thermal shock. The excellent thermal shock resistance and low thermal conductivity coupled with good chemical resistance to molten metals (particularly aluminium) result in the material fulfilling several metal contact applications in the foundry industry. They are also used in the automotive industry as an insulating liner for exhaust manifolds.

The main properties of aluminium titanate are summarized in Table 22.10.

REFERENCES

(1) Meetham, G.W. High-temperature materials. In *Ullmann's Encyclopedia of Industrial Chemistry*; Wiley-VCH Verlag: New York, 2002.

(2) UltraHard Materials USA Website; Available at: http://www.ultrahardmaterials.com.

(3) Cannon, W.R.; Gugel, E.; Leimer, G.; Woetting, G. Ceramics, advanced structural products. In *Ullmann's Encyclopedia of Industrial Chemistry*; Wiley-VCH Verlag: New York, 2002.

(4) Crystal Maker 7.0.0 Software; Crystal Maker Software Limited: Oxfordshire, UK; Available at: http://www.crystalmaker.com; using data from Inorganic Crystal Structure Database; Available at: http://www.fiz-informationsdienste.de/en/DB/icsd/index.html.

(5) Cannon, W.R. Transformation toughened ceramics for structural uses. In *Structural Ceramics*; Wachtman, J.B.; Jr., (Editor); Academic Press: Boston, MA, 1989.

(6) Freedonia Group Inc. Website; Available at: http://www.freedoniagroup.com.

(7) Business Communications Company Inc. (BCC) Website; Available at: http://www.bccresearch.com.

(8) Hudson, L.K.; Misra, C.; Perrotta, A.J.; Wefers, K.; Williams, F.S. Aluminum oxide. In *Ullmann's Encyclopedia of Industrial Chemistry*; Wiley-VCH Verlag: New York, 2002.

(9) Marketech Internacional Inc. Website; Available at: http://www.mkt-intl.com/ceramics.

(10) Wefers, K.; Misra, C. Oxides and Hydroxides of Aluminum. Alcoa Technical Paper no. 19, Pittsburgh, PA, 1987.

(11) Misra, C. Industrial alumina chemicals. *ACS Monograph 184*; The American Chemical Society: Washington, DC, 1986.

(12) Mel Chemical Website; Available at: http://www.zrchem.com.

(13) Curtis, C.E. *J. Am. Ceram. Soc.* **1947**, *30*, 180.

(14) Dambreville, A.; Phillipe, M.; Ray, A. In *The French Orthopaedic Web Journal*; Available at: http://www.maitrise-orthop.com/indexus.html.

(15) Schwetz, K.A. Boron carbide, boron nitride, and metal borides. In *Ullmann's Encyclopedia of Industrial Chemistry*; Wiley-VCH Verlag: New York, 2002.

(16) Lipp, A.; Schwetz, K.A.; Hunold, K.; *J. Eur. Ceram. Soc.* **1989**, *5*, 3.

(17) Merzhanov, A.G.; Rogachev, A.S.; A.E. Sychev. *Doklady Phys. Chem.* **1998**, *362*, 299.

(18) Liethschmidt, K. Silicon carbide. In *Ullmann's Encyclopedia of Industrial Chemistry*; Wiley-VCH Verlag: New York, 2002.

(19) Ramdsdell, L.S. *Am. Mineral.* **1947**, *32*, 64.

(20) *JANAF Thermochemical Tables, 2nd ed.*; U.S. Department of Commerce: Washington, DC, 1971.

(21) Fuchs, H. *Chem. Ing. Tech.* **1974**, *46*, 139.

(22) Frisch, B.; Thiele, W.R.; Drumm, R.; Münnich, B. *Ber. Dtsch. Keram. Ges.* **1988**, *65*, 277.

(23) Iseki, T.; Kameda, T.; Maruyama, T. *J. Mater. Sci.* **1984**, *19*, 1692.

(24) Katz, R.N. *Science* **1980**, *208*, 841.

(25) Mauri, F.; Vast, N.; Prickard, C.J. *Phys. Rev. Lett.* **2001**, *87*, 085506.

(26) Lipp, A. *Tech. Rundsch.* **1965**, *57*, 5.

(27) CERAM Website; Available at: http://www.ceram.com.

Index

Synthesis, Properties, and Applications of Oxide Nanomaterials, Edited by José A. Rodríguez and Marcos Fernández-Garcia